Jaime M. Ross, Giuseppe Coppotelli and
Lars Olson (Eds.)

Mitochondrial Dysfunction in Ageing and Diseases

MDPI

This book is a reprint of the Special Issue that appeared in the online, open access journal, *International Journal of Molecular Sciences* (ISSN 1422-0067) from 2014–2015 (available at: http://www.mdpi.com/journal/ijms/special_issues/mitochondrial-dysfunction).

Guest Editors
Jaime M. Ross, Giuseppe Coppotelli and Lars Olson
Department of Neuroscience, Karolinska Institutet
Sweden

Editorial Office
MDPI AG
St. Alban-Anlage 66
Basel, Switzerland

Publisher
Shu-Kun Lin

Managing Editor
Yong Ren

1. Edition 2016

MDPI • Basel • Beijing • Wuhan • Barcelona

ISBN 978-3-03842-251-8 (Hbk)
ISBN 978-3-03842-252-5 (PDF)

Table of Contents

List of Contributors IX

About the Guest Editors.. XVII

Giuseppe Coppotelli and Jaime M. Ross
Preface to "Mitochondrial Dysfunction in Ageing and Diseases"
Reprinted from: *Int. J. Mol. Sci.* **2016**, *17*(5), 711
http://www.mdpi.com/1422-0067/17/5/711 XIX

Naghia Ahmed, Dario Ronchi and Giacomo Pietro Comi
Genes and Pathways Involved in Adult Onset Disorders Featuring Muscle
Mitochondrial DNA Instability
Reprinted from: *Int. J. Mol. Sci.* **2015**, *16*(8), 18054–18076
http://www.mdpi.com/1422-0067/16/8/18054 1

Thierry Arnould, Sébastien Michel and Patricia Renard
Mitochondria Retrograde Signaling and the UPRmt: Where Are We in Mammals?
Reprinted from: *Int. J. Mol. Sci.* **2015**, *16*(8), 18224–18251
http://www.mdpi.com/1422-0067/16/8/18224 27

Lin Ding and Yilun Liu
Borrowing Nuclear DNA Helicases to Protect Mitochondrial DNA
Reprinted from: *Int. J. Mol. Sci.* **2015**, *16*(5), 10870–10887
http://www.mdpi.com/1422-0067/16/5/10870 59

Jaime M. Ross, Lars Olson and Giuseppe Coppotelli
Mitochondrial and Ubiquitin Proteasome System Dysfunction in Ageing and
Disease: Two Sides of the Same Coin?
Reprinted from: *Int. J. Mol. Sci.* **2015**, *16*(8), 19458–19476
http://www.mdpi.com/1422-0067/16/8/19458 79

Raffaella Crescenzo, Francesca Bianco, Arianna Mazzoli, Antonia Giacco, Giovanna Liverini and Susanna Iossa
Skeletal Muscle Mitochondrial Energetic Efficiency and Aging
Reprinted from: *Int. J. Mol. Sci.* **2015**, *16*(5), 10674–10685
http://www.mdpi.com/1422-0067/16/5/10674 101

Paola Maura Tricarico, Sergio Crovella and Fulvio Celsi
Mevalonate Pathway Blockade, Mitochondrial Dysfunction and Autophagy: A Possible Link
Reprinted from: *Int. J. Mol. Sci.* **2015**, *16*(7), 16067–16084
http://www.mdpi.com/1422-0067/16/7/16067 115

Jieying Zhang, Kunlu Wu, Xiaojuan Xiao, Jiling Liao, Qikang Hu, Huiyong Chen, Jing Liu and Xiuli An
Autophagy as a Regulatory Component of Erythropoiesis
Reprinted from: *Int. J. Mol. Sci.* **2015**, *16*(2), 4083–4094
http://www.mdpi.com/1422-0067/16/2/4083 134

Yuliya Mikhed, Andreas Daiber and Sebastian Steven
Mitochondrial Oxidative Stress, Mitochondrial DNA Damage and Their Role in Age-Related Vascular Dysfunction
Reprinted from: *Int. J. Mol. Sci.* **2015**, *16*(7), 15918–15953
http://www.mdpi.com/1422-0067/16/7/15918 147

Janina A. Vaitkus, Jared S. Farrar and Francesco S. Celi
Thyroid Hormone Mediated Modulation of Energy Expenditure
Reprinted from: *Int. J. Mol. Sci.* **2015**, *16*(7), 16158–16175
http://www.mdpi.com/1422-0067/16/7/16158 188

Francesca Forini, Nadia Ucciferri, Claudia Kusmic, Giuseppina Nicolini, Antonella Cecchettini, Silvia Rocchiccioli, Lorenzo Citti and Giorgio Iervasi
Low T3 State Is Correlated with Cardiac Mitochondrial Impairments after Ischemia Reperfusion Injury: Evidence from a Proteomic Approach
Reprinted from: *Int. J. Mol. Sci.* **2015**, *16*(11), 26687–26705
http://www.mdpi.com/1422-0067/16/11/25973 208

Francesca Forini, Giuseppina Nicolini and Giorgio Iervasi
Mitochondria as Key Targets of Cardioprotection in Cardiac Ischemic Disease:
Role of Thyroid Hormone Triiodothyronine
Reprinted from: *Int. J. Mol. Sci.* **2015**, *16*(3), 6312–6336
http://www.mdpi.com/1422-0067/16/3/6312 234

María Cecilia Cimolai, Silvia Alvarez, Christoph Bode and Heiko Bugger
Mitochondrial Mechanisms in Septic Cardiomyopathy
Reprinted from: *Int. J. Mol. Sci.* **2015**, *16*(8), 17763–17778
http://www.mdpi.com/1422-0067/16/8/17763 262

Ana A. Baburamani, Chloe Hurling, Helen Stolp, Kristina Sobotka,
Pierre Gressens, Henrik Hagberg and Claire Thornton
Mitochondrial Optic Atrophy (OPA) 1 Processing Is Altered in Response to
Neonatal Hypoxic-Ischemic Brain Injury
Reprinted from: *Int. J. Mol. Sci.* **2015**, *16*(9), 22509–22526
http://www.mdpi.com/1422-0067/16/9/22509 280

Yu Luo, Alan Hoffer, Barry Hoffer and Xin Qi
Mitochondria: A Therapeutic Target for Parkinson's Disease?
Reprinted from: *Int. J. Mol. Sci.* **2015**, *16*(9), 20704–20730
http://www.mdpi.com/1422-0067/16/9/20704 298

Seok-Jo Kim, Paul Cheresh, Renea P. Jablonski, David B. Williams and
David W. Kamp
The Role of Mitochondrial DNA in Mediating Alveolar Epithelial Cell Apoptosis
and Pulmonary Fibrosis
Reprinted from: *Int. J. Mol. Sci.* **2015**, *16*(9), 21486–21519
http://www.mdpi.com/1422-0067/16/9/21486 328

Giovanni Pagano, Annarita Aiello Talamanca, Giuseppe Castello,
Mario D. Cordero, Marco d'Ischia, Maria Nicola Gadaleta, Federico V. Pallardó,
Sandra Petrović, Luca Tiano and Adriana Zatterale
Current Experience in Testing Mitochondrial Nutrients in Disorders Featuring
Oxidative Stress and Mitochondrial Dysfunction: Rational Design of
Chemoprevention Trials
Reprinted from: *Int. J. Mol. Sci.* **2014**, *15*(11), 20169–20208
http://www.mdpi.com/1422-0067/15/11/20169 367

Josephine S. Modica-Napolitano and Volkmar Weissig
Treatment Strategies that Enhance the Efficacy and Selectivity of Mitochondria-
Targeted Anticancer Agents
Reprinted from: *Int. J. Mol. Sci.* **2015**, *16*(8), 17394–17421
http://www.mdpi.com/1422-0067/16/8/17394 411

Xiaonan Zhang, Angelo de Milito, Maria Hägg Olofsson, Joachim Gullbo,
Padraig D'Arcy and Stig Linder
Targeting Mitochondrial Function to Treat Quiescent Tumor Cells in Solid Tumors
Reprinted from: *Int. J. Mol. Sci.* **2015**, *16*(11), 27313–27326
http://www.mdpi.com/1422-0067/16/11/26020 442

Kimitoshi Kohno, Ke-Yong Wang, Mayu Takahashi, Tomoko Kurita,
Yoichiro Yoshida, Masakazu Hirakawa, Yoshikazu Harada, Akihiro Kuma,
Hiroto Izumi and Shinji Matsumoto
Mitochondrial Transcription Factor A and Mitochondrial Genome as Molecular
Targets for Cisplatin-Based Cancer Chemotherapy
Reprinted from: *Int. J. Mol. Sci.* **2015**, *16*(8), 19836–19850
http://www.mdpi.com/1422-0067/16/8/19836 462

Zhicheng Wang, Jie Wang, Rufeng Xie, Ruilai Liu and Yuan Lu
Mitochondria-Derived Reactive Oxygen Species Play an Important Role in
Doxorubicin-Induced Platelet Apoptosis
Reprinted from: *Int. J. Mol. Sci.* **2015**, *16*(5), 11087–11100
http://www.mdpi.com/1422-0067/16/5/11087 478

Ruey-Sheng Wang, Heng-Yu Chang, Shu-Huei Kao, Cheng-Heng Kao, Yi-Chen Wu, Shuyuan Yeh, Chii-Reuy Tzeng and Chawnshang Chang
Abnormal Mitochondrial Function and Impaired Granulosa Cell Differentiation in Androgen Receptor Knockout Mice
Reprinted from: *Int. J. Mol. Sci.* **2015**, *16*(5), 9831–9849
http://www.mdpi.com/1422-0067/16/5/9831 494

List of Contributors

Naghia Ahmed Neurology Unit, IRCCS Foundation Ca' Granda Ospedale Maggiore Policlinico, Dino Ferrari Centre, Department of Pathophysiology and Transplantation, Università degli Studi di Milano, via Francesco Sforza 35, Milan 20122, Italy.

Annarita Aiello Talamanca Istituto Nazionale Tumori Fondazione G. Pascale—Cancer Research Center at Mercogliano (CROM)—IRCCS, Naples I-80131, Italy.

Silvia Alvarez Institute of Biochemistry and Molecular Medicine, School of Pharmacy and Biochemistry, University of Buenos Aires-National Scientific and Technical Research Council (UBA-CONICET), Junín 956, C1113AAD Buenos Aires, Argentina.

Xiuli An College of Life Science, Zhengzhou University, Zhengzhou 450001, China; Laboratory of Membrane Biology, New York Blood Center, New York, NY 10065, USA.

Thierry Arnould Laboratory of Biochemistry and Cell Biology (URBC), Namur Research Institute for Life Sciences (NARILIS), University of Namur, Rue de Bruxelles 61, 5000 Namur, Belgium.

Ana A. Baburamani Centre for the Developing Brain, Division of Imaging Sciences and Biomedical Engineering, King's College London, St. Thomas' Hospital, SE1 7EH London, UK.

Francesca Bianco Department of Biology, University of Naples "Federico II", Napoli I-80126, Italy.

Christoph Bode Department of Cardiology and Angiology, Heart Center Freiburg University, Hugstetter Str. 55, 79106 Freiburg, Germany.

Heiko Bugger Department of Cardiology and Angiology, Heart Center Freiburg University, Hugstetter Str. 55, 79106 Freiburg, Germany.

Giuseppe Castello Istituto Nazionale Tumori Fondazione G. Pascale—Cancer Research Center at Mercogliano (CROM)—IRCCS, Naples I-80131, Italy.

Antonella Cecchettini Department of Clinical and Experimental Medicine, University of Pisa, Via Volta Pisa 56124, Italy.

Francesco S. Celi Division of Endocrinology and Metabolism, Department of Internal Medicine, Virginia Commonwealth University School of Medicine, Richmond, VA 23298, USA.

Fulvio Celsi Institute for Maternal and Child Health "Burlo Garofolo", via dell'Istria 65/1, 34137 Trieste, Italy.

Chawnshang Chang George H. Whipple Lab for Cancer Research, Departments of Pathology, Urology and Radiation Oncology, University of Rochester Medical Center, Rochester, NY 14642, USA.

Heng-Yu Chang Department of Biochemistry and Molecular Cell Biology, College of Medicine, Taipei Medical University, Taipei 110, Taiwan.

Huiyong Chen State Key Laboratory of Medical Genetics & School of Life Sciences, Central South University, Changsha 410078, China.

Paul Cheresh Division of Pulmonary & Critical Care Medicine, Northwestern University Feinberg School of Medicine, Chicago, IL 60611, USA; Department of Medicine, Division of Pulmonary and Critical Care Medicine, Jesse Brown VA Medical Center, Chicago, IL 60612, USA.

María Cecilia Cimolai Institute of Biochemistry and Molecular Medicine, School of Pharmacy and Biochemistry, University of Buenos Aires-National Scientific and Technical Research Council (UBA-CONICET), Junín 956, C1113AAD Buenos Aires, Argentina; Department of Cardiology and Angiology, Heart Center Freiburg University, Hugstetter Str. 55, 79106 Freiburg, Germany.

Lorenzo Citti Consiglio Nazionale delle Ricerche, Institute of Clinical Physiology,Via Giuseppe Moruzzi 1, Pisa 56124, Italy.

Giacomo Pietro Comi Neurology Unit, IRCCS Foundation Ca' Granda Ospedale Maggiore Policlinico, Dino Ferrari Centre, Department of Pathophysiology and Transplantation, Università degli Studi di Milano, via Francesco Sforza 35, Milan 20122, Italy.

Giuseppe Coppotelli Department of Neuroscience, Karolinska Institutet, Retzius väg 8, Stockholm 171 77, Sweden; Department of Genetics, Harvard Medical School, 77 Avenue Louis Pasteur, Boston, MA 02215, USA.

Mario D. Cordero Research Laboratory, Dental School, Universidad de Sevilla, Sevilla 41009, Spain.

Raffaella Crescenzo Department of Biology, University of Naples "Federico II", Napoli I-80126, Italy.

Sergio Crovella Department of Medicine, Surgery and Health Sciences, University of Trieste, Piazzale Europa 1, 34128 Trieste, Italy; Institute for Maternal and Child Health "Burlo Garofolo", via dell'Istria 65/1, 34137 Trieste, Italy.

Padraig D'Arcy Department of Medical and Health Sciences, Linköping University, SE-581 83 Linköping, Sweden; Department of Oncology-Pathology, Karolinska Institute, SE-171 76 Stockholm, Sweden.

Andreas Daiber 2nd Medical Clinic, Medical Center of the Johannes Gutenberg-University, Mainz 55131, Germany.

Angelo de Milito Department of Oncology-Pathology, Karolinska Institute, SE-171 76 Stockholm, Sweden.

Lin Ding Department of Radiation Biology, Beckman Research Institute, City of Hope, Duarte, CA 91010-3000, USA.

Marco d'Ischia Department of Chemical Sciences, University of Naples "Federico II", Naples I-80126, Italy.

Jared S. Farrar Division of Endocrinology and Metabolism, Department of Internal Medicine, Virginia Commonwealth University School of Medicine, Richmond, VA 23298, USA.

Francesca Forini CNR Institute of Clinical Physiology, Via G. Moruzzi 1, Pisa 56124, Italy; Consiglio Nazionale delle Ricerche, Institute of Clinical Physiology, Via Giuseppe Moruzzi 1, Pisa 56124, Italy.

Maria Nicola Gadaleta National Research Council, Institute of Biomembranes and Bioenergetics, Bari I-70126, Italy.

Antonia Giacco Department of Biology, University of Naples "Federico II", Napoli I-80126, Italy.

Pierre Gressens Centre for the Developing Brain, Division of Imaging Sciences and Biomedical Engineering, King's College London, St. Thomas' Hospital, SE1 7EH London, UK; Inserm, U 1141, 75019 Paris, France; University Paris Diderot, Sorbonne Paris Cité, UMRS 1141, 75019 Paris, France.

Joachim Gullbo Department of Immunology, Genetics and Pathology, Section of Oncology, Uppsala University, 751 85 Uppsala, Sweden.

Henrik Hagberg Centre for the Developing Brain, Division of Imaging Sciences and Biomedical Engineering, King's College London, St. Thomas' Hospital, SE1 7EH London, UK; Perinatal Center, Institute for Clinical Sciences and Physiology & Neuroscience, Sahlgrenska Academy, University of Gothenburg, 41685 Gothenburg, Sweden.

Yoshikazu Harada Department of Pathology and Cell Biology, University of Occupational and Environmental Health School of Medicine, Yahatanishi-ku, Kitakyushu-shi 807-8555, Japan.

Masakazu Hirakawa Department of Radiology, Beppu Hospital, Kyushu University, Beppu 874-0838, Japan.

Alan Hoffer Department of Neurological Surgery, Case Western Reserve University, Cleveland, OH 44106, USA.

Qikang Hu State Key Laboratory of Medical Genetics & School of Life Sciences, Central South University, Changsha 410078, China.

Chloe Hurling Centre for the Developing Brain, Division of Imaging Sciences and Biomedical Engineering, King's College London, St. Thomas' Hospital, SE1 7EH London, UK.

Giorgio Iervasi CNR Institute of Clinical Physiology, Via G. Moruzzi 1, Pisa 56124, Italy; Consiglio Nazionale delle Ricerche, Institute of Clinical Physiology, Via Giuseppe Moruzzi 1, Pisa 56124, Italy.

Susanna Iossa Department of Biology, University of Naples "Federico II", Napoli I-80126, Italy.

Hiroto Izumi Department of Occupational Pneumology, Institute of Industrial Ecological Science, University of Occupational and Environmental Health, Yahatanishi-ku, Kitakyushu-shi 807-8555, Japan.

Renea P. Jablonski Department of Medicine, Division of Pulmonary and Critical Care Medicine, Jesse Brown VA Medical Center, Chicago, IL 60612, USA; Division of Pulmonary & Critical Care Medicine, Northwestern University Feinberg School of Medicine, Chicago, IL 60611, USA.

David W. Kamp Department of Medicine, Division of Pulmonary and Critical Care Medicine, Jesse Brown VA Medical Center, Chicago, IL 60612, USA; Division of Pulmonary & Critical Care Medicine, Northwestern University Feinberg School of Medicine, Chicago, IL 60611, USA.

Cheng-Heng Kao Center of General Education, Chang Gung University, Taoyuan 333, Taiwan.

Shu-Huei Kao School of Medical Laboratory Science and Biotechnology, College of Medical Science and Technology, Taipei Medical University, Taipei 110, Taiwan.

Seok-Jo Kim Department of Medicine, Division of Pulmonary and Critical Care Medicine, Jesse Brown VA Medical Center, Chicago, IL 60612, USA; Division of Pulmonary & Critical Care Medicine, Northwestern University Feinberg School of Medicine, Chicago, IL 60611, USA.

Kimitoshi Kohno Asahi Matsumoto Hospital, Kokuramimami-ku Tsuda, Kitakyushu-shi 800-0242, Japan.

Akihiro Kuma Second Department of Internal Medicine, University of Occupational and Environmental Health School of Medicine, Yahatanishi-ku, Kitakyushu-shi 807-8555, Japan.

Tomoko Kurita Department of Gynecology, University of Occupational and Environmental Health School of Medicine, Yahatanishi-ku, Kitakyushu-shi 807-8555, Japan.

Claudia Kusmic Consiglio Nazionale delle Ricerche, Institute of Clinical Physiology, Via Giuseppe Moruzzi 1, Pisa 56124, Italy.

Jiling Liao State Key Laboratory of Medical Genetics & School of Life Sciences, Central South University, Changsha 410078, China.

Stig Linder Department of Medical and Health Sciences, Linköping University, SE-581 83 Linköping, Sweden; Department of Oncology-Pathology, Karolinska Institute, SE-171 76 Stockholm, Sweden.

Jing Liu State Key Laboratory of Medical Genetics & School of Life Sciences, Central South University, Changsha 410078, China.

Ruilai Liu Department of Laboratory Medicine, Huashan Hospital, Shanghai Medical College, Fudan University, Shanghai 200040, China.

Yilun Liu Department of Radiation Biology, Beckman Research Institute, City of Hope, Duarte, CA 91010-3000, USA.

Giovanna Liverini Department of Biology, University of Naples "Federico II", Napoli I-80126, Italy.

Yuan Lu Department of Laboratory Medicine, Huashan Hospital, Shanghai Medical College, Fudan University, Shanghai 200040, China.

Yu Luo Department of Neurological Surgery, Case Western Reserve University, Cleveland, OH 44106, USA.

Shinji Matsumoto Asahi Matsumoto Hospital, Kokuramimami-ku Tsuda, Kitakyushu-shi 800-0242, Japan.

Arianna Mazzoli Department of Biology, University of Naples "Federico II", Napoli I-80126, Italy.

Sébastien Michel Laboratory of Biochemistry and Cell Biology (URBC), Namur Research Institute for Life Sciences (NARILIS), University of Namur, Rue de Bruxelles 61, 5000 Namur, Belgium; Department of Physiology, University of Lausanne, Rue du Bugnon 7, CH-1005 Lausanne, Switzerland.

Yuliya Mikhed 2nd Medical Clinic, Medical Center of the Johannes Gutenberg-University, Mainz 55131, Germany.

Josephine S. Modica-Napolitano Department of Biology, Merrimack College, North Andover, MA 01845, USA.

Giuseppina Nicolini CNR Institute of Clinical Physiology, Via G. Moruzzi 1, Pisa 56124, Italy; Tuscany Region G. Monasterio Foundation, Via G. Moruzzi 1, Pisa 56124, Italy; Consiglio Nazionale delle Ricerche, Institute of Clinical Physiology, Via Giuseppe Moruzzi 1, Pisa 56124, Italy; Fondazione Toscana Gabriele Monasterio, Via Giuseppe Moruzzi 1, Pisa 56124, Italy.

Maria Hägg Olofsson Department of Oncology-Pathology, Karolinska Institute, SE-171 76 Stockholm, Sweden.

Lars Olson Department of Neuroscience, Karolinska Institutet, Retzius väg 8, Stockholm 171 77, Sweden.

Giovanni Pagano Istituto Nazionale Tumori Fondazione G. Pascale—Cancer Research Center at Mercogliano (CROM)—IRCCS, Naples I-80131, Italy.

Federico V. Pallardó CIBERER (Centro de Investigación Biomédica en Red de Enfermedades Raras), University of Valencia—INCLIVA, Valencia 46010, Spain.

Sandra Petrović Vinca" Institute of Nuclear Sciences, University of Belgrade, Belgrade 11001, Serbia.

Xin Qi Department of Physiology, Case Western Reserve University, Cleveland, OH 44106, USA.

Patricia Renard Laboratory of Biochemistry and Cell Biology (URBC), Namur Research Institute for Life Sciences (NARILIS), University of Namur, Rue de Bruxelles 61, 5000 Namur, Belgium.

Silvia Rocchiccioli Consiglio Nazionale delle Ricerche, Institute of Clinical Physiology, Via Giuseppe Moruzzi 1, Pisa 56124, Italy.

Dario Ronchi Neurology Unit, IRCCS Foundation Ca' Granda Ospedale Maggiore Policlinico, Dino Ferrari Centre, Department of Pathophysiology and Transplantation, Università degli Studi di Milano, via Francesco Sforza 35, Milan 20122, Italy.

Jaime M. Ross Department of Neuroscience, Karolinska Institutet, Retzius väg 8, Stockholm 171 77, Sweden; Department of Genetics, Harvard Medical School, 77 Avenue Louis Pasteur, Boston, MA 02215, USA.

Kristina Sobotka Perinatal Center, Institute for Clinical Sciences and Physiology & Neuroscience, Sahlgrenska Academy, University of Gothenburg, 41685 Gothenburg, Sweden.

Sebastian Steven Center for Thrombosis and Hemostasis; 2nd Medical Clinic, Medical Center of the Johannes Gutenberg-University, Mainz 55131, Germany.

Helen Stolp Centre for the Developing Brain, Division of Imaging Sciences and Biomedical Engineering, King's College London, St. Thomas' Hospital, SE1 7EH London, UK.

Mayu Takahashi Department of Neurosurgery, University of Occupational and Environmental Health School of Medicine, Yahatanishi-ku, Kitakyushu-shi 807-8555, Japan.

Claire Thornton Centre for the Developing Brain, Division of Imaging Sciences and Biomedical Engineering, King's College London, St. Thomas' Hospital, SE1 7EH London, UK.

Luca Tiano Biochemistry Unit, Department of Clinical and Dental Sciences, Polytechnical University of Marche, Ancona I-60131, Italy.

Paola Maura Tricarico Department of Medicine, Surgery and Health Sciences, University of Trieste, Piazzale Europa 1, 34128 Trieste, Italy.

Chii-Reuy Tzeng Graduate Institute of Clinical Medicine & Department of Obstetrics and Gynecology, College of Medicine, Taipei Medical University, Taipei 110, Taiwan.

Nadia Ucciferri Consiglio Nazionale delle Ricerche, Institute of Clinical Physiology, Via Giuseppe Moruzzi 1, Pisa 56124, Italy.

Janina A. Vaitkus Division of Endocrinology and Metabolism, Department of Internal Medicine, Virginia Commonwealth University School of Medicine, Richmond, VA 23298, USA.

Jie Wang Department of Laboratory Medicine, Huashan Hospital, Shanghai Medical College, Fudan University, Shanghai 200040, China.

Ke-Yong Wang Shared-Use Research Center, University of Occupational and Environmental Health School of Medicine, Yahatanishi-ku, Kitakyushu-shi 807-8555, Japan.

Ruey-Sheng Wang Graduate Institute of Clinical Medicine & Department of Obstetrics and Gynecology, College of Medicine, Taipei Medical University, Taipei 110, Taiwan.

Zhicheng Wang Department of Laboratory Medicine, Huashan Hospital, Shanghai Medical College, Fudan University, Shanghai 200040, China.

Volkmar Weissig Department of Pharmaceutical Sciences, Midwestern University, College of Pharmacy, Glendale, AZ 85308, USA.

David B. Williams Department of Medicine, Division of Pulmonary and Critical Care Medicine, Jesse Brown VA Medical Center, Chicago, IL 60612, USA; Division of Pulmonary & Critical Care Medicine, Northwestern University Feinberg School of Medicine, Chicago, IL 60611, USA.

Kunlu Wu State Key Laboratory of Medical Genetics & School of Life Sciences, Central South University, Changsha 410078, China.

Yi-Chen Wu Graduate Institute of Clinical Medicine & Department of Obstetrics and Gynecology, College of Medicine, Taipei Medical University, Taipei 110, Taiwan.

Xiaojuan Xiao State Key Laboratory of Medical Genetics & School of Life Sciences, Central South University, Changsha 410078, China.

Rufeng Xie Blood Engineering Laboratory, Shanghai Blood Center, Shanghai 200051, China.

Shuyuan Yeh George H. Whipple Lab for Cancer Research, Departments of Pathology, Urology and Radiation Oncology, University of Rochester Medical Center, Rochester, NY 14642, USA.

Yoichiro Yoshida Department of Gastroenterological Surgery, School of Medicine, Fukuoka University, Fukuoka 814-0180, Japan.

Adriana Zatterale Genetics Unit, Azienda Sanitaria Locale (ASL) Napoli 1 Centro, Naples I-80136, Italy.

Jieying Zhang State Key Laboratory of Medical Genetics & School of Life Sciences, Central South University, Changsha 410078, China; Red Cell Physiology Laboratory, New York Blood Center, New York, NY 10065, USA.

Xiaonan Zhang Department of Medical and Health Sciences, Linköping University, SE-581 83 Linköping, Sweden; Department of Oncology-Pathology, Karolinska Institute, SE-171 76 Stockholm, Sweden.

About the Guest Editors

Jaime M. Ross, Ph.D., a native of Southern Vermont in the USA, is currently a Senior Research Fellow at the Karolinska Institute in Stockholm, Sweden and Harvard Medical School in Boston, Massachusetts. Dr. Ross received her Ph.D. in Neuroscience from the Karolinska Institute as part of the National Institutes of Health–Karolinska Institute Graduate Partnerships Program. Her overall research interests aim to dissect the metabolic consequences and underlying mechanisms of mitochondrial dysfunction that contribute to aging and disease, including various age-related diseases such as neurodegenerative diseases. Dr. Ross uses transgenic mouse models together with cell culture to address outstanding questions in the field, and employs both in vivo and in vitro techniques. Lifestyle interventions, such as diet and exercise, and pharmaceutical treatments to counteract both aging phenotypes and mitochondrial diseases are also being investigated.

Giuseppe Coppotelli, Ph.D., is currently a Senior Research Fellow, affiliated with both the Karolinska Institute in Stockholm, Sweden and Harvard Medical School in Boston, Massachusetts. A native of Italy, Dr. Coppotelli received his Ph.D. in Molecular Endocrinology at the University of Rome "La Sapienza". His research efforts are directed toward dissecting the underlying molecular mechanisms of the aging process and discovering novel markers and treatments that could predict and ameliorate aging as well as age-associated diseases. Dr. Coppotelli aims to determine the cause of the protein homeostasis dysregulation observed in aging, focusing on the ubiquitin proteasome and autophagy lysosome systems. In particular, he is investigating the interconnectedness of protein homeostasis dysregulation with mitochondrial dysfunction, and the effect that cellular redox status, ATP/ADP, and NAD+/NADH levels have on the molecular machinery used for protein degradation. Additional research also focuses on understanding the role of dysfunctional mitochondria and dysregulation of proteostasis with inflammation.

Lars Olson, Ph.D., is a senior professor of neurobiology at the Karolinska Institute in Stockholm, Sweden. His research interests include monoamine neurons, development, nerve growth factors, regeneration, aging, transplantation in the central nervous system, models for Parkinson's disease and its treatment, models for spinal cord injury and treatment strategies, the roles of transcription factors in the nervous system, genetic risk factors for Parkinson's disease, and the Nogo signaling system. Research has been taken all the way from animal studies to clinical trials. Current focus is aging, neuro-degenerative diseases, and the role of the Nogo nerve growth inhibitory signaling system for structural synaptic plasticity and the formation of lasting memories.

Preface to "Mitochondria in Ageing and Diseases: The Super Trouper of the Cell"

Giuseppe Coppotelli and Jaime M. Ross

Reprinted from *Int. J. Mol. Sci.* Cite as: Coppotelli, G.; Ross, J.M. Mitochondria in Ageing and Diseases: The Super Trouper of the Cell. *Int. J. Mol. Sci.* **2016**, *17*, 711.

The past decade has witnessed an explosion of knowledge regarding how mitochondrial dysfunction may translate into ageing and disease phenotypes, as well as how it is modulated by genetic and lifestyle factors. In addition to energy production, mitochondria play an important role in regulating apoptosis, buffering calcium release, retrograde signaling to the nuclear genome, producing reactive oxygen species (ROS), participating in steroid synthesis, signaling to the immune system, as well as controlling the cell cycle and cell growth. Impairment of the mitochondria may be caused by mutations or deletions in nuclear or mitochondrial DNA (mtDNA). Hallmarks of mitochondrial dysfunction include decreased ATP production, decreased mitochondrial membrane potential, swollen mitochondria, damaged cristae, increased oxidative stress, and decreased mitochondrial DNA copy number.

Dysfunctional mitochondria have been implicated in ageing and in several diseases, many of which are age-related, including mitochondrial diseases, cancers, metabolic diseases and diabetes, inflammatory conditions, neuropathy, and neurodegenerative diseases such as Alzheimer's, Parkinson's, and Huntington's disease. Additionally, a possible link between mitochondrial metabolism and the ubiquitin-proteasome and autophagy-lysosome systems is emerging as a novel factor contributing to the progression of several human diseases. The purpose of the Special Issue "Mitochondrial Dysfunction in Ageing and Diseases" published in the *International Journal of Molecular Sciences* [1] was to capture reviews, perspectives, and original research articles to address the progress and current standing in the vast field of mitochondrial biology. A total of 21 papers consisting of 17 reviews and 4 articles have been published as part of the Special Issue as detailed in Table 1. Topics included range from mitochondrial function, cell signaling, and protein homeostasis to disorders and diseases where mitochondrial dysfunction are implicated, such as metabolic diseases, ageing, several age-related diseases, as well as cancer. Various therapies to counteract mitochondrial dysfunction are also discussed.

Table 1. Summary of papers in the Special Issue, arranged by topic as pertaining to mitochondrial dysfunction.

Authors	Title	Topics/Keywords	Type
Ahmed *et al.* [2]	Genes and Pathways Involved in Adult Onset Disorders Featuring Muscle Mitochondrial DNA Instability	mtDNA Maintenance; Mitochondrial Disorders	Review
Arnould *et al.* [3]	Mitochondria Retrograde Signaling and the UPRmt: Where Are We in Mammals?	Unfolded Protein Response; Cell Signaling	Review
Ding *et al.* [4]	Borrowing Nuclear DNA Helicases to Protect Mitochondrial DNA	DNA Replication and Repair; Diseases	Review
Ross *et al.* [5]	Mitochondrial and Ubiquitin Proteasome System Dysfunction in Ageing and Disease: Two Sides of the Same Coin?	Ageing and Age-Related Diseases; Ubiquitin; Proteasome	Review
Crescenzo *et al.* [6]	Skeletal Muscle Mitochondrial Energetic Efficiency and Aging	Ageing; Skeletal Muscle	Review
Tricarico *et al.* [7]	Mevalonate Pathway Blockade, Mitochondrial Dysfunction and Autophagy: A Possible Link	Cholesterol Synthesis; Autophagy; Inflammation	Review
Zhang *et al.* [8]	Autophagy as a Regulatory Component of Erythropoiesis	Autophagy; Erythroid Differentiation	Review
Mikhed *et al.* [9]	Mitochondrial Oxidative Stress, Mitochondrial DNA Damage and Their Role in Age-Related Vascular Dysfunction	Ageing; Cardiovascular Diseases; Oxidative Stress	Review
Vaitkus *et al.* [10]	Thyroid Hormone Mediated Modulation of Energy Expenditure	Thyroid Hormone; Energy Expenditure	Review
Forini *et al.* [11]	Low T3 State Is Correlated with Cardiac Mitochondrial Impairments after Ischemia Reperfusion Injury: Evidence from a Proteomic Approach	Thyroid Hormone; Cardiac Ischemia and Reperfusion; Proteomics	Article
Forini *et al.* [12]	Mitochondria as Key Targets of Cardioprotection in Cardiac Ischemic Disease: Role of Thyroid Hormone Triiodothyronine	Cardiac Ischemia; Thyroid Hormone; Cardioprotection	Review
Cimolai *et al.* [13]	Mitochondrial Mechanisms in Septic Cardiomyopathy	Septic Cardiomyopathy; Mitophagy	Review
Baburamani *et al.* [14]	Mitochondrial Optic Atrophy (OPA) 1 Processing Is Altered in Response to Neonatal Hypoxic-Ischemic Brain Injury	Hypoxia-Ischemia; Brain Injury	Article
Luo *et al.* [15]	Mitochondria: A Therapeutic Target for Parkinson's Disease?	Parkinson's Disease; Mitochondrial Dynamics	Review
Kim *et al.* [16]	The Role of Mitochondrial DNA in Mediating Alveolar Epithelial Cell Apoptosis and Pulmonary Fibrosis	Pulmonary Fibrosis; Sirtuins	Review

Table 1. *Cont.*

Authors	Title	Topics/Keywords	Type
Pagano *et al.* [17]	Current Experience in Testing Mitochondrial Nutrients in Disorders Featuring Oxidative Stress and Mitochondrial Dysfunction: Rational Design of Chemoprevention Trials	Mitochondrial Co-Factors as Therapeutics	Review
Modica-Napolitano *et al.* [18]	Treatment Strategies that Enhance the Efficacy and Selectivity of Mitochondria-Targeted Anticancer Agents	Cancer; Combination Therapy	Review
Zhang *et al.* [19]	Targeting Mitochondrial Function to Treat Quiescent Tumor Cells in Solid Tumors	Cancer; Cancer Therapies	Review
Kohno *et al.* [20]	Mitochondrial Transcription Factor A and Mitochondrial Genome as Molecular Targets for Cisplatin-Based Cancer Chemotherapy	mtTFAM; Cancer Chemotherapy	Review
Wang *et al.* [21]	Mitochondria-Derived Reactive Oxygen Species Play an Important Role in Doxorubicin-Induced Platelet Apoptosis	Chemotherapy Agent (Doxorubicin); Oxidative Stress	Article
Wang *et al.* [22]	Abnormal Mitochondrial Function and Impaired Granulosa Cell Differentiation in Androgen Receptor Knockout Mice	Androgen Receptor; Reproduction	Article

The Special Issue opens with three reviews describing aspects of the mitochondrial machinery in the context of maintaining homeostasis and disease [2–4]. Ding and Liu [4] together with Ahmed *et al.* [2] summarize the maintenance and replication of the mitochondrial genome, and discuss how DNA helicases and mtDNA instability affect integrity of the mtDNA, thus contributing to mitochondrial diseases and disorders. Mitochondrial retrograde signaling, specifically the mitochondrial unfolded protein response, involved in proteostasis is reviewed by Arnould *et al.* [3]. The next five papers review the interconnectedness of mitochondrial dysfunction and protein homeostasis in health, ageing, and diseases [5–9]. Ross *et al.* [5] discuss the interplay of mitochondrial dysfunction and impairment of the ubiquitin proteasome system in ageing and disease, and provide a hypothetical model to address the heterogeneity often described during ageing. The heterogeneity of skeletal muscle performance in ageing is examined by Crescenzo *et al.* [6], taking into account the diverse mitochondrial populations present in skeletal muscle. Tricarico *et al.* [7] focus on a possible link between mitochondrial dysfunction, defective protein prenylation, and the mevalonate pathway, crucial for cholesterol synthesis, with disease. Zhang *et al.* [8] describe under physiological and pathological conditions the modulators of autophagy that regulate erythropoiesis, a process during which mitochondria and other intracellular organelles are removed.

The intersection of mtDNA damage and oxidative stress on age-related vascular dysfunction is presented by Mikhed et al. [9], with particular focus on nicotinamide adenosine dinucleotide phosphate (NADPH) oxidases.

We received several reviews and research articles implicating mitochondria in cardiovascular diseases and ischemia as well as cerebral hypoxia-ischemia [11–14]. Interestingly, a few of these contributions highlight the role of thyroid hormone [10–12]. Vaikus et al. [10] review the diverse effects that thyroid hormone has on mitochondria and energy expenditure, including mitochondrial biogenesis and clinical correlates. Forini et al. [12] discuss possible thyroid hormone triiodothyronine (T3) supplementation to improve mitochondrial function in the context of ischemic heart disease. The same research group also present findings [11] indicating that low T3 levels are correlated with mitochondrial impairments following cardiac ischemia reperfusion injury. Mitochondrial dysfunction is also associated with septic cardiomyopathy, a complication of sepsis, which is a serious condition where the pathogenesis and underlying mechanisms remain unclear, as described by Cimolai et al. [13]. Findings discussed by Babiramani et al. [14] suggest that neonatal cerebral hypoxic-ischemia may alter mitochondrial dynamics, affecting optic atrophy 1 (OPA1). Impaired mitochondrial dynamics have also been described in neurodegenerative diseases, such as Parkinson's disease (PD). Luo et al. [15] review recent literature that support the role of compromised mitochondrial dynamics, mitophagy, and mitochondrial import in PD, and also offer a list of potential therapeutics that target mitochondria. The review by Kim et al. [16] discusses the link between mitochondrial dysfunction and alveolar epithelial cell apoptosis in contributing to age-related lung diseases, as well as how sirtuin family members may constitute therapeutic candidates. Mitochondrial co-factors, such as α-lipoic acid, carnitine, and Coenzyme Q10 have been used to treat mitochondria-associated disorders and diseases, and the results of several clinical trials using these co-factors with and without antioxidants/herbal compounds are systematically presented by Pagano et al. [17].

A few contributions regarding the involvement of mitochondria in cancer and possible therapies were also received [18–21]. Ever since the "Warburg effect" was described nearly a century ago, mitochondria have been increasingly implicated in cancer biology. Extensive research has revealed notable differences between cancerous and healthy cells, such as altered mitochondrial size, shape, metabolic profiles, membrane potential, as well as elevated levels of mtDNA mutations, mitochondrial transcription factor A (TFAM), and oxidative stress [18,20]. Using these findings to exploit mitochondria, several promising anti-cancer treatments have been developed, but have unfortunately proven to have limitations [18–21]. Thus, researchers have recently explored alternative strategies as summarized in Modica-Napolitano and Weissig [18] as well as targeting specific microenvironments within tumors as discussed by Zhang *et al.* [19]. Moreover, the mechanisms by which cancer cells develop drug-resistance are currently being investigated, as reviewed by Kohno *et al.* [20], and possible means to mitigate the side effects of anti-cancer therapies are also being studied by Wang *et al.* [21]. Collectively, these contributions focus on mitochondrial mechanisms as an avenue to reveal possible novel interventions in order to combat cancer.

Lastly, but certainly not of least importance, recent studies of the role of mitochondrial function in fertility and oocyte quality have been extensive. Research by Wang *et al.* [22] demonstrates that androgen receptor knockout mice have poor oocyte maturating rates, impaired ATP production in granulosa cell mitochondria, and impaired mitochondria biogenesis. Additional research is needed to better understand how mitochondrial function may affect fertility and fecundity in order to develop therapeutic approaches.

Overall, the 21 contributions published in the Special Issue illustrate how essential mitochondria are to overall health and success of an organism. The involvement of mitochondria in several biological disciplines, diseases, and disorders, ranging from cancer biology, metabolism, and proteostasis to neurodegenerative and cardiovascular diseases is a testament to their importance and fundamental contributions. We would like to thank all of the authors who contributed their work to the Special Issue. The main objective was to provide ample breadth and depth to depict the interconnectedness of mitochondrial function in ageing and mitochondrial-associated diseases. While the underlying mechanisms linking impaired mitochondria with the ageing process and disease states remain incompletely elucidated, the overall field of mitochondrial biology has made leaps and bounds in only the past two decades. Based on these breakthroughs, new "mito-research" platforms have emerged; for example, mitochondrial function in fertility or in stem cell niches. We remain hopeful that harnessing the power of the mitochondrial network will help us stay healthy.

Acknowledgments: The Foundation for Geriatric Diseases at Karolinska Institutet (G.C.), Karolinska Institutet Research Foundations (G.C.), Loo och Hans Ostermans Foundation for Medical Research (G.C.), the Swedish Society for Medical Research (G.C., J.M.R.), Swedish Brain Power (G.C., J.M.R.), the Swedish Research Council (537-2014-6856; J.M.R.), the Swedish Brain Foundation (J.M.R.), Swedish Lundbeck Foundation (J.M.R.), and Konung Gustaf V:s och Drottning Victorias Frimurarestiftelse (J.M.R.).

Conflicts of Interest: The authors declare no conflict of interest.

References

1. Special issue "mitochondrial dysfunction in ageing and diseases". *Int. J. Mol. Sci.* Available online: http://www.mdpi.com/journal/ijms/special_issues/mitochondrial-dysfunction (accessed on 9 May 2016).

2. Ahmed, N.; Ronchi, D.; Comi, G.P. Genes and pathways involved in adult onset disorders featuring muscle mitochondrial DNA instability. *Int. J. Mol. Sci.* **2015**, *16*, 18054–18076.

3. Arnould, T.; Michel, S.; Renard, P. Mitochondria retrograde signaling and the UPR^mt: Where are we in mammals? *Int. J. Mol. Sci.* **2015**, *16*, 18224–18251.

4. Ding, L.; Liu, Y. Borrowing nuclear DNA helicases to protect mitochondrial DNA. *Int. J. Mol. Sci.* **2015**, *16*, 10870–10887.

5. Ross, J.M.; Olson, L.; Coppotelli, G. Mitochondrial and ubiquitin proteasome system dysfunction in ageing and disease: Two sides of the same coin? *Int. J. Mol. Sci.* **2015**, *16*, 19458–19476.

6. Crescenzo, R.; Bianco, F.; Mazzoli, A.; Giacco, A.; Liverini, G.; Iossa, S. Skeletal muscle mitochondrial energetic efficiency and aging. *Int. J. Mol. Sci.* **2015**, *16*, 10674–10685.

7. Tricarico, P.M.; Crovella, S.; Celsi, F. Mevalonate pathway blockade, mitochondrial dysfunction and autophagy: A possible link. *Int. J. Mol. Sci.* **2015**, *16*, 16067–16084.

8. Zhang, J.; Wu, K.; Xiao, X.; Liao, J.; Hu, Q.; Chen, H.; Liu, J.; An, X. Autophagy as a regulatory component of erythropoiesis. *Int. J. Mol. Sci.* **2015**, *16*, 4083–4094.

9. Mikhed, Y.; Daiber, A.; Steven, S. Mitochondrial oxidative stress, mitochondrial DNA damage and their role in age-related vascular dysfunction. *Int. J. Mol. Sci.* **2015**, *16*, 15918–15953.

10. Vaitkus, J.A.; Farrar, J.S.; Celi, F.S. Thyroid hormone mediated modulation of energy expenditure. *Int. J. Mol. Sci.* **2015**, *16*, 16158–16175.

11. Forini, F.; Ucciferri, N.; Kusmic, C.; Nicolini, G.; Cecchettini, A.; Rocchiccioli, S.; Citti, L.; Iervasi, G. Low T3 state is correlated with cardiac mitochondrial impairments after ischemia reperfusion injury: Evidence from a proteomic approach. *Int. J. Mol. Sci.* **2015**, *16*, 26687–26705.

12. Forini, F.; Nicolini, G.; Iervasi, G. Mitochondria as key targets of cardioprotection in cardiac ischemic disease: Role of thyroid hormone triiodothyronine. *Int. J. Mol. Sci.* **2015**, *16*, 6312–6336.

13. Cimolai, M.C.; Alvarez, S.; Bode, C.; Bugger, H. Mitochondrial mechanisms in septic cardiomyopathy. *Int. J. Mol. Sci.* **2015**, *16*, 17763–17778.

14. Baburamani, A.A.; Hurling, C.; Stolp, H.; Sobotka, K.; Gressens, P.; Hagberg, H.; Thornton, C. Mitochondrial optic atrophy (OPA) 1 processing is altered in response to neonatal hypoxic-ischemic brain injury. *Int. J. Mol. Sci.* **2015**, *16*, 22509–22526.

15. Luo, Y.; Hoffer, A.; Hoffer, B.; Qi, X. Mitochondria: A therapeutic target for Parkinson's disease? *Int. J. Mol. Sci.* **2015**, *16*, 20704–20730.

16. Kim, S.-J.; Cheresh, P.; Jablonski, R.P.; Williams, D.B.; Kamp, D.W. The role of mitochondrial DNA in mediating alveolar epithelial cell apoptosis and pulmonary fibrosis. *Int. J. Mol. Sci.* **2015**, *16*, 21486–21519.

17. Pagano, G.; Aiello Talamanca, A.; Castello, G.; Cordero, M.D.; d'Ischia, M.; Gadaleta, M.N.; Pallardó, F.V.; Petrović, S.; Tiano, L.; Zatterale, A. Current experience in testing mitochondrial nutrients in disorders featuring oxidative stress and mitochondrial dysfunction: Rational design of chemoprevention trials. *Int. J. Mol. Sci.* **2014**, *15*, 20169–20208.

18. Modica-Napolitano, J.S.; Weissig, V. Treatment strategies that enhance the efficacy and selectivity of mitochondria-targeted anticancer agents. *Int. J. Mol. Sci.* **2015**, *16*, 17394–17421.

19. Zhang, X.; de Milito, A.; Olofsson, M.H.; Gullbo, J.; D'Arcy, P.; Linder, S. Targeting mitochondrial function to treat quiescent tumor cells in solid tumors. *Int. J. Mol. Sci.* **2015**, *16*, 27313–27326.

20. Kohno, K.; Wang, K.-Y.; Takahashi, M.; Kurita, T.; Yoshida, Y.; Hirakawa, M.; Harada, Y.; Kuma, A.; Izumi, H.; Matsumoto, S. Mitochondrial transcription factor A and mitochondrial genome as molecular targets for cisplatin-based cancer chemotherapy. *Int. J. Mol. Sci.* **2015**, *16*, 19836–19850.

21. Wang, Z.; Wang, J.; Xie, R.; Liu, R.; Lu, Y. Mitochondria-derived reactive oxygen species play an important role in Doxorubicin-induced platelet apoptosis. *Int. J. Mol. Sci.* **2015**, *16*, 11087–11100.

22. Wang, R.-S.; Chang, H.-Y.; Kao, S.-H.; Kao, C.-H.; Wu, Y.-C.; Yeh, S.; Tzeng, C.-R.; Chang, C. Abnormal mitochondrial function and impaired granulosa cell differentiation in androgen receptor knockout mice. *Int. J. Mol. Sci.* **2015**, *16*, 9831–9849.

Genes and Pathways Involved in Adult Onset Disorders Featuring Muscle Mitochondrial DNA Instability

Naghia Ahmed, Dario Ronchi and Giacomo Pietro Comi

Abstract: Replication and maintenance of mtDNA entirely relies on a set of proteins encoded by the nuclear genome, which include members of the core replicative machinery, proteins involved in the homeostasis of mitochondrial dNTPs pools or deputed to the control of mitochondrial dynamics and morphology. Mutations in their coding genes have been observed in familial and sporadic forms of pediatric and adult-onset clinical phenotypes featuring mtDNA instability. The list of defects involved in these disorders has recently expanded, including mutations in the exo-/endo-nuclease flap-processing proteins MGME1 and DNA2, supporting the notion that an enzymatic DNA repair system actively takes place in mitochondria. The results obtained in the last few years acknowledge the contribution of next-generation sequencing methods in the identification of new disease loci in small groups of patients and even single probands. Although heterogeneous, these genes can be conveniently classified according to the pathway to which they belong. The definition of the molecular and biochemical features of these pathways might be helpful for fundamental knowledge of these disorders, to accelerate genetic diagnosis of patients and the development of rational therapies. In this review, we discuss the molecular findings disclosed in adult patients with muscle pathology hallmarked by mtDNA instability.

Reprinted from *Int. J. Mol. Sci.* Cite as: Ahmed, N.; Ronchi, D.; Comi, G.P. Genes and Pathways Involved in Adult Onset Disorders Featuring Muscle Mitochondrial DNA Instability. *Int. J. Mol. Sci.* **2015**, *16*, 18054–18072.

1. Introduction

Mitochondrial disorders display heterogeneous clinical presentations in terms of age at onset, progression and symptoms. This variability reflects their complex pathogenesis, which might affect structural proteins and enzymes involved in oxidative metabolism, finally leading to the mitochondrial dysfunction observed in multiple cell types and tissues [1].

Serum abnormalities (*i.e.*, increased lactate levels) or evidence of respiratory chain impairment in affected tissues (*i.e.*, isolated or combined complex deficiency) are often indicative of inadequate mitochondrial respiration, although they are not observed in many primary mitochondrial disorders [2]. Moreover, they lack specificity or should be tested in inaccessible tissues, jeopardizing their efficacy in a clinical setting. Therefore, clinical variability and the unavailability of reliable biomarkers delay the diagnosis in early- and late-onset mitochondrial disorders.

In adult non-syndromic clinical presentations, the involvement of skeletal muscle is mainly exhibited by external ophthalmoplegia, isolated or accompanied by limb weakness [3]. In these patients, histological studies on muscle biopsy usually reveal the presence of fibers not reacting to cytochrome c oxidase staining (COX-negative fibers) or showing signs of compensatory mitochondrial proliferation (ragged-red fibers) [4]. Southern blot analysis of muscle-extracted mitochondrial DNA (mtDNA) can detect reduced mtDNA content and multiple deletions, reflecting mtDNA instability [5]. Although subtle mtDNA deletions can accumulate in the muscle of patients harboring mtDNA point mutations [6], they are largely considered hallmarks of adult disorders due to defects in nuclear genes. Therefore, the disclosure of these alterations is helpful to identify a specific subgroup of mitochondrial disorders (multiple mtDNA deletions syndromes), but the large number of imputable genes hampers the chance of a prompt molecular diagnosis, especially in sporadic cases.

In 2000, missense mutations in *SLC25A4* (encoding for the mitochondrial translocator *ANT1*) have been identified as the first molecular defects resulting in impaired mtDNA maintenance in human [7]. Since then, the hunt for molecular defects underlying multiple mtDNA deletion syndromes has demonstrated to be a very active field of research contributing to improving the diagnostic yield of these conditions. At the same time, these studies expanded our knowledge of the proteins involved in the replication and repair of the mitochondrial genome. Some of these proteins were found to have multiple, sometimes redundant activities, while others seem to have unique functions [8].

In this review, we overview the genes hosting mutations linked with adult-onset phenotypes featuring the accumulation of mtDNA deletions in skeletal muscle. The limited number of patients harboring mutations in the most recently-described genes prevents the definition of clear genotype-phenotype correlations, which are still elusive, even for the most frequent genetic defects. However, the systematic collection of clinical and molecular data of diagnosed patients might help to disclose features shared by subjects harboring mutations in the same gene or pathway.

2. Molecular Features in Adult Mitochondrial Disorders Featuring Muscle mtDNA Instability

Disturbances of mtDNA homeostasis result in clinical presentations affecting every stage of life. Infantile and pediatric forms are mostly associated with a strong reduction of mtDNA content in affected tissues; therefore, they are collectively termed mtDNA depletion syndromes [9]. The low residual mtDNA levels cannot sustain a proper respiratory chain assembly: the resulting dysfunction mainly impacts complexes containing mtDNA-encoded subunits and spares complex II, which is entirely encoded by nuclear DNA.

In adults, large-scale deletions of mtDNA spontaneously accumulate in post-mitotic tissues with ageing [6]. The position and extension of such deletions within the mitochondrial genome may be heterogeneous, and in most cases, their accumulation does not result in clinical presentations, although a biochemical defect could be observed by histochemical or enzymatic studies. These slight alterations might also be involved in muscle weakness naturally occurring with ageing. On the opposite side, in primary mitochondrial disorders due to impaired mtDNA maintenance, deletions tend to accumulate in post-mitotic tissues (*i.e.*, muscle) and result in clinical phenotypes, even in relatively young subjects [10]. These alterations are easily detectable on muscle-extracted DNA specimens using Southern blot analysis and standard or quantitative PCR protocols. The co-existence of a quantitative (depletion) and qualitative (deletions) alteration of mtDNA in the same patient is not excluded.

Both pediatric and adult-onset presentations due to mtDNA instability originate from molecular defects in nuclear genes. This is relevant for human health, since such disorders are transmitted according to Mendelian inheritance. Infantile- or juvenile-onset mtDNA depletion syndromes are associated with recessive mutations, resulting in the loss of function of the encoded enzymes. Adult presentations are genetically more heterogeneous, presenting both recessive and dominant types of transmission, these two forms co-existing even for the same gene [11] (Figure 1).

The number of genes involved in this class of disorders has rapidly increased in the last few years, mostly due to the results achieved using next-generation sequencing protocols in familial and sporadic cases. Although the encoded proteins might display multiple catalytic activities and the functional characterization is incomplete for most of them, a (provisional) classification of the genes involved in human mtDNA instability disorders can be attempted (Table 1):

(1)   Genes encoding for key proteins of the core mtDNA replication machinery: *POLG* [12], *POLG2* [13], *PEO1* [14];

(2)   Genes encoding for proteins involved in mtDNA repair and maintenance: *DNA2* [15], *MGME1* [16];

(3)   Genes encoding for proteins preserving the mitochondrial nucleotide pool: *TP* [17], *TK2* [18], *DGUOK* [19], *SLC25A4* [7], *RRM2B* [20], *SUCLA2* [21], *SUCLG1* [22], *ABAT* [23];

(4)   Genes encoding for proteins involved in mitochondrial dynamics and remodelling of mitochondrial membranes: *OPA1* [24], *MFN2* [25], *FBXL4* [26,27].

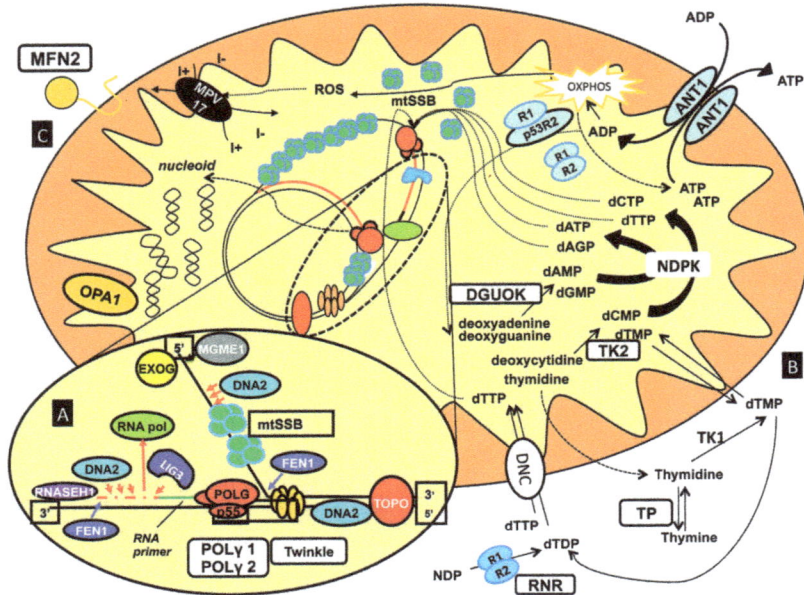

Figure 1. Schematic overview of the proteins and pathways involved in mtDNA maintenance. Zooming in on the mtDNA allows the identification of factors involved in mtDNA replication and repair (**A**) including: POLG, Twinkle, DNA2, MGME1; on the right (**B**) are the proteins assumed to affect the import and metabolism of the mitochondrial dNTP precursors; on the left (**C**) are the supposed localization of key factors ruling mitochondrial dynamics (MFN2, OPA1).

Although mutations of some genes have been so far restricted to infantile (*i.e.*, *FBXL4*) or adult (*i.e.*, *MGME1*) presentations, novel evidence supports the concept of a genetic overlap between these forms, with mutations in the same gene resulting in striking differences in age of onset, tissues affected, progression and outcome. The reason for this heterogeneity is currently unknown. The hypothesis that pediatric presentations could be associated with deleterious mutations, strongly impacting on enzyme activity of the encoded protein, while variants disclosed in adult patients could preserve higher levels of residual activity, is only partially supported by experimental data [28]. The low levels of expression and basal activity of TK2 observed in skeletal muscle might explain the tissue selectivity of TK2 dysfunction, but they are not useful to understand why age of onset and progression are so different between infantile and adult forms [29]. Moreover, the number of pediatric and adult patients sharing the same molecular defects is still low, making genotype-phenotype correlations hazardous or not obvious. It is likely that on-going projects of massive parallel sequencing in novel probands featuring mtDNA instability will further expand this genetic overlap.

Table 1. Genes involved in disorders featuring mitochondrial DNA instability.

Pathway	Gene	Locus	Encoded Protein	Transmission	Onset	mtDNA Defects	Tissues Mainly Affected	Clinical Phenotypes
mtDNA replication	POLG	15q25	DNA polymerase gamma, catalytic subunit	AD, AR	Adult	dels	muscle	PEO, MM
				AR	Infantile	depl	liver	MDS, ME, AS
					Adult	depl	cerebellum	MIRAS
	POLG2	17q	DNA polymerase gamma, accessory subunit	AD	Adult	dels	muscle	PEO
	PEO1	10q24	Twinkle	AD	Adult	dels	muscle	PEO
				AR	Infantile	depl	liver	MDS
					Infantile	depl	brain	IOSCA, ME
mtDNA repair	DNA2	10q21.3–q22.1	DNA replication helicase/nuclease 2	AD	Adult	dels	muscle	PEO, MM
	MGME1	20p11.23	Mitochondrial genome maintenance exonuclease 1	AR	Adult	dels/depl	muscle	PEO, MM
dNTPs pools maintenance	SLC25A4	4q35	Adenine nucleotide translocator	AD	Adult	dels	muscle	PEO
	TYMP	22q13	Thymidine phosphorylase	AR	Late childhood Adolescence	dels/depl	muscle	MNGIE
	TK2	16q22–q23.1	Thymidine kinase 2	AR	Early childhood	depl	muscle	MDS
				AR	Adult	dels	muscle	PEO, MM
	DGUOK	2p13	Deoxyguanosine kinase	AR	Neonatal Infantile	depl	liver/muscle	MDS
				AR	Adult	dels	muscle	PEO
	RRM2B	8q23.1	Ribonucleotide reductase M2 B	AR	Infantile	depl	muscle	MDS
				AR	Adult	depl	muscle	MNGIE
				AD	Adult	dels	muscle	PEO
	SUCLA2	13q12.2–q13.1	Succinyl-CoA ligase, beta subunit	AR	Early childhood	depl	muscle	MDS
	SUCLG1	2p11.2	Succinyl-CoA ligase, alpha subunit	AR	Neonatal Infantile	depl	muscle/liver	MDS
	ABAT	16p13.2	4-aminobutyrate aminotransferase	AR	Infantile	depl	brain/muscle	MDS
Mitochondrial dynamics	OPA1	3q28–q29	Mitochondrial dynamin-like GTPase	AD	Adult	dels	muscle	OA plus
	MFN2	1p36.22	Mitofusin 2	AR	Adult	dels	muscle	OA plus
	MPV17	2p23.2	Mpv17 mitochondrial inner membrane protein	AR	Neonatal Infantile	depl	liver	MDS
				AR	Adult	dels	brain	ME
	FBXL4	6q16.1	F-box and leucine-rich repeat (LRR) protein	AR	Neonatal Infantile	depl	brain/muscle	ME/

Abbreviations: mitochondrial DNA (mtDNA), autosomal dominant (AD), autosomal recessive (AR), multiple mtDNA deletions (dels), mtDNA depletion (depl), progressive external ophthalmoplegia (PEO), mitochondrial myopathy (MM), mtDNA depletion syndrome (MDS), Alpers' syndrome (AS), mitochondrial recessive ataxia syndrome (MIRAS), mitochondrial encephalopathy (ME), optic atrophy (OA), mitochondrial neurogastrointestinal encephalopathy (MNGIE), infantile-onset spinocerebellar ataxia (IOSCA).

The proposed classification does not include all of the genetic defects featuring muscle mtDNA instability. For example, although mutations in *SPG7*, encoding paraplegin, classically result in neurodegenerative diseases, such as hereditary spastic paraplegia [30] and optic neuropathies [31], they have been also described in patients with complex phenotypes showing prominent spastic ataxia with mitochondrial muscle pathology hallmarked by mtDNA deletions [32]. Similarly, mutations in *CHCHD10* have been originally described in familial amyotrophic lateral sclerosis with fronto-temporal dementia [33]. Muscle biopsies of affected subjects revealed respiratory chain dysfunction, COX-negative fibers and a large amount of mtDNA deletions. Following this observation, muscle mitochondrial defects, without signs of mtDNA instability, have been observed in other patients presenting pure motor neuron phenotypes [34,35] or isolated mitochondrial myopathy [36]. The role of CHCHD10 is not completely understood: some evidence indicates that it might take part in mitochondrial dynamics and cristae remodelling [37].

In the following paragraphs, we will focus on genes associated with adult presentations featuring primary accumulation of mtDNA deletions in muscle.

2.1. Genes Encoding for Members of the mtDNA Replication Machinery

The minimum replicative apparatus (replisome) of mtDNA includes the polymerase γ (POLG), the only DNA polymerase active within mitochondria of animal cells, the hexameric helicase Twinkle and the single-strand binding proteins (mtSSBPs). Other enzymatic activities, likely essential for mtDNA replication, include POLRMT, which supplies primers to start replication at the origin of the heavy strand, RNASEH1, which removes primers used during the synthesis of lagging strand, and TOP1MT, a 72-kDa topoisomerase, required to relax negative supercoils [38].

2.1.1. POLG

The mammalian mitochondrial polymerase γ is a 250-kDa heterotrimer composed of a 140-kDa catalytic α subunit (encoded by human *POLG*) and two 55-kDa accessory β subunits (encoded by human *POLG2*) [39]. It is synthesized as a precursor containing an amino-terminal leader sequence, which targets the protein towards mitochondria, where the precursor is cleaved. The catalytic subunit contains a carboxy terminal polymerase domain (*pol*) and an amino terminal 3′–5′ exonuclease domain (*exo*) with proofreading activity, separated by a linker region [40]. The intrinsic exonuclease 3′–5′ function greatly improves the replication fidelity, and it is necessary to suppress mtDNA deletions between direct repetitions [41].

Mutations in α subunits result in pediatric- and juvenile-onset presentations such as Alpers–Huttenlocher syndrome (severe infantile-onset encephalopathy with epilepsy associated with hepatic failure), SANDO (sensory ataxic neuropathy, dysarthria and ophthalmoplegia) and MIRAS (mitochondrial recessive ataxia syndrome) [42]. The analysis of muscle mtDNA reveals reduced mtDNA content and multiple deletions. *POLG* mutations also constitute the most frequent cause of familiar (dominant or recessive) and sporadic chronic external ophthalmoplegia [12]. In recessive forms, peripheral neuropathy is frequent and might occur decades before ptosis; the onset is precocious compared to dominant forms [43]. Beside muscle,

POLG mutations often strike the adult central nervous system with presentations including cognitive impairment, dementia and obsessive disorders [44]. *POLG* defects have been documented in different clinical presentations sharing ataxia [45]. Parkinsonism accompanying progressive external ophthalmoplegia (PEO), ptosis and neuropathy was also found to segregate with *POLG* mutations in a few families [44,46,47]. Therefore, *POLG* defects should be regarded as a secondary genetic cause of Parkinson's disease, with affected patients presenting earlier onset and variable response to levodopa. Few reports also described premature ovarian failure with Parkinsonism and PEO in women harboring *POLG* variants [48,49].

The A467T mutation is the most common pathogenetic substitution in *POLG*, and it is estimated to occur in 36% of the mutated alleles. Its frequency varies between 0.2% and 1% in the general European asymptomatic population [50]. Almost all of the dominant *POLG* mutations associated with PEO are mapped on the polymerase domain of the enzyme. Mutated enzymes compete against wild-type proteins for binding to the replicative fork [51]. Experiments in *S. cerevisiae* displayed a similar behavior between human *POLG* mutations and the corresponding substitutions of the orthologue *mip1*: the severity of yeast phenotypes correlated with the age of onset of human presentations [52].

2.1.2. POLG2

The p55 β (accessory) subunits of polymerase γ are encoded by *POLG2*. They constitute a homodimer that binds asymmetrically to the catalytic portion of the holoenzyme (encoded by *POLG*). The proximal subunit strengthens DNA binding, while the distal subunit facilitates the nucleotide incorporation. The disruption of the interaction between the p55 accessory subunit and the p140 catalytic part might promote the stalling of the replication fork and produce mtDNA deletions. Furthermore, POLG2 has a role in nucleoids' structure [38,53].

Few pathogenetic mutations have been so far identified in *POLG2* (G451E, G416A, c.1207_1208ins24, R369G), the patients presenting autosomal dominant ptosis and PEO with onset in the third to fourth decade, mild proximal muscle weakness with exercise intolerance and other neurologic or systemic symptoms, similarly to *POLG*-mutated PEO patients [13,54,55].

2.1.3. PEO1

Also known as *C10orf2*, *PEO1* encodes for a hexameric helicase of the RecA-type superfamily, named Twinkle. It has structural similarities with the gp4 protein of T7 phage primase/helicase. The linker region, important for subunit interactions and the establishment of the functional hexamer, is known to be a mutational hotspot for dominant PEO [14]. Apart from 5′–3′ helicase activity, Twinkle is essential for the maintenance and the regulation of mtDNA copy number [56].

Mutations that suppress Twinkle helicase activity also compromise mtDNA replication and transcription, since the progression of the replicative fork is hampered, causing the accumulation of replication intermediates [57]. The presence of a cluster of mutations in small highly-conserved regions supports the proposed negative-dominant behavior of Twinkle mutations on mtDNA maintenance and transcription [43]. As a consequence, *PEO1* mutations are inherited according to an autosomal-dominant fashion. Heterozygous mutations might also arise *de novo* in sporadic patients [58]. Fratter and colleagues studied a cohort of 33 mutated patients (26 probands) presenting either missense mutations or in-frame duplications [59].

PEO1-mutated patients show variable onset ranging from the second to the eighth decade. Clinical features might include ptosis, ophthalmoparesis, proximal hyposthenia, ataxia, peripheral neuropathy, bulbar signs, cardiomyopathy, endocrine disorders, cataract and depression or avoiding personality tracts. A syndromic presentation, including sensory neuropathy with ataxia, dysarthria and ophthalmoparesis (SANDO), can also occur [60]. As for *POLG* mutations, *PEO1* defects have been found to segregate with Parkinsonism and additional syndromic features in dominant PEO families [61,62]. Recessive mutations in *PEO1* are less frequent and result in severe pediatric presentations, including IOSCA (infantile-onset spinocerebellar ataxia), mostly occurring in Finnish patients [63], and a hepatocerebral form of mtDNA depletion syndrome [64].

2.2. Genes Encoding for Factors Involved in mtDNA Repair

The accumulation of DNA damage is thought to play a critical role in the aging process, as well as mitochondrial disorders [65]. It was thought for many years that mitochondria lacked an enzymatic DNA repair system comparable to that in the nuclear compartment. However, it is now well established that DNA repair actively takes place in mitochondria, preserving genomic integrity through oxidative DNA damage processing, base excision repair (BER) pathways and further, still uncharacterized, mechanisms [66]. It is likely that the enzymatic activities required for repairing specific mtDNA damage might be also helpful in the processing of replication intermediates, narrowing the functional border between these two groups of proteins involved in mtDNA maintenance. The BER pathway engages the recognition and removal of deaminated, oxidized and alkylated DNA bases. Following the recognition of the damaged base, a group of endo/exonucleases catalyzes the consecutive steps of this orchestrated process. The nuclear BER has been largely explored, as well as the consequences of its dysfunction for human health [67]. Mammalian mitochondria also possess BER activities, promoted by a set of enzymes presenting double localization (nuclear and mitochondrial) or exclusively targeted to the mitochondrial compartment [68]. Interestingly, clinical presentations have been recently associated with altered mtDNA maintenance due to defective mitochondrial endo-/exo-nuclease activities.

2.2.1. DNA2

This gene has been recently considered as a candidate for the molecular screening of familiar or sporadic cases with PEO and/or myopathy accompanied by muscle accumulation of mtDNA deletions [15]. Human *DNA2* encodes for a member of the nuclease/helicase family. It was found in mammalian mitochondria, where it participates in the removal of RNA primers during mtDNA replication [69], but exclusive mitochondrial localization is debated [70]. The encoded protein interacts with polymerase γ and stimulates its catalytic activity [69]. As a member of the nuclease/helicase family, human DNA2 contains conserved nuclease, ATPase and helicase domains.

Experimental studies have demonstrated a role for DNA2 in the mitochondrial long patch base excision repair pathway primed by mtDNA defects due to oxidation, alkylation and hydrolysis. In particular, the nuclease activity is involved in the processing of flap intermediates occurring during the removal of damaged bases in BER. Inside mitochondria, this task is accomplished with the partnership of the flap structure-specific endonuclease 1 (FEN1). DNA2 seems also involved in the processing of Okazaki fragments, as demonstrated in yeast [71]. The function of DNA2 as a DNA helicase is less demonstrated. In yeast, helicase activity facilitates the formation of flap intermediates *in vitro* [72], but it was found to be dispensable *in vivo* [73]. Similarly, the mutations identified in patients and localized within nuclease and ATPase domains also impair helicase activity, as shown by *in vitro* studies [15]. Therefore, the helicase activity of DNA2 is also likely involved in mtDNA maintenance. At present, Twinkle remains the only established replicative helicase inside mitochondria. However, DNA2 might partially support Twinkle, since it co-localizes with Twinkle in nucleoids and its mitochondrial recruitment is induced by *PEO1* mutations [74].

A *DNA2* homozygous out of frame truncating mutation has been identified as a genetic cause of Seckel syndrome [75], a disorder characterized by *in utero* and postnatal growth retardation, intellectual disability, microcephaly, facial dysmorphisms and, rarely, cardiac malformations. Enhanced senescence with a marked increase of damaged nuclear DNA has been observed in patients' fibroblasts, supporting the hypothesis that DNA2 repair activity is affected in this disorder [75]. While these findings acknowledge a nuclear role for human DNA2, the striking differences between pediatric and adult presentations are less clear. The heterozygous mutations disclosed in adult patients impair DNA2 activity less severely with respect to the homozygous truncating mutation linked with Seckel syndrome. Alternative nuclear and mitochondrial factors, replacing defective DNA2 and their differential expression among tissues, might also modulate the phenotype.

2.2.2. MGME1

This gene, previously known as *C20orf72*, encodes a RecB-type exonuclease belonging to the PD-(D/E)XK nuclease superfamily [76]. Cell fractioning experiments showed that the encoded polypeptide is targeted to mitochondria, where it is involved in mtDNA maintenance, promoting the turnover of the replication intermediates 7S. MGME1 cuts DNA, but not RNA or DNA-RNA hybrids; it seems to require free 5'-ends to exert its function (it does not work on circular DNA), and it has a greater affinity for single DNA than double-strand DNA *in vitro*. As in the case of DNA2, MGME1 also interacts with pol γ, therefore contributing a further 5'–3' exonuclease activity to the mitochondrial machinery assembled at the replication fork [77].

Recessive mutations within the *MGME1* coding sequence have been observed in multiple patients from two families and a sporadic patient. Clinical presentation includes external ophthalmoplegia, muscle weakness and a progressive respiratory impairment, requiring assisted ventilation [16].

2.3. Genes Encoding for Proteins Maintaining the Mitochondrial dNTP Pool

As for the nuclear genome, mtDNA also require a balanced pool of dNTPs for an effective replication (and repair). In replicative tissues, cytosolic dNTPs, used in nuclear DNA synthesis, can be also imported into mitochondria through dedicated transporters, sustaining mtDNA maintenance. In non-replicative cells, the cytosolic *de novo* synthesis of dNTPs is downregulated, and the mtDNA replication, which persists in post-mitotic cells, relies on a set of intra-mitochondrial reactions to preserve the proportions of deoxyribonucleotides (dNTPs salvage pathway) [78]. Recessive mutations in key enzymes of this pathway might impair the supply of substrates for the mitochondrial DNA polymerase, resulting in defective mtDNA synthesis or increased replicative errors and genomic instability. In quiescent cells, *de novo* dNTP synthesis still partially supports the replication of mtDNA and the repair of the nuclear genome. Therefore, other proteins involved in nucleoside transport and metabolism might be important for mtDNA maintenance, as observed for the cytosolic ribonucleotide reductase (see below) [79]. Besides the obvious role in the supply of the "building blocks" for DNA synthesis, the deregulation of nucleotide levels might also have more complex consequences, overall conditioning the mitochondrial mass in the cell [80].

2.3.1. *TYMP*

Thymidine phosphorylase (TP, encoded by *TYMP*, also known as *ECGF1*) acts as a homodimer catalyzing the phosphorylation of thymidine phosphate to thymidine and 2-deoxy-D-ribose 2-phosphate. The direct reaction is important for nucleosides catabolism, while the reverse reaction participates in the pyrimidine salvage pathway.

Recessive mutations of this gene have been identified as the cause of mitochondrial neurogastrointestinal encephalomyopathy (MNGIE) [17]. These loss-of-function defects result in the reduction (or substantial absence) of thymidine phosphorylase activity and the toxic accumulation of nucleotides in plasma [17]. The imbalance of the dNTP pool increases the rate of mtDNA point mutations, and both mtDNA depletion and deletions are observed at Southern blot analysis of muscle mtDNA. Clinical manifestations depend on residual TP activity. Since *TP* is a homodimer enzyme, the presence of a mutant allele induces 25% wild-type enzyme synthesis: asymptomatic heterozygotes have 25%–35% of enzymatic residual activity, and this threshold represents a target for therapeutic strategies, like liver transplantation, that results in being be six-fold more efficient as a TP source than bone marrow [81]. Neuroradiological studies in *TYMP*-mutated patients often show leukoencephalopathy [82]. Brain MRI has been suggested as a tool for differential diagnosis of *POLG* involvement in recessive PEO presentations [43].

2.3.2. *TK2*

This gene encodes for the thymidine phosphorylase type 2, the enzyme that catalyze the rate-limiting step of deoxypirimidine phosphorylation within the mitochondrial dNTPs salvage pathway. The gene is upregulated in non-replicating cells, where the encoded product safeguards pool availability.

TK2 deficiency has been classically associated with the myopathic form of mtDNA depletion syndrome, featuring an infantile-onset severe phenotype with motor regression and early death due to diaphragmatic paralysis [18]. Recessive mutations have been also reported in presentations resembling spinal muscular atrophy [83]. Mutations in the same gene have been more recently observed in adult phenotypes with ptosis, external ophthalmoplegia, slowly-progressive proximal muscle weakness, muscular atrophy and dysarthria [84,85]. In adult patients with PEO due to *TK2* mutations, multiple deletions, but not depletion, have been found in muscle. The low basal TK2 activity in muscles could explain why most of the clinical features related to TK2 dysfunction affect this tissue [29]. The reason for the different presentations in pediatric and adult subjects is more elusive: indeed, biochemical studies performed in fibroblasts from adult patients with mild myopathy revealed very low TK2 activity levels, not dissimilar to those observed in pediatric cases.

2.3.3. DGUOK

DGUOK encodes for the mitochondrial deoxyguanosine kinase that catalyzes the first reaction in the purine salvage pathway inside mitochondrial matrix. Recessive mutations have been described in pediatric cases showing severe encephalohepatopathy with severe reduction of liver mtDNA content and premature death, unless the patients undergo a liver transplant [19]. Recently, the massive parallel sequencing of genes encoding for established or predicted mitochondrial proteins (MitoExome) has revealed compound heterozygous mutations even in adult sporadic and familial forms of mitochondrial myopathy [28]. Muscle analysis in mutated patients disclosed mtDNA deletions with variable mtDNA content and significant deficiency of DGUOK protein levels and activity. The main clinical features of adult forms include: muscle weakness, PEO, hearing loss and bulbar signs. One of the probands reported had previously undergone liver transplant. Notably, two siblings harboring *DGUOK* mutations displayed an atypical phenotype characterized by slowly progressive, predominantly distal, upper and lower limb muscle weakness, mild dysphonia and dysphagia, similarly to SMA-like presentations due to TK2 deficiency. These findings clearly demonstrate that the effects of *DGUOK* and *TK2* mutations are not limited to those tissues that are primarily affected in terms of severity and age of involvement.

2.3.4. SLC25A4

This gene encodes for the ANT1 member of the ADP/ATP translocator family, localized in the inner mitochondrial membrane. ANT1 predominates in post-mitotic tissues, including skeletal muscle, heart and brain. It acts as a homodimeric gate-channel operating the ADP/ATP exchange between mitochondria and cytoplasm. Since ANT1 senses the adenosine concentration in the two compartments, it is a part of the signaling pathway coupling cellular energy consumption and mitochondrial respiratory chain activity [86]. Moreover, ANT1 is a structural element of the mitochondrial permeability transition pores (MPTP) and has a role in the intrinsic apoptotic pathway [87].

Heterozygous *SLC25A4* mutations have been reported in patients with PEO and ptosis with or without generalized muscle weakness [7]. Cardiac involvement, including cardiomyopathy, disarrangement of myofibers, inflammation linked to heart disease and ischemic attacks [88], has been repeatedly described. Neurosensory hypoacusia, thyroid gout and dementia without affective aspects have also been reported [43]. Mutations in the same gene also result in Sengers syndrome, a multisystemic disorder featuring congenital cataract, hypertrophic cardiomyopathy, mitochondrial myopathy and lactic acidosis. Two forms are known for Sengers syndrome: a neonatal form with poor prognosis and a late-onset presentation with lifespan observed until the third decade [89].

2.3.5. *RRM2B*

This gene maps on chromosome 8q22.1–q23.3 and encodes the minor subunit (p53R2) of ribonucleotide reductase (RNR), the tetrameric enzyme catalyzing *de novo* dNTPs synthesis from the reduction of the corresponding ribonucleosides diphosphate. *RRM2B* expression is tightly regulated by the oncosuppressor p53. In post-mitotic cells, ribonucleotide reductase contributes to mitochondrial dNTP supply [90], in parallel with the salvage pathway.

Recessive mutations within the *RRM2B* coding sequence have been associated with the infantile encephalomyopathic form of mtDNA depletion syndrome [20]. In 2009, linkage analysis was used to drive the discovery of a heterozygous non-sense *RRM2B* mutation in affected members of a big North American family of European origin presenting adult-onset dominant PEO with muscle accumulation of mtDNA deletions. The same mutation in a second family was associated with additional clinical features, including: ataxic gait, hypoacusia, reduction of tendon reflexes and psychiatric disorders [91]. Kearns–Sayre syndrome (KSS) and MNGIE are rarer manifestations. *RRM2B* mutations were observed as the third most common molecular cause of PEO in a large cohort of British adult patients, following *POLG* and *PEO1* [92]. The mutated or truncated protein p53R2 might compete against wild-type protein for the binding of heterodimeric RNR, exerting a dominant negative effect [91], a pathogenic mechanism compatible with defects disclosed in both pediatric and adult subjects.

2.4. *Genes Encoding for Protein Involved in Mitochondrial Dynamics and Remodeling*

It is now well established that intracellular mitochondria are organized in dynamic networks and physiologically undergo remodelling cycling between fission and fusion events [93]. This behavior is crucial for the turnover of aged mitochondria and their intracellular transport, but also compensates the defects accumulating in the organelle as bypass products of oxidative metabolism [94]. Together with mitochondrial protein quality control system, mitochondrial dynamics have emerged as an important field of investigation to address the pathogenesis of human neurodegeneration. Indeed, increased mitochondrial fragmentation due to the reduced ratio between fusion (*i.e.*, MFN2 and OPA1) and fission (*i.e.*, DLP1/DRP1 and FIS1) proteins has been observed in tissues and cell cultures obtained from patients with major neurodegenerative disorders, including Alzheimer's and Huntington's disease. Furthermore, Parkinson's disease has been associated with mutations in PARK2 and PINK1 proteins, which orchestrate mitochondrial protein quality control and the turnover of aged mitochondria by ubiquitination and degradation of MFN2 [95].

2.4.1. *OPA1*

The product encoded by this gene is a regulatory mitochondrial fusion protein localized in the inner membrane, showing GTPase activity [96]. OPA1 is a main regulator of cristae morphology, but also plays a key role in response to cellular cytotoxic insult generated by hyperactivation of *N*-methyl-D-aspartate (NMDA) receptors for an altered homeostasis of mitochondrial calcium. Indeed, the increased agonism of NMDA receptors induces mitochondrial fragmentation and ultrastructural defects of the inner mitochondrial membrane [97]. Conversely, *OPA1* overexpression is protective against cytotoxic glutamate response, with an increased survival of neurons and preserved mitochondrial morphology [98].

Mutations in *OPA1* are the most common cause of isolated optic atrophy with dominant inheritance (ADOA) [99,100], as well as "plus" phenotypes where optic atrophy is accompanied by syndromic features, such as neurosensory deafness and ophthalmoplegia. Muscle from *OPA1* plus patients showed signs of mitochondrial sufferance, including COX-negative fibers and accumulation of mtDNA deletions [24,101]. OPA1 defects not only impact retinal gangliar cells leading to the progressive degeneration of optic nerve, but also nervous and muscle tissues. Some studies have documented a compensatory increase of mtDNA copies in affected muscle fibers, independently of the severity of the disease [102]. The percentage of COX-negative fibers seems to be four-fold superior in the DOA plus form than in pure presentations.

Recently, *OPA1* sequencing in probands from two Italian families affected by dominant chronic PEO complicated by Parkinsonism and dementia revealed two heterozygous missense mutations affecting highly conserved amino acid positions in the GTPase domain [103]. Multiple mtDNA deletions were detected in available muscle biopsies.

2.4.2. *MFN2*

This gene encodes an outer membrane protein with GTPase activity involved in mitochondrial fusion. Molecular defects affecting *MFN2* are a major cause of the Charcot–Marie–Tooth axonal neuropathy type 2A, an autosomal dominant disease characterized by a pronounced impairment of motor and sensory neurons [104].

Mutations in *MFN2* have been recently disclosed in familial multisystemic disorder with optic atrophy beginning in early childhood, associated with axonal neuropathy and mitochondrial myopathy in adult life, a presentation resembling *OPA1* plus phenotype. Mitochondrial DNA deletions were found in muscle while patients' fibroblasts disclosed defective repair of stress-induced mtDNA damage, leading to respiratory chain impairment [25]. Notably, a detrimental effect on mtDNA replication has been also recently documented in fibroblasts from CMT2A patients, which displayed mtDNA depletion and a minimal amount of deleted molecules [105],

14

suggesting a common pathogenetic mechanism resulting from *MFN2* dysfunction, irrespective of the clinical phenotype.

2.4.3. *MPV17*

This gene encodes for an inner mitochondrial membrane protein whose function has remained obscure for several years. The encoded product is a member of the family of integral membrane proteins comprising PXMP2, MPV17, MP-L and FKSG24 (MPV17L2) in mammals. Studies on vertebrates suggested a role in nucleotide trafficking between mitochondria and cytoplasm [106]. Recently, MPV17 was identified as a non-selective ion channel, responsible for modulation of mitochondrial membrane potential, which is influenced by several conditions in the organelle, such as redox state and protein phosphorylation [107].

Mutations in *MPV17* have been firstly described in a severe juvenile-onset hepatoencephalopathy with major reduction of liver mtDNA content [108]. Coming to adult presentations, recessive *MPV17* mutations have been described in three patients presenting adult-onset multi-systemic disorders with neuropathy and leukoencephalopathy and featuring multiple mtDNA deletions in muscle [109–111].

3. Genotype-Phenotype Correlations

Mitochondrial disorders display a spectacular clinical heterogeneity. Presentations featuring muscle mtDNA instability, which constitute a tiny subgroup of the spectrum of mitochondrial disorders, do not elude this general consideration. Indeed, the overlapping of clinical features is frequent with heterogeneous genetic defects resulting in undistinguishable phenotypes (Figure 2).

Previous studies reporting a cohort of patients accumulating multiple mtDNA deletions in muscle have attempted genotype-phenotype correlations, with modest results, even considering only forms due to mutations in the same gene [43]. The elucidation of such correlations might be useful to speed molecular diagnosis. The number of candidate genes for molecular tests overcomes the resources of standard laboratories, even if the application of next-generation sequencing in a clinical setting is expected to improve the diagnostic yield, reducing the cost of genetic testing [112]. Despite these issues, some remarks emerge from previous studies. Patients harboring *POLG* mutations display severer clinical features and age-related penetrance, while *ANT1* mutations are associated with milder phenotypes. *PEO1* patients show clinical presentations with intermediate severity of symptoms [43]. Bulbar weakness, deafness and gastrointestinal symptoms are observed in *TYMP* [113] and *RRM2B* patients [92], favoring the analysis of these genes before more frequent defects, such as *POLG* and *PEO1*. Cardiac involvement, reflected by conduction defects and left hypertrophy, is more representative of ANT1 involvement rather than *RRM2B* [92].

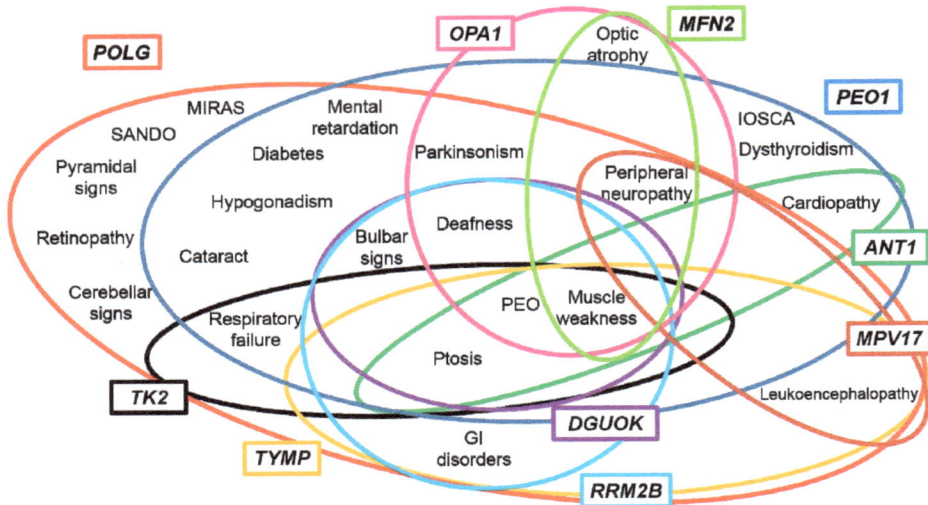

Figure 2. Representation of clinical phenotypes related to nuclear genes involved in mtDNA maintenance. Abbreviations: progressive external ophthalmoplegia (PEO), mitochondrial recessive ataxia syndrome (MIRAS), infantile-onset spinocerebellar ataxia (IOSCA), sensory ataxic neuropathy, dysarthria, and ophthalmoparesis (SANDO).

Peripheral neuropathy, even if not exclusive, is often associated with *POLG* mutations; however, it also occurs in *PEO1* cases [60]. Recently, a diagnostic flow-chart has been suggested for *POLG*-related diseases in pauci-symptomatic patients affected by peripheral neuropathy and suspected mitochondrial impairment [114]. Peripheral neuropathy has been also frequently observed in ADOA plus patients harboring *MFN2* and *OPA1* mutations [115].

4. Towards a Candidate-Pathway Approach

Despite the relevant achievements gained approaching mitochondrial patients with panel or exome sequencing technologies [116,117], the proportion of subjects without a molecular diagnosis is still high. The number of phenotypes reflecting muscle mtDNA instability has increased, while the borders between clinical presentations are getting narrow. As for other Mendelian disorders, the application of next-generation sequencing techniques in a research setting led to the discovery of novel genetic defects [112]. Overall, in only a few years, next-generation sequencing protocols disclosed more genetic defects than any other strategy (linkage studies, gene-candidate approach) in the last two decades. Nevertheless, the number of established diagnosis only slightly increased with a global improvement in the diagnostic yield of around 10%.

In the institutions where panel sequencing is routinely used for diagnostic purposes, the potential reduction of the time-to-diagnosis is challenged by the increased number of variants detected and the consequent efforts required to obtain a reliable genetic and biochemical validation. Taking into account these considerations, a detailed clinical assessment and a more accurate (or extended) genotyping performance might be important, but not decisive advancements. Conversely, the identification of clinical and molecular features shared by patients presenting defects in the same functional pathway, accompanied by quantitative measurements of deleted mtDNA molecules in muscle samples, could be used to drive the selection of genes or pathways to be addressed in molecular testing, improving the diagnosis of single patients in a short timeframe.

Acknowledgments: The financial support of Fondazione Cariplo to Dario Ronchi (Project number: 2014-1010) is gratefully acknowledged.

Conflicts of Interest: The authors declare no conflict of interest.

References

1. Schapira, A.H. Mitochondrial diseases. *Lancet* **2012**, *379*, 1825–1834.
2. Delonlay, P.; Rötig, A.; Sarnat, H.B. Respiratory chain deficiencies. *Handb. Clin. Neurol.* **2013**, *113*, 1651–1666.
3. Yu-Wai-Man, C.; Smith, F.E.; Firbank, M.J.; Guthrie, G.; Guthrie, S.; Gorman, G.S.; Taylor, R.W.; Turnbull, D.M.; Griffiths, P.G.; Blamire, A.M.; *et al.* Extraocular muscle atrophy and central nervous system involvement in chronic progressive external ophthalmoplegia. *PLoS ONE* **2013**, *8*, e75048.
4. Sundaram, C.; Meena, A.K.; Uppin, M.S.; Govindaraj, P.; Vanniarajan, A.; Thangaraj, K.; Kaul, S.; Kekunnaya, R.; Murthy, J.M. Contribution of muscle biopsy and genetics to the diagnosis of chronic progressive external opthalmoplegia of mitochondrial origin. *J. Clin. Neurosci.* **2011**, *18*, 535–538.
5. Wong, L.J.; Boles, R.G. Mitochondrial DNA analysis in clinical laboratory diagnostics. *Clin. Chim. Acta* **2005**, *354*, 1–20.
6. Zhang, C.; Baumer, A.; Maxwell, R.J.; Linnane, A.W.; Nagley, P. Multiple mitochondrial DNA deletions in an elderly human individual. *FEBS* **1992**, *297*, 34–38.
7. Kaukonen, J.; Juselius, J.K.; Tiranti, V.; Kyttälä, A.; Zeviani, M.; Comi, G.P.; Keränen, S.; Peltonen, L.; Suomalainen, A. Role of adenine nucleotide translocator 1 in mtDNA maintenaince. *Science* **2000**, *289*, 782–785.
8. Holt, I.J.; Reyes, A. Human mitochondrial DNA replication. *Cold Spring Harb. Perspect. Biol.* **2012**, *4*.
9. Finsterer, J.; Ahting, U. Mitochondrial depletion syndromes in children and adults. *Can. J. Neurol. Sci.* **2013**, *40*, 635–644.
10. Mitochondrial DNA Deletion Syndromes. Available online: http://www.ncbi.nlm.nih.gov/books/NBK1203/PubMedPMID:20301382 (accessed on 3 May 2011).

11. Spinazzola, A.; Zeviani, M. Mitochondrial diseases: A cross-talk between mitochondrial and nuclear genomes. *Adv. Exp. Med. Biol.* **2009**, *652*, 69–84.

12. Van Goethem, G.; Dermaut, B.; Löfgren, A.; Martin, J.J.; van Broeckhoven, C. Mutation of *POLG* is associated with progressive external ophthalmoplegia characterized by mtDNA deletions. *Nat. Genet.* **2001**, *28*, 211–212.

13. Longley, M.J.; Clark, S.; Yu, W.; Man, C.; Hudson, G.; Durham, S.E.; Taylor, R.W.; Nightingale, S.; Turnbull, D.M.; Copeland, W.C.; *et al.* Mutant *POLG2* disrupts DNA polymerasegamma subunits and causes progressive external ophthalmoplegia. *Am. J. Hum. Genet.* **2006**, *78*, 1026–1034.

14. Spelbrink, J.N.; Li, F.Y.; Tiranti, V.; Nikali, K.; Yuan, Q.P.; Tariq, M.; Wanrooij, S.; Garrido, N.; Comi, G.; Morandi, L.; *et al.* Human mitochondrial DNA deletions associated with mutations in the gene encoding Twinkle, a phage T7 gene 4-like protein localized in mitochondria. *Nat. Genet.* **2001**, *28*, 223–231.

15. Ronchi, D.; di Fonzo, A.; Lin, W.; Bordoni, A.; Liu, C.; Fassone, E.; Pagliarani, S.; Rizzuti, M.; Zheng, L.; Filosto, M.; *et al.* Mutations in *DNA2* link progressive myopathy to mitochondrial DNA instability. *Am. J. Hum. Genet.* **2013**, *92*, 293–300.

16. Kornblum, C.; Nicholls, T.J.; Haack, T.B.; Schöler, S.; Peeva, V.; Danhauser, K.; Hallmann, K.; Zsurka, G.; Rorbach, J.; Iuso, A.; *et al.* Loss-of-function mutations in *MGME1* impair mtDNA replication and cause multisystemic mitochondrial disease. *Nat. Genet.* **2013**, *45*, 214–249.

17. Nishino, I.; Spinazzola, A.; Papadimitriou, A.; Hammans, S.; Steiner, I.; Hahn, C.D.; Connolly, A.M.; Verloes, A.; Guimarães, J.; Maillard, I.; *et al.* Mitochondrial neurogastrointestinal encephalomyopathy: An autosomal recessive disorder due to thymidine phosphorylase mutations. *Ann. Neurol.* **2000**, *47*, 792–800.

18. Saada, A.; Shaag, A.; Mandel, H.; Nevo, Y.; Eriksson, S.; Elpeleg, O. Mutant mitochondrial thymidine kinase in mitochondrial DNA depletion myopathy. *Nat. Genet.* **2001**, *29*, 342–344.

19. Mandel, H.; Szargel, R.; Labay, V.; Elpeleg, O.; Saada, A.; Shalata, A.; Anbinder, Y.; Berkowitz, D.; Hartman, C.; Barak, M.; *et al.* The deoxyguanosine kinase gene is mutated in individuals with depleted hepatocerebral mitochondrial DNA. *Nat. Genet.* **2001**, *29*, 337–341, Erratum in: *Nat. Genet.* 29 December 2001.

20. Bourdon, A.; Minai, L.; Serre, V.; Jais, J.P.; Sarzi, E.; Aubert, S.; Chrétien, D.; de Lonlay, P.; Paquis-Flucklinger, V.; Arakawa, H.; *et al.* Mutation of *RRM2B*, encoding p53-controlled ribonucleotide reductase (p53R2), causes severe mitochondrial DNA depletion. *Nat. Genet.* **2007**, *39*, 776–780.

21. Elpeleg, O.; Miller, C.; Hershkovitz, E.; Bitner-Glindzicz, M.; Bondi-Rubinstein, G.; Rahman, S.; Pagnamenta, A.; Eshhar, S.; Saada, A. Deficiency of the ADP-forming succinyl-CoA synthase activity is associated with encephalomyopathy and mitochondrial DNA depletion. *Am. J. Hum. Genet.* **2005**, *76*, 1081–1086.

22. Ostergaard, E.; Christensen, E.; Kristensen, E.; Mogensen, B.; Duno, M.; Shoubridge, E.A.; Wibrand, F. Deficiency of the alpha subunit of succinate-coenzyme A ligase causes fatal infantile lactic acidosis with mitochondrial DNA depletion. *Am. J. Hum. Genet.* **2007**, *81*, 383–387.

23. Besse, A.; Wu, P.; Bruni, F.; Donti, T.; Graham, B.H.; Craigen, W.J.; McFarland, R.; Moretti, P.; Lalani, S.; Scott, K.L.; *et al.* The GABA transaminase, *ABAT*, is essential for mitochondrial nucleoside metabolism. *Cell Metab.* **2015**, *21*, 417–427.

24. Amati-Bonneau, P.; Valentino, M.L.; Reynier, P.; Gallardo, M.E.; Bornstein, B.; Boissière, A.; Campos, Y.; Rivera, H.; de la Aleja, J.G.; Carroccia, R.; *et al.* OPA1 mutations induce mitochondrial DNA instability and optic atrophy "plus" phenotypes. *Brain* **2008**, *131*, 338–351.

25. Rouzier, C.; Bannwarth, S.; Chaussenot, A.; Chevrollier, A.; Verschueren, A.; Bonello-Palot, N.; Fragaki, K.; Cano, A.; Pouget, J.; Pellissier, J.F.; *et al.* The *MFN2* gene is responsible for mitochondrial DNA instability and optic atrophy "plus" phenotype. *Brain* **2012**, *135*, 23–34.

26. Bonnen, P.E.; Yarham, J.W.; Besse, A.; Wu, P.; Faqeih, E.A.; Al-Asmari, A.M.; Saleh, M.A.; Eyaid, W.; Hadeel, A.; He, L.; *et al.* Mutations in *FBXL4* cause mitochondrial encephalopathy and a disorder of mitochondrial DNA maintenance. *Am. J. Hum. Genet.* **2003**, *93*, 471–481, Erratum in: *Am. J. Hum. Genet.* **2013**, *93*, 773.

27. Gai, X.; Ghezzi, D.; Johnson, M.A.; Biagosch, C.A.; Shamseldin, H.E.; Haack, T.B.; Reyes, A.; Tsukikawa, M.; Sheldon, C.A.; Srinivasan, S.; *et al.* Mutations in *FBXL4*, encoding a mitochondrial protein, cause early-onset mitochondrial encephalomyopathy. *Am. J. Hum. Genet.* **2013**, *93*, 482–495.

28. Ronchi, D.; Garone, C.; Bordoni, A.; Gutierrez Rios, P.; Calvo, S.E.; Ripolone, M.; Ranieri, M.; Rizzuti, M.; Villa, L.; Magri, F.; *et al.* Next-generation sequencing reveals *DGUOK* mutations in adult patients with mitochondrial DNA multiple deletions. *Brain* **2012**, *135*, 3404–3415.

29. Saada, A.; Shaag, A.; Elpeleg, O. mtDNA depletion myopathy: Elucidation of the tissue specificity in the mitochondrial thymidine kinase (*TK2*) deficiency. *Mol. Genet. Metab.* **2003**, *79*, 1–5.

30. Casari, G.; de Fusco, M.; Ciarmatori, S.; Zeviani, M.; Mora, M.; Fernandez, P.; de Michele, G.; Filla, A.; Cocozza, S.; Marconi, R.; *et al.* Spastic paraplegia and OXPHOS impairment caused by mutations in paraplegin, a nuclear-encoded mitochondrial metalloprotease. *Cell* **1998**, *93*, 973–983.

31. Klebe, S.; Depienne, C.; Gerber, S.; Challe, G.; Anheim, M.; Charles, P.; Fedirko, E.; Lejeune, E.; Cottineau, J.; Brusco, A.; *et al.* Spastic paraplegia gene 7 in patients with spasticity and/or optic neuropathy. *Brain* **2012**, *135*, 2980–2993.

32. Pfeffer, G.; Gorman, G.S.; Griffin, H.; Kurzawa-Akanbi, M.; Blakely, E.L.; Wilson, I.; Sitarz, K.; Moore, D.; Murphy, J.L.; Alston, C.L.; *et al.* Mutations in the SPG7 gene cause chronic progressive external ophthalmoplegia through disordered mitochondrial DNA maintenance. *Brain* **2014**, *137*, 1323–1336.

19

33. Bannwarth, S.; Ait-El-Mkadem, S.; Chaussenot, A.; Genin, E.C.; Lacas-Gervais, S.; Fragaki, K.; Berg-Alonso, L.; Yusuke Kageyama.; Serre, V.; Moore, D.G.; *et al.* A mitochondrial origin for frontotemporal dementia and amyotrophic lateral sclerosis through *CHCHD10* involvement. *Brain* **2014**, *137*, 2329–2345.

34. Ronchi, D.; Riboldi, G.; del Bo, R.; Ticozzi, N.; Scarlato, M.; Galimberti, D.; Corti, S.; Silani, V.; Bresolin, N.; Comi, G.P. *CHCHD10* mutations in Italian patients with sporadic amyotrophic lateral sclerosis. *Brain* **2015**.

35. Penttilä, S.; Jokela, M.; Bouquin, H.; Saukkonen, A.M.; Toivanen, J.; Udd, B. Late onset spinal motor neuronopathy is caused by mutation in *CHCHD10*. *Ann. Neurol.* **2015**, *77*, 163–172.

36. Ajroud-Driss, S.; Fecto, F.; Ajroud, K.; Lalani, I.; Calvo, S.E.; Mootha, V.K.; Deng, H.X.; Siddique, N.; Tahmoush, A.J.; Heiman-Patterson, T.D.; *et al.* Mutation in the novel nuclear-encoded mitochondrial protein *CHCHD10* in a family with autosomal dominant mitochondrial myopathy. *Neurogenetics* **2015**, *16*, 1–9.

37. Cogliati, S.; Frezza, C.; Soriano, M.E.; Varanita, T.; Quintana-Cabrera, R.; Corrado, M.; Cipolat, S.; Costa, V.; Casarin, A.; Gomes, L.C.; *et al.* Mitochondrial cristae shape determines respiratory chain supercomplexes assembly and respiratory efficiency. *Cell* **2013**, *155*, 160–171.

38. Wanrooij, S.; Falkenberg, M. The human mitochondrial replication fork in health and disease. *Biochim. Biophys. Acta* **2010**, *1797*, 1378–1388.

39. Spelbrink, J.N.; Toivonen, J.M.; Hakkaart, G.A.; Kurkela, J.M.; Cooper, H.M.; Lehtinen, S.K.; Lecrenier, N.; Back, J.W.; Speijer, D.; Foury, F.; *et al. In vivo* functional analysis of the human mitochondrial DNA polymerase *POLG* expressed in cultured human cells. *J. Biol. Chem.* **2000**, *275*, 24818–24828.

40. Graziewicz, M.A.; Longley, M.J.; Bienstock, R.J.; Zeviani, M.; Copeland, W.C. Structure-function defects of human mitochondrial DNA polymerase in autosomal dominant progressive external ophthalmoplegia. *Nat. Struct. Mol. Biol.* **2004**, *11*, 770–776.

41. Del Bo, R.; Bordoni, A.; Sciacco, M.; di Fonzo, A.; Galbiati, S.; Crimi, M.; Bresolin, N.; Comi, G.P. Remarkable infidelity of polymerase gammaA associated with mutations in *POLG1* exonuclease domain. *Neurology* **2003**, *61*, 903–908.

42. Isohanni, P.; Hakonen, A.H.; Euro, L.; Paetau, I.; Linnankivi, T.; Liukkonen, E.; Wallden, T.; Luostarinen, L.; Valanne, L.; Paetau, A.; *et al. POLG1* manifestations in childhood. *Neurology* **2011**, *76*, 811–815.

43. Van Goethem, G.; Martin, J.J.; van Broeckhoven, C. Progressive External Ophtalmoplegia characterized by multiple deletions of mitochondrial DNA. *Neuromol. Med.* **2003**, *3*, 129–146.

44. Delgado-Alvarado, M.; de la Riva, P.; Jiménez-Urbieta, H.; Gago, B.; Gabilondo, A.; Bornstein, B.; Rodríguez-Oroz, M.C. Parkinsonism, cognitive deficit and behavioural disturbance caused by a novel mutation in the polymerase gamma gene. *J. Neurol. Sci.* **2015**, *350*, 93–97.

45. Synofzik, M.; Srulijes, K.; Godau, J.; Berg, D.; Schöls, L. Characterizing *POLG* ataxia: Clinics, electrophysiology and imaging. *Cerebellum* **2012**, *11*, 1002–1011.

46. Mancuso, M.; Filosto, M.; Oh, S.J.; di Mauro, S. A novel polymerase gamma mutation in a family with ophthalmoplegia, neuropathy, and Parkinsonism. *Arch. Neurol.* **2004**, *61*, 1777–1779.

47. Davidzon, G.; Greene, P.; Mancuso, M.; Klos, K.J.; Ahlskog, J.E.; Hirano, M.; di Mauro, S. Early-onset familial parkinsonism due to *POLG* mutations. *Ann. Neurol.* **2006**, *59*, 859–862.

48. Luoma, P.; Melberg, A.; Rinne, J.O.; Kaukonen, J.A.; Nupponen, N.N.; Chalmers, R.M.; Oldfors, A.; Rautakorpi, I.; Peltonen, L.; Majamaa, K.; *et al.* Parkinsonism, premature menopause, and mitochondrial DNA polymerase gamma mutations: Clinical and molecular genetic study. *Lancet* **2004**, *364*, 875–882.

49. Pagnamenta, A.T.; Taanman, J.W.; Wilson, C.J.; Anderson, N.E.; Marotta, R.; Duncan, A.J.; Bitner-Glindzicz, M.; Taylor, R.W.; Laskowski, A.; Thorburn, D.R.; *et al.* Dominant inheritance of premature ovarian failure associated with mutant mitochondrial DNA polymerase gamma. *Hum. Reprod.* **2006**, *21*, 2467–2473.

50. Neeve, V.C.M.; Samuels, D.C.; Bindoff, L.A.; van den Bosch, B.; van Goethem, G.; Smeet, H.; Lombès, A.; Jardel, C.; Hirano, M.; DiMauro, S.; *et al.* What is influencing the phenotype of the common homozygous polymerase γ mutation p.Ala467Thr? *Brain* **2012**, *135*, 3614–3626.

51. Stuart, G.R.; Santos, J.H.; Strand, M.K.; van Houten, B.; Copeland, W.C. Mitochondrial and nuclear DNA defects in Saccharomyces cerevisiae with mutations in DNA polymerase gamma associated with progressive external ophthalmoplegia. *Hum. Mol. Genet.* **2006**, *15*, 363–374.

52. Stumpf, J.D.; Bailey, C.M.; Spell, D.; Stillwagon, M.; Anderson, K.S.; Copeland, W.C. *mip1* containing mutations associated with mitochondrial disease causes mutagenesis and depletion of mtDNA in Saccharomyces cerevisiae. *Hum. Mol. Genet.* **2010**, *19*, 2123–2133.

53. Di Re, M.; Sembongi, H.; He, J.; Reyes, A.; Yasukawa, T.; Martinsson, P.; Bailey, L.J.; Goffart, S.; Boyd-Kirkup, J.D.; Wong, T.S.; *et al.* The accessory subunit of mitochondrial DNA polymerase gamma determines the DNA content of mitochondrial nucleoids in human cultured cells. *Nucleic Acids Res.* **2009**, *37*, 5701–5713.

54. Walter, M.C.; Czermin, B.; Muller-Ziermann, S.; Bulst, S.; Stewart, J.D.; Hudson, G.; Schneiderat, P.; Abicht, A.; Holinski-Feder, E.; Lochmüller, H.; *et al.* Late-onset ptosis and myopathy in a patient with a heterozygous insertion in *POLG*2. *J. Neurol.* **2010**, *257*, 1517–1523.

55. Young, M.J.; Longley, M.J.; Li, F.Y.; Kasiviswanathan, R.; Wong, L.J.; Copeland, W.C. Biochemical analysis of human *POLG*2 variants associated with mitochondrial disease. *Hum. Mol. Genet.* **2011**, *20*, 3052–3066.

56. Tyynismaa, H.; Sembongi, H.; Bokori-Brown, M.; Granycome, C.; Ashley, N.; Poulton, J.; Jalanko, A.; Spelbrink, J.N.; Holt, I.J.; Suomalainen, A. Twinkle helicase is essential for mtDNA maintenance and regulates mtDNA copy number. *Hum. Mol. Genet.* **2004**, *13*, 3219–3227.

57. Milenkovic, D.; Matic, S.; Kühl, I.; Ruzzenente, B.; Freyer, C.; Jemt, E.; Park, C.B.; Falkenberg, M.; Larsson, N.G. TWINKLE is an essential mitochondrial helicase required for synthesis of nascent D-loop strands and complete mtDNA replication. *Hum. Mol. Genet.* **2013**, *22*, 1983–1993.

58. Virgilio, R.; Ronchi, D.; Hadjigeorgiou, G.M.; Bordoni, A.; Saladino, F.; Moggio, M.; Adobbati, L.; Kafetsouli, D.; Tsironi, E.; Previtali, S.; *et al.* Novel Twinkle (*PEO1*) gene mutations in Mendelian progressive external ophthalmoplegia. *J. Neurol.* **2008**, *255*, 1384–1391.

59. Fratter, C.; Gorman, G.S.; Stewart, J.D.; Buddles, M.; Smith, C.; Evans, J.; Seller, A.; Poulton, J.; Roberts, M.; Hanna, M.G.; *et al.* The clinical, histochemical, and molecular spectrum of *PEO1* (Twinkle)-linked adPEO. *Neurology* **2010**, *74*, 1619–1626.

60. Hanisch, F.; Kornhuber, M.; Alston, C.L.; Taylor, R.W.; Deschauer, M.; Zierz, S. SANDO syndrome in a cohort of 107 patients with CPEO and mitochondrial DNA deletions. *J. Neurol. Neurosurg. Psychiatry* **2014**, *86*, 630–634.

61. Baloh, R.H.; Salavaggione, E.; Milbrandt, J.; Pestronk, A. Familial parkinsonism and ophthalmoplegia from a mutation in the mitochondrial DNA helicase twinkle. *Arch. Neurol.* **2007**, *64*, 998–1000.

62. Kiferle, L.; Orsucci, D.; Mancuso, M.; Lo Gerfo, A.; Petrozzi, L.; Siciliano, G.; Ceravolo, R.; Bonuccelli, U. Twinkle mutation in an Italian family with external progressive ophthalmoplegia and parkinsonism: A case report and an update on the state of art. *Neurosci. Lett.* **2013**, *27*, 1–4.

63. Nikali, K.; Suomalainen, A.; Saharinen, J.; Kuokkanen, M.; Spelbrink, J.N.; Lönnqvist, T.; Peltonen, L. Infantile onset spinocerebellar ataxia is caused by recessive mutations in mitochondrial proteins Twinkle and Twinky. *Hum. Mol. Genet.* **2005**, *14*, 2981–2990.

64. Sarzi, E.; Goffart, S.; Serre, V.; Chrétien, D.; Slama, A.; Munnich, A.; Spelbrink, J.N.; Rötig, A. Twinkle helicase (*PEO1*) gene mutation causes mitochondrial DNA depletion. *Ann. Neurol.* **2007**, *62*, 579–587.

65. Kazachkova, N.; Ramos, A.; Santos, C.; Lima, M. Mitochondrial DNA damage patterns and aging: Revising the evidences for humans and mice. *Aging Dis.* **2013**, *4*, 337–350.

66. Alexeyev, M.; Shokolenko, I.; Wilson, G.; LeDoux, S. The maintenance of mitochondrial DNA integrity—Critical analysis and update. *Cold Spring Harb. Perspect. Biol.* **2013**, *5*.

67. Akbari, M.; Morevati, M.; Croteau, D.; Bohr, V.A. The role of DNA base excision repair in brain homeostasis and diseases. *DNA Repair* **2015**.

68. Prakash, A.; Doublié, S. Base Excision Repair in mitochondria. *J. Cell. Biochem.* **2015**, *116*, 1490–1499.

69. Zheng, L.; Zhou, M.; Guo, Z.; Lu, H.; Qian, L.; Dai, H.; Qiu, J.; Yakubovskaya, E.; Bogenhagen, D.F.; Demple, B.; *et al.* Human *DNA2* is a mitochondrial nuclease/helicase for efficient processing of DNA replication and repair intermediates. *Mol. Cell* **2008**, *32*, 325–336.

70. Duxin, J.P.; Dao, B.; Martinsson, P.; Rajala, N.; Guittat, L.; Campbell, J.L.; Spelbrink, J.N.; Stewart, S.A. Human *DNA2* is a nuclear and mitochondrial DNA maintenance protein. *Mol. Cell Biol.* **2009**, *29*, 4274–4282.

71. Gloor, J.W.; Balakrishnan, L.; Campbell, J.L.; Bambara, R.A. Biochemical analyses indicate that binding and cleavage specificities define the ordered processing of human Okazaki fragments by *DNA2* and FEN1. *Nucleic Acids Res.* **2012**, *40*, 6774–6786.

72. Shim, E.Y.; Chung, W.H.; Nicolette, M.L.; Zhang, Y.; Davis, M.; Zhu, Z.; Paull, T.T.; Ira, G.; Lee, S.E. Saccharomyces cerevisiae Mre11/Rad50/Xrs2 and Ku proteins regulate association of Exo1 and *DNA2* with DNA breaks. *EMBO J.* **2010**, *29*, 3370–3380.

73. Budd, M.E.; Campbell, J.L. *DNA2* is involved in CA strand resection and nascent lagging strand completion at native yeast telomeres. *J. Biol. Chem.* **2013**, *288*, 29414–29429.

74. Ding, L.; Liu, Y. Borrowing nuclear DNA helicases to protect mitochondrial DNA. *Int. J. Mol. Sci.* **2015**, *16*, 10870–10887.

75. Shaheen, R.; Faqeih, E.; Ansari, S. Genomic analysis of primordial dwarfism reveals novel disease genes. *Genome Res.* **2014**, *24*, 291–299.

76. Steczkiewicz, K.; Muszewska, A.; Knizewski, L.; Rychlewski, L.; Ginalski, K. Sequence, structure and functional diversity of PD-(D/E)XK phosphodiesterase superfamily. *Nucleic Acids Res.* **2012**, *40*, 7016–7045.

77. Nicholls, T.J.; Zsurka, G.; Reeva, V.; Schoeler, S.; Szczesny, R.J.; Cysewski, D.; Reyes, A.; Kornblum, C.; Sciacco, M.; Moggio, M.; *et al.* Linear mtDNA fragments and unusual mtDNA rearrangements associate with pathological deficiency of *MGME1* exonuclease. *Hum. Mol. Genet.* **2014**, *23*, 6147–6162.

78. Rampazzo, C.; Ferraro, P.; Pontarin, G.; Fabris, S.; Reichard, P.; Bianchi, V. Mitochondrial deoxyribonucleotides, pool sizes, synthesis, and regulation. *J. Biol. Chem.* **2004**, *279*, 17019–17026.

79. Pontarin, G.; Fijolek, A.; Pizzo, P.; Ferraro, P.; Rampazzo, C.; Pozzan, T.; Thelander, L.; Reichard, P.A.; Bianchi, V. Ribonucleotide reduction is a cytosolic process in mammalian cells independently of DNA damage. *Proc. Natl. Acad. Sci. USA* **2008**, *105*, 17801–17806.

80. Jenuth, J.P.; Peterson, A.C.; Shoubridge, E.A. Tissue-specific selection for different mtDNA genotypes in heteroplasmic mice. *Nat. Genet.* **1997**, *16*, 93–95.

81. Boschetti, E.; D'Alessandro, R.; Bianco, F.; Carelli, V.; Cenacchi, G.; Pinna, A.D.; del Gaudio, M.; Rinaldi, R.; Stanghellini, V.; Pironi, L.; *et al.* Liver as a source of thimidine phosphorylase replacement in mitochondrial neurogastrointestinal encephalomyopathy. *PLoS ONE* **2014**, *9*, e96692.

82. Scarpelli, M.; Ricciardi, G.K.; Beltramello, A.; Zocca, I.; Calabria, F.; Russignan, A.; Zappini, F.; Cotelli, M.S.; Padovani, A.; Tomelleri, G.; *et al.* The role of brain MRI in mitochondrial neurogastrointestinal encephalomyopathy. *Neuroradiol. J.* **2013**, *26*, 520–530.

83. Mancuso, M.; Salviati, L.; Sacconi, S.; Otaegui, D.; Camaño, P.; Marina, A.; Bacman, S.; Moraes, C.T.; Carlo, J.R.; Garcia, M.; *et al.* Mitochondrial DNA depletion: Mutations in thymidine kinase gene with myopathy and SMA. *Neurology* **2002**, *59*, 1197–1202.

84. Tyynismaa, H.; Sun, R.; Ahola-Erkkilä, S.; Almusa, H.; Pöyhönen, R.; Korpela, M.; Honkaniemi, J.; Isohanni, P.; Paetau, A.; Wang, L.; *et al.* Thymidine kinase 2 mutations in autosomal recessive progressive external ophtalmoplegia with multiple mitochondrial DNA deletions. *Hum. Mol. Genet.* **2012**, *21*, 66–75.

85. Alston, C.L.; Schaefer, A.M.; Raman, P.; Solaroli, N.; Krishnan, K.J.; Blakely, E.L.; He, L.; Craig, K.; Roberts, M.; Vyas, A.; *et al.* Late-onset respiratory failure due to *TK2* mutations causing multiple mtDNA deletions. *Neurology* **2013**, *81*, 2051–2053.

86. Liu, Y.; Chen, X.J. Adenine nucleotide translocase, mitochondrial stress, and degenerative cell death. *Oxid. Med. Cell. Longev.* **2013**, *2013*, 146860.

87. Baines, C.P.; Song, C.X.; Zheng, Y.T.; Wang, G.W.; Zhang, J.; Wang, O.L.; Guo, Y.; Bolli, R.; Cardwell, E.M.; Ping, P. Protein kinase Cepsilon interacts with and inhibits thepermeability transition pore in cardiac mitochondria. *Circ. Res.* **2003**, *92*, 873–880.

88. Dörner, A.; Schultheiss, H.P. Adenine nucleotide translocase in the focus of cardiovascular diseases. *Trends Cardiovasc. Med.* **2007**, *17*, 284–290.

89. Jordens, E.Z.; Palmieri, L.; Huizing, M.; van den Heuvel, L.P.; Sengers, R.C.; Dörner, A.; Ruitenbeek, W.; Trijbels, F.J.; Valsson, J.; Sigfusson, G.; *et al.* Adenine nucleotide translocator 1 deficiency associated with Sengers syndrome. *Ann. Neurol.* **2002**, *52*, 95–99.

90. Tanaka, H.; Arakawa, H.; Yamaguchi, T.; Shiraishi, K.; Fukuda, S.; Matsui, K.; Takei, Y.; Nakamura, Y. A ribonucleotide reductase gene involved in a p53-dependent cell-cycle checkpoint for DNA damage. *Nature* **2000**, *404*, 42–49.

91. Tyynismaa, H.; Ylikallio, E.; Patel, M.; Molnar, M.J.; Haller, R.G.; Suomalainen, A. A heterozygous truncating mutation in *RRM2B* causes autosomal-dominant progressive external ophtalmoplegia with multiple mtDNA deletions. *Am. J. Hum. Genet.* **2009**, *85*, 290–295.

92. Pitceathly, R.D.S.; Smith, C.; Fratter, C.; Alston, C.L.; He, L.; Craig, K.; Blakely, E.L.; Evans, J.C.; Taylor, J.; Shabbir, Z.; *et al.* Adults with *RRM2B*-related mitochondrial disease have distinct clinical and molecular characteristics. *Brain* **2012**, *135*, 3392–3403.

93. Ranieri, M.; Brajkovic, S.; Riboldi, G.; Ronchi, D.; Rizzo, F.; Bresolin, N.; Corti, S.; Comi, G.P. Mitochondrial fusion proteins and human diseases. *Neurol. Res. Int.* **2013**.

94. Mishra, P.; Chan, D.C. Mitochondrial dynamics and inheritance during cell division, development and disease. *Nat. Rev. Mol. Cell Biol.* **2014**, *15*, 634–646.

95. Itoh, K.; Nakamura, K.; Iijima, M.; Sesaki, H. Mitochondrial dynamics in neurodegeneration. *Trends Cell Biol.* **2013**, *23*, 64–71.

96. Belenguer, P.; Pellegrini, L. The dynamin GTPase *OPA1*, more than mitochondria? *Biochim. Biophys. Acta* **2013**, *1833*, 176–183.

97. Jahani-Asl, A.; Pilon-Larose, K.; Xu, W.; MacLaurin, J.G.; Park, D.S.; McBride, H.M.; Slack, R.S. The mitochondrial inner membrane GTPase, optic atrophy 1 (*OPA1*), restores mitochondrial morphology and promotes neuronal survival following excitotoxicity. *J. Biol. Chem.* **2011**, *286*, 4772–4782.

98. Nguyen, D.; Alavi, M.V.; Kim, K.Y.; Kang, T.; Scott, R.T.; Noh, Y.H.; Lindsey, J.D.; Wissinger, B.; Ellisman, M.H.; Weinreb, R.N.; *et al.* A new vicious cycle involvingglutamate excitotoxicity, oxidative stress and mitochondrial dynamics. *Cell Death Dis.* **2011**, *2*.

99. Alexander, C.; Votruba, M.; Pesch, U.E.; Thiselton, D.L.; Mayer, S.; Moore, A.; Rodriguez, M.; Kellner, U.; Leo-Kottler, B.; Auburger, G.; *et al.* OPA1, encoding a dynamin-related GTPase, is mutated in autosomal dominant optic atrophy linked to chromosome 3q28. *Nat. Genet.* **2000**, *26*, 211–215.

100. Delettre, C.; Lenaers, G.; Griffoin, J.M.; Gigarel, N.; Lorenzo, C.; Belenguer, P.; Pelloquin, L.; Grosgeorge, J.; Turc-Carel, C.; Perret, E.; *et al.* Nuclear gene OPA1, encoding a mitochondrial dynamin-related protein, is mutated in dominant optic atrophy. *Nat. Genet.* **2000**, *26*, 207–210.

101. Hudson, G.; Amati-Bonneau, P.; Blakely, E.L.; Stewart, J.D.; He, L.; Schaefer, A.M.; Griffiths, P.G.; Ahlqvist, K.; Suomalainen, A.; Reynier, P.; *et al.* Mutation of *OPA1* causes dominant optic atrophy with external ophthalmoplegia, ataxia, deafness and multiple mitochondrial DNA deletions: A novel disorder of mtDNA maintenance. *Brain* **2008**, *131*, 329–337.

102. Zanna, C.; Ghelli, A.; Porcelli, A.M.; Karbowski, M.; Youle, R.J.; Schimpf, S.; Wissinger, B.; Pinti, M.; Cossarizza, A.; Vidoni, S.; *et al.* OPA1 mutations associated with dominant optic atrophy impair oxidative phosphorylation and mitochondrial fusion. *Brain* **2008**, *131*, 352–367.

103. Carelli, V.; Musumeci, O.; Caporali, L.; Zanna, C.; la Morgia, C.; del Dotto, V.; Porcelli, A.M.; Rugolo, M.; Valentino, M.L.; Iommarini, L.; *et al.* Syndromic parkinsonism and dementia associated with *OPA1* missense mutations. *Ann. Neurol.* **2015**.

104. Evgrafov, O.V.; Mersiyanova, I.; Irobi., J.; van den Bosch, L.; Dierick, I.; Leung, C.L.; Schagina, O.; Verpoorten, N.; van Impe, K.; Fedotov, V.; *et al.* Mutant small heat-shock protein 27 causes axonal Charcot-Marie-Tooth disease and distal hereditary motor neuropathy. *Nat. Genet.* **2004**, *36*, 602–606.

105. Vielhaber, S.; Debska-Vielhaber, G.; Peeva, V.; Schoeler, S.; Kudin, A.P.; Minin, I.; Schreiber, S.; Dengler, R.; Kollewe, K.; Zuschratter, W.; *et al.* Mitofusin 2 mutations affect mitochondrial function by mitochondrial DNA depletion. *Acta Neuropathol.* **2013**, *125*, 245–256.

106. Löllgen, S.; Weiher, H. The role of Mpv17 protein mutations of which cause mitochondrial DNA depletion syndromes (MDDS): Lessons from homologs in different species. *Biol. Chem.* **2015**, *396*, 13–25.

107. Antonenkov, V.D.; Isomursu, A.; Mennerich, D.; Vapola, M.H.; Weiher, H.; Kietzmann, T.; Hiltunen, J.K. The human mtDNA depletion syndrome gene MPV17 encodes a non-selective channel that modulates membrane potential. *J. Biol. Chem.* **2015**, *290*, 13840–13861.

108. Spinazzola, A.; Viscomi, C.; Fernandez-Vizarra, E.; Carrara, F.; D'Adamo, P.; Calvo, S.; Marsano, R.M.; Donnini, C.; Weiher, H.; Strisciuglio, P.; *et al.* MPV17 encodes an inner mitochondrial membrane protein and is mutated in infantile hepatic mitochondrial DNA depletion. *Nat. Genet.* **2006**, *38*, 570–576.

109. Garone, C.; Rubio, J.C.; Calvo, S.E.; Naini, A.; Tanji, K.; di Mauro, S.; Mootha, V.K.; Hirano, M. MPV17 mutations causing adult-onset multisystemic disorder with multiple mitochondrial DNA deletions. *Arch. Neurol.* **2012**, *69*, 1649–1651.

110. Blakely, E.L.; Butterworth, A.; Hadden, R.D.M.; Bodi, I.; Landping, H.; McFarland, R.; Taylor, R.W. MPV17 mutation causes neuropathy and leukoencephalopathy with multiple mtDNA deletions in muscle. *Neuromuscol. Disord.* **2012**, *22*, 587–591.

111. Mendelsohn, B.A.; Mehta, N.; Hameed, B.; Pekmezci, M.; Packman, S.; Ralph, J. Adul Onset Fatal Neurohepatopathy in a Woman Caused by MPV17 Mutation. *JIMD Rep.* **2014**, *13*, 37–41.

112. Wortmann, S.B.; Koolen, D.A.; Smeitink, J.A.; van den Heuvel, L.; Rodenburg, R.J. Whole exome sequencing of suspected mitochondrial patients in clinical practice. *J. Inherit. Metab. Dis.* **2015**, *38*, 437–443.

113. Garone, C.; Tadesse, S.; Hirano, M. Clinical and genetic spectrum of mitochondrial neurogastrointestinal encephalomyopathy. *Brain* **2011**, *134*, 3326–3332.

114. Tchikviladzé, M.; Gilleron, M.; Maisonobe, T.; Galanaud, D.; Laforêt, P.; Durr, A.; Eymard, B.; Mochel, F.; Ogier, H.; Béhin, A.; *et al.* A diagnostic flow chart for *POLG*-related diseases based on signs sensitivity and specificity. *J. Neurol. Neurosurg. Psychiatry* **2015**, *86*, 646–654.

115. Pareyson, D.; Saveri, P.; Sagnelli, A.; Piscosquito, G. Mitochondrial dynamics and inherited peripheral nerve diseases. *Neurosci. Lett.* **2015**, *596*, 66–77.

116. Lieber, D.S.; Calvo, S.E.; Shanahan, K.; Slate, N.G.; Liu, S.; Hershman, S.G.; Gold, N.B.; Chapman, B.A.; Thorburn, D.R.; Berry, G.T.; *et al.* Targeted exome sequencing of suspected mitochondrial disorders. *Neurology* **2013**, *80*, 1762–1770.

117. Calvo, S.E.; Compton, A.G.; Hershman, S.G.; Lim, S.C.; Lieber, D.S.; Tucker, E.J.; Laskowski, A.; Garone, C.; Liu, S.; Jaffe, D.B.; *et al.* Molecular diagnosis of infantile mitochondrial disease with targeted next-generation sequencing. *Sci. Transl. Med.* **2012**, *4*.

Mitochondria Retrograde Signaling and the UPR^mt: Where Are We in Mammals?

Thierry Arnould, Sébastien Michel and Patricia Renard

Abstract: Mitochondrial unfolded protein response is a form of retrograde signaling that contributes to ensuring the maintenance of quality control of mitochondria, allowing functional integrity of the mitochondrial proteome. When misfolded proteins or unassembled complexes accumulate beyond the folding capacity, it leads to alteration of proteostasis, damages, and organelle/cell dysfunction. Extensively studied for the ER, it was recently reported that this kind of signaling for mitochondrion would also be able to communicate with the nucleus in response to impaired proteostasis. The mitochondrial unfolded protein response (UPR^mt) is activated in response to different types and levels of stress, especially in conditions where unfolded or misfolded mitochondrial proteins accumulate and aggregate. A specific UPR^mt could thus be initiated to boost folding and degradation capacity in response to unfolded and aggregated protein accumulation. Although first described in mammals, the UPR^mt was mainly studied in *Caenorhabditis elegans*, and accumulating evidence suggests that mechanisms triggered in response to a UPR^mt might be different in *C. elegans* and mammals. In this review, we discuss and integrate recent data from the literature to address whether the UPR^mt is relevant to mitochondrial homeostasis in mammals and to analyze the putative role of integrated stress response (ISR) activation in response to the inhibition of mtDNA expression and/or accumulation of mitochondrial mis/unfolded proteins.

Reprinted from *Int. J. Mol. Sci.* Cite as: Arnould, T.; Michel, S.; Renard, P. Mitochondria Retrograde Signaling and the UPR^mt: Where Are We in Mammals? *Int. J. Mol. Sci.* **2015**, *16*, 18224–18247.

1. Introduction

Mitochondrial, multifunctional, and dynamic organelles (resulting from fusion and fission events of mitochondria fragments, according to cell type and conditions) are linked to pathologies far beyond the *sensu stricto* "mitochondrial diseases" because they regulate metabolism via the Krebs cycle and oxidative phosphorylation (OXPHOS). Linked to their bioenergetics, mitochondria that contain more than 1000 different proteins/peptides, synthesized from both mitochondrial and nuclear genomes in a coordinated manner [1,2] also participate in synthesis (steroids, amino acids, nucleotides), the biogenesis of iron-sulfur centers, calcium homeostasis and redox status [3], control epigenetics marks in the nuclear genome, and constitute an integration platform for signaling and cell death/survival signals, connecting the organelle to apoptosis and autophagy [4–7]. The continued organelle maintenance can be seen as a balance between the biogenesis of the organelle and the mechanisms that provide quality control (involved in the remodeling and mitophagy) that guarantees cell homeostasis and function [8,9]. This function is maintained by the participation of several chaperones, antioxidant enzymes (SOD2, PRDX3, PRDX5, GPX1: human nomenclature according to Uniprot), and quality control proteases that promote protein folding and stability on the mitochondria while performing the degradation of un- or mis-folded proteins that accumulate [10]. Essential for this control is the molecular communication between the mitochondria and nucleus, in which ATP, calcium, and reactive oxygen species have been described to play major roles [1,11–14]. Indeed, upon organelle dysfunction, which can be caused by many events, such as mtDNA depletion, mutations, deletions, oxidative stress, aggregation of misfolded proteins, or dramatic changes in morphology and dynamics, a mitochondrial-to-nucleus communication, known as "retrograde signaling," triggers an orchestrated expression of nuclear genes in an attempt to relieve/resolve the stress and/or to compensate the defect [15]. The importance of these stress response pathways can be highlighted by the various pathophysiological developments associated with their impairment [15]. The aim of the current review is to highlight recent developments in the role of mitochondrial unfolded protein response (UPR^{mt}) in the field of mitochondrial dysfunction and putative connection with other retrograde signaling pathways, with a particular focus on differences between mechanisms retrieved in mammals and other organisms.

1.1. Mitochondrial Retrograde Responses

Several signaling pathways have been described in the mitochondria-to-nucleus communication observed in response to organelle stress and dysfunction. This mitochondrial stress response can be perceived as an attempt to compensate for the metabolic defect by stimulating several biological processes including mitochondria biogenesis. The transcriptional regulation of mitochondrial biogenesis is mediated by a set of transcription factors such as NRF1 and NRF2/GA-binding protein subunit β-1 (nuclear respiratory factors 1 and 2), cyclicAMP-responsive element binding protein1 (CREB1), steroid hormone receptor ERR1 (ERRα), and PPARG (PPARγ). The activity of these transcription factors is coordinated by members of the PGC-1 co-activator family [16]. The retrograde communication often relies on some of these transcriptional regulators to modulate gene expression to adapt mitochondria function in response to the original cue and promote recovery and function to resolve the stress. We now discuss the signaling pathways that regulate the expression, localization, and activity of these transcription factors. Notably, if the following pathways are usually associated with the regulation of mitochondrial protein abundance, lipid content might also be regulated and controlled by retrograde signaling. As with the absence of mtDNA in HeLa cells, mitochondrial DNA absence sensitive (MIDAS) factor increases mitochondrial mass by regulating the expression of enzymes involved in cardiolipin synthesis [17].

Retrograde communication in response to mitochondrial stress comes from seminal studies from Butow's group in the budding yeast *Saccharomyces cerevisiae* depleted of mtDNA [18,19], which showed that electron transport chain (ETC) deficiency leads to the transcription of genes associated with glutamate metabolism (reviewed in [20]). In yeast, there are three retrograde response genes (RTG): Rtg1p and Rtg3p are transcription factors, forming a dimer that translocates from the cytosol to the nucleus to regulate gene expression, while Rtg2p would act/behave as a sensor of mitochondrial stress [21]. Although mammalian orthologs of these proteins have not been found, similar signaling pathways do exist in mammals, as detailed in the next section.

1.1.1. Retrograde Signaling Involving Increased Cytosolic Calcium Concentration

Pioneering works on retrograde signaling in mammalian cells have been described in mtDNA-depleted cells through the pharmacological inhibition of ETC complexes or disruption of the mitochondrial membrane potential and focused on calcium in mouse C2C12 myocytes and human pulmonary carcinoma A549 cells. Avadhani's group showed that mitochondrial dysfunction associated with membrane depolarization leads to increased cytosolic calcium concentration and triggers nuclear factor of activated T-cells (NF-AT) and activating transcription factor 2 (ATF2) translocation to the nucleus in a calcineurin (Cn)- and Mitogen-activated protein kinase 8 (MAPK8)/JNK1-dependent manner, respectively [11,22]. The list of transcriptional regulators activated by calcium, either by a phosphorylation in a JNK1- or calcium/calmodulin-dependent protein kinase type IV (CaMKIV)-dependent manner, or by a dephosphorylation mediated by calcineurin in response to mitochondrial dysfunction has now expanded and includes the transcription factors CREB1, nuclear factor of κ light polypeptide gene enhancer in B-cells 1 (NF-κB), cellular tumor antigen p53 (p53), myocyte-specific enhancer factor 2A (MEF-2), and co-activators peroxisome proliferator-activated receptor γ coactivator 1-α (PGC-1α) and heterogenous nuclear ribonucleoprotein A2 (hnRNPA2) [13,22–25]. Altogether, data about calcium-dependent mitochondrial retrograde response represent a good example of functional compensation after organelle stress. Overexpressed nuclear genes include proteins associated with calcium homeostasis, such as calsequestrin and calreticulin, OXPHOS subunits for stabilizing energy production, and chloride intracellular channel protein 4 (Clic 4), which contribute to mitochondrial membrane potential [26–28].

1.1.2. Retrograde Response Associated with ROS Production and Signaling

Reactive oxygen species (ROS) are by-products of ETC complexes, and while they represent important physiological second messengers, their excessive production can lead to protein, lipid, and DNA damages and, ultimately, cell death. When antioxidant defenses are overwhelmed, a transcriptional program regulated by the NRF2/GA-binding protein subunit β-1 is initiated [29]. Indeed, excessive production of ROS triggers the stabilization and translocation of the factor to the nucleus, where the transcription of target genes containing an antioxidant-response element (ARE) in their promoter is increased, including NRF1, which regulates mitochondrial gene expression [29,30]. It has also been suggested that NRF1 might be directly regulated by redox signaling in phosphatidylinositol 3-kinase (PI3K) and RAC-α serine/threonine-protein kinase/Akt-dependent manners to regulate mtDNA transcription and replication by increasing mitochondrial transcription factor A (TFAM) protein expression [31]. PGC-1α is also activated by AMP-activated protein kinase (AMPK) in response to mitochondrial ROS to promote organelle biogenesis [32]. As for calcium, ROS have been the focus of numerous studies on retrograde communication, and additional transcription factors, such as NF-κB, p53, and AP-1, are activated by oxidants [33–35]. Surprisingly, while several studies have demonstrated enhanced biogenesis in response to mitochondrial stress [33,36–38], recent biological analysis of retrograde signaling by Chae and collaborators demonstrated the role of ROS in the down-regulation of OXPHOS enzymes [39]. These authors reported that, among the 72 transcription factors, for which activity was modified in response to mtDNA mutation, the abundance of RXRα Retinoic acid receptor RXR α) decreased after ROS-dependent JNK1 activation. As a direct consequence, the interaction between RXRα and PGC-1α was reduced, which attenuated the expression of gene encoding of OXPHOS enzymes as well as mitochondrial ribosomal proteins—two phenomena that further exacerbate mitochondrial dysfunction [39].

1.1.3. Energy Deprivation and Retrograde Response

The most obvious retrograde signaling might be the stress response associated with energy deprivation. AMPK is a sentinel activated in response to an increase in the AMP/ATP ratio, which triggers allosteric activation and the serine/threonine-protein kinase STK11 (LKB1)-dependent Thr172 phosphorylation of AMPK. Once activated, AMPK increases ATP production by inducing mitochondrial biogenesis, thus stimulating fatty acid β-oxidation and glycolysis, and shuts down energy consuming processes such as cell growth and lipid and protein synthesis [40–43]. AMPK regulates PGC-1α activity and abundance as well as its binding activity to transcription factors, such as NRF1, at the promoter of nuclear gene encoding mitochondrial proteins [36,42]. Moreover, AMPK is also well-known for regulating the serine/threonine-protein kinase mammalian target of rapamycin (mTOR) through control of tuberous sclerosis 2 (Tuberin) activity [44]. In mammals, two complexes have been described: mTORC1 and mTORC2. While mTORC1 regulates processes, such as transcription, translation, autophagy, and metabolism, mTORC2 is mainly associated with cell survival, proliferation, and metabolism. The activities of mTOR complexes are negatively regulated by Hamartin/Tuberin (tuberous sclerosis 1/2) via phosphorylation. In response to mitochondrial stress and energy deprivation, AMPK decreases mTOR activity and downstream processes, such as protein synthesis, by a mechanism involving either direct phosphorylation of the raptor or the activating phosphorylation of Tuberin [41,43,44].

Finally, the $NAD^+/NADH$ ratio and acetyl-coenzyme A (acetyl-CoA) are important mitochondrial coenzymes and metabolites, respectively, which are associated with retrograde response [45]. However, the reduction of NAD^+ into NADH during the Krebs cycle and its oxidation by the NADH dehydrogenase complex represent the basis of mitochondrial respiration. Because NAD^+ is also a substrate used by enzymes, such as poly[ADP-ribose] polymerase-1 (PARP-1) and Sirtuins, the $NAD^+/NADH$ ratio, which not only results from the redox status of the molecule, but also a balance between its synthesis and its degradation, might modulate gene expression [46]. Conversely, acetyl-CoA, resulting from pyruvate metabolism and fatty acid β-oxidation, is a substrate for lysine acetylation by acetyltransferases that controls the acetylation status of the transcription factors of the forkhead box protein O family members [47], histone acetylation, chromatin remodeling, and thus gene expression [48,49]. Yet, even if both $NAD^+/NADH$ and acetyl-CoA are associated with the activity of post-translational modifying enzymes, and thus regulation of signaling pathways or gene expression, their direct involvement in retrograde signaling has not been extensively characterized to date.

1.1.4. Beyond Regulation by Transcription Factors

Beyond this level of retrograde communication, which regulates nuclear gene expression by the activation of transcription factors, it has been shown that the mtDNA copy number can also lead to epigenetic modifications. It is known that ATP, acetyl-CoA concentration, $NAD^+/NADH$ ratio, and SAM (s-adenosylmethionine) are cofactors and metabolites whose concentrations can be modulated by mitochondria. Several of them, such as acetyl-CoA and SAM, are important acetyl- or methyl-group donors for DNA, RNA, and protein modifications that could also control gene expression and contribute to adaptation to mitochondria dysfunction [50]. These molecules can thus be seen as messengers for retrograde signaling responses [50].

Indeed, it has been demonstrated that mitochondrial depletion leads to increased promoter methylation of nuclear genes, which can be reverted, at least partly, by mtDNA repletion [7]. In addition, in conditions of mitochondrial stress, mitochondria-to-nucleus signaling is also know to confer a long-term adaptive response to the cell, which can protect it from future insults related or not to the initial stress. This concept, called mitochondrial hormesis or mitohormesis, provides a long-term increase in cytoprotective function after mild and transient mitochondrial stress such as ROS production. Mitohormesis has been particularly highlighted in the context of increased lifespan in various organisms such as yeast, worms, flies, and mice [51]. Moreover, it has been suggested that mitochondrial stress can generate a non-cell-autonomous response, meaning the activation of a stress response in a cell/tissue that is not affected by the initial triggered stress event. For instance, in response to OXPHOS deficiency, muscle fibers secrete the cytokine fibroblast growth factor 21 (FGF21) in the bloodstream, which results in chronic lipid recruitment and mobilization from adipose tissue and triggers ketogenesis in the liver [52,53]. It seems that skeletal muscle increases FGF21 expression in mitochondrial disorders to compensate for metabolic insufficiency by activating the mTOR Transcriptional repressor protein YY1-PGC-1α pathway [54]. Durieux and co-workers have described another example in *C. elegans*, in which stressed mitochondria, via electron chain manipulations in key tissues, release soluble signals (the so-called mitokines), which trigger expression of cytoprotective genes in distal tissues and have a beneficial effect on longevity of the worm [55]. Mitokines have also been unveiled in the context of a recent retrograde response, which we discuss in the next sections of this review: the mitochondrial unfolded protein response (UPR^{mt}). Following the example of the endoplasmic reticulum unfolded protein response (UPR^{er}), when damaged and/or unfolded proteins accumulate within mitochondria, this stress response coordinates expression of chaperones and proteases in a positive feedback loop and results in a first attempt to resolve the stress and bring cytoprotection [56].

1.2. The Mitochondrial Unfolded Protein Response (UPR^{mt})

Protein homeostasis relies on the equilibrium between the number of unfolded proteins and the folding capacity of a compartment. The mitochondria possess their own arsenal of chaperones and proteases. When unfolded, misfolded, or unassembled proteins accumulate beyond folding capacity, this leads to damage and organelle/cell dysfunction. As seen previously, mitochondrion is able to communicate with the nucleus in response to organelle dysfunction and bioenergetic impairment. As with the endoplasmic reticulum, for which the different signaling pathways and branches of the unfolded protein response are well described [57], and affecting the regulation of autophagy and the biology of mitochondria [58,59], a specific UPR^{mt} could also be initiated to boost folding and degradation capacity in response to unfolded and aggregated protein accumulations in mitochondria. Thus, mitochondrial protein homeostasis—or proteostasis—is preserved by retrograde communication to coordinate transcriptional activation of nuclear-encoded mitochondrial chaperones and proteases. Although its existence was initially reported in mammals almost 20 years ago [60], later studies related to the UPR^{mt} mainly focused in *Caenorhabditis elegans*, with mechanisms mainly revealed in the model organism and linked to aging, healthy lifespan, and longevity [61–63]. The UPR^{mt} is now considered as a mechanism that allows synchronization of nuclear and mitochondrial genomes [64].

1.2.1. The UPR^{mt} in *C. elegans*

Stress Models for the Study of UPR^{mt}

The worm *C. elegans* has been a useful model to identify stresses that activate a UPR^{mt} and to decipher signaling from proteotoxic stress to expression of stress-responsive genes. Most of the time, the induction of the UPR^{mt} can be easily assayed in worms that express green fluorescent protein (GFP) under the control of the promoter that drives the expression of gene encoding chaperones and orthologues of the mammalian mtHSP70 and heat shock 60 kDa protein 1 (HSPD1), respectively [61–63]. Ethidium bromide, an inhibitor of the mitochondrial DNA polymerase γ (POLG) was the first UPR^{mt} stressor described in *C. elegans* [65]. Because mitochondrial ETC complexes rely on a precise stoichiometry between nuclear- and mtDNA-encoded proteins, it is reasonable to think that the inhibition of mitochondrial transcription and replication by ethidium bromide, as transcriptions in mitochondria, clearly depend on mtDNA replication [66], which will lead to a protein imbalance and an increase in unassembled components. Similar mechanisms were later used to induce a UPR^{mt} with other molecules. On the one hand, decreasing mtDNA-encoded proteins through interference with mitochondrial translation, either with inhibitors, such as doxycycline, or with siRNA, allowing the silencing of mitoribosomal proteins, triggers a UPR^{mt} [46]. On the other hand, the stimulation of organelle biogenesis, by NAD$^+$ supplementation or mTOR inhibition by rapamycin, has similar effects [67]. Additional stresses, which induce a UPR^{mt}, following impaired assembly of mitochondrial complexes, defective protein folding, or processing, have also been described in a genome-wide RNAi screen [65].

Signaling Pathway of the UPR^{mt}

Ten years of research using the UPR^{mt} reporter in worms and large-scale experiments with siRNA were needed to shed light on a currently well-accepted model of retrograde signaling that followed mitochondrial proteotoxic stress in *C. elegans*. Briefly, accumulating unfolded proteins within mitochondria are degraded by the ATP-dependent Clp protease proteolytic subunit (ClpP) into small peptides, which are actively exported across the inner mitochondrial membrane (IMM) by HAF-1, a matrix peptide exporter belonging to the ATP-binding cassette (ABC) transporters [68,69]. Passive diffusion then brings peptides through the outer mitochondrial membrane (OMM) to the cytosol that triggers nuclear translocation of activating transcription factor associated with stress-1 (ATFS-1), a transcription factor that orchestrates expression of mitochondrial chaperones and proteases as well as other genes associated with ROS detoxification, mitochondrial protein import, and glycolysis [70]. Experiments in which the expression of ClpP and HAF-1 was silenced confirmed that both proteins were essential for the nuclear translocation of ATFS-1 and the activation of a UPRmt [68,69]. The mechanism that controls the cellular localization of ATFS-1 was recently discovered [70]. In addition to its nuclear localization sequence (NLS), the transcription factor also contains a mitochondrial targeting sequence (MTS). Under basal conditions, ATFS-1 is constitutively imported in mitochondria and degraded by the Lon protease [71]. However, upon mitochondrial proteotoxic stress conditions, mitochondrial import efficiency decreases, and a fraction of ATFS-1 proteins accumulates in the cytosol and translocates to the nucleus. Very recently, Haynes and colleagues showed that the ATFS-1 transcription factor not only induces chaperones, OXPHOS assembly factor, and glycolysis genes, but also directly regulates OXPHOS gene promoters by limiting the accumulation of OXPHOS transcripts. Altogether, ATFS-1 coordinates the abundance of transcripts involved in OXPHOS expression and assembly factors to the protein-folding capacity of mitochondria [72]. The importance of the efficiency of the import machinery of mitochondrial proteins in the re-localization of ATFS-1 is supported by the fact that the disruption of mitochondrial import in response to the silencing of the mitochondrial import inner membrane translocase subunit Tim23 is sufficient to trigger ATFS-1 accumulation in the nucleus. It can also activate the GFP reporter construct driven by the promoter of HSPD1 encoding HSP60 [70], a mitochondrial matrix chaperonin crucial for the folding and assembly of newly imported mitochondrial proteins [73].

Interestingly, Rainbolt and colleagues obtained similar results and showed an increase in the expression of HSP60-GFP in response to the silencing of Tim17, a response that can be prevented in ATFS-1 but not HAF-1 mutant worms [74]. In addition, these authors noted that silencing of either Tim23 or Tim17 conferred increased resistance to paraquat-induced oxidative stress, although the protection was rather modest in mammalian cells. However, increased resistance to oxidative stress was not disrupted in the ATFS-1 mutant worm, thereby questioning the active participation of the UPRmt in this protection. Finally, while the reduced protein import was able to induce stress-responsive genes independently of mitochondrial matrix stress (HAF-1 mutant), in the context of a UPRmt, both the reduced mitochondrial protein import and the ATFS-1 nuclear translocation relied on HAF-1 and peptides efflux; however, the precise mechanism and tight regulation are still unclear [70].

In addition to ATFS-1, in *C. elegans*, the transcription factors DVE-1 and ubiquitin-like protein 5 (UBL-5) appear to be involved in the transcriptional regulation of the UPR^{mt}. During the onset of mitochondrial stress, both proteins redistribute into the nucleus and form a complex that binds the promoter of gene encoding chaperones such as mtHSP70 and HSPD1. As for ATFS-1, ClpP is also required for DVE-1 nuclear translocation, but HAF-1 would be dispensable [68]. While these proteins might help in chromatin remodeling to facilitate ATFS-1 access to the promoters of the specific target genes of the UPR^{mt}, future research on the importance of the contribution of DVE-1/UBL-5 complex to the UPR^{mt} is still needed. In addition, it is also very likely that the activation of a UPR^{mt} could be accompanied by signals that trigger and connect the activation of other cell signaling pathways.

Crosstalk between the UPR^{mt} and Other Stress Pathways in *C. elegans*

The UPR^{mt} relies on more than the over-expression of chaperones and proteases to maintain mitochondria proteostasis; several genes that are directly or indirectly controlled by ATFS-1 have been identified by comparing the relative abundance of transcripts between wild type and ATFS-1 mutant worms raised either under basal or UPR^{mt}-inducing conditions. In these conditions, among 685 genes differentially expressed in response to the UPR^{mt}, only 391 genes (encoding important components of ROS detoxification and antioxidant enzymes, glycolytic enzymes, and proteins of the mitochondrial protein import machinery) seemed to be dependent on ATFS-1, suggesting the activation of other pathways [70].

Other studies described the UPR^{mt} as interconnected cell signaling with other stress-activated pathways such as the integrative stress response (ISR) [75] and the antioxidant response [67]. Baker and co-workers demonstrated that, during mitochondrial stress, eIF2α is phosphorylated by GCN2 in a ROS-dependent manner, attenuating the protein synthesis within the cytosol. This is similar to what happens in the well-characterized UPR^{er} [75]. These two pathways are complementary and maintain the activity of mitochondria. These authors show that, in a worm genetic model of UPR^{mt}, RNAi-mediated silencing of eukaryotic translation initiation factor 2-α kinase 4/GCN2-like protein (GCN2) is associated with the accumulation of carbonylated proteins and decreased oxygen consumption, despite enhanced activity of the mtHSP70-GFP reporter plasmid. Conversely, upon mitochondrial stress, the inhibition of the UPR^{mt} increases the load on the ISR, as demonstrated by an increased eIF2α phosphorylation. In addition, concomitant alterations of both pathways result in growth defects, even in the absence of any stress [75].

In addition, boosting the NAD$^+$ level not only activates Sirtuins and, especially, mitochondrial Sirtuin 3 [76], but also induces mitonuclear protein imbalances and activates UPR^{mt} [67]. Importantly, these authors revealed that, in mice and *C. elegans*, in time-course experiments, treatments with nicotinamide riboside (NR), a NAD$^+$ precursor led to a protective response in a two-step process. First, during the early phase response (one day of treatment), the UPR^{mt} was activated, with the associated increase in ClpP and HSP6 expression. Then, after three days of treatment with NAD$^+$ booster/NR, with the UPR^{mt} markers still overexpressed, an antioxidant response was also triggered, as demonstrated by enhanced activation of the promoter of the gene encoding SOD3 and nuclear localization of the transcription factor daf-16 (FOXO3A orthologue in mammals). Interestingly, it appears that both pathways might be interconnected; the induction of Extracellular superoxide dismutase [Cu/Zn] (SOD3) was dependent on UBL-5, a transcriptional regulator activated during the UPR^{mt} [67].

1.2.2. The UPR^{mt} Models in Mammals

In mammals, two different models of UPR^{mt} have been described (Figure 1). While Hoogenraad's group described a DDIT3 (DNA damage-inducible transcript 3 protein)/CHOP-10-dependent UPR^{mt} induced by the overexpression of a truncated form of a mitochondrial matrix enzyme [77], Germain and colleagues reported a CHOP-10-independent UPR^{mt} model in which protein aggregates accumulated within the internal membrane space (IMS) when a mutant (catalytically inactive enzyme) form of endonuclease G (N174A) was overexpressed in breast adenocarcinoma MCF-7 cells [78].

Accumulation of Unfolded Proteins within the Mitochondrial Matrix

Pioneering work by Hoogenraad and collaborators first showed the selective induction of mitochondrial chaperones HSPD1 and HSPE1 in response to complete mitochondrial genome depletion ($\rho°$) in rat hepatoma H4 cells [60]. Later on, they set up a UPR^{mt} model based on the overexpression of a truncated form of the mitochondrial ornithine carbamoyltransferase (OTC) enzyme (OTCΔ). The truncated form of the enzyme, while correctly localized in the organelle matrix, formed aggregates associated with the insoluble fraction after detergent extraction [79]. First clues of a UPR^{mt} have been described in the monkey kidney COS-7 cell line, as aggregated protein elimination correlates with the up-regulation of HSPD1 and ClpP. Therefore, OTCΔ abundance decreases when chaperones and proteases are overexpressed, yet abundance of the wild type OTC remains stable. In addition, co-immunoprecipitation experiments have also shown that HSPD1 and ClpP were bound to OTCΔ but not to its wild type counterpart [79].

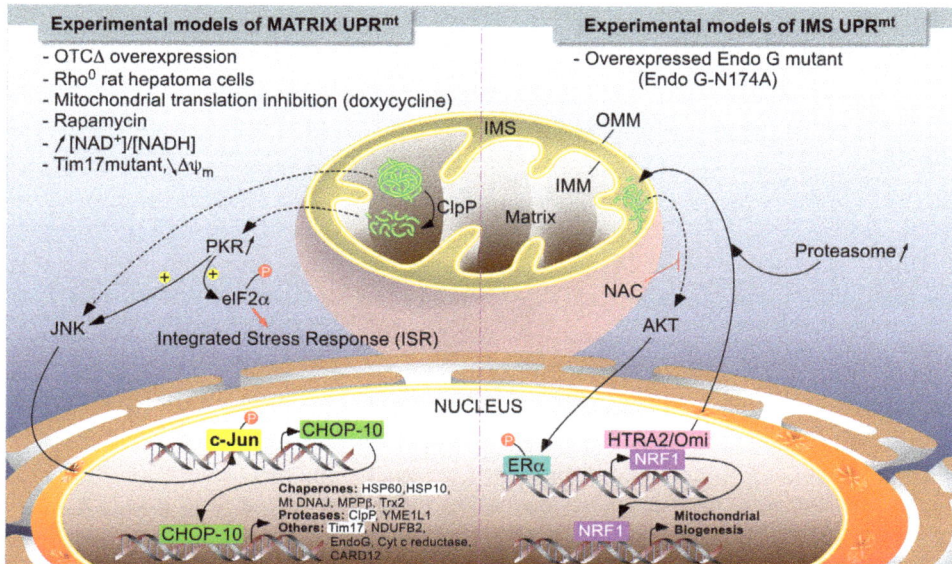

Figure 1. The UPR^mt models in mammalians. Two major independent models of UPR^mt have been described. When unfolded proteins accumulate in the mitochondrial matrix, they are first cleaved by ClpP proteases. Peptides exit mitochondria by unknown mechanisms and trigger a signaling pathway, leading to the activation of c-Jun N-terminal kinase (JNK) and PKR, which phosphorylate as c-Jun (part of AP-1 transcription factor) and eIF2α, respectively. The phosphorylation of eIF2α can also be mediated by GCN2 in response to mitochondrial translation inhibition (not illustrated), and it turns on the integrated stress response (see Figure 2). PKR could also activate JNK. The activation of c-Jun triggers CHOP-10 expression. In turn, the transcription factor regulates the expression of stress-resolving genes such as mitochondrial proteases and chaperones (left side of the chart). When unfolded proteins accumulate in the inter-membrane space, the UPR^mt is mediated by the activation of ERα, triggered by an AKT-dependent phosphorylation. The activation of AKT would then be mediated by oxidative stress, as inhibited by *N*-acetylcysteine (NAC). Mitochondrial stress resolves by an increase in the expression of HTRA2/Omi protease and enhanced proteasome activity, while the biogenesis of mitochondria is enhanced and under the control of NRF1, a target gene of ERα. The highlighted acronyms represent genes for which an endogenous differential expression (RT-qPCR or Western blot analysis) in response to UPR^mt has been experimentally demonstrated to be dependent on CHOP-10; this is in addition to analysis using reporter constructs. The non-highlighted genes refer to genes controlled by CHOP-10 in response to a UPR^mt, but only demonstrated using reporter constructs.

Hoogenraad and colleagues also used reporter plasmids to highlight the induction of several genes that encoded mitochondrial proteins in response to OTCΔ aggregation such as chaperones (HSP60, HSP10, DnaJ homolog subfamily B member 1/HSP40, mitochondrial-processing peptidase subunit β, mitochondrial thioredoxin/MTRX), proteases (ClpP, ATP-dependent zinc metalloprotease YME1L1), and other mitochondrial proteins (Tim17, NADH dehydrogenase (ubiquinone) 1β subcomplex subunit 2/NDUFB2, caspase recruitment domain-containing protein 12/CARD12, endonuclease G, cytochrome C reductase) [80]. However, it is important to emphasize that increased expression of most of these target genes using artificial reporters has not been confirmed for endogenous-related genes [81] (personal unpublished data)—an observation that could be explained by differential epigenetic regulation and chromatin remodeling between a naked DNA promoter in the reporter construct and the DNA of authentic endogenous promoters [81]. In addition, the response seems to be specific; no change has been detected in the abundance for the UPR^{er} gene markers (BiP/GRP-78/78 kDa glucose-regulated protein, endoplasmin/GRP-94/94 kDa glucose-regulated protein), for cytosolic chaperones (Heat shock cognate 71 kDa protein, heat shock-related 70 kDa protein 2), and the requirement for mitochondrial localization. Demonstrated as the induction of nuclear gene encoding, mitochondrial proteins were not observed when cells were transfected with OTCΔ lacking the mitochondrial signal peptide [77,79].

Since these works, few advances have been made in the comprehension of mechanisms by which a UPR^{mt} is set up in mammals. However, the pathological relevance of the protein kinase RNA-activated (PKR) and the UPR^{mt} emerged in two murine models of colitis as well as in patients with inflammatory bowel diseases (IBDs) [82]. The expression of PKR increases in Mode K intestinal epithelial cells that overexpress the truncated form of OTC (OTCΔ). Silencing experiments pointed out that PKR might play a role in the early phase of the UPR^{mt}; the silencing of the kinase expression using siRNA prevents the phosphorylation of c-Jun and the induction of CHOP-10 [82]. In addition, these authors showed a PKR-dependent phosphorylation of eIF2α, suggesting the concomitant induction of ISR and UPR^{mt}, as previously described in *C. elegans* [82]. Various strategies were used to disrupt the stoichiometric equilibrium between nuclear- and mitochondrial-encoded ETC (Electron Transport Chain) subunits to trigger a UPR^{mt}. Among them, doxycycline, an inhibitor of mitochondrial translation, triggered the activation of a UPR^{mt} activation in AML12 cells (a mouse hepatic cell line), as demonstrated by an increased expression of HSPD1 protein and the activation of a reporter gene driven by the ClpP promoter [45]. Similar results were obtained when mitochondrial protein abundance was increased by either increasing the NAD$^+$ concentration with two precursors of its synthesis (nicotinamide mononucleotide (NAM) and nicotinamide riboside (NR)) known to trigger the expression of several nuclear gene encoding mitochondrial proteins in an NAD-dependent protein deacetylase Sirtuin-1 (SIRT1)-dependent manner [83] or by boosting mitochondrial biogenesis by mTOR inhibition with rapamycin [67]. In primary murine hepatocytes, it was also confirmed that an antioxidant response mediated by superoxide dismutase 2 (SOD2) was associated with the UPR^{mt}, and that both responses were dependent on the Sirtuin-1 deacetylase [67]. Interestingly, at least one other Sirtuin is involved in the resolution of mitochondrial proteotoxic stress. Indeed, NAD-dependent protein deacetylase Sirtuin-3 (SIRT3) has been shown to be activated in response to a large panel of mitochondrial stresses known to induce the accumulation of protein aggregates and/or ROS, including antimycin A, rotenone, an inhibitor of heat shock protein HSP90 and the overexpression of mutated EndoG (ΔMLS-Endo G-His), mutated SOD1, and OTCΔ. These mitochondrial stresses activated an antioxidant response and mitophagy in a SIRT3-dependent manner, but the possible SIRT3-dependency of the UPR^{mt} (evaluated by CHOP-10 and HSP60 mRNA levels) was not reported. Interestingly, the proteotoxic stress resulted in a heterogeneous mitochondrial phenotype, a fraction of the mitochondria under severe stress. The cell viability under proteotoxic stress was compromised in the absence of SIRT3. Altogether, this set of data suggests that cells undergoing proteotoxic stress present a heterogeneous adaptive response with the induction of a UPR^{mt} to resolve the stress in mildly affected mitochondria and a SIRT3-orchestrated mitophagy to remove irreversibly damaged organelles [84].

Despite the requirement for HAF-1 to observe the activation of a UPR^{mt} in *C. elegans*, no clear mammalian orthologue of this protein has been identified thus far. However, Rainbolt and colleagues confirmed that, in mammalian cells, the expression of the worm Tim17 mutant and the reduction of mitochondrial imported induce UPR^{mt}. In addition, the knock down of Tim17 in HEK293 cells resulted in a slight increase in HSPD1 and YME1L1 mRNA abundance [74]. Interestingly, these authors proposed a stress pathway based on the phosphorylation of eIF2α and the attenuation of translation that leads to a down-regulation of Tim17 abundance, which is further enhanced by an increased YME1L1-dependent degradation of the protein. Ultimately, a lower abundance of Tim17 and, as a consequence, the reduced mitochondrial protein import might thus facilitate the activation of the UPR^{mt} [74].

As for HAF-1, no mammalian orthologue of ATFS-1 has been identified to date. Detailed mechanisms as well as several actors of the UPR^{mt} are thus still likely missing in mammals, especially regarding the effectors involved in the transcriptional regulation of target genes. However, both DVE-1 and UBL5 have mammalian orthologues: SATB2 and UBL5, respectively [68]. Interestingly, SATB2 is a global chromatin organizer, and this protein is also able to form a complex with UBL5; however, no evidence of involvement of SATB2/UBL5 in the mammalian UPR^{mt} has been experimentally demonstrated thus far [68].

The analysis of the HSPD1/HSPE1 bidirectional promoter revealed the existence of a CHOP-10-C/EBPβ binding element crucial for the activation of the UPR^{mt}-responsive genes. In addition, the abundance of these two transcription factors increased upon the overexpression of OTCΔ [79,80]. As in *C. elegans*, the current mammalian model of the UPR^{mt}-induced gene expression would be a two-step process. It would firstly require the expression of transcription factors, such as CHOP-10, which would in turn activate the expression of gene encoding of mitochondrial chaperones/chaperonins and proteases.

In terms of upstream signaling pathways, it seems that the phosphorylation of c-Jun by JNK2 and/or PKR is necessary to activate the expression of these transcription factors [77,82]. Phosphorylated c-Jun would bind to an AP-1 binding site within the CHOP-10 promoter, allowing the regulation of a UPR^{mt}-specific gene expression. These genes do not contain the well-described endoplasmic reticulum stress response element (ERSE) required for the classic genes regulated during the UPR^{er} [80]. However, considering the multiple cell responses involving CHOP-10 and the putative 3522 promoters of nuclear genes predicted to contain the binding consensus element for this transcription factor, it is likely that additional factors are required to specifically regulate gene expression during the UPR^{mt} [80]. The sequence alignment of the 1000 bp promoter sequences of the UPR^{mt}-induced genes revealed two well-conserved elements (except for the HSPD1/HSPE1 bidirectional promoter) surrounding the CHOP-10 binding site at a constant distance, named MURE1/2 (Mitochondrial unfolded protein response element 1/2) [80]. The requirement of these regulatory elements was demonstrated by reporter plasmid experiments as mutations within MURE1 or MURE2 to reduce the transcriptional activity of the UPR^{mt}-target genes triggered by the overexpression of OTCΔ [80]. However, the relevance of the MURE1-CHOP-10-MURE2 element might be questionable because (i) it has not been confirmed since then; (ii) this motif was widespread in the genome and was not restricted to the UPR^{mt} responsive genes (personal unpublished data); and (iii) the ligands putatively associated with these two regulatory elements have not yet been identified.

IMS-Associated mtUPR

In addition to the UPR^{mt} initiated by events occurring in the mitochondrial matrix, Germain and Papa described another UPR^{mt} model triggered by mitochondrial IMS protein aggregate accumulations, which caused the overexpression of a mutant form of endonuclease G (EndoG-N174A) in breast cancer MCF-7 cells, a condition that increases HTRA2/OMI protein abundance (a mitochondria-located serine protease that can be released by mitochondria during apoptosis) [85], NRF1 expression, and enhanced proteasome activity [78]. To contrast with OTCΔ-induced UPR^{mt}, the IMS stress response is not dependent on CHOP-10 expression, but rather relies on the ligand-independent activation of estrogen receptor α (ERα). It has been clearly shown that the overexpression of EndoG-N174A induces the RAC-α serine/threonine-protein kinase/AKT/PKB-dependent phosphorylation of ERα phosphorylation (Ser167). Additionally, AKT activation is dependent on ROS production triggered by the IMS stress as a treatment, with the antioxidant N-acetyl cysteine (NAC), completely abolishing the activation of ERα [78]. More recently, this research group deciphered events associated with IMS stress in triple negative cancer cells (lacking ERα) and highlighted the involvement of the mitochondrial SIRT3 in the regulation of the antioxidant response and mitophagy [84]. Strikingly, while these authors had previously shown that EndoG-N174A overexpression in MCF-7 cells did not induce a cell signaling response comparable to the one observed in response to the accumulation of truncated OTC in the mitochondrial matrix, in ERα-deficient breast cancer cells, they recently revealed that IMS stress also increases the expression of CHOP-10 and HSPD1 [84].

In conclusion, it seems that different retrograde responses that are dependent on the mitochondrial compartment and location at the origin of the stress could be initiated. Indeed, as we have seen, different quality control factors have been described, depending on the organelle compartment, such as ClpP and HSPD1, for matrix proteins or serine protease HTRA2/Omi and the proteasome for IMS proteins. Thus, because initiating stress and responding effectors to recover protein homeostasis are different between the matrix and IMS stress, this may explain the existence of independent signaling pathways such as those described with CHOP-10 and OTCΔ, or ERα and EndoG-N174A. In a recent study, we showed that, the transcription factor CHOP-10 was systematically overexpressed when the expression of mtDNA was inhibited by different means and in different cell types, while the UPR^{mt}-related gene markers HSPD1 and ClpP were not induced [86]. Furthermore, increased expression of this transcription factor correlated with the activation of another stress-responsive pathway called the integrated stress response (ISR) [86].

1.3. The Integrated Stress Response (ISR)

Although the integrated stress response (ISR) is not usually described as a typical retrograde response, several studies reported its activation in response to mitochondrial dysfunction [86–90]. The next paragraph presents the main features of ISR as well as its crosstalk with mitochondrial stress because it is a complementary pathway to the UPR^{mt}.

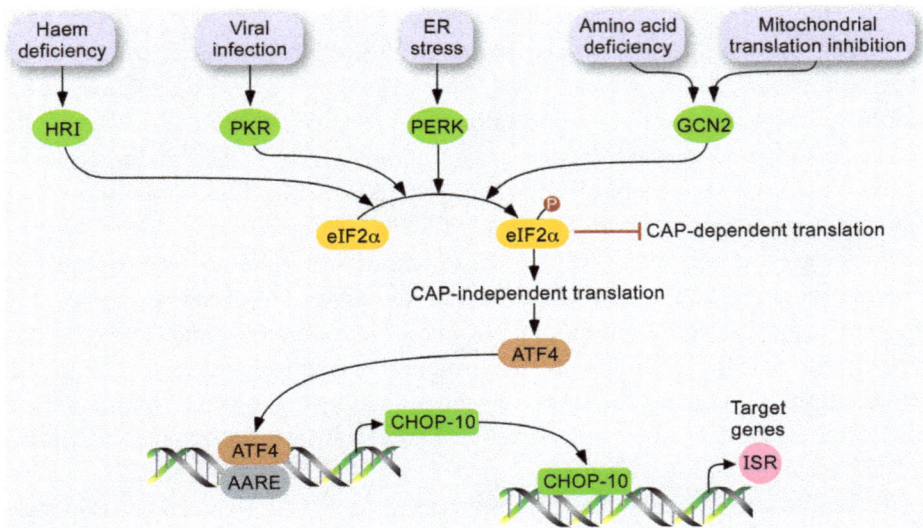

Figure 2. Hypothetical model for stress response to mitochondrial dysfunction. In response to various stresses, several kinases, such as HRI, PKR, PERK, and GCN2, are activated and converge to phosphorylate the translation initiation factor eIF2α. In turn, the phosphorylation of eIF2α has at least two consequences. While the cytosolic translation is globally attenuated as it inhibits CAP-dependent translation, mRNAs containing a uORF are preferentially translated, such as the transcription factor ATF4, which controls the expression of CHOP-10. The profit of this GCN2-eIF2α-ATF4 pathway might be stress attenuation gained by relieving the load of proteins imported into the mitochondria and increasing the expression of integrated stress responsive genes.

The ISR is a stress adaptive pathway conserved by evolution as recovered from yeast to human. In mammals, molecular effectors of this signaling pathway (Figure 2) are activated during ER stress, amino acid depletion [91], virus infection, oxidative stress, heme deprivation, or UV irradiation [92].

These stressors are sensed by four different kinases: Eukaryotic translation initiation factor 2-α kinase 1/heme-regulated inhibitor (HRI), PKR, RNA-dependent protein kinase-like endoplasmic reticulum kinase (PERK), and general control non-derepressible 2 (GCN2) that converge to the phosphorylation of eIF2α eukaryotic translation Initiation Factor 2α at serine 51 [93]. Phosphorylation of eIF2α is a central event of the ISR; it integrates upstream signals to reduce the ATP-consuming events, such as cytosolic translation, and to regulate the specific expression of a set of nuclear genes. Phosphorylated eIF2α (P-eIF2α) inhibits recycling of its GTP-bound form (active) by eIF2B, which is required for delivery of the methionyl-tRNA to the ribosome during translation initiation [94]. In addition, P-eIF2α also elicits preferential translation of mRNAs containing an upstream open reading frame (uORF) in their 5'-leader sequence such as the bZIP transcription factor activating transcription factor (4ATF4). Under unstressed conditions, translation initiation occurs at the start codon of the first uORF (uORF1) within the 5'-leader of ATF4 mRNA. While the ribosome scans the sequence, after translation termination at the uORF1, the ribosome does not dissociate from the mRNA and resumes scanning to reach uORF2. When eIF2α is not phosphorylated, it is rapidly recycled in its eIF2α-GTP form—a reaction that allows rapid re-initiation of translation at the site. However, the inhibitory uORF2 is out of frame from the ATF4 coding region, and the ribosome dissociates without translation of ATF4. When cells are under stressful conditions, the availability of eIF2α-GTP decreases and after termination at uORF1, the ribosome will scan through uORF2 because there is not enough eIF2α-GTP to re-initiate translation in a rapid manner. As the ribosome scans to reach the coding sequence of ATF4, eIF2α-GTP is sufficiently recycled to initiate translation at the start codon [92].

ISR also regulates gene expression by an interconnected set of transcription factors, such as ATF3 and CHOP-10, which are both induced by ATF4. Together with other bZIP transcription factors, they regulate many other downstream genes, such as homocysteine-responsive endoplasmic reticulum-resident ubiquitin-like domain member 1 (HERPUD1) and tribbles homolog 3 (TRIB3), which first help to either resolve the stress or, alternatively, to induce apoptosis if the stress is too severe or persists for too long [91]. The protein encoded by HERPUD1 is a protective protein that participates in the endoplasmic reticulum-associated degradation (ERAD) pathway during ER stress [95]. Interestingly, these authors showed that HERPUD1 might be neuroprotective during ER stress by stabilization of Ca^{2+} storage and maintenance of mitochondria function [95]. The tribbles homolog 3 protein is a pseudo-kinase because, despite having a region of high homology with kinase domains, it lacks a consensus sequence for protein phosphorylation. So far, TRIB3 is known to directly interact with different proteins, such as AKT, C/EBPs (CCAAT/enhancer binding proteins), and ATF4/CHOP-10, thereby regulating the biological process and responses such as insulin signaling, autophagy, adipogenesis, and induction of apoptosis, respectively [96].

The ISR is interconnected with several pathways, and some effectors are shared by other signaling such as the UPR^{er} or the UPR^{mt} [82]. In addition, the GCN2 kinase can be activated by amino acid depletion, glucose depletion, UV irradiation, proteasome inhibition, and oxidative stress [93]. The mechanism of activation upon amino acid starvation is well described. Under normal conditions, the kinase is kept inactive by auto-inhibitory molecular interactions. Upon binding of unloaded tRNA to the histidyl-tRNA synthetase-like domain, an allosteric rearrangement creates an open conformation of the protein, and GCN2 is activated by auto-phosphorylation (Thr898 and Thr903) within the activation loop [97].

This kinase is a good example to illustrate divergent signaling pathways of the cell responses following eIF2α phosphorylation. Indeed, while the signaling pathway downstream the P-eIF2α triggered by the UV-irradiation enhances the activation of NF-κB and cell survival, the phosphorylation of eIF2α in response to the inhibition of proteasome leads to the expression of CHOP-10 [98,99]. As mentioned, CHOP-10 is a pleiotropic transcription factor that can form heterodimers with other members of the bZIP family transcription factors, such as C/EBPs and ATF4, to function either as a trans-activator or a trans-repressor of target genes [100–102]. Moreover, while CHOP-10 has been associated with apoptosis in many different conditions, paradoxically, recent studies have revealed that it can also promote cell survival during ER stress and amino acid starvation by regulating the expression of several autophagy-related genes such as p62, ATG5, ATG7, and ATG10 [103,104]. The resulting pro- or anti-apoptotic response is most likely dependent on the integration and balance of multiple signals that are activated in parallel with the cell nuclear background, together with the type/origin, nature, severity, and duration of the stress.

In addition to this coordinated expression of chaperones and proteases, when the defect is too severe, the next level of quality control is the degradation of the entire organelle by specific autophagy called mitophagy [105]. We have seen clearly that mitochondria display several lines of quality control mechanisms: mitochondria-specific chaperones and proteases protect against misfolded proteins at the molecular level, and fission/fusion and mitophagy segregate and eliminate damage at the organelle level [9]. Therefore, if an increase in unfolded proteins in mitochondria activates a UPR^{mt} to increase chaperone production, the mitochondrial serine/threonine-protein kinase PINK1 and the E3 ubiquitin-protein ligase parkin, whose mutations cause familial Parkinson's disease, remove depolarized mitochondria through mitophagy [106]. This research group also showed that both phenomenon were linked; the expression of unfolded proteins in the matrix causes the accumulation of PINK1 on energetically healthy mitochondria, resulting in mitochondrial translocation of Parkin and mitophagy and subsequent reduction of unfolded protein load, a response that is strongly enhanced by the knockdown of the mitochondrial Lon protease homolog [105].

1.4. What Is the (Patho)Physiological Relevance of a UPR^{mt} in Mammals?

1.4.1. The UPR^{mt} and Neurodegenerative Diseases

The accumulation of unfolded or aggregated proteins is a hallmark of a number of neurodegenerative diseases, including Alzheimer's disease (AD) and Parkinson's disease (PD), although it is currently unclear if these protein aggregates participate or are the consequence of such pathologies. For instance, extracellular amyloid β deposits are characteristic of AD, but they are also present in mitochondria [107]. Amyloid β precursor protein is cleaved by the mitochondrial HTRA2/Omi serine protease [108] and the latter has been shown to delay the aggregation of the amyloid β 1–42 peptide [109]. Although HTRA2 is a target gene of IMS-induced UPR^{mt} (Figure 1), a putative role of the UPR^{mt} in this pathology has not been demonstrated clearly. Regarding PD, accumulation of α-synuclein has been demonstrated in the mitochondria of post-mortem brains of PD patients as well as in cell models of PD [110,111]. Interestingly, a higher level of misfolded respiratory proteins was detected in post-mortem brains of PD patients, which correlates with an elevated abundance of HSP60 [112], suggesting the occurrence of a UPR^{mt}. However, the UPR^{mt} is still poorly characterized in PD models, contrary to mitophagy, because PINK and/or Parkin deficiencies are strongly associated with PD [113]. Clearly, investigating the relationship between the UPR^{mt} and neurodegenerative diseases would be of great interest. In addition, the development of new tools to better monitor the UPR^{mt} *in vivo* in individual cells would be welcome, as the mitochondrial proteotoxic stress generates a heterogeneous phenotype associated with a UPR^{mt} and/or mitophagy [84].

1.4.2. The UPR^{mt} in Aging

One of the pioneering studies reporting a link between aging and the UPR^{mt} was the silencing of cco-1, a subunit of Cytochrome c oxidase, in neuronal tissue of *C. elegans*, which increases lifespan and induces a UPR^{mt} not only in neurons, but also in the intestine [55], suggesting an inter-tissue signaling that results in a UPR^{mt} activation. Since then, the UPR^{mt} has been implicated in increased lifespan in worms, flies, and mice, highlighting the importance of maintaining mitochondrial proteostasis for longevity. This aspect will not be developed here; several reviews have been recently devoted to UPR^{mt} and aging [56,62,114].

1.4.3. The UPR^{mt} and Stem Cells

Recently, Mohrin and co-workers highlighted a SIRT7-NRF1 regulatory axis in the UPR^{mt} and linked the UPR^{mt} to hematopoietic stem cells' (HSC) quiescence and aging [115]. They first demonstrated that the NAD$^+$-dependent deacetylase SIRT7 is induced in response to a UPR^{mt}. SIRT7 directly binds to the transcription factor NRF1, specifically on the promoter of gene encoding mitochondrial ribosomal proteins and mitochondrial translation factors, to reduce, indirectly, their expression. SIRT7 induction thus contributes to attenuating the mitochondrial translation, thereby decreasing the proteotoxic stress imposed on the organelle. In addition, SIRT7 deficiency results in a constitutive UPR^{mt}. Because stem cells undergo an oxidative metabolic shift when entering into differentiation, it is not surprising that SIRT7, by limiting mitochondria biogenesis and the UPR^{mt}, contributes to HSCs' quiescence and prevents differentiation. Furthermore, the activation of a UPR^{mt} observed in aged HSCs has been linked to SIRT7 decreasing in these cells, contributing to their decline. Interestingly, reintroduction of SIRT7 in aged HSCs reduces the UPR^{mt} and improves their regenerative capacities. Altogether, these authors concluded that the UPR^{mt} might be an aged-sensitive metabolic checkpoint to regulate HSCs' renewal properties.

1.4.4. The UPR^{mt} in Innate Immunity

Several bacterial pathogens target the mitochondria, disrupting essential functions, such as calcium and redox homeostasis, or mitochondrial morphology [116]. However, the impact of pathogens on the UPR^{mt} is still poorly described, especially in mammals. In *C. elegans*, several bacteria strains, and particularly *P. aeruginosa*, activate a UPR^{mt}, as shown by a specific chaperone induction in the intestine [117]. Because the worm is more susceptible to *P. aeruginosa* exposure when ATFS-1 expression is down-regulated, the UPR^{mt} might be protective against bacterial infection. In addition, the knockdown of an mtDNA-encoded ATP synthase subunit not only activates the UPR^{mt}, but also the expression of innate immune genes and makes the worm more resistant to *P. aeruginosa* exposure. Finally, these authors showed that human cells undergoing a UPR^{mt} also induce the transcription of antimicrobial peptides (reviewed in [113]). In line with these observations, PKR-mediated UPR^{mt} has been highlighted in two murine models of colitis, as well as in patients with inflammatory bowel diseases under inflammatory conditions [82]. The UPR^{mt} might thus represent one of the mechanisms that senses bacterial pathogens in the intestinal mucosae and adapts the cell functions and immune response (reviewed in [118]). However, it is still currently unclear whether PKR-dependent induction of a UPR^{mt} in colitis models is detrimental or beneficial due to controversial data found in the literature [82,119].

Importantly, the sensitivity to the UPR^{mt} may vary according to different parameters, including tissue-specificity and stress intensity. This has been demonstrated by Dogan and co-workers, who studied a mouse model that lacked a protein directly involved in mitochondrial translation, such as the gene DARS2 encoding the mitochondrial aspartate-tRNA ligase, specifically in heart and skeletal muscle tissues [120]. While both tissues presented a comparable mitochondrial respiratory deficiency, the strategies developed to cope with this proteotoxic stress were different, depending on the tissue. DARS2-deficient hearts from 6-week-old mice presented a mitochondrial cardiopathy, with an increase in mitochondria biogenesis and activation of a UPR^{mt}, accompanied by reduced autophagy markers. However, 3-week-old mice showed no sign of cardiopathy, suggesting that the mitochondrial stress imposed by the conditional loss of DARS2 had not reached a sufficient level to trigger these stress responses. Conversely, these stress responses were not observed in the skeletal muscles of 6-week-old mice despite a 60% to 80% decrease in the activity of respiratory chain complexes, presumably because a low turnover of mitochondrial transcripts was the preferred strategy to cope with mitochondrial stress in this tissue [120].

2. Conclusions

2.1. The UPR^{mt} versus ISR or Both in Response to a Mitochondrial Stress?

While the number of papers reporting on the UPR^{mt} is ever growing, it is interesting to note that this stress response is still poorly defined in terms of target genes and signaling pathways, especially in mammalians. This is probably reinforced by the fact that some researchers tend to designate UPR^{mt} as any kind of retrograde response triggered by a mitochondrial stress, even if the stress has not been associated with the accumulation of unfolded protein aggregates. Although the pioneering work performed by the group of Hoogenraad [79] established a large list of target genes specifically induced by the UPR^{mt} (and not by the UPR^{er}), these interesting findings were mainly obtained with reporter assays. Only a few of these target genes were shown later to be induced at the endogenous level by one or several of the mitochondria stressing conditions listed in Figure 1. These endogenous UPR^{mt} targets were CHOP-10 [79], HSP60 [78,79,82,115], ClpP [115], and YME1L1 [74], although the last three UPR^{mt} markers were generally induced in a modest manner. However, it is interesting to point out that other types of mitochondrial stresses, such as the inhibition of mitochondrial translation or the depletion of mtDNA [86], MELAS and NARP mtDNA point mutations [88], inhibition of the respiratory chain by rotenone and antimycin A [89], or loss of HTRA2 expression [90] do not induce HSP60 and ClpP expression, but rather activate the integrated stress response (ISR) characterized by an attenuation of the cytosolic translation achieved through the phosphorylation of eIF2α and by the induction of another set of target genes including CHOP-10, TRIB3, and HERPUD1 [86]. Importantly, Rainbolt and co-workers revealed that arsenite, an ETC inhibitor, triggered the ISR, with a transient phosphorylation of eIF2α, accompanied by the induction of HSP60 and Tim17 at the transcript level, but not at the protein level [74]. Tim17 protein abundance strongly decreased in response to arsenite due to both a decreased biogenesis and an increased YME1L1-dependent degradation, thereby reducing the mitochondrial protein import. Interestingly, depletion of Tim17 by RNA interference not only reduced mitochondrial import, but also modestly increased (1.5-fold) the transcript level of HSP60 and YME1L1 [74]. Altogether, this suggests a two-wave response to the arsenite-induced mitochondrial stress: the first wave activated the ISR, with a transient phosphorylation of eIF2α that reduces Tim17 synthesis. The lower mitochondrial import capacity associated with reduced Tim17 abundance in turn induced the transcription of some UPR^{mt} target genes. It is important to know whether cooperation between the ISR and the UPR^{mt} activation could be generalized to other types of mitochondrial stresses, which still needs to be addressed, although Haller's group has already shown that the UPR^{mt} observed in human intestinal epithelial cells overexpressing OTCΔ is dependent on PKR-mediated ISR [82]. Researchers interested in mitochondria-induced stress responses should therefore be encouraged to monitor the kinetics—systematically—of both the UPR^{mt} and the ISR markers, including the efficiency of the mitochondrial import capacity.

It is important to emphasize the fact that, while the ISR and the UPR^{mt} were first discriminated based on the source of mitochondrial stress generated (depletion of mtDNA *versus* accumulation of misfolded proteins) in experimental models, it is most likely that a certain level of overlap exists between these two signaling pathways, especially as several effectors are common to both cell responses. One can easily imagine that mtDNA depletion and mutations affecting ETC that are known to trigger ISR could also eventually lead to ROS-dependent oxidation of mitochondrial proteins and lipids, causing defects in folding or misassembling of mitochondrial protein complexes, initiating a UPR^{mt}.

2.2. Future Directions

Important questions remain unanswered regarding the UPR^{mt}, as well as mitochondria-induced ISR. First, while the link between a mitochondria proteotoxic stress and the cytosolic UPR^{mt} cascade is relatively clear in *C. elegans*, it remains poorly understood in mammals. Does the mammalian UPR^{mt} signaling depend on peptide efflux? If so, what are the peptide transporters and the primary cytosolic targets?

Second, the attenuation of cap-dependent translation, achieved through phosphorylation of eIF2α, is a shared and common feature between the UPR^{er}, the UPR^{mt}, and the ISR (Figures 1 and 2). These three pathways are powerful mechanisms for alleviating the proteotoxic stress imposed on organelles or cytosol. However, an intriguing feature is the kinetics of eIF2α phosphorylation: while transient in the case of a UPR^{er}, it is sustained (up to 30–48 h) in the case of a UPR^{mt} triggered by ΔOTC overexpression [82], and in the case of ISR, it is induced by the inhibition of mitochondrial translation [86]. In response to a UPR^{er}, it is known that the induction of encoding the protein phosphatase 1 regulatory subunit 15A/Growth arrest and DNA damage-inducible protein (PPP1R15A/GADD34) contributes to dephosphorylation of the phosphorylated form of eIF2α and thus resumes cap-dependent translation. One might wonder how cells deal with a reduced cap-dependent translation for as long as 2–3 days, a phenomenon accompanied by cell growth arrest in some cell types, but not all of them (Sébastien Michel, University of Namur, Belgium. Personal communication, 2014). During this period, several ISR-target genes are induced. Even if the presence of internal ribosome entry sites (IRES) has been described for at least some of them, one cannot exclude a specific regulation at the translation level on these ISR-responsive transcripts, following the mechanism of RNA regulon control [121]. In line with this hypothesis, a recent multi-layered "omics" dissection of mitochondrial activity in the liver proteome from 40 strains of the BXD mouse genetic reference population revealed that six members of the UPR^{mt} pathway are coordinately regulated; this occurs more strongly at the protein level than at the transcript level [122].

Third, the specificity of CHOP-10-dependent cell responses needs to be investigated. Indeed, CHOP-10 is induced in response to number of cell stress conditions, such as nutrient deprivation, viral infection, hypoxia, or accumulation of protein aggregates in organelles such as the endoplasmic reticulum and mitochondria. This transcription factor is undoubtedly a key actor for coordinating the adequate cell response that ranges from apoptosis, the UPR^{er}, and the UPR^{mt} to the ISR. In addition, CHOP-10 was also shown to reduce the proliferation of intestinal cells in models of acute and chronic colitis [123]. However, the molecular determinants that confer the specificity of CHOP-10-dependent stress-response are still largely unclear and might involve specific post-transcriptional modifications of CHOP-10 as well as the activation of additional transcriptional regulators.

Fourth, while inducing chaperones and proteases to restore proteostasis makes sense, the biological significance of some ISR target genes in the context of a mitochondrial proteotoxic stress remains undetermined. For instance, it has been shown that the induction of TRIB3 participates in a negative feedback loop to repress most of the ATF4-dependent ISR target genes [124]. However, the biological role of other ISR-induced genes, and particularly the induction of several aminoacyl-tRNA synthetases in response to mitochondrial stresses [88,124], still requires analysis. Briefly, in all eukaryotes, aminoacyl-tRNA synthetases are sequestrated in large multimeric complexes, where they ensure their primary role of tRNA aminoacylation. It has been shown that upon release from these complexes, they can carry out alternative functions. When yeasts are submitted to the diauxic transition from anaerobic glycolysis to oxidative respiration, two aminoacyl-tRNA synthetases, cERS and cMRS, dissociate from their cytosolic anchor complex, relocating to the mitochondria and the nucleus, respectively. While cERS stimulates mitochondrial translation, cMRS triggers the transcription of nuclear gene encoding of ATP synthase subunits, enabling synchronization of the expression of mitochondria-encoded and nuclear-encoded subunits of the complex V [125]. The fact that non-canonical functions have been recently highlighted for two tRNA synthetases in the yeast could pave the way for new researches needed in the future to fully understand signaling and effectors of the UPR^{mt} in mammalians.

Acknowledgments: Sebastien Michel was a recipient of a doctoral fellowship from the Fonds pour la Recherche dans l'Industrie et l'Agriculture (FRIA, Brussels, Belgium). The authors also thank the Belgian Association for Muscular Diseases for their support (ABMM, La Louvière, Belgium, grant 2013-08) and Michel Savels for the Figure layout.

Author Contributions: Thierry Arnould initiated the writing and mainly contributed to the writing of the mitochondrial retrograde signaling and UPR^{mt}. Sébastien Michel was involved in the different aspects of UPR^{mt} signaling in various species and Patricia Renard contributed to the physiological relevance of UPR^{mt} and drafted the figures. All authors contributed actively to the writing and the revision of the review.

Conflicts of Interest: The authors declare no conflict of interest.

References

1. Goffart, S.; Wiesner, R.J. Regulation and co-ordination of nuclear gene expression during mitochondrial biogenesis. *Exp. Physiol.* **2003**, *88*, 33–40.

2. Ryan, M.T.; Hoogenraad, N.J. Mitochondrial-nuclear communications. *Annu. Rev. Biochem.* **2007**, *76*, 701–722.

3. Collins, Y.; Chouchani, E.T.; James, A.M.; Menger, K.E.; Cocheme, H.M.; Murphy, M.P. Mitochondrial redox signalling at a glance. *J. Cell Sci.* **2012**, *125*, 801–806.

4. Lackner, L.L. Shaping the dynamic mitochondrial network. *BMC Biol.* **2014**, *12*, 35.

5. Van Vliet, A.R.; Verfaillie, T.; Agostinis, P. New functions of mitochondria associated membranes in cellular signaling. *Biochim. Biophys. Acta* **2014**, *1843*, 2253–2262.

6. Milane, L.; Trivedi, M.; Singh, A.; Talekar, M.; Amiji, M. Mitochondrial biology, targets, and drug delivery. *J. Control. Release* **2015**, *207*, 40–58.

7. Smiraglia, D.J.; Kulawiec, M.; Bistulfi, G.L.; Gupta, S.G.; Singh, K.K. A novel role for mitochondria in regulating epigenetic modification in the nucleus. *Cancer Biol. Ther.* **2008**, *7*, 1182–1190.

8. Kotiadis, V.N.; Duchen, M.R.; Osellame, L.D. Mitochondrial quality control and communications with the nucleus are important in maintaining mitochondrial function and cell health. *Biochim. Biophys. Acta* **2014**, *1840*, 1254–1265.

9. Michel, S.; Wanet, A.; de Pauw, A.; Rommelaere, G.; Arnould, T.; Renard, P. Crosstalk between mitochondrial (dys)function and mitochondrial abundance. *J. Cell. Physiol.* **2012**, *227*, 2297–2310.

10. Rugarli, E.I.; Langer, T. Mitochondrial quality control: A matter of life and death for neurons. *EMBO J.* **2012**, *31*, 1336–1349.

11. Biswas, G.; Adebanjo, O.A.; Freedman, B.D.; Anandatheerthavarada, H.K.; Vijayasarathy, C.; Zaidi, M.; Kotlikoff, M.; Avadhani, N.G. Retrograde Ca^{2+} signaling in C2C12 skeletal myocytes in response to mitochondrial genetic and metabolic stress: A novel mode of inter-organelle crosstalk. *EMBO J.* **1999**, *18*, 522–533.

12. Amuthan, G.; Biswas, G.; Ananadatheerthavarada, H.K.; Vijayasarathy, C.; Shephard, H.M.; Avadhani, N.G. Mitochondrial stress-induced calcium signaling, phenotypic changes and invasive behavior in human lung carcinoma A549 cells. *Oncogene* **2002**, *21*, 7839–7849.

13. Arnould, T.; Vankoningsloo, S.; Renard, P.; Houbion, A.; Ninane, N.; Demazy, C.; Remacle, J.; Raes, M. CREB activation induced by mitochondrial dysfunction is a new signaling pathway that impairs cell proliferation. *EMBO J.* **2002**, *21*, 53–63.

14. Whelan, S.P.; Zuckerbraun, B.S. Mitochondrial signaling: Forwards, backwards, and in between. *Oxid. Med. Cell. Longev.* **2013**, *2013*, 351613.

15. Cagin, U.; Enriquez, J.A. The complex crosstalk between mitochondria and the nucleus: What goes in between? *Int. J. Biochem. Cell Biol.* **2015**, *63*, 10–15.

16. Hock, M.B.; Kralli, A. Transcriptional control of mitochondrial biogenesis and function. *Annu. Rev. Physiol.* **2009**, *71*, 177–203.

17. Nakashima-Kamimura, N.; Asoh, S.; Ishibashi, Y.; Mukai, Y.; Shidara, Y.; Oda, H.; Munakata, K.; Goto, Y.; Ohta, S. MIDAS/GPP34, a nuclear gene product, regulates total mitochondrial mass in response to mitochondrial dysfunction. *J. Cell Sci.* **2005**, *118*, 5357–5367.

18. Parikh, V.S.; Morgan, M.M.; Scott, R.; Clements, L.S.; Butow, R.A. The mitochondrial genotype can influence nuclear gene expression in yeast. *Science* **1987**, *235*, 576–580.

19. Parikh, V.S.; Conrad-Webb, H.; Docherty, R.; Butow, R.A. Interaction between the yeast mitochondrial and nuclear genomes influences the abundance of novel transcripts derived from the spacer region of the nuclear ribosomal DNA repeat. *Mol. Cell. Biol.* **1989**, *9*, 1897–1907.

20. Jazwinski, S.M. The retrograde response: When mitochondrial quality control is not enough. *Biochim. Biophys. Acta* **2013**, *1833*, 400–409.

21. Liao, X.; Butow, R.A. *RTG1* and *RTG2*: Two yeast genes required for a novel path of communication from mitochondria to the nucleus. *Cell* **1993**, *72*, 61–71.

22. Biswas, G.; Anandatheerthavarada, H.K.; Zaidi, M.; Avadhani, N.G. Mitochondria to nucleus stress signaling: A distinctive mechanism of NF-κB/Rel activation through calcineurin-mediated inactivation of IκBβ. *J. Cell Biol.* **2003**, *161*, 507–519.

23. Vankoningsloo, S.; de Pauw, A.; Houbion, A.; Tejerina, S.; Demazy, C.; de Longueville, F.; Bertholet, V.; Renard, P.; Remacle, J.; Holvoet, P.; *et al.* CREB activation induced by mitochondrial dysfunction triggers triglyceride accumulation in 3T3-L1 preadipocytes. *J. Cell Sci.* **2006**, *119*, 1266–1282.

24. Guha, M.; Pan, H.; Fang, J.K.; Avadhani, N.G. Heterogeneous nuclear ribonucleoprotein A2 is a common transcriptional coactivator in the nuclear transcription response to mitochondrial respiratory stress. *Mol. Biol. Cell* **2009**, *20*, 4107–4119.

25. Acharya, P.; Engel, J.C.; Correia, M.A. Hepatic CYP3A suppression by high concentrations of proteasomal inhibitors: A consequence of endoplasmic reticulum (ER) stress induction, activation of RNA-dependent protein kinase-like ER-bound eukaryotic initiation factor 2α (eIF2α)-kinase (PERK) and general control nonderepressible-2 eIF2α kinase (GCN2), and global translational shutoff. *Mol. Pharmacol.* **2009**, *76*, 503–515.

26. Arnould, T.; Mercy, L.; Houbion, A.; Vankoningsloo, S.; Renard, P.; Pascal, T.; Ninane, N.; Demazy, C.; Raes, M. mtCLIC is up-regulated and maintains a mitochondrial membrane potential in mtDNA-depleted L929 cells. *FASEB J.* **2003**, *17*, 2145–2147.

27. Rohas, L.M.; St-Pierre, J.; Uldry, M.; Jager, S.; Handschin, C.; Spiegelman, B.M. A fundamental system of cellular energy homeostasis regulated by PGC-1α. *Proc. Natl. Acad. Sci. USA* **2007**, *104*, 7933–7938.

28. Guha, M.; Tang, W.; Sondheimer, N.; Avadhani, N.G. Role of calcineurin, hnRNPA2 and Akt in mitochondrial respiratory stress-mediated transcription activation of nuclear gene targets. *Biochim. Biophys. Acta* **2010**, *1797*, 1055–1065.

29. Itoh, K.; Wakabayashi, N.; Katoh, Y.; Ishii, T.; Igarashi, K.; Engel, J.D.; Yamamoto, M. Keap1 represses nuclear activation of antioxidant responsive elements by Nrf2 through binding to the amino-terminal Neh2 domain. *Genes Dev.* **1999**, *13*, 76–86.

30. Lerner, C.; Bitto, A.; Pulliam, D.; Nacarelli, T.; Konigsberg, M.; van Remmen, H.; Torres, C.; Sell, C. Reduced mammalian target of rapamycin activity facilitates mitochondrial retrograde signaling and increases life span in normal human fibroblasts. *Aging Cell* **2013**, *12*, 966–977.

31. Piantadosi, C.A.; Suliman, H.B. Mitochondrial transcription factor a induction by redox activation of nuclear respiratory factor 1. *J. Biol. Chem.* **2006**, *281*, 324–333.

32. Favre, C.; Zhdanov, A.; Leahy, M.; Papkovsky, D.; O'Connor, R. Mitochondrial pyrimidine nucleotide carrier (PNC1) regulates mitochondrial biogenesis and the invasive phenotype of cancer cells. *Oncogene* **2010**, *29*, 3964–3976.

33. Perez-de-Arce, K.; Foncea, R.; Leighton, F. Reactive oxygen species mediates homocysteine-induced mitochondrial biogenesis in human endothelial cells: Modulation by antioxidants. *Biochem. Biophys. Res. Commun.* **2005**, *338*, 1103–1109.

34. Storz, P.; Doppler, H.; Toker, A. Protein kinase D mediates mitochondrion-to-nucleus signaling and detoxification from mitochondrial reactive oxygen species. *Mol. Cell. Biol.* **2005**, *25*, 8520–8530.

35. Liu, B.; Chen, Y.; St Clair, D.K. ROS and p53: A versatile partnership. *Free Radic. Biol. Med.* **2008**, *44*, 1529–1535.

36. Bergeron, R.; Ren, J.M.; Cadman, K.S.; Moore, I.K.; Perret, P.; Pypaert, M.; Young, L.H.; Semenkovich, C.F.; Shulman, G.I. Chronic activation of AMP kinase results in NRF-1 activation and mitochondrial biogenesis. *Am. J. Physiol. Endocrinol. Metab.* **2001**, *281*, E1340–E1346.

37. Holmuhamedov, E.; Jahangir, A.; Bienengraeber, M.; Lewis, L.D.; Terzic, A. Deletion of mtDNA disrupts mitochondrial function and structure, but not biogenesis. *Mitochondrion* **2003**, *3*, 13–19.

38. Mercy, L.; Pauw, A.; Payen, L.; Tejerina, S.; Houbion, A.; Demazy, C.; Raes, M.; Renard, P.; Arnould, T. Mitochondrial biogenesis in mtDNA-depleted cells involves a Ca^{2+}-dependent pathway and a reduced mitochondrial protein import. *FEBS J.* **2005**, *272*, 5031–5055.

39. Chae, S.; Ahn, B.Y.; Byun, K.; Cho, Y.M.; Yu, M.H.; Lee, B.; Hwang, D.; Park, K.S. A systems approach for decoding mitochondrial retrograde signaling pathways. *Sci. Signal.* **2013**, *6*, rs4.

40. Hardie, D.G. The AMP-activated protein kinase pathway—New players upstream and downstream. *J. Cell Sci.* **2004**, *117*, 5479–5487.

41. Bolster, D.R.; Crozier, S.J.; Kimball, S.R.; Jefferson, L.S. AMP-activated protein kinase suppresses protein synthesis in rat skeletal muscle through down-regulated mammalian target of rapamycin (mTOR) signaling. *J. Biol. Chem.* **2002**, *277*, 23977–23980.

42. Jager, S.; Handschin, C.; St-Pierre, J.; Spiegelman, B.M. AMP-activated protein kinase (AMPK) action in skeletal muscle via direct phosphorylation of PGC-1α. *Proc. Natl. Acad. Sci. USA* **2007**, *104*, 12017–12022.

43. Krause, U.; Bertrand, L.; Hue, L. Control of p70 ribosomal protein S6 kinase and acetyl-CoA carboxylase by AMP-activated protein kinase and protein phosphatases in isolated hepatocytes. *Eur. J. Biochem.* **2002**, *269*, 3751–3759.

44. Inoki, K.; Zhu, T.; Guan, K.L. TSC2 mediates cellular energy response to control cell growth and survival. *Cell* **2003**, *115*, 577–590.

45. Houtkooper, R.H.; Mouchiroud, L.; Ryu, D.; Moullan, N.; Katsyuba, E.; Knott, G.; Williams, R.W.; Auwerx, J. Mitonuclear protein imbalance as a conserved longevity mechanism. *Nature* **2013**, *497*, 451–457.

46. Houtkooper, R.H.; Canto, C.; Wanders, R.J.; Auwerx, J. The secret life of NAD$^+$: An old metabolite controlling new metabolic signaling pathways. *Endocr. Rev.* **2010**, *31*, 194–223.

47. Daitoku, H.; Sakamaki, J.; Fukamizu, A. Regulation of FoxO transcription factors by acetylation and protein–protein interactions. *Biochim. Biophys. Acta* **2011**, *1813*, 1954–1960.

48. Madiraju, P.; Pande, S.V.; Prentki, M.; Madiraju, S.R. Mitochondrial acetylcarnitine provides acetyl groups for nuclear histone acetylation. *Epigenetics* **2009**, *4*, 399–403.

49. Spange, S.; Wagner, T.; Heinzel, T.; Kramer, O.H. Acetylation of non-histone proteins modulates cellular signalling at multiple levels. *Int. J. Biochem. Cell Biol.* **2009**, *41*, 185–198.

50. Wallace, D.C.; Fan, W. Energetics, epigenetics, mitochondrial genetics. *Mitochondrion* **2010**, *10*, 12–31.

51. Yun, J.; Finkel, T. Mitohormesis. *Cell Metab.* **2014**, *19*, 757–766.

52. Nunnari, J.; Suomalainen, A. Mitochondria: In sickness and in health. *Cell* **2012**, *148*, 1145–1159.

53. Ribas, F.; Villarroya, J.; Hondares, E.; Giralt, M.; Villarroya, F. FGF21 expression and release in muscle cells: Involvement of MyoD and regulation by mitochondria-driven signalling. *Biochem. J.* **2014**, *463*, 191–199.

54. Ji, K.; Zheng, J.; Lv, J.; Xu, J.; Ji, X.; Luo, Y.B.; Li, W.; Zhao, Y.; Yan, C. Skeletal muscle increase FGF21 expression in mitochondrial disorder to compensate for the energy metabolic insufficiency by activating mTOR-YY1-PGC1α pathway. *Free Radic. Biol. Med.* **2015**, *84*, 161–170.

55. Durieux, J.; Wolff, S.; Dillin, A. The cell-non-autonomous nature of electron transport chain-mediated longevity. *Cell* **2011**, *144*, 79–91.

56. Schulz, A.M.; Haynes, C.M. UPR^{mt}-mediated cytoprotection and organismal aging. *Biochim. Biophys. Acta* **2015**.

57. Chakrabarti, A.; Chen, A.W.; Varner, J.D. A review of the mammalian unfolded protein response. *Biotechnol. Bioeng.* **2011**, *108*, 2777–2793.

58. Vannuvel, K.; Renard, P.; Raes, M.; Arnould, T. Functional and morphological impact of ER stress on mitochondria. *J. Cell. Physiol.* **2013**, *228*, 1802–1818.

59. Senft, D.; Ronai, Z.A. UPR, autophagy, and mitochondria crosstalk underlies the ER stress response. *Trends Biochem. Sci.* **2015**, *40*, 141–148.

60. Martinus, R.D.; Garth, G.P.; Webster, T.L.; Cartwright, P.; Naylor, D.J.; Hoj, P.B.; Hoogenraad, N.J. Selective induction of mitochondrial chaperones in response to loss of the mitochondrial genome. *Eur. J. Biochem.* **1996**, *240*, 98–103.

61. Bennett, C.F.; Kaeberlein, M. The mitochondrial unfolded protein response and increased longevity: Cause, consequence, or correlation? *Exp. Gerontol.* **2014**, *56*, 142–146.

62. Jensen, M.B.; Jasper, H. Mitochondrial proteostasis in the control of aging and longevity. *Cell Metab.* **2014**, *20*, 214–225.

63. Jovaisaite, V.; Mouchiroud, L.; Auwerx, J. The mitochondrial unfolded protein response, a conserved stress response pathway with implications in health and disease. *J. Exp. Biol.* **2014**, *217*, 137–143.

64. Jovaisaite, V.; Auwerx, J. The mitochondrial unfolded protein response-synchronizing genomes. *Curr. Opin. Cell Biol.* **2015**, *33*, 74–81.

65. Yoneda, T.; Benedetti, C.; Urano, F.; Clark, S.G.; Harding, H.P.; Ron, D. Compartment-specific perturbation of protein handling activates genes encoding mitochondrial chaperones. *J. Cell Sci.* **2004**, *117*, 4055–4066.

66. Taanman, J.W. The mitochondrial genome: Structure, transcription, translation and replication. *Biochim. Biophys. Acta* **1999**, *1410*, 103–123.

67. Mouchiroud, L.; Houtkooper, R.H.; Moullan, N.; Katsyuba, E.; Ryu, D.; Canto, C.; Mottis, A.; Jo, Y.S.; Viswanathan, M.; Schoonjans, K.; *et al.* The NAD$^+$/sirtuin pathway modulates longevity through activation of mitochondrial UPR and FoxO signaling. *Cell* **2013**, *154*, 430–441.

68. Haynes, C.M.; Petrova, K.; Benedetti, C.; Yang, Y.; Ron, D. ClpP mediates activation of a mitochondrial unfolded protein response in *C. elegans. Dev. Cell* **2007**, *13*, 467–480.

69. Haynes, C.M.; Yang, Y.; Blais, S.P.; Neubert, T.A.; Ron, D. The matrix peptide exporter HAF-1 signals a mitochondrial UPR by activating the transcription factor ZC376.7 in *C. Elegans. Mol. Cell* **2010**, *37*, 529–540.

70. Nargund, A.M.; Pellegrino, M.W.; Fiorese, C.J.; Baker, B.M.; Haynes, C.M. Mitochondrial import efficiency of ATFS-1 regulates mitochondrial UPR activation. *Science* **2012**, *337*, 587–590.

71. Harbauer, A.B.; Zahedi, R.P.; Sickmann, A.; Pfanner, N.; Meisinger, C. The protein import machinery of mitochondria—A regulatory hub in metabolism, stress, and disease. *Cell Metab.* **2014**, *19*, 357–372.

72. Nargund, A.M.; Fiorese, C.J.; Pellegrino, M.W.; Deng, P.; Haynes, C.M. Mitochondrial and nuclear accumulation of the transcription factor ATFS-1 promotes OXPHOS recovery during the UPR^{mt}. *Mol. Cell* **2015**, *58*, 123–133.

73. Cheng, M.Y.; Hartl, F.U.; Horwich, A.L. The mitochondrial chaperonin HSP60 is required for its own assembly. *Nature* **1990**, *348*, 455–458.

74. Rainbolt, T.K.; Atanassova, N.; Genereux, J.C.; Wiseman, R.L. Stress-regulated translational attenuation adapts mitochondrial protein import through Tim17A degradation. *Cell Metab.* **2013**, *18*, 908–919.

75. Baker, B.M.; Nargund, A.M.; Sun, T.; Haynes, C.M. Protective coupling of mitochondrial function and protein synthesis via the eIF2α kinase GCN-2. *PLoS Genet.* **2012**, *8*, e1002760.

76. Kincaid, B.; Bossy-Wetzel, E. Forever young: SIRT3 a shield against mitochondrial meltdown, aging, and neurodegeneration. *Front. Aging Neurosci.* **2013**, *5*, 48.

77. Horibe, T.; Hoogenraad, N.J. The *chop* gene contains an element for the positive regulation of the mitochondrial unfolded protein response. *PLoS ONE* **2007**, *2*, e835.

78. Papa, L.; Germain, D. Estrogen receptor mediates a distinct mitochondrial unfolded protein response. *J. Cell Sci.* **2011**, *124*, 1396–1402.

79. Zhao, Q.; Wang, J.; Levichkin, I.V.; Stasinopoulos, S.; Ryan, M.T.; Hoogenraad, N.J. A mitochondrial specific stress response in mammalian cells. *EMBO J.* **2002**, *21*, 4411–4419.

80. Aldridge, J.E.; Horibe, T.; Hoogenraad, N.J. Discovery of genes activated by the mitochondrial unfolded protein response (mtUPR) and cognate promoter elements. *PLoS ONE* **2007**, *2*, e874.

81. Cairns, B.R. The logic of chromatin architecture and remodelling at promoters. *Nature* **2009**, *461*, 193–198.

82. Rath, E.; Berger, E.; Messlik, A.; Nunes, T.; Liu, B.; Kim, S.C.; Hoogenraad, N.; Sans, M.; Sartor, R.B.; Haller, D. Induction of dsRNA-activated protein kinase links mitochondrial unfolded protein response to the pathogenesis of intestinal inflammation. *Gut* **2012**, *61*, 1269–1278.

83. Brenmoehl, J.; Hoeflich, A. Dual control of mitochondrial biogenesis by sirtuin 1 and sirtuin 3. *Mitochondrion* **2013**, *13*, 755–761.

84. Papa, L.; Germain, D. SirT3 regulates the mitochondrial unfolded protein response. *Mol. Cell. Biol.* **2014**, *34*, 699–710.

85. Hegde, R.; Srinivasula, S.M.; Zhang, Z.; Wassell, R.; Mukattash, R.; Cilenti, L.; DuBois, G.; Lazebnik, Y.; Zervos, A.S.; Fernandes-Alnemri, T.; *et al.* Identification of Omi/HtrA2 as a mitochondrial apoptotic serine protease that disrupts inhibitor of apoptosis protein-caspase interaction. *J. Biol. Chem.* **2002**, *277*, 432–438.

86. Michel, S.; Canonne, M.; Arnould, T.; Renard, P. Inhibition of mitochondrial genome expression triggers the activation of chop-10 by a cell signaling dependent on the integrated stress response but not the mitochondrial unfolded protein response. *Mitochondrion* **2015**, *21*, 58–68.

87. Cortopassi, G.; Danielson, S.; Alemi, M.; Zhan, S.S.; Tong, W.; Carelli, V.; Martinuzzi, A.; Marzuki, S.; Majamaa, K.; Wong, A. Mitochondrial disease activates transcripts of the unfolded protein response and cell cycle and inhibits vesicular secretion and oligodendrocyte-specific transcripts. *Mitochondrion* **2006**, *6*, 161–175.

88. Fujita, Y.; Ito, M.; Nozawa, Y.; Yoneda, M.; Oshida, Y.; Tanaka, M. CHOP (C/EBP homologous protein) and ASNS (asparagine synthetase) induction in cybrid cells harboring MELAS and NARP mitochondrial DNA mutations. *Mitochondrion* **2007**, *7*, 80–88.

89. Ishikawa, F.; Akimoto, T.; Yamamoto, H.; Araki, Y.; Yoshie, T.; Mori, K.; Hayashi, H.; Nose, K.; Shibanuma, M. Gene expression profiling identifies a role for CHOP during inhibition of the mitochondrial respiratory chain. *J. Biochem.* **2009**, *146*, 123–132.

90. Moisoi, N.; Klupsch, K.; Fedele, V.; East, P.; Sharma, S.; Renton, A.; Plun-Favreau, H.; Edwards, R.E.; Teismann, P.; Esposti, M.D.; *et al.* Mitochondrial dysfunction triggered by loss of HtrA2 results in the activation of a brain-specific transcriptional stress response. *Cell Death Differ.* **2009**, *16*, 449–464.

91. Harding, H.P.; Zhang, Y.; Zeng, H.; Novoa, I.; Lu, P.D.; Calfon, M.; Sadri, N.; Yun, C.; Popko, B.; Paules, R.; *et al.* An integrated stress response regulates amino acid metabolism and resistance to oxidative stress. *Mol. Cell* **2003**, *11*, 619–633.

92. Wek, R.C.; Jiang, H.Y.; Anthony, T.G. Coping with stress: eIF2 kinases and translational control. *Biochem. Soc. Trans.* **2006**, *34*, 7–11.

93. Donnelly, N.; Gorman, A.M.; Gupta, S.; Samali, A. The eIF2α kinases: Their structures and functions. *Cell Mol. Life Sci.* **2013**, *70*, 3493–3511.

94. Baird, T.D.; Wek, R.C. Eukaryotic initiation factor 2 phosphorylation and translational control in metabolism. *Adv. Nutr.* **2012**, *3*, 307–321.

95. Chan, S.L.; Fu, W.; Zhang, P.; Cheng, A.; Lee, J.; Kokame, K.; Mattson, M.P. Herp stabilizes neuronal Ca^{2+} homeostasis and mitochondrial function during endoplasmic reticulum stress. *J. Biol. Chem.* **2004**, *279*, 28733–28743.

96. Prudente, S.; Sesti, G.; Pandolfi, A.; Andreozzi, F.; Consoli, A.; Trischitta, V. The mammalian tribbles homolog TRIB3, glucose homeostasis, and cardiovascular diseases. *Endocr. Rev.* **2012**, *33*, 526–546.

97. Padyana, A.K.; Qiu, H.; Roll-Mecak, A.; Hinnebusch, A.G.; Burley, S.K. Structural basis for autoinhibition and mutational activation of eukaryotic initiation factor 2α protein kinase GCN2. *J. Biol. Chem.* **2005**, *280*, 29289–29299.

98. Jiang, H.Y.; Wek, R.C. GCN2 phosphorylation of eIF2α activates NF-κB in response to UV irradiation. *Biochem. J.* **2005**, *385*, 371–380.

99. Jiang, H.Y.; Wek, R.C. Phosphorylation of the α-subunit of the eukaryotic initiation factor-2 (eIF2α) reduces protein synthesis and enhances apoptosis in response to proteasome inhibition. *J. Biol. Chem.* **2005**, *280*, 14189–14202.

100. Ron, D.; Habener, J.F. CHOP, a novel developmentally regulated nuclear protein that dimerizes with transcription factors C/EBP and LAP and functions as a dominant-negative inhibitor of gene transcription. *Genes Dev.* **1992**, *6*, 439–453.

101. Ohoka, N.; Yoshii, S.; Hattori, T.; Onozaki, K.; Hayashi, H. TRB3, a novel ER stress-inducible gene, is induced via ATF4-CHOP pathway and is involved in cell death. *EMBO J.* **2005**, *24*, 1243–1255.

102. Su, N.; Kilberg, M.S. C/EBP homology protein (CHOP) interacts with activating transcription factor 4 (ATF4) and negatively regulates the stress-dependent induction of the asparagine synthetase gene. *J. Biol. Chem.* **2008**, *283*, 35106–35117.

103. B'Chir, W.; Maurin, A.C.; Carraro, V.; Averous, J.; Jousse, C.; Muranishi, Y.; Parry, L.; Stepien, G.; Fafournoux, P.; Bruhat, A. The eIF2α/ATF4 pathway is essential for stress-induced autophagy gene expression. *Nucleic Acids Res.* **2013**, *41*, 7683–7699.

104. B'Chir, W.; Chaveroux, C.; Carraro, V.; Averous, J.; Maurin, A.C.; Jousse, C.; Muranishi, Y.; Parry, L.; Fafournoux, P.; Bruhat, A. Dual role for CHOP in the crosstalk between autophagy and apoptosis to determine cell fate in response to amino acid deprivation. *Cell Signal.* **2014**, *26*, 1385–1391.

105. Jin, S.M.; Youle, R.J. The accumulation of misfolded proteins in the mitochondrial matrix is sensed by PINK1 to induce PARK2/Parkin-mediated mitophagy of polarized mitochondria. *Autophagy* **2013**, *9*, 1750–1757.

106. Jin, S.M.; Lazarou, M.; Wang, C.; Kane, L.A.; Narendra, D.P.; Youle, R.J. Mitochondrial membrane potential regulates PINK1 import and proteolytic destabilization by PARL. *J. Cell Biol.* **2010**, *191*, 933–942.

107. Caspersen, C.; Wang, N.; Yao, J.; Sosunov, A.; Chen, X.; Lustbader, J.W.; Xu, H.W.; Stern, D.; McKhann, G.; Yan, S.D. Mitochondrial Aβ: A potential focal point for neuronal metabolic dysfunction in Alzheimer's disease. *FASEB J.* **2005**, *19*, 2040–2041.

108. Park, H.J.; Kim, S.S.; Seong, Y.M.; Kim, K.H.; Goo, H.G.; Yoon, E.J.; Min, D.S.; Kang, S.; Rhim, H. β-Amyloid precursor protein is a direct cleavage target of HtrA2 serine protease. Implications for the physiological function of HtrA2 in the mitochondria. *J. Biol. Chem.* **2006**, *281*, 34277–34287.

109. Kooistra, J.; Milojevic, J.; Melacini, G.; Ortega, J. A new function of human HtrA2 as an amyloid-β oligomerization inhibitor. *J. Alzheimer's Dis.* **2009**, *17*, 281–294.

110. Devi, L.; Raghavendran, V.; Prabhu, B.M.; Avadhani, N.G.; Anandatheerthavarada, H.K. Mitochondrial import and accumulation of α-synuclein impair complex I in human dopaminergic neuronal cultures and Parkinson disease brain. *J. Biol. Chem.* **2008**, *283*, 9089–9100.

111. Shavali, S.; Brown-Borg, H.M.; Ebadi, M.; Porter, J. Mitochondrial localization of α-synuclein protein in α-synuclein overexpressing cells. *Neurosci. Lett.* **2008**, *439*, 125–128.

112. De Castro, I.P.; Costa, A.C.; Lam, D.; Tufi, R.; Fedele, V.; Moisoi, N.; Dinsdale, D.; Deas, E.; Loh, S.H.; Martins, L.M. Genetic analysis of mitochondrial protein misfolding in *Drosophila melanogaster*. *Cell Death Differ.* **2012**, *19*, 1308–1316.

113. Pellegrino, M.W.; Haynes, C.M. Mitophagy and the mitochondrial unfolded protein response in neurodegeneration and bacterial infection. *BMC Biol.* **2015**, *13*, 22.

114. Hill, S.; van Remmen, H. Mitochondrial stress signaling in longevity: A new role for mitochondrial function in aging. *Redox Biol.* **2014**, *2*, 936–944.

115. Mohrin, M.; Shin, J.; Liu, Y.; Brown, K.; Luo, H.; Xi, Y.; Haynes, C.M.; Chen, D. Stem cell aging. A mitochondrial UPR-mediated metabolic checkpoint regulates hematopoietic stem cell aging. *Science* **2015**, *347*, 1374–1377.

116. Lobet, E.; Letesson, J.J.; Arnould, T. Mitochondria: A target for bacteria. *Biochem. Pharmacol.* **2015**, *94*, 173–185.

117. Pellegrino, M.W.; Nargund, A.M.; Kirienko, N.V.; Gillis, R.; Fiorese, C.J.; Haynes, C.M. Mitochondrial UPR-regulated innate immunity provides resistance to pathogen infection. *Nature* **2014**, *516*, 414–417.

118. Rath, E.; Haller, D. Unfolded protein responses in the intestinal epithelium: Sensors for the microbial and metabolic environment. *J. Clin. Gastroenterol.* **2012**, *46*, S3–S5.

119. Cao, S.S.; Song, B.; Kaufman, R.J. PKR protects colonic epithelium against colitis through the unfolded protein response and prosurvival signaling. *Inflamm. Bowel Dis.* **2012**, *18*, 1735–1742.

120. Dogan, S.A.; Pujol, C.; Maiti, P.; Kukat, A.; Wang, S.; Hermans, S.; Senft, K.; Wibom, R.; Rugarli, E.I.; Trifunovic, A. Tissue-specific loss of DARS2 activates stress responses independently of respiratory chain deficiency in the heart. *Cell Metab.* **2014**, *19*, 458–469.

121. Blackinton, J.G.; Keene, J.D. Post-transcriptional RNA regulons affecting cell cycle and proliferation. *Semin. Cell Dev. Biol.* **2014**, *34*, 44–54.

122. Wu, Y.; Williams, E.G.; Dubuis, S.; Mottis, A.; Jovaisaite, V.; Houten, S.M.; Argmann, C.A.; Faridi, P.; Wolski, W.; Kutalik, Z.; *et al.* Multilayered genetic and omics dissection of mitochondrial activity in a mouse reference population. *Cell* **2014**, *158*, 1415–1430.

123. Waldschmitt, N.; Berger, E.; Rath, E.; Sartor, R.B.; Weigmann, B.; Heikenwalder, M.; Gerhard, M.; Janssen, K.P.; Haller, D. C/EBP homologous protein inhibits tissue repair in response to gut injury and is inversely regulated with chronic inflammation. *Mucosal Immunol.* **2014**, *7*, 1452–1466.

124. Jousse, C.; Deval, C.; Maurin, A.C.; Parry, L.; Cherasse, Y.; Chaveroux, C.; Lefloch, R.; Lenormand, P.; Bruhat, A.; Fafournoux, P. TRB3 inhibits the transcriptional activation of stress-regulated genes by a negative feedback on the ATF4 pathway. *J. Biol. Chem.* **2007**, *282*, 15851–15861.

125. Frechin, M.; Enkler, L.; Tetaud, E.; Laporte, D.; Senger, B.; Blancard, C.; Hammann, P.; Bader, G.; Clauder-Munster, S.; Steinmetz, L.M.; *et al.* Expression of nuclear and mitochondrial genes encoding ATP synthase is synchronized by disassembly of a multisynthetase complex. *Mol. Cell* **2014**, *56*, 763–776.

Borrowing Nuclear DNA Helicases to Protect Mitochondrial DNA

Lin Ding and Yilun Liu

Abstract: In normal cells, mitochondria are the primary organelles that generate energy, which is critical for cellular metabolism. Mitochondrial dysfunction, caused by mitochondrial DNA (mtDNA) mutations or an abnormal mtDNA copy number, is linked to a range of human diseases, including Alzheimer's disease, premature aging and cancer. mtDNA resides in the mitochondrial lumen, and its duplication requires the mtDNA replicative helicase, Twinkle. In addition to Twinkle, many DNA helicases, which are encoded by the nuclear genome and are crucial for nuclear genome integrity, are transported into the mitochondrion to also function in mtDNA replication and repair. To date, these helicases include RecQ-like helicase 4 (RECQ4), petite integration frequency 1 (PIF1), DNA replication helicase/nuclease 2 (DNA2) and suppressor of var1 3-like protein 1 (SUV3). Although the nuclear functions of some of these DNA helicases have been extensively studied, the regulation of their mitochondrial transport and the mechanisms by which they contribute to mtDNA synthesis and maintenance remain largely unknown. In this review, we attempt to summarize recent research progress on the role of mammalian DNA helicases in mitochondrial genome maintenance and the effects on mitochondria-associated diseases.

Reprinted from *Int. J. Mol. Sci.* Cite as: Ding, L.; Liu, Y. Borrowing Nuclear DNA Helicases to Protect Mitochondrial DNA. *Int. J. Mol. Sci.* **2015**, *16*, 10870–10887.

1. Introduction

The mitochondrion, once an autonomous free-living Proteobacterium, became a part of the eukaryotic cell through endosymbiosis approximately two billion years ago [1]. A symbiotic relationship was established, and now, mitochondria not only serve as the powerhouses of the cell by generating adenosine triphosphate (ATP) via oxidative phosphorylation, but also regulate cellular metabolism through synthesizing heme and steroids, supplying reactive oxygen species (ROS), establishing the membrane potential and controlling calcium and apoptotic signaling [2]. Human mitochondria are maternally inherited organelles, which reside in the cytoplasm. The mitochondrial architecture consists of an outer membrane, an inner membrane, an intermembrane space and the matrix or lumen (Figure 1). The mitochondrial number per cell differs from one cell type to another, and each mitochondrion contains multiple copies of the mitochondrial DNA (mtDNA), ranging from one to 15 copies per mitochondrion [3,4]. mtDNA copy number per cell

59

also varies among different tissues due to the tissue-specific epigenetic regulation of the expression of mtDNA replication polymerase γ (Pol γ) [5]. The human mtDNA resides in the lumen and attaches to the inner membrane [6]. The mtDNA forms a small circle, which consists of 16,569 base pairs that encode two rRNA genes, 22 tRNA genes and 13 protein-encoding genes that produce parts of the electron transport chain and ATP Synthase complexes.

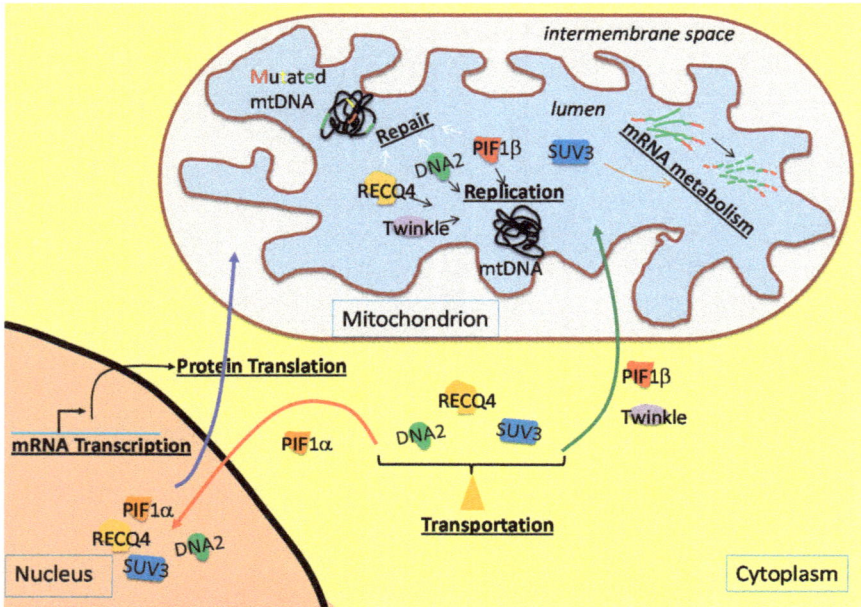

Figure 1. Schematic diagram of the production and the cellular localization of the DNA helicases (Twinkle, purple; RecQ-like helicase 4 (RECQ4), yellow; DNA replication helicase/nuclease 2 (DNA2), green; petite integration frequency 1 (PIF1), red; suppressor of var1 3-like protein 1 (SUV3), blue) that function in the mitochondrion. These DNA helicases are encoded in the nuclear genome, produced in the cytoplasm and transported into the mitochondrial lumen. With the exception of Twinkle, other DNA helicases, including RECQ4, DNA2, PIF1 and SUV3, are transported into the mitochondrial lumen or nucleus depending on the molecular cue. In the mitochondrion, these helicases participate in DNA replication and repair, as well as mRNA metabolism, in order to maintain mtDNA stability.

mtDNA is thought to be duplicated through either strand-displacement replication or RNA incorporation throughout the lagging strand [7]. Interestingly, mtDNA sequences are highly polymorphic, even within an individual. This is due to the fact that somatic mutations in mtDNA, as a result of replication errors,

ROS exposure and aging, make mtDNA sequences different from each other, even within the same cell (heteroplasmic), rather than genetically identical (homoplasmic). To safeguard a healthy population of mitochondria in a cell, mitochondria are constantly dividing (fission) and rejoining (fusion). However, should a pathogenic somatic mutation be introduced into the mtDNA genome, the entire mitochondrial population could be affected. Therefore, to maintain mtDNA stability, it is crucial to ensure faithful mtDNA synthesis. In addition, mitochondria employ several DNA repair pathways to restore DNA integrity in response to damage or replication errors [8]. Failure to do so causes mitochondrial morphological changes [9], which may lead to mitochondrial dysfunction, a phenomenon that has been linked to Alzheimer's disease and premature aging [10]. Moreover, recent studies have shown that changes in mtDNA copy number are often associated with human cancers [11,12].

DNA synthesis and DNA repair are sophisticated processes that involve multi-protein complexes. Due to the small size of the mtDNA genome and the limited number of genes it encodes, mitochondria have adapted a mechanism to "borrow" enzymes encoded in the nuclear genome for many of its functions, including mtDNA synthesis and repair. For example, DNA helicases are ATPases that break the hydrogen bonds between DNA base pairs and transiently convert double-stranded DNA (dsDNA) into single-stranded DNA (ssDNA), the latter of which can serve as the template for DNA synthesis or allow the repair of damaged bases or nucleotides. These DNA helicases are transcribed in the nucleus, synthesized in the cytoplasm and imported into the mitochondrial compartment. Mitochondrial transport occurs primarily through either the presequence pathway or the carrier pathway. Both pathways involve interactions with the translocase of the outer membrane (TOM) and the translocase of the inner membrane (TIM) protein complexes, though the protein subunits are different for each pathway [13–15]. The presequence pathway targets the precursor protein to the lumen, where a mitochondrial targeting signal (MTS) located at the N-terminus of the precursor protein is then cleaved by a mitochondrial processing peptidase. The precursor proteins that also express hydrophobic sorting signals are either inserted into the inner membrane or released into the inter membrane space. The carrier pathway usually targets the mitochondrial proteins to the inner membrane, and these precursors have a non-cleavable internal targeting signal (ITS) and form complexes with cytosolic chaperones to prevent aggregation. Nonetheless, there are exceptions to these rules; some proteins, such as the tumor suppressor p53, can also be targeted to the mitochondrion via protein-protein interactions [16].

In mammalian cells, mtDNA replication is promoted by the replicative DNA helicase Twinkle, which is encoded by the *C10orf2* gene in the nucleus. In addition to Twinkle, there are many DNA helicases that contribute to mammalian mtDNA integrity. Interestingly, unlike Twinkle, which is known to exclusively function in the mitochondrion, many of these DNA helicases not only are expressed from the nuclear genome, but also are involved in nuclear DNA replication and repair (Figure 1). This raises several questions. How do these DNA helicases balance their distribution and function in the nucleus and mitochondrion? What triggers the translocation of these helicases between different cellular compartments? How do Twinkle and other helicases collaborate in mtDNA replication and repair? In this review, we summarize recent findings on how these nuclear-encoding DNA helicases contribute to mtDNA integrity and associated diseases, and we will try to shine light on future studies in this active field.

2. Mitochondrial Replicative Helicase, Twinkle

Twinkle (for the T7 gp4-like protein with intramitochondrial nucleoid localization or PEO1) was first identified based on a sequence homology search as T7 gene 4 primase/helicase in 2001 [17]. Although Twinkle is conserved in many eukaryotes, such as the mouse, *Drosophila* and zebra fish, it has no orthologs in yeast [18]. It is possible that other yeast helicases compensate for its role in mtDNA replication. Twinkle is essential for embryonic development in mammalian systems, and it is known to unwind mtDNA for mtDNA synthesis by Pol γ [19]. Immunofluorescence microscopy has revealed that Twinkle proteins form punctate foci within mitochondria and colocalize with mitochondrial nucleoids [17], which are aggregates containing mtDNA and proteins that enact mitochondrial genome maintenance and transcription [20,21]. These foci resemble twinkling stars [17]. Human Twinkle, a 684 amino-acid (aa)-long polypeptide with a molecular weight of 77 kDa, oligomerizes to form a hexamer and exhibits 5'-3' helicase activity due to the conserved superfamily 4 (SF4) helicase domain located at its *C*-terminus (Figure 2) [22]. In addition to the conserved SF4 domain, Twinkle also contains a 42-aa MTS for mitochondrial targeting and a non-functional *N*-terminal primase-like domain that connects to the SF4 domain by a linker domain (Figure 2) [22,23]. The linker region is important for the hexamerization of Twinkle and its DNA helicase activity [24]. Recent studies have also found that Twinkle exhibits DNA annealing activity, indicating a possible involvement of Twinkle in recombination-mediated replication initiation or the fork regression pathway of DNA repair [25]. Interestingly, an alternatively-spliced product, Twinky, lacks part of the *C*-terminus, exists as monomers and has no enzymatic activity [23]. The function of Twinky remains unclear, as it cannot localize to the mitochondrial nucleoids [17] nor associate with Twinkle [23], despite the fact that Twinky contains the proposed MTS at the *N*-terminus (Figure 2). This suggests that the unique *C*-terminus of Twinkle may contain an additional sequence that is also important for its mitochondrial localization. Recombinant human Twinkle, combined with Pol γ purified from insect cells, is sufficient to form the minimal mammalian mtDNA replisome [26]. The hexameric Twinkle ring can efficiently bind to the single-stranded region of a closed circular DNA without a helicase loader and support DNA synthesis by Pol γ through the duplex region [26]. The helicase activity of Twinkle is stimulated by mitochondrial single-stranded DNA-binding protein (mtSSB) [22,27].

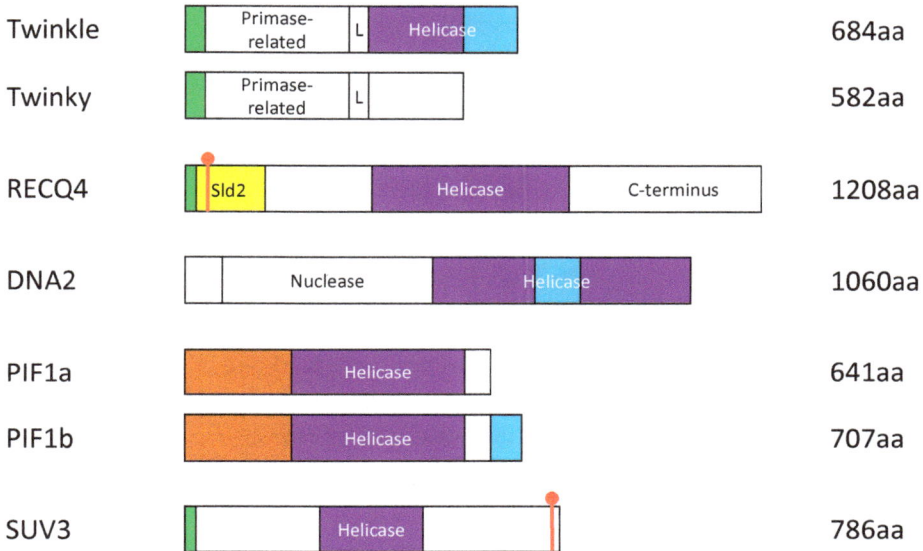

Figure 2. Schematic diagram of the protein domains and alternatively-spliced variants of the human DNA helicases that have known functions in the mitochondrion. Green: mitochondrial targeting sequence (MTS). Red: nuclear localization signal (NLS). Purple: helicase domain. Blue: non-MTS sequence required for mitochondrial localization. Brown: arginine-rich region where potential NLSs reside. Yellow: unique sld2-like domain. L = linker region.

Given the essential role of Twinkle in mtDNA synthesis, mtDNA stability is greatly influenced by the Twinkle expression level in cells. For example, overexpression of the wild-type Twinkle is associated with increased mtDNA copy number in skeletal muscle in mice and reduced ROS-induced mtDNA mutations [28,29], whereas depletion of the Twinkle protein by small interfering RNA (siRNA) leads to a significant decrease in mtDNA copy number [30]. Furthermore, increasing evidence has linked a set of mutations, which change the stability and enzymatic activity of Twinkle [31], to a wide range of diseases [32], such as mitochondrial myopathy [33] and autosomal dominant progressive external ophthalmoplegia (adPEO) [34]. Individuals suffering from adPEO bear multiple deletions in their mitochondrial genome and exhibit multiple symptoms, including muscle weakening, hearing loss, nerve damage and Parkinsonism [35,36].

3. The Involvement of the Nuclear DNA Helicases

3.1. RecQ-Like Helicase 4 (RECQ4)

The gene that encodes RecQ-like helicase 4 (RECQ4) was first identified and cloned based on its limited sequence homology to the highly-conserved RECQ family of superfamily 2 (SF2) DNA helicases [37]. RECQ4 mutations were later identified in patients suffering from Rothmund-Thomson syndrome (RTS), Baller-Gerold syndrome and RAPADILINO (RAdial hypo-/aplasia, PAtellae hypo-/aplasia and cleft or highly arched PAlate, DIarrhea and DIslocated joints, LIttle size and LImb malformation, NOse slender and NOrmal intelligence) syndrome, with phenotypes ranging from premature aging to cancer predisposition [38]. *In vitro*, purified recombinant RECQ4 proteins exist as multimeric proteins and unwind DNA in a 3'–5' direction [39]. Interestingly, RECQ4 not only unwinds DNA, but also exhibits strong DNA annealing activity [40]. In addition to the conserved SF2 helicase domain, the vertebrate RECQ4 contains a unique Sld2-like *N*-terminus (Figure 2) that resembles the essential yeast DNA replication initiation factor Sld2 [41]. Researchers have shown that RECQ4 forms a chromatin-specific complex via this Sld2-like *N*-terminal domain with the MCM2-7 replicative helicase complex and participates in nuclear DNA replication initiation [42–47]. This function explains why *recq4* knockout results in embryonic lethality in mice. Furthermore, the expression of a RECQ4 fragment containing only the Sld2-like *N*-terminal domain is sufficient to support embryonic development [48,49]. The *C*-terminus of RECQ4, which is highly conserved among vertebrates, contains a putative RecQ-*C*-terminal domain (RQC) [50]. Although this *C*-terminal domain is not required for unperturbed DNA replication, a recent study suggests that it is crucial for replication elongation when cells are exposed to ionizing radiation [51]. It has been demonstrated with other members of the RECQ family helicases that the RQC domain is important for the DNA unwinding activity [52]. Therefore, it is possible that the helicase activity of RECQ4 is involved in stabilizing or repairing the damaged replication forks. Because many disease-associated RECQ4 mutations disrupt the conserved *C*-terminal domain [38], understanding the potential function of RECQ4 in replication fork stability in response to ionizing radiation may provide important insight into the pathogenicity of these diseases.

In addition to affecting nuclear DNA replication, RECQ4 expression level also affects mtDNA copy number [53]. Consistent with this, RECQ4 also localizes to the mitochondrion [16,53–56], and the existence of a MTS within the first 20 aa has been proposed [16]. That said, whether RECQ4 is targeted to the mitochondrion via the conventional MTS remains to be validated. Given that RECQ4 interacts with the nuclear DNA replicative helicase complex and plays a critical role in nuclear DNA replication [42–47], it is possible that RECQ4 might have a similar role in mtDNA synthesis. Indeed, in a recent study from our laboratory, we reported a weak interaction between RECQ4 and the mitochondrial replicative helicase Twinkle that can be detected in human whole-cell extracts [56]. Perhaps most surprisingly, we found that this interaction between RECQ4 and Twinkle was significantly enhanced in human cells carrying the most common lymphoma-prone RECQ4 mutation: c.1390+2 delT. This mutation produces RECQ4 polypeptides lacking Ala420-Ala463 residues immediately upstream of the conserved helicase domain [56]. As a consequence, there is increased mtDNA synthesis, leading to an increase in the mtDNA copy number and mitochondrial dysfunction in these cells. Clearly, residues Ala420–Ala463, which are missing in this cancer-prone RECQ4 mutant, have an important inhibitory role in mtDNA synthesis, and we further elucidated how this regulation works. We found that residues Ala420–Ala463 of RECQ4 are required for the interaction with p32, and this interaction negatively regulates RECQ4 mitochondrial localization [56]. p32, which resides in both the mitochondrion and the nucleus [56,57], is involved in regulating mitochondrial innate immunity [58], energy production [57,59] and mitochondrial protein translocation [60]. Cells expressing RECQ4 mutants that lack these 44 aa show defective RECQ4-p32 interactions, increased RECQ4 mutant proteins in the mitochondrion and decreased nuclear RECQ4, suggesting that the excess of RECQ4 molecules in the mitochondrion likely results from increased nuclear-mitochondrial transport [56]. Therefore, this work presents a model for the mechanism used by cells to balance the distribution of RECQ4 in the nucleus and mitochondrion via direct protein-protein interaction.

Although mitochondrial localization of RECQ4 is restricted by p32, RECQ4 itself has also been suggested to function as a positive regulator of the mitochondrial transport of the p53 tumor suppressor via a direct protein–protein interaction [16]. Interestingly, mitochondrial localization of p53 can be blocked by the chaperone protein nucleophosmin (NPM) [61], which was found to also interact with RECQ4 in the nucleoplasm [56]. Although the domain of RECQ4 that interacts with NPM remains to be determined, it is tempting to speculate that NPM inhibits the mitochondrial transport of p53 via its interaction with RECQ4 in the nucleus. In summary, RECQ4 is a dynamic interacting protein, and its protein-protein interactions not only govern the rate of nuclear and mitochondrial DNA synthesis, but also regulate its cellular localization.

3.2. DNA Replication Helicase/Nuclease 2 (DNA2)

The DNA replication helicase/nuclease 2 (*DNA2*) gene was first isolated from a genetic screen in budding yeast [62], and the human *DNA2* gene was later identified based on its sequence homology to the yeast counterpart [63]. Human DNA2, a 120-kDa polypeptide, has two independent functional domains: the *N*-terminal nuclease and the *C*-terminal helicase domain (Figure 2). DNA2 is highly conserved among the eukaryotes. As such, expression of either the human or *Xenopus laevis* DNA2 complements the temperature sensitivity of a *DNA2* (*DNA2-1*) mutation in budding yeast [64]. However, the specificity of DNA2 enzymatic activity may vary across species due to the widely divergent sequences of the distal *N*-terminal regions [65,66]. DNA2 proteins purified from human cells and insect cells show that the nuclease domain has both 5'–3' and 3'–5' nuclease activities [65,67], whereas the helicase unwinds dsDNA that contains a 5' flap as a tail [67]. Data from yeast studies suggest that the 5'–3' helicase activity of DNA2 facilitates the production of a 5' flap structure, a substrate of DNA2 nuclease activity *in vitro* [68]. Nonetheless, the yeast helicase activity *in vivo* is dispensable for cell growth under normal conditions [69,70]. It remains to be determined if this is also the case in higher eukaryotes and if other DNA helicases can compensate for the DNA2 helicase activity.

In the nucleus, DNA2 interacts with proliferating cell nuclear antigen, also known as PCNA, a protein that is important for replication processivity and prevents the accumulation of DNA double-strand breaks (DSBs) during replication [71]. In addition, DNA2 interacts with the Fanconi anemia complementation group D2 (FANCD2) protein and functions in the FANCD2-dependent interstrand crosslink repair pathway [72]. Cells with depleted DNA2 show increased DSBs [71], internuclear chromatin bridges [73] and increased sensitivity to interstrand crosslinking agents due to a reduced homologous recombination frequency [72]. Furthermore, DNA2 participates in long-range DNA resection, in concert with the Werner syndrome ATP-dependent helicase (WRN) and the Bloom syndrome protein (BLM), in DSB repair [74–76]. DNA2 also stimulates BLM helicase activity [75]. Recently, DNA2 was also implicated in telomere maintenance based on its ability to cleave G-quadruplex DNA, and heterozygous DNA2 knockout mice were found to be prone to telomeric DNA damage and aneuploidy [77].

Although DNA2 can localize to the nucleus and play a role in nuclear DNA repair, immunofluorescence microscopy data suggest that the majority of the DNA2 molecules are found in the mitochondrion [73,78]. DNA2 does not contain a classical MTS/ITS, but its localization to the mitochondrion requires the sequence located within 734 and 829 aa [78]. It remains unclear how cells regulate the distribution of DNA2 in the mitochondrion and the nucleus in response to either the cell cycle or DNA damage. DNA2 interacts with and stimulates Pol γ in the mitochondrion and is thought to also function in concert with flap structure-specific endonuclease 1 (FEN1) to process 5'-flap intermediates and participate in repairing oxidative lesions in mtDNA by long-range base excision repair [78]. Indeed, DNA2 proteins colocalize with mtDNA nucleoids and Twinkle and through an unknown mechanism; this localization increases in cells carrying some of the adPEO-associated Twinkle mutations [73]. Interestingly, point mutations in the *DNA2* gene itself have also been linked to adPEO, and these patients show progressive myopathy with mitochondrial dysfunction [79]. Importantly, one mutation located within the helicase domain altered the DNA unwinding efficiency [79], suggesting that the helicase activity of DNA2 has an important role in mtDNA maintenance in humans. Therefore, similar to Twinkle, DNA2 is important for maintaining healthy mitochondrial DNA and preventing related diseases.

3.3. Petite Integration Frequency 1 (PIF1)

PIF1, which stands for petite integration frequency 1, is conserved in both budding yeast and humans [80–83]. PIF1 is a member of the superfamily 1 (SF1) helicase family and has 5'–3' DNA unwinding activity (Figure 2) [84–86]. Similar to RECQ4 and DNA2, PIF1 localizes to both the nucleus and the mitochondrion [83]. However, unlike RECQ4 and DNA2, PIF1 mitochondrial localization in human cells is regulated by alternative splicing, which produces α and β isoforms [83]. Both the α and β isoforms contain the intact helicase domain and the *N*-terminus (Figure 2), which has arginine-rich nuclear localization signals [83] and is important for the interaction with ssDNA [84].

The PIF1 α isoform consists of 641 aa and has a short distal *C*-terminus. This isoform localizes to the nucleus [83], and PIF1 function in the nucleus has been extensively demonstrated [83,85,87,88]. The expression of PIF1 is cell cycle regulated, and the downregulation of PIF1 leads to cell cycle delay [81,83]. Both yeast and human PIF1 bind DNA and promote DNA replication through interaction with G-quadruplex DNA regions [86,87,89–92]. This activity is important for maintaining telomere integrity and for resolving stalled replication forks [85]. Reduction of PIF1α by siRNA knockdown decreases cancer cell survival, but has no impact on non-malignant cells [93], and this is likely due to its role in restarting stalled replication forks [85,94].

The PIF1 β isoform (707 aa) has a long distal *C*-terminus with a lipocalin motif (protein secretion signal; Figure 2). This *C*-terminal region results from alternative splicing and is not present in the α isoform. PIF1β is expected to have similar biochemical properties, compared to PIF1α, as they contain the same helicase domain. However, unlike the α isoform, the majority of this β isoform localizes to the mitochondrion, with some residual nuclear signal [83]. Evidence from yeast studies suggests that PIF1 may associate with mtDNA and mitochondrial inner membranes [95] and contribute to reducing DSBs in mtDNA [96]. Furthermore, it is required for repairing UV- and ethidium bromide-damaged mtDNA [80]. In addition, Twinkle, which cannot efficiently unwind G-quadruplex DNA [97], may rely on PIF1 helicase activity to remove G-quadruplexes, which could potentially lead to mtDNA deletions. Nonetheless, it is unknown how the distal *C*-terminus, which is unique to PIF1β, promotes its mitochondrial localization and how PIF1β protects mtDNA from DSBs. Interestingly, deletion of *PIF1* rescued the lethal phenotype of DNA2 in budding yeast, suggesting that PIF1 and DNA2 may be involved in similar, but non-redundant pathways in the mitochondrion [98].

3.4. Suppressor of Var1 3-Like Protein 1 (SUV3)

SUV3, a member of the DExH-box helicase family, was first identified in budding yeast as the suppressor of var1 (the small subunit of mitochondrial ribosomal protein) [99], and the gene was later found to be conserved in humans [100]. *SUV3* knockout mice are embryonic lethal, whereas heterozygous mice have shortened lifespan and develop tumors at multiple sites, due to a reduced mtDNA copy number and an elevated number of mtDNA mutations [101]. Reduced SUV3 expression was observed in human breast tumor samples [101]. Nonetheless, unlike RECQ4, PIF1 and DNA2 helicases, the effect of SUV3 deficiency on mtDNA copy number and stability is likely indirect. For example, in the mitochondrion, SUV3 forms a complex with polynucleotide phosphorylase (PNPase) to function in mtRNA degradation [102]. Indeed, analysis using purified recombinant human SUV3 proteins demonstrated that SUV3 is an active ATPase and capable of unwinding not only DNA, but also RNA in a 3'–5' direction [102–104]. This SUV3-PNPase complex transiently associates with the mitochondrial polyadenylation polymerase when the inorganic phosphate level is low in the mitochondrial lumen [105]. The three-component complex is capable of regulating the length of the RNA poly(A) tail. Consistent with this, siRNA knockdown leads to an increase in the amount of mtRNA with shorter poly(A) tails, a reduction in mtDNA copy number [106] and an increase in the rate of apoptosis [107]. In addition, expression of a mutant defective in the ATPase function leads to an abnormally high level of mtRNA, due to the slow mRNA turnover rate [108]. Although it remains unclear how a defect in mtRNA degradation contributes to mtDNA instability in SUV3-deficient cells, it is possible that the abnormal level of mtRNA imposes cellular stress, leading to overproduction of ROS and mtDNA damage.

Early studies suggest that SUV3 localizes to the lumen of the mitochondrion, presumably through cleavage of an MTS localized at the distal *N*-terminus (Figure 2) [103,107]. However, recent studies provide evidence that SUV3 also localizes to the nucleus with a potential nuclear localization signal located between residues 777 and 781 at the *C*-terminus [104,107]. In the nucleus, SUV3 interacts with nuclear DNA replication and repair factors, such as the RECQ helicases BLM and WRN [109], as well as replication protein A (RPA) and FEN1 [104]. Therefore, it is possible that, at least in humans, SUV3 is a key player in nuclear genome maintenance due to its participation in DNA damage repair, whereas it maintains mitochondrial genome integrity by participating in mtRNA metabolism. The reason why cells utilize an mtRNA helicase in nuclear DNA damage repair remains unknown. Interestingly, in mammalian cells, there is an increase in the degradation of mtRNA, but not cytoplasmic RNAs, to protect cells in response to oxidative stress [110]. It is possible that the involvement of SUV3 in nuclear DNA repair provides a mechanism for cells to "sense" oxidative DNA damage and induce mtRNA degradation. Therefore, identifying the molecular switch that balances the localization and the two distinct functions of SUV3 might reveal a novel crosstalk between the nucleus and the mitochondrion in response to DNA damage.

4. Conclusions

Given that mitochondria provide the vital ATP energy source needed by diverse cellular processes that support the development of an organism, it is not surprising that abnormal mtDNA copy number and mitochondrial dysfunction have been correlated with a decline in tissue maintenance and regeneration. Tissue degeneration may contribute to some of the symptoms, such as muscle weakening, hearing loss, nerve damage and Parkinsonism observed in the adPEO1 patients [35,36]. Growing evidence also suggests a close association between mitochondrial dysfunction and age-related bone diseases. For example, osteoporosis is a result of the loss of bone mass and is one of the common symptoms associated with aging [12,111]. Studies in mice indicate that increased apoptosis in osteoblasts, due to the accumulation of ROS generated by damaged mitochondria, is one of the main causes of bone loss [112]. Interestingly, RTS patients with RECQ4 mutations show abnormal bone development and osteoporosis at an early age [113]. Therefore, it is possible that these RTS-associated RECQ4 mutations lead to mitochondrial dysfunction and contribute to the premature aging phenotypes.

In addition to their association with tissue degeneration and developmental defects, mitochondria have recently gained attention for their potential use both as diagnostic tools and as therapeutic targets for cancer treatment [11]. Variations in mtDNA copy number are observed in many cancers and correlate with tumor aggressiveness and survival outcome. For example, mtDNA copy number is significantly elevated in various types of lymphoma, including Burkitt lymphoma and non-Hodgkin lymphoma [114–117]. In addition, highly invasive osteosarcoma cells contain enlarged mitochondria and larger amounts of mtDNA, and inhibiting replication of mtDNA in these cells also effectively slows down tumor growth [118–120]. Because mtDNA copy number correlates with cell growth [121], deregulated mtDNA synthesis could be a risk factor that contributes to cancer pathogenesis or that sustains cancer cell growth. Therefore, reducing aberrant mtDNA synthesis in cancer by targeting enzymes involved in mtDNA synthesis or mtDNA repair may be an effective strategy for controlling tumor progression. It would be of great interest for future studies to explore the possibility that the DNA helicases we have summarized here may be cancer drug targets or biomarkers for cancer diagnosis and prevention.

Acknowledgments: We thank Nancy Linford and Keely Walker for their comments and expert editing of this manuscript. Yilun Liu was supported by funding from the National Cancer Institute (R01 CA151245).

Author Contributions: Lin Ding and Yilun Liu wrote the manuscript.

Conflicts of Interest: The authors declare no conflict of interest.

References

1. Kurland, C.G.; Andersson, S.G. Origin and evolution of the mitochondrial proteome. *Microb. Mol. Biol. Rev.* **2000**, *64*, 786–820.
2. Nunnari, J.; Suomalainen, A. Mitochondria: In sickness and in health. *Cell* **2012**, *148*, 1145–1159.
3. Iborra, F.J.; Kimura, H.; Cook, P.R. The functional organization of mitochondrial genomes in human cells. *BMC Biol.* **2004**, *2*, 9.
4. Satoh, M.; Kuroiwa, T. Organization of multiple nucleoids and DNA molecules in mitochondria of a human cell. *Exp. Cell Res.* **1991**, *196*, 137–140.
5. Kelly, R.D.; Mahmud, A.; McKenzie, M.; Trounce, I.A.; St John, J.C. Mitochondrial DNA copy number is regulated in a tissue specific manner by DNA methylation of the nuclear-encoded DNA polymerase gamma A. *Nucleic Acids Res.* **2012**, *40*, 10124–10138.
6. Albring, M.; Griffith, J.; Attardi, G. Association of a protein structure of probable membrane derivation with HeLa cell mitochondrial DNA near its origin of replication. *Proc. Natl. Acad. Sci. USA* **1977**, *74*, 1348–1352.
7. Holt, I.J.; Reyes, A. Human mitochondrial DNA replication. *Cold Spring Harb. Perspect. Biol.* **2012**, *4*, a012971.
8. Kazak, L.; Reyes, A.; Holt, I.J. Minimizing the damage: Repair pathways keep mitochondrial DNA intact. *Nat. Rev. Mol. Cell Biol.* **2012**, *13*, 659–671.
9. Gilkerson, R.W.; Margineantu, D.H.; Capaldi, R.A.; Selker, J.M. Mitochondrial DNA depletion causes morphological changes in the mitochondrial reticulum of cultured human cells. *FEBS Lett.* **2000**, *474*, 1–4.
10. Schapira, A.H. Mitochondrial disease. *Lancet* **2006**, *368*, 70–82.
11. Yu, M. Generation, function and diagnostic value of mitochondrial DNA copy number alterations in human cancers. *Life Sci.* **2011**, *89*, 65–71.
12. Clay Montier, L.L.; Deng, J.J.; Bai, Y. Number matters: Control of mammalian mitochondrial DNA copy number. *J. Genet. Genomics* **2009**, *36*, 125–131.
13. Becker, T.; Bottinger, L.; Pfanner, N. Mitochondrial protein import: From transport pathways to an integrated network. *Trends Biochem. Sci.* **2012**, *37*, 85–91.
14. Schmidt, O.; Pfanner, N.; Meisinger, C. Mitochondrial protein import: From proteomics to functional mechanisms. *Nat. Rev. Mol. Cell Biol.* **2010**, *11*, 655–667.

15. Gebert, N.; Ryan, M.T.; Pfanner, N.; Wiedemann, N.; Stojanovski, D. Mitochondrial protein import machineries and lipids: A functional connection. *Biochim. Biophys. Acta* **2011**, *1808*, 1002–1011.

16. De, S.; Kumari, J.; Mudgal, R.; Modi, P.; Gupta, S.; Futami, K.; Goto, H.; Lindor, N.M.; Furuichi, Y.; Mohanty, D.; *et al.* RECQL4 is essential for the transport of p53 to mitochondria in normal human cells in the absence of exogenous stress. *J. Cell Sci.* **2012**, *125*, 2509–2522.

17. Spelbrink, J.N.; Li, F.Y.; Tiranti, V.; Nikali, K.; Yuan, Q.P.; Tariq, M.; Wanrooij, S.; Garrido, N.; Comi, G.; Morandi, L.; *et al.* Human mitochondrial DNA deletions associated with mutations in the gene encoding Twinkle, a phage T7 gene 4-like protein localized in mitochondria. *Nat. Genet.* **2001**, *28*, 223–231.

18. Westermann, B. Mitochondrial inheritance in yeast. *Biochim. Biophys. Acta* **2014**, *1837*, 1039–1046.

19. Milenkovic, D.; Matic, S.; Kuhl, I.; Ruzzenente, B.; Freyer, C.; Jemt, E.; Park, C.B.; Falkenberg, M.; Larsson, N.G. TWINKLE is an essential mitochondrial helicase required for synthesis of nascent D-loop strands and complete mtDNA replication. *Hum. Mol. Genet.* **2013**, *22*, 1983–1993.

20. Gilkerson, R.; Bravo, L.; Garcia, I.; Gaytan, N.; Herrera, A.; Maldonado, A.; Quintanilla, B. The mitochondrial nucleoid: Integrating mitochondrial DNA into cellular homeostasis. *Cold Spring Harb. Perspect. Biol.* **2013**, *5*, a011080.

21. Bogenhagen, D.F.; Rousseau, D.; Burke, S. The layered structure of human mitochondrial DNA nucleoids. *J. Biol. Chem.* **2008**, *283*, 3665–3675.

22. Korhonen, J.A.; Gaspari, M.; Falkenberg, M. TWINKLE has 5'→3' DNA helicase activity and is specifically stimulated by mitochondrial single-stranded DNA-binding protein. *J. Biol. Chem.* **2003**, *278*, 48627–48632.

23. Farge, G.; Holmlund, T.; Khvorostova, J.; Rofougaran, R.; Hofer, A.; Falkenberg, M. The N-terminal domain of TWINKLE contributes to single-stranded DNA binding and DNA helicase activities. *Nucleic Acids Res.* **2008**, *36*, 393–403.

24. Korhonen, J.A.; Pande, V.; Holmlund, T.; Farge, G.; Pham, X.H.; Nilsson, L.; Falkenberg, M. Structure–function defects of the TWINKLE linker region in progressive external ophthalmoplegia. *J. Mol. Biol.* **2008**, *377*, 691–705.

25. Sen, D.; Nandakumar, D.; Tang, G.Q.; Patel, S.S. Human mitochondrial DNA helicase TWINKLE is both an unwinding and annealing helicase. *J. Biol. Chem.* **2012**, *287*, 14545–14556.

26. Jemt, E.; Farge, G.; Backstrom, S.; Holmlund, T.; Gustafsson, C.M.; Falkenberg, M. The mitochondrial DNA helicase TWINKLE can assemble on a closed circular template and support initiation of DNA synthesis. *Nucleic Acids Res.* **2011**, *39*, 9238–9249.

27. Korhonen, J.A.; Pham, X.H.; Pellegrini, M.; Falkenberg, M. Reconstitution of a minimal mtDNA replisome *in vitro.*. *EMBO J.* **2004**, *23*, 2423–2429.

28. Ylikallio, E.; Tyynismaa, H.; Tsutsui, H.; Ide, T.; Suomalainen, A. High mitochondrial DNA copy number has detrimental effects in mice. *Hum. Mol. Genet.* **2010**, *19*, 2695–2705.

71

29. Pohjoismaki, J.L.; Williams, S.L.; Boettger, T.; Goffart, S.; Kim, J.; Suomalainen, A.; Moraes, C.T.; Braun, T. Over-expression of Twinkle-helicase protects cardiomyocytes from genotoxic stress caused by reactive oxygen species. *Proc. Natl. Acad. Sci. USA* **2013**, *110*, 19408–19013.

30. Tyynismaa, H.; Sembongi, H.; Bokori-Brown, M.; Granycome, C.; Ashley, N.; Poulton, J.; Jalanko, A.; Spelbrink, J.N.; Holt, I.J.; Suomalainen, A. Twinkle helicase is essential for mtDNA maintenance and regulates mtDNA copy number. *Hum. Mol. Genet.* **2004**, *13*, 3219–3227.

31. Longley, M.J.; Humble, M.M.; Sharief, F.S.; Copeland, W.C. Disease variants of the human mitochondrial DNA helicase encoded by C10orf2 differentially alter protein stability, nucleotide hydrolysis, and helicase activity. *J. Biol. Chem.* **2010**, *285*, 29690–29702.

32. Greaves, L.C.; Reeve, A.K.; Taylor, R.W.; Turnbull, D.M. Mitochondrial DNA and disease. *J. Pathol.* **2012**, *226*, 274–286.

33. Arenas, J.; Briem, E.; Dahl, H.; Hutchison, W.; Lewis, S.; Martin, M.A.; Spelbrink, H.; Tiranti, V.; Jacobs, H.; Zeviani, M. The V368i mutation in Twinkle does not segregate with adPEO. *Ann. Neurol.* **2003**, *53*, 278.

34. Li, F.Y.; Tariq, M.; Croxen, R.; Morten, K.; Squier, W.; Newsom-Davis, J.; Beeson, D.; Larsson, C. Mapping of autosomal dominant progressive external ophthalmoplegia to a 7-cM critical region on 10q24. *Neurology* **1999**, *53*, 1265–1271.

35. Zeviani, M.; Servidei, S.; Gellera, C.; Bertini, E.; DiMauro, S.; DiDonato, S. An autosomal dominant disorder with multiple deletions of mitochondrial DNA starting at the D-loop region. *Nature* **1989**, *339*, 309–311.

36. Moslemi, A.R.; Melberg, A.; Holme, E.; Oldfors, A. Autosomal dominant progressive external ophthalmoplegia: Distribution of multiple mitochondrial DNA deletions. *Neurology* **1999**, *53*, 79–84.

37. Kitao, S.; Ohsugi, I.; Ichikawa, K.; Goto, M.; Furuichi, Y.; Shimamoto, A. Cloning of two new human helicase genes of the RecQ family: Biological significance of multiple species in higher eukaryotes. *Genomics* **1998**, *54*, 443–452.

38. Liu, Y. Rothmund-Thomson syndrome helicase, RECQ4: On the crossroad between DNA replication and repair. *DNA Repair. (Amst.)* **2010**, *9*, 325–330.

39. Suzuki, T.; Kohno, T.; Ishimi, Y. DNA helicase activity in purified human RECQL4 protein. *J. Biochem.* **2009**, *146*, 327–335.

40. Xu, X.; Liu, Y. Dual DNA unwinding activities of the Rothmund-Thomson syndrome protein, RECQ4. *EMBO J.* **2009**, *28*, 568–577.

41. Kamimura, Y.; Masumoto, H.; Sugino, A.; Araki, H. Sld2, which interacts with Dpb11 in Saccharomyces cerevisiae, is required for chromosomal DNA replication. *Mol. Cell. Biol.* **1998**, *18*, 6102–6109.

42. Im, J.S.; Ki, S.H.; Farina, A.; Jung, D.S.; Hurwitz, J.; Lee, J.K. Assembly of the Cdc45-Mcm2-7-GINS complex in human cells requires the Ctf4/And-1, RecQL4, and Mcm10 proteins. *Proc. Natl. Acad. Sci. USA* **2009**, *106*, 15628–15632.

43. Xu, X.; Rochette, P.J.; Feyissa, E.A.; Su, T.V.; Liu, Y. MCM10 mediates RECQ4 association with MCM2-7 helicase complex during DNA replication. *EMBO J.* **2009**, *28*, 3005–3014.

44. Thangavel, S.; Mendoza-Maldonado, R.; Tissino, E.; Sidorova, J.M.; Yin, J.; Wang, W.; Monnat, R.J., Jr.; Falaschi, A.; Vindigni, A. Human RECQ1 and RECQ4 helicases play distinct roles in DNA replication initiation. *Mol. Cell. Biol.* **2010**, *30*, 1382–1396.

45. Im, J.S.; Park, S.Y.; Cho, W.H.; Bae, S.H.; Hurwitz, J.; Lee, J.K. RecQL4 is required for the association of Mcm10 and Ctf4 with replication origins in human cells. *Cell Cycle* **2015**, *14*, 1001–1009.

46. Sangrithi, M.N.; Bernal, J.A.; Madine, M.; Philpott, A.; Lee, J.; Dunphy, W.G.; Venkitaraman, A.R. Initiation of DNA replication requires the RECQL4 protein mutated in Rothmund-Thomson syndrome. *Cell* **2005**, *121*, 887–898.

47. Matsuno, K.; Kumano, M.; Kubota, Y.; Hashimoto, Y.; Takisawa, H. The *N*-terminal noncatalytic region of Xenopus RecQ4 is required for chromatin binding of DNA polymerase α in the initiation of DNA replication. *Mol. Cell. Biol.* **2006**, *26*, 4843–4852.

48. Ichikawa, K.; Noda, T.; Furuichi, Y. Preparation of the gene targeted knockout mice for human premature aging diseases, Werner syndrome, and Rothmund-Thomson syndrome caused by the mutation of DNA helicases. *Nihon Yakurigaku Zasshi* **2002**, *119*, 219–226.

49. Mann, M.B.; Hodges, C.A.; Barnes, E.; Vogel, H.; Hassold, T.J.; Luo, G. Defective sister-chromatid cohesion, aneuploidy and cancer predisposition in a mouse model of type II Rothmund-Thomson syndrome. *Hum. Mol. Genet.* **2005**, *14*, 813–825.

50. Marino, F.; Vindigni, A.; Onesti, S. Bioinformatic analysis of RecQ4 helicases reveals the presence of a RQC domain and a Zn knuckle. *Biophys. Chem.* **2013**, *177*, 34–39.

51. Kohzaki, M.; Chiourea, M.; Versini, G.; Adachi, N.; Takeda, S.; Gagos, S.; Halazonetis, T.D. The helicase domain and C-terminus of human RecQL4 facilitate replication elongation on DNA templates damaged by ionizing radiation. *Carcinogenesis* **2012**, *33*, 1203–1210.

52. Lucic, B.; Zhang, Y.; King, O.; Mendoza-Maldonado, R.; Berti, M.; Niesen, F.H.; Burgess-Brown, N.A.; Pike, A.C.; Cooper, C.D.; Gileadi, O.; *et al.* A prominent beta-hairpin structure in the winged-helix domain of RECQ1 is required for DNA unwinding and oligomer formation. *Nucleic Acids Res.* **2011**, *39*, 1703–1717.

53. Chi, Z.; Nie, L.; Peng, Z.; Yang, Q.; Yang, K.; Tao, J.; Mi, Y.; Fang, X.; Balajee, A.S.; Zhao, Y. RecQL4 cytoplasmic localization: Implications in mitochondrial DNA oxidative damage repair. *Int. J. Biochem. Cell Biol.* **2012**, *44*, 1942–1951.

54. Croteau, D.L.; Rossi, M.L.; Canugovi, C.; Tian, J.; Sykora, P.; Ramamoorthy, M.; Wang, Z.M.; Singh, D.K.; Akbari, M.; Kasiviswanathan, R.; *et al.* RECQL4 localizes to mitochondria and preserves mitochondrial DNA integrity. *Aging Cell* **2012**, *11*, 456–466.

55. Gupta, S.; De, S.; Srivastava, V.; Hussain, M.; Kumari, J.; Muniyappa, K.; Sengupta, S. RECQL4 and p53 potentiate the activity of polymerase gamma and maintain the integrity of the human mitochondrial genome. *Carcinogenesis* **2014**, *35*, 34–45.

56. Wang, J.T.; Xu, X.; Alontaga, A.Y.; Chen, Y.; Liu, Y. Impaired p32 regulation caused by the lymphoma-prone RECQ4 mutation drives mitochondrial dysfunction. *Cell Rep.* **2014**, *7*, 848–858.

57. Muta, T.; Kang, D.; Kitajima, S.; Fujiwara, T.; Hamasaki, N. p32 protein, a splicing factor 2-associated protein, is localized in mitochondrial matrix and is functionally important in maintaining oxidative phosphorylation. *J. Biol. Chem.* **1997**, *272*, 24363–24370.

58. West, A.P.; Shadel, G.S.; Ghosh, S. Mitochondria in innate immune responses. *Nat. Rev. Immunol.* **2011**, *11*, 389–402.

59. Fogal, V.; Richardson, A.D.; Karmali, P.P.; Scheffler, I.E.; Smith, J.W.; Ruoslahti, E. Mitochondrial p32 protein is a critical regulator of tumor metabolism via maintenance of oxidative phosphorylation. *Mol. Cell. Biol.* **2010**, *30*, 1303–1318.

60. Itahana, K.; Zhang, Y. Mitochondrial p32 is a critical mediator of ARF-induced apoptosis. *Cancer Cell* **2008**, *13*, 542–553.

61. Dhar, S.K.; St Clair, D.K. Nucleophosmin blocks mitochondrial localization of p53 and apoptosis. *J. Biol. Chem.* **2009**, *284*, 16409–16418.

62. Dumas, L.B.; Lussky, J.P.; McFarland, E.J.; Shampay, J. New temperature-sensitive mutants of *Saccharomyces cerevisiae* affecting DNA replication. *Mol. Gen. Genet.* **1982**, *187*, 42–46.

63. Eki, T.; Okumura, K.; Shiratori, A.; Abe, M.; Nogami, M.; Taguchi, H.; Shibata, T.; Murakami, Y.; Hanaoka, F. Assignment of the closest human homologue (DNA2L:KIAA0083) of the yeast DNA2 helicase gene to chromosome band 10q21.3-q22.1. *Genomics* **1996**, *37*, 408–410.

64. Imamura, O.; Campbell, J.L. The human Bloom syndrome gene suppresses the DNA replication and repair defects of yeast DNA2 mutants. *Proc. Natl. Acad. Sci. USA* **2003**, *100*, 8193–8198.

65. Kim, J.H.; Kim, H.D.; Ryu, G.H.; Kim, D.H.; Hurwitz, J.; Seo, Y.S. Isolation of human DNA2 endonuclease and characterization of its enzymatic properties. *Nucleic Acids Res.* **2006**, *34*, 1854–1864.

66. Lee, C.H.; Lee, M.; Kang, H.J.; Kim, D.H.; Kang, Y.H.; Bae, S.H.; Seo, Y.S. The *N*-terminal 45-kDa domain of DNA2 endonuclease/helicase targets the enzyme to secondary structure DNA. *J. Biol. Chem.* **2013**, *288*, 9468–9481.

67. Masuda-Sasa, T.; Imamura, O.; Campbell, J.L. Biochemical analysis of human DNA2. *Nucleic Acids Res.* **2006**, *34*, 1865–1875.

68. Bae, S.H.; Kim, D.W.; Kim, J.; Kim, J.H.; Kim, D.H.; Kim, H.D.; Kang, H.Y.; Seo, Y.S. Coupling of DNA helicase and endonuclease activities of yeast DNA2 facilitates Okazaki fragment processing. *J. Biol. Chem.* **2002**, *277*, 26632–26641.

69. Hu, J.; Sun, L.; Shen, F.; Chen, Y.; Hua, Y.; Liu, Y.; Zhang, M.; Hu, Y.; Wang, Q.; Xu, W.; *et al.* The intra-S phase checkpoint targets DNA2 to prevent stalled replication forks from reversing. *Cell* **2012**, *149*, 1221–1232.

70. Formosa, T.; Nittis, T. DNA2 mutants reveal interactions with DNA polymerase α and Ctf4, a Pol α accessory factor, and show that full DNA2 helicase activity is not essential for growth. *Genetics* **1999**, *151*, 1459–1470.

71. Peng, G.; Dai, H.; Zhang, W.; Hsieh, H.J.; Pan, M.R.; Park, Y.Y.; Tsai, R.Y.; Bedrosian, I.; Lee, J.S.; Ira, G.; *et al.* Human nuclease/helicase DNA2 alleviates replication stress by promoting DNA end resection. *Cancer Res.* **2012**, *72*, 2802–2813.

72. Karanja, K.K.; Cox, S.W.; Duxin, J.P.; Stewart, S.A.; Campbell, J.L. DNA2 and EXO1 in replication-coupled, homology-directed repair and in the interplay between HDR and the FA/BRCA network. *Cell Cycle* **2012**, *11*, 3983–3996.

73. Duxin, J.P.; Dao, B.; Martinsson, P.; Rajala, N.; Guittat, L.; Campbell, J.L.; Spelbrink, J.N.; Stewart, S.A. Human DNA2 is a nuclear and mitochondrial DNA maintenance protein. *Mol. Cell. Biol.* **2009**, *29*, 4274–4282.

74. Nimonkar, A.V.; Genschel, J.; Kinoshita, E.; Polaczek, P.; Campbell, J.L.; Wyman, C.; Modrich, P.; Kowalczykowski, S.C. BLM-DNA2-RPA-MRN and EXO1-BLM-RPA-MRN constitute two DNA end resection machineries for human DNA break repair. *Genes Dev.* **2011**, *25*, 350–362.

75. Daley, J.M.; Chiba, T.; Xue, X.; Niu, H.; Sung, P. Multifaceted role of the Topo IIIα-RMI1-RMI2 complex and DNA2 in the BLM-dependent pathway of DNA break end resection. *Nucleic Acids Res.* **2014**, *42*, 11083–11091.

76. Sturzenegger, A.; Burdova, K.; Kanagaraj, R.; Levikova, M.; Pinto, C.; Cejka, P.; Janscak, P. DNA2 cooperates with the WRN and BLM RecQ helicases to mediate long-range DNA end resection in human cells. *J. Biol. Chem.* **2014**, *289*, 27314–27326.

77. Lin, W.; Sampathi, S.; Dai, H.; Liu, C.; Zhou, M.; Hu, J.; Huang, Q.; Campbell, J.; Shin-Ya, K.; Zheng, L.; *et al.* Mammalian DNA2 helicase/nuclease cleaves G-quadruplex DNA and is required for telomere integrity. *EMBO J.* **2013**, *32*, 1425–1439.

78. Zheng, L.; Zhou, M.; Guo, Z.; Lu, H.; Qian, L.; Dai, H.; Qiu, J.; Yakubovskaya, E.; Bogenhagen, D.F.; Demple, B.; *et al.* Human DNA2 is a mitochondrial nuclease/helicase for efficient processing of DNA replication and repair intermediates. *Mol. Cell* **2008**, *32*, 325–336.

79. Ronchi, D.; Di Fonzo, A.; Lin, W.; Bordoni, A.; Liu, C.; Fassone, E.; Pagliarani, S.; Rizzuti, M.; Zheng, L.; Filosto, M.; *et al.* Mutations in DNA2 link progressive myopathy to mitochondrial DNA instability. *Am. J. Hum. Genet.* **2013**, *92*, 293–300.

80. Foury, F.; Kolodynski, J. PIF mutation blocks recombination between mitochondrial ρ$^+$ and ρ$^-$ genomes having tandemly arrayed repeat units in Saccharomyces cerevisiae. *Proc. Natl. Acad. Sci. USA* **1983**, *80*, 5345–5349.

81. Mateyak, M.K.; Zakian, V.A. Human PIF helicase is cell cycle regulated and associates with telomerase. *Cell Cycle* **2006**, *5*, 2796–2804.

82. Zhang, D.H.; Zhou, B.; Huang, Y.; Xu, L.X.; Zhou, J.Q. The human PIF1 helicase, a potential Escherichia coli RecD homologue, inhibits telomerase activity. *Nucleic Acids Res.* **2006**, *34*, 1393–1404.

83. Futami, K.; Shimamoto, A.; Furuichi, Y. Mitochondrial and nuclear localization of human PIF1 helicase. *Biol. Pharm. Bull.* **2007**, *30*, 1685–1692.

84. Gu, Y.; Masuda, Y.; Kamiya, K. Biochemical analysis of human PIF1 helicase and functions of its N-terminal domain. *Nucleic Acids Res.* **2008**, *36*, 6295–6308.

85. George, T.; Wen, Q.; Griffiths, R.; Ganesh, A.; Meuth, M.; Sanders, C.M. Human PIF1 helicase unwinds synthetic DNA structures resembling stalled DNA replication forks. *Nucleic Acids Res.* **2009**, *37*, 6491–6502.

86. Sanders, C.M. Human PIF1 helicase is a G-quadruplex DNA-binding protein with G-quadruplex DNA-unwinding activity. *Biochem. J.* **2010**, *430*, 119–128.

87. Bochman, M.L.; Sabouri, N.; Zakian, V.A. Unwinding the functions of the PIF1 family helicases. *DNA Repair (Amst.)* **2010**, *9*, 237–249.

88. Gu, Y.; Wang, J.; Li, S.; Kamiya, K.; Chen, X.; Zhou, P. Determination of the biochemical properties of full-length human PIF1 ATPase. *Prion* **2013**, *7*, 341–347.

89. Duan, X.L.; Liu, N.N.; Yang, Y.T.; Li, H.H.; Li, M.; Dou, S.X.; Xi, X. G-quadruplexes significantly stimulate PIF1 helicase-catalyzed duplex DNA unwinding. *J. Biol. Chem.* **2015**, *290*, 7722–7735.

90. Hou, X.M.; Wu, W.Q.; Duan, X.L.; Liu, N.N.; Li, H.H.; Fu, J.; Dou, S.X.; Li, M.; Xi, X.G. Molecular mechanism of G-quadruplex unwinding helicase: Sequential and repetitive unfolding of G-quadruplex by PIF1 helicase. *Biochem. J.* **2015**, *466*, 189–199.

91. Zhou, R.; Zhang, J.; Bochman, M.L.; Zakian, V.A.; Ha, T. Periodic DNA patrolling underlies diverse functions of PIF1 on R-loops and G-rich DNA. *Elife* **2014**, *3*, e02190.

92. Byrd, A.K.; Raney, K.D. A parallel quadruplex DNA is bound. Tightly but unfolded slowly by PIF1 helicase. *J. Biol. Chem.* **2015**, *290*, 6482–6494.

93. Gagou, M.E.; Ganesh, A.; Thompson, R.; Phear, G.; Sanders, C.; Meuth, M. Suppression of apoptosis by PIF1 helicase in human tumor cells. *Cancer Res.* **2011**, *71*, 4998–5008.

94. Gagou, M.E.; Ganesh, A.; Phear, G.; Robinson, D.; Petermann, E.; Cox, A.; Meuth, M. Human PIF1 helicase supports DNA replication and cell growth under oncogenic-stress. *Oncotarget* **2014**, *5*, 11381–11398.

95. Cheng, X.; Ivessa, A.S. Association of the yeast DNA helicase PIF1p with mitochondrial membranes and mitochondrial DNA. *Eur. J. Cell Biol.* **2010**, *89*, 742–747.

96. Cheng, X.; Dunaway, S.; Ivessa, A.S. The role of Pif1p, a DNA helicase in *Saccharomyces cerevisiae*, in maintaining mitochondrial DNA. *Mitochondrion* **2007**, *7*, 211–222.

97. Bharti, S.K.; Sommers, J.A.; Zhou, J.; Kaplan, D.L.; Spelbrink, J.N.; Mergny, J.L.; Brosh, R.M., Jr. DNA sequences proximal to human mitochondrial DNA deletion breakpoints prevalent in human disease form G-quadruplexes, a class of DNA structures inefficiently unwound by the mitochondrial replicative Twinkle helicase. *J. Biol. Chem.* **2014**, *289*, 29975–29993.

98. Budd, M.E.; Reis, C.C.; Smith, S.; Myung, K.; Campbell, J.L. Evidence suggesting that PIF1 helicase functions in DNA replication with the DNA2 helicase/nuclease and DNA polymerase delta. *Mol. Cell. Biol.* **2006**, *26*, 2490–2500.

99. Butow, R.A.; Zhu, H.; Perlman, P.; Conrad-Webb, H. The role of a conserved dodecamer sequence in yeast mitochondrial gene expression. *Genome* **1989**, *31*, 757–760.

100. Dmochowska, A.; Kalita, K.; Krawczyk, M.; Golik, P.; Mroczek, K.; Lazowska, J.; Stepien, P.P.; Bartnik, E. A human putative SUV3-like RNA helicase is conserved between Rhodobacter and all eukaryotes. *Acta Biochim. Pol.* **1999**, *46*, 155–162.

101. Chen, P.L.; Chen, C.F.; Chen, Y.; Guo, X.E.; Huang, C.K.; Shew, J.Y.; Reddick, R.L.; Wallace, D.C.; Lee, W.H. Mitochondrial genome instability resulting from SUV3 haploinsufficiency leads to tumorigenesis and shortened lifespan. *Oncogene* **2013**, *32*, 1193–1201.

102. Wang, D.D.; Shu, Z.; Lieser, S.A.; Chen, P.L.; Lee, W.H. Human mitochondrial SUV3 and polynucleotide phosphorylase form a 330-kDa heteropentamer to cooperatively degrade double-stranded RNA with a 3'-to-5' directionality. *J. Biol. Chem.* **2009**, *284*, 20812–20821.

103. Minczuk, M.; Piwowarski, J.; Papworth, M.A.; Awiszus, K.; Schalinski, S.; Dziembowski, A.; Dmochowska, A.; Bartnik, E.; Tokatlidis, K.; Stepien, P.P.; *et al.* Localisation of the human hSUV3p helicase in the mitochondrial matrix and its preferential unwinding of dsDNA. *Nucleic Acids Res.* **2002**, *30*, 5074–5086.

104. Veno, S.T.; Kulikowicz, T.; Pestana, C.; Stepien, P.P.; Stevnsner, T.; Bohr, V.A. The human SUV3 helicase interacts with replication protein A and flap endonuclease 1 in the nucleus. *Biochem. J.* **2011**, *440*, 293–300.

105. Wang, D.D.; Guo, X.E.; Modrek, A.S.; Chen, C.F.; Chen, P.L.; Lee, W.H. Helicase SUV3, polynucleotide phosphorylase, and mitochondrial polyadenylation polymerase form a transient complex to modulate mitochondrial mRNA polyadenylated tail lengths in response to energetic changes. *J. Biol. Chem.* **2014**, *289*, 16727–16735.

106. Khidr, L.; Wu, G.; Davila, A.; Procaccio, V.; Wallace, D.; Lee, W.H. Role of SUV3 helicase in maintaining mitochondrial homeostasis in human cells. *J. Biol. Chem.* **2008**, *283*, 27064–27073.

107. Szczesny, R.J.; Obriot, H.; Paczkowska, A.; Jedrzejczak, R.; Dmochowska, A.; Bartnik, E.; Formstecher, P.; Polakowska, R.; Stepien, P.P. Down-regulation of human RNA/DNA helicase SUV3 induces apoptosis by a caspase- and AIF-dependent pathway. *Biol. Cell* **2007**, *99*, 323–332.

108. Szczesny, R.J.; Borowski, L.S.; Brzezniak, L.K.; Dmochowska, A.; Gewartowski, K.; Bartnik, E.; Stepien, P.P. Human mitochondrial RNA turnover caught in flagranti: Involvement of hSUV3p helicase in RNA surveillance. *Nucleic Acids Res.* **2010**, *38*, 279–298.

109. Pereira, M.; Mason, P.; Szczesny, R.J.; Maddukuri, L.; Dziwura, S.; Jedrzejczak, R.; Paul, E.; Wojcik, A.; Dybczynska, L.; Tudek, B.; *et al.* Interaction of human SUV3 RNA/DNA helicase with BLM helicase: Loss of the *SUV3* gene results in mouse embryonic lethality. *Mech. Ageing Dev.* **2007**, *128*, 609–617.

110. Crawford, D.R.; Wang, Y.; Schools, G.P.; Kochheiser, J.; Davies, K.J. Down-regulation of mammalian mitochondrial RNAs during oxidative stress. *Free Radic. Biol. Med.* **1997**, *22*, 551–559.

111. Desler, C.; Marcker, M.L.; Singh, K.K.; Rasmussen, L.J. The importance of mitochondrial DNA in aging and cancer. *J. Aging Res.* **2011**, *2011*, 407536.

112. Almeida, M. Aging and oxidative stress: A new look at old bone. *IBMS BoneKEy* **2010**, *7*, 340–352.

113. Siitonen, H.A.; Sotkasiira, J.; Biervliet, M.; Benmansour, A.; Capri, Y.; Cormier-Daire, V.; Crandall, B.; Hannula-Jouppi, K.; Hennekam, R.; Herzog, D.; *et al.* The mutation spectrum in RECQL4 diseases. *Eur. J. Hum. Genet.* **2009**, *17*, 151–158.

114. Carew, J.S.; Nawrocki, S.T.; Xu, R.H.; Dunner, K.; McConkey, D.J.; Wierda, W.G.; Keating, M.J.; Huang, P. Increased mitochondrial biogenesis in primary leukemia cells: The role of endogenous nitric oxide and impact on sensitivity to fludarabine. *Leukemia* **2004**, *18*, 1934–1940.

115. Lan, Q.; Lim, U.; Liu, C.S.; Weinstein, S.J.; Chanock, S.; Bonner, M.R.; Virtamo, J.; Albanes, D.; Rothman, N. A prospective study of mitochondrial DNA copy number and risk of non-Hodgkin lymphoma. *Blood* **2008**, *112*, 4247–4249.

116. D'Souza, A.D.; Parikh, N.; Kaech, S.M.; Shadel, G.S. Convergence of multiple signaling pathways is required to coordinately up-regulate mtDNA and mitochondrial biogenesis during T cell activation. *Mitochondrion* **2007**, *7*, 374–385.

117. Jeon, J.P.; Shim, S.M.; Nam, H.Y.; Baik, S.Y.; Kim, J.W.; Han, B.G. Copy number increase of 1p36.33 and mitochondrial genome amplification in Epstein-Barr virus-transformed lymphoblastoid cell lines. *Cancer Genet. Cytogenet.* **2007**, *173*, 122–130.

118. Shapovalov, Y.; Hoffman, D.; Zuch, D.; de Mesy Bentley, K.L.; Eliseev, R.A. Mitochondrial dysfunction in cancer cells due to aberrant mitochondrial replication. *J. Biol. Chem.* **2011**, *286*, 22331–22338.

119. Akiyama, T.; Dass, C.R.; Choong, P.F. Novel therapeutic strategy for osteosarcoma targeting osteoclast differentiation, bone-resorbing activity, and apoptosis pathway. *Mol. Cancer Ther.* **2008**, *7*, 3461–3469.

120. Miyazaki, T.; Mori, S.; Shigemoto, K.; Larsson, N.; Nakamura, T.; Kato, S.; Nakashima, T.; Takayanagi, H.; Tanaka, S. Maintenance of mitochondrial DNA copy number is essential for osteoclast survival. *Arthritis Res. Ther.* **2012**, *14*, 40.

121. Jeng, J.Y.; Yeh, T.S.; Lee, J.W.; Lin, S.H.; Fong, T.H.; Hsieh, R.H. Maintenance of mitochondrial DNA copy number and expression are essential for preservation of mitochondrial function and cell growth. *J. Cell. Biochem.* **2008**, *103*, 347–357.

Mitochondrial and Ubiquitin Proteasome System Dysfunction in Ageing and Disease: Two Sides of the Same Coin?

Jaime M. Ross, Lars Olson and Giuseppe Coppotelli

Abstract: Mitochondrial dysfunction and impairment of the ubiquitin proteasome system have been described as two hallmarks of the ageing process. Additionally, both systems have been implicated in the etiopathogenesis of many age-related diseases, particularly neurodegenerative disorders, such as Alzheimer's and Parkinson's disease. Interestingly, these two systems are closely interconnected, with the ubiquitin proteasome system maintaining mitochondrial homeostasis by regulating organelle dynamics, the proteome, and mitophagy, and mitochondrial dysfunction impairing cellular protein homeostasis by oxidative damage. Here, we review the current literature and argue that the interplay of the two systems should be considered in order to better understand the cellular dysfunction observed in ageing and age-related diseases. Such an approach may provide valuable insights into molecular mechanisms underlying the ageing process, and further discovery of treatments to counteract ageing and its associated diseases. Furthermore, we provide a hypothetical model for the heterogeneity described among individuals during ageing.

Reprinted from *Int. J. Mol. Sci.* Cite as: Ross, J.M.; Olson, L.; Coppotelli, G. Mitochondrial and Ubiquitin Proteasome System Dysfunction in Ageing and Disease: Two Sides of the Same Coin? *Int. J. Mol. Sci.* **2015**, *16*, 19458–19473.

1. Introduction

An increase in the average age of the world population has heightened the interest in ageing research in order to find treatments to improve health in old age. However, despite vast scientific efforts, the mechanisms that regulate ageing remain poorly understood. Outstanding questions include when the process starts and how it proceeds, why different species age at different rates, and why even individuals within the same species age differently. Ageing is a complex process, including genetic and environmental factors, both with stochastic components, all concurring and integrating in a manner difficult to predict. In a recent review, López-Otín and colleagues underlined nine hallmarks of ageing: genomic instability, telomere attrition, epigenetic alterations, loss of proteostasis, deregulated nutrient-sensing, mitochondrial dysfunction, cellular senescence, stem cell exhaustion, and altered intercellular communication [1]. Notably, such putative hallmarks are not isolated cellular processes but are highly interconnected. In order to properly understand the ageing process and to identify therapies to combat ageing, the role and interconnectedness of the putative hallmarks must be further dissected.

Impairment of the ubiquitin proteasome system (UPS) and mitochondrial dysfunction are two hallmarks of ageing and both have been implicated in a plethora of ageing-associated diseases, such as Alzheimer's and Parkinson's disease and certain cancers [1–6]. UPS is part of the "proteostasis network" (PN), and together with the autophagy lysosome system (ALS) and the molecular chaperone network contribute to maintaining cellular protein homeostasis by removing unwanted or damaged proteins that could aggregate and become toxic for the cell [7–10]. Mitochondria are the main source of energy production, generating ATP through oxidative phosphorylation (OXPHOS), and are also involved in many other important cellular processes, such as calcium buffering, apoptosis, steroid synthesis, and reactive oxygen species (ROS) production [11–13]. Although mitochondria are equipped with several mechanisms to quench free radicals, they are still subject to oxidative damage and thus rely on the UPS along with other quality control mechanisms to remove damaged mitochondrial proteins. Hence, an efficient UPS is crucial to preserve healthy mitochondria, and *vice versa*, healthy mitochondria are needed to maintain an efficient UPS system, since excessive ROS production could not only overflow the proteasome by increasing the amount of damaged proteins to be removed, but could also oxidize and damage the proteasomal subunits themselves and thereby decrease their catalytic activities. Once either mitochondrial dysfunction or proteasomal impairment develops, a vicious cycle may start, leading to progressive failure of both systems. Here, we summarize current knowledge of the interplay between the two systems, underlining how they affect each other in health, ageing, and disease, as well as how therapies targeting one deficiency might also benefit the other.

2. The Ubiquitin Proteasome System

The discovery of the ubiquitin-mediated protein degradation system earned Aaron Ciechanover, Avram Hershko, and Irwin Rose the 2004 Nobel Prize in Chemistry. Before uncovering the UPS, protein degradation was thought to occur mainly in the lysosome, an organelle filled with hydrolytic enzymes with an optimal proteolytic activity at a low pH [14]. Proteasome-mediated protein degradation differs from lysosomal-mediated proteolysis by operating at a neutral pH, mainly degrading short-lived proteins, taking place in a protein complex, and by not involving intracellular compartmentalization. The conjugation of a polyubiquitin chain is an essential step to target unwanted or damaged proteins for proteasomal degradation [9]. Proteasome activity generates small peptides that are further digested into amino acids by the abundant cytosolic endopeptidases and aminopeptidases, while lysosomal degradation directly produces single amino acids [15]. The UPS is a highly selective system and operates in both nuclear and cytoplasmic compartments. Conversely, lysosomes are present only in the cytoplasm and are able to remove a wide range of substrates, ranging from a single protein delivered to it via chaperone-mediated autophagy (CMA) to large aggregates and whole organelles (e.g., mitochondria) engulfed via macroautophagy [16,17].

Ubiquitin [Ub] is a 76 amino acid ≈ 8 kDa protein that is highly conserved among Eukaryota [18,19]. Protein ubiquitination is an ATP dependent process that occurs through a three-step sequential enzymatic cascade performed by the ubiquitin-activating enzyme (E1), ubiquitin-conjugating enzyme (E2), and ubiquitin ligase (E3). The result generates an isopeptidyl bond between ubiquitin at glycine 76 and either the ε-amino group of an internal lysine residue on the protein substrate or its amino terminus. Subsequently, multiple rounds of ubiquitination extend the ubiquitin chain by adding more ubiquitins on one of the seven internal lysine residues (Lys 6, 11, 27, 29, 33, 48 and 63) of the previously added ubiquitin, which generates polyubiquitin chains with different linkages (e.g., K48, K63, *etc.*) [20]. The length and type of the ubiquitin chain determine the fate of the ubiquitinated protein; the K48-linked polyubiquitin chain is the main signal that targets substrates for 26S proteasome degradation, while other types of linkages have been shown to play a role in receptor signaling, endocytosis, transcription, DNA repair, and autophagy [21]. The E3 ligase enzyme confers specificity to the ubiquitination system by recognizing the target's substrate; indeed, while there is one type of ubiquitin-activating E1 enzyme (ubiquitin-like modifier-activating enzyme 1, UBA1) present in all cells and a second E1 type (Ubiquitin-activating Enzyme 1-like 2, UBE1L2) with seemingly more tissue specificity [22], there are about 30 E2 enzymes and more than 600 members of the E3 family. E3 ligase enzymes can be grouped into two classes: those that are homologous to the E6-AP carboxyl terminus (HECT) and the really interesting new gene [RING] ligases. The two classes differ not only in their structure but also in the way they catalyze the last step of ubiquitination. The HECT ligases accept the activated ubiquitin from an E2 enzyme on a cysteine residue in the active domain and then transfer it to the substrate, whereas the RING ligases act as scaffold proteins by bringing together an E2 conjugating enzyme and the substrate [23].

Ubiquitination is a reversible post-translational modification, and a family of proteases, the deubiquitinating enzymes (DUBs), can remove ubiquitin from substrates, thereby regulating the ubiquitination process and recycling ubiquitin. DUBs are highly specific and have been grouped into five subfamilies: Ub carboxyl-terminal hydrolases (UCH), Ub-specific proteases (Usp), ovarian tumor like proeases, JAB1/MPN/Mov34 (JAMM/MPN) metalloproteases, and the Machado–Jakob disease proteases. Removal of ubiquitin adducts from the substrate is a critical step for proteasomal degradation [24,25].

The 26S proteasome is a multi-subunit holoenzyme of ≈2.5 MDa, with two distinct subdomains, a 20S core particle (CP) and, in the classical conformation, either one or two 19S (PA 700) regulatory particles (RP) on either side of the CP. The CP is a barrel-shaped complex made by two α- and two β-rings, each containing seven subunits (α_{1-7} and β_{1-7}), and arranged with two β-rings in the middle and two α-rings on either side. The proteolytic activity is carried out by three β subunits (β1, β2, β5), each with different amino acid specificity, caspase-, trypsin-, and chymotrypsin-like activity, respectively [26]. The α subunits seem to have a regulatory function, allowing only unfolded substrates access to the inner chamber, where the proteolytic activities are located, thus avoiding non-specific degradation of cellular proteins. Ubiquitinated substrates are docked and unfolded by the 19S RP, which can be functionally subdivided into a base and a lid [27]. The base consists of six AAA-ATPase rings (Rpt1-6) and three non-ATPase subunits (Rpn1, Rpn2, Rpn13), while another subunit, Rpn10, seems to associate with the base and the lid after their assembly. The AAA-ATPases use energy to unfold the substrate and translocate it through the central pore of the 20S chamber, while two of the non-ATPase subunits (Rpn10, Rpn13) serve as ubiquitin receptors [28–32]. The lid has more than nine proteins, including the deubiquinating enzyme Rpn11, which is essential for efficient substrate degradation [33]. Other regulatory particles have also been described, such as 11S (PA 28) and PA 200, with different functions and activations as compared to the 19S RP. The 11S RP is involved in the immune-proteasome and is regulated by γ-interferon, whereas PA 200 RP is only present in the nucleus, although little is known about its specific function [26].

3. Mitochondria

The endosymbiotic origin of mitochondria explains some of the unique biological aspects of these organelles [34], which form a dynamic network, often referred to as the mitochondrial network [35]. Mitochondria are regulated by fusion and fission, processes that are crucial to maintain functional mitochondria and energetic homeostasis. These processes, for example, enable small mitochondria to move along the cytoskeleton and relocate to areas where energy delivery is needed, such as the presynaptic terminals of an axon. In mammals, several proteins have been implicated in the regulation of fusion and fission of mitochondria. Mitofusin-1 and -2 (MFN1, MFN2) together with the optic atrophy 1 protein (OPA1) are required for mitochondrial fusion, while dynamin-related protein 1 (DRP1) is indispensable for fission [36,37]. All mitochondria contain two lipid bi-layers, an outer membrane (OMM) and an inner membrane (IMM), leading to the intermembrane space (IMS), chemically equivalent to the cytoplasm, and the matrix, an internal space that contains enzymes important for fatty acid oxidation as well as for the tricarboxylic acid (TCA), or Krebs cycle, as well as mtDNA. The IMM is highly impermeable, and by folding in a convoluted manner, forms the *cristae*, a large surface area where the respiratory chain (RC) complexes I–V are located (Figure 1).

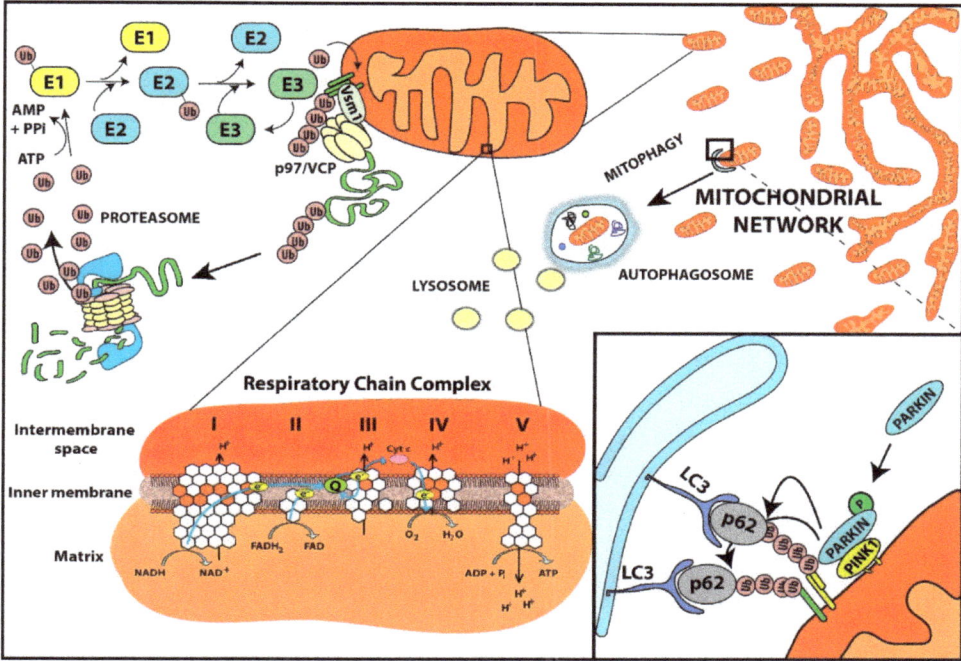

Figure 1. UPS and mitochondrial quality control. Polyubiquitination of mitochondrial proteins by the catalyzed reaction of E1, E2 and E3 enzymes in this depiction leads to the recruitment of the p97/VCP complex to the mitochondrial outer membrane (**upper left**). p97/VCP can extract a ubiquitinated protein in an ATP-dependent process that facilitates its proteasomal degradation. The UPS is also needed for the autophagic degradation of damaged mitochondria, a process known as mitophagy. Loss of mitochondrial membrane polarization stabilizes PINK1, which relocalizes to the outer membrane where it recruits and activates the E3 ligase PARKIN by phosphorylation. Once activated, PARKIN ubiquitinates several mitochondrial proteins, which flag the mitochondria for autophagic degradation (**lower right**). A schematic representation of the mitochondrial respiratory chain (complexes I, II, III, IV and V) is shown, with nuclear-encoded subunits depicted as white hexagons and the mitochondrial-encoded subunits as orange (**lower left**).

Mitochondria are the only organelles that contain their own DNA. In humans, mitochondrial DNA (mtDNA) is a circular molecule that encodes 13 proteins, all of which are involved in OXPHOS, 22 transfer RNA species (tRNAs), and two ribosomal RNA types (16S, 12S). Each cell can contain several hundred copies of mtDNA (10^3–10^4 copies per cell) depending on the energy demand of the tissue, the differentiation stage of the cell, hormonal balance, and exercise level [38,39]. The vast majority of the ≈1000 mitochondrial proteins are encoded by nuclear genes [40], synthetized in the cytoplasm, and imported into the mitochondria in an unfolded state. During this process, cellular and mitochondrial chaperones (mtHSP70, mtHSP60, mtHSP10, *etc.*) assist the folding of imported proteins to ensure that they reach their destination to execute their function [41,42]. Mitochondria are the main source of reactive oxygen species (ROS), a natural by-product of OXPHOS. If not properly regulated, ROS can be extremely harmful to DNA, lipids and proteins, especially matrix proteins, which are not accessible by the cellular quality control machinery. In this regard, mitochondria possess their own quality control system consisting of several proteases, such as Lon, ClpXP, *i*-AAA, and *m*-AAA, to ensure that damaged or unfolded proteins that cannot be rescued and refolded by the mitochondrial chaperons are turned-over, thereby avoiding toxicity. Several reviews have been published on this topic [43–45]. The UPS is also an integral component of the mitochondrial protein quality control system, and mediates degradation not only of outer membrane embedded proteins, but also matrix proteins, implicating the existence of retro-translocation mechanisms of proteins from the mitochondrial matrix to the cytoplasm for proteasomal degradation [46].

4. Role of the Ubiquitin Proteasome System in Mitochondrial Protein Quality Control

The involvement of the UPS in the quality control of mitochondrial proteins started to emerge after several studies found components of the UPS in the mitochondria as well as ubiquitination of numerous mitochondrial proteins. In an early study conducted in yeast, the SCF ubiquitin ligase complex subunit Mdm30 (mitochondrial distribution and morphology protein 30) was shown to affect mitochondrial shape by regulating the steady-state level of Fzo1, an ortholog of mammalian mitofusin-1 and -2; thus connecting the ubiquitin proteasome system with mitochondria [47]. While attempting to determine the mitochondrial proteome of *Saccharomyces cerevisiae*, numerous E3 ligases and DUBs were found to be associated with the mitochondrial compartment [48]. In another study, the purification of total ubiquitinated proteins from mouse heart expressing 8xHis/Flag-Ubiquitin (HisF-Ub) under the α-myosin heavy chain (α-MHC) promoter, led to the finding that 38% of all ubiquitinated proteins were mitochondrial and found in all compartments, including the matrix [49]. One possible explanation for such findings could be that nuclear encoded mitochondrial proteins that are not properly folded during translation are directly targeted for degradation. In this regard, it has been estimated that one third of all synthetized proteins are defective ribosomal products (DRiPs), due to errors in transcription and/or translation, and are turned-over by the proteasome before reaching their final destination [50]. However, an interesting alternative possibility has been proposed: the existence of a mechanism to retro-translocate mitochondrial proteins into the cytosol for degradation, akin to the endoplasmic reticulum-associated degradation (ERAD) pathway, and thus named the mitochondria-associated degradation (MAD) system, also referred to as the outer mitochondrial membrane-associated degradation (OMMAD) system [51–53].

In support of the MAD process, it has been shown that colon cancer cells (COLO 205) treated with inhibitors of the chaperone protein, heat shock protein 90 (HSP90), undergo apoptotic cell death preceded by dramatic changes in the mitochondrial compartment [54]. The most prominent change was an accumulation of mitochondrial proteins due to an increase in protein half-life, as determined by ^{35}S-methionine/cysteine pulse-chase. The authors found that one protein in particular, oligomycin-sensitivity-conferring protein (OSCP), which is a component of the mitochondrial membrane ATP synthase (F1F0-ATP synthase or complex V) and located in the IMM, was ubiquitinated and degraded by the proteasome in an HSP90-dependent manner [54]. Additionally, a role for ubiquitination and proteasome degradation has been described for the mitochondrial uncoupling protein 1 and 2 (UCP1, 2) as well as for the endonuclease G (endoG) protein [55–57]. Similarly with what has been described in the ERAD pathway, the Cdc48/p97 complex (cdc48: cell division control protein 48) seems to be required for the extraction of mitochondrial proteins in the MAD system [58]. In fact, it has been shown in yeast treated with mitochondrial stressors that the cytoplasmic protein Vms1 (valosin-containing protein (VCP)/Cdc48-associated mitochondrial stress-responsive 1) re-localizes to mitochondria and recruits the Cdc48/p97–Npl4 (Npl4: nuclear protein localization protein 4) complex (Figure 1) [52]. Interestingly, Vms1 overexpression in yeast has been shown to counteract the mitochondrial damage and cell death induced by the expression of UBB+1, a frame-shift variant of ubiquitin B, which is associated with Alzheimer's disease [59]. Complex p97, known as VCP in mammals and Cdc48 in yeast, belongs to the ATPases associated with diverse cellular activities (AAA+) protein family, and is a barrel-shaped hexameric complex that uses ATP to unfold and extract proteins from membranes and protein complexes [23]. Notably, Cdc48/VCP mutations have been shown to induce a decrease in mitochondrial membrane potential and to increase mitochondrial oxygen consumption leading to mitochondrial damage and cell death both in yeast and human-derived fibroblasts [60,61].

Among the numerous E3 ligases associated with mitochondria, PARKIN is by far the most studied. Mutations in the PARK2 locus, where the *PARKIN* gene is located, were initially associated with autosomal recessive juvenile Parkinson's disease (AR-JP) [62]. Further studies have contributed to understanding the function of PARKIN and the possible mechanism by which it might promote disease [63].

PARKIN has been described as an hybrid E3 ubiquitin ligase that possesses both RING and HECT E3 ligase characteristics [64]. Upon mitochondrial depolarization, the self-inactivated enzyme is thought to be recruited to the mitochondrial membrane where it is phosphorylated and activated by PTEN-induced putative kinase 1 (PINK1) [65]. PINK1 is constantly imported and degraded in healthy mitochondria; however, when perturbations of mitochondrial homeostasis affect the mitochondrial membrane potential, PINK1 escapes degradation and accumulates on the outer membrane. There, it recruits and activates PARKIN by phosphorylating the Ser65 residue of the PARKIN ubiquitin-like domain; however, its full activation also requires the phosphorylation of Ser65 on the ubiquitin molecule [66–68]. Once activated, PARKIN induces the removal of depolarized mitochondria by mitophagy through a poorly understood mechanism, which requires the poly-ubiquitination of several other outer membrane mitochondrial proteins, including MFN1 and 2, Mitochondrial Rho GTPase (RHOT)-1 and 2, and voltage-dependent anion channel (VDAC)-1, 2, and 3 [69–72]. Notably, up-regulation of Parkin in *Drosophila* resulted in increased mean and maximal lifespan, and was associated with reduced protein aggregation and improved mitochondrial activity in aged flies [73]. Although the PINK1/PARKIN pathway has been shown to be involved in the removal of depolarized mitochondria induced by stressors, such as carbonyl cyanide 3-chlorophenylhydrazone (CCCP), an uncoupler of oxidative phosphorylation, its involvement in the physiological removal of mitochondria seems to be nonessential, as demonstrated by the absence of striking phenotypes in Parkin and Pink1 knockout mice, thus suggesting the presence of additional mechanisms for the removal of mitochondria, independent of the PINK1/PARKIN pathway (reviewed in [74]). In fact, a study from our group showed that dysfunctional mitochondria in a mouse model for Parkinson's disease generated by knocking out the mitochondrial transcription factor A (TFAM) in dopaminergic neurons, did not recruit PARKIN. Neither removal of defective mitochondria nor the neurodegenerative phenotype was affected by the absence of PARKIN in these mice [75].

Another RING/E3 ubiquitin ligase that seems to regulate mitochondrial dynamics is MITOL/MARCH-V (mitochondrial ubiquitin ligase), a membrane protein located in the OMM where it interacts with and ubiquitinates several substrates [76]. One such substrate is Drp1, which is degraded upon MITOL-mediated ubiquitination; thus, MITOL might affect mitochondrial fission by regulating Drp1 levels [77,78]. Furthermore, MITOL seems to be involved in the ubiquitination and degradation of misfolded proteins located in mitochondria, such as a mutated form of superoxide dismutase 1 (SOD1), an antioxidant enzyme that has been implicated in amyotrophic lateral sclerosis (ALS, or Lou Gehrig's disease) [79]. Additionally, several DUBs have been localized to mitochondria, such as ataxin-3, a deubiquitinating enzyme that is associated with Machado-Joseph disease and seems to interact with PARKIN in order to counteract self-ubiquitination [80].

Taken together, these studies support a central role for the UPS in the maintenance of mitochondrial homeostasis by regulating organelle dynamics (fission and fusion), the proteome, and mitophagy. Thus, it is not surprising that disturbances affecting UPS activity might also have an effect on mitochondrial function. With that said, studies also support that the converse is also true.

5. Effect of Mitochondrial Dysfunction on the Ubiquitin Proteasome System

Evidence that mitochondrial dysfunction might affect proteasomal activity has been reported in different systems, including yeast, *C. elegans*, and mammalian cells. It has been shown that inhibition of OXPHOS in rat-derived cortical neurons also affects proteasomal activity and protein ubiquitination [81]. Two recent reports have helped to shed light on the possible molecular mechanisms underlying such an effect [82,83]. Stimulation of ROS production in a respiration-deficient yeast mutant ($\Delta fzo1$) was shown to induce proteasome disassembly, with the complete detachment of the 20S CP and 19S RPs, similar to what was observed in yeast and mammalian cells treated with either hydrogen peroxide (H_2O_2) or antimycin A, a cytochrome *c* reductase inhibitor. Proteasome disassembly was associated with proteasomal substrate accumulation and was reversed upon treatment with antioxidants or dithiothreitol (DTT), a strong reducing agent [82]. Comparable results were obtained in a different study, using a short-lived ubiquitin fused protein expressed in *C. elegans* as a reporter, to screen for factors involved in regulating protein turnover. Screening revealed reporter accumulation in two worm mutants carrying mutations in proteins involved in mitochondrial processes: IVD-1 and ACS-19. IVD-1 is the ortholog of a human mitochondrial enzyme (isovaleryl-CoA dehydrogenase) involved in the leucin catabolism pathway, while ACS-19 is predicted to be the ortholog of a human enzyme (ACSS2, acetyl-CoA synthetase) involved in fatty acid metabolism in the mitochondrial matrix. In both cases, the effect of mitochondrial dysfunction on proteasomal function was due to an increase in ROS production, which was prevented by treatment with the antioxidant *N*-acetylcysteine (NAC) [83].

ROS is a group of potentially harmful compounds that can damage all cellular components, including proteins, DNA, and lipids. Oxidation can affect protein structure, thus impairing function, and might also render proteins prone to aggregation, which could result in toxicity. The complete disassembly of the proteasome, resulting in an increase of 20S CPs, could be a protective mechanism to counteract a temporary rise in oxidative damage. It has been shown that 20S CP is more resistant to oxidative damage, compared to 19S RP, and is able to bind and degrade mis-folded oxidized proteins without the need for ubiquitination and ATP expenditure [84–86]. Thus, a temporary disassembly of the proteasome holoenzyme together with an up-regulation of an antioxidant stress response, heat shock proteins, and autophagic flux could be seen as part of a cellular strategy to counteract an acute increase in oxidative damage. Hence, through the uncapping of the 20S CP, cells might redirect the degradation capability of the proteasome from the removal of ubiquitinated substrates to the removal of oxidized proteins. However, since oxidative stress is a hallmark of ageing and age-related diseases, chronic exposure to oxidative stress could result in proteasome disassembly, which could further aggravate these conditions (Figure 2).

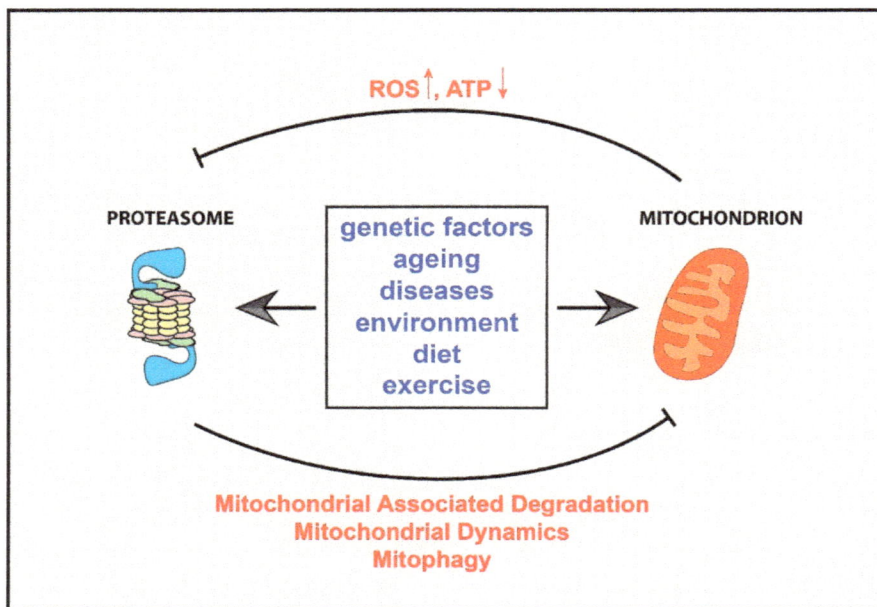

Figure 2. UPS and mitochondrial cross-talk. Several factors, including genes, environment, age, diseases, diet, and exercise can either positively or negatively affect UPS activity and mitochondrial function. Impairment of one of the two systems can then drive the malfunctioning of the other and result in a vicious cycle. A decrease in cellular ATP levels and an increase in ROS production can impair proteasomal function by affecting protein ubiquitination and proteasome assembly and stability, while a decrease in UPS activity could impair mitochondrial function by affecting mitochondrial dynamics, mitophagy, and the removal of damaged mitochondrial proteins.

ATP depletion is another mechanism through which mitochondrial dysfunction might affect proteasomal activity. ATP is required for both protein ubiquitination [87] and proteasome assembly and stability [88–90]. Intracellular ATP levels have been shown to regulate proteasomal activity both *in vitro* and in cultured cells [91], and manipulation of intracellular ATP levels by inhibition of complex I has been shown to decrease proteasomal activity in primary mesencephalic cell cultures, an effect which was counteracted by increasing the glucose concentration in the cellular medium [92].

6. The "Mitochondrion—Ubiquitin Proteasome System Axis" in Ageing and Age-Related Diseases

The UPS and mitochondria are two systems among several reportedly affected by ageing; an accumulation of mis-folded proteins and oxidative stress have been denoted as two features of the ageing process. A decline in UPS activity has been shown in yeast (*Saccharomyces cerevisiae*) [93], fly (*Drosophila melanogaster*) [94], rodents [95–97], and also in human-derived dermal fibroblasts [98]. Conversely, it has been shown that proteasome activation by genetic manipulation in different models can ameliorate the ageing process and also increase lifespan (reviewed in [99]). Several possibilities have been proposed to explain the UPS decline associated with ageing, including down-regulation and/or modification of proteasomal subunits, disassembly of the holoenzyme, an increase in substrates and aggregates that could clog the proteasome, and reduction in ATP levels, which could impair the overall process of protein ubiquitination and unfolding [100]. As mentioned, an increase in oxidative damage is a major contributor to the UPS decline, and with OXPHOS as the main source of ROS production, mitochondria have thus been suspected to play a central role in the ageing process. Based on this notion, Denham Harman proposed the "Free Radical Theory of Aging" (FRTA) in 1956, suggesting that ageing is driven by the accumulation of oxidative damage to cellular structures over time [101]. It has been proposed that accumulation of mtDNA mutations could be a possible cause of the mitochondrial dysfunction described in ageing, and in this regard data from different groups, including ours, have shown a cause-effect relationship between increased mtDNA mutational load and ageing phenotypes [102–110]. However, it has also been argued that the level of mtDNA mutations observed in normally aged tissues is much less than the threshold needed to cause respiratory chain dysfunction [111,112]. Thus, another possibility for the age-associated decline in mitochondrial function could be a loss in protein homeostasis due to the impairment of the UPS and/or autophagic systems.

As described, the UPS and mitochondria systems are tightly interdependent, and once a vicious cycle of dysfunction starts it is difficult to identify which one was the trigger (Figure 2). This is demonstrated in neurodegenerative diseases, such as Alzheimer's disease (AD) and Parkinson's disease (PD), with ageing consistently implicated as the major risk factor. In both diseases, it has been seemingly difficult to isolate UPS impairment from dysfunctional mitochondria, and *vice versa*, in order to understand the contribution of each system in disease onset and progression. PD is a neurodegenerative disorder that arises from the loss of dopaminergic neurons, mainly in substantia nigra, and is characterized by resting tremor, bradykinesia, and muscle rigidity. The discovery of Lewy bodies in neurons, aggregates containing α-synuclein, ubiquitinated proteins, and components of the UPS, strongly implicated the proteasome in the pathogenesis of the disease [113]. However, other studies have reported a compelling correlation between mitochondrial dysfunction and PD, and mouse models mimicking the disease have been generated by genetically impairing mitochondrial function in dopaminergic neurons [114] or by using toxins that affect mitochondria, such as 1-methyl-4-phenyl-1,2,3,6-tetrahydropyridine (MPTP) [115]. In all likelihood, PD will turn out to be several different diseases characterized by different etiologies, although only partially different phenotypes. AD patients exhibit gross brain atrophy, with both neuronal and synaptic loss, accumulation of amyloid plaques containing amyloid β peptides, and intracellular neurofibrillary tangles of phosphorylated Tau protein [116]. The involvement of the UPS in AD has been postulated based on studies demonstrating a decrease in proteasome activity associated with AD and the presence of ubiquitin and UPS components in the plaques [117]. As similarly shown with PD, another body of literature has focused on mitochondrial dysfunction as representing the major etiopathogenesis of AD [118]. Taking both perspectives into consideration, perhaps these two interconnected systems should be regarded as the "Mitochondrion-UPS Axis" when trying to understand and dissect the cellular dysfunction observed in ageing and age-related diseases. That is, UPS impairment and mitochondrial dysfunction could be two sides of the same coin in that either system cannot be separated from the other since they affect each other in a vicious cycle (Figure 2).

In order to explain the differences observed among individuals during ageing, we propose a model that takes into consideration the decline in both mitochondrial function and UPS activity over time. We speculate that the point of interception between the two systems might represent the age at which cellular dysfunction begins (Figure 3). While both systems decline with age in a dependent manner, the shape of each curve will vary slightly between individuals, due to the compounded effects of an individual's genetic background, environmental stressors (*i.e.*, toxins, smoking), diet, and exercise. Taking these factors into account, the age of cellular dysfunction onset for a given person could start decades earlier as compared to another, leading to the ageing heterogeneity of the human population.

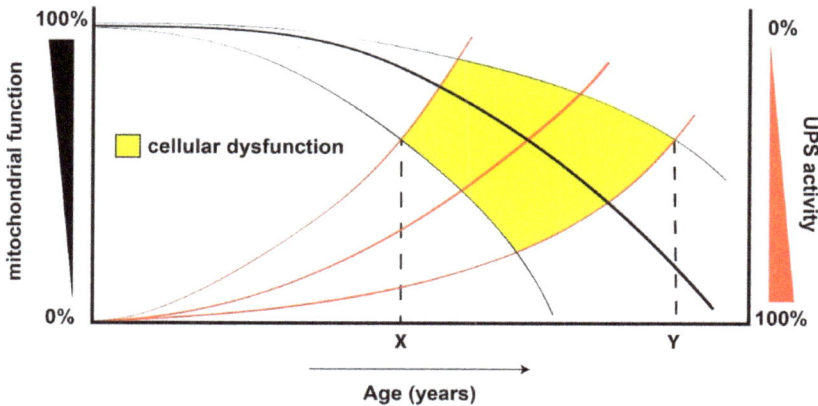

Figure 3. Hypothetical model to explain the heterogeneity of the ageing process among individuals taking into consideration changes in both UPS and mitochondrial function. A theoretical model to explain the idiosyncratic differences observed during ageing by taking into consideration the variation of both UPS activity and mitochondrial function over time. Both systems decline with age in a dependent manner, but the slope of the curve may vary between individuals, depending on factors such as genetics, environment, diet, and exercise, as depicted by the thin lines. The point of interception of the two curves hypothetically represents the age of onset of cellular dysfunction, defined as any point in time when cellular homeostasis is perturbed. Thus, two individuals, each following different extremities of mitochondrial dysfunction and UPS decline, might experience the onset of cellular dysfunction at different ages (X and Y), which could be decades apart from each other.

7. Conclusions and Future Prospects

The last century has witnessed a considerable increase of life expectancy due to better living conditions and medical advancements in the cure and prevention of many once fatal diseases. Several compounds, such as resveratrol, metformin, and rapamycin, have shown potential in improving overall health and lifespan in experimental organisms. Finding drugs to combat ageing might therefore not be just fantasy, but actually feasible [119,120]. However, the only currently known proven interventions shown to improve ageing phenotypes in humans are a hypocaloric diet and exercise [121,122]. Therefore, understanding the underlying molecular mechanism of the ageing process is the *condicio sine qua non* for developing any promising therapeutic intervention to slow the ageing process. In this regard, we suggest that dissecting the "Mitochondrion-UPS Axis" may help in the search for drugs to counteract ageing and age-related diseases.

91

Acknowledgments: The Swedish Research Council (537-2014-6856; JMR, K2012-62X-03185-42-4; LO), the Swedish Brain Foundation (Jaime M. Ross), Swedish Lundbeck Foundation (Jaime M. Ross), Swedish Brain Power (Jaime M. Ross, Lars Olson, Giuseppe Coppotelli), the Swedish Society for Medical Research (Giuseppe Coppotelli), the Foundation for Geriatric Diseases at Karolinska Institutet (Giuseppe Coppotelli), Karolinska Institutet Research Foundations (Giuseppe Coppotelli), Loo och Hans Östermans Foundation for Medical Research (Giuseppe Coppotelli), ERC Advanced Investigator grant (322744; Lars Olson), and the Karolinska Distinguished Professor Award (Lars Olson).

Conflicts of Interest: The authors declare no conflict of interest.

Abbreviations

AD = Alzheimer's Disease; DUBs = Deubiquitinating enzymes; HECT = Homologous to the E6-AP Carboxyl Terminus; IMM = Inner Mitochondrial Membrane; IMS = Intermembrane Space; RC = Respiratory Chain; MAD = Mithochondrial Associated Degradation; OMM = Outer Mitochondrial Membrane; OXPHOS = Oxidative Phosphorylation; PD = Parkinson's Disease; RING = Really Interesting New Gene; ROS = Reactive Oxygen Species; Ub = Ubiquitin; UPS = Ubiquitin Proteasome System.

References

1. López-Otín, C.; Blasco, M.A.; Partridge, L.; Serrano, M.; Kroemer, G. The hallmarks of aging. *Cell* **2013**, *153*, 1194–1217.
2. Wang, X.; Wang, W.; Li, L.; Perry, G.; Lee, H.-G.; Zhu, X. Oxidative stress and mitochondrial dysfunction in Alzheimer's disease. *Biochim. Biophys. Acta* **2014**, *1842*, 1240–1247.
3. Morimoto, R.I. Proteotoxic stress and inducible chaperone networks in neurodegenerative disease and aging. *Genes Dev.* **2008**, *22*, 1427–1438.
4. Keller, J.N.; Hanni, K.B.; Markesbery, W.R. Possible involvement of proteasome inhibition in aging: Implications for oxidative stress. *Mech. Ageing Dev.* **2000**, *113*, 61–70.
5. Gautier, C.A.; Corti, O.; Brice, A. Mitochondrial dysfunctions in Parkinson's disease. *Rev. Neurol.* **2014**, *170*, 339–343.
6. Büeler, H. Impaired mitochondrial dynamics and function in the pathogenesis of Parkinson's disease. *Exp. Neurol.* **2009**, *218*, 235–246.
7. Jung, T.; Catalgol, B.; Grune, T. The proteasomal system. *Mol. Asp. Med.* **2009**, *30*, 191–296.
8. Hershko, A.; Ciechanover, A. The ubiquitin system. *Annu. Rev. Biochem.* **1998**, *67*, 425–479.
9. Glickman, M.H.; Ciechanover, A. The ubiquitin-proteasome proteolytic pathway: Destruction for the sake of construction. *Physiol. Rev.* **2002**, *82*, 373–428.
10. Hochstrasser, M. Ubiquitin-dependent protein degradation. *Annu. Rev. Genet.* **1996**, *30*, 405–439.
11. Kim, A. A panoramic overview of mitochondria and mitochondrial redox biology. *Toxicol. Res.* **2014**, *30*, 221–234.

12. Mishra, P.; Chan, D.C. Mitochondrial dynamics and inheritance during cell division, development and disease. *Nat. Rev. Mol. Cell Biol.* **2014**, *15*, 634–646.

13. Green, D.R.; Galluzzi, L.; Kroemer, G. Cell biology. Metabolic control of cell death. *Science* **2014**, *345*, 1250256.

14. De Duve, C.; Pressman, B.C.; Gianetto, R.; Wattiaux, R.; Appelmans, F. Tissue fractionation studies. 6. Intracellular distribution patterns of enzymes in rat-liver tissue. *Biochem. J.* **1955**, *60*, 604–617.

15. Baraibar, M.A.; Friguet, B. Changes of the proteasomal system during the aging process. *Prog. Mol. Biol. Transl. Sci.* **2012**, *109*, 249–275.

16. Kaushik, S.; Cuervo, A.M. Chaperone-mediated autophagy: A unique way to enter the lysosome world. *Trends Cell Biol.* **2012**, *22*, 407–417.

17. Mizushima, N. Autophagy: Process and function. *Genes Dev.* **2007**, *21*, 2861–2873.

18. Ciechanover, A. Proteolysis: From the lysosome to ubiquitin and the proteasome. *Nat. Rev. Mol. Cell Biol.* **2005**, *6*, 79–87.

19. Pickart, C.M.; Eddins, M.J. Ubiquitin: Structures, functions, mechanisms. *Biochim. Biophys. Acta* **2004**, *1695*, 55–72.

20. Komander, D.; Rape, M. The ubiquitin code. *Annu. Rev. Biochem.* **2012**, *81*, 203–229.

21. Finley, D. Recognition and processing of ubiquitin-protein conjugates by the proteasome. *Annu. Rev. Biochem.* **2009**, *78*, 477–513.

22. Pelzer, C.; Kassner, I.; Matentzoglu, K.; Singh, R.K.; Wollscheid, H.P.; Scheffner, M.; Schmidtke, G.; Groettrup, M. UBE1L2, a novel E1 enzyme specific for ubiquitin. *J. Biol. Chem.* **2007**, *282*, 23010–23014.

23. Metzger, M.B.; Hristova, V.A.; Weissman, A.M. HECT and RING finger families of E3 ubiquitin ligases at a glance. *J. Cell Sci.* **2012**, *125*, 531–537.

24. Amerik, A.Y.; Hochstrasser, M. Mechanism and function of deubiquitinating enzymes. *Biochim. Biophys. Acta* **2004**, *1695*, 189–207.

25. Reyes-Turcu, F.E.; Ventii, K.H.; Wilkinson, K.D. Regulation and cellular roles of ubiquitin-specific deubiquitinating enzymes. *Annu. Rev. Biochem.* **2009**, *78*, 363–397.

26. Jung, T.; Grune, T. Structure of the proteasome. *Prog. Mol. Biol. Transl. Sci.* **2012**, *109*, 1–39.

27. Lander, G.C.; Estrin, E.; Matyskiela, M.E.; Bashore, C.; Nogales, E.; Martin, A. Complete subunit architecture of the proteasome regulatory particle. *Nature* **2012**, *482*, 186–191.

28. Van Nocker, S.; Sadis, S.; Rubin, D.M.; Glickman, M.; Fu, H.; Coux, O.; Wefes, I.; Finley, D.; Vierstra, R.D. The multiubiquitin-chain-binding protein Mcb1 is a component of the 26S proteasome in *Saccharomyces cerevisiae* and plays a nonessential, substrate-specific role in protein turnover. *Mol. Cell. Biol.* **1996**, *16*, 6020–6028.

29. Deveraux, Q.; Ustrell, V.; Pickart, C.; Rechsteiner, M. A 26 S protease subunit that binds ubiquitin conjugates. *J. Biol. Chem.* **1994**, *269*, 7059–7061.

30. Husnjak, K.; Elsasser, S.; Zhang, N.; Chen, X.; Randles, L.; Shi, Y.; Hofmann, K.; Walters, K.J.; Finley, D.; Dikic, I. Proteasome subunit Rpn13 is a novel ubiquitin receptor. *Nature* **2008**, *453*, 481–488.

31. Braun, B.C.; Glickman, M.; Kraft, R.; Dahlmann, B.; Kloetzel, P.M.; Finley, D.; Schmidt, M. The base of the proteasome regulatory particle exhibits chaperone-like activity. *Nat. Cell Biol.* **1999**, *1*, 221–226.

32. Liu, C.W.; Li, X.; Thompson, D.; Wooding, K.; Chang, T.; Tang, Z.; Yu, H.; Thomas, P.J.; DeMartino, G.N. ATP binding and ATP hydrolysis play distinct roles in the function of 26S proteasome. *Mol. Cell* **2006**, *24*, 39–50.

33. Verma, R.; Aravind, L.; Oania, R.; McDonald, W.H.; Yates, J.R.; Koonin, E.V.; Deshaies, R.J. Role of Rpn11 metalloprotease in deubiquitination and degradation by the 26S proteasome. *Science* **2002**, *298*, 611–615.

34. Müller, M.; Mentel, M.; van Hellemond, J.J.; Henze, K.; Woehle, C.; Gould, S.B.; Yu, R.Y.; van der Giezen, M.; Tielens, A.G.; Martin, W.F. Biochemistry and evolution of anaerobic energy metabolism in eukaryotes. *Microbiol. Mol. Biol. Rev.* **2012**, *76*, 444–495.

35. Rafelski, S.M. Mitochondrial network morphology: Building an integrative, geometrical view. *BMC Biol.* **2013**.

36. Dhingra, R.; Kirshenbaum, L.A. Regulation of mitochondrial dynamics and cell fate. *Circ. J.* **2014**, *78*, 803–810.

37. Van der Bliek, A.M.; Shen, Q.; Kawajiri, S. Mechanisms of mitochondrial fission and fusion. *Cold Spring Harb. Perspect. Biol.* **2013**, *5*.

38. Kuznetsov, A.V.; Hermann, M.; Saks, V.; Hengster, P.; Margreiter, R. The cell-type specificity of mitochondrial dynamics. *Int. J. Biochem. Cell Biol.* **2009**, *41*, 1928–1939.

39. Kuznetsov, A.V.; Margreiter, R. Heterogeneity of mitochondria and mitochondrial function within cells as another level of mitochondrial complexity. *Int. J. Mol. Sci.* **2009**, *10*, 1911–1929.

40. Pagliarini, D.J.; Calvo, S.E.; Chang, B.; Sheth, S.A.; Vafai, S.B.; Ong, S.E.; Walford, G.A.; Sugiana, C.; Boneh, A.; Chen, W.K. A mitochondrial protein compendium elucidates complex I disease biology. *Cell* **2008**, *134*, 112–123.

41. Chacinska, A.; Koehler, C.M.; Milenkovic, D.; Lithgow, T.; Pfanner, N. Importing mitochondrial proteins: Machineries and mechanisms. *Cell* **2009**, *138*, 628–644.

42. Baker, M.J.; Frazier, A.E.; Gulbis, J.M.; Ryan, M.T. Mitochondrial protein-import machinery: Correlating structure with function. *Trends Cell Biol.* **2007**, *17*, 456–464.

43. Baker, M.J.; Palmer, C.S.; Stojanovski, D. Mitochondrial protein quality control in health and disease. *Br. J. Pharmacol.* **2014**, *171*, 1870–1889.

44. Baker, M.J.; Tatsuta, T.; Langer, T. Quality control of mitochondrial proteostasis. *Cold Spring Harb. Perspect. Biol.* **2011**, *3*.

45. Voos, W.; Ward, L.A.; Truscott, K.N. The role of AAA+ proteases in mitochondrial protein biogenesis, homeostasis and activity control. *Subcell Biochem.* **2013**, *66*, 223–263.

46. Taylor, E.B.; Rutter, J. Mitochondrial quality control by the ubiquitin-proteasome system. *Biochem. Soc. Trans.* **2011**, *39*, 1509–1513.

47. Fritz, S.; Weinbach, N.; Westermann, B. Mdm30 is an F-box protein required for maintenance of fusion-competent mitochondria in yeast. *Mol. Biol. Cell* **2003**, *14*, 2303–2313.

48. Sickmann, A.; Reinders, J.; Wagner, Y.; Joppich, C.; Zahedi, R.; Meyer, H.E.; Schönfisch, B.; Perschil, I.; Chacinska, A.; Guiard, B. The proteome of *Saccharomyces cerevisiae* mitochondria. *Proc. Natl. Acad. Sci. USA* **2003**, *100*, 13207–13212.

49. Jeon, H.B.; Choi, E.S.; Yoon, J.H.; Hwang, J.H.; Chang, J.W.; Lee, E.K.; Choi, H.W.; Park, Z.Y.; Yoo, Y.J. A proteomics approach to identify the ubiquitinated proteins in mouse heart. *Biochem. Biophys. Res. Commun.* **2007**, *357*, 731–736.

50. Schubert, U.; Antón, L.C.; Gibbs, J.; Norbury, C.C.; Yewdell, J.W.; Bennink, J.R. Rapid degradation of a large fraction of newly synthesized proteins by proteasomes. *Nature* **2000**, *404*, 770–774.

51. Chatenay-Lapointe, M.; Shadel, G.S. Stressed-out mitochondria get MAD. *Cell Metab.* **2010**, *12*, 559–560.

52. Heo, J.-M.; Livnat-Levanon, N.; Taylor, E.B.; Jones, K.T.; Dephoure, N.; Ring, J.; Xie, J.; Brodsky, J.L.; Madeo, F.; Gygi, S.P. A stress-responsive system for mitochondrial protein degradation. *Mol. Cell* **2010**, *40*, 465–480.

53. Neutzner, A.; Youle, R.J.; Karbowski, M. Outer mitochondrial membrane protein degradation by the proteasome. *Novartis Found. Symp.* **2007**, *287*, 4–14.

54. Margineantu, D.H.; Emerson, C.B.; Diaz, D.; Hockenbery, D.M. Hsp90 inhibition decreases mitochondrial protein turnover. *PLoS ONE* **2007**, *2*, e1066.

55. Azzu, V.; Brand, M.D. Degradation of an intramitochondrial protein by the cytosolic proteasome. *J. Cell Sci.* **2010**, *123*, 578–585.

56. Clarke, K.J.; Adams, A.E.; Manzke, L.H.; Pearson, T.W.; Borchers, C.H.; Porter, R.K. A role for ubiquitinylation and the cytosolic proteasome in turnover of mitochondrial uncoupling protein 1 (UCP1). *Biochim. Biophys. Acta* **2012**, *1817*, 1759–1767.

57. Radke, S.; Chander, H.; Schäfer, P.; Meiss, G.; Krüger, R.; Schulz, J.B.; Germain, D. Mitochondrial protein quality control by the proteasome involves ubiquitination and the protease Omi. *J. Biol. Chem.* **2008**, *283*, 12681–12685.

58. Xu, S.; Peng, G.; Wang, Y.; Fang, S.; Karbowski, M. The AAA-ATPase p97 is essential for outer mitochondrial membrane protein turnover. *Mol. Biol. Cell.* **2011**, *22*, 291–300.

59. Braun, R.J.; Sommer, C.; Leibiger, C.; Gentier, R.J.G.; Dumit, V.I.; Paduch, K.; Eisenberg, T.; Habernig, L.; Trausinger, G.; Magnes, C. Accumulation of basic amino acids at mitochondria dictates the cytotoxicity of aberrant ubiquitin. *Cell Rep.* **2015**.

60. Bartolome, F.; Wu, H.C.; Burchell, V.S.; Preza, E.; Wray, S.; Mahoney, C.J.; Fox, N.C.; Calvo, A.; Canosa, A.; Moglia, C.; *et al.* Pathogenic *VCP* mutations induce mitochondrial uncoupling and reduced ATP levels. *Neuron* **2013**, *78*, 57–64.

61. Braun, R.J.; Zischka, H.; Madeo, F.; Eisenberg, T.; Wissing, S.; Büttner, S.; Engelhardt, S.M.; Büringer, D.; Ueffing, M. Crucial mitochondrial impairment upon CDC48 mutation in apoptotic yeast. *J. Biol. Chem.* **2006**, *281*, 25757–25767.

62. Kitada, T.; Asakawa, S.; Hattori, N.; Matsumine, H.; Yamamura, Y.; Minoshima, S.; Yokochi, M.; Mizuno, Y.; Shimizu, N. Mutations in the *parkin* gene cause autosomal recessive juvenile parkinsonism. *Nature* **1998**, *392*, 605–608.

63. Koyano, F.; Matsuda, N. Molecular mechanisms underlying PINK1 and Parkin catalyzed ubiquitylation of substrates on damaged mitochondria. *Biochim. Biophys. Acta* **2015**.

64. Wenzel, D.M.; Lissounov, A.; Brzovic, P.S.; Klevit, R.E. UBCH7 reactivity profile reveals parkin and HHARI to be RING/HECT hybrids. *Nature* **2011**, *474*, 105–108.

65. Matsuda, N.; Sato, S.; Shiba, K.; Okatsu, K.; Saisho, K.; Gautier, C.A.; Sou, Y.S.; Saiki, S.; Kawajiri, S.; Sato, F.; Kimura, M.; *et al.* PINK1 stabilized by mitochondrial depolarization recruits Parkin to damaged mitochondria and activates latent Parkin for mitophagy. *J. Cell Biol.* **2010**, *189*, 211–221.

66. Kane, L.A.; Lazarou, M.; Fogel, A.I.; Li, Y.; Yamano, K.; Sarraf, S.A.; Banerjee, S.; Youle, R.J. PINK1 phosphorylates ubiquitin to activate Parkin E3 ubiquitin ligase activity. *J. Cell Biol.* **2014**, *205*, 143–153.

67. Koyano, F.; Okatsu, K.; Kosako, H.; Tamura, Y.; Go, E.; Kimura, M.; Kimura, Y.; Tsuchiya, H.; Yoshihara, H.; Hirokawa, T.; *et al.* Ubiquitin is phosphorylated by PINK1 to activate parkin. *Nature* **2014**, *510*, 162–166.

68. Kazlauskaite, A.; Kondapalli, C.; Gourlay, R.; Campbell, D.G.; Ritorto, M.S.; Hofmann, K.; Alessi, D.R.; Knebel, A.; Trost, M.; Muqit, M.M.K. Parkin is activated by PINK1-dependent phosphorylation of ubiquitin at Ser65. *Biochem. J.* **2014**, *460*, 127–139.

69. Geisler, S.; Holmström, K.M.; Skujat, D.; Fiesel, F.C.; Rothfuss, O.C.; Kahle, P.J.; Springer, W. PINK1/Parkin-mediated mitophagy is dependent on VDAC1 and p62/SQSTM1. *Nat. Cell Biol.* **2010**, *12*, 119–131.

70. Glauser, L.; Sonnay, S.; Stafa, K.; Moore, D.J. Parkin promotes the ubiquitination and degradation of the mitochondrial fusion factor mitofusin 1. *J. Neurochem.* **2011**, *118*, 636–645.

71. Narendra, D.; Tanaka, A.; Suen, D.-F.; Youle, R.J. Parkin is recruited selectively to impaired mitochondria and promotes their autophagy. *J. Cell Biol.* **2008**, *183*, 795–803.

72. Sarraf, S.A.; Raman, M.; Guarani-Pereira, V.; Sowa, M.E.; Huttlin, E.L.; Gygi, S.P.; Harper, J.W. Landscape of the PARKIN-dependent ubiquitylome in response to mitochondrial depolarization. *Nature* **2013**, *496*, 372–376.

73. Rana, A.; Rera, M.; Walker, D.W. Parkin overexpression during aging reduces proteotoxicity, alters mitochondrial dynamics, and extends lifespan. *Proc. Natl. Acad. Sci. USA* **2013**, *110*, 8638–8643.

74. Melrose, H.L.; Lincoln, S.J.; Tyndall, G.M.; Farrer, M.J. Parkinson's disease: A rethink of rodent models. *Exp. Brain Res.* **2006**, *173*, 196–204.

75. Sterky, F.H.; Lee, S.; Wibom, R.; Olson, L.; Larsson, N.G. Impaired mitochondrial transport and Parkin-independent degeneration of respiratory chain-deficient dopamine neurons *in vivo*. *Proc. Natl. Acad. Sci. USA* **2011**, *108*, 12937–12942.

76. Nagashima, S.; Tokuyama, T.; Yonashiro, R.; Inatome, R.; Yanagi, S. Roles of mitochondrial ubiquitin ligase MITOL/MARCH5 in mitochondrial dynamics and diseases. *J. Biochem.* **2014**, *155*, 273–279.

77. Karbowski, M.; Neutzner, A.; Youle, R.J. The mitochondrial E3 ubiquitin ligase MARCH5 is required for Drp1 dependent mitochondrial division. *J. Cell Biol.* **2007**, *178*, 71–84.

78. Yonashiro, R.; Ishido, S.; Kyo, S.; Fukuda, T.; Goto, E.; Matsuki, Y.; Ohmura-Hoshino, M.; Sada, K.; Hotta, H.; Yamamura, H.; *et al.* A novel mitochondrial ubiquitin ligase plays a critical role in mitochondrial dynamics. *EMBO J.* **2006**, *25*, 3618–3626.

79. Yonashiro, R.; Sugiura, A.; Miyachi, M.; Fukuda, T.; Matsushita, N.; Inatome, R.; Ogata, Y.; Suzuki, T.; Dohmae, N.; Yanagi, S. Mitochondrial ubiquitin ligase MITOL ubiquitinates mutant SOD1 and attenuates mutant SOD1-induced reactive oxygen species generation. *Mol. Biol. Cell* **2009**, *20*, 4524–4530.

80. Durcan, T.M.; Kontogiannea, M.; Thorarinsdottir, T.; Fallon, L.; Williams, A.J.; Djarmati, A.; Fantaneanu, T.; Paulson, H.L.; Fon, E.A. The Machado–Joseph disease-associated mutant form of ataxin-3 regulates parkin ubiquitination and stability. *Hum. Mol. Genet.* **2011**, *20*, 141–154.

81. Huang, Q.; Wang, H.; Perry, S.W.; Figueiredo-Pereira, M.E. Negative regulation of 26S proteasome stability via calpain-mediated cleavage of Rpn10 upon mitochondrial dysfunction in neurons. *J. Biol. Chem.* **2013**, *288*, 12161–12174.

82. Livnat-Levanon, N.; Kevei, É.; Kleifeld, O.; Krutauz, D.; Segref, A.; Rinaldi, T.; Erpapazoglou, Z.; Cohen, M.; Reis, N.; Hoppe, T.; *et al.* Reversible 26S proteasome disassembly upon mitochondrial stress. *Cell Rep.* **2014**, *7*, 1371–1380.

83. Segref, A.; Kevei, É.; Pokrzywa, W.; Schmeisser, K.; Mansfeld, J.; Livnat-Levanon, N.; Ensenauer, R.; Glickman, M.H.; Ristow, M.; Hoppe, T. Pathogenesis of human mitochondrial diseases is modulated by reduced activity of the ubiquitin/proteasome system. *Cell Metab.* **2014**, *19*, 642–652.

84. Grune, T.; Reinheckel, T.; Davies, K.J. Degradation of oxidized proteins in K562 human hematopoietic cells by proteasome. *J. Biol. Chem.* **1996**, *271*, 15504–15509.

85. Grune, T.; Merker, K.; Sandig, G.; Davies, K.J.A. Selective degradation of oxidatively modified protein substrates by the proteasome. *Biochem. Biophys. Res. Commun.* **2003**, *305*, 709–718.

86. Shringarpure, R.; Grune, T.; Mehlhase, J.; Davies, K.J.A. Ubiquitin conjugation is not required for the degradation of oxidized proteins by proteasome. *J. Biol. Chem.* **2003**, *278*, 311–318.

87. Hershko, A.; Heller, H.; Elias, S.; Ciechanover, A. Components of ubiquitin-protein ligase system. Resolution, affinity purification, and role in protein breakdown. *J. Biol. Chem.* **1983**, *258*, 8206–8214.

88. Eytan, E.; Ganoth, D.; Armon, T.; Hershko, A. ATP-dependent incorporation of 20S protease into the 26S complex that degrades proteins conjugated to ubiquitin. *Proc. Natl. Acad. Sci. USA* **1989**, *86*, 7751–7755.

89. Dahlmann, B.; Kuehn, L.; Reinauer, H. Studies on the activation by ATP of the 26 S proteasome complex from rat skeletal muscle. *Biochem. J.* **1995**, *309*, 195–202.

90. Kleijnen, M.F.; Roelofs, J.; Park, S.; Hathaway, N.A.; Glickman, M.; King, R.W.; Finley, D. Stability of the proteasome can be regulated allosterically through engagement of its proteolytic active sites. *Nat. Struct. Mol. Biol.* **2007**, *14*, 1180–1188.

91. Huang, H.; Zhang, X.; Li, S.; Liu, N.; Lian, W.; McDowell, E.; Zhou, P.; Zhao, C.; Guo, H.; Zhang, C.; *et al.* Physiological levels of ATP negatively regulate proteasome function. *Cell Res.* **2010**, *20*, 1372–1385.

92. Höglinger, G.U.; Carrard, G.; Michel, P.P.; Medja, F.; Lombès, A.; Ruberg, M.; Friguet, B.; Hirsch, E.C. Dysfunction of mitochondrial complex I and the proteasome: Interactions between two biochemical deficits in a cellular model of Parkinson's disease. *J. Neurochem.* **2003**, *86*, 1297–1307.

93. Chen, Q.; Thorpe, J.; Ding, Q.; El-Amouri, I.S.; Keller, J.N. Proteasome synthesis and assembly are required for survival during stationary phase. *Free Radic. Biol. Med.* **2004**, *37*, 859–868.

94. Vernace, V.A.; Arnaud, L.; Schmidt-Glenewinkel, T.; Figueiredo-Pereira, M.E. Aging perturbs 26S proteasome assembly in *Drosophila melanogaster*. *FASEB J.* **2007**, *21*, 2672–2682.

95. Keller, J.N.; Huang, F.F.; Markesbery, W.R. Decreased levels of proteasome activity and proteasome expression in aging spinal cord. *Neuroscience* **2000**, *98*, 149–156.

96. Dasuri, K.; Zhang, L.; Ebenezer, P.; Liu, Y.; Fernandez-Kim, S.O.; Keller, J.N. Aging and dietary restriction alter proteasome biogenesis and composition in the brain and liver. *Mech. Ageing Dev.* **2009**, *130*, 777–783.

97. Ferrington, D.A.; Husom, A.D.; Thompson, L.V. Altered proteasome structure, function, and oxidation in aged muscle. *FASEB J.* **2005**, *19*, 644–646.

98. Hwang, J.S.; Hwang, J.S.; Chang, I.; Kim, S. Age-associated decrease in proteasome content and activities in human dermal fibroblasts: Restoration of normal level of proteasome subunits reduces aging markers in fibroblasts from elderly persons. *J. Gerontol. A Biol. Sci. Med. Sci.* **2007**, *62*, 490–499.

99. Chondrogianni, N.; Voutetakis, K.; Kapetanou, M.; Delitsikou, V.; Papaevgeniou, N.; Sakellari, M.; Lefaki, M.; Filippopoulou, K.; Gonos, E.S. Proteasome activation: An innovative promising approach for delaying aging and retarding age-related diseases. *Ageing Res. Rev.* **2015**, *23*, 37–55.

100. Vernace, V.A.; Schmidt-Glenewinkel, T.; Figueiredo-Pereira, M.E. Aging and regulated protein degradation: Who has the UPPer hand? *Aging Cell.* **2007**, *6*, 599–606.

101. Harman, D. Aging: A theory based on free radical and radiation chemistry. *J. Gerontol.* **1956**, *11*, 298–300.

102. Schwarze, S.R.; Lee, C.M.; Chung, S.S.; Roecker, E.B.; Weindruch, R.; Aiken, J.M. High levels of mitochondrial DNA deletions in skeletal muscle of old rhesus monkeys. *Mech. Ageing Dev.* **1995**, *83*, 91–101.

103. Khaidakov, M.; Heflich, R.H.; Manjanatha, M.G.; Myers, M.B.; Aidoo, A. Accumulation of point mutations in mitochondrial DNA of aging mice. *Mutat. Res.* **2003**, *526*, 1–7.

104. Corral-Debrinski, M.; Horton, T.; Lott, M.T.; Shoffner, J.M.; Beal, M.F.; Wallace, D.C. Mitochondrial DNA deletions in human brain: Regional variability and increase with advanced age. *Nat. Genet.* **1992**, *2*, 324–329.

105. Soong, N.W.; Hinton, D.R.; Cortopassi, G.; Arnheim, N. Mosaicism for a specific somatic mitochondrial DNA mutation in adult human brain. *Nat. Genet.* **1992**, *2*, 318–323.

106. Ross, J.M.; Stewart, J.B.; Hagström, E.; Brené, S.; Mourier, A.; Coppotelli, G.; Freyer, C.; Lagouge, M.; Hoffer, B.J.; Olson, L.; *et al.* Germline mitochondrial DNA mutations aggravate ageing and can impair brain development. *Nature* **2013**, *501*, 412–415.

107. Ross, J.M.; Öberg, J.; Brené, S.; Coppotelli, G.; Terzioglu, M.; Pernold, K.; Goiny, M.; Sitnikov, R.; Kehr, J.; Trifunovic, A.; *et al.* High brain lactate is a hallmark of aging and caused by a shift in the lactate dehydrogenase A/B ratio. *Proc. Natl. Acad. Sci. USA* **2010**, *107*, 20087–20092.

108. Ross, J.M.; Coppotelli, G.; Hoffer, B.J.; Olson, L. Maternally transmitted mitochondrial DNA mutations can reduce lifespan. *Sci. Rep.* **2014**.

109. Trifunovic, A.; Wredenberg, A.; Falkenberg, M.; Spelbrink, J.N.; Rovio, A.T.; Bruder, C.E.; Bohlooly, Y.M.; Gidlöf, S.; Oldfors, A.; Wibom, R.; *et al.* Premature ageing in mice expressing defective mitochondrial DNA polymerase. *Nature* **2004**, *429*, 417–423.

110. Kujoth, G.C.; Hiona, A.; Pugh, T.D.; Someya, S.; Panzer, K.; Wohlgemuth, S.E.; Hofer, T.; Seo, A.Y.; Sullivan, R.; Jobling, W.A.; *et al.* Mitochondrial DNA mutations, oxidative stress, and apoptosis in mammalian aging. *Science* **2005**, *309*, 481–484.

111. Larsson, N.G.; Oldfors, A. Mitochondrial myopathies. *Acta Physiol. Scand.* **2001**, *171*, 385–393.

112. Cottrell, D.A.; Turnbull, D.M. Mitochondria and ageing. *Curr. Opin. Clin. Nutr. Metab. Care* **2000**, *3*, 473–478.

113. Cook, C.; Petrucelli, L. A critical evaluation of the ubiquitin-proteasome system in Parkinson's disease. *Biochim. Biophys. Acta* **2009**, *1792*, 664–675.

114. Ekstrand, M.I.; Terzioglu, M.; Galter, D.; Zhu, S.; Hofstetter, C.; Lindqvist, E.; Thams, S.; Bergstrand, A.; Hansson, F.S.; Trifunovic, A.; *et al.* Progressive parkinsonism in mice with respiratory-chain-deficient dopamine neurons. *Proc. Natl. Acad. Sci. USA* **2007**, *104*, 1325–1330.

115. Schmidt, N.; Ferger, B. Neurochemical findings in the MPTP model of Parkinson's disease. *J. Neural. Transm.* **2001**, *108*, 1263–1282.

116. Harrington, C.R. The molecular pathology of Alzheimer's disease. *Neuroimaging Clin. N. Am.* **2012**, *22*, 11–22.

117. Riederer, B.M.; Leuba, G.; Vernay, A.; Riederer, I.M. The role of the ubiquitin proteasome system in Alzheimer's disease. *Exp. Biol. Med.* **2011**, *236*, 268–276.

118. Friedland-Leuner, K.; Stockburger, C.; Denzer, I.; Eckert, G.P.; Müller, W.E. Mitochondrial dysfunction: Cause and consequence of Alzheimer's disease. *Prog. Mol. Biol. Transl. Sci.* **2014**, *127*, 183–210.

119. Argyropoulou, A.; Aligiannis, N.; Trougakos, I.P.; Skaltsounis, A.L. Natural compounds with anti-ageing activity. *Nat. Prod. Rep.* **2013**, *30*, 1412–1437.

120. De Cabo, R.; Carmona-Gutierrez, D.; Bernier, M.; Hall, M.N.; Madeo, F. The search for antiaging interventions: From elixirs to fasting regimens. *Cell* **2014**, *157*, 1515–1526.

121. Mizushima, S.; Moriguchi, E.H.; Ishikawa, P.; Hekman, P.; Nara, Y.; Mimura, G.; Moriguchi, Y.; Yamori, Y. Fish intake and cardiovascular risk among middle-aged Japanese in Japan and Brazil. *J. Cardiovasc. Risk* **1997**, *4*, 191–199.

122. Huffman, K.M.; Slentz, C.A.; Bateman, L.A.; Thompson, D.; Muehlbauer, M.J.; Bain, J.R.; Stevens, R.D.; Wenner, B.R.; Kraus, V.B.; Newgard, C.B.; *et al.* Exercise-induced changes in metabolic intermediates, hormones, and inflammatory markers associated with improvements in insulin sensitivity. *Diabetes Care* **2011**, *34*, 174–176.

Skeletal Muscle Mitochondrial Energetic Efficiency and Aging

Raffaella Crescenzo, Francesca Bianco, Arianna Mazzoli, Antonia Giacco, Giovanna Liverini and Susanna Iossa

Abstract: Aging is associated with a progressive loss of maximal cell functionality, and mitochondria are considered a key factor in aging process, since they determine the ATP availability in the cells. Mitochondrial performance during aging in skeletal muscle is reported to be either decreased or unchanged. This heterogeneity of results could partly be due to the method used to assess mitochondrial performance. In addition, in skeletal muscle the mitochondrial population is heterogeneous, composed of subsarcolemmal and intermyofibrillar mitochondria. Therefore, the purpose of the present review is to summarize the results obtained on the functionality of the above mitochondrial populations during aging, taking into account that the mitochondrial performance depends on organelle number, organelle activity, and energetic efficiency of the mitochondrial machinery in synthesizing ATP from the oxidation of fuels.

Reprinted from *Int. J. Mol. Sci.* Cite as: Crescenzo, R.; Bianco, F.; Mazzoli, A.; Giacco, A.; Liverini, G.; Iossa, S. Skeletal Muscle Mitochondrial Energetic Efficiency and Aging. *Int. J. Mol. Sci.* **2015**, *16*, 10674–10683.

1. Introduction

Increasing age leads to a decline in cell functionality, generally termed as "aging" [1]. All tissue and organs of the body are involved in the phenomenon of aging, but the extent of the cellular impairment is greatly variable, since post-mitotic tissues are the most sensitive targets of the aging process [2]. Among the latter, skeletal muscle tissue is profoundly affected during aging, and its functional decline is characterized by a progressive atrophy, that becomes more severe with advancing age and that from a certain point onwards can lead to mobility impairment, increased risk of falls, and physical frailty [3,4]. The loss of function in skeletal muscle is also responsible for the development of age-associated insulin resistance and related metabolic disturbances [5–9]. For these reasons, understanding the mechanisms underlying aging in skeletal muscle is fundamental for promotion of health and mobility in the elderly. To this end, the most used animal model of human aging is represented by ad libitum fed caged rodents, that exhibit a sedentary lifestyle and unrestricted diet, and that develop an age-induced spontaneous obesity and insulin resistance already at 4 months of age [7].

Although the loss of skeletal muscle protein mass during aging could at least partially explain the decline in muscle performance, a role for mitochondria has been postulated in the aging process [10,11]. In fact, mitochondria have a major role in energetic homeostasis by determining ATP availability in the cells. A decrease in mitochondrial function could therefore cause an inability to meet cellular ATP demands, so that skeletal muscle cells lose their capacity to adapt to physiological stress imposed across the entire lifespan [12]. In addition, dysfunctional mitochondria could contribute to the development of age-induced insulin resistance [13], since mitochondrial oxidative capacity has been considered a good predictor of insulin sensitivity [14].

One important issue that should be taken into account when studying mitochondria in skeletal muscle is that the mitochondrial population is heterogeneous, composed of mitochondria located either beneath the sarcolemmal membrane (subsarcolemmal, SS) or between the myofibrils (intermyofibrillar, IMF) [15] (Figure 1). Since these two mitochondrial populations exhibit different energetic characteristics and can be differently affected by physiological stimuli [16], it is important that both are separately studied. Therefore, the purpose of the present review is to summarize the results obtained on the functionality of the above mitochondrial populations during aging, taking into account that the mitochondrial performance depends on organelle number, organelle activity, and energetic efficiency of the mitochondrial machinery in synthesizing ATP from the oxidation of fuels. A search in PubMed of relevant articles was conducted, by using query "skeletal muscle mitochondria and aging", "subsarcolemmal mitochondria and aging", and "intermyofibrillar mitochondria and aging", with the inclusion of related articles by the same groups.

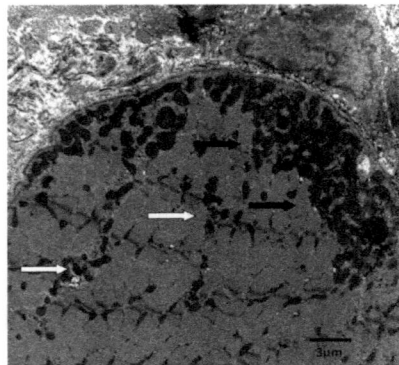

Figure 1. Heterogeneous mitochondrial populations in skeletal muscle cells. Scale bar = 3 μm. White arrows = Intermyofibrillar mitochondria; Black arrows = Subsarcolemmal mitochondria.

2. Age-Related Changes in Mitochondria

Although many studies have addressed the issue of mitochondrial performance attenuation during aging in skeletal muscle, the obtained results are contradictory, with some papers reporting impairment of mitochondria with increasing age [17–23] and others showing no age-induced change [24,25], underscoring the complexity of understanding in this area.

The discrepancy between the published data can partly be explained by differences in experimental approach. However, even when comparing the results obtained with similar methodological approach, divergent outcomes are evident. For example, among the studies that have measured the activity of enzymatic complexes, such as citrate synthase and electron transport chain complexes, to obtain indirect insights into the energy producing (respiratory) capacity of the mitochondria, some of them have reported an age-dependent decrease in aging muscle [26–28] while other studies found no variation [29] or reported a variable response in different muscles [30]. An alternative approach used to study mitochondrial alteration in aging muscle is assessing total mitochondrial content. Again, some studies reported that mitochondrial content is reduced in aging muscle [31,32] and others found no change [21,33–35]. There is however consensus on the finding that aging skeletal muscle has a blunted capacity for generation of new mitochondria in response to both endurance exercise training [36] and chronic electrical stimulation [37,38].

To our knowledge, studies carried out specifically on mitochondria located beneath the sarcolemmal membrane (SS) and between the myofibrils (IMF) from skeletal muscle during aging are scarce. In a pioneering research, Farrar *et al.* [39] found that ADP-stimulated respiration did not change with increasing age in IMF and SS mitochondria, while aging decreased the amount of IMF proteins. Chabi *et al.* [32] studied senescent (3 years) rats and have found that in SS and IMF mitochondria the capacity for ATP production was reduced, as a result of diminished mitochondrial content per gram of muscle. Drew *et al.* [40] found a decrease in ATP production between 12 and 26 months of age in SS mitochondria from gastrocnemius muscle in Fisher rats. Very few studies specifically addressed the differential regulation of SS and IMF mitochondria with aging in humans. These studies evaluated the SS and IMF mitochondrial content of skeletal muscle in young and old men and reported no age-dependent change in both mitochondrial populations [34,41]. From all the above results, it emerges a differential effect of aging on SS and IMF mitochondria, at least in animal models. More studies on humans are needed to validate the differential effect of aging on the two mitochondrial populations. In addition, studies on changes in mitochondrial function induced by aging or other physiological stimuli should be carried out on the two different mitochondrial populations existing in skeletal muscle.

3. Mitochondrial Energetic Efficiency during Aging

From the analysis of the published results on the differential effect of aging on IMF and SS mitochondria, no clear picture can be obtained, partly because of differences in the experimental approach used to evaluate mitochondrial function. The mitochondrial performance depends on organelle number, organelle activity, and energetic efficiency of the mitochondrial machinery in synthesizing ATP from the oxidation of fuels. In addition, it is well known that the amount of fuels oxidized by the cell is dictated mainly by ATP turnover rather than by mitochondrial oxidative capacity and therefore a decrease in mitochondrial capacity and/or number becomes more important when cells increase their metabolic activity, *i.e.*, during contraction, while an increased mitochondrial efficiency still alter the amount of oxidized fuels also at rest (Figure 2). The efficiency with which dietary calories are converted to ATP is determined by the degree of coupling of oxidative phosphorylation. If the respiratory chain is highly efficient at pumping protons out of the mitochondrial inner membrane, and the ATP synthesis is highly efficient at converting the proton flow through its proton channel into ATP (from ADP), then the mitochondria will generate maximum ATP and minimum heat per calorie. In contrast, if the efficiency of proton pumping is reduced and/or more protons are required to make each ATP molecule, then each calorie will yield less ATP.

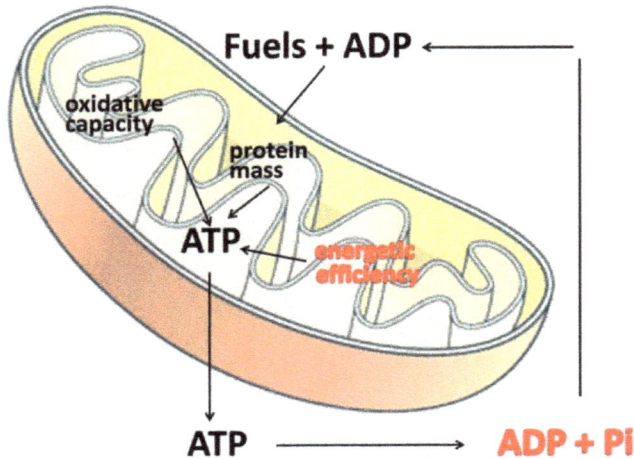

Figure 2. Factors affecting cellular fuel oxidation. The amount of burned fuels mainly depends on mitochondrial energetic efficiency and ATP turnover (in red). Mitochondrial mass and oxidative capacity are less important because mitochondria are thought to have a much greater capacity to generate ATP than what is usually required [42].

The main point of regulation of the oxidative phosphorylation efficiency [43] is represented by the degree of coupling between oxygen consumption and ATP synthesis, which is always lower than 1 and can vary according to the metabolic needs of the cell [44]. Among the factors, which affect mitochondrial degree of coupling, an important role is played by the permeability of the mitochondrial inner membrane to H^+ ions (leak). It is now well known that mitochondrial inner membrane exhibit a basal proton leak pathway, whose contribution to the basal metabolic rate in rats has been estimated to be about 20%–25% [45]. In addition, it is well known that fatty acids can act as natural uncouplers of oxidative phosphorylation, by generating a fatty acid-dependent proton leak pathway [46–48], which is a function of the amount of unbound fatty acids in the cell.

The issue of mitochondrial efficiency has been explored in humans *in vivo*, and it has been found that the effect of aging is fiber type-dependent. In fact, in mildly uncoupled fiber types (*i.e.*, tibialis anterior) no age effect is evident, while a substantial uncoupling takes place with aging in well coupled fiber types (*i.e.*, dorsal interosseous) [49]. Due to the *in vivo* conditions, these studies do not distinguish between SS and IMF mitochondria. Since aging has been show to selectively affect IMF but not SS mitochondria in heart [50], and taking into account the above considerations, a study was carried out on the putative changes in mitochondrial performance and efficiency during aging in SS and IMF mitochondria. In the transition from young adulthood (60 days) to adulthood (180 days), SS and IMF skeletal muscle mitochondria displayed an increase in the degree of coupling and efficiency, as well as a decreased fatty acid dependent proton leak [51]. The above modifications in mitochondrial performance occurred concomitantly with an increase in whole body lipids and plasma non-esterified fatty acids (NEFA) [51], suggesting a link between skeletal muscle increased mitochondrial efficiency and metabolic impairments.

These results were extended to the evaluation of how the progression of aging affects skeletal muscle mitochondrial function by measuring mitochondrial respiratory capacity and proton leak in SS and IMF mitochondria from adult (six months) and old (two years) rats [52]. A significant decrease in oxidative capacity was found in skeletal muscle homogenates, as well as in SS and IMF mitochondria from old rats. Oxidative capacity measured in the homogenate reflects the product of mitochondrial mass and activity, while oxidative capacity in isolated organelles depends only on mitochondrial activity, and therefore the similar age-induced decrease in oxidative capacity in homogenates and isolated mitochondria found in old rats could be indicative of a lack of changes in mitochondrial mass. A decreased mitochondrial mass has been found in senescent rats (36 months old) [32], thus suggesting that an impairment of mitochondrial biogenesis occurs in late aging and/or it takes place selectively in specific muscles, such as gastrocnemius, whose mitochondrial mass has been found decreased in old rats [20,53].

Both mitochondrial populations from old rats exhibited a significant decrease in proton leak [52], suggesting that with increasing age the efficiency of oxidative phosphorylation increases both in SS and IMF mitochondria. Similar results have been obtained *in vivo* in aged rat skeletal muscle, where a trend for a higher coupling efficiency was found [20]. Skeletal muscle cells in sedentary laboratory rats operate in conditions of low ADP availability, near to state 4 (when no ADP is available), with a high contribution of proton leak to total oxygen consumed by mitochondria [54], and therefore the decreased proton leak found in SS and IMF mitochondria from old rats is physiologically relevant. When mitochondria are more efficient, less substrates are oxidized to obtain ATP. Therefore, the increased mitochondrial coupling in skeletal muscle could contribute to the decreased energy expenditure that is evident even after the decrease in lean mass has been taken into account and that is at the basis of age-induced obesity, since skeletal muscle energy metabolism accounts for about 30% of whole body energy expenditure in resting conditions [45].

One interesting question is: What could be the impact of the increased mitochondrial coupling on ATP yield? Although it is very difficult to calculate the theoretical difference in ATP yield per calorie in highly efficient *vs.* inefficient proton pumping, a rough estimate could be obtained using a published estimate of the control value of proton leak on P/O ratio in skeletal muscle [55]. This value is reported to be -0.72 [55], so we can calculate that the 40% decrease in proton leak found in old rats [52] would result in a 29% increase in P/O ratio, and therefore in the amount of ATP obtained per unit of oxygen consumed. However, when mitochondria are more coupled, ATP is produced at a slower rate [44,56], that could be unable to meet cellular energy demands, especially during skeletal muscle contraction. In fact, in elderly subjects it has been found that a lower speed of ATP production is associated with higher fatigability [57].

Another deleterious consequence of reduced substrate burning in skeletal muscle could be intracellular triglyceride accumulation and lipotoxicity, since NEFA serum levels are significantly higher in older rats. In fact, in conditions of increased plasma NEFA, more fatty acids enter in the cells and if they are poorly oxidized, they accumulate intracellularly in myocytes mainly as long-chain fatty acyl-CoA, monoacylglcyerol, diacylglycerol, phosphatidic acid, triacylglycerol and ceramide. Among these fatty acid derivatives, high intramyocellular levels of diacylglycerol and ceramides are directly associated with insulin resistance [58]. The metabolic implications of increased mitochondrial energetic efficiency are summarized in Figure 3.

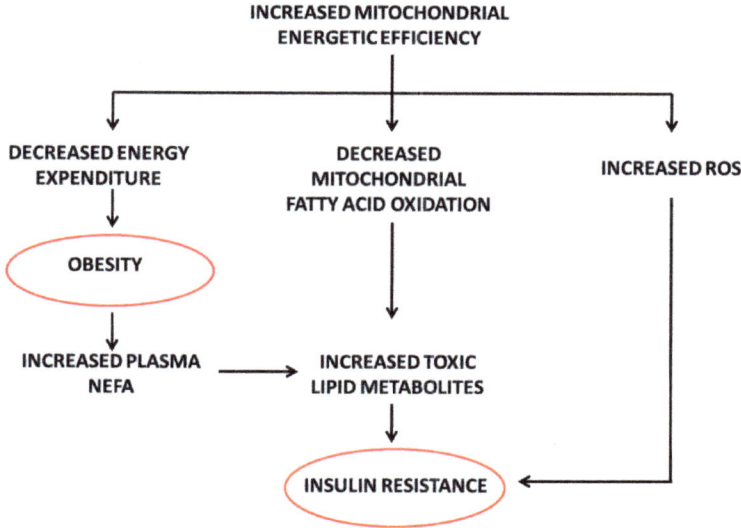

Figure 3. Summary of the metabolic implications of the increased mitochondrial energetic efficiency in skeletal muscle with aging. The red circles identify the pathological outcomes.

4. Oxidative Stress in Aging Mitochondria

Reactive oxygen species (ROS) production increases when mitochondrial potential is higher [59,60], and therefore increased mitochondrial energetic efficiency could induce a condition of increased oxidative stress. In fact, one of the postulated physiological roles for the uncoupling effect of fatty acids is to maintain mitochondrial membrane potential below the critical threshold for ROS production, especially *in situ*ations of low rates of ATP turnover, such as in resting skeletal muscle [61]. It has been proposed that ROS generation during the normal oxidative activity of mitochondria leads to damage of lipids, proteins and DNA, especially in postmitotic tissues, such as skeletal muscle, and that this oxidative damage is at the basis of the biological phenomenon of cellular aging [10,11]. Therefore, the increased coupling of mitochondrial oxidative phosphorylation found in 180 day-old rats [51] could led to the oxidative damage of skeletal muscle cell, detectable later in life. However, the increased uncoupling protein 3 (UCP3) protein content found in SS and IMF mitochondria from 180 days old rats could be involved in the protection from oxidative damage [62,63]. In fact, it has been proposed that UCP3 translocates fatty acid peroxides from the inner to the outer membrane leaflet, thus preserving macromolecules from being oxidized by very aggressive fatty acid peroxides [62,63], while its uncoupling effect is considered very low, due to its low content in skeletal muscle mitochondria [64]. Therefore, the up-regulation of UCP3 during aging could buffer oxidative damage, that otherwise could be even higher.

In 2 year-old rats, an increase in the degree of lipid peroxidation was found only in SS mitochondria, although the decrease in proton leak is the same in both mitochondrial populations [52], while Chabi *et al.* [32] studied senescent (3 years) rats and found that ROS production was enhanced in SS and IMF mitochondria. The differential susceptibility to oxidative stress of SS and IMF mitochondria could depend on their different content of UCP3. In fact, a significant increase in UCP3 content was evident in SS and IMF mitochondria from old rats, but it was more marked in IMF mitochondria (about 10-fold) than in SS mitochondria (about 5-fold). Therefore, the lower increase in UCP3 content in SS mitochondria is probably the cause of the higher oxidative damage found in this mitochondrial population, while IMF mitochondria seem more protected by oxidative damage by the marked up-regulation of UCP3, so preserving the capacity to produce ATP for muscle contraction. One could speculate that, as aging proceeds, SS mitochondria undergo progressive oxidative damage with loss of functional activity. In fact, in senescent (30–36 month-old) animals, an increase in mitochondrial proton leak [65], a decrease in mitochondrial coupling [66] or a decrease in mitochondrial membrane potential in SS but not in IMF mitochondria has been found [32]. In conclusion, the increased coupling of SS mitochondria causes an increase of the oxidative damage in this mitochondrial population, that is located beneath the plasma membrane and provides ATP for membrane transports and signal transduction pathways [67]. On the other hand, IMF mitochondria, that provide ATP for muscle contraction, seems to be more protected from oxidative damage, and could thus increase their oxidative capacity and density in response to endurance training even in old age [68].

A decreased UCP3 content has been found in skeletal muscle mitochondria from old rats [31,69,70], but these results have been obtained using Fischer 344 rats, a rat strain that gain weight only moderately with age compared with other strains (*i.e.*, Sprague-Dawley, Wistar, Long Evans) [8]. Therefore, it can be hypothesized that the regulation of UCP3 with aging in skeletal muscle mitochondria is strain-dependent. Therefore, it is possible that age-induced oxidative damage in skeletal muscle and age-induced obesity are intimately linked. Further studies on the degree of obesity and oxidative damage induced by aging in different strains and species are needed to substantiate the hypothesis.

5. Conclusions

In the rat model of human obesity the progression of aging is accompanied by an increased efficiency of SS and IMF mitochondria but an increased oxidative damage occurs only in the SS population. Therefore, a differential susceptibility of SS and IMF mitochondria to aging-induced damage emerges, although more studies on humans are needed to validate the differential effect of aging on the two mitochondrial populations. These observations also indicate that studies on changes in mitochondrial function induced by aging or other physiological stimuli should be carried out on the two different mitochondrial populations existing in skeletal muscle.

Acknowledgments: The work was supported by University "Federico II" of Naples.

Author Contributions: All authors contributed to the writing of this review.

Conflicts of Interest: The authors declare no conflict of interest.

References

1. Figueiredo, P.A.; Mota, M.P.; Appell, H.J.; Duarte, J.A. The role of mitochondria in aging of skeletal muscle. *Biogerontology* **2008**, *9*, 67–84.

2. Kwong, L.K.; Sohal, R.S. Age-related changes in activities of mitochondrial electron transport complexes in various tissues of the mouse. *Arch. Biochem. Biophys.* **2000**, *373*, 16–22.

3. Cruz-Jentoft, J.; Landi, F.; Topinkova, E.; Michel, J.P. Understanding sarcopenia as a geriatric syndrome. *Curr. Opin. Clin. Nutr. Metab. Care* **2010**, *13*, 1–7.

4. Marzetti, E.; Calvani, R.; Cesari, M.; Buford, T.W.; Lorenzi, M.; Behnke, B.J.; Leeuwenburgh, C. Mitochondrial dysfunction and sarcopenia of aging: From signaling pathways to clinical trials. *Int. J. Biochem. Cell Biol.* **2013**, *45*, 2288–2301.

5. Shimokata, H.; Tobin, J.; Muller, D.C.; Elahi, D.; Coon, P.J.; Andres, R. Studies in the distribution of body fat: 1. Effects of age, sex and obesity. *J. Gerontol.* **1989**, *44*, M66–M73.

6. Reaven, G.M.; Reaven, E.P. Age, glucose intolerance, and non-insulin dependent diabetes mellitus. *J. Am. Geriatr. Soc.* **1985**, 286–290.

7. Barzilai, N.; Rossetti, L. Relationship between changes in body composition and insulin responsiveness in models of the aging rat. *Am. J. Physiol.* **1995**, *269*, E591–E597.

8. Larkin, L.M.; Reynolds, T.H.; Supiano, M.A.; Kahn, B.B.; Halter, J.B. Effect of aging and obesity on insulin responsiveness and glut-4 glucose transporter content in skeletal muscle of Fisher 344 × Brown Norway rats. *J. Gerontol.* **2001**, *56*, B486–B492.

9. Iossa, S.; Mollica, M.P.; Lionetti, L.; Crescenzo, R.; Botta, M.; Barletta, A.; Liverini, G. Acetyl-L-carnitine supplementation differently influences nutrient partitioning, serum leptin concentration and skeletal muscle mitochondrial respiration in young and old rats. *J. Nutr.* **2002**, *132*, 636–642.

10. Lenaz, G. Role of mitochondria in oxidative stress and ageing. *Biochim. Biophyis. Acta* **1998**, *1366*, 53–67.

11. Lenaz, G.; D'Aurelio, M.; Merlo Pich, M.; Genova, M.L.; Ventura, B.; Bovina, C.; Formiggini, G.; Parenti Castelli, G. Mitochondrial bioenergetics in aging. *Biochim. Biophys. Acta* **2000**, *1459*, 397–404.

12. Shigenaga, M.K.; Hagen, T.M.; Ames, B.N. Oxidative damage and mitochondrial decay in aging. *Proc. Natl. Acad. Sci. USA* **1994**, *91*, 10771–10778.

13. Petersen, K.F.; Befroy, D.; Dufour, S.; Dziura, J.; Arijan, C.; Rothman, D.L.; DiPietro, L.; Cline, G.W.; Shulman, G.I. Mitochondrial disfunction in the elderly: Possible role in insulin resistance. *Science* **2003**, *300*, 1140–1142.

14. Bruce, C.R.; Anderson, M.J.; Carey, A.L.; Newman, D.G.; Bonen, A.; Kriketos, A.D.; Cooney, G.J.; Hawley, J.A. Muscle oxidative capacity is a better predictor of insulin sensitivity than lipid status. *J. Clin. Endocrinol. Metab.* **2003**, *88*, 5444–5451.

15. Drew, B.; Leeuwenburgh, C. Ageing and subcellular distribution of mitochondria: Role of mitochondrial DNA deletions and energy production. *Acta Physiol. Scand.* **2004**, *182*, 333–341.

16. Mollica, M.P.; Lionetti, L.; Crescenzo, R.; D'Andrea, E.; Ferraro, M.; Liverini, G.; Iossa, S. Heterogeneous bioenergetic behaviour of subsarcolemmal and intermyofibrillar mitochondria in fed and fasted rats. *Cell. Mol. Life Sci.* **2006**, *63*, 358–366.

17. Figueiredo, P.A.; Powers, S.K.; Ferreira, R.M.; Appell, H.J.; Duarte, J.A. Aging impairs skeletal muscle mitochondrial bioenergetic function. *J. Gerontol. A Biol. Sci. Med. Sci.* **2009**, *64*, 21–33.

18. Trounce, I.; Byrne, E.; Marzuki, S. Decline in skeletal muscle mitochondrial respiratory chain function: Possible factor in ageing. *Lancet* **1989**, *1*, 637–639.

19. Conley, K.E.; Jubrias, S.A.; Esselman, P.C. Oxidative capacity and aging in human muscle. *J. Physiol.* **2000**, *526*, 203–210.

20. Gouspillou, G.; Bourdel-Marchasson, I.; Rouland, R.; Calmettes, G.; Biran, M.; Deschodt-Arsac, V.; Miraux, S.; Thiaudiere, E.; Pasdois, P.; Detaille, D.; *et al.* Mitochondrial energetics is impaired *in vivo* in aged skeletal muscle. *Aging Cell* **2014**, 39–48.

21. Gouspillou, G.; Sgarioto, N.; Kapchinsky, S.; Purves-Smith, F.; Norris, B.; Pion, C.H.; Barbat-Artigas, S.; Lemieux, F.; Taivassalo, T.; Morais, J.A.; *et al.* Increased sensitivity to mitochondrial permeability transition and myonuclear translocation of endonuclease G in atrophied muscle of physically active older humans. *FASEB J.* **2014**, *28*, 1621–1633.

22. Gouspillou, G.; Bourdel-Marchasson, I.; Rouland, R.; Calmettes, G.; Franconi, J.M.; Deschodt-Arsac, V.; Diolez, P. Alteration of mitochondrial oxidative phosphorylation in aged skeletal muscle involves modification of adenine nucleotide translocator. *Biochim. Biophys. Acta* **2010**, *1797*, 143–151.

23. Picard, M.; Ritchie, D.; Thomas, M.M.; Wright, K.J.; Hepple, R.T. Alterations in intrinsic mitochondrial function with aging are fiber type-specific and do not explain differential atrophy between muscles. *Aging Cell* **2011**, *10*, 1047–1055.

24. Rasmussen, U.F.; Krustrup, P.; Kjaer, M.; Rasmussen, H.N. Experimental evidence against the mitochondrial theory of aging. A study on isolated human skeletal muscle mitochondria. *Exp. Gerontol.* **2003**, *38*, 877–886.

25. Rasmussen, U.F.; Krustrup, P.; Kjaer, M.; Rasmussen, H.N. Human skeletal muscle mitochondrial metabolism in youth and senescence: No signs of functional changes of ATP formation and mitochondrial capacity. *Pflug. Arch.* **2003**, *446*, 270–278.

26. Desai, V.G.; Weindruch, R.; Hart, R.W.; Feuers, R.J. Influences of age and dietary restriction on gastrocnemius electron transport system activities in mice. *Arch. Biochem. Biophys.* **1996**, *333*, 145–151.

27. Hagen, J.L.; Krause, D.J.; Baker, D.J.; Fu, M.; Tarnopolsky, M.A.; Hepple, R.T. Skeletal muscle aging in F344BN F1-hybrid rats: I. Mitochondrial dysfunction contributes to the age-associated reduction in VO_{2max}. *J. Gerontol.* **2004**, *59*, 1099–1110.

28. Lombardi, A.; Silvestri, E.; Cioffi, F.; Senese, R.; Lanni, A.; Goglia, F.; de Lange, P.; Moreno, M. Defining the trascriptomic and proteomic profiles of rat ageing skeletal muscle by the use of a cDNA array, 2D- and Blue native-PAGE approach. *J. Proteomics* **2009**, *72*, 708–721.

29. Barrientos, A.; Casademont, J.; Rotig, A.; Miro, O.; Urbano-Marquez, A.; Rustin, P.; Cardellach, F. Absence of relationship between the level of electron transport chain activities and aging in human skeletal muscle. *Biochem. Biophys. Res. Commun.* **1996**, *229*, 536–539.

30. Lyons, C.N.; Mathieu-Costello, O.; Moyes, C.D. Regulation of skeletal muscle mitochondrial content during aging. *J. Gerontol.* **2006**, *61*, 3–13.

31. Kerner, J.; Turkaly, P.J.; Minkler, P.E.; Hoppel, C.L. Aging skeletal muscle mitochondria in the rat: Decreased uncoupling protein-3 content. *Am. J. Physiol.* **2001**, *281*, E1054–E1062.

32. Chabi, B.; Ljubicic, V.; Menzies, K.J.; Huang, J.H.; Saleem, A.; Hood, D.A. Mitochondrial function and apoptotic susceptibility in aging skeletal muscle. *Aging Cell* **2008**, *7*, 2–12.

33. Mathieu-Costello, O.; Ju, Y.; Trejo-Morales, M.; Cui, L. Greater capillary-fiber interface per fiber mitochondrial volume in skeletal muscles of old rats. *J. Appl. Physiol.* **2005**, *99*, 281–289.

34. Callahan, D.M.; Bedrin, N.G.; Subramanian, M.; Berking, J.; Ades, P.A.; Toth, M.J.; Miller, M.S. Age-related structural alterations in human skeletal muscle fibers and mitochondria are sex specific: Relationship to single-fiber function. *J. Appl. Physiol.* **2014**, *116*, 1582–1592.

35. Konopka, R.; Suer, M.K.; Wolff, C.A.; Harber, M.P. Markers of human skeletal muscle mitochondrial biogenesis and quality control: Effects of age and aerobic exercise training. *J. Gerontol.* **2014**, *69*, 371–378.

36. Betik, C.; Thomas, M.M.; Wright, K.J.; Riel, C.D.; Hepple, R.T. Exercise training from late middle age until senescence does not attenuate the declines in skeletal muscle aerobic function. *Am. J. Physiol.* **2009**, *297*, R744–R755.

37. Ljubicic, V.; Hood, D.A. Diminished contraction-induced intracellular signaling towards mitochondrial biogenesis in aged skeletal muscle. *Aging Cell* **2009**, *8*, 394–404.

38. Ljubicic, V.; Joseph, A.M.; Adhihetty, P.J.; Huang, J.H.; Saleem, A.; Uguccioni, G.; Hood, D.A. Molecular basis for an attenuated mitochondrial adaptive plasticity in aged skeletal muscle. *Aging* **2009**, *1*, 818–830.

39. Farrar, R.P.; Martin, T.P.; Ardies, C.M. The interaction of aging and endurance exercise upon the mitochondrial function of skeletal muscle. *J. Gerontol.* **1981**, *36*, 642–647.

40. Drew, B.; Phaneuf, S.; Dirks, A.; Selman, C.; Gredilla, R.; Lezza, A.; Barja, G.; Leeuwenburgh, C. Effects of aging and caloric restriction on mitochondrial energy production in gastrocnemius muscle and heart. *Am. J. Physiol.* **2003**, *284*, R474–R480.

41. Nielsen, J.; Suetaa, C.; Hvid, L.G.; Schroder, H.D.; Aagaard, P.; Orteblad, N. Subcellular localization-dependent decrements in skeletal muscle glycogen and mitochondria content following short-term disuse in young and old men. *Am. J. Physiol.* **2010**, *299*, E1053–E1060.

42. Holloszy, J.O. Skeletal muscle "mitochondrial deficiency" does not mediate insulin resistance. *Am. J. Clin. Nutr.* **2009**, *89*, 463S–466S.

43. Kadenbach, B. Intrinsic and extrinsic uncoupling of oxidative phosphorylation. *Biochim. Biophys. Acta* **2003**, *1604*, 77–94.

44. Stucki, J.W. The optimal efficiency and the economic degrees of coupling of oxidative phosphorylation. *Eur. J. Biochem.* **1980**, *109*, 269–283.

45. Rolfe, D.F.S.; Brown, G.C. Cellular energy utilisation and molecular origin of standard metabolic rate in mammals. *Physiol. Rev.* **1997**, *77*, 731–758.

46. Skulachev, V.P. Fatty acid circuit as a physiological mechanism of uncoupling of oxidative phosphorylation. *FEBS Lett.* **1991**, *294*, 158–162.

47. Jezek, P.; Engstova, H.; Zackova, M.; Vercesi, A.E.; Costa, A.D.T.; Arruda, P.; Garlid, K.D. Fatty acid cycling mechanism and mitochondrial uncoupling proteins. *Biochim. Biophys. Acta* **1998**, *1365*, 319–327.

48. Soboll, S.; Grundel, S.; Schwabe, U.; Scholtz, R. Influence of fatty acids on energy metabolism. 2. Kinetics of changes in metabolic rates and changes in subcellular adenine nucleotide contents and pH gradients following addition of octanoate and oleate in perfused rat liver. *Eur. J. Biochem.* **1984**, *141*, 231–236.

49. Amara, C.E.; Shankland, E.G.; Jubrias, S.A.; Marcinek, D.J.; Kushmerick, M.J.; Conley, K.E. Mild mitochondrial uncoupling impacts cellular aging in human muscles *in vivo*. *Proc. Natl. Acad. Sci. USA* **2007**, *104*, 1057–1062.

50. Fannin, S.W.; Lesnefsky, E.J.; Slabe, T.J.; Hassan, M.O.; Hoppel, C.L. Aging selectively decreases oxidative capacity in rat heart interfibrillar mitochondria. *Arch. Biochem. Biophys.* **1999**, *372*, 399–407.

51. Iossa, S.; Mollica, M.P.; Lionetti, L.; Crescenzo, R.; Tasso, R.; Liverini, G. A possible link between skeletal muscle mitochondrial efficiency and age-induced insulin resistance. *Diabetes* **2004**, *53*, 2861–2866.

52. Crescenzo, R.; Bianco, F.; Mazzoli, A.; Giacco, A.; Liverini, G.; Iossa, S. Alterations in proton leak, oxidative status and uncoupling protein 3 content in skeletal muscle subsarcolemmal and intermyofibrillar mitochondria in old rats. *BMC Geriatr.* **2014**, *14*, 79.

53. Koltai, E.; Hart, N.; Taylor, A.W.; Goto, S.; Ngo, J.K.; Davies, K.J.A.; Radak, Z. Age-associated declines in mitochondrial biogenesis and protein quality control factors are minimized by exercise training. *Am. J. Physiol.* **2012**, *303*, R127–R134.

54. Hafner, R.P.; Brown, G.C.; Brand, M.D. Analysis of the control of respiration rate, phosphorylation rate, proton leak rate and protonmotive force in isolated mitochondria using the "top-down" approach of metabolic control theory. *Eur. J. Biochem.* **1990**, *188*, 313–319.

55. Brand, M.D.; Chien, L.F.; Ainscow, E.K.; Rolfe, D.F.S.; Porter, R.K. The causes and functions of mitochondrial proton leak. *Biochim. Biophys. Acta* **1994**, *1187*, 132–139.

56. Cairns, C.B.; Walther, J.; Harken, A.H.; Banerjee, A. Mitochondrial oxidative phosphorylation thermodynamic efficiencies reflect physiological organ roles. *Am. J. Physiol.* **1998**, *274*, R1376–R1383.

57. Santanasto, A.J.; Glynn, N.W.; Jubrias, S.A.; Conley, K.E.; Boudreau, R.M.; Amati, F.; Mackey, D.C.; Simonsick, E.M.; Strotmeyer, E.S.; Coen, P.M.; *et al.* Skeletal muscle mitochondrial function and fatigability in older adults. *J. Gerontol. A Biol. Sci. Med. Sci.* **2015**.

58. Stannard, S.R.; Johnson, N.A. Insulin resistance and elevated triglyceride in muscle: More important for survival than thrifty genes? *J. Physiol.* **2003**, *554*, 595–607.

59. Korshunov, S.S.; Skulachev, V.P.; Starkov, A.A. High protonic potential actuates a mechanism of production of reactive oxygen species in mitochondria. *FEBS Lett.* **1997**, *416*, 15–18.

60. Mailloux, R.J.; Harper, M.E. Uncoupling proteins and the control of mitochondrial reactive oxygen species production. *Free Radic. Biol. Med.* **2011**, *51*, 1106–1115.

61. Brand, M.D. Uncoupling to survive? The role of mitochondrial inefficiency in ageing. *Exp. Gerontol.* **2000**, *35*, 811–820.

62. Lombardi, A.; Busiello, R.A.; Napolitano, L.; Cioffi, F.; Moreno, M.; de Lange, P.; Silvestri, E.; Lanni, A.; Goglia, F. UCP3 translocates lipid hydroperoxide and mediates lipid hydroperoxide-dependent mitochondrial uncoupling. *J. Biol. Chem.* **2010**, *285*, 16599–16605.

63. Goglia, F.; Skulachev, V.P. A function for novel uncoupling proteins: Antioxidant defence of mitochondrial matrix by translocating fatty acid peroxides from the inner to the outer membrane leaflet. *FASEB J.* **2003**, *17*, 1585–1591.

64. Divakaruni, A.S.; Brand, M.D. The regulation and physiology of mitochondrial proton leak. *Physiology* **2011**, *26*, 192–205.

65. Lal, S.B.; Ramsey, J.J.; Monemdjou, S.; Weindruch, R.; Harper, M.E. Effects of caloric restriction on skeletal muscle mitochondrial proton leak in aging rats. *J. Gerontol.* **2001**, *56*, B116–B122.

66. Marcinek, D.J.; Schenkman, K.A.; Ciesielski, W.A.; Lee, D.; Conley, K.E. Reduced mitochondrial coupling *in vivo* alters cellular energetic in aged mouse skeletal muscle. *J. Physiol.* **2005**, *569*, 467–473.

67. Hood, D. Plasticity in skeletal, cardiac, and smooth muscle: Contractile activity-induced mitochondrial biogenesis in skeletal muscle. *J. Appl. Physiol.* **2001**, *90*, 1137–1157.

68. Beyer, R.E.; Starnes, J.W.; Edington, D.W.; Lipton, R.J.; Compton, R.T.; Kwasman, M.A. Exercise-induced reversal of age-related declines of oxidative reactions, mitochondrial yield, and flavins in skeletal muscle of the rat. *Mech. Ageing Dev.* **1984**, *24*, 309–323.

69. Bevilacqua, L.; Ramsey, J.J.; Hagopian, K.; Weindruch, R.; Harper, M.E. Long-term caloric restriction increases UCP3 content but decreases proton leak and reactive oxygen species production in rat skeletal muscle mitochondria. *Am. J. Physiol.* **2005**, *289*, E429–E438.

70. Barazzoni, R.; Nair, K.S. Changes in uncoupling protein-2 and -3 expression in aging rat skeletal muscle, liver, and heart. *Am. J. Physiol.* **2001**, *280*, E413–E419.

Mevalonate Pathway Blockade, Mitochondrial Dysfunction and Autophagy: A Possible Link

Paola Maura Tricarico, Sergio Crovella and Fulvio Celsi

Abstract: The mevalonate pathway, crucial for cholesterol synthesis, plays a key role in multiple cellular processes. Deregulation of this pathway is also correlated with diminished protein prenylation, an important post-translational modification necessary to localize certain proteins, such as small GTPases, to membranes. Mevalonate pathway blockade has been linked to mitochondrial dysfunction: especially involving lower mitochondrial membrane potential and increased release of pro-apoptotic factors in cytosol. Furthermore a severe reduction of protein prenylation has also been associated with defective autophagy, possibly causing inflammasome activation and subsequent cell death. So, it is tempting to hypothesize a mechanism in which defective autophagy fails to remove damaged mitochondria, resulting in increased cell death. This mechanism could play a significant role in Mevalonate Kinase Deficiency, an autoinflammatory disease characterized by a defect in Mevalonate Kinase, a key enzyme of the mevalonate pathway. Patients carrying mutations in the *MVK* gene, encoding this enzyme, show increased inflammation and lower protein prenylation levels. This review aims at analysing the correlation between mevalonate pathway defects, mitochondrial dysfunction and defective autophagy, as well as inflammation, using Mevalonate Kinase Deficiency as a model to clarify the current pathogenetic hypothesis as the basis of the disease.

Reprinted from *Int. J. Mol. Sci.* Cite as: Tricarico, P.M.; Crovella, S.; Celsi, F. Mevalonate Pathway Blockade, Mitochondrial Dysfunction and Autophagy: A Possible Link. *Int. J. Mol. Sci.* **2015**, *16*, 16067–16081.

1. Mevalonate Pathway

The mevalonate pathway, fundamental for cholesterol synthesis, is one of the most important metabolic networks in the cell; it provides essential cell constituents, such as cholesterol, and some of its branches produce key metabolites, such as geranylgeranyl pyrophosphate and farnesyl pyrophosphate, necessary for normal cell metabolism.

The first step of the mevalonate pathway is the synthesis of 3-hydroxy-3-methylglutaryl-CoA (HMG-CoA) from three molecules of acetyl-CoA, firstly by a condensation reaction forming acetoacetyl-CoA through acetoacetyl-CoA thiolase (EC 2.3.1.9) and subsequently through a second condensation between acetoacetyl-CoA and a third acetyl-CoA molecule catalysed by HMG-CoA synthase (EC 2.3.3.10) (Figure 1a, 1). In the second step, HMG-CoA is reduced to mevalonate acid by NADPH, a reaction catalysed by the HMG-CoA reductase (HMGR) enzyme (EC 1.1.1.88 and EC 1.1.1.34). HMGR is the rate-limiting enzyme for the mevalonate pathway and is one of the most finely regulated enzymes [1]. Regulation begins at the transcriptional level; if cholesterol or other sterol isoprenoids are in shortage, sterol regulatory element binding proteins (SREBP) are activated and they bind to sterol regulatory elements (SREs) present on the HMGR promoter, increasing its transcription [2,3]. Cholesterol also regulates the degradation of HMGR, promoting its association with gp78, an ubiquitin-E3 ligase that directs the enzyme towards proteasome 26s. HMGR is also regulated at the post-translational level, by phosphorylation mediated through AMP-activated protein kinase (AMPK). This enzyme is sensitive to the AMP:ATP ratio, and is activated by increased AMP concentration, thus in casea of metabolic stress, it deactivates HMGR, reducing cellular metabolism [4] (Figure 1a, 2).

The third key enzyme of the mevalonate pathway is the one responsible for converting mevalonic acid into mevalonate-5-phosphate, a key pathway intermediate. Mevalonate kinase (EC 2.7.1.36) (MVK) catalyses this conversion, using ATP as a phosphate donor and energy source. Furthermore, this enzyme is finely regulated, firstly at transcriptional level in a similar manner of HMGR: SREs are present at the MVK promoter and increases its transcription upon cholesterol shortage [5]. In addition, MVK presents feedback inhibition from some of the mevalonate pathway substrates, specifically geranylgeranyl pyrophosphate and farnesyl pyrophosphate, demonstrating that non-sterol isoprenoid could have a key role in regulation of this enzyme [6] (Figure 1a, 3).

In the fourth step of the mevalonate pathway, Mevalonate-5-phosphate is then converted into mevalonate-5-pyrophosphate by phosphomevalonate kinase (EC 2.7.4.2), using again ATP as phosphate and energy donor. Differently from MVK, this enzyme does not present feedback inhibition from its products [6]. However, various compounds were recently found to be inhibitors for this enzyme, suggesting novel mechanisms to inhibit the mevalonate pathway [7] and more interestingly, phosphomevalonate kinase appears also to be regulated by cholesterol shortage, as for mevalonate kinase and HMGR [8]. Indeed, cholesterol shortage induces increases in all the three first enzymes of the mevalonate pathway, thus guaranteeing a continued supply of this key membrane component (Figure 1a, 4).

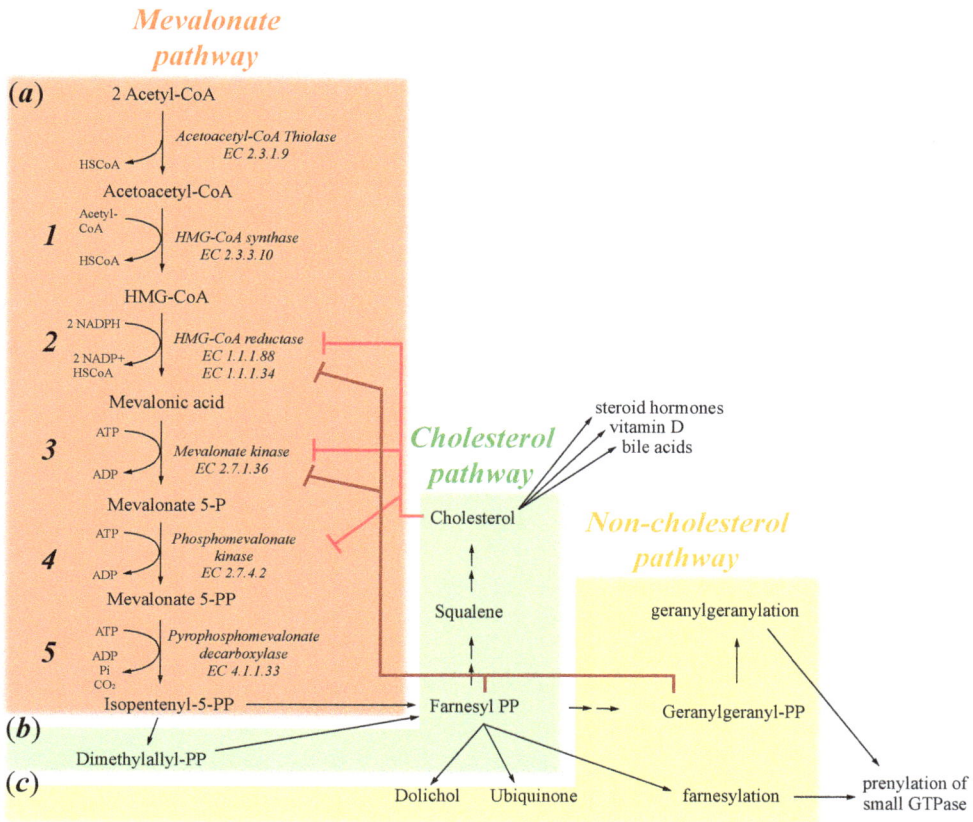

Figure 1. Schematic representation of the mevalonate pathway divided into: (**a**) The mevalonate pathway that produces mevalonate 5-PP and then isppententenyl 5-PP; (**b**) The cholesterol pathway that produces cholesterol, which in turn induces the formation of steroid hormones, vitamin D and bile acids; and (**c**) The non-cholesterol pathway important for the production of farnesyl-PP and geranylgeranyl-PP that induces respectively farnesylation and geranylgeranylation of small GTPase.

The fifth enzyme of the mevalonate pathway is pyrophosphomevalonate decarboxylase or diphosphomevalonate decarboxylase (EC 4.1.1.33). It converts mevalonate-5-pyrophosphate into isopentenyl-5-pyrophosphate (IPP), the final product of mevalonate pathway and the starting substrate for successive biosyntheticals reactions, especially cholesterol and isoprenoid production. This enzyme performs two key reactions: firstly, it phosphorylates mevalonate-5-pyrophosphate generating an intermediate product that, secondly, it is dephosphorylated and decarboxylated, obtaining thus IPP as a final product (Figure 1a, 5).

The mevalonate pathway, as described above, generates a key intermediate for cholesterol production, a fundamental constituent of cell membranes. Moreover, cholesterol is also converted to steroid hormones, regulating different cellular pathways, vitamin D and bile acid production (Figure 1b). IPP is the first step also in other, non-cholesterol, reactions; it is important for the production of farnesyl-pyrophosphate (FPP). FPP is converted in dolichols, used to assemble carbohydrate chains in glycoproteins, or in ubiquinones (or coenzyme Q10), electron transporters in mitochondria; or it is used to farnesylate or geranylate proteins, thus targeting them to cell membranes (Figure 1c). In summary, the mevalonate pathway is responsible for numerous cellular processes and the key enzymes described above undergo different regulation to maintain a constant supply of IPP.

2. Exogenous Mevalonate Pathway Blockade

The principal compounds that induce exogenous mevalonate pathway blockade are the statins, which are a class of compounds that act as competitive inhibitors of 3-hydroxy-3-methylglutaryl coenzyme-A (HMG-CoA) reductase, a key enzyme of the mevalonate pathway, which converts HMG-CoA into mevalonic acid. Statins, in general, are able to bind to a portion of HMG-CoA binding site, thus blocking the access of this substrate to the active site of the enzyme; effectively reducing the rate of mevalonate productions [9,10].

In 1976 Endo and coauthors discovered the first statin, isolated from *Penicillium citrinium* [11]. Subsequently, over the last two decades, several statins have been identified and classified in several ways. The most commonly used classification divides them into statins produced by fungi (such as Lovastatin, Simvastatin) and statins synthetically made (such as Atorvastatin, Fluvastatin).

All statins share a conserved HMG-like moiety covalently linked to a more or less extended hydrophobic group.

By blocking HMG-CoA reductase, statins induce a decrease in cholesterol level and simultaneously other by-products of the mevalonate pathway such as farnesyl pyrophosphate (FPP), geranylgeranyl pyrophosphate (GGPP), dolichols and coenzyme Q10 [12,13]. As reviewed in Winter-Vann and Casey (2005), inhibition of HMG-CoA reductase has a pleiotropic effect, due to the different affinities of key enzymes in the mevalonate pathway. FPP, the main metabolite in this pathway, could be converted to cholesterol through squalene synthase and this enzyme has a Km for the substrate of about 2 μM. GGPP synthase, instead, could convert FPP to GGPP, with a Km of 1 μM; GGPP is attached to different proteins (the majority of which pertain to the Rab family) to ensure their correct localization. On the other hand, protein farnesyl trasferase (FTase) uses FPP to attach a farnesyl group to specific proteins, such as the family of small GTPase proteins (Ras and Rho GTPases), with a Km of 5 nM. Therefore, inhibition of HMG-CoA reductase lowers FPP levels and the

118

first consequence is a reduction in cholesterol levels; following that, GGPP levels are reduced, causing mislocalization and loss of activity of specific proteins. Instead, due to the high affinity of FTase towards FPP, farnesylation levels of key cellular enzymes remain stable [14].

Indeed, a widely adopted view considers the pleiotropic effects of statins independent of lowering cholesterol levels, but rather connected to a lack of these prenylated proteins [12]. In the last few years there has been an increase in interest of these pleiotropic effects, because of their possible main responsibility for statin anti-cancer and immunomodulatory effects [15–18].

For all these reasons, the role of statins are debatable, and there are many studies describing statins as drugs for treatment of a variety of disease such as hypercholesterolemia, cancer, cardiovascular diseases, inflammatory diseases [19–24].

Furthermore, statins are used as a pharmacological compound to biochemically reproduce some features of Mevalonate Kinase Deficiency (MKD)—a pathology characterized by a defect in a key enzyme of mevalonate pathway [13,25,26]. In some studies, mevalonate pathway blockade, obtained in neuronal and monocytic cell lines by statin (Lovastatin) administration, induces an increase of apoptosis correlated to mitochondrial damage [27–29].

Also, Van der Burgh and co-workers have recently demonstrated that mevalonate pathway blockade, obtained in monocytic cell line by statin (Simvastatin) administration, produces mitochondrial damage and autophagy impairment, related to a decrease in protein prenylation levels [25,30].

2.1. Mitochondrial Dysfunction and Statin

Mevalonate pathway blockade, obtained by treatment with statins, has been linked to mitochondrial dysfunction, specifically by lowering mitochondrial membrane potential and increasing release of pro-apoptotic factors.

Usually, mitochondrial dysfunction is associated with intrinsic apoptosis, also known as the mitochondrial apoptotic pathway. This pathway is characterized by activation of caspase-9 and -3, and inhibition or activation of anti- or pro-apoptotic Bcl-2 family members. Furthermore, mitochondrial membrane potential decreases, causing release of pro-apoptotic factors, oxidative stress and then cell death [31].

In a biochemical MKD model, obtained by Lovastatin treatment in neuroblastoma cell lines, we observed mitochondrial dysfunction correlated to increased intrinsic apoptosis, also confirmed by activation of caspase-3 and -9 [27,28]; furthermore, in monocyte cell lines, we observed a similar increase in oxidative stress [29].

Mitochondrial dysfunction, caused by statins, could be related to oxidative stress, shortage of prenylated proteins or both. In fact, it was observed that the block of mevalonate pathway, obtained by statin (Simvastin) treatment in endothelial

119

cancer cell lines, resulted in G1 cell cycle arrest, apoptosis, DNA damage and cellular stress [32].

Another study showed that simvastatin, in lung cancer cells, inhibited the proliferation and significantly increased oxidative stress, in particular augmenting reactive oxygen species (ROS) production and the activity of total superoxide dismutase (SOD) and in particular the mitochondrial form, superoxide dismutase 2 (SOD2) [33].

Strong oxidative stress, which induces mitochondrial dysfunction, could be due to the action of statins on the mevalonate pathway, decreasing coenzyme Q10 and dolichol levels, considered as anti-oxidants defense systems.

Coenzyme Q10 is a product of the mevalonate pathway and is an important electron transporter of the mitochondrial respiratory chain. A decrease in coenzyme Q10 levels, caused by mevalonate pathway blockade, could result in an abnormal mitochondrial respiratory function causing mitochondrial and oxidative damage [34].

Dolichol, a polyprenol compound, is an important free-radical scavenger in cell membranes [35]. Ciosek and co-workers observed a significant decrease in dolichol levels after Lovastatin administration in *in vivo* models [36]; a lack of this compound might cause oxidative stress and mitochondrial damage [13,37].

Nevertheless, mitochondrial dysfunction caused by statins could also be related to a decrease in prenylated protein levels; indeed, statins treatment could lead to a reduction in cholesterol level, and also in farnesyl pyrophosphate (FPP) and in geranylgeranyl pyrophosphate (GGPP). Xia and co-workers have demonstrated that apoptosis induced by Lovastatin treatment, in human AML cells is connected to a decrease in GGPP and, to a lesser extent, related to a FPP decrease [38].

Agarwal and co-workers also observed the close correlation between decrease in prenylated proteins levels, apoptosis and mitochondrial damage in Lovastatin-treated colon cancer cells. The treatment caused a decrease in expression of anti-apoptotic protein Bcl-2 and an increase of pro-apoptotic protein such as Bax; the subsequent addition of GGPP prevented Lovastatin apoptosis, confirming a key role of prenylated proteins levels [39].

Recently, Van der Burgh and co-workers have demonstrated that, in simvastatin-treated cells, mitochondria clearance is reduced, with lower oxygen consumption and glycolysis rate. These conditions suggested that accumulation of damaged mitochondria could be the trigger for NACHT, LRR and PYD domains-containing protein 3 (NALP3) inflammasome activation. The authors also speculate that prenylated proteins could be the main mediators of statin adverse effects [25,30].

Further confirmations that statins treatment could impair mitochondrial activity come from two recent studies done using a completely different model, *C. elegans*. In the first paper, the authors show that animals resistant to statin treatment have an increased mitochondrial unfolded protein response (UPRmt) which, they speculate, could lower protein turnover and thus lessening the need for protein prenylation [40]. In the second paper, the authors demonstrate that statin abrogates the *C. elegans* ability to sense mitochondrial damage and that this ability could be partially rescued through GGPP subministration [41]. Taken together, these results demonstrate once more the key role of protein prenylation in mitochondrial homeostasis.

A mitochondrial-specific effect of statin, probably mediated through lowering prenylated proteins, suggests that this class of compound could be considered as anti-cancer drugs. However, further studies are necessary to completely clarify the variety of effects of these drugs and at present, only hypotheses have been raised to explain actions of statins on cellular survival.

2.2. Autophagy and Statins

Mevalonate pathway blockade has been linked to defective autophagy, possibly causing inflammasome activation and subsequent cell death. Autophagy (macroautophagy) is the main catabolic mechanism involved in the turnover of cytoplasmic components and selective removal of damaged or redundant organelles (such as mitochondria, peroxisomes and endoplasmic reticulum), through the lysosome machinery. Initial steps include the formation of phagophore or pre-autophagosome, an isolated membrane able to elongate and forming the autophagosome, a double-membrane compartment that sequesters the cytoplasmic materials. Subsequently, the fusion of autophagosome with lysosome forms the autolysosome, where the captured material is degraded [42] (Figure 2).

Autophagy involves numerous molecular mediators called autophagy-related (ATG) proteins; among these, prenylation of proteins appears to be one of the key regulation mechanisms [43]. Specifically, different small GTPase are indicated as ATG proteins: Rabs, Rheb, RalB. Rabs are crucial proteins for developing subdomains on membranes to facilitate maturation; recent studies have shown that some Rabs are essential for autophagy. The best known Rabs at the present time are: Rab1, able to regulate the autophagosome formation; Rab5, an early endosome protein, responsible for autophagosome membrane elongation and autophagosome formation, regulating Beclin1-Vps34-Atg6 class III PI3-kinase complex; Rab7, a late endosome protein, responsible for autophagosome maturation, promoting the microtubule plus-end-directed transport and facilitating fusion of autophagosome with lysosome; Rab24, normally present in reticular distribution around the nucleus, is important for autophagosome maturation; Rab32, generally present in mitochondria, regulates the membrane trafficking, and is required for autophagosome formation; Rab33 modulates autophagosome formation [44,45] (Figure 2).

121

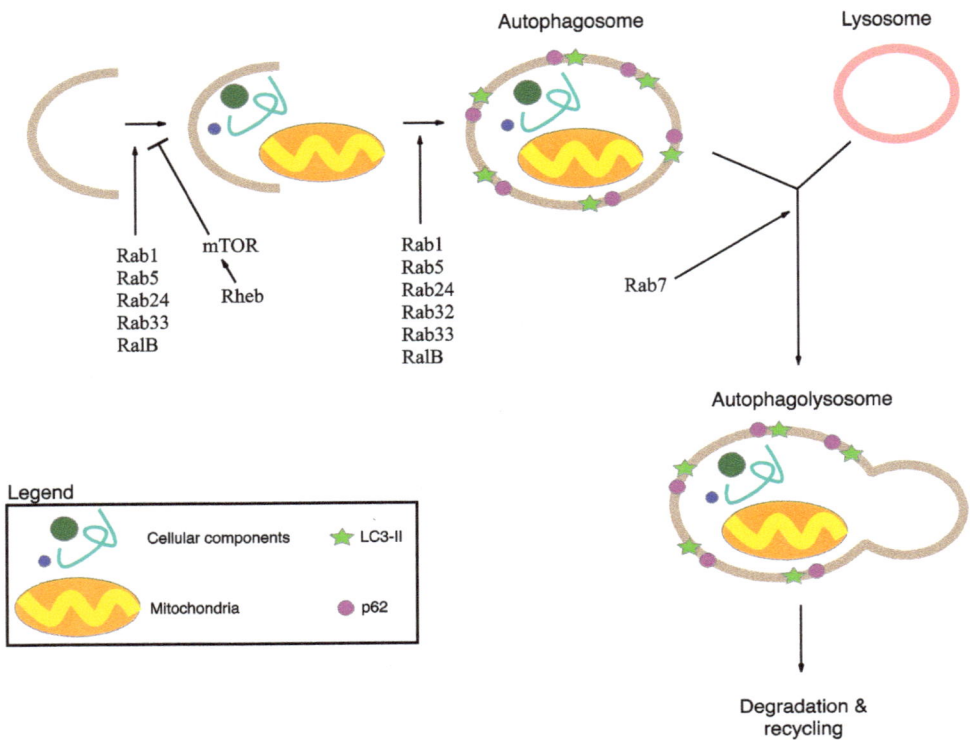

Figure 2. Schematic representation of macroautophagy mechanism and its main actors. Macroautophagy delivers cellular components and damaged or redundant organelles (such as mitochondria), to the lysosome through the intermediary of a double membrane-bound vesicle, referred to as an autophagosome. Autophagy is initiated by the formation of the isolation membrane that induces the formation of autophagosome. Subsequently, the autophagosome fuses with the lysosome to form an autolysosome. Finally all the material is degraded in the autophagolysosome and recycled. The p62 protein interacts with damaged proteins in the cells, and the complexes are then selectively tied to the autophagosome through LC3-II. Rabs, Rheb and RalB are autophagy-related protein important for the regulation of macroautophagy mechanism.

Ras homology enriched in brain (Rheb) directly binds and selectively activates the multiprotein complex 1 of mammalian target of rapamycin (mTORC1) [46] which is composed of mTOR, a negative regulator of autophagy, and mLST8 [47,48]. mTOR activity inhibits mammalian autophagy and indeed a recent study demonstrated that autophagy impairment is correlated to mTORC1 hyperactivation in β-cell [49] (Figure 2).

RalB localizes in nascent autophagosome and is activated due to nutrient deprivation and, thanks to the binding to its effector Exo84, induces the assembly of active Beclin1-VPS34 and ULK1. The resulting complex induces the isolation of pre-autophagosomal membrane and maturation of autophagosomes [50,51] (Figure 2).

All these proteins require prenylation for their activation, either farnesylated or geranylgeranylated; indeed statins, blocking the mevalonate pathway and causing a decrease in prenylated proteins levels, could play a regulatory role in autophagy. Nevertheless, the statin effects in autophagy remain poorly understood.

Recently, Van der Burgh and co-workers have observed defective autophagy in statin-treated monocytes, correlated to damaged mitochondria and NALP3 inflammasome activation [25]. The same authors have subsequently demonstrated that statin treatment increases levels of unprenylated RhoA, which in turn activates protein kinase B (PKB) possibly playing a role in statin-induced autophagy blockade [30].

On the contrary, other studies show that treatment with statins induces an increase in autophagy levels, and for this reason, statins could be considered as anti-cancer drugs. Indeed, a study showed that statins, such as Cerivastatin, Pitavastatin and Fluvastatin, are the most potent autophagy inducing agents in human cancer cells; the authors, however, did not analyze levels of p62, thus the possibility that autophagy is increased but subsequentrly blocked was not examined [52].

Another study demonstrated that statin induces autophagy through depleting cellular levels of geranylgeranyl diphosphonate (GGPP), independently of the decreased activity in two major small G proteins, Rheb and Ras [53]. However, the authors did not examine all autophagic pathways, thus the observations are not conclusive and further investigations are needed.

Lastly, Wei and co-workers observed that simvastatin inhibits the Rac1-mTOR pathway and thereby increases autophagy, in coronary arterial myocytes [54].

All these results show that the role of statin in autophagy is related to GGPP and prenylated proteins levels, thus being important actors in this mechanism. Indeed, cellular differences in GGPP and prenylated proteins levels could explain the contradictory findings in the studies described above. However, further studies are necessary to clarify the mechanism of action and the molecular targets involved in statin-modulated autophagy.

3. Endogenous Mevalonate Pathway Blockade

The mevalonate pathway could also be blocked by enzymatic defects due to mutations in genes involved in this pathway. In particular, a rare disease involving mutations on *MVK* gene (12q24.11) has been described: Mevalonate Kinase

Deficiency (MKD). Currently, 82 mutations of the *MVK* gene have been reported in the Human Gene Mutation Database [55].

MKD possesses two distinct phenotypes: a milder one, also called Hyper IgD Syndrome (HIDS; OMIM#260920), in which the patients suffer recurrent fevers, have skin rashes, hepatosplenomegalia and generally a sustained inflammatory response; and a severe, rarer, one, called Mevalonic Aciduria (MA; OMIM #610377), characterized by the involvement of the Central Nervous System, with cerebellar ataxia, psychomotor retardation and also, as in HIDS, recurrent fever attacks [56].

Residual MVK enzymatic activity marks the boundary between HIDS and MA, with MA patients having less than 1% activity, while HIDS between 1% and 7% of activity [57]. Initially, disease pathogenesis was thought to derive from low cholesterol levels, being MVK a central enzyme in the mevalonate pathway. However, patients, either with HIDS or MA, showed normal cholesterol levels, probably due to dietary intake. Subsequently, accumulation of Mevalonic acid was indicated as responsible for the MKD phenotype. Still, a small clinical trial, involving two MA patients, using statin (Lovastatin) to reduce Mevalonic acid, resulted in worsening of the symptoms [58]; on the contrary, Simvastatin appeared to be beneficial for treating HIDS patients [56]. Thus, the hypothesis pointing to Mevalonic acid levels as causative for MKD does not explain completely the disease's manifestations. Celec and Behuliak in 2008 hypothesized that MVK dysfunctions could diminish non-steroid isoprenoids, causing oxidative stress and ultimately leading to chronic hyperinflammation [13].

The shortage of isoprenoids, correlated to a severe reduction in protein prenylation, in particular of geranylgeranyl pyrophosphate (GGPP), has been linked with the activation of caspase-1 and thereby with the production of IL-1β [59,60].

In particular, the IL-1 family is strongly supposed to play a fundamental role in MKD inflammatory processes, indeed, several biological therapies have successfully targeted these molecules [61–63].

In a previous work it has been shown that, in monocytes from MKD patients, a key component of the inflammation machinery is NACHT, LRR and PYD domains-containing protein 3 (NALP3) [64]; NALP3 interacts with another protein, pyrin domain (PYD) of apoptosis-associated speck-like protein containing a CARD domain (ASC), constituting the inflammasome platform. The CARD domain recruits pro-caspase-1, which self-cleaves into active caspase-1 and then converts pro-IL-1β to active IL-1β, activating one of the main pathways of inflammation [65,66].

However it remains an open question how isoprenoid shortage, as in MKD and in presence of the biochemical block, activates NALP3 and the inflammasome pathway.

Recently Van der Burgh and co-workers have demonstrated that, in MKD patients' cells and in statin-treated monocytes, autophagy is impaired and specifically mitochondria clearance is slowed, suggesting that accumulation of damaged mitochondria could be the trigger to NALP3 inflammasome activation [25]. The same group has recently reported that statin treatment increases levels of unprenylated Ras homolog gene family member A (RhoA), which in turn activates PKB, representing then the hypothetical starting point for autophagy. The authors also demonstrated that levels of unprenylated RhoA correlate with IL-1b release, thus partially confirming this hypothetical link [25,30].

4. Autophagy, Inflammation and Damaged Mitochondria

An emerging concept in recent years tightly associates autophagy with inflammation regulation mechanisms. Autophagy can regulate different aspects of the immune response: it is involved in the degradation of bacteria/virus engulfed by the cell, it regulates Pattern Recognition Receptors (PRRs) and can act as their effector, it can process Antigens for MHC presentation and finally autophagy can regulate inflammasome activation and secretion of different cytokines [67]. The first three mechanisms have been extensively described by Deretic (2013) [68], and will be briefly discussed here.

Autophagy can be envisaged as a mechanism to clear the cytosol from invading intracellular pathogens, either bacteria or viruses, which are engulfed in autophagic membranes and targeted to lysosomes to be degraded. This process is facilitated by sequestosome 1/p62-like receptors (SLRs) that recognize pathogens and facilitate their encasement in autophagosomes, possessing an LC3 interacting region.

Furthermore, autophagy could be an effector for PRRs, degrading targets marked by toll-like receptors (TLRs) or it can help in delivering ligands to TLRs, as for TLR7 [68].

Antigen processing for MHC II presentation represents another important mechanism regulated by autophagy. Indeed, autophagy is crucial for viral immunosurvelliance and is also inhibited by HIV-1 by upregulating mTOR in dendritic cells [69,70]. Moreover, autophagy is required for positive selection of naïve T cells in thymus, whereas knocking-out key autophagy proteins results in autoimmune syndromes [71].

How autophagy regulates inflammasome activation is currently a subject of numerous studies and a general consensus has not been yet reached. However, two hypotheses are at the moment explored: a first one suggesting that autophagy regulates processing of IL-1b and other pro-inflammatory molecules; the second proposing that autophagy removes damaged mitochondria, thus dampening NALP3 activation. It is possible that those two processes are not mutually exclusive and act in parallel to maintain the inflammatory status in "inactive" condition.

Direct regulation of IL-1b processing by inflammasomes has been demonstrated in macrophages and *in vivo* where treatment with rapamycin (autophagy inducer) decreases its secreted and circulating levels [72]. Further confirmation came from the work of Shi and co-authors (2012), in which they demonstrate how inflammasome activation induces autophagy and autophagosomes formation containing inflammasomes components such as ASC or absent in melanoma 2 (AIM2), in a self-limiting process. IL-1b does not possess a "canonical" secretory signature and it is translated in the cytosol, via polyribosomes linked to the cytoskeleton [73–75]. Later it was demonstrated that secretion of IL-1b is dependent on its localization in lysosome-associated vesicles and agents that regulates autophagy can modulate its release [72,76]. These works clearly show the regulation of IL-1b processing by autophagy.

Other studies put autophagy upstream of inflammasome activation. Reactive oxygen species (ROS) can activate NALP3 inflammasome and damaged mitochondria are the main source for ROS. Mitophagy, a specialized form of autophagy, constantly removes damaged mitochondria, thus lowering ROS levels. If cells are challenged with 3-methyladenine, a blocker of mitophagy, NALP3 activation is increased and redistributed near mitochondria-ER contact points, working as a sensor for mitochondrial damage [77].

Moreover, mitochondrial DNA (mtDNA) can work as NALP3 activation inducer, thus sensing mitochondrial damage and 3-MA could increase NALP3 activation caused by mtDNA [78,79]. These data show a possible pipeline: decreased mitophagy leading to increased ROS and mtDNA in cytosol, leading to increased NALP3 activation.

The two mechanisms described above are not necessarily self-exclusive. It is then possible that autophagy machinery collaborates in dampening inflammation and a disturbance in mitochondria homeostasis could lead to exacerbating inflammation.

In MKD is it then possible that a disturbance in autophagy mechanisms causes damage in mitochondria, impairing their recycling and thus increasing ROS levels; this, in turn, increases activation of inflammation. This chain of events still remains to be verified; however exploring this mechanism could represent a novel strategy to fight this debilitating disease.

5. Conclusive Remarks

MKD is an orphan drug disease, so many efforts are being made in search of potential targets for novel treatments tailored to prevent, or at least to diminish, apoptosis in MKD patients. Several studies reported *in vitro* administration of natural and synthetic isoprenoids to restore the mevalonate pathway in cell cultures (both models and patients' derived monocytes) treated with statins to biochemically mimic the genetic defect.

As described above, a possible link exists between defective protein prenylation and mitochondrial dysfunction, supposedly made by autophagy (Figure 3).

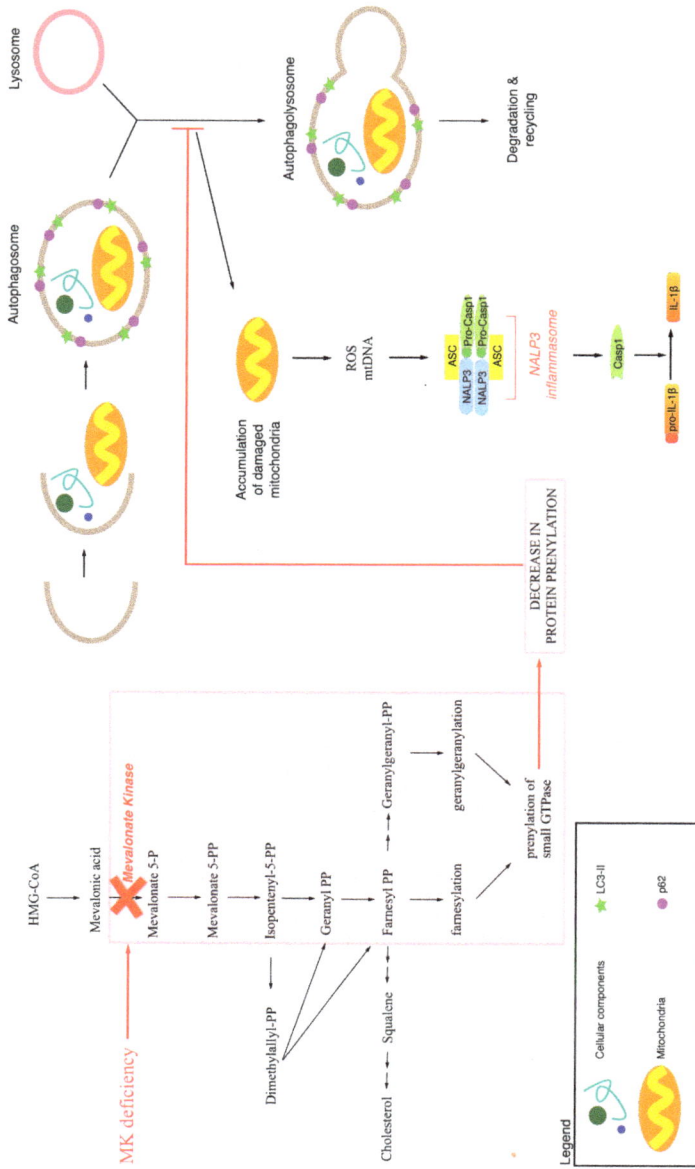

Figure 3. Schematic representation of a possibile link between defective protein prenylation, mitochondrial dysfunction and autophagy. Mevalonate Kinase Deficiency is characterized by a block of the mevalonate pathway induces by mutation in a gene that encodes for Mevalonate Kinase. Blockade of the mevalonate pathway induces decrease in protein prenylation that could alter the macroautophagy mechanism and in particular mitochondrial degradation and recycling. Accumulation of damaged mitochondria induces ROS production and mtDNA release. All these events are important for the activation of NALP3 inflammasome that cleaves and activates IL-1b.

Furthermore, it remains to be determined how autophagy is impaired in MKD. Protein prenylation seems to be one of the regulation mechanisms involved in autophagy. Compounds able to restore protein prenylation could be considered as potential therapy to tackle MKD; unfortunaly such compounds are at the present time, not actively researched. Other strategies, such as modulation of farnesyl protein transferase could be exploited to fight MKD.

For all these reasons, our review aimed at recalling the attention of the scientific community on another possible mechanism at the basis of MKD pathogenesis, and intends to point out that autophagy should be considered when trying to design novel therapeutic strategies to fight cell death in MKD patients.

Acknowledgments: This study was supported by a grant from the Institute for Maternal and Child Health—IRCCS "Burlo Garofolo"—Trieste, Italy (RC 42/2011).

Author Contributions: Paola Maura Tricarico wrote "Exogenous Mevalonate pathway blockade"; Sergio Crovella wrote "Endogenous mevalonate pathway blockade" and "Conclusive remarks"; Fulvio Celsi wrote "Mevalonate pathway" and "Autophagy, inflammation and damaged mitochondria". All authors revised and assembled the paper.

Conflicts of Interest: The authors declare no conflict of interest.

References

1.   Goldstein, J.L.; Brown, M.S. Regulation of the mevalonate pathway. *Nature* **1990**, *343*, 425–430.
2.   Horton, J.D. Sterol regulatory element-binding proteins: Transcriptional activators of lipid synthesis. *Biochem. Soc. Trans.* **2002**, *30*, 1091–1095.
3.   Weber, L.W.; Boll, M.; Stampfl, A. Maintaining cholesterol homeostasis: Sterol regulatory element-binding proteins. *World J. Gastroenterol.* **2004**, *10*, 3081–3087.
4.   Burg, J.S.; Espenshade, P.J. Regulation of HMG-CoA reductase in mammals and yeast. *Progress Lipid Res.* **2011**, *50*, 403–410.
5.   Murphy, C.; Murray, A.M.; Meaney, S.; Gåfvels, M. Regulation by SREBP-2 defines a potential link between isoprenoid and adenosylcobalamin metabolism. *Biochem. Biophys. Res. Commun.* **2007**, *355*, 359–364.
6.   Hinson, D.D.; Chambliss, K.L.; Toth, M.J.; Tanaka, R.D.; Gibson, K.M. Post-translational regulation of mevalonate kinase by intermediates of the cholesterol and nonsterol isoprene biosynthetic pathways. *J. Lipid Res.* **1997**, *38*, 2216–2223.
7.   Boonsri, P.; Neumann, T.S.; Olson, A.L.; Cai, S.; Herdendorf, T.J.; Miziorko, H.M.; Hannongbua, S.; Sem, D.S. Molecular docking and NMR binding studies to identify novel inhibitors of human phosphomevalonate kinase. *Biochem. Biophys. Res. Commun.* **2013**, *430*, 313–319.
8.   Olivier, L.M.; Chambliss, K.L.; Gibson, K.M.; Krisans, S.K. Characterization of phosphomevalonate kinase: Chromosomal localization, regulation, and subcellular targeting. *J. Lipid Res.* **1999**, *40*, 672–679.

9. Istvan, E.S.; Deisenhofer, J. Structural mechanism for statin inhibition of HMG-CoA reductase. *Science* **2001**, *292*, 1160–1164.

10. Corsini, A.; Maggi, F.M.; Catapano, A.L. Pharmacology of competitive inhibitors of HMG-CoA reductase. *Pharmacol. Res.* **1995**, *31*, 9–27.

11. Endo, A.; Kuroda, M.; Tsujita, Y. ML-236A, ML-236B, and ML-236C, new inhibitors of cholesterogenesis produced by Penicillium citrinium. *J. Antibiot. (Tokyo)* **1976**, *29*, 1346–1348.

12. Alegret, M.; Silvestre, J.S. Pleiotropic effects of statins and related pharmacological experimental approaches. *Methods Find Exp. Clin. Pharmacol.* **2006**, *28*, 627–656.

13. Celec, P.; Behuliak, M. The lack of non-steroid isoprenoids causes oxidative stress in patients with mevalonic aciduria. *Med. Hypotheses* **2008**, *70*, 938–940.

14. Winter-Vann, A.M.; Casey, P.J. Post-prenylation-processing enzymes as new targets in oncogenesis. *Nat. Rev. Cancer* **2005**, *5*, 405–412.

15. Wong, W.W.; Dimitroulakos, J.; Minden, M.D.; Penn, L.Z. HMG-CoA reductase inhibitors and the malignant cell: The statin family of drugs as triggers of tumor-specific apoptosis. *Leukemia* **2002**, *16*, 508–519.

16. Sassano, A.; Platanias, L.C. Statins in tumor suppression. *Cancer Lett.* **2008**, *260*, 11–19.

17. Osmak, M. Statins and cancer: Current and future prospects. *Cancer Lett.* **2012**, *324*, 1–12.

18. Blum, A.; Shamburek, R. The pleiotropic effects of statins on endothelial function, vascular inflammation, immunomodulation and thrombogenesis. *Atherosclerosis* **2009**, *203*, 325–330.

19. Olsson, A.G.; Istad, H.; Luurila, O.; Ose, L.; Stender, S.; Tuomilehto, J.; Wiklund, O.; Southworth, H.; Pears, J.; Wilpshaar, J.W.; *et al.* Effects of rosuvastatin and atorvastatin compared over 52 weeks of treatment in patients with hypercholesterolemia. *Am. Heart J.* **2002**, *144*, 1044–1051.

20. Cardwell, C.R.; Mc Menamin, Ú.; Hughes, C.M.; Murray, L.J. Statin use and survival from lung cancer: A population-based cohort study. *Cancer Epidemiol. Biomarkers Prev.* **2015**, *24*, 833–841.

21. Shepherd, J.; Blauw, G.J.; Murphy, M.B.; Bollen, E.L.; Buckley, B.M.; Cobbe, S.M.; Ford, I.; Gaw, A.; Hyland, M.; Jukema, J.W.; *et al.*; PROSPER study group Pravastatin in elderly individuals at risk of vascular disease (PROSPER): A randomised controlled trial. *Lancet* **2002**, *23*, 360, 1623–1630.

22. Dursun, S.; Çuhadar, S.; Köseoğlu, M.; Atay, A.; Aktaş, S.G. The anti-inflammatory and antioxidant effects of pravastatin and nebivolol in rat aorta. *Anadolu Kardiyol. Derg.* **2014**, *14*, 229–233.

23. Leung, B.P.; Sattar, N.; Crilly, A.; Prach, M.; McCarey, D.W.; Payne, H.; Madhok, R.; Campbell, C.; Gracie, J.A.; Liew, F.Y.; *et al.* A novel anti-inflammatory role for simvastatin in inflammatory arthritis. *J. Immunol.* **2003**, *170*, 1524–1530.

24. Barsante, M.M.; Roffê, E.; Yokoro, C.M.; Tafuri, W.L.; Souza, D.G.; Pinho, V.; Castro, M.S.; Teixeira, M.M. Anti-inflammatory and analgesic effects of atorvastatin in a rat model of adjuvant-induced arthritis. *Eur. J. Pharmacol.* **2005**, *516*, 282–289.

25. Van der Burgh, R.; Nijhuis, L.; Pervolaraki, K.; Compeer, E.; Jongeneel, L.H.; van Gijn, M.; Coffer, P.J.; Murphy, M.P.; Mastroberardino, P.G.; Frenkel, J.; *et al.* Defects in mitochondrial clearance predispose human monocytes to interleukin-1β hypersecretion. *J. Biol. Chem.* **2014**, *289*, 5000–5012.

26. Kuijk, L.M.; Beekman, J.M.; Koster, J.; Waterham, H.R.; Frenkel, J.; Coffer, P.J. HMG-CoA reductase inhibition induces IL-1β release through Rac1/PI3K/PKB-dependent caspase-1 activation. *Blood* **2008**, *112*, 3563–3573.

27. Marcuzzi, A.; Tricarico, P.M.; Piscianz, E.; Kleiner, G.; Brumatti, L.V.; Crovella, S. Lovastatin induces apoptosis through the mitochondrial pathway in an undifferentiated SH-SY5Y neuroblastoma cell line. *Cell Death Dis.* **2013**, *4*, e585.

28. Marcuzzi, A.; Zanin, V.; Piscianz, E.; Tricarico, P.M.; Vuch, J.; Girardelli, M.; Monasta, L.; Bianco, A.M.; Crovella, S. Lovastatin-induced apoptosis is modulated by geranylgeraniol in a neuroblastoma cell line. *Int. J. Dev. Neurosci.* **2012**, *30*, 451–456.

29. Tricarico, P.M.; Kleiner, G.; Valencic, E.; Campisciano, G.; Girardelli, M.; Crovella, S.; Knowles, A.; Marcuzzi, A. Block of the mevalonate pathway triggers oxidative and inflammatory molecular mechanisms modulated by exogenous isoprenoid compounds. *Int. J. Mol. Sci.* **2014**, *15*, 6843–6856.

30. Van der Burgh, R.; Pervolaraki, K.; Turkenburg, M.; Waterham, H.R.; Frenkel, J.; Boes, M. Unprenylated RhoA contributes to IL-1β hypersecretion in mevalonate kinase deficiency model through stimulation of Rac1 activity. *J. Biol. Chem.* **2014**, *289*, 27757–27765.

31. Tricarico, P.M.; Marcuzzi, A.; Piscianz, E.; Monasta, L.; Crovella, S.; Kleiner, G. Mevalonate kinase deficiency and neuroinflammation: Balance between apoptosis and pyroptosis. *Int. J. Mol. Sci.* **2013**, *14*, 23274–23288.

32. Schointuch, M.N.; Gilliam, T.P.; Stine, J.E.; Han, X.; Zhou, C.; Gehrig, P.A.; Kim, K.; Bae-Jump, V.L. Simvastatin, an HMG-CoA reductase inhibitor, exhibits anti-metastatic and anti-tumorigenic effects in endometrial cancer. *Gynecol. Oncol.* **2014**, *134*, 346–355.

33. Li, Y.; Fu, J.; Yuan, X.; Hu, C. Simvastatin inhibits the proliferation of A549 lung cancer cells through oxidative stress and up-regulation of SOD2. *Pharmazie* **2014**, *69*, 610–614.

34. Tavintharan, S.; Ong, C.N.; Jeyaseelan, K.; Sivakumar, M.; Lim, S.C.; Sum, C.F. Reduced mitochondrial coenzyme Q10 levels in HepG2 cells treated with high-dose simvastatin: A possible role in statin-induced hepatotoxicity? *Toxicol. Appl. Pharmacol.* **2007**, *223*, 173–179.

35. Bergamini, E.; Bizzarri, R.; Cavallini, G.; Cerbai, B.; Chiellini, E.; Donati, A.; Gori, Z.; Manfrini, A.; Parentini, I.; Signori, F.; *et al.* Ageing and oxidative stress: A role for dolichol in the antioxidant machinery of cell membranes? *J. Alzheimers Dis.* **2004**, *6*, 129–135.

36. Ciosek, C.P.; Magnin, D.R.; Harrity, T.W.; Logan, J.V.; Dickson, J.K.; Gordon, E.M.; Hamilton, K.A.; Jolibois, K.G.; Kunselman, L.K.; Lawrence, R.M.; *et al.* Lipophilic 1,1-bisphosphonates are potent squalene synthase inhibitors and orally active cholesterol lowering agents *in vivo*. *J. Biol. Chem.* **1993**, *268*, 24832–24837.

37. Sirvent, P.; Mercier, J.; Lacampagne, A. New insights into mechanisms of statin-associated myotoxicity. *Curr. Opin. Pharmacol.* **2008**, *8*, 333–338.

38. Xia, Z.; Tan, M.M.; Wong, W.W.; Dimitroulakos, J.; Minden, M.D.; Penn, L.Z. Blocking protein geranylgeranylation is essential for lovastatin-induced apoptosis of human acute myeloid leukemia cells. *Leukemia (Baltimore)* **2001**, *15*, 1398–1407.

39. Agarwal, B.; Bhendwal, S.; Halmos, B.; Moss, S.F.; Ramey, W.G.; Holt, P.R. Lovastatin augments apoptosis induced by chemotherapeutic agents in colon cancer cells. *Clin. Cancer Res.* **1999**, *5*, 2223–2229.

40. Rauthan, M.; Ranji, P.; Aguilera Pradenas, N.; Pitot, C.; Pilon, M. The mitochondrial unfolded protein response activator ATFS-1 protects cells from inhibition of the mevalonate pathway. *Proc. Natl. Acad. Sci. USA* **2013**, *110*, 5981–5986.

41. Liu, Y.; Samuel, B.S.; Breen, P.C.; Ruvkun, G. Caenorhabditis elegans pathways that surveil and defend mitochondria. *Nature* **2014**, *508*, 406–410.

42. Levine, B.; Kroemer, G. Autophagy in the pathogenesis of disease. *Cell* **2008**, *132*, 27–42.

43. Longatti, A.; Tooze, S.A. Vesicular trafficking and autophagosome formation. *Cell Death Differ.* **2009**, *16*, 956–965.

44. Zhu, Y.; Casey, P.J.; Kumar, A.P.; Pervaiz, S. Deciphering the signaling networks underlying simvastatin-induced apoptosis in human cancer cells: Evidence for non-canonical activation of RhoA and Rac1 GTPases. *Cell Death Dis.* **2013**, *4*, e568.

45. Hutagalung, A.H.; Novick, P.J. Role of Rab GTPases in membrane traffic and cell physiology. *Physiol. Rev.* **2011**, *91*, 119–149.

46. Sciarretta, S.; Zhai, P.; Shao, D.; Maejima, Y.; Robbins, J.; Volpe, M.; Condorelli, G.; Sadoshima, J. Rheb is a critical regulator of autophagy during myocardial ischemia: Pathophysiological implications in obesity and metabolic syndrome. *Circulation* **2012**, *125*, 1134–1146.

47. Ravikumar, B.; Futter, M.; Jahreiss, L.; Korolchuk, V.I.; Lichtenberg, M.; Luo, S.; Massey, D.C.; Menzies, F.M.; Narayanan, U.; Renna, M.; *et al.* Mammalian macroautophagy at a glance. *J. Cell Sci.* **2009**, *122*, 1707–1711.

48. Hall, M.N. mTOR-what does it do? *Transplant. Proc.* **2008**, *40*, S5–S8.

49. Bartolomé, A.; Kimura-Koyanagi, M.; Asahara, S.; Guillén, C.; Inoue, H.; Teruyama, K.; Shimizu, S.; Kanno, A.; García-Aguilar, A.; Koike, M.; Uchiyama, Y.; *et al.* Pancreatic β-cell failure mediated by mTORC1 hyperactivity and autophagic impairment. *Diabetes* **2014**, *63*, 2996–3008.

50. Bodemann, B.O.; Orvedahl, A.; Cheng, T.; Ram, R.R.; Ou, Y.H.; Formstecher, E.; Maiti, M.; Hazelett, C.C.; Wauson, E.M.; Balakireva, M.; Camonis, J.H.; *et al.* RalB and the exocyst mediate the cellular starvation response by direct activation of autophagosome assembly. *Cell* **2011**, *144*, 253–267.

51. Bento, C.F.; Puri, C.; Moreau, K.; Rubinsztein, D.C. The role of membrane-trafficking small GTPases in the regulation of autophagy. *J. Cell Sci.* **2013**, *126*, 1059–1069.

52. Jiang, P.; Mukthavaram, R.; Chao, Y.; Nomura, N.; Bharati, I.S.; Fogal, V.; Pastorino, S.; Teng, D.; Cong, X.; Pingle, S.C.; Kapoor, S.; *et al. In vitro* and *in vivo* anticancer effects of mevalonate pathway modulation on human cancer cells. *Br. J. Cancer* **2014**, *111*, 1562–1571.

53. Araki, M.; Maeda, M.; Motojima, K. Hydrophobic statins induce autophagy and cell death in human rhabdomyosarcoma cells by depleting geranylgeranyl diphosphate. *Eur. J. Pharmacol.* **2012**, *674*, 95–103.

54. Wei, Y.M.; Li, X.; Xu, M.; Abais, J.M.; Chen, Y.; Riebling, C.R.; Boini, K.M.; Li, P.L.; Zhang, Y. Enhancement of autophagy by simvastatin through inhibition of Rac1-mTOR signaling pathway in coronary arterial myocytes. *Cell Physiol. Biochem.* **2013**, *31*, 925–937.

55. Stenson, P.D.; Mort, M.; Ball, E.V.; Shaw, K.; Phillips, A.; Cooper, D.N. The Human Gene Mutation Database: Building a comprehensive mutation repository for clinical and molecular genetics, diagnostic testing and personalized genomic medicine. *Hum. Genet.* **2014**, *133*, 1–9.

56. Simon, A.; Kremer, H.P.; Wevers, R.A.; Scheffer, H.; de Jong, J.G.; van der Meer, J.W.; Drenth, J.P. Mevalonate kinase deficiency: Evidence for a phenotypic continuum. *Neurology* **2004**, *62*, 994–997.

57. Hoffmann, G.F.; Charpentier, C.; Mayatepek, E.; Mancini, J.; Leichsenring, M.; Gibson, K.M.; Divry, P.; Hrebicek, M.; Lehnert, W.; Sartor, K.; *et al.* Clinical and biochemical phenotype in 11 patients with mevalonic aciduria. *Pediatrics* **1993**, *91*, 915–921.

58. Haas, D.; Hoffmann, G.F. Mevalonate kinase deficiencies: From mevalonic aciduria to hyperimmunoglobulinemia D syndrome. *Orphanet. J. Rare Dis.* **2006**, *1*, 13.

59. Mandey, S.H.; Kuijk, L.M.; Frenkel, J.; Waterham, H.R. A role for geranylgeranylation in interleukin-1β secretion. *Arthritis Rheum.* **2006**, *54*, 3690–3695.

60. Kuijk, L.M.; Mandey, S.H.; Schellens, I.; Waterham, H.R.; Rijkers, G.T.; Coffer, P.J.; Frenkel, J. Statin synergizes with LPS to induce IL-1β release by THP-1 cells through activation of caspase-1. *Mol. Immunol.* **2008**, *45*, 2158–2165.

61. Cailliez, M.; Garaix, F.; Rousset-Rouvière, C.; Bruno, D.; Kone-Paut, I.; Sarles, J.; Chabrol, B.; Tsimaratos, M. Anakinra is safe and effective in controlling hyperimmunoglobulinaemia D syndrome-associated febrile crisis. *J. Inherit. Metab. Dis.* **2006**, *29*, 763.

62. Bodar, E.J.; Kuijk, L.M.; Drenth, J.P.; van der Meer, J.W.; Simon, A.; Frenkel, J. On-demand anakinra treatment is effective in mevalonate kinase deficiency. *Ann. Rheum. Dis.* **2011**, *70*, 2155–2158.

63. Galeotti, C.; Meinzer, U.; Quartier, P.; Rossi-Semerano, L.; Bader-Meunier, B.; Pillet, P.; Koné-Paut, I. Efficacy of interleukin-1-targeting drugs in mevalonate kinase deficiency. *Rheumatology (Oxford)* **2012**, *51*, 1855–1859.

64. Pontillo, A.; Paoluzzi, E.; Crovella, S. The inhibition of mevalonate pathway induces upregulation of NALP3 expression: New insight in the pathogenesis of mevalonate kinase deficiency. *Eur. J. Hum. Genet.* **2010**, *18*, 844–847.

65. Lamkanfi, M.; Dixit, V.M. The inflammasomes. *PLoS Pathog.* **2009**, *5*, e1000510.

66. Martinon, F.; Mayor, A.; Tschopp, J. The inflammasomes: Guardians of the body. *Annu. Rev. Immunol.* **2009**, *27*, 229–265.

67. Deretic, V.; Saitoh, T.; Akira, S. Autophagy in infection, inflammation and immunity. *Nat. Rev. Immunol.* **2013**, *13*, 722–737.

68. Lee, H.K.; Lund, J.M.; Ramanathan, B.; Mizushima, N.; Iwasaki, A. Autophagy-Dependent Viral Recognition by Plasmacytoid Dendritic Cells. *Science* **2007**, *315*, 1398–1401.

69. Paludan, C.; Schmid, D.; Landthaler, M.; Vockerodt, M.; Kube, D.; Tuschl, T.; Münz, C. Endogenous MHC class II processing of a viral nuclear antigen after autophagy. *Science* **2005**, *307*, 593–596.

70. Blanchet, F.P.; Moris, A.; Nikolic, D.S.; Lehmann, M.; Cardinaud, S.; Stalder, R.; Garcia, E.; Dinkins, C.; Leuba, F.; Wu, L.; *et al.* Human immunodeficiency virus-1 inhibition of immunoamphisomes in dendritic cells impairs early innate and adaptive immune responses. *Immunity* **2010**, *32*, 654–669.

71. Nedjic, J.; Aichinger, M.; Emmerich, J.; Mizushima, N.; Klein, L. Autophagy in thymic epithelium shapes the T-cell repertoire and is essential for tolerance. *Nature* **2008**, *455*, 396–400.

72. Harris, J.; Hartman, M.; Roche, C.; Zeng, S.G.; O'Shea, A.; Sharp, F.A.; Lambe, E.M.; Creagh, E.M.; Golenbock, D.T.; Tschopp, J.; *et al.* Autophagy Controls IL-1 Secretion by Targeting Pro-IL-1 for Degradation. *J. Biol. Chem.* **2011**, *286*, 9587–9597.

73. Auron, P.E.; Webb, A.C.; Rosenwasser, L.J.; Mucci, S.F.; Rich, A.; Wolff, S.M.; Dinarello, C.A. Nucleotide sequence of human monocyte interleukin 1 precursor cDNA. *Proc. Natl. Acad. Sci. USA* **1984**, *81*, 7907–7911.

74. Matsushima, K.; Taguchi, M.; Kovacs, E.J.; Young, H.A.; Oppenheim, J.J. Intracellular localization of human monocyte associated interleukin 1 (IL 1) activity and release of biologically active IL 1 from monocytes by trypsin and plasmin. *J. Immunol.* **1986**, *136*, 2883–2891.

75. Shi, C.S.; Shenderov, K.; Huang, N.N.; Kabat, J.; Abu-Asab, M.; Fitzgerald, K.A.; Sher, A.; Kehrl, J.H. Activation of autophagy by inflammatory signals limits IL-1β production by targeting ubiquitinated inflammasomes for destruction. *Nat. Immunol.* **2012**, *13*, 255–263.

76. Andrei, C.; Dazzi, C.; Lotti, L.; Torrisi, M.R.; Chimini, G.; Rubartelli, A. The secretory route of the leaderless protein interleukin 1β involves exocytosis of endolysosome-related vesicles. *Mol. Biol. Cell* **1999**, *10*, 1463–1475.

77. Zhou, R.; Yazdi, A.S.; Menu, P.; Tschopp, J. A role for mitochondria in NLRP3 inflammasome activation. *Nature* **2011**, *469*, 221–225.

78. Shimada, K.; Crother, T.R.; Karlin, J.; Dagvadorj, J.; Chiba, N.; Chen, S.; Ramanujan, V.K.; Wolf, A.J.; Vergnes, L.; *et al.* Oxidized Mitochondrial DNA Activates the NLRP3 Inflammasome during Apoptosis. *Immunity* **2012**, *36*, 401–414.

79. Ding, Z.; Liu, S.; Wang, X.; Khaidakov, M.; Dai, Y.; Mehta, J.L. Oxidant stress in mitochondrial DNA damage, autophagy and inflammation in atherosclerosis. *Sci. Rep.* **2013**, *3*, 1077.

Autophagy as a Regulatory Component of Erythropoiesis

Jieying Zhang, Kunlu Wu, Xiaojuan Xiao, Jiling Liao, Qikang Hu,
Huiyong Chen, Jing Liu and Xiuli An

Abstract: Autophagy is a process that leads to the degradation of unnecessary or dysfunctional cellular components and long-lived protein aggregates. Erythropoiesis is a branch of hematopoietic differentiation by which mature red blood cells (RBCs) are generated from multi-potential hematopoietic stem cells (HSCs). Autophagy plays a critical role in the elimination of mitochondria, ribosomes and other organelles during erythroid terminal differentiation. Here, the modulators of autophagy that regulate erythroid differentiation were summarized, including autophagy-related (Atg) genes, the B-cell lymphoma 2 (Bcl-2) family member Bcl-2/adenovirus E1B 19 kDa interacting protein 3-like (Nix/Binp3L), transcription factors globin transcription factor 1 (GATA1) and forkhead box O3 (FoxO3), intermediary factor KRAB-associated protein1 (KAP1), and other modulators, such as focal adhesion kinase family-interacting protein of 200-kDa (FIP200), Ca^{2+} and 15-lipoxygenase. Understanding the modulators of autophagy in erythropoiesis will benefit the autophagy research field and facilitate the prevention and treatment of autophagy-related red blood cell disorders.

Reprinted from *Int. J. Mol. Sci.* Cite as: Zhang, J.; Wu, K.; Xiao, X.; Liao, J.; Hu, Q.; Chen, H.; Liu, J.; An, X. Autophagy as a Regulatory Component of Erythropoiesis. *Int. J. Mol. Sci.* **2015**, *16*, 4083–4092.

1. Introduction

Erythropoiesis is a continuous and dynamic process by which erythrocytes are generated from multipotent hematopoietic stem cells (HSCs). Erythropoiesis is mainly divided into two stages, early erythroid progenitor proliferation and terminal erythroid differentiation. HSCs proliferate and differentiate into the earliest erythroid progenitors: burst-forming-unit erythroid (BFU-E) cells, and then, colony-forming-unit erythroid (CFU-E) cells. Subsequently, terminal erythroid differentiation starts with proerythroblasts, which undergo three mitoses to produce basophilic, polychromatic, and orthochromatic erythroblasts. Eventually, orthochromatic erythroblasts expel their nucleus and become reticulocytes, which subsequently become mature erythrocytes [1]. Noticeable changes in cellular composition and structure occur during terminal erythroid differentiation, including the filling of the cells with abundant hemoglobin and the clearance of all intracellular organelles from the cells, such as mitochondria and ribosomes [2].

A study performed in 1962 revealed an abundance of membranous structures in murine hepatocytes following treatment with glucagon and found that the mitochondria are degraded by lysosomes [3]. Deter and de Duve first proposed the biological concept of autophagy in an international forum [4]. Autophagy is a key cellular catabolic pathway that can be divided into macroautophagy, microautophagy and chaperone-mediated autophagy, according to the various enveloped substances and transport methods. Macroautophagy is comprised of two types: selective and non-selective. Selective autophagy includes mitophagy and pexophagy, and non-selective autophagy plays an important role in cell starvation [5,6].

Lemasters *et al.* [7] formally proposed the concept of mitophagy in 2005. This group observed that decreasing mitochondrial membrane potentials and the opening of the conductance permeability transition pores of the mitochondrial inner membrane cause mitophagy. The mitochondrion is the powerhouse of the cell, providing almost all energy for cellular activities and generating reactive oxygen species (ROS). ROS may cause damage to mitochondria, releasing apoptosis-inducing factors and leading to cell death. As the major site of biosynthesis (hemoglobin and lipid), the mitochondrion participates in the regulation of intracellular calcium. Therefore, the timely sequestration of damaged mitochondria is important for the normal growth of cells and the maintenance of a stable cellular environment [8,9].

Mitochondrial clearance from reticulocytes occurs through a special process that is regulated by multi-domain autophagy-related protein. The programmed removal of the mitochondria that occurs in reticulocytes represents a physiological model for studying the molecular mechanisms involved in mitophagy [10]. A hemin-induced human myeloid leukemia cell line (K562) has been shown to possess the capacity for erythroid differentiation *in vitro*. Multi-vesicular bodies and autophagy have been observed during K562 cell erythroid maturation [11]. In this review, we summarize the relevant modulators of autophagy involved in the regulation of erythroid differentiation under physiological and pathological conditions.

2. Autophagy Regulators and Erythroid Maturation

The targeted deletion of genes related to autophagy has been shown to cause anemia, indicating the presence of defective erythrocyte maturation and impaired mitophagy during terminal erythroid differentiation. The reported autophagy-related regulators that act during erythropoiesis have been summarized in Table 1.

Table 1. Autophagy-related modulators in erythropoiesis.

Modulators	Interactions with Other Molecules or Targets	Functions	References
Atg1/Ulk1	Atg13, Hsp90-Cdc37	Regulation of mitochondrial and ribosomal clearance	[12–14]
Atg4	-	Fusion of autophagosomes with lysosomes	[15]
Atg7	Atg5	Regulation of mitochondrial removal	[16–21]
Nix/Bnip3L	LC3, Atg8, miRNA	Modulation of mitochondrial clearance and autophagosome formation	[22–27]
GATA1	FoxO3, LC3-I	Direct regulation of autophagy genes	[28–33]
KRAB/KAP1-miRNA	Nix/Bnip3L, Ulk1	Participation in cascade controlling mitophagy	[34]
FIP200	Ulk1, Atg13	Essential autophagy gene in hematopoietic cells	[35,36]
Ca^{2+} and 15-lipoxygenase	-	Ca^{2+} promotes binding of 15-lipoxygenase to modulate the clearance of mitochondria	[37–39]

2.1. Autophagy-Related Gene (Atg) Family

Many autophagy-related genes have been identified that are critical for selective and/or nonselective autophagic regulatory mechanisms [40]. Atg1 (Ulk1), Atg13 and Atg17 are serine-threonine kinase complexes that regulate the cell cycle and cell growth and proliferation; E1-like enzyme Atg7 can activate Atg12 and conjugate Atg5 and E2-like protein Atg10 to form the preautophagosomal structure. Atg7 also mediates the conjugation of Atg12 to Atg5 and of Atg8 to phosphatidylethanolamine (PE), which participates in the extension of autophagy vesicles [41,42]. Among these genes, Ulk1, Atg4 and Atg7 are reported to play important roles during erythropoiesis.

2.2. Uncoordinated 51-Like Autophagy Activating Kinase 1 (Ulk1)

Ulk1, which is a homolog of yeast Atg1, is critical for mitochondrial and ribosomal clearance during erythroid terminal differentiation. The number of reticulocytes, mean cell volume (MCV), mean corpuscular hemoglobin level (MCH), and relative distribution width (RDW) of mature erythroid cells are increased in $Ulk1^{-/-}$ mice. $Ulk1^{-/-}$ reticulocytes exhibit the delayed removal of mitochondria, ribosomes and other organelles *in vitro*, and this defect is overcome via treatment with carbonyl cyanide 3-chlorophenylhydrazone (CCCP), which is a mitochondrial uncoupler that produces ROS and causes membrane depolarization [12,13]. Ulk1 interacts with the Hsp90-Cdc37 complex to promote its stability and activation. In addition, this interaction is conducive to Ulk1-directed phosphorylation and the recruitment of Atg13 to damaged mitochondria. As a Hsp90 antagonist, 17-allylamino-17-demethoxygeldanamycin (17AAG) is the synthetic derivative of geldanamycin, that can inhibit ATP binding and hydrolysis, and block the formation of chaperone complexes. When differentiating erythroid cells are treated with 2.5 μM 17AAG, they display significantly decreased Ulk1 protein levels, but Ulk1 mRNA levels are not affected. Although this treatment does not affect reticulocyte maturation, it notably reduces reticulocytes harboring mitochondria containing autophagosomes. Hsp90-Cdc37, Ulk1 and Atg13 are all required for mitophagy during erythroid differentiation [14].

2.3. Autophagy-Related 4 (Atg4)

Autophagy is induced in polychromatic erythroblasts, and autophagosomes remain abundant until enucleation, which stimulates the expression of Atg4 family members (Atg4A and Atg4D) and Atg8. The quantitative electron microscope assay has shown that compared to wild-type, fewer autophagosomes are assembled in Atg4 cysteine mutant Atg4B (C74A)-expressing progenitor cells, suggesting that the roles of Atg4 family members (particularly Atg4B) are important for autophagosome fusion during the differentiation of human erythroblasts [15].

2.4. Autophagy-Related 7 (Atg7)

Atg7 plays a critical role in mitochondrial autophagy in the mammalian hematopoietic system and has a unique pro-apoptotic effect on lysosome dysfunction. A previous study has shown that Atg7 is essential for the self-renewal, proliferation and normal functioning of HSCs [16]. According to this study, Vav-Atg7$^{-/-}$ mice showed reductions in hematopoietic stem cells and progenitors of multiple lineages. Furthermore, Atg7-deficient Lin$^-$Sca1$^+$c-Kit$^+$ (LSK) cells accumulate mitochondria and ROS, causing DNA damage, which suggests that mitophagy is important to the regulation of HSCs maintenance. In Atg7$^{-/-}$ erythroid cells, the mitochondria are targeted to form autophagosomes, but autophagosome elongation is impaired, and mitochondrion engulfment is inhibited [16,17]. It has been shown that half of Atg7$^{-/-}$ fetal liver cell-transplanted mice die, and the surviving mice display anemia, reticulocytosis, and lymphopenia [17]. Vav-Atg7$^{-/-}$ mice generated using Atg7 Flox/Flox and Vav-iCre mice have been reported to show severe anemia and shortened lifespans [18]. Additionally, the transferrin receptor is up-regulated, and mitochondrial loss is initiated in Ter119$^+$/CD71$^-$ elevated cells in the bone marrow of

Vav-Atg7$^{-/-}$ mice. The loss of Atg7-mediated mitophagy in Atg7$^{-/-}$ erythroblasts leads to the accumulation of damaged mitochondria with the increased formation of isolation membranes, resulting in cell death. In the absence of Atg7, mitochondrial proteins are selectively removed by mitophagy, but proteins associated with the endoplasmic reticulum and ribosomes are unaffected [18,19]. An additional study has shown that the number of mitochondria and mitochondrial ROS in developing red blood cells (RBCs) are increased in Vav-Atg7$^{-/-}$ mice, and the developing RBCs display phosphatidylserine at their surfaces and undergo caspase 3-mediated apoptosis [20]. A recent study showed that when Atg7 was deleted from erythroid progenitors of wild-type and mtDNA-mutator mice, the genetic disruption of autophagy did not cause anemia in wild-type mice but accelerated the mitochondrial respiration decline and induced macrocytic anemia in the mtDNA-mutator mice [21].

2.5. Bcl-2 Family: Bcl-2/Adenovirus E1B 19 kDa Interacting Protein 3-Like (Nix/Binp3L)

The Bcl-2 family is known to play a key role in apoptosis, and its function as a regulator of autophagy has also received increasing interest. Bcl-2 functions as an antiautophagy protein via interacting with the conserved autophagy protein Beclin 1 [43]. Nix (also named Bnip3L), which is a mitochondrial outer membrane protein, is the BH3-only member of the Bcl-2 family that inhibits the proliferation of tumor cell lines [44,45]. Nix activity is mediated through the minimal essential region (MER) in its cytoplasmic domain. The mutation of the central leucine residue of MER causes loss of Nix activity and deters rescue of mitochondrial clearance in reticulocytes [46]. Nix is also a selective receptor that combines with mammalian Atg8 homologs, including microtubule-associated protein light chain 3 (LC3/GABARAP) and ubiquitin-like modifiers, which are indispensable for the maturation of phagophores and autophagosomes [22]. Through its N-terminal LC3-interacting region, Nix can recruit GABARAP-L1 to depolarize impaired mitochondria. LC3, which is a mammalian homolog of Atg8, has unmodified (LC3-I) and lipid-modified (LC3-II) forms [23].

In erythroid cells, Nix is upregulated during reticulocyte maturation [47], which is essential for mitochondrial membrane potential dispersion and autophagosome formation. Nix promotes the conversion of LC3-I to LC3-II. In addition, the elimination of the Nix: LC3/GABARAP interaction delays mitochondrial clearance in erythrocytes [24,25]. A previous study has shown that the reticulocytes of Nix$^{-/-}$ mice exhibit markedly abnormal mitochondrial residues. Nix$^{-/-}$ mice display hemolytic anemia and erythroid hyperplasia and increased levels of caspase activation and phosphatidylserine due to the increased production of ROS [25,26]. Nix may function through the ancillary release of cytochrome c or the interaction with other mitochondrial effector molecules. In the absence of Nix, mitochondria are not incorporated into autophagosomes in a timely manner for clearance, leading to an erythroid maturation defect. The Nix-dependent clearance of mitochondria has also been detected in human K562 cells that have been induced to undergo erythroid lineage maturation [26]. Nix may be activated to signal into mitochondria to dissipate their mitochondrial transmembrane potential ($\Delta\Psi m$) during erythroid cells maturation. Nix interacts with other molecules in mitochondria leading to selective sequestration of mitochondria into autophagosomes [27].

2.6. Transcription Factors and KAP1

Transcription factors are regarded as additional essential elements that are in autophagy during erythropoiesis and include erythroid-specific genes, such as GATA1, which play a critical role in erythroid differentiation [28,29]. It has been found that GATA1 directly upregulates the transcription of genes encoding autophagy-related components, such as LC3B and its homologs. In murine erythroid cells, GATA1 activates autophagy-related genes, increasing their expression levels during human erythropoiesis [30]. The forkhead protein FoxO3 is required for the GATA1-mediated induction of LC3 and the formation of autophagosomes in erythroid cells [31,32]. Recently, McIver *et al.* found that GATA-1/FoxO3 could repress the expression of Exosc8, a pivotal component of the exosome complex. When downregulated in primary erythroid precursor cells, Exosc8 could induce erythroid cell maturation [33]. GATA1 establishes a dependent pathway to activate the formation of LC3 and autophagosomes for mitochondrial clearance during erythropoiesis.

KRAB-associated protein 1 (KAP1), which is also named tripartite motif containing 28 (TRIM28), transcription intermediary factor 1 β (TIF1β) or KRAB-interacting protein 1 (KRIP-1), is a transcriptional intermediary factor that acts as a scaffold in transcription complexes. The KRAB/KAP1-miRNA regulatory cascade controls mitophagy during human erythropoiesis [34]. KAP1-depleted erythroblasts exhibit erythrocyte maturation defects and accumulate mitochondria. A luciferase reporter assay performed using mouse erythroleukemia (MEL) cells has shown that miR-351 targets the Nix 3'-UTR. Overexpression of miR-351 inhibits erythroid differentiation and causes mitochondrial accumulation in MEL cells [34]. In human erythroleukemia (HEL) cells, knockdown of Kap1 also leads to the impairment of erythroid differentiation, increased mitochondria and the blockage of autophagy effectors, including Nix. Additionally, hsa-miR-125a-5p expression was increased in KAP1-depleted HEL cells. When hsa-miR-125a-5p is over-expressed, the downregulation of Nix and increased numbers of mitochondria are also observed [34]. Therefore, multi-factorial molecules interact with miRNAs to form a regulatory network during mitophagy for the control of erythropoiesis.

2.7. Other Modulators: FIP200, Ca2+ and 15-Lipoxygenase

FIP200 (200-kDa focal adhesion kinase family-interacting protein) is known to play an essential role in mammalian autophagy and diverse cellular functions. Several studies have shown that FIP200 is an important part of the Ulk1-Atg13-FIP200 complex in autophagosome formation [35]. The deletion of FIP200 results in increased HSCs cycling, the loss of HSCs reconstituting capacities, aberrant myeloid expansion and the blocking of erythroid maturation. Furthermore, FIP200-null HSCs exhibit abnormal accumulation of mitochondria and have increased ROS levels. These studies suggest that FIP200 is a key regulator of fetal HSCs and plays a potential role in autophagy for the maintenance of fetal hematopoiesis [35,36].

Lipoxygenase is the key enzyme involved in unsaturated fatty acid metabolism, and it can translate arachidonic acid, linoleic acid and other fatty acids into their bioactive metabolites, affecting cell structure, metabolism and signal transduction. 15-lipoxygenase acts in response to oxidative damage and modulates the clearance of mitochondria during reticulocyte maturation [37]. A previous study has indicated that 15-lipoxygenase sediments with the mitochondrial fraction in rabbit reticulocytes [38]. Ca^{2+} promotes the binding of 15-lipoxygenase to reticulocyte mitochondria and stimulates the lipid peroxidation of mitochondrial lipids and free linoleic acid. Therefore, Ca^{2+} is important for regulating the 15-lipoxygenase-mediated degradation of mitochondria in reticulocytes [39].

Mammalian autophagy signaling pathways are complicated and the mTOR-dependent pathway is the most prominent. AMPK (AMP-activated protein kinase) and mTOR (mammalian target of rapamycin) regulate autophagy through the direct phosphorylation of Ulk1. In nutrient-deficient conditions, AMPK promotes autophagy through the activation of Ulk1, and conversely, mTOR activity prevents Ulk1 activation under normal situations [48]. The implicated roles of the intracellular autophagy pathway in erythroid cell differentiation and maturation are summarized in Figure 1.

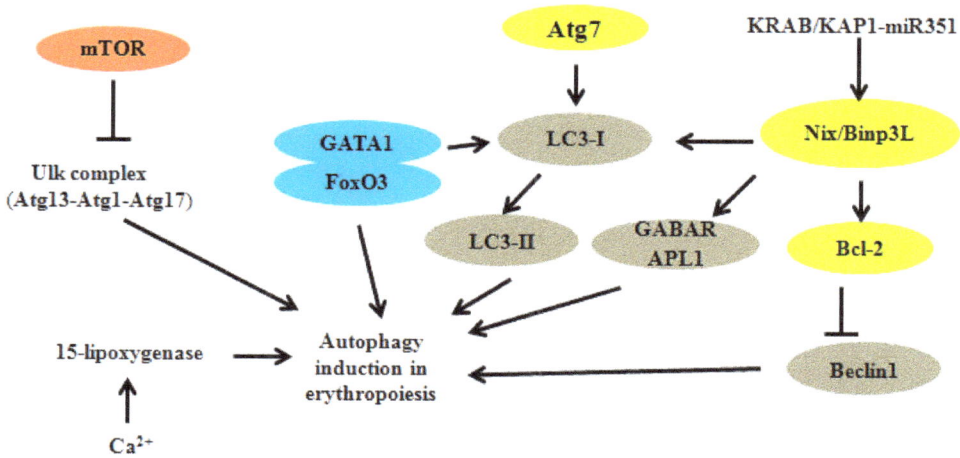

Figure 1. Autophagy-related factors are involved in the regulation of signal pathways in erythroid cells. The mTOR pathway is an important pathway that directly modulates the Ulk1 complex, and the inhibition of mTOR represses autophagy-related processes. Atg7 and Nix/Bnip3L are required for the removal of mitochondria, inducing the conversion of LC3-I to its lipid-modified form, LC3-II, to promote autophagy. miRNAs can regulate the expressions of key transcriptional components, and Ca^{2+} promotes the binding of 15-lipoxygenase to reticulocyte mitochondria.

3. Autophagy and β-Thalassemia

Several recent reports have indicated the key role of autophagy in red cell disorders, including β-thalassemia and myelodysplasia syndrome (MDS) [49–51]. When cultured CD34[+] erythroid progenitor cells from peripheral blood obtained from normal and β-thalassemia patients are induced to erythroid differentiation, autophagy is increased in the erythroblasts from the β-thalassemia patients compared with the normal erythroblasts, and this increase is mediated by the high levels of Ca^{2+} in the β-thalassemia erythroblasts. Normal erythroblasts show increased apoptosis following treatment with L-asparagine, which is an autophagy inhibitor, but this is not observed in erythroblasts from patients with β-thalassemia. Furthermore, reduced Ca^{2+} levels cause decreases in both autophagy and apoptosis. The high levels of autophagy may contribute to the increased apoptosis, leading to anemia and ineffective erythropoiesis in erythroblasts from β-thalassemia patients [51]. Notably, it has also been demonstrated that the occurrence of autophagy and early differentiation are linked in hESCs [52].

4. Perspectives

Autophagy is important in maintaining a cellular homeostatic environment. Autophagy has been implicated in many diseases, including various cancers, central nervous system (CNS)-related disorders, neurodegenerative disorders and heart disease in addition to aging. Autophagy studies are developing rapidly in the field of biology, especially in the erythroid research field. If autophagy is impaired during erythroid differentiation, erythroid maturation will be deficient. In this review, we summarize the relevant modulators of autophagy involved in the regulation of erythroid differentiation, which imply that autophagy is an important and complex process during erythroid differentiation. Understanding the modulators of autophagy in normal and pathologic erythropoiesis may facilitate the prevention and treatment of red blood cell-related disorders.

Acknowledgments: This work was supported by the grants from New Century Excellent Talents in University (NCET-11-0518), Doctoral Fund of Ministry of Education of China (No. 20120162110054) and the National Natural Science Foundation of China (No. 81270576, 81470362 and 31101686).

Author Contributions: Jing Liu, Xiuli An and Jieying Zhang designed the paper; Jieying Zhang, Kunlu Wu, Xiaojuan Xiao, Jiling Liao, Qikang Hu, Huiyong Chen, Jing Liu and Xiuli An participated in collecting and analyzing the data; Jieying Zhang and Jing Liu wrote the paper; and Jing Liu, Xiuli An and Jieying Zhang revised the paper.

Conflicts of Interest: The authors declare no conflict of interest.

Abbreviation

17AAG	17-allylamino-17-demethoxygeldanamycin
AMPK	AMP-activated protein kinase
Atg	autophagy-related
Bcl-2	B-cell lymphoma 2
BFU-E	burst-forming-unit erythroid
CCCP	carbonyl cyanide 3-chlorophenylhydrazone
Cdc37	cell division cycle 37
CFU-E	colony-forming-unit erythroid
CNS	central nervous system
FIP200	focal adhesion kinase family-interacting protein of 200-kDa
FoxO3	forkhead box O3
GABARAP	gamma aminobutyric acid A receptor-associated protein
GATA1	globin transcription factor
HEL	human erythroleukemia
HSCs	hematopoietic stem cells
Hsp90	90 kDa heat shock protein

KRAB	Krueppel-associated box
KAP-1	KRAB-associated protein 1
KRIP-1	KRAB-interacting protein 1
LC3	light chain 3
LSK	$Lin^-Sca1^+c\text{-}Kit^+$
MCV	mean cell volume
MCH	mean corpuscular hemoglobin
MDS	myelodysplasia syndrome
MEL	mouse erythroleukemia
MER	minimal essential region
mTOR	mammalian target of rapamycin
Nix/Binp3L	Bcl-2/adenovirus E1B 19 kDa interacting protein 3-like
PE	phosphatidylethanolamine
RBCs	red blood cells
ROS	reactive oxygen species
RDW	relative distribution width
TRIM28	tripartite motif containing 28
TIF1β	transcription intermediary factor 1 beta
Ulk1	uncoordinated 51-like autophagy activating kinase 1

References

1. Stephenson, J.R.; Axelrad, A.A.; McLeod, D.L.; Shreeve, M.M. Induction of colonies of hemoglobin-synthesizing cells by erythropoietin *in vitro*. *Proc. Natl. Acad. Sci. USA* **1971**, *68*, 1542–1546.

2. Géminard, C.; de Gassart, A.; Vidal, M. Reticulocyte maturation: Mitoptosis and exosome release. *Biocell* **2002**, *26*, 205–215.

3. Ashford, T.P.; Porter, K.R. Cytoplasmic components in hepatic cell lysosomes. *J. Cell Biol.* **1962**, *12*, 198–202.

4. Deter, R.L.; DeDuve, C. Influence of glucagon, an inducer of cellular autophagy on some physical properties of rat liver lysosomes. *J. Cell Biol.* **1967**, *33*, 437–449.

5. Klionsky, D.J.; Emr, S.D. Autophagy as a regulated pathway of cellular degradation. *Science* **2000**, *290*, 1717–1721.

6. Komatsu, M.; Ichimura, Y. Selective autophagy regulates various cellular functions. *Genes Cells* **2010**, *10*, 923–933.

7. Lemasters, J.J. Selective mitochondrial autophagy, or mitophagy, as a targeted defense against oxidative stress, mitochondrial dysfunction, and aging. *Rejuvenation Res.* **2005**, *8*, 3–5.

8. Scheffler, I.E. A century of mitochondrial research: Achievements and perspectives. *Mitochondrion* **2001**, *1*, 3–31.

9. Duchen, M.R. Mitochondria in health and disease: Perspectives on a new mitochondrial biology. *Mol. Asp. Med.* **2004**, *25*, 365–451.

10. Zhang, J.; Ney, P.A. Reticulocyte mitophagy: Monitoring mitochondrial clearance in a mammalian model. *Autophagy* **2010**, *6*, 405–408.

11. Fader, C.M.; Colombo, M.I. Multivesicular bodies and autophagy in erythrocyte maturation. *Autophagy* **2006**, *2*, 122–125.

12. Chan, E.Y.; Kir, S.; Tooze, S.A. siRNA screening of the kinome identifies ULK1 as a multidomain modulator of autophagy. *J. Biol. Chem.* **2007**, *282*, 25464–25474.

13. Kundu, M.; Lindsten, T.; Yang, C.Y.; Wu, J.M.; Zhao, F.P.; Zhang, J.; Selak, M.A.; Ney, P.A.; Thompson, C.B. Ulk1 plays a critical role in the autophagic clearance of mitochondria and ribosomes during reticulocyte maturation. *Blood* **2008**, *112*, 1493–1502.

14. Joo, J.H.; Dorsey, F.C.; Joshi, A.; Hennessy-Walters, K.M.; Rose, K.L.; McCastlain, K.; Zhang, J.; Iyengar, R.; Jung, C.H.; Suen, D.F.; *et al.* Hsp90-Cdc37 chaperone complex regulates Ulk1- and Atg13-mediated mitophagy. *Mol. Cell* **2011**, *43*, 572–585.

15. Betin, V.M.; Singleton, B.K.; Parsons, S.F.; Anstee, D.J.; Lane, J.D. Autophagy facilitates organelle clearance during differentiation of human erythroblasts: Evidence for a role for Atg4 paralogs during autophagosome maturation. *Autophagy* **2013**, *9*, 881–893.

16. Mortensen, M.; Soilleux, E.J.; Djordjevic, G.; Tripp, R.; Lutteropp, M.; Sadighi-Akha, E.; Stranks, A.J.; Glanville, J.; Knight, S.; Jacobsen, S.E.; *et al.* The autophagy protein Atg7 is essential for hematopoietic stem cell maintenance. *J. Exp. Med.* **2011**, *208*, 455–467.

17. Zhang, J.; Randall, M.S.; Loyd, M.R.; Dorsey, F.C.; Kundu, M.; Cleveland, J.L.; Ney, P.A. Mitochondrial clearance is regulated by Atg7-dependent and -independent mechanisms during reticulocyte maturation. *Blood* **2009**, *114*, 157–164.

18. Komatsu, M.; Waguri, S.; Ueno, T.; Iwata, J.; Murata, S.; Tanida, I.; Ezaki, J.; Mizushima, N.; Ohsumi, Y.; Uchiyama, Y.; *et al.* Impairment of starvation-induced and constitutive autophagy in Atg7-deficient mice. *J. Cell Biol.* **2005**, *169*, 425–434.

19. Mortensen, M.; Ferguson, D.J.; Edelmann, M.; Kessler, B.; Morten, K.J.; Komatsu, M.; Simon, A.K. Loss of autophagy in erythroid cells leads to defective removal of mitochondria and severe anemia *in vivo. Proc. Natl. Acad. Sci. USA* **2010**, *107*, 832–837.

20. Mortensen, M.; Simon, A.K. Nonredundant role of Atg7 in mitochondrial clearance during erythroid development. *Autophagy* **2010**, *6*, 423–425.

21. Li-Harms, X.; Milasta, S.; Lynch, J.; Wright, C.; Joshi, A.; Iyengar, R.; Neale, G.; Wang, X.; Wang, Y.D.; Prolla, T.A.; *et al.* Mito-protective autophagy is impaired in erythroid cells of aged mtDNA-mutator mice. *Blood* **2015**, *125*, 162–174.

22. Novak, I.; Kirkin, V.; McEwan, D.G.; Zhang, J.; Wild, P.; Rozenknop, A.; Rogov, V.; Löhr, F.; Popovic, D.; Occhipinti, A.; *et al.* Nix is a selective autophagy receptor for mitochondrial clearance. *EMBO Rep.* **2010**, *11*, 45–51.

23. Schwarten, M.; Mohrluder, J.; Ma, P.; Stoldt, M.; Thielmann, Y.; Stangler, T.; Hersch, N.; Hoffmann, B.; Merkel, R.; Willbold, D.; *et al.* Nix directly binds to GABARAP: A possible crosstalk between apoptosisand autophagy. *Autophagy* **2009**, *5*, 690–698.

24. Schweers, R.L.; Zhang, J.; Randall, M.S.; Loyd, M.R.; Li, W.; Dorsey, F.C.; Kundu, M.; Opferman, J.T.; Cleveland, J.L.; Miller, J.L.; *et al.* NIX is required for programmed mitochondrial clearance during reticulocyte maturation. *Proc. Natl. Acad. Sci. USA* **2007**, *104*, 19500–19505.

25. Zhang, J.; Ney, P.A. Nix induces mitochondrial autophagy in reticulocytes. *Autophagy* **2008**, *4*, 354–356.

26. Sandoval, H.; Thiagarajan, P.; Dasgupta, S.K.; Schumacher, A.; Prchal, J.T.; Chen, M.; Wang, J. Essential role for Nix in autophagic maturation of erythroid cells. *Nature* **2008**, *454*, 232–235.

27. Chen, M.; Sandoval, H.; Wang, J. Selective mitochondrial autophagy during erythroid maturation. *Autophagy* **2008**, *4*, 926–928.

28. Welch, J.J.; Watts, J.A.; Vakoc, C.R.; Yao, Y.; Wang, H.; Hardison, R.C.; Blobel, G.A.; Chodosh, L.A.; Weiss, M.J. Gobal regulation of erythroid gene expression by transcription factor GATA-1. *Blood* **2004**, *104*, 3136–3147.

29. Layon, M.E.; Layon, C.J.; West, R.J.; Lowrey, C.H. Expression of GATA-1 in a non-hematopoietic cell line induces β-globin locus control region chromatin structure remode ling and an erythroid pattern of gene expression. *J. Mol. Biol.* **2007**, *366*, 737–744.

30. Kang, Y.A.; Sanalkumar, R.; O'Geen, H.; Linnemann, A.K.; Chang, C.J.; Bouhassira, E.E.; Farnham, P.J.; Keles, S.; Bresnick, E.H. Autophagy driven by a master regulator of hematopoiesis. *Mol. Cell. Biol.* **2012**, *32*, 226–239.

31. Spitali, P.; Grumati, P.; Hiller, M.; Chrisam, M.; Aartsma-Rus, A.; Bonaldo, P. Autophagy is impaired in the tibialis anterior of dystrophin null mice. *PLoS Curr.* **2013**, *5*, 1–9.

32. Bakker, W.J.; van Dijk, T.B.; Parren-van Amelsvoort, M.; Kolbus, A.; Yamamoto, K.; Steinlein, P.; Verhaak, R.G.; Mak, T.W.; Beug, H.; Löwenberg, B.; *et al.* Differential regulation of FoxO3a target genes in erythropoiesis. *Mol. Cell. Biol.* **2007**, *27*, 3839–3854.

33. McIver, S.C.; Kang, Y.A.; DeVilbiss, A.W.; O'Driscoll, C.A.; Ouellette, J.N.; Pope, N.J.; Camprecios, G.; Chang, C.J.; Yang, D.; Bouhassira, E.E.; *et al.* The exosome complex establishes a barricade to erythroid maturation. *Blood* **2014**, *124*, 2285–2297.

34. Barde, I.; Rauwel, B.; Marin-Florez, R.M.; Corsinotti, A.; Laurenti, E.; Verp, S.; Offner, S.; Marquis, J.; Kapopoulou, A.; Vanicek, J.; *et al.* A KRAB/KAP1-miRNA cascade regulates erythropoiesis through stage-specific control of mitophagy. *Science* **2013**, *340*, 350–353.

35. Jung, C.H.; Jun, C.B.; Ro, S.H.; Kim, Y.M.; Otto, N.M.; Cao, J.; Kundu, M.; Kim, D.H. ULK-Atg13-FIP200 complexes mediate mTOR signaling to the autophagy machinery. *Mol. Biol. Cell* **2009**, *20*, 1992–2003.

36. Liu, F.; Lee, J.Y.; Wei, H.; Tanabe, O.; Engel, J.D.; Morrison, S.J.; Guan, J.L. FIP200 is required for the cell-autonomous maintenance of fetal hematopoietic stem cells. *Blood* **2010**, *116*, 4806–4814.

37. Kuhn, H.; Belkner, J.; Wiesner, R. Subcellular distribution of lipoxygenase products in rabbit reticulocyte membranes. *Eur. J. Biochem.* **1990**, *191*, 221–227.

38. Vijayvergiya, C.; DeAngelis, D.; Walther, M.; Kühn, H.; Duvoisin, R.M.; Smith, D.H.; Wiedmann, M. High-level expression of rabbit 15-lipoxygenase induces collapse of the mitochondrial pH gradient in cell culture. *Biochemistry* **2004**, *43*, 15296–15302.

39. Watson, A.; Doherty, F.J. Calcium promotes membrane association of reticulocyte 15-lipoxygenase. *Biochem. J.* **1994**, *298*, 377–383.

40. Xie, Z.; Klionsky, D.J. Autophagosome formation: Core machinery and adaptations. *Nat. Cell Biol.* **2007**, *9*, 1102–1109.

41. Geng, J.; Klionsky, D.J. The Atg8 and Atg12 ubiquitin-like conjugation systems in macroautophagy: Protein modifications: Beyond the usual suspects' review series. *EMBO Rep.* **2008**, *9*, 859–864.

42. Walls, K.C.; Ghosh, A.P.; Franklin, A.V.; Klocke, B.J.; Ballestas, M.; Shacka, J.J.; Zhang, J.; Roth, K.A. Lysosome dysfunction triggers Atg7-dependent neural apoptosis. *J. Biol. Chem.* **2010**, *285*, 10497–10507.

43. Pattingre, S.; Tassa, A.; Qu, X.; Garuti, R.; Liang, X.H.; Mizushima, N.; Packer, M.; Schneider, M.D.; Levine, B. Bcl-2 antiapoptotic proteins inhibit Beclin 1-dependent autophagy. *Cell* **2005**, *122*, 927–939.

44. Boyd, J.M.; Malstrom, S.; Subramanian, T.; Venkatesh, L.K.; Schaeper, U.; Elangovan, B.; D'Sa-Eipper, C.; Chinnadurai, G. Adenovirus E1B 19 kDa and Bcl-2 proteins interact with a common set of cellular proteins. *Cell* **1994**, *79*, 341–351.

45. Matsushima, M.; Fujiwara, T.; Takahashi, E.; Minaguchi, T.; Eguchi, Y.; Tsujimoto, Y.; Suzumori, K.; Nakamura, Y. Isolation, mapping, and functional analysis of a novel human cDNA (BNIP3L) encoding a protein homologous to human NIP3. *Genes Chromosomes Cancer* **1998**, *21*, 230–235.

46. Zhang, J.; Loyd, M.R.; Randall, M.S.; Waddell, M.B.; Kriwacki, R.W.; Ney, P.A. A short linear motif in BNIP3L (Nix) mediates mitochondrial clearance in reticulocytes. *Autophagy* **2012**, *8*, 1325–1332.

47. Aerbajinai, W.; Giattina, M.; Lee, Y.T.; Raffeld, M.; Miller, J.L. The proapoptotic factor Nix is coexpressed with Bcl-xL during terminal erythroid differentiation. *Blood* **2003**, *102*, 712–717.

48. Kim, J.; Kundu, M.; Viollet, B.; Guan, K.L. AMPK and mTOR regulate autophagy through direct phosphorylation of Ulk1. *Nat. Cell Biol.* **2011**, *13*, 132–141.

49. Watson, A.S.; Mortensen, M.; Simon, A.K. Autophagy in the pathogenesis of myelodysplastic syndromeand acute myeloid leukemia. *Cell Cycle* **2011**, *10*, 1719–1725.

50. Fang, J.; Rhyasen, G.; Bolanos, L.; Rasch, C.; Varney, M.; Wunderlich, M.; Goyama, S.; Jansen, G.; Cloos, J.; Rigolino, C.; *et al.* Cytotoxic effects of bortezomib in myelodysplastic syndrome/acute myeloidleukemia depend on autophagy-mediated lysosomal degradation of TRAF6 and repression of PSMA1. *Blood* **2012**, *120*, 858–867.

51. Lithanatudom, P.; Wannatung, T.; Leecharoenkiat, A.; Svasti, S.; Fucharoen, S.; Smith, D.R. Enhanced activation of autophagy in β-thalassemia/Hb E erythroblasts during erythropoiesis. *Ann. Hematol.* **2011**, *90*, 747–758.

52. Tra, T.; Gong, L.; Kao, L.P.; Li, X.L.; Grandela, C.; Devenish, R.J.; Wolvetang, E.; Prescott, M. Autophagy in human embryonic stem cells. *PLoS ONE* **2011**, *6*, e27485.

Mitochondrial Oxidative Stress, Mitochondrial DNA Damage and Their Role in Age-Related Vascular Dysfunction

Yuliya Mikhed, Andreas Daiber and Sebastian Steven

Abstract: The prevalence of cardiovascular diseases is significantly increased in the older population. Risk factors and predictors of future cardiovascular events such as hypertension, atherosclerosis, or diabetes are observed with higher frequency in elderly individuals. A major determinant of vascular aging is endothelial dysfunction, characterized by impaired endothelium-dependent signaling processes. Increased production of reactive oxygen species (ROS) leads to oxidative stress, loss of nitric oxide ($^{\bullet}$NO) signaling, loss of endothelial barrier function and infiltration of leukocytes to the vascular wall, explaining the low-grade inflammation characteristic for the aged vasculature. We here discuss the importance of different sources of ROS for vascular aging and their contribution to the increased cardiovascular risk in the elderly population with special emphasis on mitochondrial ROS formation and oxidative damage of mitochondrial DNA. Also the interaction (crosstalk) of mitochondria with nicotinamide adenosine dinucleotide phosphate (NADPH) oxidases is highlighted. Current concepts of vascular aging, consequences for the development of cardiovascular events and the particular role of ROS are evaluated on the basis of cell culture experiments, animal studies and clinical trials. Present data point to a more important role of oxidative stress for the maximal healthspan (healthy aging) than for the maximal lifespan.

Reprinted from *Int. J. Mol. Sci.* Cite as: Mikhed, Y.; Daiber, A.; Steven, S. Mitochondrial Oxidative Stress, Mitochondrial DNA Damage and Their Role in Age-Related Vascular Dysfunction. *Int. J. Mol. Sci.* **2015**, *16*, 15918–15948.

1. Introduction

Demographic change is an emerging issue in the Western world. The proportion of people older than 65 years will dramatically increase within the next decades [1]. Besides its negative effects on the costs for retirement funds, an increasing average age will amplify the economic burden for healthcare costs in these countries. Cardiovascular diseases (CVD) are a main cause of morbidity and mortality in elderly people and their incidence is closely correlated with age [2] (Figure 1A). The term "vascular aging" outlines all changes in vessels, which are associated with aging. Smooth muscle cells and endothelial cells are involved in these changes during vascular aging. Progressive aging leads to arterial stiffness and endothelial

147

dysfunction, which is known to correlate with future cardiovascular events in humans [3]. Furthermore, aged vessels are more prone to develop atherosclerotic lesions, vascular injury, impaired angiogenesis and calcification [4]. Consequently, the incidence and frequency of cardiovascular diseases such as atherosclerosis and its late complications such as coronary artery disease or stroke, increase substantially with age [5]. However, endothelial dysfunction in the elderly is not only associated with CVD, but also with other disorders related to aging, such as erectile dysfunction, renal dysfunction, Alzheimer's disease or retinopathy [6–9]. Since the CVD burden is predicted to increase dramatically in Western societies and the knowledge about vascular aging is limited, there is urgent need for research in this field in order to reduce morbidity and mortality in the aging population. Within the last years, scientists identified three key players in the vascular aging process: nitric oxide signaling, oxidative stress and inflammation [10]. It should be noted, that these players do not stand alone as they affect and influence each other. Especially pathophysiological convergence of different organ diseases with associated comorbidities increases at higher age and represents another important field of research that needs to be addressed in the future [11–13].

Nitric oxide ($^\bullet$NO) is essential for a functional endothelium and diminished $^\bullet$NO bioavailability leads to endothelial dysfunction [14,15]. In aged vessels the bioavailability of $^\bullet$NO is reduced, whereas production of reactive oxygen species (ROS) is increased [10,16,17]. It is not only the reaction of $^\bullet$NO with superoxide anion ($O_2^{\bullet-}$), leading in turn to production of peroxynitrite ($ONOO^-$) [17], which reduces $^\bullet$NO bioavailability. Also dysregulation of the endothelial nitric oxide synthase (eNOS), known as eNOS uncoupling, results in impaired $^\bullet$NO release from the endothelium and leads instead to increased superoxide production [18]. The mechanisms of this uncoupling process are complex and multiple. They include decreased availability of the eNOS substrate L-arginine or the cofactor tetrahydrobiopterin (BH4), but also phosphorylation state (Ser1177, Thr495, Tyr657) or S-glutathionlyation of the protein (for review see [19]). All of them play an important role for the coupling state of the enzyme and many of them were identified in the vascular aging process [16,20–22]. Imbalance of $^\bullet$NO bioavailability by ROS is not only induced by eNOS itself. Increased oxidative stress from mitochondria and other enzymatic sources are observed in aged animals and affect the coupling state of eNOS [23]. This observation points to a strong correlation between aging, oxidative stress, and as a consequence of imbalanced $^\bullet$NO bioavailability, the development of endothelial dysfunction (Figure 1B) [24]. The impact of vascular oxidative stress on endothelial function and the predictive value of this parameter was previously shown by a large clinical trial demonstrating better cardiovascular prognosis in patients with lower burden of vascular oxidative stress (less pronounced effect of vitamin C infusion on flow-mediated dilation) (Figure 1C) [25].

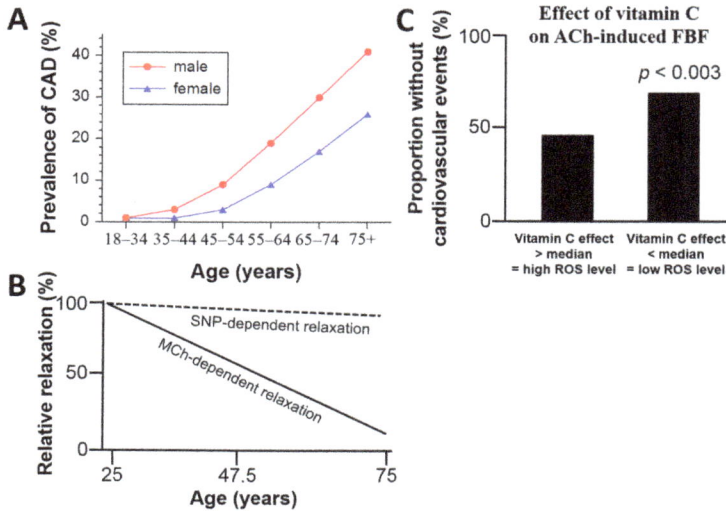

Figure 1. (**A**) Prevalence of coronary artery diseases (CAD) increases with progressing age and gender in Germany. Drawn from results of the Detect Study [26,27] and figure adopted from [28] with permission of the publisher. © Springer Science+Business Media, LLC 2010; (**B**) Correlation between age and endothelium-dependent (methacholine (MCh), solid line) and -independent (sodium nitroprusside (SNP), broken line) relaxation. Healthy individuals with an age of 25–70 years were tested for MCh- and SNP-dependent vasodilation. Endothelium-dependent relaxation was impaired with age ($r = 0.81$, $p < 0.001$, r is the correlation coefficient), whereas endothelium-independent relaxation was decreased only by trend in older individuals ($r = 0.1$, not significant). According to [24]; and (**C**) Results from Kaplan-Meier-analysis for the cardiovascular event rate in two cohorts of patients displaying either pronounced or weak effect of vitamin C on endothelial function (measured by forearm plethysmography after infusion of acetylcholine (ACh)) over a time period of more than 6 years. The take-home message is: Higher levels of vascular oxidative stress (free radicals) are associated with a more pronounced beneficial effect of the radical scavenger vitamin C on endothelial function and an increased cardiovascular event rate. FBF means forearm blood flow. According to Heitzer *et al.*, Circulation 2001 [25].

There is good evidence that mitochondria represent a major source of ROS in aging tissues [29,30]. Mitochondrial DNA damage accumulates in the aging cell leading to mitochondrial dysfunction [31] and aging-related cardiovascular and neurodegenerative disease [10,32]. The present review will discuss the impact of mitochondrial oxidative stress and mitochondrial DNA damage on vascular dysfunction in the aging process.

2. The Cardiovascular System

The cardiovascular system is a closed network containing arteries, veins and capillaries. The center of this network is the heart. Transportation is one of the most important functions of the human cardiovascular system. By every heartbeat nutrients, oxygen, carbon dioxide and hormones are distributed to all parts of the body. Furthermore, the cardiovascular system is involved in host defense by the inflammatory system and hemostasis by the coagulation system. As the interface between blood and vessel wall, the endothelium plays a crucial role as a specialized monolayered squamous epithelium that lines the interior surface of blood vessels. Preserving the blood barrier function and thereby preventing adhesion of immune cells is a defense against infiltration of immune cells such as monocytes into lesion-prone areas of the endothelium, an essential step in the development of atherosclerotic plaques [33,34]. There are over 2 billion heartbeats in one human life and every heartbeat is associated with increased sheer stress and elongation of the vessel. The endothelial cell layer has to control the vascular tone under all circumstances by nitric oxide ($^{\bullet}$NO), endothelium-derived hyperpolarizing factor, prostacyclin or natriuretic peptides. Furthermore, these mediators released by the endothelium have antiaggregatory properties and suppress thrombus formation, vascular stenosis [35] and cardiac hypertrophy. On the other hand, the endothelium acts synergistically with a regulatory system, which consists of vasoconstrictors such as catecholamines and other vasoactive peptides (*i.e.*, angiotensin, vasopressin, endothelin).

The aging endothelium is more and more unable to fulfill all these tasks. In elderly people a significant impairment of endothelium-dependent relaxation (endothelial dysfunction) can be found [36,37]. This dysfunction promotes thrombosis, vasoconstriction, leukocyte infiltration and cell proliferation in the vessel wall. Thus, endothelial dysfunction in aging is an early predictor for the development of atherosclerosis, hypertension and future cardiovascular events. Besides this interaction during the aging process, this correlation was also proven by a meta-analysis of 23 studies [3], which nicely demonstrates flow-mediated dilation (FMD) of the brachial artery as a prognostic marker for cardiovascular events. Although this study cannot prove endothelial dysfunction as the cause of increased cardiovascular morbidity, it demonstrates that endothelial dysfunction is a precursor of cardiovascular disease. Not only are macrovessels, like aorta or coronary arteries, affected by aging-dependent endothelial dysfunction and oxidative stress but the microcirculation (resistance vessels) are especially affected by vascular aging (for review see [38]). Studies of Mayhan *et al.* demonstrated impaired eNOS-dependent reactivity of cerebral arterioles, which was associated with increased oxidative stress [39]. Similar evidence for endothelial dysfunction could be found for retinal vessels during the aging process [40] and its contribution to neurodegenerative disease is very likely [41,42] Our group and many others revealed impaired •NO signaling, vascular inflammation and oxidative stress as key players in the pathogenesis of aging dependent endothelial dysfunction (for review see [28]).

3. Aging and Oxidative Stress

As early as in 1954, Harman expressed for the first time the free radical theory of aging [43]. This idea was based on the observations, that aging is a universal phenomenon, and its contributing factors must be present in every living organism. His first hypothesis emphasized the importance of the hydroxyl radical, as well as molecular oxygen in the aging process [44]. Later, this concept was extended to mitochondria which are the most abundant cellular source of ROS. Mitochondrial ROS formation probably contributes to the high mutation rate of the mitochondrial genome. In general, assembly of the respiratory chain components requires the contribution of two spatially separated genomes, the nuclear DNA and the maternally inherited mtDNA [45]. Malfunctioning of the mitochondrial genome is directly correlated with impaired mitochondrial physiology and depleted ATP-synthesis, which are accompanied by enhanced ROS formation and increased apoptosis [29]. Age-dependent impairment of vascular redox regulation is demonstrated by the bioavailability of another free radical species –•NO. Nitric oxide is not only involved in vasodilation, but also in vascular smooth muscle cell proliferation, inhibition of platelet aggregation and several others [46]. It has been postulated that •NO is gradually reduced with age and might serve as an applicable biomarker for age-dependent endothelial dysfunction. The prevailing paradigm is that an age-dependent increase in superoxide rapidly consumes •NO, consequently reduces its endothelial levels and thereby leads to impaired vasorelaxation [24,47].

Oxidative stress burden usually correlates with cellular thiol levels or vice versa cellular thiol/disulfide ratio is a well-accepted indicator of the redox state of a cell. Therefore, thiols and thiol-dependent enzymes were in the focus of oxidative stress and aging-related research. Cellular thiols possess significant antioxidant effects and affect the organismal healthspan. Glutathione peroxidases (GPx) belong to the class of enzymes responsible for the removal of H_2O_2 from the intracellular compartments. *GPx* deficiency leads to increased levels of oxidative stress, pronounced vascular dysfunction [16] and increased senescence of fibroblasts [48]. Even though genetic depletion of either *GPx-1* or *GPx-4* has no effect on the lifespan of the experimental animals, their effect on the process of healthy aging cannot be disputed [16]. Thioredoxins (Trx) are another class of antioxidant enzymes that can directly react with peroxides and eliminate damage caused by peroxides via reduction of disulfides and methionine sulfoxides [49]. Complete knock-out of the mitochondrial Trx isoform ($Trx2^{-/-}$) is embryonically lethal and partial knock-out ($Trx2^{+/-}$) mice show elevated levels of lipid peroxides, oxidized nucleobases and proteins [50]. Although

Trx2+/− mice exhibited reduction in their lifespan by trend, a further increase of the significance power would require higher number of animals. On the other hand, genetic knock-in of the cytosolic isoform of thioredoxin, *Trx1*, showed considerable increase in mice longevity and stronger resistance to oxidative stress inducers, like UV-light or ischemia/reperfusion, further supporting the previous notion of the importance of antioxidant enzymes [51].

The impact of antioxidant defense enzymes on aging-related cardiovascular complications is well established and has been previously demonstrated for the mitochondrial superoxide dismutase 2 (SOD2) [52], the cytosolic superoxide dismutase (SOD1) [47,53], the extracellular superoxide dismutase (ecSOD), and the thioredoxin-1 protein (Trx) [49]. Considering the fact that superoxide is the major contributing factor to the endothelial dysfunction in the aging vasculature further investigations of the antioxidant systems have been conducted in order to understand why these defense mechanisms are not able to combat increasing levels of the oxidative stress. On the example of SOD2, it has been shown that in aging vessels, MnSOD has been heavily nitrated, most probably by peroxynitrite, as indicated by increased staining for 3-nitrotyrosine [54]. Inhibition of this protective enzyme results in the activation of the vicious cycle of increased oxidative burden. On the other hand, no direct correlation between lifespan and deletion of or overexpression of most antioxidant enzymes (*SOD2+/−* or transgenic overexpression of *SOD2* (*SOD2tg*), *GPx-1−/−*, *GPx-4−/−* or *MsrA−/−*, transgenic overexpression of *SOD1* (*SOD1tg*), transgenic overexpression of catalase (*catalasetg*)) could affect the longevity [55]. Only *SOD1−/−* mice and mice with double gene ablation combinations showed reduced life expectancy [55,56]. It is worth to mention that mice completely deficient in *SOD2* show lethality at the embryo stage or a few weeks after birth, once again stressing the importance of these antioxidant enzymes in the normal physiology [57,58]. Of note, overexpression of *Trx1* increased the lifespan and stress resistance [51]. Although there is only a limited role for oxidative stress as a direct determinant for accelerated aging or decreased lifespan [55,56,59], there is substantial evidence for a contribution of oxidative stress to detrimental effects on physiological organ function during the aging process preventing healthy aging [60–63].

The observation that antioxidant enzymes have a significant effect on the healthspan of animals during normal aging (e.g., indicated by decreased aging-associated cardiovascular complications during the sunset years) and also on the resistance to stress conditions is of high clinical importance [51]. Previously, we demonstrated that genetic deletion of the mitochondrial antioxidant proteins aldehyde dehydrogenase-2 (*ALDH-2*) and manganese superoxide dismutase (*Mn-SOD*) leads to vascular dysfunction and mitochondrial oxidative stress with increasing age [23]. These data support the concept that oxidative stress in general and mitochondrial ROS formation in particular, despite not playing a key role for the lifespan, have significant impact on the quality of aging, the healthspan [60–63].

4. Vascular Function, Oxidative Stress and •NO Bioavailability in Aging

Endothelial dysfunction was found in several animal models of hypertension or atherosclerosis, both representing important cardiovascular risk factors (for review see [19]). Furthermore, patients with endothelial dysfunction display a higher burden of oxidative stress and have increased risk factors for cardiovascular disease and events (see Figure 1). Endothelial dysfunction and most cardiovascular disease are characterized by increased levels of ROS formation due to an imbalance between pro-oxidative enzymes (xanthine oxidase, NADPH oxidase, uncoupled eNOS or enzymes of mitochondrial respiration) and antioxidant enzymes (Cu, Zn-SOD, Mn-SOD and extracellular SOD), resulting in a deviation of cellular redox environment from the normal [64]. A similar pattern of vascular dysregulation can be found in aging tissues (for review see [28]) and was first described in 1956 by Harman *et al.* ("free radical theory of aging").

Irreversible oxidations and accumulation of oxidized biological macromolecules (e.g., DNA mutations, oxidized proteins) appear in biological systems, which are suffering from chronic oxidative stress. Besides these long-term consequences, ROS interfere rapidly with nitric oxide (•NO) signaling. The accepted concept for reduced •NO bioavailability is the reaction of superoxide with •NO under formation of peroxynitrite ($ONOO^-$) [65]. Not only is reduced bioavailability of the important vasodilator •NO problematic for the vascular system, $ONOO^-$ itself has the ability to disturb the enzymatic function of proteins by nitration of tyrosine residues and oxidation of cysteine-thiol-groups [66–68]. Among others, the mitochondrial isoform of superoxide dismutase, Mn-SOD, is affected by nitration and becomes inactivated which further reduces antioxidant capacity of the cell [69].

The •NO producing enzyme eNOS itself is highly susceptible to damage by increased oxidative stress [15]. Tetrahydrobiopterin (BH4), a cofactor of eNOS, can be oxidized by $ONOO^-$ and the latter can potentially uncouple eNOS [70,71]. BH4 is a redox cofactor of eNOS and regulates the catalytic activity. In aged animals reduced vascular BH4 levels were shown [72], but the results in the literature are controversial [73]. Nevertheless, pharmacological supplementation of BH4 improves endothelial function in aged humans compared to young subjects [22]. This shortage of cofactor leads to a conformational change in eNOS resulting in uncoupling. Besides BH4, eNOS has several other redox switches that may lead to dysfunction/uncoupling in a ROS-dependent fashion: it can be S-glutathionylated [74], inhibited by asymmetric dimethylarginine (ADMA) and phosphorylated in a protein kinase C (PKC) or protein tyrosine kinase (PYK-2) dependent manner at Thr495 or Tyr657 [75]. Likewise, the zinc-sulfur-complex at the dimer-binding-interface can be oxidatively disrupted [76].

Uncoupling of eNOS switches the enzyme from good to evil [77]. In the uncoupled state, eNOS is generating ROS, which further oxidize BH4 and reduce •NO bioavailability [71]. This vicious circle is an established concept and part of the pathogenesis of endothelial dysfunction in aged vessels [24,36]. Several groups reported on increased eNOS expression levels in aging, which might be a counter-regulatory effect to compensate for decreased •NO bioavailability. In contrast, other groups found no change of eNOS expression in aging, but observed decreased Akt-dependent phosphorylation of eNOS at Ser1177 with increasing age as a potential explanation for an impaired endothelial dysfunction in the elderly [78]. Our group just recently provided evidence for S-glutathionylation and adverse phosphorylation of eNOS at Thr495 and Tyr657 by redox-sensitive PKC and PYK-2, respectively, as important mechanisms in the process of aging-induced vascular dysfunction [16].

NADPH oxidase (Nox) is a major source of ROS in the cardiovascular system [79,80]. Isoforms 1, 2, 4 and 5 are significantly expressed in heart and vessels. Nox2 and Nox4 are known to be upregulated in vascular tissue of aged mice [81]. In addition, these enzymes are regulated by tumor necrosis factor-α (TNF-α), which is known to be elevated in aged animals and humans [82,83]. The cytokine TNF-α plays an important role in many inflammatory disorders and vascular dysfunction is closely linked to inflammatory processes [84]. In fact, administration of TNF-α can promote oxidative stress by activation of Nox, endothelial dysfunction, endothelial apoptosis, and upregulation of proatherogenic inflammatory mediators, like inducible nitric oxide synthase (iNOS) and adhesion molecules [85,86]. Furthermore, TNF-α stimulates mitochondrial superoxide production in human retinal endothelial cells [87]. Chronic TNF-α inhibition improves flow-mediated arterial dilation in resistance arteries of aged animals, while reducing iNOS and intercellular adhesion molecule-1 (ICAM-1) expression [88]. All the mentioned effects are similar to functional alterations of the aged vascular endothelium. Not only TNF-α, but also interleukins (IL-1β, IL-6, IL-17) and C-reactive protein (CRP) are elevated in aging independent to other risk factors (e.g., smoking) [89]. Since, infiltrating leukocytes contribute to increased oxidative stress and reduced •NO bioavailability in the vessel wall [90,91], cytokine release and chemoattraction of leukocytes by the endothelium are important in the pathogenesis of aging-mediated endothelial dysfunction.

5. Aging, Mitochondrial Oxidative Stress, Mitochondrial DNA Damage and Endothelial Dysfunction

In 1972, Harman modified his "free radical theory of aging" to specify the role of mitochondria [92]. He tried to explain why exogenous supplementation of antioxidants to rodents could not improve their lifespan. His explanation was that these antioxidants do not reach the mitochondrion. He proposed that mitochondria are both the primary origin and target of oxidative stress.

Recently, we demonstrated in two different knock-out models with increased mitochondrial ROS ($ALDH\text{-}2^{-/-}$, $MnSOD^{-/-}$ mice), that mitochondrial ROS formation and oxidative mitochondrial DNA (mtDNA) lesions as well as vascular dysfunction are increasing with age [23] (Figure 2). According to our data, endothelial dysfunction was clearly correlated with mitochondrial oxidative stress. The increase of mitochondrial ROS was more dependent on aging, then on the presence or absence of antioxidant proteins. The correlation between mtROS and mtDNA strand breaks, led us to speculate that the mitochondrial DNA damage could induce even more mtROS. Since the mitochondrial DNA mainly encodes for proteins of the mitochondrial respiration chain, one could assume that impaired mtDNA translation leads to mitochondrial uncoupling with secondary increase in mtROS formation.

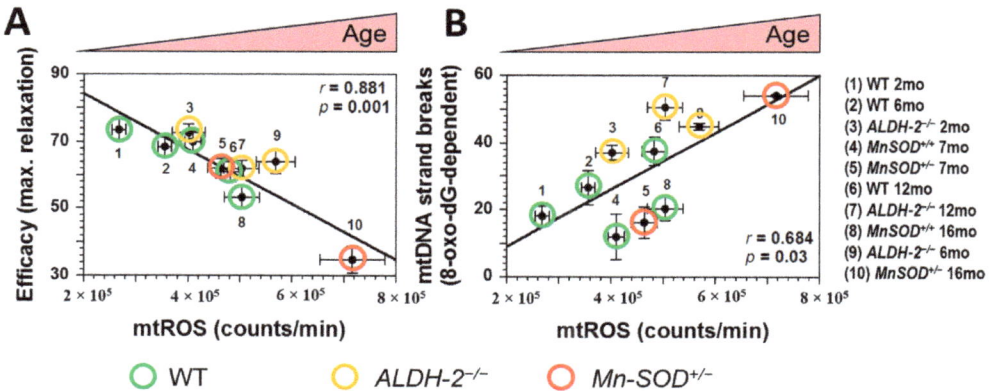

Figure 2. Correlations between mitochondrial oxidative stress (mtROS), mitochondrial DNA (mtDNA) damage and vascular (endothelial) function (ACh-induced maximal relaxation). (**A**) mtROS formation was plotted for all age-groups and mouse strains *versus* the corresponding maximal efficacy in response to acetylcholine (ACh); (**B**) mtROS was plotted for all age-groups and mouse strains *versus* the corresponding mtDNA damage. ROS were measured using L-012 (100 µM) enhanced chemiluminescence in isolated cardiac mitochondria upon stimulation with succinate (5 mM). r is the correlation coefficient. The groups are: 1 = B6 WT, 2mo; 2 = B6 WT, 6mo; 3 = $ALDH\text{-}2^{-/-}$, 2mo; 4 = $MnSOD^{+/+}$, 7mo; 5 = $MnSOD^{+/-}$, 7mo; 6 = WT B6, 12mo; 7 = $ALDH\text{-}2^{-/-}$, 12mo; 8 = $MnSOD^{+/+}$, 16mo; 9 = $ALDH\text{-}2^{-/-}$, 6mo; 10 = $MnSOD^{+/-}$, 16mo. The age of measured groups increases from the left to the right. Adopted from Wenzel *et al.*, Cardiovasc. Res. 2008 [23]. With permission of the European Society of Cardiology. All rights reserved. © The Author and Oxford University Press 2008.

Previous reports highlighted increased mitochondrial and systemic oxidative stress in mice with genetic deficiency in glutathione peroxidase-1 (*GPx-1*) [48]. In addition, *GPx-1* deficiency showed synergistic negative effects on vascular function in the setting of diabetes [93], hyperlipidemia [94], and hypertension [95]. Moreover, increased senescence was reported for fibroblasts from $GPx-1^{-/-}$ mice [48]. Most importantly, a correlation between cardiovascular risk and GPx-1 activity in blood cells was previously reported conferring high clinical relevance to the expression/activity of GPx-1 [96] and again supporting the concept that antioxidant enzymes and oxidative stress might contribute significantly to the healthspan and comorbidity of the elderly [60–63].

Recently, we demonstrated for the first time that aging per se leads to eNOS dysfunction and eNOS uncoupling via increased adverse phosphorylation and *S*-glutathionylation of the enzyme (Figure 3B) [16]. We also established that *GPx-1* deficiency resulted in a phenotype of endothelial and vascular dysfunction, which was substantially potentiated by the aging process (Figure 3A). By using oxidative stress-prone $GPx-1^{-/-}$ mice (a model representing decreased break-down of cellular hydrogen peroxide) in a study of the aging process, we can provide a strong mechanistic link between oxidative stress, eNOS dysfunction and vascular dysfunction in aging animals (Figure 3). Most importantly, •NO bioavailability was also significantly decreased in aged $GPx-1^{-/-}$ mice (Figure 4) supporting a dysregulation of eNOS and/or increased oxidative degradation of •NO during the aging process in general and in animals with decreased antioxidant defense in particular.

As a proof of endothelial and vascular dysfunction, we showed that both, endothelium-dependent and -independent relaxation was impaired in aged $GPx-1^{-/-}$ mice [16]. Altered eNOS function by inactivating or uncoupling phosphorylation, PKC-dependent at Thr495 [99,100] or PYK-2-dependent at Tyr657 [101], and *S*-glutathionylation [74] leading to diminished •NO bioavailability are plausible explanations for this phenotype. The deregulatory modifications of eNOS were also translated to increased uncoupling of the enzyme as envisaged by endothelial superoxide formation, which increased with age, was more pronounced in the $GPx-1^{-/-}$ group and nicely correlated with *S*-glutathionylation as a marker of uncoupled eNOS (Figure 3B) [16]. Since the •NO target enzyme soluble guanylyl cyclase (sGC) is also subject to oxidative inactivation (*S*-oxidation, *S*-nitrosation, heme-oxidation) [102–108], it might be speculated that the aging process will lead to an inactivation or at least desensitization of the enzyme. At least decreased expression of sGC subunits have been reported in tissues of old animals [109–111]. Future studies with sGC activators will prove whether apo-sGC or heme-oxidized sGC play a role for aging-induced vascular dysfunction.

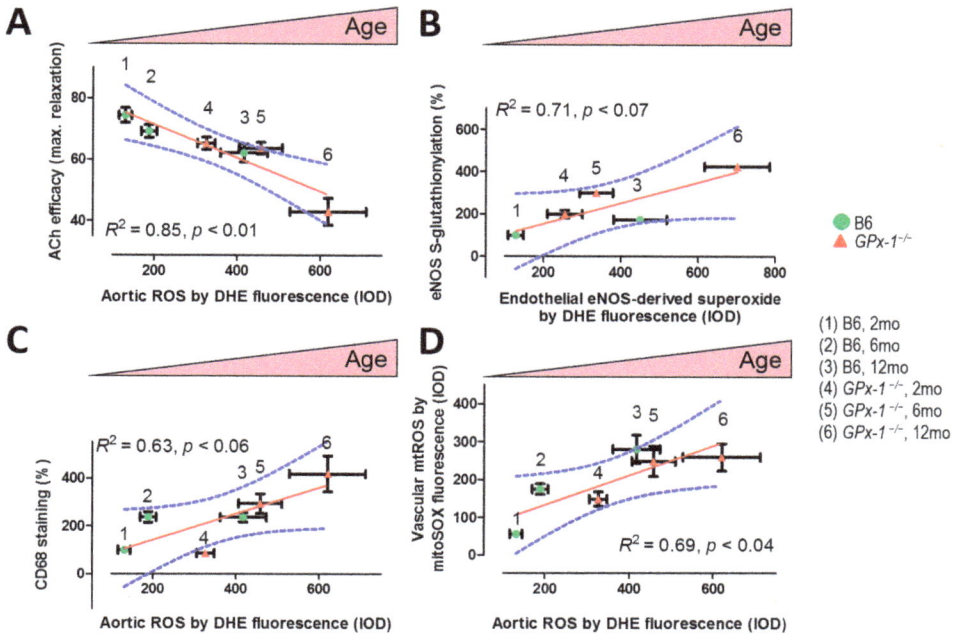

Figure 3. (**A**) Correlation between endothelium (ACh)-dependent relaxation (isometric tension measurement in isolated aortic ring segments) and aortic ROS formation (DHE staining of the aortic wall). Linear regression: $p < 0.01$, $R^2 = 0.85$; (**B**) Correlation between eNOS uncoupling marker (S-glutathionylated eNOS) and endothelial (eNOS-derived) superoxide formation (endothelial DHE staining). Linear regression: $p < 0.07$, $R^2 = 0.71$; (**C**) Correlation between inflammation (CD68 staining) and aortic ROS formation (DHE staining of the aortic wall). Linear regression: $p < 0.06$, $R^2 = 0.63$; and (**D**) Correlation between mitochondrial ROS formation (mitoSOX staining) and aortic ROS formation (DHE staining of the aortic wall). Linear regression: $p < 0.04$, $R^2 = 0.69$. Linear regressions were performed with GraphPad Prism 6 for Windows (version 6.02). All data were collected from our previous work [16]. Each data point was based on measurement of 4–10 animals. B6 means C57/BL6 wild type control; GPx-$1^{-/-}$ means glutathione peroxidase-1 knockout mice on C57/BL6 background. The age of measured groups increases from the left to the right. The solid red lines are simple linear regression fits to the data. Blue lines are the 95% confidence intervals on the estimated means. With permission of Wolters Kluwer Health, Inc. Copyright © 2014, Wolters Kluwer Health.

In cultured endothelial cells we demonstrated that GPx-1 silencing increased adhesion of leukocytes, which may contribute to the observed endothelial/vascular dysfunction (e.g., by increased oxidative breakdown of •NO and/or impairment of the •NO-cGMP signaling cascade by infiltrated leukocytes) [16]. Furthermore, we observed an appreciable increase in cardiovascular oxidative stress and mild vascular remodeling, as detected by Sirius red and Masson's trichrome staining (indicative for increased fibrosis of the intima and thus a decrease in intima/media ratio) [16].

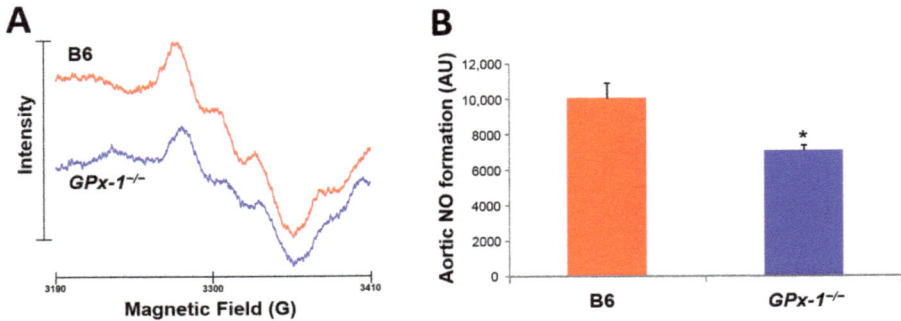

Figure 4. Nitric oxide formation in isolated aortic ring segments in old (12 mo) wild type (WT) and *GPx-1* knockout mice. •NO was measured in aortic ring segments (1 thoracic aorta for each measurement) upon stimulation with calcium ionophore (A23187, 10 μM) for 60 min at 37 °C in the presence of freshly prepared lipophilic spin trap Fe(II)(DETC)$_2$. •NO bound to the spin trap generates a stable paramagnetic nitrosyl-iron species that yield a typical triplet signal when measured by electron spin resonance (EPR) spectroscopy in liquid nitrogen. The detailed method was published in [97,98] and samples were used from a published study [16]. (**A**) Representative spectra and (**B**) quantification of signal intensity. Data are mean ± SEM of 9 mice per group. *, $p < 0.05$ *versus* wild type.

GPx-1 deficiency has been demonstrated to increase the susceptibility of cultured endothelial cells to lipopolysaccharide (LPS) by enforcing TLR4/CD14 signaling [112]. In conductance vessels, sustained overproduction of vasodilators (e.g., •NO by iNOS) may reduce the responsiveness of the vasculature to these messengers because of a desensitization of the •NO/cGMP pathway [113]. Indeed, increased iNOS expression and activity has been demonstrated for selenium-deficient RAW cell macrophages [114] and selenium is the precursor for selenocysteine synthesis forming the active site of GPx-1. iNOS-derived •NO formation could also provide the basis for extensive protein tyrosine nitration as reported for old mice in general and *GPx-1* deficient mice in particular [16]. In summary, the progressing phenotype of low-grade inflammation in GPx-1 deficient mice during the aging process was nicely reflected by the correlation of the marker of inflammation (CD68 staining) with the overall vascular ROS formation (dihydroethidine staining (DHE staining)) in dependence of age and antioxidant defense state of the animals (Figure 3C) [16]. Since global vascular ROS formation was nicely correlated with mitochondrial ROS formation (Figure 3D), and all other parameters were linked to global vascular ROS formation, one can assume that mtROS formation has significant impact on eNOS dysregulation/uncoupling, vascular function and low-grade inflammation during the aging process [16]. This assumption is in good accordance to the reports on mtROS-driven activation of the inflammasome and expression of proinflammatory cytokines [115–118].

We observed a moderate but consistent decline in reduced thiol groups in aged $GPx-1^{-/-}$ mice as compared to only a minor tendency of this decline in the aged B6 wild type mice [16]. Overall, the majority of literature supports a trend of decrease in reduced thiols during the aging process, which could affect the S-glutathionylation pattern and accordingly the coupling state of eNOS. Smith $et~al.$ showed in 2006 that the decline in endothelial GSH may contribute to a change of eNOS phosphorylation pattern (decline in P-Ser1176 and increase in P-Thr494) that was associated with a loss of vascular •NO bioavailability, increased proinflammatory cytokines and impaired endothelium-dependent vasodilation [119]. Recent work by Crabtree and coworkers even described an interplay of BH4 deficiency and eNOS S-glutathionylation in cells with diminished GTP-cyclohydrolase-1 expression providing a functional link between these two routes of eNOS uncoupling [120] that could be of relevance for the aging process as well.

According to our own previous data, mitochondrial oxidative stress increases with age and is a strong trigger of age-related endothelial/vascular dysfunction (Figure 2A) [23,28]. Using two genetic mouse models with ablated mitochondrial aldehyde dehydrogenase ($ALDH-2^{-/-}$) or mitochondrial superoxide dismutase ($MnSOD^{+/-}$), both important antioxidant enzymes, we could show that increased mitochondrial oxidative stress is associated with augmented oxidative mtDNA lesions (Figure 2B). Outcome from studies in genetic animal models with increased mitochondrial ROS formation (e.g., $MnSOD$- or $Trx-2$-deficiency) strongly supports an important link between cellular aging and mitochondrial dysfunction. Of note, overexpression of mitochondria-targeted catalase enhanced protection of mitochondria from ROS-induced damage and extended life span in mice [121].

Mitochondria represent an important source of reactive oxygen species, caused by electron leakage in the respiratory chain that results in univalent reduction of oxygen into $O_2^{•-}$. The steady state concentration of superoxide in the mitochondrial matrix is about 5- to 10-fold higher than that in the cytosolic and nuclear spaces. These apparently high mitochondrial superoxide formation rates correlate well with the reported mitochondrial oxidative DNA lesions being 10- to 20-fold higher in mitochondrial compared to nuclear DNA [29]. Cadenas and Davies proposed that susceptibility of mtDNA to oxidative damage may be ascribed to a combination of factors besides the higher superoxide formation rate in the mitochondrial matrix: unlike nuclear DNA, mitochondrial DNA lacks protective histones and enjoys only a relatively low DNA repair activity [29]. Therefore, mitochondrial 8-oxo-deoxyguanosine (8-oxo-dG) DNA lesion could represent an interesting marker for the burden of oxidative stress during the aging process [122]. According to Sastre *et al.* "mitochondrial oxidative stress should be considered a hallmark of cellular aging" [123]. The impact of mitochondrial ROS production on longevity may involve direct signal transduction pathways that are sensitive to oxidative stress, and indirect pathways related to the accumulation of oxidative damage to mitochondrial DNA, proteins, and lipids. Of note, the majority of mtDNA encodes for proteins of the mitochondrial respiratory chain and accumulation of mtDNA lesions might contribute to further uncoupling of mitochondrial electron flow at the expense of oxygen reduction to water but instead favor the formation of superoxide [124].

The free radical hypothesis of aging highlights that reactive oxygen species are responsible for the accumulation of altered biological macromolecules such as DNA, over an organism's lifespan [31]. Nucleic acid, in particular mitochondrial DNA (mtDNA), is regarded as a highly susceptible target for oxidant-induced mutations and deletions, which causes progressive deterioration of mitochondrial function over time (Figure 5). mtDNA deterioration belongs to a destructive cycle in which mitochondrial dysfunction further increases oxidative burden resulting in loss of cellular functions and finally apoptosis and necrosis. One of the major oxidative modifications of the mtDNA is 8-oxo-deoxyguanosine (8-oxo-dG) [125,126]. 8-oxo-dG is a mutagenic lesion and its accumulation is directly correlated with the development of pathological processes [127]. The correlation of lifespan with oxidant-induced mtDNA damage was demonstrated for the first time by Barja and co-workers [122]. These authors showed that in short-lived animals 8-oxo-dG content in nuclear and mitochondrial DNA was increased in cardiac tissue (Figure 6). These findings could be attributed to higher burden of oxidative stress in these short-lived animals (due to higher metabolic rate, less efficient antioxidant defense and/or less efficient DNA repair machinery). However, when brain tissue was investigated, accumulation of 8-oxo-dG in nuclear DNA was more pronounced in the long-lived animals (data not shown) [122]. This important study demonstrates that accumulation of oxidative DNA damage cannot per se be assumed for more living years among all different animal species, but each of them obviously has distinct kinetics of formation and repair of DNA damage, which, on top of this, depends on the specific tissue used for the quantification.

This assumption was later expanded in genetically modified mice with a proofreading-deficient mitochondrial polymerase-γ (Polγ). These mice accumulated severe mtDNA mutations, leading to mitochondrial dysfunction, increase in apoptosis and premature aging [31,128]. One of the more recent studies clearly depicted that a transgenic mouse with cardiac tissue-specific overexpression of mutated human Polγ [129] developed early aging symptoms. Elevated ROS generation and severe cardiomyopathy, typical for the "mtDNA-mutator mouse", was also observed in these animals. Our current knowledge of maternally inherited human diseases [130], e.g., DAD-syndrome (Leu-UUR tRNA = diabetes mellitus and deafness), MELAS- (mitochondrial encephalomyopathy, lactic acidosis, and stroke-like syndrome) [131–133] or KSS- (Kearns-Sayre syndrome) [131,133], highlight the importance of mtDNA. Various mutations or deletions, especially in tissues with high oxygen and energy consumption such as the myocardium, increase the rate of apoptosis, free radical species formation, leading to the functional impairment of the specific organ [131,134]. Therefore, mutated mtDNA is commonly regarded as a major contributor to vascular aging and various cardiovascular disorders [131,134].

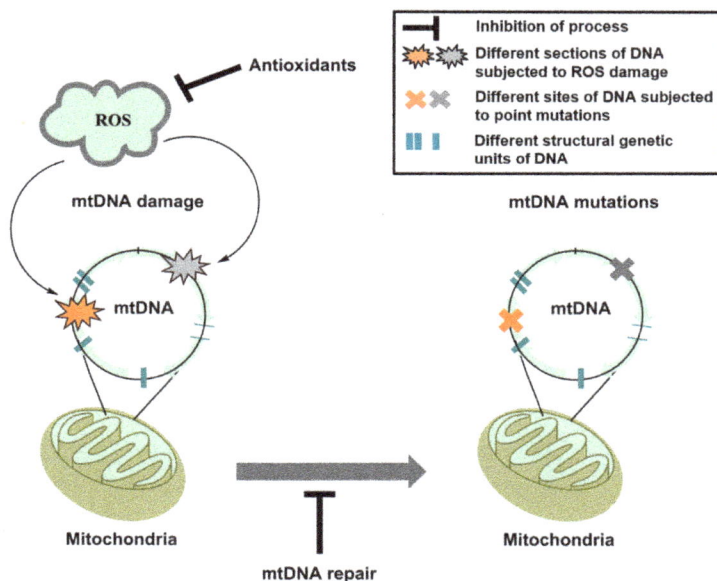

Figure 5. Schematic representation of mitochondrial DNA (mtDNA) damage by reactive oxygen species (ROS) leading to mtDNA mutations. Deleterious activity of ROS can be prevented by the administration of antioxidants. Damaged mtDNA can be repaired by the appropriate mtDNA repair pathway.

Considering that mtDNA damage accumulation with age is already widely accepted, more profound molecular explanations were provided during the last years. It has been recently shown that the observed mtDNA mutations might be mainly ascribed to errors in the replication process, to unrepaired damages, or to the spontaneous deamination of the nucleobases, indicating effects of ROS only as concomitant [135]. Investigations conducted by Itsara *et al.* [136] showed that malfunction of the replication process leads to a high accumulation of mtDNA mutations. This observation might be based on the fact that mtDNA propagates throughout the lifespan of the somatic cell, providing higher chances for the occurrence of mtDNA damage [137]. A typical mtDNA modification introduced by mitochondrial ROS is 8-hydroxy-deoxyguanosine (8-oxo-dG) [125,126]. This type of damage results in G:C to T:A transversions after DNA replication [138]. Nonetheless, this concept has not been confirmed by experimental studies, detecting only 10% of all mutations to be G:C to T:A transversions, having no correlation with the age of the animals. The most prevailing type of detected mutation was G:C to A:T transitions, attributed to the imperfect activity of mtDNA polymerase γ.

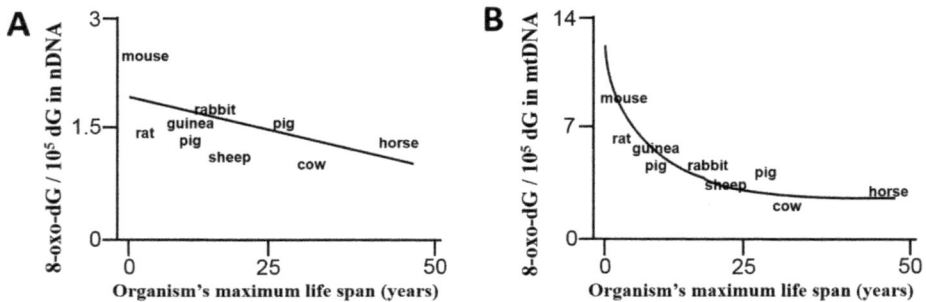

Figure 6. Correlation of an organism's life span with the content of 8-oxo-dG in nuclear (**A**) or mitochondrial (**B**) DNA in cardiac tissue. Surprisingly, oxidative DNA lesions do not per se accumulate with increasing number of living years but show distinct kinetics among different animal species, tissues and probably the underlying burden of oxidative stress, efficiency of the antioxidant system and DNA repair machinery. According to Barja et al. [122].

Another study by Kennedy et al. [139], utilizing a more clinically relevant experimental model of aging human brain, showed that mtDNA mutations during aging consist mostly of transition mutations (G:C to A:T), as depicted by ultra-sensitive detection technique and Duplex Sequencing methodology. These mtDNA mutations showed a nice correlation with the malfunction of Poly or deamination of cytidine resulting in uracil and adenosine deamination, yielding inosine as the major mutagenic contributors in mtDNA.

Several recent papers that were summarized in the review by Cha et al. [140], highlight the direct correlation of mitochondrial DNA mutations with neurodegenerative diseases, in particular Huntington's, Parkinson's, Alzheimer's diseases and amyotrophic lateral sclerosis (ALS). The incidence of these neurodegenerative disease increases with higher age [32]. In the setting of Alzheimer's diseases (AD) it has been shown that post-mortem brain tissues of AD patients have elevated levels of degraded mtDNA [141] and defective base excision repair [142]. Another group was able to show that patients suffering from AD have an increased number of cytochrome C oxidase deficient neurons [143], supporting the above mentioned concept that accumulating mtDNA damage during the aging process leads to impaired biosynthesis of the respiratory chain proteins.

6. Aging and Mitochondrial DNA Repair

Not only the particular damage of the mtDNA molecule itself, but also the lifetime of mtDNA lesions has profound effects on the progression and characteristics of the aging phenotype. The accumulation of mtDNA damage is not only determined by the higher mtROS formation rates but also by the activity or expression levels of mitochondrial DNA repair systems. Therefore, the following part of the review concentrates on the impact of aging on the DNA repair machinery.

Taking into account the highly oxidizing environment of the mitochondria, malfunctioning of the Polγ and peculiarity of the mtDNA replication process would ultimately lead to the accumulation of mtDNA lesions. During mtDNA replication, the lagging strand is kept in the single-stranded condition for a prolonged period of time, and thereby the chances are increased for impairment of the replication. For this reason, mitochondrial protein machinery evolutionary created a multi-layered DNA repair system that is able to resolve such complex challenges [144].

One of the most extensively studied DNA repair pathways in mitochondria is base excision repair (BER) [145]. Its enzymatic pathways are identical to the ones that take place in the nucleus, starting with the modified base identification by specific glycosylase, followed by its excision, insertion of the correct nucleotide and finishing with the DNA strand ligation [146]. The family of glycosylases involved in mitochondrial BER consists of several members: OGG1 that is able to remove the most common and the most mutagenic modification introduced to the mtDNA—8-oxo-dG [147]; UNG1, which is responsible for the removal of uracil, resulting from the spontaneous deamination of the cytosine [148]; NTH1-dichotomous enzyme, possessing glycosylase and AP lyase activities, that excises pyrimidine lesions (ThyGly, FapyG and FapyA) [149]; NEIL1 and NEIL2 that are also involved in the removal of FapyG and FapyA [150]; MYH, enzyme that detects adenine placed opposite to 8-oxo-dG [151]. After specific glycosylases have identified and extracted the damaged base, the abasic sites are resolved by APE endonuclease that prepares the DNA molecule for the Polγ mediated repair process by cleaving 3'-hydroxyl and 5'-deoxyribose-5-phosphate sites [152]. Afterwards, the newly synthesized section of DNA is ligated to the rest of the molecule by DNA ligase III.

In case that the incorrect base has not been identified by BER and the replication process actually resulted in the formation of a mismatch, the adequate mismatch repair pathway (MMR) can be activated. There has been a lot of dispute in the field about the existence of this repair mechanism, since mitochondria doesn't possess classical players of MMR, like MSH3, MSH6 or MLH1 [153]. Though recently, with the help of specific mass spectrometry analysis, a novel MMR repair protein has been identified. Y-box binding protein (YB-1) has been shown to efficiently bind DNA containing mismatches [154]. This finding was further supported by identifying 3′ and 5′-exonuclease activity of YB-1 on single stranded DNA and weak endonuclease activity on double-stranded DNA. Depletion of this enzyme by siRNA showed increased levels of mtDNA mutagenesis, confirming the presence of a unique mitochondrial MMR pathway. In contrast to the nucleus, mitochondrion doesn't possess a nucleotide excision repair. This limitation leads to the accumulation of unrepaired bulky pyrimidine dimers, as observed upon UV irradiation.

Enzymatic reactants of the homologous recombination (HR) or non-homologous end joining (NHEJ) that are quite abundantly involved in the DNA repair inside the nucleus have not been precisely identified in the mitochondria. So far, only products of their reactivity have been detected by super-resolution and transmission electron microscopy. Among them are the following ones: 1. mtDNA deletions, which are attributed to the exo- and endonucleases activity; 2. Holiday junctions, detected in the mtDNA of the human heart and brain [155]; 3. Formation of diverse brunched structures and multiple-way junctions. All three of them are indicative of the active homologous recombination process [156]. The identification of the novel single strand annealing enzyme Mgm101 further supports this notion [157].

Recent findings on the expression levels of DNA repair systems in the matrix of aging mitochondria are quite specific for the different DNA repair systems [158]. For example, it has been shown that incidents of the homologous recombination become more abundant with the progress of aging, being strongly correlated with linear increase in the mtDNA damage [159]. Alternatively, the specific DNA glycosylase, OGG1, has been shown to be down-regulated with the onset of aging, augmenting the deleterious effects of the oxidative stress burden in the mitochondrial environment and accelerating the process of cellular senescence [160].

7. Crosstalk between Mitochondrial and NADPH Oxidase-Generated Reactive Oxygen and Nitrogen Species and Impact on Endothelial Function

Primary mtROS not only cause direct oxidative damage to cellular structures but may also contribute to the activation of secondary ROS sources such as the NADPH oxidases. In a feed-forward fashion, this crosstalk initiates a vicious cycle that finally may lead to eNOS dysfunction/uncoupling and impairment of endothelial function [75]. Previously, we reported on the crosstalk between mtROS and cytosolic ROS in the model of nitroglycerin-induced nitrate tolerance that is associated with increased mitochondrial oxidative stress. Nitrate tolerance is an excellent model of vascular dysfunction and oxidative stress in general [161] and mitochondrial oxidative stress and dysfunction in particular [162–164]. Two distinct ROS sources could be identified: endothelial dysfunction was linked to activation of NADPH oxidases, whereas vascular dysfunction was associated with mitochondrial oxidative stress [165]. Cyclosporine A, the mitochondrial permeability transition pore inhibitor, blocked this interaction (crosstalk) between mitochondrial and NADPH-dependent ROS formation and selectively improved endothelial dysfunction, whereas nitrate tolerance was not affected. In support of this observation, $gp91phox^{-/-}$ and $p47phox^{-/-}$ mice developed tolerance under nitroglycerin treatment but no endothelial dysfunction. In contrast, endothelial dysfunction and tolerance was improved by the respiratory complex I inhibitor rotenone (preventing complex I-derived ROS formation by reverse electron transfer [166]). Likewise, administration of the K_{ATP} opener diazoxide caused a nitrate tolerance-like phenomenon in control animals, whereas tolerance phenomenon was improved by the K_{ATP} inhibitor glibenclamide.

Very similar effects of these compounds have been recently demonstrated in animal and cell culture models of angiotensin-II induced hypertension [167,168]. NADPH oxidase-driven activation of mitochondrial ROS formation via K_{ATP} channels (acting as a redox switch) and changes in the membrane potential was previously proposed [169,170] and later confirmed [167]. The proposed mechanism for mtROS-driven activation of NADPH oxidase is redox-sensitive stimulation of protein kinases such as PKC and cSrc in a calcium-dependent process [97,166,171]. Synergistic formation of mtROS and NADPH oxidase derived ROS lead to uncoupling of eNOS, nitration of prostacyclin synthase and desensitization sGC [19]. Recently, we showed that this interaction (crosstalk) plays an essential role in angiotensin-II induced hypertension, vascular oxidative stress and endothelial dysfunction (envisaged by eNOS S-glutathionylation) [97]. In support of this proposed mechanism, increased oxidative stress per se is able to activate NADPH oxidase in a positive feedback fashion [172].

We and others propose a similar crosstalk between mitochondrial and NADPH oxidase-dependent ROS formation in the aging vasculature leading to age-related vascular dysfunction (Figure 7) [23,28,97,173]. This proposal is based on the finding that mtROS formation increases with age (and is higher in $MnSOD^{+/-}$ mice) and endothelial function is impaired with age (to a higher extent in $MnSOD^{+/-}$ mice) [23]. Of note, we and others also showed that mtROS participate in the activation of immune cells via stimulation of the phagocytic NADPH oxidase (Nox2) [97,174]. mtROS have also been reported to participate in the activation of the inflammasome and trigger the expression of proinflammatory cytokines [115–118]. In summary these data support a proinflammatory role of mtROS for progression of low-grade inflammation in the aging vasculature.

165

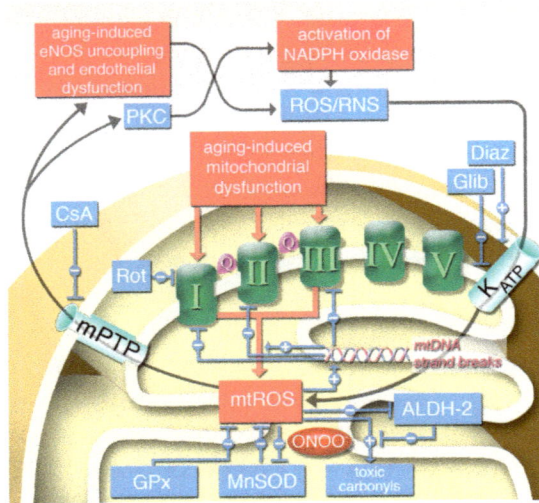

Figure 7. Hypothetic scheme of aging-induced vascular dysfunction and the role of mitochondria in this process. Aging-induced mitochondrial dysfunction triggers mitochondrial reactive oxygen species (mtROS) formation from respiratory complexes I, II, and III (Q = ubiquinone), whereas respiratory complexes IV and V were not reported to contribute to mtROS formation directly. Break-down of mtROS is catalyzed by glutathione peroxidase (GPx, for H_2O_2) or manganese superoxide dismutase (MnSOD), the latter is in turn inhibited by mitochondrial peroxynitrite ($ONOO^-$) formation. mtROS increase the levels of toxic aldehydes and inhibit the mitochondrial aldehyde dehydrogenase (ALDH-2), the detoxifying enzyme of those aldehydes. Increase in mtROS and toxic aldehydes also leads to mtDNA strand breaks which leads to augmented dysfunction in respiratory chain complexes and further increase in mtROS since mtDNA encodes mainly for those respiratory complexes. mtROS also activates mitochondrial permeability transition pore (mPTP), which upon opening releases mtROS to the cytosol leading to protein kinase C (PKC)-dependent NADPH oxidase activation, eNOS uncoupling and finally to endothelial dysfunction [165]. Cytosolic reactive oxygen and nitrogen species (ROS/RNS) in turn were demonstrated to activate K_{ATP} channels, which causes alterations in mitochondrial membrane potential (C) and further augments mtROS levels [167]. Effects of rotenone (Rot), cyclosporine A (CsA), diazoxide (Diaz) and glibenclamide (Glib) have been recently demonstrated in related models of vascular dysfunction and oxidative stress, nitroglycerin-induced nitrate tolerance and angiotensin-II triggered hypertension [165,167]. + means promotion of pathways; − means inhibition of pathways. Adopted from Wenzel *et al.*, Cardiovasc. Res. 2008 [23]. With permission of the European Society of Cardiology. All rights reserved. © The Author and Oxford University Press 2008.

8. Emerging Concepts of Aging

Moosmann and coworkers have published an interesting concept, based on the observation that intramembrane accumulation of methionine exhibits antioxidant and cytoprotective properties in living cells [175]. Their findings highlight methionine as an evolutionarily selected antioxidant of the respiratory chain complexes. On top of that they provide evidence that methionine is able to redox-cycle between oxidized and reduced forms, and in such a way becomes a vital member of the antioxidant defense system. Knockout mice not expressing methionine sulfoxide reductase enzyme are characterized by a decreased lifespan and several other pathologies [176,177]. Methionine sulfoxide reductases are expressed in endothelial cells [178], are potentially involved in the prevention of endothelial dysfunction via regulation of RUNX2 transcription factor activity and biological function in endothelial cells [179]. The functional polymorphism (rs10903323 G/A) in methionine sulfoxide reductase A shows an association with the increased risk of coronary artery disease in Chinese population [180]. Moreover, vascular smooth muscle cells were protected against oxidative damage by cytosolic overexpression of methionine sulfoxide reductase A [181]. A meta-examination of 248 animal species genome sequences with known maximum lifespan, including mammals, birds, fish, insects, and helminths demonstrated that the frequency of cysteine encoded by mitochondrial DNA is a specific marker of aerobic longevity [182]: long-lived organisms synthesize respiratory chain complexes with low abundance of cysteine. These results provide distinct (indirect) support for the free radical theory of aging.

Another explanation for worsening of the physical condition with age is based on a profound role of the mitochondrial enzyme p66Shc as an adaptor protein which is implicated in mitochondrial reactive oxygen species generation and translation of oxidative signals into apoptosis [183]. Mice deficient in $p66Shc^{-/-}$ gene produce decreased quantities of intracellular oxidants and display 30% longer life span. In order to elucidate the function of p66Shc and its possible implication in age-associated cardiovascular diseases, a series of studies were initiated. Extensive research revealed that $p66Shc^{-/-}$ mice are protected from age-dependent endothelial dysfunction [184] as well as age-related risk factors such as diabetes and hypercholesterolemia. The review of Camici *et al.* focused on deciphering the novel role of the p66Shc adaptor protein and its involvement in the age-associated cardiovascular disease and pathophysiology of aging. One of the major findings of these authors is that p66Shc N terminus is activated through reversible tetramerization by forming two disulfide bonds, as a result of which it forms a redox module responsible for apoptosis initiation that is tightly associated with senescence [185]. Other systems that are involved in p66Shc regulation are two antioxidant enzymatic classes—glutathione and thioredoxin enzymes that can inactivate p66Shc through a reduction mechanism. Consequently, this forms a thiol-based redox sensor system and initiates apoptosis once cellular protection systems cannot ameliorate cellular stress anymore. Protein kinase C β and prolyl isomerase 1 effectively regulates the mitochondrial effects of the longevity-associated stimulator-p66Shc [186]. Recently, highly specialized signaling pathways leading to mitochondrial import of p66Shc that are responsible for its proapoptotic activity upon oxidative stress were analyzed [187]. In contrast, advantages for diabetic cardiovascular complications [188] and age-associated endothelial •NO formation [189] were reported for genetic deletion of *p66Shc*. Last, but not least, it was validated that a p53-p66Shc signaling pathway controls intracellular redox status and contributes to increased levels of oxidative DNA damage [190] and mitochondrial oxidative stress [191]. The important role of p66Shc for vascular function is supported by recent findings that p66Shc expression correlates with the prognosis of stroke patients and post-ischemic inhibition of p66Shc reduces ischemia/reperfusion brain injury [192].

Another emerging concept in the field of aging is based on the contribution of epigenetic pathways. Epigenetic mechanisms that are involved in the aging process include alterations of the DNA methylation status, modifications of the histone tails (mainly acetylation and methylation) and changes in the expression of non-coding RNAs [193]. With increasing age, global DNA demethylation processes have been detected, leading to a DNA hypomethylation condition. This has been mostly observed on the CpG islands, which constitute regions of the DNA consisting for more than 50% of cytosine and guanine repeats [194]. On the other hand, there is a lot of information regarding sequestered, locus specific DNA hypermethylation. These nuclear regions are called senescence-associated heterochromatin foci (SAHF) [195]. The presence of such DNA segments is of high interest since they are at variance with the generally accepted fact that aging is associated with a more relaxed state of chromatin, called euchromatin [196].

One of the major epigenetic hallmarks of aging is the increased acetylation level of the histone tails. Such state is mostly attributed to the decreased level of histone deacetylases, in particular NAD^+-dependent sirtuins [197]. The most disputed histone modification that is associated with aging is the methylation status of lysine terminal ends. Recent data support a trend of increasing tri- and di-methylation of lysine 4 and lysine 36 on the histone variant 3 (H3K4me2/3 and H3K36me3), which are considered to be gene activating marks, during the aging process. Likewise, a trend of decreasing methylation levels of lysine 27 and lysine 9 on the histone variant 3 (H3K27me3, H3K9me2/3), which are known to be gene repressing marks, was reported for the aging process. These observations support a relaxed chromatin state, which is transcriptionally active [196]. Such epigenetic and genetic alterations of the coding nucleic acid molecules might at least partially explain the increased inflammation levels due to the transcriptional gene activation that are associated with the aging process.

The field of non-coding RNAs, especially in the context of aging is currently one of the least discovered. The class of non-coding RNAs consists of microRNA, long non-coding RNAs and piwiRNA. The role of non-coding RNAs in the senescence mechanisms has been much appreciated in the form of biomarkers [198]. The novel microRNA—miR-34a, was documented as an aging marker in several tissues. Boon and colleagues have shown that miR-34a is induced in a cardiac aging model, particularly in cardiac tissue. miR-34a involvement has been also shown in acute myocardial infarction (MI). miR-34a inhibition reduces cell death and fibrosis associated with post MI conditions [199]. Investigations of Boon *et al.* clearly highlighted miR-34a and its target PNUTS as a key molecule regulating heart contractile function during the aging process by inducing DNA damage responses and telomere shortening.

Finally, the aging process is closely linked to immunosenescence and low-grade inflammation contributing to higher prevalence of metabolic syndrome, diabetes and associated cardiovascular complications [200]. Since the risk factors for cardiovascular diseases and diabetes show a clear overlap, especially at the level of inflammation, the latter could represent a key player for aging-associated disorders and increased comorbidity in the elderly [11–13]. Therefore, targeting of the chronic inflammatory phenotype in the elderly could represent a promising strategy to normalize the increased prevalence of comorbidities in the elderly and increase their healthspan [201]. mtROS formation is augmented with increasing age [28] and mtROS can lead to the activation of immune cells and their phagocytic NADPH oxidase, induce the inflammasome and trigger cytokine release [97,115–118,174]. Therefore, targeting mitochondrial oxidative stress in the elderly could represent another strategy to suppress the chronic inflammatory phenotype and adverse effects of the aging process. This was recently shown in an animal model for metformin-dependent AMP-activated protein kinase activation that suppressed oxidative damage and chronic inflammation in old animals and increased their lifespan and healthspan [202].

9. Clinical Impact

The link between endothelial dysfunction, oxidative stress and aging has not only been defined in animal models. Hypertension, dyslipidemia and atherosclerosis are precursors of cardiovascular events, like stroke or myocardial infarction [203,204]. Oxidative stress and reduced •NO bioavailability impair vascular protective function of the endothelium [205]. In patients, non-invasive flow-mediated dilation (FMD) of the forearm brachial artery or plethysmography (acetylcholine-dependent vasodilation) are used to determine endothelial function *in vivo* [24] and are a predictor for cardiovascular events [3]. In patients with well-defined risk factors such as dyslipidemia or smoking, a significantly impaired acetylcholine-triggered endothelium-dependent vasodilation was observed and both risk factors are known to be associated with oxidative stress. Endothelial dysfunction of coronary arteries is directly associated with an increased risk of myocardial infarction [206]. Furthermore, Heitzer *et al.* demonstrated that patients who showed improved endothelial function after vitamin C infusion had a worse prognosis for cardiovascular events as compared to patients with low vitamin C effects (Figure 1C) [25]. These findings indicate that oxidative stress is not only a key player in the pathogenesis of endothelial dysfunction *in vivo*, but furthermore also reflects the prognosis of patients with established coronary artery disease. The concept of an increased burden of oxidative stress is broadly accepted, but nevertheless results from clinical trials investigating antioxidant vitamins like B, C, E and folic acid are disappointing (for review see [207]) and for vitamin E even increased mortality and numbers of heart failure were found (HOPE and HOPE-TOO). These findings prove what Harman already postulated in 1975 with his modified "free radical theory of aging" – that site-specific formation of ROS (e.g., in mitochondria) might be the key to understand the obvious discrepancy between association of most cardiovascular disease with oxidative stress but failure of antioxidant therapy so far. The reasons for the failure of large clinical trials on antioxidant therapy to display profound beneficial effects could be, among others, that this secondary prevention is applied to patients with already irreversible tissue/organ damage. In addition, secondary antioxidant prevention does not specifically target the defective defense and repair mechanisms (e.g., it could interfere with important intrinsic protective mechanisms such as ischemic preconditioning). Finally, the concentrations of the antioxidants reached at the sites of ROS formation could be too low (e.g., compliance was not controlled by measurement of plasma levels of the administrated antioxidants) [208]. The controversial and even contradictory results from antioxidant clinical trials or experiments manipulating antioxidants by pharmacological (or genetic approaches) suggest that aging is a complex and multifaceted process that cannot be explained by a single theory.

In humans, the incidence of hypertension, diabetes and atherosclerosis correlates with age (Figure 1A) [26,27]. In parallel, endothelial dysfunction manifests with age (Figure 1B) [24], together with oxidative stress from mitochondria and other enzymatic sources [209,210]. Aging is an independent risk factor for cardiovascular disease, which is mainly caused by endothelial dysfunction due to oxidative stress and low-grade inflammation [24,211]. Recent data provided evidence for changes in high-density lipoprotein composition and function in dependence of age [212]. Therefore, the quality of high density lipoprotein (HDL) may be another future target for therapeutic intervention in and/or diagnosis of age-related cardiovascular disease [213] because HDL quality decreases with age and this will negatively influence the endothelial function. Recently, the nitrovasodilator pentaerithrityl tetranitrate was demonstrated to provide heritable blood pressure lowering effects in hypertensive rats associated with enhanced H3K27ac and H3K4me3 and transcriptional activation of cardioprotective genes such as *eNOS*, *SOD2*, *GPx-1*, and *HO-1* [214]. Drugs like pentaerithrityl tetranitrate could represent the prototype of future epigenetic drugs that could also improve the healthspan in general and cardiovascular aging in particular. Mitochondria-targeted antioxidants could represent another promising therapeutic strategy to combat the "side effects" of the aging process. Mito-quinone improved age-related endothelial dysfunction in mice [215]. Since aging cannot be stopped like smoking [216] and unspecific antioxidant treatment using vitamins is not effective, aging-induced mitochondrial ROS formation may become a target for diagnosis and treatment of cardiovascular disease in the elderly.

10. Perspective

In the present review, we have provided strong evidence from our own and others studies that mitochondrial oxidative stress plays a key determinant for aging-induced impairment of cellular signaling and as a consequence cell death. There is a large body of evidence for the association of cellular aging with mitochondrial dysfunction based on genetic animal models with increased mitochondrial reactive oxygen species (ROS) formation (e.g., $MnSOD^{+/-}$ or *Trx-2*-deficiency). Interestingly, overexpression of catalase, another enzyme crucially involved in the antioxidant defense, enhanced protection of mitochondria from ROS, led to an extended life span in mice [121]. The increase in life span was much more pronounced when catalase was targeted to mitochondria as compared to overexpression in peroxisomes or in the nucleus. There is also evidence for an interaction (crosstalk) between mitochondrial oxidative stress and cytosolic sources of oxidative stress providing a direct link between aging and vascular dysfunction. Therefore, a better understanding of the role of mitochondria in the aging process may lead to specifically designed therapies to interfere with mitochondrial dysfunction and to delay the aging process for longevity. Since regulation of the vascular tone largely depends on a redox-balance in favor of a more reductive milieu, increased oxidative stress impairs vascular function and leads to endothelial dysfunction, atherosclerosis and other cardiovascular complications. Therefore, therapeutic intervention at the level of mitochondrial dysfunction would not only beneficially influence the aging process but also most kinds of cardiovascular diseases. The hyperacetylation state of the histone tails and impaired DNA repair capacity in aged tissues imply a therapeutic modulation of the aging process by epigenetic drugs and improved DNA repair. Since cardiovascular diseases are the main reason for mortality in the Western world and their prevalence increases with age, development of therapeutic interventions not only promises a prolonged healthspan (maybe even increased lifespan) for a large part of the world population but also represents an appreciable pharmaceutical market.

Acknowledgments: The present work was supported by continuous funding by Stufe1 and NMFZ programs of the Johannes Gutenberg-University Mainz and University Medical Center Mainz as well as Mainzer Herz Foundation (Andreas Daiber). Sebastian Steven holds a Virchow-Fellowship from the Center of Thrombosis and Hemostasis (Mainz, Germany) funded by the Federal Ministry of Education and Research (BMBF 01EO1003). Yuliya Mikhed holds a stipend from the International PhD Program on the "Dynamics of Gene Regulation, Epigenetics and DNA Damage Response" from the Institute of Molecular Biology gGmbH, (Mainz, Germany) funded by the Boehringer Ingelheim Foundation. All authors of this review were supported by the European Cooperation in Science and Technology (COST Action BM1203/EU-ROS).

Conflicts of Interest: The authors declare no conflict of interest.

References

1. Kelly, D.T. Paul dudley white international lecture. Our future society. A global challenge. *Circulation* **1997**, *95*, 2459–2464.

2. Lakatta, E.G.; Levy, D. Arterial and cardiac aging: Major shareholders in cardiovascular disease enterprises: Part I: Aging arteries: A "set up" for vascular disease. *Circulation* **2003**, *107*, 139–146.

3. Ras, R.T.; Streppel, M.T.; Draijer, R.; Zock, P.L. Flow-mediated dilation and cardiovascular risk prediction: A systematic review with meta-analysis. *Int. J. Cardiol.* **2013**, *168*, 344–351.

4. Herrera, M.D.; Mingorance, C.; Rodriguez-Rodriguez, R.; Alvarez de Sotomayor, M. Endothelial dysfunction and aging: An update. *Ageing Res. Rev.* **2010**, *9*, 142–152.

5. Bischoff, B.; Silber, S.; Richartz, B.M.; Pieper, L.; Klotsche, J.; Wittchen, H.U. Inadequate medical treatment of patients with coronary artery disease by primary care physicians in germany. *Clin. Res. Cardiol.* **2006**, *95*, 405–412.

6. Burnett, A.L. The role of nitric oxide in erectile dysfunction: Implications for medical therapy. *J. Clin. Hypertens.* **2006**, *8*, 53–62.

7. Csiszar, A.; Toth, J.; Peti-Peterdi, J.; Ungvari, Z. The aging kidney: Role of endothelial oxidative stress and inflammation. *Acta Physiol. Hung.* **2007**, *94*, 107–115.

8. Price, J.M.; Hellermann, A.; Hellermann, G.; Sutton, E.T. Aging enhances vascular dysfunction induced by the alzheimer's peptide β-amyloid. *Neurol. Res.* **2004**, *26*, 305–311.

9. Coleman, H.R.; Chan, C.C.; Ferris, F.L., III; Chew, E.Y. Age-related macular degeneration. *Lancet* **2008**, *372*, 1835–1845.

10. El Assar, M.; Angulo, J.; Rodriguez-Manas, L. Oxidative stress and vascular inflammation in aging. *Free Radic. Biol. Med.* **2013**, *65*, 380–401.

11. Cesari, M.; Onder, G.; Russo, A.; Zamboni, V.; Barillaro, C.; Ferrucci, L.; Pahor, M.; Bernabei, R.; Landi, F. Comorbidity and physical function: Results from the aging and longevity study in the Sirente geographic area (iLSIRENTE study). *Gerontology* **2006**, *52*, 24–32.

12. Yancik, R.; Ershler, W.; Satariano, W.; Hazzard, W.; Cohen, H.J.; Ferrucci, L. Report of the national institute on aging task force on comorbidity. *J. Gerontol. Ser. A* **2007**, *62*, 275–280.

13. Wieland, G.D. From bedside to bench: Research in comorbidity and aging. *Sci. Aging Knowl. Environ.* **2005**, *2005*, pe29.

14. Munzel, T.; Daiber, A.; Mulsch, A. Explaining the phenomenon of nitrate tolerance. *Circ. Res.* **2005**, *97*, 618–628.

15. Munzel, T.; Daiber, A.; Ullrich, V.; Mulsch, A. Vascular consequences of endothelial nitric oxide synthase uncoupling for the activity and expression of the soluble guanylyl cyclase and the cGMP-dependent protein kinase. *Arterioscler. Thromb. Vasc. Biol.* **2005**, *25*, 1551–1557.

16. Oelze, M.; Kroller-Schon, S.; Steven, S.; Lubos, E.; Doppler, C.; Hausding, M.; Tobias, S.; Brochhausen, C.; Li, H.; Torzewski, M.; *et al.* Glutathione peroxidase-1 deficiency potentiates dysregulatory modifications of endothelial nitric oxide synthase and vascular dysfunction in aging. *Hypertension* **2014**, *63*, 390–396.

17. Van der Loo, B.; Labugger, R.; Skepper, J.N.; Bachschmid, M.; Kilo, J.; Powell, J.M.; Palacios-Callender, M.; Erusalimsky, J.D.; Quaschning, T.; Malinski, T.; *et al.* Enhanced peroxynitrite formation is associated with vascular aging. *J. Exp. Med.* **2000**, *192*, 1731–1744.

18. Forstermann, U.; Sessa, W.C. Nitric oxide synthases: Regulation and function. *Eur. Heart J.* **2012**, *33*, 829–837.

19. Daiber, A.; Oelze, M.; Daub, S.; Steven, S.; Schuff, A.; Kroller-Schon, S.; Hausding, M.; Wenzel, P.; Schulz, E.; Gori, T.; *et al.* Vascular redox signaling, redox switches in endothelial nitric oxide synthase and endothelial dysfunction. In *Systems Biology of Free Radicals and Antioxidants*; Laher, I., Ed.; Springer-Verlag: Berlin, Germany; Heidelberg, Germany, 2014; pp. 1177–1211.

20. Donato, A.J.; Gano, L.B.; Eskurza, I.; Silver, A.E.; Gates, P.E.; Jablonski, K.; Seals, D.R. Vascular endothelial dysfunction with aging: Endothelin-1 and endothelial nitric oxide synthase. *Am. J. Phys. Heart Circ. Physiol.* **2009**, *297*, H425–H432.

21. Donato, A.J.; Magerko, K.A.; Lawson, B.R.; Durrant, J.R.; Lesniewski, L.A.; Seals, D.R. SIRT-1 and vascular endothelial dysfunction with ageing in mice and humans. *J. Physiol.* **2011**, *589*, 4545–4554.

22. Higashi, Y.; Sasaki, S.; Nakagawa, K.; Kimura, M.; Noma, K.; Hara, K.; Jitsuiki, D.; Goto, C.; Oshima, T.; Chayama, K.; *et al.* Tetrahydrobiopterin improves aging-related impairment of endothelium-dependent vasodilation through increase in nitric oxide production. *Atherosclerosis* **2006**, *186*, 390–395.

23. Wenzel, P.; Schuhmacher, S.; Kienhofer, J.; Muller, J.; Hortmann, M.; Oelze, M.; Schulz, E.; Treiber, N.; Kawamoto, T.; Scharffetter-Kochanek, K.; *et al.* Manganese superoxide dismutase and aldehyde dehydrogenase deficiency increase mitochondrial oxidative stress and aggravate age-dependent vascular dysfunction. *Cardiovasc. Res.* **2008**, *80*, 280–289.

24. Gerhard, M.; Roddy, M.A.; Creager, S.J.; Creager, M.A. Aging progressively impairs endothelium-dependent vasodilation in forearm resistance vessels of humans. *Hypertension* **1996**, *27*, 849–853.

25. Heitzer, T.; Schlinzig, T.; Krohn, K.; Meinertz, T.; Munzel, T. Endothelial dysfunction, oxidative stress, and risk of cardiovascular events in patients with coronary artery disease. *Circulation* **2001**, *104*, 2673–2678.

26. Savji, N.; Rockman, C.B.; Skolnick, A.H.; Guo, Y.; Adelman, M.A.; Riles, T.; Berger, J.S. Association between advanced age and vascular disease in different arterial territories: A population database of over 3.6 million subjects. *J. Am. Coll. Cardiol.* **2013**, *61*, 1736–1743.

27. Ong, K.L.; Cheung, B.M.; Man, Y.B.; Lau, C.P.; Lam, K.S. Prevalence, awareness, treatment, and control of hypertension among united states adults 1999–2004. *Hypertension* **2007**, *49*, 69–75.

28. Daiber, A.; Kienhoefer, J.; Zee, R.; Ullrich, V.; van der Loo, B.; Bachschmid, M. The role of mitochondrial reactive oxygen species formation for age-induced vascular dysfunction. In *Aging and Age-Related Disorders*; Bondy, S.C., Maiese, K., Eds.; Humana Press: Clifton, NJ, USA, 2010; pp. 237–257.

29. Cadenas, E.; Davies, K.J. Mitochondrial free radical generation, oxidative stress, and aging. *Free Radic. Biol. Med.* **2000**, *29*, 222–230.

30. Lenaz, G.; Bovina, C.; D'Aurelio, M.; Fato, R.; Formiggini, G.; Genova, M.L.; Giuliano, G.; Pich, M.M.; Paolucci, U.; Castelli, G.P.; *et al.* Role of mitochondria in oxidative stress and aging. *Ann. N. Y. Acad. Sci.* **2002**, *959*, 199–213.

31. Kujoth, G.C.; Hiona, A.; Pugh, T.D.; Someya, S.; Panzer, K.; Wohlgemuth, S.E.; Hofer, T.; Seo, A.Y.; Sullivan, R.; Jobling, W.A.; *et al.* Mitochondrial DNA mutations, oxidative stress, and apoptosis in mammalian aging. *Science* **2005**, *309*, 481–484.

32. Yin, F.; Boveris, A.; Cadenas, E. Mitochondrial energy metabolism and redox signaling in brain aging and neurodegeneration. *Antioxid. Redox Signal.* **2014**, *20*, 353–371.

33. Cheng, Z.J.; Vapaatalo, H.; Mervaala, E. Angiotensin II and vascular inflammation. *Med. Sci. Monit.* **2005**, *11*, RA194–RA205.

34. Lau, D.; Baldus, S. Myeloperoxidase and its contributory role in inflammatory vascular disease. *Pharmacol. Ther.* **2006**, *111*, 16–26.

35. Willerson, J.T.; Golino, P.; Eidt, J.; Campbell, W.B.; Buja, L.M. Specific platelet mediators and unstable coronary artery lesions. Experimental evidence and potential clinical implications. *Circulation* **1989**, *80*, 198–205.

36. Seals, D.R.; Jablonski, K.L.; Donato, A.J. Aging and vascular endothelial function in humans. *Clin. Sci.* **2011**, *120*, 357–375.

37. Tanaka, H.; Dinenno, F.A.; Seals, D.R. Age-related increase in femoral intima-media thickness in healthy humans. *Arterioscler. Thromb. Vasc. Biol.* **2000**, *20*, 2172.

38. Crimi, E.; Ignarro, L.J.; Napoli, C. Microcirculation and oxidative stress. *Free Radic. Res.* **2007**, *41*, 1364–1375.

39. Mayhan, W.G.; Arrick, D.M.; Sharpe, G.M.; Sun, H. Age-related alterations in reactivity of cerebral arterioles: Role of oxidative stress. *Microcirculation* **2008**, *15*, 225–236.

40. Militante, J.; Lombardini, J.B. Age-related retinal degeneration in animal models of aging: Possible involvement of taurine deficiency and oxidative stress. *Neurochem. Res.* **2004**, *29*, 151–160.

41. Fischer, R.; Maier, O. Interrelation of oxidative stress and inflammation in neurodegenerative disease: Role of TNF. *Oxidative Med. Cell. Longev.* **2015**, *2015*, 610813.

42. Blasiak, J.; Petrovski, G.; Vereb, Z.; Facsko, A.; Kaarniranta, K. Oxidative stress, hypoxia, and autophagy in the neovascular processes of age-related macular degeneration. *BioMed. Res. Int.* **2014**, *2014*, 768026.

43. Harman, D. Aging: A theory based on free radical and radiation chemistry. *J. Gerontol.* **1956**, *11*, 298–300.

44. Waters, W.A. Some recent developments in the chemistry of free radicals. *J. Chem. Soc.* **1946**, 409–415.

45. Rogell, B.; Dean, R.; Lemos, B.; Dowling, D.K. Mito-nuclear interactions as drivers of gene movement on and off the X-chromosome. *BMC Genomics* **2014**, *15*, 330.

46. Thomas, D.D.; Ridnour, L.A.; Isenberg, J.S.; Flores-Santana, W.; Switzer, C.H.; Donzelli, S.; Hussain, P.; Vecoli, C.; Paolocci, N.; Ambs, S.; *et al.* The chemical biology of nitric oxide: Implications in cellular signaling. *Free Radic. Biol. Med.* **2008**, *45*, 18–31.

47. Van der Loo, B.; Bachschmid, M.; Skepper, J.N.; Labugger, R.; Schildknecht, S.; Hahn, R.; Mussig, E.; Gygi, D.; Luscher, T.F. Age-associated cellular relocation of Sod 1 as a self-defense is a futile mechanism to prevent vascular aging. *Biochem. Biophys. Res. Commun.* **2006**, *344*, 972–980.

48. De Haan, J.B.; Bladier, C.; Lotfi-Miri, M.; Taylor, J.; Hutchinson, P.; Crack, P.J.; Hertzog, P.; Kola, I. Fibroblasts derived from Gpx1 knockout mice display senescent-like features and are susceptible to H_2O_2-mediated cell death. *Free Radic. Biol. Med.* **2004**, *36*, 53–64.

49. Altschmied, J.; Haendeler, J. Thioredoxin-1 and endothelial cell aging: Role in cardiovascular diseases. *Antioxid. Redox Signal.* **2009**, *11*, 1733–1740.

50. Go, Y.M.; Jones, D.P. Redox control systems in the nucleus: Mechanisms and functions. *Antioxid. Redox Signal.* **2010**, *13*, 489–509.

51. Salmon, A.B.; Richardson, A.; Perez, V.I. Update on the oxidative stress theory of aging: Does oxidative stress play a role in aging or healthy aging? *Free Radic. Biol. Med.* **2010**, *48*, 642–655.

52. Brown, K.A.; Didion, S.P.; Andresen, J.J.; Faraci, F.M. Effect of aging, MnSOD deficiency, and genetic background on endothelial function: Evidence for MnSOD haploinsufficiency. *Arterioscler. Thromb. Vasc. Biol.* **2007**, *27*, 1941–1946.

53. Didion, S.P.; Kinzenbaw, D.A.; Schrader, L.I.; Faraci, F.M. Heterozygous CuZn superoxide dismutase deficiency produces a vascular phenotype with aging. *Hypertension* **2006**, *48*, 1072–1079.

54. Goldstein, S.; Czapski, G.; Lind, J.; Merenyi, G. Tyrosine nitration by simultaneous generation of $^\bullet NO$ and $O^\bullet_2$ under physiological conditions. How the radicals do the job. *J. Biol. Chem.* **2000**, *275*, 3031–3036.

55. Perez, V.I.; Bokov, A.; van Remmen, H.; Mele, J.; Ran, Q.; Ikeno, Y.; Richardson, A. Is the oxidative stress theory of aging dead? *Biochim. Biophys. Acta* **2009**, *1790*, 1005–1014.

56. Muller, F.L.; Lustgarten, M.S.; Jang, Y.; Richardson, A.; van Remmen, H. Trends in oxidative aging theories. *Free Radic. Biol. Med.* **2007**, *43*, 477–503.

57. Lebovitz, R.M.; Zhang, H.; Vogel, H.; Cartwright, J., Jr.; Dionne, L.; Lu, N.; Huang, S.; Matzuk, M.M. Neurodegeneration, myocardial injury, and perinatal death in mitochondrial superoxide dismutase-deficient mice. *Proc. Natl. Acad. Sci. USA* **1996**, *93*, 9782–9787.

58. Li, Y.; Huang, T.T.; Carlson, E.J.; Melov, S.; Ursell, P.C.; Olson, J.L.; Noble, L.J.; Yoshimura, M.P.; Berger, C.; Chan, P.H.; *et al.* Dilated cardiomyopathy and neonatal lethality in mutant mice lacking manganese superoxide dismutase. *Nat. Genet.* **1995**, *11*, 376–381.

59. Jang, Y.C.; Remmen, H.V. The mitochondrial theory of aging: Insight from transgenic and knockout mouse models. *Exp. Gerontol.* **2009**, *44*, 256–260.

60. Dai, D.F.; Chiao, Y.A.; Marcinek, D.J.; Szeto, H.H.; Rabinovitch, P.S. Mitochondrial oxidative stress in aging and healthspan. *Longev. Healthspan* **2014**, *3*, 6.

61. Hamilton, R.T.; Walsh, M.E.; van Remmen, H. Mouse models of oxidative stress indicate a role for modulating healthy aging. *J. Clin. Exp. Pathol.* **2012**.

62. Berry, A.; Cirulli, F. The *p66Shc* gene paves the way for healthspan: Evolutionary and mechanistic perspectives. *Neurosci. Biobehav. Rev.* **2013**, *37*, 790–802.

63. Wanagat, J.; Dai, D.F.; Rabinovitch, P. Mitochondrial oxidative stress and mammalian healthspan. *Mech. Ageing Dev.* **2010**, *131*, 527–535.

64. Sies, H. Oxidative stress: A concept in redox biology and medicine. *Redox Biol.* **2015**, *4*, 180–183.

65. Beckman, J.S.; Koppenol, W.H. Nitric oxide, superoxide, and peroxynitrite: The good, the bad, and ugly. *Am. J. Physiol.* **1996**, *271*, C1424–C1437.

66. Daiber, A.; Bachschmid, M. Enzyme inhibition by peroxynitrite-mediated tyrosine nitration and thiol oxidation. *Curr. Enzym. Inhib.* **2007**, *3*, 103–117.

67. Beckman, J.S. Protein tyrosine nitration and peroxynitrite. *FASEB J.* **2002**, *16*, 1144.

68. Quijano, C.; Alvarez, B.; Gatti, R.M.; Augusto, O.; Radi, R. Pathways of peroxynitrite oxidation of thiol groups. *Biochem. J.* **1997**, *322*, 167–173.

69. MacMillan-Crow, L.A.; Crow, J.P.; Kerby, J.D.; Beckman, J.S.; Thompson, J.A. Nitration and inactivation of manganese superoxide dismutase in chronic rejection of human renal allografts. *Proc. Natl. Acad. Sci. USA* **1996**, *93*, 11853–11858.

70. Kuzkaya, N.; Weissmann, N.; Harrison, D.G.; Dikalov, S. Interactions of peroxynitrite, tetrahydrobiopterin, ascorbic acid, and thiols: Implications for uncoupling endothelial nitric-oxide synthase. *J. Biol. Chem.* **2003**, *278*, 22546–22554.

71. Schulz, E.; Jansen, T.; Wenzel, P.; Daiber, A.; Munzel, T. Nitric oxide, tetrahydrobiopterin, oxidative stress, and endothelial dysfunction in hypertension. *Antioxid. Redox Signal.* **2008**, *10*, 1115–1126.

72. Yoshida, Y.I.; Eda, S.; Masada, M. Alterations of tetrahydrobiopterin biosynthesis and pteridine levels in mouse tissues during growth and aging. *Brain Dev.* **2000**, *22*, S45–S49.

73. Blackwell, K.A.; Sorenson, J.P.; Richardson, D.M.; Smith, L.A.; Suda, O.; Nath, K.; Katusic, Z.S. Mechanisms of aging-induced impairment of endothelium-dependent relaxation: Role of tetrahydrobiopterin. *Am. J. Physiol. Heart Circ. Physiol.* **2004**, *287*, H2448–H2453.

74. Chen, C.A.; Wang, T.Y.; Varadharaj, S.; Reyes, L.A.; Hemann, C.; Talukder, M.A.; Chen, Y.R.; Druhan, L.J.; Zweier, J.L. *S*-glutathionylation uncouples eNOS and regulates its cellular and vascular function. *Nature* **2010**, *468*, 1115–1118.

75. Schulz, E.; Wenzel, P.; Munzel, T.; Daiber, A. Mitochondrial redox signaling: Interaction of mitochondrial reactive oxygen species with other sources of oxidative stress. *Antioxid. Redox Signal.* **2014**, *20*, 308–324.

76. Zou, M.H.; Shi, C.; Cohen, R.A. Oxidation of the zinc-thiolate complex and uncoupling of endothelial nitric oxide synthase by peroxynitrite. *J. Clin. Investig.* **2002**, *109*, 817–826.

77. Forstermann, U.; Munzel, T. Endothelial nitric oxide synthase in vascular disease: From marvel to menace. *Circulation* **2006**, *113*, 1708–1714.

78. Soucy, K.G.; Ryoo, S.; Benjo, A.; Lim, H.K.; Gupta, G.; Sohi, J.S.; Elser, J.; Aon, M.A.; Nyhan, D.; Shoukas, A.A.; *et al.* Impaired shear stress-induced nitric oxide production through decreased NOS phosphorylation contributes to age-related vascular stiffness. *J. Appl. Physiol.* **2006**, *101*, 1751–1759.

79. Cave, A.C.; Brewer, A.C.; Narayanapanicker, A.; Ray, R.; Grieve, D.J.; Walker, S.; Shah, A.M. Nadph oxidases in cardiovascular health and disease. *Antioxid. Redox signal.* **2006**, *8*, 691–728.

80. Griendling, K.K.; Sorescu, D.; Ushio-Fukai, M. NAD(P)H oxidase: Role in cardiovascular biology and disease. *Circ. Res.* **2000**, *86*, 494–501.

81. Paneni, F.; Osto, E.; Costantino, S.; Mateescu, B.; Briand, S.; Coppolino, G.; Perna, E.; Mocharla, P.; Akhmedov, A.; Kubant, R.; *et al.* Deletion of the activated protein-1 transcription factor JunD induces oxidative stress and accelerates age-related endothelial dysfunction. *Circulation* **2013**, *127*, 1229–1240.

82. Roubenoff, R.; Harris, T.B.; Abad, L.W.; Wilson, P.W.; Dallal, G.E.; Dinarello, C.A. Monocyte cytokine production in an elderly population: Effect of age and inflammation. *J. Gerontol. Ser. A* **1998**, *53*, M20–M26.

83. Moe, K.T.; Aulia, S.; Jiang, F.; Chua, Y.L.; Koh, T.H.; Wong, M.C.; Dusting, G.J. Differential upregulation of Nox homologues of NADPH oxidase by tumor necrosis factor-α in human aortic smooth muscle and embryonic kidney cells. *J. Cell. Mol. Med.* **2006**, *10*, 231–239.

84. Karbach, S.; Wenzel, P.; Waisman, A.; Munzel, T.; Daiber, A. eNOS uncoupling in cardiovascular diseases—The role of oxidative stress and inflammation. *Curr. Pharm. Des.* **2014**, *20*, 3579–3594.

85. Nandi, J.; Saud, B.; Zinkievich, J.M.; Yang, Z.J.; Levine, R.A. TNF-α modulates INOS expression in an experimental rat model of indomethacin-induced jejunoileitis. *Mol. Cell. Biochem.* **2010**, *336*, 17–24.

86. Ungvari, Z.; Csiszar, A.; Edwards, J.G.; Kaminski, P.M.; Wolin, M.S.; Kaley, G.; Koller, A. Increased superoxide production in coronary arteries in hyperhomocysteinemia: Role of tumor necrosis factor-α, NAD(P)H oxidase, and inducible nitric oxide synthase. *Arterioscler. Thromb. Vasc. Biol.* **2003**, *23*, 418–424.

87. Busik, J.V.; Mohr, S.; Grant, M.B. Hyperglycemia-induced reactive oxygen species toxicity to endothelial cells is dependent on paracrine mediators. *Diabetes* **2008**, *57*, 1952–1965.

88. Csiszar, A.; Labinskyy, N.; Smith, K.; Rivera, A.; Orosz, Z.; Ungvari, Z. Vasculoprotective effects of anti-tumor necrosis factor-α treatment in aging. *Am. J. Pathol.* **2007**, *170*, 388–398.

89. Ferrucci, L.; Corsi, A.; Lauretani, F.; Bandinelli, S.; Bartali, B.; Taub, D.D.; Guralnik, J.M.; Longo, D.L. The origins of age-related proinflammatory state. *Blood* **2005**, *105*, 2294–2299.

90. Wenzel, P.; Knorr, M.; Kossmann, S.; Stratmann, J.; Hausding, M.; Schuhmacher, S.; Karbach, S.H.; Schwenk, M.; Yogev, N.; Schulz, E.; *et al.* Lysozyme M-positive monocytes mediate angiotensin II-induced arterial hypertension and vascular dysfunction. *Circulation* **2011**, *124*, 1370–1381.

91. Guzik, T.J.; Hoch, N.E.; Brown, K.A.; McCann, L.A.; Rahman, A.; Dikalov, S.; Goronzy, J.; Weyand, C.; Harrison, D.G. Role of the T cell in the genesis of angiotensin II induced hypertension and vascular dysfunction. *J. Exp. Med.* **2007**, *204*, 2449–2460.

92. Harman, D. The biologic clock: The mitochondria? *J. Am. Geriatr. Soc.* **1972**, *20*, 145–147.

93. Lewis, P.; Stefanovic, N.; Pete, J.; Calkin, A.C.; Giunti, S.; Thallas-Bonke, V.; Jandeleit-Dahm, K.A.; Allen, T.J.; Kola, I.; Cooper, M.E.; *et al.* Lack of the antioxidant enzyme glutathione peroxidase-1 accelerates atherosclerosis in diabetic apolipoprotein e-deficient mice. *Circulation* **2007**, *115*, 2178–2187.

94. Forgione, M.A.; Cap, A.; Liao, R.; Moldovan, N.I.; Eberhardt, R.T.; Lim, C.C.; Jones, J.; Goldschmidt-Clermont, P.J.; Loscalzo, J. Heterozygous cellular glutathione peroxidase deficiency in the mouse: Abnormalities in vascular and cardiac function and structure. *Circulation* **2002**, *106*, 1154–1158.

95. Chrissobolis, S.; Didion, S.P.; Kinzenbaw, D.A.; Schrader, L.I.; Dayal, S.; Lentz, S.R.; Faraci, F.M. Glutathione peroxidase-1 plays a major role in protecting against angiotensin II-induced vascular dysfunction. *Hypertension* **2008**, *51*, 872–877.

96. Blankenberg, S.; Rupprecht, H.J.; Bickel, C.; Torzewski, M.; Hafner, G.; Tiret, L.; Smieja, M.; Cambien, F.; Meyer, J.; Lackner, K.J. Glutathione peroxidase 1 activity and cardiovascular events in patients with coronary artery disease. *N. Engl. J. Med.* **2003**, *349*, 1605–1613.

97. Kroller-Schon, S.; Steven, S.; Kossmann, S.; Scholz, A.; Daub, S.; Oelze, M.; Xia, N.; Hausding, M.; Mikhed, Y.; Zinssius, E.; *et al.* Molecular mechanisms of the crosstalk between mitochondria and NADPH oxidase through reactive oxygen species-studies in white blood cells and in animal models. *Antioxid. Redox Signal.* **2014**, *20*, 247–266.

98. Hausding, M.; Jurk, K.; Daub, S.; Kroller-Schon, S.; Stein, J.; Schwenk, M.; Oelze, M.; Mikhed, Y.; Kerahrodi, J.G.; Kossmann, S.; *et al.* CD40L contributes to angiotensin II-induced pro-thrombotic state, vascular inflammation, oxidative stress and endothelial dysfunction. *Basic Res. Cardiol.* **2013**, *108*, 386.

99. Fleming, I.; Fisslthaler, B.; Dimmeler, S.; Kemp, B.E.; Busse, R. Phosphorylation of Thr^{495} regulates Ca^{2+}/calmodulin-dependent endothelial nitric oxide synthase activity. *Circ. Res.* **2001**, *88*, E68–E75.

100. Lin, M.I.; Fulton, D.; Babbitt, R.; Fleming, I.; Busse, R.; Pritchard, K.A., Jr.; Sessa, W.C. Phosphorylation of threonine 497 in endothelial nitric-oxide synthase coordinates the coupling of L-arginine metabolism to efficient nitric oxide production. *J. Biol. Chem.* **2003**, *278*, 44719–44726.

101. Loot, A.E.; Schreiber, J.G.; Fisslthaler, B.; Fleming, I. Angiotensin ii impairs endothelial function via tyrosine phosphorylation of the endothelial nitric oxide synthase. *J. Exp. Med.* **2009**, *206*, 2889–2896.

102. Brune, B.; Schmidt, K.U.; Ullrich, V. Activation of soluble guanylate cyclase by carbon monoxide and inhibition by superoxide anion. *Eur. J. Biochem.* **1990**, *192*, 683–688.

103. Weber, M.; Lauer, N.; Mulsch, A.; Kojda, G. The effect of peroxynitrite on the catalytic activity of soluble guanylyl cyclase. *Free Radic. Biol. Med.* **2001**, *31*, 1360–1367.

104. Artz, J.D.; Schmidt, B.; McCracken, J.L.; Marletta, M.A. Effects of nitroglycerin on soluble guanylate cyclase: Implications for nitrate tolerance. *J. Biol. Chem.* **2002**, *277*, 18253–18256.

105. Crassous, P.A.; Couloubaly, S.; Huang, C.; Zhou, Z.; Baskaran, P.; Kim, D.D.; Papapetropoulos, A.; Fioramonti, X.; Duran, W.N.; Beuve, A. Soluble guanylyl cyclase is a target of angiotensin II-induced nitrosative stress in a hypertensive rat model. *Am. J. Physiol. Heart Circ. Physiol.* **2012**, *303*, H597–H604.

106. Mayer, B.; Kleschyov, A.L.; Stessel, H.; Russwurm, M.; Munzel, T.; Koesling, D.; Schmidt, K. Inactivation of soluble guanylate cyclase by stoichiometric S-nitrosation. *Mol. Pharmacol.* **2009**, *75*, 886–891.

107. Sayed, N.; Kim, D.D.; Fioramonti, X.; Iwahashi, T.; Duran, W.N.; Beuve, A. Nitroglycerin-induced S-nitrosylation and desensitization of soluble guanylyl cyclase contribute to nitrate tolerance. *Circ. Res.* **2008**, *103*, 606–614.

108. Stasch, J.P.; Schmidt, P.M.; Nedvetsky, P.I.; Nedvetskaya, T.Y.; H S, A.K.; Meurer, S.; Deile, M.; Taye, A.; Knorr, A.; Lapp, H.; *et al.* Targeting the heme-oxidized nitric oxide receptor for selective vasodilatation of diseased blood vessels. *J. Clin. Investig.* **2006**, *116*, 2552–2561.

109. Chen, L.; Daum, G.; Fischer, J.W.; Hawkins, S.; Bochaton-Piallat, M.L.; Gabbiani, G.; Clowes, A.W. Loss of expression of the β subunit of soluble guanylyl cyclase prevents nitric oxide-mediated inhibition of DNA synthesis in smooth muscle cells of old rats. *Circ. Res.* **2000**, *86*, 520–525.

110. Ruetten, H.; Zabel, U.; Linz, W.; Schmidt, H.H. Downregulation of soluble guanylyl cyclase in young and aging spontaneously hypertensive rats. *Circ. Res.* **1999**, *85*, 534–541.

111. Kloss, S.; Bouloumie, A.; Mulsch, A. Aging and chronic hypertension decrease expression of rat aortic soluble guanylyl cyclase. *Hypertension* **2000**, *35*, 43–47.

112. Lubos, E.; Mahoney, C.E.; Leopold, J.A.; Zhang, Y.Y.; Loscalzo, J.; Handy, D.E. Glutathione peroxidase-1 modulates lipopolysaccharide-induced adhesion molecule expression in endothelial cells by altering CD14 expression. *FASEB J.* **2010**, *24*, 2525–2532.

113. Kessler, P.; Bauersachs, J.; Busse, R.; Schini-Kerth, V.B. Inhibition of inducible nitric oxide synthase restores endothelium-dependent relaxations in proinflammatory mediator-induced blood vessels. *Arterioscler. Thromb. Vasc. Biol.* **1997**, *17*, 1746–1755.

114. Prabhu, K.S.; Zamamiri-Davis, F.; Stewart, J.B.; Thompson, J.T.; Sordillo, L.M.; Reddy, C.C. Selenium deficiency increases the expression of inducible nitric oxide synthase in RAW 264.7 macrophages: Role of nuclear factor-κB in up-regulation. *Biochem. J.* **2002**, *366*, 203–209.

115. Bulua, A.C.; Simon, A.; Maddipati, R.; Pelletier, M.; Park, H.; Kim, K.Y.; Sack, M.N.; Kastner, D.L.; Siegel, R.M. Mitochondrial reactive oxygen species promote production of proinflammatory cytokines and are elevated in TNFR1-associated periodic syndrome (TRAPS). *J. Exp. Med.* **2011**, *208*, 519–533.

116. West, A.P.; Brodsky, I.E.; Rahner, C.; Woo, D.K.; Erdjument-Bromage, H.; Tempst, P.; Walsh, M.C.; Choi, Y.; Shadel, G.S.; Ghosh, S. TLR signalling augments macrophage bactericidal activity through mitochondrial ROS. *Nature* **2011**, *472*, 476–480.

117. Zhou, R.; Yazdi, A.S.; Menu, P.; Tschopp, J. A role for mitochondria in NLRP3 inflammasome activation. *Nature* **2011**, *469*, 221–225.

118. Zhou, R.; Tardivel, A.; Thorens, B.; Choi, I.; Tschopp, J. Thioredoxin-interacting protein links oxidative stress to inflammasome activation. *Nat. Immunol.* **2010**, *11*, 136–140.

119. Smith, A.R.; Visioli, F.; Frei, B.; Hagen, T.M. Age-related changes in endothelial nitric oxide synthase phosphorylation and nitric oxide dependent vasodilation: Evidence for a novel mechanism involving sphingomyelinase and ceramide-activated phosphatase 2A. *Aging Cell* **2006**, *5*, 391–400.

120. Crabtree, M.J.; Brixey, R.; Batchelor, H.; Hale, A.B.; Channon, K.M. Integrated redox sensor and effector functions for tetrahydrobiopterin- and glutathionylation-dependent endothelial nitric-oxide synthase uncoupling. *J. Biol. Chem.* **2013**, *288*, 561–569.

121. Schriner, S.E.; Linford, N.J.; Martin, G.M.; Treuting, P.; Ogburn, C.E.; Emond, M.; Coskun, P.E.; Ladiges, W.; Wolf, N.; van Remmen, H.; *et al.* Extension of murine life span by overexpression of catalase targeted to mitochondria. *Science* **2005**, *308*, 1909–1911.

122. Barja, G.; Herrero, A. Oxidative damage to mitochondrial DNA is inversely related to maximum life span in the heart and brain of mammals. *FASEB J.* **2000**, *14*, 312–318.

123. Sastre, J.; Pallardo, F.V.; Vina, J. The role of mitochondrial oxidative stress in aging. *Free Radic. Biol. Med.* **2003**, *35*, 1–8.

124. Madamanchi, N.R.; Runge, M.S. Mitochondrial dysfunction in atherosclerosis. *Circ. Res.* **2007**, *100*, 460–473.

125. De Souza-Pinto, N.C.; Eide, L.; Hogue, B.A.; Thybo, T.; Stevnsner, T.; Seeberg, E.; Klungland, A.; Bohr, V.A. Repair of 8-oxodeoxyguanosine lesions in mitochondrial DNA depends on the oxoguanine DNA glycosylase (OGG1) gene and 8-oxoguanine accumulates in the mitochondrial dna of OGG1-defective mice. *Cancer Res.* **2001**, *61*, 5378–5381.

126. De Souza-Pinto, N.C.; Hogue, B.A.; Bohr, V.A. DNA repair and aging in mouse liver: 8-oxodG glycosylase activity increase in mitochondrial but not in nuclear extracts. *Free Radic. Biol. Med.* **2001**, *30*, 916–923.

127. Souza-Pinto, N.C.; Croteau, D.L.; Hudson, E.K.; Hansford, R.G.; Bohr, V.A. Age-associated increase in 8-oxo-deoxyguanosine glycosylase/ap lyase activity in rat mitochondria. *Nucleic Acids Res.* **1999**, *27*, 1935–1942.

128. Trifunovic, A.; Wredenberg, A.; Falkenberg, M.; Spelbrink, J.N.; Rovio, A.T.; Bruder, C.E.; Bohlooly, Y.M.; Gidlof, S.; Oldfors, A.; Wibom, R.; *et al.* Premature ageing in mice expressing defective mitochondrial DNA polymerase. *Nature* **2004**, *429*, 417–423.

129. Lewis, W.; Day, B.J.; Kohler, J.J.; Hosseini, S.H.; Chan, S.S.; Green, E.C.; Haase, C.P.; Keebaugh, E.S.; Long, R.; Ludaway, T.; *et al.* Decreased mtDNA, oxidative stress, cardiomyopathy, and death from transgenic cardiac targeted human mutant polymerase γ. *Lab. Investig.* **2007**, *87*, 326–335.

130. Finsterer, J. Overview on visceral manifestations of mitochondrial disorders. *Neth. J. Med.* **2006**, *64*, 61–71.

131. Anan, R.; Nakagawa, M.; Miyata, M.; Higuchi, I.; Nakao, S.; Suehara, M.; Osame, M.; Tanaka, H. Cardiac involvement in mitochondrial diseases. A study on 17 patients with documented mitochondrial DNA defects. *Circulation* **1995**, *91*, 955–961.

132. Pinsky, D.J.; Oz, M.C.; Koga, S.; Taha, Z.; Broekman, M.J.; Marcus, A.J.; Liao, H.; Naka, Y.; Brett, J.; Cannon, P.J.; *et al.* Cardiac preservation is enhanced in a heterotopic rat transplant model by supplementing the nitric oxide pathway. *J. Clin. Investig.* **1994**, *93*, 2291–2297.

133. Zeviani, M.; di Donato, S. Mitochondrial disorders. *Brain* **2004**, *127*, 2153–2172.

134. Ballinger, S.W.; Patterson, C.; Knight-Lozano, C.A.; Burow, D.L.; Conklin, C.A.; Hu, Z.; Reuf, J.; Horaist, C.; Lebovitz, R.; Hunter, G.C.; *et al.* Mitochondrial integrity and function in atherogenesis. *Circulation* **2002**, *106*, 544–549.

135. Sevini, F.; Giuliani, C.; Vianello, D.; Giampieri, E.; Santoro, A.; Biondi, F.; Garagnani, P.; Passarino, G.; Luiselli, D.; Capri, M.; *et al.* mtDNA mutations in human aging and longevity: Controversies and new perspectives opened by high-throughput technologies. *Exp. Gerontol.* **2014**, *56*, 234–244.

136. Itsara, L.S.; Kennedy, S.R.; Fox, E.J.; Yu, S.; Hewitt, J.J.; Sanchez-Contreras, M.; Cardozo-Pelaez, F.; Pallanck, L.J. Oxidative stress is not a major contributor to somatic mitochondrial DNA mutations. *PLoS Genet.* **2014**, *10*, e1003974.

137. Larsson, N.G. Somatic mitochondrial DNA mutations in mammalian aging. *Annu. Rev. Biochem.* **2010**, *79*, 683–706.

138. De Bont, R.; van Larebeke, N. Endogenous DNA damage in humans: A review of quantitative data. *Mutagenesis* **2004**, *19*, 169–185.

139. Kennedy, S.R.; Salk, J.J.; Schmitt, M.W.; Loeb, L.A. Ultra-sensitive sequencing reveals an age-related increase in somatic mitochondrial mutations that are inconsistent with oxidative damage. *PLoS Genet.* **2013**, *9*, e1003794.

140. Cha, M.Y.; Kim, D.K.; Mook-Jung, I. The role of mitochondrial DNA mutation on neurodegenerative diseases. *Exp. Mol. Med.* **2015**, *47*, e150.

141. Reddy, P.H. Amyloid β, mitochondrial structural and functional dynamics in alzheimer's disease. *Exp. Neurol.* **2009**, *218*, 286–292.

142. Canugovi, C.; Shamanna, R.A.; Croteau, D.L.; Bohr, V.A. Base excision DNA repair levels in mitochondrial lysates of alzheimer's disease. *Neurobiol. Aging* **2014**, *35*, 1293–1300.

143. Krishnan, K.J.; Ratnaike, T.E.; de Gruyter, H.L.; Jaros, E.; Turnbull, D.M. Mitochondrial DNA deletions cause the biochemical defect observed in alzheimer's disease. *Neurobiol. Aging* **2012**, *33*, 2210–2214.

144. Muftuoglu, M.; Mori, M.P.; de Souza-Pinto, N.C. Formation and repair of oxidative damage in the mitochondrial DNA. *Mitochondrion* **2014**, *17*, 164–181.

145. Liu, P.; Demple, B. DNA repair in mammalian mitochondria: Much more than we thought? *Environ. Mol. Mutagen.* **2010**, *51*, 417–426.

146. Bogenhagen, D.F. Repair of mtDNA in vertebrates. *Am. J. Hum. Genet.* **1999**, *64*, 1276–1281.

147. Nishioka, K.; Ohtsubo, T.; Oda, H.; Fujiwara, T.; Kang, D.; Sugimachi, K.; Nakabeppu, Y. Expression and differential intracellular localization of two major forms of human 8-oxoguanine DNA glycosylase encoded by alternatively spliced OGG1 mRNAs. *Mol. Biol. Cell* **1999**, *10*, 1637–1652.

148. Nilsen, H.; Otterlei, M.; Haug, T.; Solum, K.; Nagelhus, T.A.; Skorpen, F.; Krokan, H.E. Nuclear and mitochondrial uracil-DNA glycosylases are generated by alternative splicing and transcription from different positions in the UNG gene. *Nucleic Acids Res.* **1997**, *25*, 750–755.

149. Ikeda, S.; Kohmoto, T.; Tabata, R.; Seki, Y. Differential intracellular localization of the human and mouse endonuclease III homologs and analysis of the sorting signals. *DNA Repair* **2002**, *1*, 847–854.

150. Hu, J.; de Souza-Pinto, N.C.; Haraguchi, K.; Hogue, B.A.; Jaruga, P.; Greenberg, M.M.; Dizdaroglu, M.; Bohr, V.A. Repair of formamidopyrimidines in DNA involves different glycosylases: Role of the OGG1, NTH1, and NEIL1 enzymes. *J. Biol. Chem.* **2005**, *280*, 40544–40551.

151. Ohtsubo, T.; Nishioka, K.; Imaiso, Y.; Iwai, S.; Shimokawa, H.; Oda, H.; Fujiwara, T.; Nakabeppu, Y. Identification of human muty homolog (hMYH) as a repair enzyme for 2-hydroxyadenine in DNA and detection of multiple forms of hMYH located in nuclei and mitochondria. *Nucleic Acids Res.* **2000**, *28*, 1355–1364.

152. Park, J.S.; Kim, H.L.; Kim, Y.J.; Weon, J.I.; Sung, M.K.; Chung, H.W.; Seo, Y.R. Human AP endonuclease 1: A potential marker for the prediction of environmental carcinogenesis risk. *Oxidative Med. Cell. Longev.* **2014**, *2014*, 730301.

153. Mason, P.A.; Matheson, E.C.; Hall, A.G.; Lightowlers, R.N. Mismatch repair activity in mammalian mitochondria. *Nucleic Acids Res.* **2003**, *31*, 1052–1058.

154. De Souza-Pinto, N.C.; Mason, P.A.; Hashiguchi, K.; Weissman, L.; Tian, J.; Guay, D.; Lebel, M.; Stevnsner, T.V.; Rasmussen, L.J.; Bohr, V.A. Novel DNA mismatch-repair activity involving YB-1 in human mitochondria. *DNA Repair* **2009**, *8*, 704–719.

155. Pohjoismaki, J.L.; Goffart, S.; Tyynismaa, H.; Willcox, S.; Ide, T.; Kang, D.; Suomalainen, A.; Karhunen, P.J.; Griffith, J.D.; Holt, I.J.; *et al.* Human heart mitochondrial DNA is organized in complex catenated networks containing abundant four-way junctions and replication forks. *J. Biol. Chem.* **2009**, *284*, 21446–21457.

156. Chen, X.J. Mechanism of homologous recombination and implications for aging-related deletions in mitochondrial DNA. *Microbiol. Mol. Biol. Rev.* **2013**, *77*, 476–496.

157. Chen, X.J.; Guan, M.X.; Clark-Walker, G.D. MGM101, a nuclear gene involved in maintenance of the mitochondrial genome in saccharomyces cerevisiae. *Nucleic Acids Res.* **1993**, *21*, 3473–3477.

158. Gredilla, R.; Garm, C.; Stevnsner, T. Nuclear and mitochondrial DNA repair in selected eukaryotic aging model systems. *Oxidative Med. Cell. Longev.* **2012**, *2012*, 282438.

159. Bender, A.; Krishnan, K.J.; Morris, C.M.; Taylor, G.A.; Reeve, A.K.; Perry, R.H.; Jaros, E.; Hersheson, J.S.; Betts, J.; Klopstock, T.; *et al.* High levels of mitochondrial DNA deletions in substantia nigra neurons in aging and Parkinson disease. *Nat. Genet.* **2006**, *38*, 515–517.

160. Chen, D.; Cao, G.; Hastings, T.; Feng, Y.; Pei, W.; O'Horo, C.; Chen, J. Age-dependent decline of DNA repair activity for oxidative lesions in rat brain mitochondria. *J. Neurochem.* **2002**, *81*, 1273–1284.

161. Daiber, A.; Oelze, M.; Wenzel, P.; Wickramanayake, J.M.; Schuhmacher, S.; Jansen, T.; Lackner, K.J.; Torzewski, M.; Munzel, T. Nitrate tolerance as a model of vascular dysfunction: Roles for mitochondrial aldehyde dehydrogenase and mitochondrial oxidative stress. *Pharmacol. Rep.* **2009**, *61*, 33–48.

162. Daiber, A.; Oelze, M.; Coldewey, M.; Bachschmid, M.; Wenzel, P.; Sydow, K.; Wendt, M.; Kleschyov, A.L.; Stalleicken, D.; Ullrich, V.; *et al.* Oxidative stress and mitochondrial aldehyde dehydrogenase activity: A comparison of pentaerythritol tetranitrate with other organic nitrates. *Mol. Pharmacol.* **2004**, *66*, 1372–1382.

163. Sydow, K.; Daiber, A.; Oelze, M.; Chen, Z.; August, M.; Wendt, M.; Ullrich, V.; Mulsch, A.; Schulz, E.; Keaney, J.F., Jr.; *et al.* Central role of mitochondrial aldehyde dehydrogenase and reactive oxygen species in nitroglycerin tolerance and cross-tolerance. *J. Clin. Investig.* **2004**, *113*, 482–489.

164. Esplugues, J.V.; Rocha, M.; Nunez, C.; Bosca, I.; Ibiza, S.; Herance, J.R.; Ortega, A.; Serrador, J.M.; D'Ocon, P.; Victor, V.M. Complex I dysfunction and tolerance to nitroglycerin: An approach based on mitochondrial-targeted antioxidants. *Circ. Res.* **2006**, *99*, 1067–1075.

165. Wenzel, P.; Mollnau, H.; Oelze, M.; Schulz, E.; Wickramanayake, J.M.; Muller, J.; Schuhmacher, S.; Hortmann, M.; Baldus, S.; Gori, T.; *et al.* First evidence for a crosstalk between mitochondrial and nadph oxidase-derived reactive oxygen species in nitroglycerin-triggered vascular dysfunction. *Antioxid. Redox Signal.* **2008**, *10*, 1435–1447.

166. Dikalov, S.I.; Nazarewicz, R.R.; Bikineyeva, A.; Hilenski, L.; Lassegue, B.; Griendling, K.K.; Harrison, D.G.; Dikalova, A.E. Nox2-induced production of mitochondrial superoxide in angiotensin II-mediated endothelial oxidative stress and hypertension. *Antioxid. Redox Signal.* **2014**, *20*, 281–294.

167. Doughan, A.K.; Harrison, D.G.; Dikalov, S.I. Molecular mechanisms of angiotensin II-mediated mitochondrial dysfunction: Linking mitochondrial oxidative damage and vascular endothelial dysfunction. *Circ. Res.* **2008**, *102*, 488–496.

168. Nazarewicz, R.R.; Dikalova, A.E.; Bikineyeva, A.; Dikalov, S.I. Nox2 as a potential target of mitochondrial superoxide and its role in endothelial oxidative stress. *Am. J. Physiol. Heart Circ. Physiol.* **2013**, *305*, H1131–H1140.

169. Brandes, R.P. Triggering mitochondrial radical release: A new function for NADPH oxidases. *Hypertension* **2005**, *45*, 847–848.

170. Kimura, S.; Zhang, G.X.; Nishiyama, A.; Shokoji, T.; Yao, L.; Fan, Y.Y.; Rahman, M.; Abe, Y. Mitochondria-derived reactive oxygen species and vascular MAP kinases: Comparison of angiotensin II and diazoxide. *Hypertension* **2005**, *45*, 438–444.

171. Dikalova, A.E.; Bikineyeva, A.T.; Budzyn, K.; Nazarewicz, R.R.; McCann, L.; Lewis, W.; Harrison, D.G.; Dikalov, S.I. Therapeutic targeting of mitochondrial superoxide in hypertension. *Circ. Res.* **2010**, *107*, 106–116.

172. Fukui, T.; Ishizaka, N.; Rajagopalan, S.; Laursen, J.B.; Capers, Q.T.; Taylor, W.R.; Harrison, D.G.; de Leon, H.; Wilcox, J.N.; Griendling, K.K. p22phox mRNA expression and NADPH oxidase activity are increased in aortas from hypertensive rats. *Circ. Res.* **1997**, *80*, 45–51.

173. Cheresh, P.; Kim, S.J.; Tulasiram, S.; Kamp, D.W. Oxidative stress and pulmonary fibrosis. *Biochim. Biophys. Acta* **2013**, *1832*, 1028–1040.

174. Nazarewicz, R.R.; Dikalov, S.I. Mitochondrial ROS in the prohypertensive immune response. *Am. J. Physiol. Regul. Integr. Comp. Physiol.* **2013**, *305*, R98–R100.

175. Bender, A.; Hajieva, P.; Moosmann, B. Adaptive antioxidant methionine accumulation in respiratory chain complexes explains the use of a deviant genetic code in mitochondria. *Proc. Natl. Acad. Sci. USA* **2008**, *105*, 16496–16501.

176. Moskovitz, J.; Bar-Noy, S.; Williams, W.M.; Requena, J.; Berlett, B.S.; Stadtman, E.R. Methionine sulfoxide reductase (MsrA) is a regulator of antioxidant defense and lifespan in mammals. *Proc. Natl. Acad. Sci. USA* **2001**, *98*, 12920–12925.

177. Stadtman, E.R.; Moskovitz, J.; Berlett, B.S.; Levine, R.L. Cyclic oxidation and reduction of protein methionine residues is an important antioxidant mechanism. *Mol. Cell. Biochem.* **2002**, *234–235*, 3–9.

178. Taungjaruwinai, W.M.; Bhawan, J.; Keady, M.; Thiele, J.J. Differential expression of the antioxidant repair enzyme methionine sulfoxide reductase (MSRA and MSRB) in human skin. *Am. J. Dermatopathol.* **2009**, *31*, 427–431.

179. Mochin, M.T.; Underwood, K.F.; Cooper, B.; McLenithan, J.C.; Pierce, A.D.; Nalvarte, C.; Arbiser, J.; Karlsson, A.I.; Moise, A.R.; Moskovitz, J.; *et al.* Hyperglycemia and redox status regulate RUNX2 DNA-binding and an angiogenic phenotype in endothelial cells. *Microvasc. Res.* **2015**, *97*, 55–64.

180. Gu, H.; Chen, W.; Yin, J.; Chen, S.; Zhang, J.; Gong, J. Methionine sulfoxide reductase A rs10903323 G/A polymorphism is associated with increased risk of coronary artery disease in a chinese population. *Clin. Biochem.* **2013**, *46*, 1668–1672.
181. Haenold, R.; Wassef, R.; Brot, N.; Neugebauer, S.; Leipold, E.; Heinemann, S.H.; Hoshi, T. Protection of vascular smooth muscle cells by over-expressed methionine sulphoxide reductase A: Role of intracellular localization and substrate availability. *Free Radic. Res.* **2008**, *42*, 978–988.
182. Moosmann, B.; Behl, C. Mitochondrially encoded cysteine predicts animal lifespan. *Aging Cell* **2008**, *7*, 32–46.
183. Camici, G.G.; Cosentino, F.; Tanner, F.C.; Luscher, T.F. The role of p66Shc deletion in age-associated arterial dysfunction and disease states. *J. Appl. Physiol.* **2008**, *105*, 1628–1631.
184. Francia, P.; delli Gatti, C.; Bachschmid, M.; Martin-Padura, I.; Savoia, C.; Migliaccio, E.; Pelicci, P.G.; Schiavoni, M.; Luscher, T.F.; Volpe, M.; *et al.* Deletion of p66Shc gene protects against age-related endothelial dysfunction. *Circulation* **2004**, *110*, 2889–2895.
185. Gertz, M.; Fischer, F.; Wolters, D.; Steegborn, C. Activation of the lifespan regulator p66Shc through reversible disulfide bond formation. *Proc. Natl. Acad. Sci. USA* **2008**, *105*, 5705–5709.
186. Pinton, P.; Rimessi, A.; Marchi, S.; Orsini, F.; Migliaccio, E.; Giorgio, M.; Contursi, C.; Minucci, S.; Mantovani, F.; Wieckowski, M.R.; *et al.* Protein kinase C β and prolyl isomerase 1 regulate mitochondrial effects of the life-span determinant p66Shc. *Science* **2007**, *315*, 659–663.
187. Pinton, P.; Rizzuto, R. P66Shc, oxidative stress and aging: Importing a lifespan determinant into mitochondria. *Cell. Cycle* **2008**, *7*, 304–308.
188. Rota, M.; LeCapitaine, N.; Hosoda, T.; Boni, A.; De Angelis, A.; Padin-Iruegas, M.E.; Esposito, G.; Vitale, S.; Urbanek, K.; Casarsa, C.; *et al.* Diabetes promotes cardiac stem cell aging and heart failure, which are prevented by deletion of the p66Shc gene. *Circ. Res.* **2006**, *99*, 42–52.
189. Yamamori, T.; White, A.R.; Mattagajasingh, I.; Khanday, F.A.; Haile, A.; Qi, B.; Jeon, B.H.; Bugayenko, A.; Kasuno, K.; Berkowitz, D.E.; *et al.* p66Shc regulates endothelial no production and endothelium-dependent vasorelaxation: Implications for age-associated vascular dysfunction. *J. Mol. Cell. Cardiol.* **2005**, *39*, 992–995.
190. Trinei, M.; Giorgio, M.; Cicalese, A.; Barozzi, S.; Ventura, A.; Migliaccio, E.; Milia, E.; Padura, I.M.; Raker, V.A.; Maccarana, M.; *et al.* A p53-p66Shc signalling pathway controls intracellular redox status, levels of oxidation-damaged DNA and oxidative stress-induced apoptosis. *Oncogene* **2002**, *21*, 3872–3878.
191. Di Lisa, F.; Kaludercic, N.; Carpi, A.; Menabo, R.; Giorgio, M. Mitochondrial pathways for ROS formation and myocardial injury: The relevance of p66Shc and monoamine oxidase. *Basic Res. Cardiol.* **2009**, *104*, 131–139.

192. Spescha, R.D.; Klohs, J.; Semerano, A.; Giacalone, G.; Derungs, R.S.; Reiner, M.F.; Rodriguez Gutierrez, D.; Mendez-Carmona, N.; Glanzmann, M.; Savarese, G.; et al. Post-ischaemic silencing of p66Shc reduces ischaemia/reperfusion brain injury and its expression correlates to clinical outcome in stroke. *Eur. Heart J.* **2015**, *36*, 1590–1600.

193. Moskalev, A.A.; Aliper, A.M.; Smit-McBride, Z.; Buzdin, A.; Zhavoronkov, A. Genetics and epigenetics of aging and longevity. *Cell. Cycle* **2014**, *13*, 1063–1077.

194. Barbot, W.; Dupressoir, A.; Lazar, V.; Heidmann, T. Epigenetic regulation of an IAP retrotransposon in the aging mouse: Progressive demethylation and de-silencing of the element by its repetitive induction. *Nucleic Acids Res.* **2002**, *30*, 2365–2373.

195. Narita, M.; Nunez, S.; Heard, E.; Narita, M.; Lin, A.W.; Hearn, S.A.; Spector, D.L.; Hannon, G.J.; Lowe, S.W. Rb-mediated heterochromatin formation and silencing of E2F target genes during cellular senescence. *Cell.* **2003**, *113*, 703–716.

196. Tsurumi, A.; Li, W.X. Global heterochromatin loss: A unifying theory of aging? *Epigenetics* **2012**, *7*, 680–688.

197. McCauley, B.S.; Dang, W. Histone methylation and aging: Lessons learned from model systems. *Biochim. Biophys. Acta* **2014**, *1839*, 1454–1462.

198. Bilsland, A.E.; Revie, J.; Keith, W. Microrna and senescence: The senectome, integration and distributed control. *Crit. Rev. Oncog.* **2013**, *18*, 373–390.

199. Boon, R.A.; Iekushi, K.; Lechner, S.; Seeger, T.; Fischer, A.; Heydt, S.; Kaluza, D.; Treguer, K.; Carmona, G.; Bonauer, A.; et al. MicroRNA-34a regulates cardiac ageing and function. *Nature* **2013**, *495*, 107–110.

200. Guarner, V.; Rubio-Ruiz, M.E. Low-grade systemic inflammation connects aging, metabolic syndrome and cardiovascular disease. *Interdiscip. Top. Gerontol.* **2015**, *40*, 99–106.

201. Howcroft, T.K.; Campisi, J.; Louis, G.B.; Smith, M.T.; Wise, B.; Wyss-Coray, T.; Augustine, A.D.; McElhaney, J.E.; Kohanski, R.; Sierra, F. The role of inflammation in age-related disease. *Aging* **2013**, *5*, 84–93.

202. Martin-Montalvo, A.; Mercken, E.M.; Mitchell, S.J.; Palacios, H.H.; Mote, P.L.; Scheibye-Knudsen, M.; Gomes, A.P.; Ward, T.M.; Minor, R.K.; Blouin, M.J.; et al. Metformin improves healthspan and lifespan in mice. *Nat. Commun.* **2013**, *4*, 2192.

203. Wilson, P.W. Established risk factors and coronary artery disease: The framingham study. *Am. J. Hypertens.* **1994**, *7*, 7S–12S.

204. Munzel, T.; Sinning, C.; Post, F.; Warnholtz, A.; Schulz, E. Pathophysiology, diagnosis and prognostic implications of endothelial dysfunction. *Ann. Med.* **2008**, *40*, 180–196.

205. Munzel, T.; Gori, T.; Bruno, R.M.; Taddei, S. Is oxidative stress a therapeutic target in cardiovascular disease? *Eur. Heart J.* **2010**, *31*, 2741–2748.

206. Schachinger, V.; Britten, M.B.; Zeiher, A.M. Prognostic impact of coronary vasodilator dysfunction on adverse long-term outcome of coronary heart disease. *Circulation* **2000**, *101*, 1899–1906.

207. Gori, T.; Munzel, T. Oxidative stress and endothelial dysfunction: Therapeutic implications. *Ann. Med.* **2011**, *43*, 259–272.

208. Chen, A.F.; Chen, D.D.; Daiber, A.; Faraci, F.M.; Li, H.; Rembold, C.M.; Laher, I. Free radical biology of the cardiovascular system. *Clin. Sci.* **2012**, *123*, 73–91.

209. Kimura, Y.; Matsumoto, M.; Den, Y.B.; Iwai, K.; Munehira, J.; Hattori, H.; Hoshino, T.; Yamada, K.; Kawanishi, K.; Tsuchiya, H. Impaired endothelial function in hypertensive elderly patients evaluated by high resolution ultrasonography. *Can. J. Cardiol.* **1999**, *15*, 563–568.

210. Wray, D.W.; Nishiyama, S.K.; Harris, R.A.; Zhao, J.; McDaniel, J.; Fjeldstad, A.S.; Witman, M.A.; Ives, S.J.; Barrett-O'Keefe, Z.; Richardson, R.S. Acute reversal of endothelial dysfunction in the elderly after antioxidant consumption. *Hypertension* **2012**, *59*, 818–824.

211. Jousilahti, P.; Vartiainen, E.; Tuomilehto, J.; Puska, P. Sex, age, cardiovascular risk factors, and coronary heart disease: A prospective follow-up study of 14 786 middle-aged men and women in Finland. *Circulation* **1999**, *99*, 1165–1172.

212. Holzer, M.; Trieb, M.; Konya, V.; Wadsack, C.; Heinemann, A.; Marsche, G. Aging affects high-density lipoprotein composition and function. *Biochim. Biophys. Acta* **2013**, *1831*, 1442–1448.

213. Besler, C.; Heinrich, K.; Riwanto, M.; Luscher, T.F.; Landmesser, U. High-density lipoprotein-mediated anti-atherosclerotic and endothelial-protective effects: A potential novel therapeutic target in cardiovascular disease. *Curr. Pharm. Des.* **2010**, *16*, 1480–1493.

214. Wu, Z.; Siuda, D.; Xia, N.; Reifenberg, G.; Daiber, A.; Munzel, T.; Forstermann, U.; Li, H. Maternal treatment of spontaneously hypertensive rats with pentaerythritol tetranitrate reduces blood pressure in female offspring. *Hypertension* **2015**, *65*, 232–237.

215. Gioscia-Ryan, R.A.; LaRocca, T.J.; Sindler, A.L.; Zigler, M.C.; Murphy, M.P.; Seals, D.R. Mitochondria-targeted antioxidant (MitoQ) ameliorates age-related arterial endothelial dysfunction in mice. *J. Physiol.* **2014**, *592*, 2549–2561.

216. Klipstein-Grobusch, K.; Geleijnse, J.M.; den Breeijen, J.H.; Boeing, H.; Hofman, A.; Grobbee, D.E.; Witteman, J.C. Dietary antioxidants and risk of myocardial infarction in the elderly: The rotterdam study. *Am. J. Clin. Nutr.* **1999**, *69*, 261–266.

Thyroid Hormone Mediated Modulation of Energy Expenditure

Janina A. Vaitkus, Jared S. Farrar and Francesco S. Celi

Abstract: Thyroid hormone (TH) has diverse effects on mitochondria and energy expenditure (EE), generating great interest and research effort into understanding and harnessing these actions for the amelioration and treatment of metabolic disorders, such as obesity and diabetes. Direct effects on ATP utilization are a result of TH's actions on metabolic cycles and increased cell membrane ion permeability. However, the majority of TH induced EE is thought to be a result of indirect effects, which, in turn, increase capacity for EE. This review discusses the direct actions of TH on EE, and places special emphasis on the indirect actions of TH, which include mitochondrial biogenesis and reduced metabolic efficiency through mitochondrial uncoupling mechanisms. TH analogs and the metabolic actions of T2 are also discussed in the context of targeted modulation of EE. Finally, clinical correlates of TH actions on metabolism are briefly presented.

Reprinted from *Int. J. Mol. Sci.* Cite as: Vaitkus, J.A.; Farrar, J.S.; Celi, F.S. Thyroid Hormone Mediated Modulation of Energy Expenditure. *Int. J. Mol. Sci.* **2015**, *16*, 16158–16171.

1. Introduction

The maintenance of life is dependent on the metabolism of substrates in the form of carbohydrates, fats, and proteins to provide energy, and in the form of ATP to assure cell integrity and functions. Although in humans the day-to-day variations in energy flux are dramatic, over time, the dynamic equilibrium between energy intake (EI) and energy expenditure (EE) is remarkable. Indeed, a small but sustained imbalance between EE and EI can lead to dramatic and severe clinical presentations, such as obesity or cachexia, both of which represent life-limiting conditions [1,2]. A variety of biochemical pathways are involved in energy metabolism, but in its broadest sense, the common requirement is chemical energy. Basal EE, otherwise defined as resting energy expenditure (REE), is the energy required to maintain basic cell and organ functions. Total EE (TEE) is defined as REE plus the energy consumed during activity (activity EE (AEE)) and diet-induced thermogenesis (DIT), the energy used to metabolize substrates above and beyond the requirements of intestinal tract mobility and absorption [3]. It is important to note that TEE is not static, as REE, AEE, and DIT are all variable and modifiable by a variety of factors. While there are several modulators of REE, and therefore overall EE, the focus of this review will be on thyroid hormone (TH) and its mechanisms of action, particularly on mitochondria. Following the complex integration of various afferent metabolic signals to the hypothalamus [4], TH releasing hormone (TRH) prompts the pituitary gland to secrete thyroid-stimulating hormone (TSH), which in turn activates the thyroid gland to produce and secrete TH [5]. In humans, this is mostly in the form of tetraiodothyronine (also referred to as thyroxine, T4), and to some degree, triiodothyronine (T3) [5]. T4 is then converted into T3 by deiodinase enzymes [5,6], which allow for time- and tissue-specific pre-receptor modulation of the hormonal signal. Most T4 and T3 are bound to thyroxine binding globulin (TBG) and other carriers in circulation, and only unbound or "free" TH exerts biological effects [7]. For the purposes of this review, TH will refer to T3 and T4, while other forms, referred to as TH analogs and "non-classical" THs, will be discussed later.

The critical role of TH in EE modulation has been known for more than a century, starting with the groundbreaking work of Magnus-Levy in 1895 (summarized in [8]). However, each specific mechanism, and in particular their regulatory systems, have yet to be fully elucidated. This review will discuss the developments in knowledge in this area, specifically regarding TH's role in modulating EE.

2. Direct Effects

Direct effects refer to TH actions that inherently cause an increase in ATP utilization. In general, these actions can be further classified into those that are related to metabolic cycles, and those that are related to ion leaks.

2.1. Metabolic Cycles

Metabolic cycles, also referred to as substrate or futile cycles, are the combination of two or more reactions which act in a cyclical manner; for a two reaction cycle, the reactions operate in reverse under the control of separate enzymes [9]. In the process of these reactions occurring, ATP is utilized, yet no product is consumed due to the cyclical nature of the products and reactants (hence the designation as a *futile cycle*). Examples of these cycles on the enzymatic level include hexokinase/glucose-6-phosphatase, phosphofructokinase/fructose 1,6-diphosphatase [9], and pyruvate kinase/malic enzyme [10]. Broadly then, futile cycles include such processes as glycolysis/gluconeogenesis, lipolysis (also referred to as fatty acid oxidation)/lipogenesis, and protein turnover, among others [9,11,12]. TH action promotes substrate cycling (reviewed by [9–11,13]). Interestingly, Grant and colleagues demonstrated that this increase in cycling results in a reduction in reactive oxygen species (ROS) formation in states of over nutrition [13]. Therefore, TH, by promoting "futile" cycles, plays an important role as an antioxidant in addition to increasing EE. With respect to TEE, however, the EE fraction affected by TH action on metabolic cycles is low compared to other mechanisms discussed later in this review [14,15].

189

2.2. Ion Leaks

A similar yet distinct target of TH activity is an increase in ion leakage, resulting from TH-induced increased cellular membrane permeability to ions. Consequently, a new ion gradient is established, and cells act to re-establish the desired ion concentrations across the membrane of interest at the cost of increased ATP utilization. Two of the most widely studied and understood ion leaks which are induced by TH and lead to futile ion cycling are the Na^+/K^+ ATPase and the sarco/endoplasmic reticulum Ca^{2+} ATPase (SERCA) (see Figure 1, orange components). TH action increases both Na^+ influx and K^+ efflux into/out of cell plasma membranes, which not only results in increased Na^+/K^+ ATPase activity [16], but also increased expression and insertion of these Na^+/K^+ ATPases into the plasma membrane [17–20]. While not as widely discussed, the Ca^{2+} ATPase on the plasma membrane of erythrocytes has also demonstrated regulation and activity modulation by TH [21], supporting the notion that TH exerts non-genomic effects [22] aside from its well-documented transcriptional action (which will be discussed later). TH also mediates leakage of Ca^{2+} from the sarcoplasmic/endoplasmic reticulum (SR/ER) into the cytosol [11], and induces increased expression of ryanodine receptors, which in turn further increase Ca^{2+} efflux out of the SR/ER into the cytosol [23]. Since Ca^{2+} is an extremely important signaling ion and second messenger used by cells, its leakage has the potential to undermine cell survival. In order to restore homeostasis, the cell compensates by increasing Ca^{2+} influx back into the SR/ER via TH-induced expression of SERCA [6,9,24]. Similar to metabolic cycles described above, futile ion cycling has been estimated to play a less substantial role in TH-dependent increases in EE [14,18].

3. Indirect Effects

While direct effects have been demonstrated to be important in TH-induced EE, the majority of the thermogenesis induced by TH can be attributed to indirect effects [9]. Indirect effects result in an increased capacity for EE through non-genomic pathways and mitochondrial biogenesis, and also a reduction in metabolic efficiency at the stage of ATP production, by activating uncoupling mechanisms.

3.1. Non-Genomic Pathways

TH participates in diverse non-genomic actions which can be initiated at the plasma membrane, in the cytoplasm, or in the mitochondria [7]. These recently discovered non-genomic actions of TH are important for the coordination of normal growth and metabolism, and include regulation of ion channels and oxidative phosphorylation [25]. The principal mediators of non-genomic TH actions on metabolism are the protein kinase signaling cascades [26]. A few examples of non-genomic TH actions are reported below, with comprehensive reviews available elsewhere [6,27]. In an example of plasma membrane TH signaling, T3 binding to the plasma membrane integrin $\alpha V\beta 3$ was found to activate the phosphatidylinositol-4,5-bisphosphate 3-kinase (PI3K) pathway, leading to thyroid hormone receptor-α1 (TRα1) receptor shuttling from the cytoplasm to the nucleus (see Figure 1, pink components) and induction of hypoxia-inducible factor 1-α (HIF1α) gene expression [28]. Non-genomic TH actions on the cardiovascular system also involve protein-kinase-dependent signaling cascades, which include protein kinase A (PKA), protein kinase C (PKC), PI3K, and mitogen-activated protein kinase (MAPK), with changes in ion channel and pump activities [29]. Other non-genomic actions of TH have been linked to AMP-activated protein kinase (AMPK) and Akt/protein kinase B [30–32]. T3 and T2 activate AMPK, a particularly important energy sensor in the cell, resulting in increased fatty acid oxidation, mitochondrial biogenesis, and glucose transporter type 4 (GLUT4) translocation [33–35]. Collectively, the non-genomic effects of TH on ion channels and protein kinase signaling cascades may account for a significant component of TH-mediated EE.

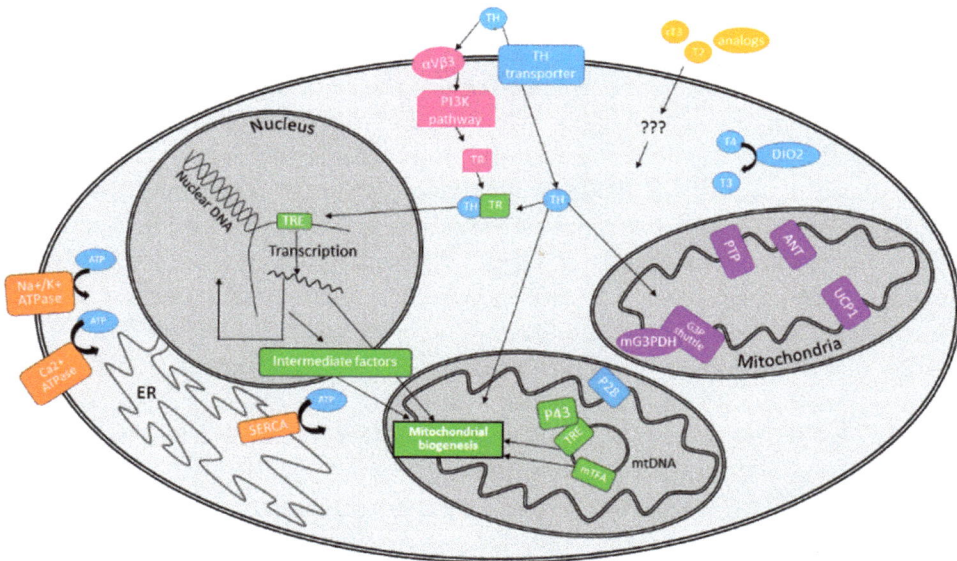

Figure 1. Summary of the mechanisms by which thyroid hormone (TH) modulates energy expenditure (EE) on the cellular level. Orange: Ion leaks. Pink: Non-genomic pathways. Green: Mitochondrial biogenesis resulting from nuclear, intermediate, and mitochondrial-specific pathways. Purple: Uncoupling mechanisms. Yellow: rT3, T2, TH analogs. Blue: TH, ATP, and intermediate steps in TH metabolism and signaling.

3.2. Mitochondrial Biogenesis

Of the roughly 1500 mitochondrial genes, the vast majority are housed within the nuclear genome, while the remainder are in the mitochondrial genome [36,37]. In 1992, Wiesner and colleagues demonstrated that the mechanisms of regulation for these two genomes are distinct [38]. TH exerts some of its thermogenic effects by stimulating mitochondrial biogenesis, which has substantial EE implications. Of note, the elevated oxidative capacity due to an increase in the number of mitochondria is not synonymous with an increase in baseline EE, but rather reflects the potential for expansion of respiration in response to an increased demand (such as muscle contraction or adaptive thermogenic response activation) [39].

TH-dependent mitochondrial biogenesis occurs via three mechanisms discussed below: (1) action on nuclear TH receptors; (2) activation of mitochondrial transcription; and (3) expression and activation of intermediate factors that span both the nucleus and the mitochondria (see Figure 1, green components).

3.2.1. Nuclear

In mammals, two genes, *c-ErbAα* and *c-ErbAβ*, lead to the production of TH receptors (TRs) (reviewed in [40]). TRα1, TRα2, and TRα3 are the protein products of *c-ErbAα*, yet only the TRα1 isoform binds TH and is functionally relevant [41]. TRβ1 and TRβ2, both of which bind TH, are the products of *c-ErbAβ* [42]. TR isoforms are tissue specific, developmentally regulated, and may have distinct functions [43]. All functional TR isoforms contain multiple functional domains, which include a DNA-binding domain (DBD) and a carboxyl-terminal ligand-binding domain (LBD) [7]. The DBD is highly conserved and interacts with specific DNA segments known as TH response elements, or TREs [7]. Thus, TRs are nuclear receptors which modulate gene expression specifically and locally through binding of circulating TH. TRs can exist as monomers, homodimers, and heterodimers; as heterodimers, they can interact with retinoid X receptor (RXR) or retinoic acid receptor (RAR) [44,45]. Through their LBD, TR can also interact with coactivators and corepressors, further modulating TH activity in a tissue specific manner [46]. TH nuclear actions modulate the activities of other transcription factors and coactivators (see Section 3.2.3 below) which are important in metabolic control and the regulation of mitochondrial DNA replication and transcription [47–49]. TH also promotes mitochondrial biogenesis through the induction of nuclear encoded mitochondrial genes such as cytochrome c, cytochrome c oxidase subunit IV, and cytochrome c subunit VIIIa [50]. Other TR interacting proteins and TR functions are reviewed extensively elsewhere [6].

3.2.2. Mitochondrial

In addition to the effects described above, TH exerts actions in/on mitochondria [51]. Aside from the nuclear genomic-based pathway of mitochondrial biogenesis, TH also induces mitochondrial genome transcription [25]. TH promotes mitochondrial genome transcription via two distinct mechanisms: directly by binding within the mitochondria to activate transcription machinery, and indirectly by binding to TR nuclear receptors which induce the expression of intermediate factors, which then go on to mitochondria and induce mitochondrial genome-specific gene expression (reviewed by [25] and discussed further in Section 3.2.3 below).

It is important to recognize that direct TH action on mitochondria is not sufficient *per se* to promote mitochondrial biogenesis, since the vast majority of the mitochondrial proteome is encoded by and regulated within the cell's nuclear genome and cytoplasm [36,37]. Still, there is evidence of direct TH action on the mitochondrial genome. Truncated forms of TRα1, p43 (mitochondrial matrix T3-binding protein) and p28 (inner mitochondrial membrane T3 binding protein), have been isolated in the mitochondrial matrix and inner mitochondrial membrane, respectively [52]. This was a novel and exciting finding, since prior to this discovery there was no knowledge of a non-nuclear TR. Subsequently, Casas and colleagues [53] demonstrated that p43 is indeed restricted to the mitochondria, and that it has similar ligand binding affinity to TRα1, indicating that p43 is the receptor which drives TH mediated transcription of the mitochondrial genome [54,55]. p43 translocates into the mitochondria via fusion to a cytosolic protein [56], and once within the mitochondrial matrix, TH binding to p43 results in p43 interaction with the mitochondrial genome via TREs located in the D loop of the heavy strand [6] to initiate transcription. This mechanism explains the observation of an increased mRNA/rRNA ratio within the mitochondria after exposure to TH [57].

3.2.3. Intermediate Factors

TH also induces mitochondrial biogenesis by bridging nuclear and mitochondrial transcription. This "bridge" is formed by a TH-dependent increase in nuclear expression of a variety of intermediate factors, which can then act on the nucleus, generating a positive feedback loop to either induce nuclear transcription, or to act on the mitochondria to induce mitochondrial transcription [25]. In an extensive review on this topic, Weitzel and Iwen distinguish two distinct classes of intermediate factors: Transcription factors and coactivators [25]. The expression of mitochondrial transcription factor A (mTFA, also referred to as TFAM) is directly regulated by TH, and modulates *in vivo* mitochondrial transcription [58]. Nuclear respiratory factors 1 and 2 (NRF1, NRF2) are transcription factors with multifaceted actions leading to stimulation of mitochondrial biogenesis ([25], and [59] for extensive review). While these intermediate factors function as transcription factors, others function as coactivators of transcription. An example of this class is represented by steroid hormone receptor coactivator 1 (SRC-1), whose action as a coactivator of TH modulates white and brown adipose tissue (BAT) energy balance [60]. Peroxisome proliferator-activated receptor gamma coactivator-1 (PGC-1, both α and β isoforms) are also transcriptionally regulated by TH [25,61] and play a pivotal role in the oxidative capacity of skeletal muscle and BAT (see below). For many metabolism-related genes which are regulated by TH, a putative TRE has yet to be found, further supporting a role for intermediate factors in TH metabolic control [48].

3.3. Uncoupling Mechanisms within the Mitochondria

While mitochondrial biogenesis increases the capacity for EE, uncoupling mechanisms manipulate and decrease the efficiency of ATP production within the cell, thereby increasing EE. TH has been demonstrated to play a role in these mechanisms (see Figure 1, purple components), as discussed below.

3.3.1. Uncoupling Proteins

Non-shivering thermogenesis consists of the direct conversion of chemical energy into heat, allowing for a rapid and efficient adaptation to changes in environmental temperature. This ultimately contributed to the evolutionary success of mammals, as it expands the ability to survive in hostile climates [62]. The biochemical hallmark of non-shivering thermogenesis is represented by uncoupling oxidative phosphorylation in the mitochondria, particularly in brown adipose tissue (BAT) [63]. This is accomplished by uncoupling protein-1 (UCP1), which renders the inner membrane of the mitochondria permeable to electrons [64]. This allows for the dissipation of chemical energy as heat, shunting the production of ATP away from the respiration complexes and therefore increasing EE. TH plays an important role in modulating this process. UCP1 transcription is positively regulated by a TRE [65], which therefore implicates TH in this energy-expending activity. Interestingly, in BAT, the intracellular concentration of T3 is relatively independent from the circulating levels of TH, and it is regulated by type 2 deiodinase (DIO2) [66]. DIO2 is driven by the β-adrenergic cyclic AMP (cAMP) signaling cascade [67], which promotes an increase in intracellular conversion of the prohormone T4 into T3, the ligand for the TH receptor. This signal pathway ultimately assures a time- and tissue-specific modulation of TH action relatively independent of circulating TH levels [66], with obvious effects on EE [68].

In addition to UCP1, which is the hallmark of brown adipose tissue transcriptome signature, other structurally-related proteins with putative uncoupling properties have been described in other tissues. UCP2 and UCP3 are the most well studied and their transcription is induced by TH [69,70]. UCP3, which is predominantly expressed in skeletal muscle, has been associated with TH-induced modulation of

REE [71] and fatty acid peroxide-induced mitochondrial uncoupling [72]. Additional actions of uncoupling proteins are reviewed elsewhere [9,73].

3.3.2. PCG-1α

While TH action directly stimulates EE in the mitochondria by promoting the uncoupling of substrate oxidation from ADP phosphorylation, TH also augments the overall capacity for non-shivering thermogenesis and therefore EE by positively regulating the transcription of PGC-1α, the master regulator of brown and "beige" adipocyte differentiation and mitochondria proliferation [74]. PGC-1α is also an important modulator of EE in muscle, where it promotes the switch from glycolytic function toward oxidative metabolism [75]. Interestingly, PGC-1α also plays a role in modulating the relative ratio between the transcriptionally active isoform of the TH receptor (TRα1) and the "inactive" TRα2 isoform devoid of the ligand binding domain, thereby generating a sort of intracellular negative feedback [76].

3.3.3. Mitochondrial Permeability Transition Pore

Mitochondrial uncoupling by T3 is driven by gating of the mitochondrial permeability transition pore (PTP) [77]. Previous studies have shown that mitochondrial PTP opening is exquisitely sensitive to mitochondrial Ca^{2+} [78], which is classically increased in states of cell stress [79]. Prolonged opening of the PTP results in mitochondrial depolarization and swelling, and if PTP conductance is sufficiently elevated, mitochondrial rupture will ensue with release of pro-apoptotic proteins and programmed cell death [80]. Interestingly, in addition to its historic role in apoptosis, recent evidence has emerged to implicate PTP in TH-mediated EE. Yehuda-Shnaidman et al. found that mitochondrial uncoupling by T3 required activation of the endoplasmic reticulum inositol 1,4,5-triphosphate receptor 1 (IP(3)R1), suggesting an upstream role for IP(3)R1 in the action of T3 on EE [77]. This study indicated a novel target for TH-dependent mitochondrial EE and the potential for targeting future TH analogs to this pathway. While much research is still necessary in this area, it is possible that IP(3)R1 may result in increased PTP opening, uncoupling, and therefore EE. For a more extensive discussion of the mitochondrial PTP and its role in TH induced EE, please see a recent review by Yehuda-Shnaidman and colleagues [9].

3.3.4. ANT

The mitochondrial adenosine diphosphate/adenosine triphosphate (ADP/ATP) translocase, or ANT, forms a gated pore in the inner mitochondrial membrane, allowing ADP to flow into the mitochondrial matrix and ATP in the opposite direction towards the cytoplasm [81]. ANT serves an important role in oxidative phosphorylation by controlling the flow of ADP substrate into the mitochondria, which is subsequently phosphorylated to ATP. As an important regulator of mitochondrial EE, ANT and cytosolic and mitochondrial ADP/ATP ratios were an early focus of studies into TH stimulated EE [82,83]. Indeed, in 1985, Seitz and colleagues demonstrated that T3 could rapidly increase mitochondrial respiration, ATP regeneration, and the activity of ANT in rat liver [82]. T3 stimulation of ANT was later confirmed and more expansively studied in rat liver mitochondrial isolates [84]. Mowbray and colleagues proposed a model in which T3 caused covalent modification of ANT, promoting a conformation with elevated ADP and cation flux [85]. This study directly linked T3 to mitochondrial uncoupling and provided evidence for the role of TH in shunting substrate towards heat generation in the mitochondria instead of ATP production. Brand *et al.* later demonstrated that basal proton conductance in the mitochondria of mice lacking ANT1 was half that of wild-type controls; firmly establishing the role of ANT in mitochondrial basal uncoupling [86] and therefore EE. Finally, ANT may serve an important role in long-term adaptive thermogenesis. In their study, Ukropec *et al.* found that mice lacking UCP1 were able to induce ANT1/2 and other proteins to compensate for long-term cold exposure [87]. Taken together, these data suggest an important role for ANT in the uncoupling of mitochondrial respiration.

3.3.5. Glycerol-3-Phosphate Shuttle

In order for the electron transport chain to produce ATP, reducing equivalents must also be present in the inner mitochondrial matrix, in addition to ADP as described above. Two mechanisms that allow for this are the malate-aspartate shuttle and the glycerol-3-phosphate (G3P) shuttle [11]. These shuttles differ in the resultant nucleotides which they provide to the electron transport chain within the mitochondria; the malate-aspartate shuttle provides NADH, while the G3P shuttle provides $FADH_2$ [9]. This seemingly minute difference has substantial implications with respect to energy balance, as subsequent oxidative phosphorylation of NADH results in the synthesis of 3 ATP, compared with only 2 ATP for a $FADH_2$ molecule (reviewed in [9]). In this sense, the G3P shuttle is less metabolically efficient, and therefore, if its action is upregulated, it can function as an energy dissipation mechanism. Indeed, TH regulates the G3P shuttle at the level of FADH-dependent mitochondrial glycerol-3-phosphate dehydrogenase (mG3PDH) [9]. mG3PDH is located on the outer side of the mitochondrial inner membrane and allows for the conversion of G3P into dihydroxyacetone phosphate (DHAP) [88]. In this conversion, $FADH_2$ is formed and shuttled into complex II of the electron transport chain. Silva and colleagues studied a transgenic *mG3PDH−/−* mouse model and found significantly higher levels of TH ([89], and reviewed in [11]). This evidence suggests a clear role for TH in thermogenesis created by the G3P shuttle. However, total oxygen consumption was not reduced as drastically as expected (only a 7%–10%

reduction in the transgenic $mG3PDH-/-$ mouse compared to controls) [89]. This suggests that compensatory mechanisms exist to lessen the reduction in EE when mG3PDH is not present.

4. TH Analogs and Non-Classical THs

4.1. TH Analogs

The diverse effects of TH on metabolism prompted researchers to study its use as a potential therapeutic for obesity and dyslipidemia. However, supra-physiologic TH levels cause a toxic state, and their systemic effects such as tachycardia, bone loss, muscle wasting, and neuropsychiatric disturbances prevent therapeutic use [90]. For these reasons, supplementing TH in euthyroid individuals for the treatment of obesity was abandoned. A logical development from research on TH actions has been the isolation and synthesis of TH derivatives with favorable side effect profiles, or "ideal" target-tissue distribution, to exploit beneficial metabolic effects while minimizing toxicity and systemic adverse effects. Newer TH derivatives have been developed with tissue and TRβ specificity (reviewed in [91,92]) (see Figure 1, yellow components). By focusing on TRβ selectivity, the adverse cardiac effects of TH have been reduced due to the low expression of TRβ receptors in the heart [93]. Tissue specificity has focused on the actions of TH in the liver, in part because synthetic TH derivatives could be made with high first-pass metabolism in the liver and greatly lowered serum concentrations [92]. The synthetic TH analog GC-1 (sobetirome) has been shown to prevent or reduce hepatosteatosis in a rat model [94] and can reduce serum triglyceride and cholesterol levels without significant side-effects on heart rate [95]. Additionally, GC-1 has been shown to increase EE and prevent fat accumulation in female rats [96].

4.2. Non-Classical THs

In addition to the "classic" THs T4 and T3, other naturally occurring "non-classical" THs may have physiological actions or be exploited therapeutically in the modulation of EE (see Figure 1, yellow components). The mechanisms of action of non-classical THs, which include 3,3′,5′-triiodothyronine (rT3), thyronamines (TAMs), and 3,5-diiodothyronine (T2) have been recently reviewed in detail elsewhere [8,97–99]. In this review, we will briefly discuss the metabolic actions of T2. T2 is found at picomolar serum concentrations in humans [100], and at similar concentrations, T2 is able to stimulate oxygen consumption in the isolated perfused livers of hypothyroid rats [101]. T2 has also been shown to directly and rapidly stimulate mitochondrial activity [102] and elevate resting EE in rats [103]. Subsequently, it was demonstrated that T2 can prevent high fat diet-induced hepatosteatosis and obesity in rats by stimulating mitochondrial uncoupling and decreasing ATP synthesis [104,105]. Furthermore, T2 can treat obesity and hepatosteatosis [106] and prevent high fat diet-induced insulin resistance in rats [107]. Finally, recent experimental evidence indicates that T2 is able to activate BAT-dependent thermogenesis and enhance mitochondrial respiration in hypothyroid rats [108]. In an attempt towards translating experimental findings to humans, Antonelli *et al.* administered T2 to healthy, euthyroid subjects and monitored changes in body weight, resting metabolic rate (RMR) and thyroid function [109]. Compared to baseline, T2-treated subjects had a significant elevation in RMR, reduced body weight, and normal thyroid and cardiac function, while no changes in any of these metrics were observed in the placebo group. Within the limitation of a very small proof-of-concept trial, this study further supports the potential of T2 to therapeutically increase RMR and reduce body weight.

5. Clinical Correlates

The recent discovery of naturally occurring mutations in the *TRα* gene [110] has provided the opportunity to assess *in vivo* the differential effects of TH signaling by comparing and contrasting the effects of TH receptor α and β mutations on energy metabolism. The human phenotype of resistance to TH (RTH) secondary to mutations in the *TRβ* gene is commonly characterized by a combination of hyper- and hypothyroid hormonal signaling at different end-organ tissues, with an overall increase in EE [111]. Conversely, the recently described syndrome of RTH secondary to *TRα* mutations is characterized by increased adiposity and decreased EE [112], in keeping with the predominance of *TRα* in high energy demanding tissues such as myocardium. Interestingly, while both isoforms are present in BAT [113], *TRβ* is the prevalent isoform, playing a critical role in the adaptive thermogenic response [114]. The data therefore strongly suggest that the modulatory activity of lipolysis and EE by *TRα* is primarily due to indirect effects, rather than direct action on the mitochondria. Interestingly, an association between polymorphisms in the *TRα* locus and increased body mass index has been reported, supporting the role of this isoform in energy metabolism [115]. From a clinical standpoint, these findings suggest that the development of a receptor isoform or tissue-specific TH agonist may represent a viable strategy to modulate end-organ targets or pathways with precision, without generating undesirable side effects.

6. Conclusions and Final Remarks

TH has pleiotropic effects on mitochondria and energy expenditure. The modulation of TH's actions is critical in the delivery of time and tissue specific signaling. The effects of TH in increasing energy expenditure via modulation of the adaptive thermogenesis response, coupled with the ability of increasing respiratory capacity by regulating mitochondrial biogenesis, are augmented by the increase in TH's non-mitochondrial effects on futile cycles and ion transport. Finally, the opportunity to selectively modulate TH effects represents a promising therapeutic target for the amelioration of a wide range of metabolic disorders.

Acknowledgments: The authors are grateful to Bin Ni, Ph.D. for his comments and constructive criticisms.

Author Contributions: Janina A. Vaitkus performed the primary literature search, wrote the first draft of the manuscript, and contributed to the subsequent revisions; Jared S. Farrar contributed to the primary literature search and to revisions; Francesco S. Celi designed the structure of the manuscript, supervised the literature search, and contributed to the subsequent revisions.

Conflicts of Interest: The authors declare no conflict of interest.

References

1. Rosen, E.D.; Spiegelman, B.M. Adipocytes as regulators of energy balance and glucose homeostasis. *Nature* **2006**, *444*, 847–853.

2. De Vos-Geelen, J.; Fearon, K.C.; Schols, A.M. The energy balance in cancer cachexia revisited. *Curr. Opin. Clin. Nutr. Metab. Care* **2014**, *17*, 509–514.

3. Haugen, H.A.; Chan, L.N.; Li, F. Indirect calorimetry: A practical guide for clinicians. *Nutr. Clin. Pract.* **2007**, *22*, 377–388.

4. Sotelo-Rivera, I.; Jaimes-Hoy, L.; Cote-Velez, A.; Espinoza-Ayala, C.; Charli, J.L.; Joseph-Bravo, P. An acute injection of corticosterone increases thyrotrophin-releasing hormone expression in the paraventricular nucleus of the hypothalamus but interferes with the rapid hypothalamus pituitary thyroid axis response to cold in male rats. *J. Neuroendocrinol.* **2014**, *26*, 861–869.

5. Medici, M.; Visser, W.E.; Visser, T.J.; Peeters, R.P. Genetic determination of the hypothalamic-pituitary-thyroid axis: Where do we stand? *Endocr. Rev.* **2015**, *36*, 214–244.

6. Cheng, S.Y.; Leonard, J.L.; Davis, P.J. Molecular aspects of thyroid hormone actions. *Endocr. Rev.* **2010**, *31*, 139–170.

7. Cioffi, F.; Senese, R.; Lanni, A.; Goglia, F. Thyroid hormones and mitochondria: With a brief look at derivatives and analogues. *Mol. Cell. Endocrinol.* **2013**, *379*, 51–61.

8. Goglia, F. The effects of 3,5-diiodothyronine on energy balance. *Front. Physiol.* **2014**, *5*.

9. Yehuda-Shnaidman, E.; Kalderon, B.; Bar-Tana, J. Thyroid hormone, thyromimetics, and metabolic efficiency. *Endocr. Rev.* **2014**, *35*, 35–58.

10. Petersen, K.F.; Cline, G.W.; Blair, J.B.; Shulman, G.I. Substrate cycling between pyruvate and oxaloacetate in awake normal and 3,3′-5-triiodo-L-thyronine-treated rats. *Am. J. Physiol.* **1994**, *267*, E273–E277.

11. Silva, J.E. Thermogenic mechanisms and their hormonal regulation. *Physiol. Rev.* **2006**, *86*, 435–464.

12. Newsholme, E.A.; Parry-Billings, M. Some evidence for the existence of substrate cycles and their utility *in vivo. Biochem. J.* **1992**, *285*, 340–341.

13. Grant, N. The role of triiodothyronine-induced substrate cycles in the hepatic response to overnutrition: Thyroid hormone as an antioxidant. *Med. Hypotheses.* **2007**, *68*, 641–649.

14. Freake, H.C.; Oppenheimer, J.H. Thermogenesis and thyroid function. *Annu. Rev. Nutr.* **1995**, *15*, 263–291.

15. Oppenheimer, J.H.; Schwartz, H.L.; Lane, J.T.; Thompson, M.P. Functional relationship of thyroid hormone-induced lipogenesis, lipolysis, and thermogenesis in the rat. *J. Clin. Investig.* **1991**, *87*, 125–132.

16. Haber, R.S.; Ismail-Beigi, F.; Loeb, J.N. Time course of Na, K transport and other metabolic responses to thyroid hormone in clone 9 cells. *Endocrinology* **1988**, *123*, 238–247.

17. Lei, J.; Nowbar, S.; Mariash, C.N.; Ingbar, D.H. Thyroid hormone stimulates Na-K-ATPase activity and its plasma membrane insertion in rat alveolar epithelial cells. *Am. J. Physiol. Lung Cell. Mol. Physiol.* **2003**, *285*, L762–L772.

18. Lei, J.; Mariash, C.N.; Ingbar, D.H. 3,3′,5-triiodo-L-thyronine up-regulation of Na, K-ATPase activity and cell surface expression in alveolar epithelial cells is src kinase- and phosphoinositide 3-kinase-dependent. *J. Biol. Chem.* **2004**, *279*, 47589–47600.

19. Gick, G.G.; Ismail-Beigi, F.; Edelman, I.S. Thyroidal regulation of rat renal and hepatic Na, K-ATPase gene expression. *J. Biol. Chem.* **1988**, *263*, 16610–16618.

20. Gick, G.G.; Ismail-Beigi, F. Thyroid hormone induction of Na^+-K^+-ATPase and its mrnas in a rat liver cell line. *Am. J. Physiol.* **1990**, *258*, C544–C551.

21. Segal, J.; Hardiman, J.; Ingbar, S.H. Stimulation of calcium-atpase activity by 3,5,3′-*tri*-iodothyronine in rat thymocyte plasma membranes. A possible role in the modulation of cellular calcium concentration. *Biochem. J.* **1989**, *261*, 749–754.

22. Vicinanza, R.; Coppotelli, G.; Malacrino, C.; Nardo, T.; Buchetti, B.; Lenti, L.; Celi, F.S.; Scarpa, S. Oxidized low-density lipoproteins impair endothelial function by inhibiting non-genomic action of thyroid hormone-mediated nitric oxide production in human endothelial cells. *Thyroid* **2013**, *23*, 231–238.

23. Jiang, M.; Xu, A.; Tokmakejian, S.; Narayanan, N. Thyroid hormone-induced overexpression of functional ryanodine receptors in the rabbit heart. *Am. J. Physiol. Heart Circ. Physiol.* **2000**, *278*, H1429–H1438.

24. Kahaly, G.J.; Dillmann, W.H. Thyroid hormone action in the heart. *Endocr. Rev.* **2005**, *26*, 704–728.

25. Weitzel, J.M.; Iwen, K.A. Coordination of mitochondrial biogenesis by thyroid hormone. *Mol. Cell. Endocrinol.* **2011**, *342*, 1–7.

26. Bassett, J.H.; Harvey, C.B.; Williams, G.R. Mechanisms of thyroid hormone receptor-specific nuclear and extra nuclear actions. *Mol. Cell. Endocrinol.* **2003**, *213*, 1–11.

27. Moeller, L.C.; Broecker-Preuss, M. Transcriptional regulation by nonclassical action of thyroid hormone. *Thyroid Res.* **2011**, *4* (Suppl. 1).

28. Lin, H.Y.; Sun, M.; Tang, H.Y.; Lin, C.; Luidens, M.K.; Mousa, S.A.; Incerpi, S.; Drusano, G.L.; Davis, F.B.; Davis, P.J. L-thyroxine *vs.* 3,5,3′-triiodo-L-thyronine and cell proliferation: Activation of mitogen-activated protein kinase and phosphatidylinositol 3-kinase. *Am. J. Physiol. Cell Physiol.* **2009**, *296*, C980–C991.

29. Axelband, F.; Dias, J.; Ferrao, F.M.; Einicker-Lamas, M. Nongenomic signaling pathways triggered by thyroid hormones and their metabolite 3-iodothyronamine on the cardiovascular system. *J. Cell. Physiol.* **2011**, *226*, 21–28.

30. Irrcher, I.; Walkinshaw, D.R.; Sheehan, T.E.; Hood, D.A. Thyroid hormone (T3) rapidly activates p38 and ampk in skeletal muscle *in vivo*. *J. Appl. Physiol.* **2008**, *104*, 178–185.

31. Moeller, L.C.; Dumitrescu, A.M.; Refetoff, S. Cytosolic action of thyroid hormone leads to induction of hypoxia-inducible factor-1 α and glycolytic genes. *Mol. Endocrinol.* **2005**, *19*, 2955–2963.

32. De Lange, P.; Senese, R.; Cioffi, F.; Moreno, M.; Lombardi, A.; Silvestri, E.; Goglia, F.; Lanni, A. Rapid activation by 3,5,3′-L-triiodothyronine of adenosine 5′-monophosphate-activated protein kinase/acetyl-coenzyme a carboxylase and akt/protein kinase b signaling pathways: Relation to changes in fuel metabolism and myosin heavy-chain protein content in rat gastrocnemius muscle *in vivo*. *Endocrinology* **2008**, *149*, 6462–6470.

33. Canto, C.; Auwerx, J. Amp-activated protein kinase and its downstream transcriptional pathways. *Cell. Mol. Life Sci.* **2010**, *67*, 3407–3423.

34. Krueger, J.J.; Ning, X.H.; Argo, B.M.; Hyyti, O.; Portman, M.A. Triidothyronine and epinephrine rapidly modify myocardial substrate selection: A 13C isotopomer analysis. *Am. J. Physiol. Endocrinol. Metab.* **2001**, *281*, E983–E990.

35. Lombardi, A.; de Lange, P.; Silvestri, E.; Busiello, R.A.; Lanni, A.; Goglia, F.; Moreno, M. 3,5-Diiodo-L-thyronine rapidly enhances mitochondrial fatty acid oxidation rate and thermogenesis in rat skeletal muscle: AMP-activated protein kinase involvement. *Am. J. Physiol. Endocrinol. Metab.* **2009**, *296*, E497–E502.

36. Anderson, S.; Bankier, A.T.; Barrell, B.G.; de Bruijn, M.H.; Coulson, A.R.; Drouin, J.; Eperon, I.C.; Nierlich, D.P.; Roe, B.A.; Sanger, F.; *et al.* Sequence and organization of the human mitochondrial genome. *Nature* **1981**, *290*, 457–465.

37. Lopez, M.F.; Kristal, B.S.; Chernokalskaya, E.; Lazarev, A.; Shestopalov, A.I.; Bogdanova, A.; Robinson, M. High-throughput profiling of the mitochondrial proteome using affinity fractionation and automation. *Electrophoresis* **2000**, *21*, 3427–3440.

38. Wiesner, R.J.; Kurowski, T.T.; Zak, R. Regulation by thyroid hormone of nuclear and mitochondrial genes encoding subunits of cytochrome-c oxidase in rat liver and skeletal muscle. *Mol. Endocrinol.* **1992**, *6*, 1458–1467.

39. Holloszy, J.O. Skeletal muscle "mitochondrial deficiency" does not mediate insulin resistance. *Am. J. Clin. Nutr.* **2009**, *89*, 463S–466S.

40. Lazar, M.A. Thyroid hormone receptors: Multiple forms, multiple possibilities. *Endocr. Rev.* **1993**, *14*, 184–193.

41. Mitsuhashi, T.; Tennyson, G.E.; Nikodem, V.M. Alternative splicing generates messages encoding rat c-erbA proteins that do not bind thyroid hormone. *Proc. Natl. Acad. Sci. USA* **1988**, *85*, 5804–5808.

42. Williams, G.R. Cloning and characterization of two novel thyroid hormone receptor β isoforms. *Mol. Cell. Biol.* **2000**, *20*, 8329–8342.

43. Cioffi, F.; Lanni, A.; Goglia, F. Thyroid hormones, mitochondrial bioenergetics and lipid handling. *Curr. Opin. Endocrinol. Diabetes Obes.* **2010**, *17*, 402–407.

44. Kakizawa, T.; Miyamoto, T.; Kaneko, A.; Yajima, H.; Ichikawa, K.; Hashizume, K. Ligand-dependent heterodimerization of thyroid hormone receptor and retinoid x receptor. *J. Biol. Chem.* **1997**, *272*, 23799–23804.

45. Lee, S.; Privalsky, M.L. Heterodimers of retinoic acid receptors and thyroid hormone receptors display unique combinatorial regulatory properties. *Mol. Endocrinol.* **2005**, *19*, 863–878.

46. Crunkhorn, S.; Patti, M.E. Links between thyroid hormone action, oxidative metabolism, and diabetes risk? *Thyroid* **2008**, *18*, 227–237.

47. McClure, T.D.; Young, M.E.; Taegtmeyer, H.; Ning, X.H.; Buroker, N.E.; Lopez-Guisa, J.; Portman, M.A. Thyroid hormone interacts with PPARα and PGC-1 during mitochondrial maturation in sheep heart. *Am. J. Physiol. Heart Circ. Physiol.* **2005**, *289*, H2258–H2264.

48. Weitzel, J.M.; Hamann, S.; Jauk, M.; Lacey, M.; Filbry, A.; Radtke, C.; Iwen, K.A.; Kutz, S.; Harneit, A.; Lizardi, P.M.; *et al.* Hepatic gene expression patterns in thyroid hormone-treated hypothyroid rats. *J. Mol. Endocrinol.* **2003**, *31*, 291–303.

49. Weitzel, J.M.; Iwen, K.A.; Seitz, H.J. Regulation of mitochondrial biogenesis by thyroid hormone. *Exp. Physiol.* **2003**, *88*, 121–128.

50. Lee, J.Y.; Takahashi, N.; Yasubuchi, M.; Kim, Y.I.; Hashizaki, H.; Kim, M.J.; Sakamoto, T.; Goto, T.; Kawada, T. Triiodothyronine induces UPC-1 expression and mitochondrial biogenesis in human adipocytes. *Am. J. Physiol. Cell Physiol.* **2012**, *302*, C463–C472.

51. Psarra, A.M.; Solakidi, S.; Sekeris, C.E. The mitochondrion as a primary site of action of steroid and thyroid hormones: Presence and action of steroid and thyroid hormone receptors in mitochondria of animal cells. *Mol. Cell. Endocrinol.* **2006**, *246*, 21–33.

52. Wrutniak, C.; Cassar-Malek, I.; Marchal, S.; Rascle, A.; Heusser, S.; Keller, J.M.; Flechon, J.; Dauca, M.; Samarut, J.; Ghysdael, J.; *et al.* A 43-kDa protein related to c-ERb A α1 is located in the mitochondrial matrix of rat liver. *J. Biol. Chem.* **1995**, *270*, 16347–16354.

53. Casas, F.; Rochard, P.; Rodier, A.; Cassar-Malek, I.; Marchal-Victorion, S.; Wiesner, R.J.; Cabello, G.; Wrutniak, C. A variant form of the nuclear triiodothyronine receptor c-ERb A α1 plays a direct role in regulation of mitochondrial rna synthesis. *Mol. Cell. Biol.* **1999**, *19*, 7913–7924.

54. Casas, F.; Pessemesse, L.; Grandemange, S.; Seyer, P.; Baris, O.; Gueguen, N.; Ramonatxo, C.; Perrin, F.; Fouret, G.; Lepourry, L.; *et al.* Overexpression of the mitochondrial T3 receptor induces skeletal muscle atrophy during aging. *PLoS ONE* **2009**, *4*, e5631.

55. Pessemesse, L.; Lepourry, L.; Bouton, K.; Levin, J.; Cabello, G.; Wrutniak-Cabello, C.; Casas, F. p28, a truncated form of TRα1 regulates mitochondrial physiology. *FEBS Lett.* **2014**, *588*, 4037–4043.

56. Carazo, A.; Levin, J.; Casas, F.; Seyer, P.; Grandemange, S.; Busson, M.; Pessemesse, L.; Wrutniak-Cabello, C.; Cabello, G. Protein sequences involved in the mitochondrial import of the 3,5,3′-L-triiodothyronine receptor p43. *J. Cell. Physiol.* **2012**, *227*, 3768–3777.

57. Enriquez, J.A.; Fernandez-Silva, P.; Garrido-Perez, N.; Lopez-Perez, M.J.; Perez-Martos, A.; Montoya, J. Direct regulation of mitochondrial rna synthesis by thyroid hormone. *Mol. Cell. Biol.* **1999**, *19*, 657–670.

58. Garstka, H.L.; Facke, M.; Escribano, J.R.; Wiesner, R.J. Stoichiometry of mitochondrial transcripts and regulation of gene expression by mitochondrial transcription factor A. *Biochem. Biophys. Res. Commun.* **1994**, *200*, 619–626.

59. Scarpulla, R.C. Transcriptional paradigms in mammalian mitochondrial biogenesis and function. *Physiol. Rev.* **2008**, *88*, 611–638.

60. Picard, F.; Gehin, M.; Annicotte, J.; Rocchi, S.; Champy, M.F.; O'Malley, B.W.; Chambon, P.; Auwerx, J. SRC-1 and TIF-2 control energy balance between white and brown adipose tissues. *Cell* **2002**, *111*, 931–941.

61. Wu, Z.; Puigserver, P.; Andersson, U.; Zhang, C.; Adelmant, G.; Mootha, V.; Troy, A.; Cinti, S.; Lowell, B.; Scarpulla, R.C.; *et al.* Mechanisms controlling mitochondrial biogenesis and respiration through the thermogenic coactivator PGC-1. *Cell* **1999**, *98*, 115–124.

62. Oelkrug, R.; Polymeropoulos, E.T.; Jastroch, M. Brown adipose tissue: Physiological function and evolutionary significance. *J. Comp. Physiol.* **2015**, 1–20.

63. Cannon, B.; Hedin, A.; Nedergaard, J. Exclusive occurrence of thermogenin antigen in brown adipose tissue. *FEBS Lett.* **1982**, *150*, 129–132.

64. Lowell, B.B.; Spiegelman, B.M. Towards a molecular understanding of adaptive thermogenesis. *Nature* **2000**, *404*, 652–660.

65. Rabelo, R.; Schifman, A.; Rubio, A.; Sheng, X.; Silva, J.E. Delineation of thyroid hormone-responsive sequences within a critical enhancer in the rat uncoupling protein gene. *Endocrinology* **1995**, *136*, 1003–1013.

66. Silva, J.E.; Larsen, P.R. Adrenergic activation of triiodothyronine production in brown adipose tissue. *Nature* **1983**, *305*, 712–713.

67. Canettieri, G.; Celi, F.S.; Baccheschi, G.; Salvatori, L.; Andreoli, M.; Centanni, M. Isolation of human type 2 deiodinase gene promoter and characterization of a functional cyclic adenosine monophosphate response element. *Endocrinology* **2000**, *141*, 1804–1813.

68. Celi, F.S. Brown adipose tissue—When it pays to be inefficient. *N. Engl. J. Med.* **2009**, *360*, 1553–1556.

69. Larkin, S.; Mull, E.; Miao, W.; Pittner, R.; Albrandt, K.; Moore, C.; Young, A.; Denaro, M.; Beaumont, K. Regulation of the third member of the uncoupling protein family, UCP3, by cold and thyroid hormone. *Biochem. Biophys. Res. Commun.* **1997**, *240*, 222–227.

70. Masaki, T.; Yoshimatsu, H.; Kakuma, T.; Hidaka, S.; Kurokawa, M.; Sakata, T. Enhanced expression of uncoupling protein 2 gene in rat white adipose tissue and skeletal muscle following chronic treatment with thyroid hormone. *FEBS Lett.* **1997**, *418*, 323–326.

71. De Lange, P.; Lanni, A.; Beneduce, L.; Moreno, M.; Lombardi, A.; Silvestri, E.; Goglia, F. Uncoupling protein-3 is a molecular determinant for the regulation of resting metabolic rate by thyroid hormone. *Endocrinology* **2001**, *142*, 3414–3420.

72. Lombardi, A.; Busiello, R.A.; Napolitano, L.; Cioffi, F.; Moreno, M.; de Lange, P.; Silvestri, E.; Lanni, A.; Goglia, F. UCP3 translocates lipid hydroperoxide and mediates lipid hydroperoxide-dependent mitochondrial uncoupling. *J. Biol. Chem.* **2010**, *285*, 16599–16605.

73. Lanni, A.; Moreno, M.; Lombardi, A.; Goglia, F. Thyroid hormone and uncoupling proteins. *FEBS Lett.* **2003**, *543*, 5–10.

74. Wulf, A.; Harneit, A.; Kroger, M.; Kebenko, M.; Wetzel, M.G.; Weitzel, J.M. T3-mediated expression of PGC-1α via a far upstream located thyroid hormone response element. *Mol. Cell. Endocrinol.* **2008**, *287*, 90–95.

75. Rodgers, J.T.; Lerin, C.; Gerhart-Hines, Z.; Puigserver, P. Metabolic adaptations through the PGC-1α and sirt1 pathways. *FEBS Lett.* **2008**, *582*, 46–53.

76. Thijssen-Timmer, D.C.; Schiphorst, M.P.; Kwakkel, J.; Emter, R.; Kralli, A.; Wiersinga, W.M.; Bakker, O. PGC-1α regulates the isoform mrna ratio of the alternatively spliced thyroid hormone receptor α transcript. *J. Mol. Endocrinol.* **2006**, *37*, 251–257.

77. Yehuda-Shnaidman, E.; Kalderon, B.; Azazmeh, N.; Bar-Tana, J. Gating of the mitochondrial permeability transition pore by thyroid hormone. *FASEB J.* **2010**, *24*, 93–104.

78. Bernardi, P. Mitochondrial transport of cations: Channels, exchangers, and permeability transition. *Physiol. Rev.* **1999**, *79*, 1127–1155.

79. Rasola, A.; Bernardi, P. Mitochondrial permeability transition in Ca^{2+}-dependent apoptosis and necrosis. *Cell Calcium* **2011**, *50*, 222–233.

80. Crompton, M. The mitochondrial permeability transition pore and its role in cell death. *Biochem. J.* **1999**, *341*, 233–249.

81. Neckelmann, N.; Li, K.; Wade, R.P.; Shuster, R.; Wallace, D.C. cDNA sequence of a human skeletal muscle ADP/ATP translocator: Lack of a leader peptide, divergence from a fibroblast translocator cDNA, and coevolution with mitochondrial DNA genes. *Proc. Natl. Acad. Sci. USA* **1987**, *84*, 7580–7584.

82. Seitz, H.J.; Muller, M.J.; Soboll, S. Rapid thyroid-hormone effect on mitochondrial and cytosolic ATP/ADP ratios in the intact liver cell. *Biochem. J.* **1985**, *227*, 149–153.

83. Seitz, H.J.; Tiedgen, M.; Tarnowski, W. Regulation of hepatic phosphoenolpyruvate carboxykinase (GTP). Role of dietary proteins and amino acids *in vivo* and in the isolated perfused rat liver. *Biochim. Biophys. Acta* **1980**, *632*, 473–482.

84. Verhoeven, A.J.; Kamer, P.; Groen, A.K.; Tager, J.M. Effects of thyroid hormone on mitochondrial oxidative phosphorylation. *Biochem. J.* **1985**, *226*, 183–192.

85. Mowbray, J.; Hardy, D.L. Direct thyroid hormone signalling via ADP-ribosylation controls mitochondrial nucleotide transport and membrane leakiness by changing the conformation of the adenine nucleotide transporter. *FEBS Lett.* **1996**, *394*, 61–65.

86. Brand, M.D.; Pakay, J.L.; Ocloo, A.; Kokoszka, J.; Wallace, D.C.; Brookes, P.S.; Cornwall, E.J. The basal proton conductance of mitochondria depends on adenine nucleotide translocase content. *Biochem. J.* **2005**, *392*, 353–362.

87. Ukropec, J.; Anunciado, R.P.; Ravussin, Y.; Hulver, M.W.; Kozak, L.P. UCP1-independent thermogenesis in white adipose tissue of cold-acclimated $Ucp1^{-/-}$ mice. *J. Biol. Chem.* **2006**, *281*, 31894–31908.

88. Hagopian, K.; Ramsey, J.J.; Weindruch, R. Enzymes of glycerol and glyceraldehyde metabolism in mouse liver: Effects of caloric restriction and age on activities. *Biosci. Rep.* **2008**, *28*, 107–115.

89. Alfadda, A.; DosSantos, R.A.; Stepanyan, Z.; Marrif, H.; Silva, J.E. Mice with deletion of the mitochondrial glycerol-3-phosphate dehydrogenase gene exhibit a thrifty phenotype: Effect of gender. *Am. J. Physiol. Regul. Integr. Comp. Physiol.* **2004**, *287*, R147–R156.

90. Burch, H.B.; Wartofsky, L. Life-threatening thyrotoxicosis. Thyroid storm. *Endocrinol. Metab. Clin. N. Am.* **1993**, *22*, 263–277.

91. Moreno, M.; de Lange, P.; Lombardi, A.; Silvestri, E.; Lanni, A.; Goglia, F. Metabolic effects of thyroid hormone derivatives. *Thyroid* **2008**, *18*, 239–253.

92. Baxter, J.D.; Webb, P. Thyroid hormone mimetics: Potential applications in atherosclerosis, obesity and type 2 diabetes. *Nat. Rev. Drug Discov.* **2009**, *8*, 308–320.

93. Grover, G.J.; Mellstrom, K.; Ye, L.; Malm, J.; Li, Y.L.; Bladh, L.G.; Sleph, P.G.; Smith, M.A.; George, R.; Vennstrom, B.; *et al.* Selective thyroid hormone receptor-β activation: A strateg—y for reduction of weight, cholesterol, and lipoprotein (a) with reduced cardiovascular liability. *Proc. Natl. Acad. Sci. USA* **2003**, *100*, 10067–10072.

94. Perra, A.; Simbula, G.; Simbula, M.; Pibiri, M.; Kowalik, M.A.; Sulas, P.; Cocco, M.T.; Ledda-Columbano, G.M.; Columbano, A. Thyroid hormone (T3) and TRβ agonist GC-1 inhibit/reverse nonalcoholic fatty liver in rats. *FASEB J.* **2008**, *22*, 2981–2989.

95. Trost, S.U.; Swanson, E.; Gloss, B.; Wang-Iverson, D.B.; Zhang, H.; Volodarsky, T.; Grover, G.J.; Baxter, J.D.; Chiellini, G.; Scanlan, T.S.; *et al.* The thyroid hormone receptor-β selective agonist GC-1 differentially affects plasma lipids and cardiac activity. *Endocrinology* **2000**, *141*, 3057–3064.

96. Villicev, C.M.; Freitas, F.R.; Aoki, M.S.; Taffarel, C.; Scanlan, T.S.; Moriscot, A.S.; Ribeiro, M.O.; Bianco, A.C.; Gouveia, C.H. Thyroid hormone receptor β-specific agonist GC-1 increases energy expenditure and prevents fat-mass accumulation in rats. *J. Endocrinol.* **2007**, *193*, 21–29.

97. Coppola, M.; Glinni, D.; Moreno, M.; Cioffi, F.; Silvestri, E.; Goglia, F. Thyroid hormone analogues and derivatives: Actions in fatty liver. *World J. Hepatol.* **2014**, *6*, 114–129.

98. Senese, R.; Cioffi, F.; de Lange, P.; Goglia, F.; Lanni, A. Thyroid: Biological actions of "nonclassical" thyroid hormones. *J. Endocrinol.* **2014**, *221*, R1–R12.

99. Piehl, S.; Hoefig, C.S.; Scanlan, T.S.; Kohrle, J. Thyronamines—Past, present, and future. *Endocr. Rev.* **2011**, *32*, 64–80.

100. Pinna, G.; Meinhold, H.; Hiedra, L.; Thoma, R.; Hoell, T.; Graf, K.J.; Stoltenburg-Didinger, G.; Eravci, M.; Prengel, H.; Brodel, O.; *et al.* Elevated 3,5-diiodothyronine concentrations in the sera of patients with nonthyroidal illnesses and brain tumors. *J. Clin. Endocrinol. Metab.* **1997**, *82*, 1535–1542.

101. Horst, C.; Rokos, H.; Seitz, H.J. Rapid stimulation of hepatic oxygen consumption by 3,5-*di*-iodo-L-thyronine. *Biochem. J.* **1989**, *261*, 945–950.

102. Lombardi, A.; Lanni, A.; Moreno, M.; Brand, M.D.; Goglia, F. Effect of 3,5-di-iodo-L-thyronine on the mitochondrial energy-transduction apparatus. *Biochem. J.* **1998**, *330*, 521–526.

103. Moreno, M.; Lanni, A.; Lombardi, A.; Goglia, F. How the thyroid controls metabolism in the rat: Different roles for triiodothyronine and diiodothyronines. *J. Physiol.* **1997**, *505*, 529–538.

104. Lanni, A.; Moreno, M.; Lombardi, A.; de Lange, P.; Silvestri, E.; Ragni, M.; Farina, P.; Baccari, G.C.; Fallahi, P.; Antonelli, A.; *et al.* 3,5-Diiodo-L-thyronine powerfully reduces adiposity in rats by increasing the burning of fats. *FASEB J.* **2005**, *19*, 1552–1554.

105. Grasselli, E.; Canesi, L.; Voci, A.; de Matteis, R.; Demori, I.; Fugassa, E.; Vergani, L. Effects of 3,5-diiodo-L-thyronine administration on the liver of high fat diet-fed rats. *Exp. Biol. Med.* **2008**, *233*, 549–557.

106. Mollica, M.P.; Lionetti, L.; Moreno, M.; Lombardi, A.; de Lange, P.; Antonelli, A.; Lanni, A.; Cavaliere, G.; Barletta, A.; Goglia, F. 3,5-Diiodo-L-thyronine, by modulating mitochondrial functions, reverses hepatic fat accumulation in rats fed a high-fat diet. *J. Hepatol.* **2009**, *51*, 363–370.

107. Moreno, M.; Silvestri, E.; de Matteis, R.; de Lange, P.; Lombardi, A.; Glinni, D.; Senese, R.; Cioffi, F.; Salzano, A.M.; Scaloni, A.; *et al.* 3,5-Diiodo-L-thyronine prevents high-fat-diet-induced insulin resistance in rat skeletal muscle through metabolic and structural adaptations. *FASEB J.* **2011**, *25*, 3312–3324.

108. Lombardi, A.; Senese, R.; de Matteis, R.; Busiello, R.A.; Cioffi, F.; Goglia, F.; Lanni, A. 3,5-Diiodo-L-thyronine activates brown adipose tissue thermogenesis in hypothyroid rats. *PLoS ONE* **2015**, *10*, e0116498.

109. Antonelli, A.; Fallahi, P.; Ferrari, S.M.; di Domenicantonio, A.; Moreno, M.; Lanni, A.; Goglia, F. 3,5-Diiodo-L-thyronine increases resting metabolic rate and reduces body weight without undesirable side effects. *J. Biol. Regul. Homeost. Agents* **2011**, *25*, 655–660.

110. Bochukova, E.; Schoenmakers, N.; Agostini, M.; Schoenmakers, E.; Rajanayagam, O.; Keogh, J.M.; Henning, E.; Reinemund, J.; Gevers, E.; Sarri, M.; *et al.* A mutation in the thyroid hormone receptor α gene. *N. Engl. J. Med.* **2012**, *366*, 243–249.

111. Moran, C.; Schoenmakers, N.; Agostini, M.; Schoenmakers, E.; Offiah, A.; Kydd, A.; Kahaly, G.; Mohr-Kahaly, S.; Rajanayagam, O.; Lyons, G.; *et al.* An adult female with resistance to thyroid hormone mediated by defective thyroid hormone receptor α. *J. Clin. Endocrinol. Metab.* **2013**, *98*, 4254–4261.

112. Mitchell, C.S.; Savage, D.B.; Dufour, S.; Schoenmakers, N.; Murgatroyd, P.; Befroy, D.; Halsall, D.; Northcott, S.; Raymond-Barker, P.; Curran, S.; *et al.* Resistance to thyroid hormone is associated with raised energy expenditure, muscle mitochondrial uncoupling, and hyperphagia. *J. Clin. Investig.* **2010**, *120*, 1345–1354.

113. Tuca, A.; Giralt, M.; Villarroya, F.; Vinas, O.; Mampel, T.; Iglesias, R. Ontogeny of thyroid hormone receptors and c-erbA expression during brown adipose tissue development: Evidence of fetal acquisition of the mature thyroid status. *Endocrinology* **1993**, *132*, 1913–1920.

114. Martinez de Mena, R.; Scanlan, T.S.; Obregon, M.J. The T3 receptor β isoform regulates UCP1 and D2 deiodinase in rat brown adipocytes. *Endocrinology* **2010**, *151*, 5074–5083.

115. Fernandez-Real, J.M.; Corella, D.; Goumidi, L.; Mercader, J.M.; Valdes, S.; Rojo Martinez, G.; Ortega, F.; Martinez-Larrad, M.T.; Gomez-Zumaquero, J.M.; Salas-Salvado, J.; *et al.* Thyroid hormone receptor α gene variants increase the risk of developing obesity and show gene-diet interactions. *Int. J. Obes.* **2013**, *37*, 1499–1505.

Low T3 State Is Correlated with Cardiac Mitochondrial Impairments after Ischemia Reperfusion Injury: Evidence from a Proteomic Approach

Francesca Forini, Nadia Ucciferri, Claudia Kusmic, Giuseppina Nicolini, Antonella Cecchettini, Silvia Rocchiccioli, Lorenzo Citti and Giorgio Iervasi

Abstract: Mitochondria are major determinants of cell fate in ischemia/reperfusion injury (IR) and common effectors of cardio-protective strategies in cardiac ischemic disease. Thyroid hormone homeostasis critically affects mitochondrial function and energy production. Since a low T3 state (LT3S) is frequently observed in the post infarction setting, the study was aimed to investigate the relationship between 72 h post IR T3 levels and both the cardiac function and the mitochondrial proteome in a rat model of IR. The low T3 group exhibits the most compromised cardiac performance along with the worst mitochondrial activity. Accordingly, our results show a different remodeling of the mitochondrial proteome in the presence or absence of a LT3S, with alterations in groups of proteins that play a key role in energy metabolism, quality control and regulation of cell death pathways. Overall, our findings highlight a relationship between LT3S in the early post IR and poor cardiac and mitochondrial outcomes, and suggest a potential implication of thyroid hormone in the cardio-protection and tissue remodeling in ischemic disease.

Reprinted from *Int. J. Mol. Sci.* Cite as: Forini, F.; Ucciferri, N.; Kusmic, C.; Nicolini, G.; Cecchettini, A.; Rocchiccioli, S.; Citti, L.; Iervasi, G. Low T3 State Is Correlated with Cardiac Mitochondrial Impairments after Ischemia Reperfusion Injury: Evidence from a Proteomic Approach. *Int. J. Mol. Sci.* **2015**, *16*, 26687–26705.

1. Introduction

Acute myocardial infarction (AMI) leading to ischemic heart disease is a major cause of death worldwide. Although timely reperfusion effectively reduces short-term mortality, restoration of blood flow also leads to ischemia/reperfusion (IR) injury which in the long run prompts adverse cardiac remodeling. Thus, limiting cardiac damage in the early stages of the wound healing process is a critical step to improve patient prognosis.

Mitochondrial dysfunction is a key pathogenic event in cardiac IR and disease progression [1–3]. As a consequence, preserving mitochondrial function is essential to limit myocardial damage in ischemic heart disease. The main mechanisms for post IR mitochondrial dysfunction include impairment in electron transport chain (ETC) complex activities [4,5], defects in supermolecular assembly of ETC complexes [6–8], impaired ionic homeostasis and formation of reactive oxygen species (ROS) [9], and impaired tricarboxylic acid (TCA) cycle anaplerosis, [10,11]. In light of this complex scenario, protein profiling has emerged as a powerful tool allowing for simultaneous measurement of the levels of many mitochondrial proteins in a single analysis [12–14]. Moreover, analyses of the mitochondrial proteome prove useful not only to expand our knowledge of mitochondrial function in physio/pathological states, but also to unveil potential strategies for therapeutic intervention [15].

Triiodothyronine (T3), the biologically active thyroid hormone (TH), is considered a major regulator of mitochondrial activity [16–18]. The breakdown of thyroid system homeostasis is associated with bioenergetic remodelling of cardiac mitochondria which leads to severe alterations in the biochemistry, structure and contractility of cardiac muscle [17]. Clinical evidence shows post IR declines of T3, known as low T3 state (LT3S) [19–21] that represent a strong independent prognostic predictor of death and major adverse cardiac events [22]. Accordingly LT3S correction exerts cardioprotective actions both in the clinical arena and in animal models of post ischemic cardiac diseases [23–25]. Since mitochondria are critical effectors of the T3 cardioprotective signaling [26,27], characterization of the mitochondrial proteome remodeling in the post IR-LT3S model may reveal candidate proteins and pathways related to this signaling.

The main objective of the present study is to compare the post IR rat mitochondrial proteome in the early phase of wound healing between normal and low T3 states. Differentially expressed proteins were assessed in comparison to sham operated animals. Our data indicate that mitochondrial proteomic alteration and dysfunction are mainly associated with a post IR decrease of T3 levels.

2. Results

2.1. Validation of the Post IR Low-T3 State (LT3S) Model

We have previously observed that in our AMI model, cardiac TH concentrations closely reflect those of the circulating free TH [27,28]. Therefore, in the present study the circulating levels of free T3 (FT3) and free T4 (FT4) were assessed at baseline and 3 days (3d) post-surgery to track the myocardial TH status. As summarized in Table 1, sham operation did not significantly change serum TH concentrations. No significant alterations of 3d FT3 or FT4 were measured even in IR-NT3 group, while a significant reduction of FT3 concentration was observed three days post-surgery in IR-LT3S rats. This difference should not be attributed to different degree of LV damage since all ischemia-injured rats exhibited the same damage score as well as comparable percentage of area at risk (AAR) (Table 1). These findings confirmed the previous observation about the occurrence of a L-T3S in a subset of animals in the early post-ischemic setting [27]. In accordance with clinical data, the L-T3S condition was observed in about 30% of ischemia-reperfused rats with matching degrees of damaged myocardium, thus suggesting that the Low-T3S rat may represent a useful preclinical model.

2.2. Post IR Myocardial Functional Parameters and Mitochondrial Activity

As a second step, we asked if the observed FT3 decrease in the L-T3S group could be associated with an impaired recovery of post-ischemic cardiac function and chamber geometry. Although both groups of infarcted rats showed similar alterations of the LV fractional shortening (FS), and the end systolic LV diameter (Figure 1A), only the IR-LT3 group exhibited a significant reduction of the systolic anterior wall thickening (SAWT) with respect to both sham and IR-NT3 group (Figure 1A) suggesting that a post IR L-T3S is associated with a significant reduction in regional contractility of LV within the AAR. To strengthen the hypothesis of a relationship between the variation of T3 levels (three days after IR with respect to the basal level) and the cardiac functional parameter SAWT, a non linear regression (sigmoid model) was applied to derive the EC_{50}, *i.e.*, the delta T3 concentration that provokes a response half way between the minimal response (bottom point) and the maximal response (top point) (Figure 1B). The derived EC_{50} was equal to -0.39, a value that falls between LT3S and NT3 groups, and may be considered a clear cutoff of the delta T3 level.

Table 1. Serum free thyroid hormone and LV damage index.

TH Level (pg/mL)	Sham		IR-NT3		IR-LT3S	
	Mean ± SEM	Median (IQR)	Mean ± SEM	Median (IQR)	Mean ± SEM	Median (IQR)
FT3 basal	3.2 ± 0.3	3.6 (2.7–3.8)	3.2 ± 0.3	3.2 (2.6–3.9)	3.5 ± 0.3	3.4 (3.3–3.6)
FT3 final	3.5 ± 0.2	3.8 (3.1–3.8)	3.3 ± 0.3	3.6 (2.6–3.6)	2.2 ± 0.3 *,#,†	2.3 (1.9–2.4) §,&,+
FT4 basal	12.4 ± 0.6	13.4 (11.6–13.7)	13.0 ± 0.7	13.3 (12.9–13.7)	12.1 ± 1.0	13.7 (13.5–14.3)
FT4 final	12.1 ± 1.5	12.6 (10.9–13.8)	14.3 ± 1.8	14.3 (12.7–16.3)	13.3 ± 1.6	10.4 (10.1–10.8)
Damage Index	Sham		IR-NT3		IR-LT3S	
	–		Mean ± SEM		Mean ± SEM	
Arrhythmic severity score	NA		3.3 ± 0.4		3.4 ± 0.3	
Area at risk (% of LV)	NA		48 ± 3		47 ± 5	

Free T3 (FT3) and free T4 (FT4) were measured before and 72 h post-IR; $n = 5$. Left columns report mean values ± SEM: * $p = 0.006$ *vs.* respective basal levels; # $p = 0.004$ *vs.* sham-operated; † $p = 0.009$ *vs.* IR-NT3. Right columns report median values and interquartile range (IQR), along with the corresponding non parametric analysis: § $p = 0.04$ *vs.* respective basal levels; & $p = 0.009$ *vs.* sham-operated; + $p = 0.014$ *vs.* IR-NT3. Arrhythmic score was calculated based on ECG data obtained from ischemia to 30 min of reperfusion. Area at risk measured at 72 h post IR is expressed as % of LV. NA = Not Applicable.

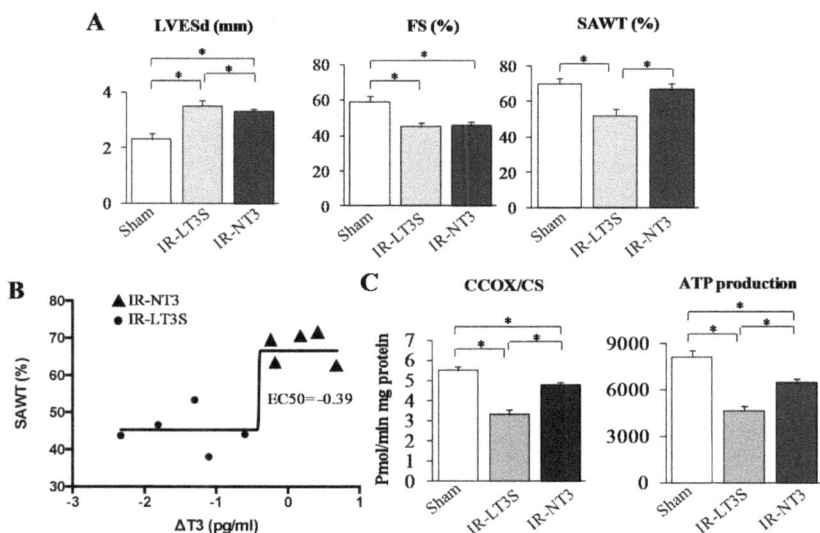

Figure 1. Effect of L-T3S on cardiac functional recovery and mitochondrial activity 72 h post IR. (**A**) LV end systolic diameter (LVESd) (**left** panel), fractional shortening (FS) (**medium** panel) and systolic anterior wall thickening (SAWT) (**right** panel); (**B**) Non linear fitting (sigmoid) relating the ΔFT3 values (difference between 3 days post IR and basal) to systolic anterior wall thickening (SAWT) in IR-LT3 and IR-NT3; and (**C**) mitochondrial function assessed in the LV area at risk (AAR): citrate synthase (CS) normalized cytochrome c oxidase (CcOx) activity (**left** panel) and ATP production (**right** panel). Data are expressed as mean ± SE; n = 5 in each group; * $p \leqslant 0.005$.

To evaluate if this alteration can be associated with a greater degree of mitochondrial impairment in the AAR of L-T3S rats, we next determined cytochrome c oxidase activity and ATP production. As shown in Figure 1C, both ischemia injured groups showed reduced citrate synthase-normalized cytochrome c activity, as well as reduced rate of ATP production, but the lowest level were in any case assessed in the L-T3S rats. These findings indicate that a decreased post IR T3 level is associated with poorer mitochondrial activity and energy production.

2.3. Mitochondrial Proteome

A proteomic study was then performed to assess if the physiological and biochemical differences observed between IR-LT3S and IR-NT3 rats might be related to quantitative changes in the cardiac mitochondrial proteome. To this end, mitochondrial protein profiling from sham, IR-NT3 and IR-LT3S rats were obtained. The principal mitochondrial proteins were identified, as shown in the Supplementary Material (Figure S1). Multiple comparisons were performed to identify differentially expressed proteins. Of the total 546 identified proteins, 138 mitochondrial proteins exhibited significant changes and were grouped according to their function using the published literature and Uniprot database (Nucleic Acids Res. 43:D204-D212, 2015). Figure 2A shows the percentage representation of different protein groups/functions (clusters) significantly changed between IR-LT3S and IR-NT3. Twenty-five percent of altered proteins are implicated either in mitochondrial quality control (21%) or in cell death (4%). It is particularly notable that the remaining 75% belongs to functional groups that are involved in ATP synthesis.

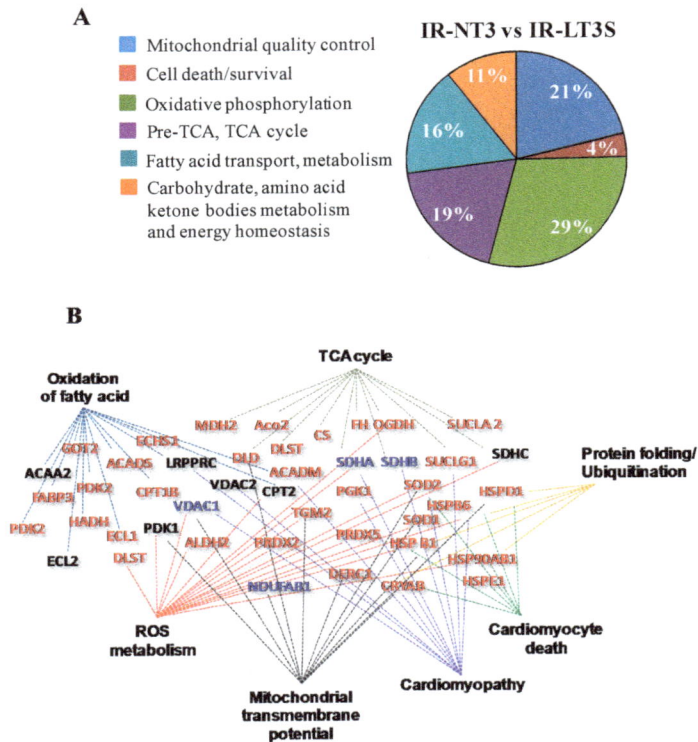

Figure 2. Mitochondrial proteomic analysis obtained at 72 h post IR. (**A**) Pie chart showing percentage of differentially expressed proteins grouped according to their function in IR-NT3 *vs.* IR-LT3; and (**B**) clustering of differentially expressed proteins in IR-LT3S *vs.* IR-NT3 generated by IPA software. Networks related to diseases, functions and canonical pathways were generated based on the information stored in IPA Knowledge base. Network nodes are named by correspondent Gene Codes. The color assigned to node name indicate the level of proteins expression: red for up-regulated, blue for down-regulated and black for no change in IR-NT3 *vs.* IR-LT3S respectively. Gene acronyms are listed in the abbreviation list.

Ingenuity Pathway Analysis (IPA, http://www.ingenuity.com/products/pathways_analysis.html, Qiagen, Venlo, Holland) was used to confirm the functional protein grouping of differentially expressed proteins in IR-NT3 *vs.* IR-LT3S and to relate them to disease. As shown in Figure 2B, the protein clusters play critical roles in mitochondrial activity and dysfunction, and in disease etiopathology (cardiomyopathy). Selected proteins from each functional group are reported in Figures 3–6 and described below (Tables S1–S5 for the complete list).

2.4. Mitochondrial Quality Control and Cell Death

IR induced a significant upregulation of stress-responsive proteins (Figure 3). Notably, IR-NT3 rats exhibited the highest level of heat shock proteins (HSP), including HSP27, HSP71, HSP90 and α-crystallin (Cryab) along with a greater increase of DNA-repair-associated proteins (40s ribosomal protein S3, Rps3; and O-acetyl-ADP-ribose deacetylase, Macrod1) (Figure 3). By considering ROS scavenging system, six antioxidant enzymes were found to be upregulated in IR-NT3 group *vs.* both sham and IR-LT3S groups, including isoforms of aldehyde deidrogenase (Aldh6a and Aldh2), peroxiredoxines (Prdx2 and Prdx5) and superoxide dismutases (Sod1 and Sod2) (Figure 3). On the contrary, in the IR-LT3S group the level of the antioxidant enzymes was comparable to the sham group (Figure 3).

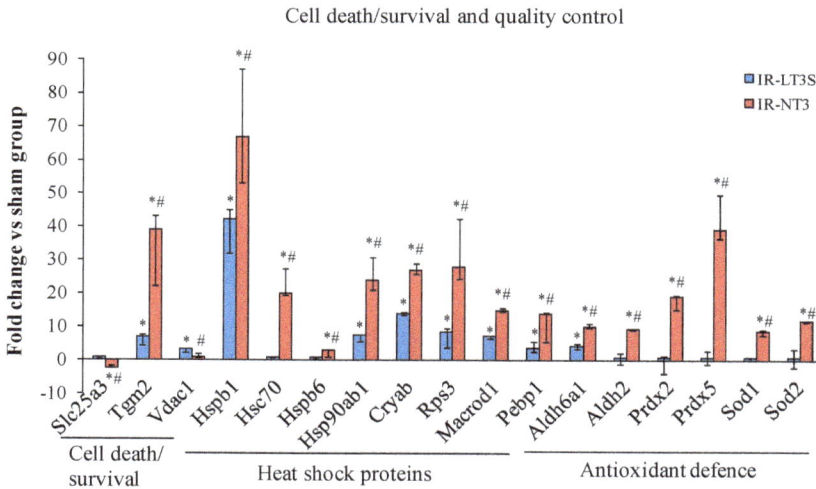

Figure 3. Differentially expressed proteins involved in cell death and mitochondrial quality control in response to stress. Data are expressed as median and interquartile range. * *p vs.* sham < 0.017; # *p vs.* IR-LT3S < 0.017. Protein acronyms are listed in the abbreviation list.

IR also differently affected proteins involved in mitochondrial-mediated cell death/survival processes. In particular, in the IR-NT3 group the phosphate carrier (Slc25a3) was downregulated *versus* both the sham and the IR-LT3S. This protein remained unchanged in the IR-LT3S rats with respect to the sham group. The increase of voltage dependent anion selective channel protein 1 (VDAC1) observed in IR-LT3S was prevented in IR-NT3 group. The protective transglutaminase 2 (TGM2) was upregulated in both IR groups, with the highest values shown by IR-NT3 rats (Figure 3).

Overall these data suggest that mitochondria of IR-NT3 rats possess higher protein quality control and greater defensive capacity to face IR injury.

2.5. Cellular Energy Metabolism

In accordance with well known post IR cardiac metabolic impairment, protein expression in IR rats was altered at crucial points in cellular energy metabolism.

Pre-TCA and TCA cycle enzyme isoforms were upregulated in response to IR, but the highest levels were measured in the IR-NT3 rats (Figure 4). The main modulated proteins are involved in: (1) anaplerotic reaction and malate/aspartate shuttle (aspartate aminotransferase, Got2); (2) regulation of pyruvate entry through TCA (pyruvate dehydrogenase alpha 1/1, Pdha1/1; and pyruvate dehydrogenase kinase 1 and 2, Pdk1 and Pdk2); (3) TCA cycle (citrate synthase, CS; aconitate hydratase Aco2; isocitrate dehydrogenase (NADP) Idh2; and fumarate hydratase, Fh); and (4) substrate level phosphorylation (E2 and E3 component of the 2-oxoglutarate dehydrogenase Dlst and Dld; and succinyl-CoA synthetase, Sucla 2 and Suclg1) (Figure 4).

Similarly to pre-TCA and TCA, IR induced an increase of protein subunits involved in fatty acid metabolism (Figure 5). The highest levels were found in IR-NT3 rats and the most highly expressed isoforms were involved in: (1) β oxidation and lipid biosynthesis (2,4-dienoyl CoA reductase 1, Decr1; acetyl-coenzyme A dehydrogenase medium and short chain, Acadm and Acads; $\delta(3,5)$-$\delta(2,4)$-dienoyl-CoA isomerase, Ech1; enoyl-CoA hydratase, Echs1; hydroxyacyl-coenzyme A dehydrogenase, Hadh; and acyl-CoA synthetase family member 2, Acsf2); (2) regulation of intracellular levels of acyl-CoAs free fatty acids and CoASH (Protein Acot 13, LOC683884); and (3) regulation of fatty acid transport and fatty acid substrate utilization (fatty acid-binding protein, Fabp3; and acyl-CoA thioesterase 2, Acot2) (Figure 5).

IR influenced also the content of several enzymes regulating other metabolic reactions (Figure 6). As for TCA and fatty acid oxidation, the highest levels were measured in IR-NT3 group. Among them, of particular interest for the post IR functional recovery are those involved in lactate metabolism and clearance, glycolysis and glycogenolysis (L-lactate dehydrogenase B chain, Ldhb; pyruvate kinase, MOR4B8; phosphoglycerate kinase 1, Pgk1; and glycogen phosphorylase,

Pigb) (Figure 6). Finally, IR-NT3 rats exhibited increased levels of the mitochondrial creatine kinase isoform (CKmt2), a protein that was more severely decreased in IR-LT3S group and that plays a critical role in energy homeostasis. (Figure 6).

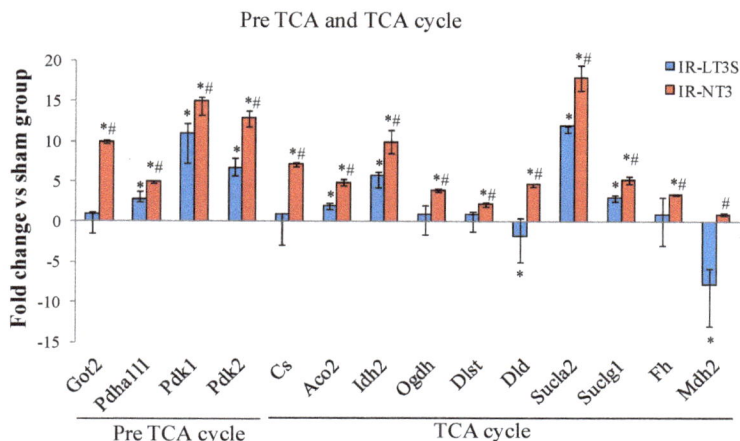

Figure 4. Differentially expressed proteins involved in TCA cycle and pre TCA cycle. Data are expressed as median and interquartile range. * p $vs.$ sham < 0.017; # p $vs.$ IR-LT3S < 0.017. Protein acronyms are listed in the abbreviation list.

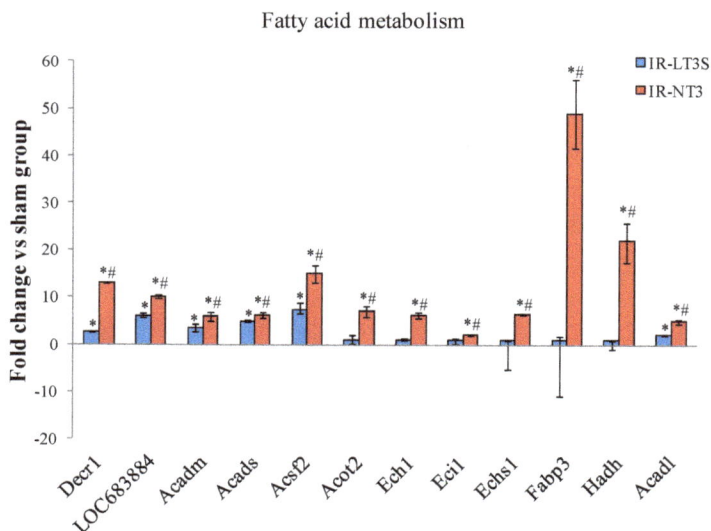

Figure 5. Differentially expressed proteins involved fatty acid metabolism. Data are expressed as median and interquartile range. * p $vs.$ sham < 0.017; # p $vs.$ IR-LT3S < 0.017. Protein acronyms are listed in the abbreviation list.

Other metabolic processes

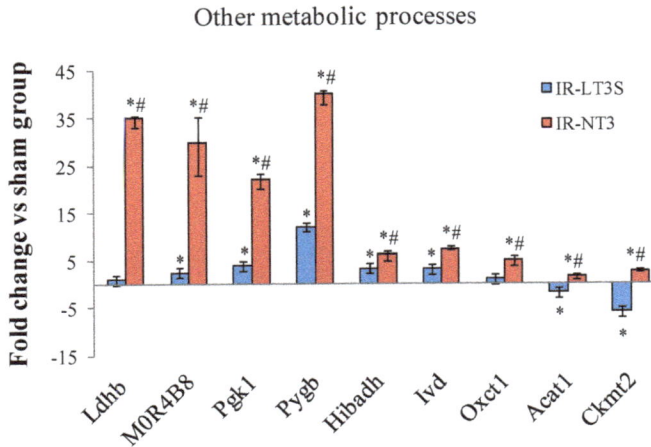

Figure 6. Differentially expressed proteins involved in other cellular energy metabolic processes. * *p vs.* sham < 0.017; # *p vs.* IR-LT3S < 0.017. Protein acronyms are listed in the abbreviation list.

In line with ATP production data, these findings suggest that, with respect to IR-LT3S rats, IR-NT3 rats present a greater ability to provide metabolic intermediates for TCA and to oxidize FFA, glucose and glycogen in order to sustain greater ATP production.

3. Discussion

The main results of the present study is that the recovery of the pre-ischemic FT3 levels in the early period of the post IR wound healing process, is associated with better cardiac functional recovery and lower mitochondrial activity impairments in the injured LV myocardium. Moreover, proteomic profiling in IR-NT3 and IR-LT3S rats revealed a different modulation of mitochondrial proteins critically involved in the regulation of mitochondrial activity and cardiomyocyte survival, as suggested by IPA (see Figure 2B). The main proteomic results are discussed below according to the functional protein grouping.

3.1. Mitochondrial Quality Control and Mitochondrial-Mediated Cell Death

3.1.1. Mitochondrial Quality Control

Mitochondrial reactive oxygen species (ROS) production following IR leads to extensive damage to different types of mitochondrial molecules resulting in mitochondrial dysfunction [29]. Different strategies of mitochondrial quality control have evolved to counteract the adverse effects resulting from oxidative stress [30]. Among these processes, repair of damaged molecules, refolding of misfolded proteins, ROS scavenging and removal of excessively damaged mitochondria play a key role.

Protein folding and repair are regulated by specialized proteins, termed chaperones, which include heat-shock proteins (HSPs) [31]. The protective functions of HSP70, HSP90, HSP 27 and α-crystallin B against IR injury was extensively investigated in previous studies using transgenic animals and isolated cardiac myocyte-derived cells [32–34]. Similarly, overexpression of the small HSPs α-crystallin B and HSP27 diminished the reversible damage after simulated or myocardial ischemia [35–37].

Accordingly, in our study, the post IR retention of physiological T3 levels was associated with a higher content of HSP71, HSP90, HSP27 and α-crystallin B, which was paralleled by preserved mitochondrial function.

Mitochondrial DNA is another target of IR. Mitochondrial DNA damage plays a key role in post IR disease progression, which highlights the importance of efficient repair machinery [38]. In our model of IR, two proteins implicated in DNA damage repair namely 40s ribosomal protein S3 (rpS3) and O-acetyl-ADP-ribose deacetylase (MACROD1) were found to be up-regulated in IR-NT3 group with respect to IR-LT3S. These proteins act synergistically through different modes of action to afford cardioprotection [39–41].

Antioxidant defenses are another important class of mitochondrial quality control molecules that were differentially modulated by post IR FT3 levels. Among them, Aldh2 protects the heart against ischemic injury through detoxification of toxic aldehyde and a differential regulation of autophagy [42,43], while peroxiredoxins and superoxide dismutases play a key cardioprotective role through oxide detoxification [44–47].

We propose that the higher level of quality control proteins observed in the IR-NT3 group promotes better repair of post IR mitochondrial damage, which is essential for preserved mitochondrial activity.

3.1.2. Cell Fate

IR-induced mitochondrial impairments favor the formation of mitochondrial pores that open in a process known as mitochondrial permeability transition leading to apoptosis and necrosis. Inhibition of this process is cardioprotective in both patients and animal models [48,49]. In our study, the two mitochondrial permeability transition activators (namely, the phosphate carrier and Vdac1) were less expressed in the IR-NT3 group than in IR-LT3S. These data are in line with our previous observations [27] that LT3S correction by T3 replacement in the post IR setting limits mitochondrial membrane depolarization and cell death and reinforces the hypothesis of a key role of post-IR T3 levels in cardiac recovery.

When mitochondrial injuries overwhelm molecular repair capacity and antioxidant defenses, removal of damaged mitochondria through mitophagy is a protective strategy to avoid cell death. Our proteomic profiling showed in the IR-LT3 group the highest level of tissue tranglutaminase 2 (Tgm2), a cardioprotective effector that participates in the maintenance of the intact mitochondrial respiratory function and in the clearance of damaged mitochondria [50,51].

3.2. Oxidative Phosphorylation

The oxidative phosphorylation system (OXPHOS) in the mitochondrial inner membrane carries out the central biological process of cardiac energy metabolism. Thus, the alteration of major OXPHOS proteins is responsible for modifying all the cardiac energy metabolism and performance. As expected, we found in both injured groups a significant dysregulation of all ETC and ATP synthesis complexes, the lowest levels being measured in IR-NT3 group. The apparently contradictory result of lower OXPHOS protein levels in the presence of better preserved functional recovery and mitochondrial activity observed in the IR-NT3 group, may have several explanations. First, with the exception of complex II, all ETC complexes can associate themselves in supercomplexes, known as respirasomes, that organize electron flux to optimize the use of available substrates [52]. Mitochondrial defects can arise from supramolecular assembly rather than from the individual components of the ETC [6]. We speculate that the higher level of quality control proteins in IR-NT3 rats guarantees the assembly of intact, correctly-folded mitochondrial components in more functional macromolecular complexes, which may explain higher ATP production in the presence of lower ETC protein content. Second, post-translational modifications (PTMs) have emerged as powerful regulators of mitochondrial function and in particular of the mitochondria-encoded subunit 1 of the complex IV [53–55]. We might speculate that different post IR T3 levels may have induced different post-translational modifications (PTMs) in ETC complexes. Our proteomic approach was not intended to analyze PTMs, further dedicated studies are needed to explore this critical issue.

3.3. Pre TCA, TCA Cycle

The tricarboxylic acid cycle (TCA) forms a major metabolic hub and as such it is involved in many disease states involving energetic imbalance. In hypoxic conditions such as in the post IR setting, when OXPHOS is impaired, the TCA supplies high-energy phosphates through matrix substrate-level phosphorylation catalyzed by the succinyl coenzyme A synthetase [56,57].

It has been demonstrated that reduction of glycogen turnover and depletion of TCA substrates contributes to impaired contractile function of ischemia/reperfused myocardium, and that TCA intermediates, along with essential substrates, such as glucose, lactate, and pyruvate, are necessary to ensure functional recovery with reperfusion. [58,59]. A central role in substrate level phosphorylation is played by α-ketoglutarate dehydrogenase, that supply succinyl-CoA to succinyl coenzyme A synthetase, and by malate-aspartate shuttle, that is involved in anaplerotic supply of substrate to TCA [56].

In our study the decreased OXPHOS protein content following the ischemic injury was counterbalanced by an increase in pre-TCA, and TCA cycle enzymes, as well as in glycolytic enzymes. This tendency was more evident in the presence of preserved post IR FT3 and in particular regarded the enzymes involved in anaplerotic reaction and in the substrate level phosphorylation. These data suggest that the preserved TH level may play a key role in favoring mitochondrial anaerobic production of ATP. Our data are in agreement with previous findings in multiple animal models showing that T3 supplementation modulates pyruvate entry into the TCA, thereby providing the energy support for improved cardiac function after reperfusion [60,61]. We speculate that the effects of preserved physiological levels of T3 on α-ketoglutarate dehydrogenase might improve post IR cardiac efficiency. Indeed, supply of α-ketoglutarate during blood cardioplegia attenuated ischemic injury in patients undergoing coronary operations [62,63]. Similarly, T3 administration in patients with L-T3S induced by cardiopulmonary bypass improved postoperative ventricular function, reduced the need for treatment with inotropic agents and mechanical devices, and decreased the incidence of myocardial ischemia [64].

3.4. Fatty Acids Metabolism

In healthy hearts, >70% of the cardiac energy is accounted for by oxidation of fatty acids (FAs) and the remainder by glucose oxidation. However, the heart changes its substrate preference from FAs towards glucose as remodeling develops in response to diverse stresses including IR [65]. This metabolic shift may be an adaptive mechanism under acute stress condition such as IR, because it lowers oxygen consumption. On the other hand, glucose oxidation yields far less ATP than FAs. Insufficient ATP production is likely to increase the susceptibility of post-infarct hearts to cardiomyocyte death and contractile dysfunction [66]. Accordingly, reversal of metabolic shift in post IR remodeling markedly improved contractile function [67].

Here we report that retention of physiological post IR T3 levels in the early post IR phase is associated with the upregulation of several proteins involved in FA oxidation.

If confirmed through mechanistic inferences, these data might support and extend to cardiovascular pathologies, such as IR, the notion that changes of TH levels affect the myocardial mitochondrial bioenergetic capacity [68].

3.5. Study Limitations and Concluding Remarks

The results, obtained in a small population of animals, are associative and, in the absence of an intervention group (for example ischemia/reperfusion animals treated with T3), at present we are unable to infer any cause and effect relationship neither between post IR T3 level and cardiac/mitochondrial function, nor between proteomic remodeling and cardiac impairment. Moreover, the assumption that circulating T3 levels are a substitute of myocardial T3 content derives from data of previous studies [27,28] rather than a direct measurement in the present study.

Nevertheless, our data clearly indicate specific changes in mitochondrial protein expression in relation to different post IR circulating T3 level. Retention of physiological T3 concentration is associated with the upregulation of proteins with functional relevance in rescue of mitochondrial integrity and in optimization of substrate utilization. These differences along with the better recovery of post IR cardiac function and mitochondrial activity in the NT3 rats prompt us to speculate that a condition of L-T3S in the early setting of the post IR wound healing might affect mitochondrial function and contribute to adverse remodeling.

4. Material and Methods

4.1. Animal Procedure

The study was performed in accordance with the European Directive (2010/63/UE) and the Italian law (D.L 26/2014), and the protocol was approved by the Animal Care Committee of the Italian Ministry of Health (Endorsement n.240/2011-B, 9 November 2011). All surgery was performed under anaesthesia, and all efforts were made to minimize suffering. The study design is depicted in the flow chart reported in Figure 7. A total of 24 male Wistar rats were used in the study. Five rats were assigned in the sham group and 19 underwent IR. Out of 19 animals subjected to LAD occlusion, two died during surgery due to irreversible ventricular fibrillation following the IR protocol.

Myocardial infarction and reperfusion was produced by 30 min ligation of the left descending coronary artery (LAD) followed by reperfusion of adult male Wistar rats 12–15 weeks old and weighing 310 ± 3 g using a technique described in detail elsewhere [27]. A standard limb D1–D3 electrocardiogram (ECG) was continuously monitored during surgery up to 60 min.

In all cases ischemia was confirmed by ST segment elevation in the ECG and visually assessed regional cardiac cyanosis. In addition, occurrence of arrhythmias both during ischemia and at reperfusion was also recorded: arrhythmias were classified according to the Lambeth Conventions. A scoring system was used to classify their severity as previously described [69].

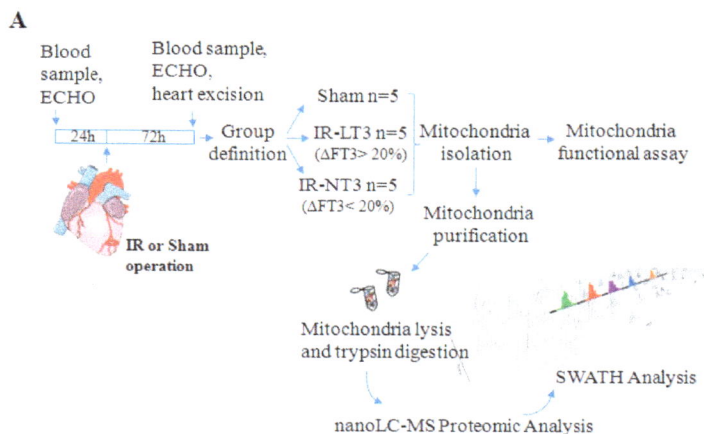

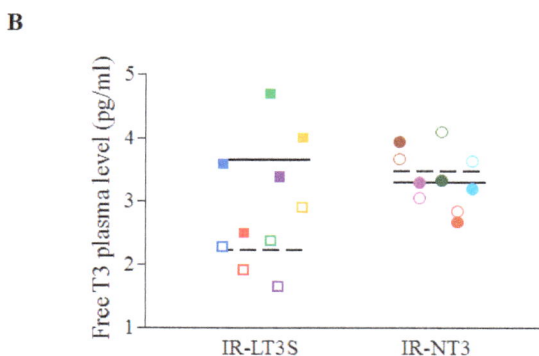

Figure 7. (**A**) Flow chart showing the study design. Seventy-two hours post IR, five rats developed a low T3 state (IR-LT3S) while 12 maintained normal FT3 plasma levels (IR-NT3 group). All five rats of the IR-LT3S group were used, and for a balanced statistical analysis between groups of equal numbers, five were randomly enrolled out of the 12 IR-NT3 rats; (**B**) Scatter-plot showing FT3 plasma levels in the IR-LT3S group (squares) and NT-3 group (circles) at baseline (filled symbols) and three days post IR (empty symbols). Different colors represent different rats. Continuous and dashed lines indicate the mean values before and after IR respectively.

The severity score of ventricular arrhythmias was used as index of ischemia damage and a score level equal or greater than three was adopted as inclusion criteria in the study. A group of sham-operated rats was used as control. Three days after surgery hearts were arrested in diastole by a lethal KCl injection. Cardiac tissue samples were obtained from the core of the ischemic reperfused region (area at risk, AAR) of LV as describe elsewhere [27]. In sham-operated animals tissues were harvested from corresponding regions. Samples from each area were immediately

221

processed for mitochondria isolation. Rats undergone I/R were allocated to two groups according to the percent decrease of T3 circulating levels measured at 72 h after surgery against the basal value. In particular, a reduction ⩾20% of the basal level of T3 measured 3 days after surgery was used as cut-off value to enroll animals in the low T3-state group (IR-LT3S). All the other rats were included in the normal T3 group (IR-NT3).

4.2. Echocardiography Study

Echocardiographic studies were performed three days after infarction with a portable ultrasound system (MyLab 25, Esaote SpA, Genova, Italy) equipped with a high frequency linear transducer (LA523, 12.5 MHz). Images were obtained from the sedated animal, from the left parasternal view. Short-axis two-dimensional view of the left ventricle (LV) was taken at the level of papillary muscles to obtain M-mode recording. Anterior (ischemia-reperfused) and posterior (viable) end-diastolic and end-systolic wall thicknesses, systolic wall thickening, and LV internal dimensions were measured following the American Society of Echocardiography guidelines. Parameters were calculated as mean of the measures obtained in three consecutive cardiac cycles. The global LV systolic function was expressed as fractional shortening (FS%).

4.3. Morphometric Analysis to Determine the Area at Risk

The area at risk (AAR) was determined as previously described [27]. Briefly, three days after surgery, the LAD was re-tied with the suture left in its original position and 1 mL 1% Evan's blue (Sigma Aldrich, Saint Louis, MO, USA) was injected in the inferior cava vein to identify the myocardial AAR as unstained. Next, the heart, arrested in diastole, was excised and cut in transversal and parallel slices about 2 mm thick. LV area and the AAR, (expressed as percentage of the LV) from each slice were measured using Image J software (open source image processing program).

4.4. Serum and Tissue Thyroid Hormone Levels

Two mL of blood were collected from the femoral vein both 3–5 days before (basal) and 3 days after LAD occlusion (terminal endpoint) in anesthetized animals. Serum free T3 (FT3) and free T4 (FT4) were assayed as previously described [27].

4.5. Mitochondria Isolation

Mitochondria were purified from LV fresh tissue according to the manufacturer's protocol (MITO-ISO1; Sigma Aldrich, Saint Louis, MO, USA) and as previously described [26]. Briefly, cardiac tissue was homogenized in buffer containing 10 mM 4-(2-Hydroxyethyl)piperazine-1-ethanesulfonic acid, N-(2-Hydroxyethyl)piperazine-

N'-(2-ethanesulfonic acid) (HEPES, Sigma Aldrich), 200 mM mannitol (Sigma Aldrich), 70 mM sucrose and 1mM Ethylene glycol-bis(2-aminoethylether)-N,N,N',N'-tetraacetic acid(EGTA) (PH 7.5, Sigma Aldrich) and centrifuged at $2000\times g$ at 4 °C for 5 min. The supernatant was collected and centrifuged at $11,000\times g$ at 4 °C for 20 min. The pellet was suspended in storage buffer at pH 7.5 containing 10 mM HEPES, 250 mM sucrose (Sigma Aldrich), 1 mM ATP (Sigma Aldrich), 0.08 mM ADP (Sigma Aldrich), 5 mM sodium succinate (Sigma Aldrich), 2 mM K_2HPO_4 (Sigma Aldrich) and 1 mM DTT (Sigma Aldrich) and stored at -80 °C until use. An aliquot of the suspended pellet was assayed for protein content with the Pierce bicinchoninic protein assay kit (Sigma Aldrich).

4.6. Mitochondrial Enzyme Activity Assays

Mitochondrial function was expressed as the ratio between the activity of the cytochrome c oxidase-1 (CcOX-1) and that of citrate synthase (CS) to normalize for mitochondrial mass. Enzyme activity was assessed in 1mL cuvette by using spectrophotometric assay kit according to the manufacturer's protocols (CYTOC-OX1 and CS0720, Sigma Aldrich) and as previously described [27]. All assays were performed in triplicate.

4.7. Measurements of ATP Production in Isolated Mitochondria

ATP synthesis rates were measured in mitochondrial fractions with the ATP Determination Kit A22066 (Thermo Scientific, Waltham, MA, USA) as previously described [27]. The assays were performed in triplicate in 96 well plate in a volume of 150 µL containing10 µg mitochondria protein, 0.25 M sucrose, 50 mM HEPES, 2 mM $MgCl_2$, 1 mM EGTA, 10 mM KH_2PO_4, 1 mM pyruvate, 1 mM malate. 1 mM ATP-free ADP and a solution of 0.5 mM luciferin and 0.25 µg/mL luciferase were added with the injectors integrated in the plate reader (Infinite M200 PRO, TECAN, Männedorf, Switzerland). The slope of luminescence increase was determined in the first 48 s after injection of luciferase reagent and ADP, and it was converted to ATP concentration using a standard curve according to the manufacturer's instruction.

4.8. Proteomics Sample Pre-Processing

For proteomic analyses the crude mitochondrial fraction was further purified on Percoll gradient as previously described with some modification [70]. Briefly, mitochondria were resuspended in 1 mL of 15% Percoll and layered on preformed gradient consisting of 22% Percoll (3 mL) layered over 50% Percoll (1 mL). Following centrifugation of the gradient at $90,000\times g$ for 40 min, the mitochondrial fraction that accumulated at the lower interface (between the 50% and 22% Percoll layers) was collected and diluted with PBS (1:8). After 2 washing step in PBS, the pellet was lysed in TRIS HCl 5 mM pH = 8.1, acetonitrile 10% (Romil, Cambridge,

UK) and protease inhibitor and sonicated in ice. Protein concentration was determined by bicinchoninic acid assay (Pierce, Thermo Scientific, Waltham, MA, USA). About 20 μg of protein were diluted in Ammonium Bicarbonate 25 mM, reduced with dithiothreitol DTT 5 mM at 80 °C for 30 min and alkylated using iodoacetamide 10 mM at 37 °C for 20 min. Digestion was performed incubating overnight with 1:100 trypsin (Roche, Basel, Switzerland):substrate at 37 °C. Peptides solution was then loaded on a C18 cartridge in order to eliminate debris and filtered with 0.22 μm filter. Peptide mix was diluted to 100 μL by 2% ACN/0.1% FA.

4.9. nanoLC-MS/MS SWATH-Based Analysis

Chromatographic separation of peptides was performed using a nano-HPLC system (Eksigent, ABSciex, Washington, DC, USA). The loading pump pre-concentrated the sample in a pre-column cartridge (PepMap-100 C18 5 μm 100 A, 0.1 × 20 mm, Thermo Scientific, Waltham, MA, USA) and then separated in a C18 PepMap-100 column (3 μm, 75 μm × 250 mm, Thermo Scientific) at a flow rate of 300 nL·min^{-1}. Runs were performed with eluent A (Ultrapure water, 0.1% Formic acid) under 60 min linear gradient from 5% to 40% of eluent B (Acetonitrile, 0.1% Formic acid) followed by 10 min of a purge step and 20 min re-equilibration step. Peptides eluted from chromatography were directly processed using 5600 TripleTOF™ mass spectrometer (ABSciex) equipped with a DuoSpray™ ion source (ABSciex). Data were acquired using the new Sequential Window Acquisition of all THeoretical Mass Spectra (SWATH™) method for shotgun data independent MRM quantification. For library, MS/MS data were processed with ProteinPilot™ Software (ABSciex). The false discovery rate (FDR) analysis was done using the integrated tools in ProteinPilot software (ABSciex, Washington, DC, USA) and a confidence level of 95% was set.

The label free statistical comparative analysis was performed using PeakView4.5 Software (ABSciex) with MS/MS(ALL) with SWATH™ Acquisition MicroApp 2.0 and MarkerViewTM (ABSciex). Retention time alignment was obtained using selected peptides from top score protein. Processing settings were: 7 peptides per protein, 7 transitions per peptide, 92% peptide confidence (according to Paragon algorithm result) and 5% FDR; XIC (Extracted-Ion Chromatogram) options: extraction window 10 min, width 50 ppm and 0.1 Da. Normalization of the sample content was done using a global normalization of profiles (based on total protein content). Principal Component Analysis (PCA) was performed in order to evidence groupings among the data set. SWATH strategy generate time-resolved fragment ion spectra sufficiently specific to confidently identify query peptides which are quantified with a consistency and accuracy comparable with that of selected reaction monitoring, the gold standard proteomic quantification method [71]. This means that differential expression analysis of SWATH data allows the profiling of disease-related proteomes with a high degree of reproducibility and confidence providing self-validated data [72].

4.10. Statistical Analysis

Results are given as mean + SEM unless otherwise stated. Since the post IR circulating free T3 (FT3) levels have been used to allocate rats to different groups, we verified the distribution of this parameter for normality (Kolmogorov-Smirnov) before inferential statistic analysis. Differences among the three groups of rats were analyzed by a one-way ANOVA followed by a Bonferroni test once normality had been proven (Kolmogorov-Smirnov test). Differences were considered statistically significant at a value of $p < 0.05$. For proteomic analyses, a non-parametric test (Kruskall-Wallis) was run considering all the three groups. Thereafter, a Mann-Whitney U-test (adjusting the α-level by Bonferroni inequality) was used to check differences between groups two by two (differences were considered statistically significant at a value of $p < 0.017$). The significant proteins resulting from the Mann-Whitney U-test comparison between IR-NT3 and IR-LT3 ($n = 82$) were used as input dataset for IPA and networks were created for the most significant linked diseases and functions. Additionally, proteins involved in TCA cycle pathway were shown (Figure 2B).

Supplementary Materials: Supplementary materials can be found at http://www.mdpi.com/1422-0067/16/11/25973/s1.

Acknowledgments: We thank Valeria Siciliano for skillful assistance in statistical analysis. We thank Nicole Di Lascio for assistance in the model fitting. This work was funded by the Tuscany Region Research Grant (DGR 1157/2011) "Study of the molecular, biochemical and metabolic mechanisms involved in the cardioprotective effect of T3".

Author Contributions: Francesca Forini conceived and designed the study, contributed data analysis and drafted the paper; Nadia Ucciferri performed the proteomic experiments and critically revised the manuscript; Claudia Kusmic contributed the animal model, performed the ECO and ECG measurements and critically revised the manuscript; Giuseppina Nicolini performed the experiments and critically revised the manuscript; Antonella Cecchettini performed the experiments and critically revised the manuscript; Silvia Rocchiccioli contributed IPA analysis tool and critically revised the manuscript; Lorenzo Citti supervised the study as senior scientist of the proteomic laboratory and Giorgio Iervasi conceived and supervised the study as principal investigator of the funding grant.

Conflicts of Interest: The authors declare no conflict of interest.

Abbreviations

Acaa2 = 3-ketoacyl-CoA thiolase.

Acadl = Long-chain specific acyl-CoA dehydrogenase.

Acadm = Acetyl-Coenzyme A dehydrogenase, medium chain.

Acads = Acetyl-Coenzyme A dehydrogenase, short chain.

Acat1 = Acetyl-CoA acetyltransferase.

Aco2 = Aconitate hydratase.

Acot2 = Acyl-CoA thioesterase 2.

Acsf2 = Acyl-CoA synthetase family member 2.

Aldh2 = Aldehyde dehydrogenase.

Aldh6a1 = Aldehyde dehydrogenase family 6, isoform CRA.

Ckmt2 = Creatine kinase mitochondrial.

Cpt1b = Carnitine O-palmitoyltransferase 1.

Cryab = α-crystallin B chain.

Cs = Citrate synthase.

Decr1 = 2,4-dienoyl CoA reductase 1.

Dld = Dihydrolipoyl dehydrogenase.

Dlst = Dihydrolipoamide S-succinyltransferase.

Ech1 = $\delta(3,5)$-$\delta(2,4)$-dienoyl-CoA isomerase.

Echs1 = Enoyl-CoA hydratase.

Eci1 = Dodecenoyl-Coenzyme A δ isomerase (3,2 trans-enoyl-Coenzyme A isomerase).

Ecl1 = Extender of the chronological lifespan protein 1.

Ecl2 = Extender of the chronological lifespan protein 2.

Fabp3 = Fatty acid-binding protein.

Fh = Fumarate hydratase 1.

Got2 = Aspartate aminotransferase.

Hadh = Hydroxyacyl-coenzyme A dehydrogenase.

Hibadh = 3-hydroxyisobutyrate dehydrogenase.

Hsc70 = Heat shock cognate 71 kDa.

Hspb6 = Heat shock protein α-crystallin-related-B6.

Hsp90ab1 = Heat shock protein HSP 90-α class B, member 1.

Hspb1 = Heat shock 27 kDa protein 1.

Hspd1 = Heat shock protein 60 kDa.

Hspe1 = Heat shock 10kDa protein 1.

Idh2 = Isocitrate dehydrogenase (NADP).

Ivd = Isovaleryl-CoA dehydrogenase.

Ldhb = L-lactate dehydrogenase B chain.

LOC683884 = Protein Acot13.

Lrpprc = Leucine-rich PPR.

Ndufab1 = NADH dehydrogenase (ubiquinone) 1, α/β subcomplex, 1.

Macrod1 = O-acetyl-ADP-ribose deacetylase MACROD1.

Mdh2 = Malate dehydrogenase.

MOR4B8 = Pyruvate kinase.

Ogdh = 2-oxoglutarate dehydrogenase.

Oxct1 = Succinyl-CoA:3-ketoacid coenzyme A transferase 1.

Pdha1/1 = Protein Pdha1/1.

Pdk1 = Pyruvate dehydrogenase kinase 1.

Pdk2 = Pyruvate dehydrogenase kinase 1.

Pepb1 = Phosphatidylethanolamine-binding protein 1.

Pgk1 = Phosphoglycerate kinase 1.

Prdx2 = Peroxiredoxin-2.

Prdx5 = Peroxiredoxin-5.

Pygb = Glycogen phosphorylase.

Rps3 = 40 S ribosomal protein S3.

Sdha = Succinate dehydrogenase (ubiquinone) flavoprotein subunit.

Sdhb = Succinate dehydrogenase (ubiquinone) iron-sulfur subunit.

Sdhc = Succinate dehydrogenase complex, subunit C, integral membrane protein.

Slc25a3 = Phosphate carrier protein.

Sod1 = Superoxide dismutase [Cu-Zn].

Sod2 = Superoxide dismutase [Cu-Zn].

Sucla2 = succinyl-CoA synthetase.

Suclg1 = Succinyl-CoA synthetase (ADP/GDP-forming) subunit α.

Tgm2 = transglutaminase 2.

Vdac1 = Voltage-dependent anion-selective channel protein 1.

Vdac2 = Voltage-dependent anion-selective channel protein 2.

References

1.   Whelan, R.S.; Kaplinskiy, V.; Kitsis, R.N. Cell death in the pathogenesis of heart disease mechanisms and significance. *Annu. Rev. Phys.* **2010**, *72*, 19–44.

2.   Baines, C.P. The cardiac mitochondrion: Nexus of stress. *Annu. Rev. Phys.* **2010**, *72*, 61–80.

3.   Marín-García, J.; Goldenthal, M.J. Mitochondrial centrality in heart failure. *Heart Fail. Rev.* **2008**, *13*, 137–150.

4.   Chen, Q.; Moghaddas, S.; Hoppel, C.L.; Lesnefsky, E.J. Ischemic defects in the electron transport chain increase the production of reactive oxygen species from isolated rat heart mitochondria. *Am. J. Physiol. Cell Physiol.* **2008**, *294*, C460–C466.

5. Paradies, G.; Petrosillo, G.; Pistolese, M.; di Venosa, N.; Federici, A.; Ruggiero, F.M. Decrease in mitochondrial complex I activity in ischemic/reperfused rat heart: Involvement of reactive oxygen species and cardiolipin. *Circ. Res.* **2004**, *94*, 53–59.

6. Rosca, M.G.; Vazquez, E.J.; Kerner, J.; Parland, W.; Chandler, M.P.; Stanley, W.; Sabbah, H.N.; Hoppel, C.L. Cardiac mitochondria in heart failure: Decrease in respirasomes and oxidative phosphorylation. *Cardiovasc. Res.* **2008**, *80*, 30–39.

7. Lenaz, G.; Baracca, A.; Barbero, G.; Bergamini, C.; Dalmonte, M.E.; del Sole, M.; Faccioli, M.; Falasca, A.; Fato, R.; Genova, M.L.; *et al.* Mitochondrial respiratory chain super-complex I–III in physiology and pathology. *Biochim. Biophys. Acta* **2010**, *1797*, 633–640.

8. Lenaz, G.; Genova, M.L. Structure and organization of mitochondrial respiratory complexes: A new understanding of an old subject. *Antioxid. Redox Signal.* **2010**, *12*, 961–1008.

9. Ambrosio, G.; Zweier, J.L.; Duilio, C.; Kuppusamy, P.; Santoro, G.; Elia, P.P.; Tritto, I.; Cirillo, P.; Condorelli, M.; Chiariello, M.; *et al.* Evidence that mitochondrial respiration is a source of potentially toxic oxygen free radicals in intact rabbit hearts subjected to ischemia and reflow. *J. Biol. Chem.* **1993**, *268*, 18532–18541.

10. Turer, A.T.; Stevens, R.D.; Bain, J.R.; Muehlbauer, M.J.; van der Westhuizen, J.; Mathew, J.P.; Schwinn, D.A.; Glower, D.D.; Newgard, C.B.; Podgoreanu, M.V. Metabolomic profiling reveals distinct patterns of myocardial substrate use in humans withcoronary artery disease or left ventricular dysfunction during surgical ischemia/reperfusion. *Circulation* **2009**, *119*, 1736–1746.

11. Russell, R.R., III; Taegtmeyer, H. Changes in citric acid cycle flux and anaplerosis antedate the functional decline in isolated rat hearts utilizing acetoacetate. *J. Clin. Investig.* **1991**, *87*, 384–390.

12. Bugger, H.; Schwarzer, M.; Chen, D.; Schrepper, A.; Amorim, P.A.; Schoepe, M.; Nguyen, T.D.; Mohr, F.W.; Khalimonchuk, O.; Weimer, B.C. Doenst Proteomic remodeling of mitochondrial oxidative pathways in pressure overload-induced heart failure. *Cardiovasc. Res.* **2010**, *85*, 376–384.

13. Mootha, V.K.; Bunkenborg, J.; Olsen, J.V.; Hjerrild, M.; Wisniewski, J.R.; Stahl, E.; Bolouri, M.S.; Ray, H.N.; Sihag, S.; Kamal, M.; *et al.* Integrated analysis of protein composition, tissue diversity, and gene regulation in mouse mitochondria. *Cell* **2003**, *115*, 629–640.

14. Birner, C.; Dietl, A.; Deutzmann, R.; Schröder, J.; Schmid, P.; Jungbauer, C.; Resch, M.; Endemann, D.; Stark, K.; Riegger, G.; *et al.* Proteomic profiling implies mitochondrial dysfunction in tachycardia-induced heart failure. *J. Card. Fail.* **2012**, *18*, 660–673.

15. Ouzounian, M.; Lee, D.S.; Gramolini, A.O.; Emili, A.; Fukuoka, M.; Liu, P.P. Predict, prevent and personalize: Genomic and proteomic approaches to cardiovascular medicine. *Can. J. Cardiol.* **2007**, *23*, 28A–33A.

16. Wrutniak-Cabello, C.; Casas, F.; Cabello, G. Thyroid hormone action in mitochondria. *J. Mol. Endocrinol.* **2001**, *26*, 67–77.

17. Goldenthal, M.J.; Ananthakrishnan, R.; Marín-García, J. Nuclear-mitochondrial cross-talk in cardiomyocyte T3 signaling: A time-course analysis. *J. Mol. Cell. Cardiol.* **2005**, *39*, 319–326.

18. Marín-García, J. Thyroid hormone and myocardial mitochondrial biogenesis. *Vascul. Pharmacol.* **2010**, *52*, 120–130.

19. Hamilton, M.A.; Stevenson, L.W.; Luu, M.; Walden, J.A. Altered thyroid hormone metabolism in advanced heart failure. *J. Am. Coll. Cardiol.* **1990**, *16*, 91–95.

20. Wiersinga, W.M.; Lie, K.I.; Touber, J.L. Thyroid hormones in acute myocardial infarction. *Clin. Endocrinol. (Oxf.)* **1981**, *14*, 367–374.

21. Friberg, L.; Drvota, V.; Bjelak, A.H.; Eggertsen, G.; Ahnve, S. Association between increased levels of reverse triiodothyronine and mortality after acute myocardial infarction. *Am. J. Med.* **2001**, *111*, 699–703.

22. Iervasi, G.; Pingitore, A.; Landi, P.; Raciti, M.; Ripoli, A.; Scarlattini, M.; L'Abbate, A.; Donato, L. Low-T3 syndrome: A strong prognostic predictor of death in patients with heart disease. *Circulation* **2003**, *107*, 708–713.

23. Pingitore, A.; Galli, E.; Barison, A.; Iervasi, A.; Scarlattini, M.; Nucci, D.; L'abbate, A.; Mariotti, R.; Iervasi, G. Acute effects of triiodothyronine (T3) replacement therapy in patients with chronic heart failure and low-T3 syndrome: A randomized, placebo-controlled study. *J. Clin. Endocrinol. Metab.* **2008**, *93*, 1351–1358.

24. Pantos, C.; Mourouzis, I.; Markakis, K.; Tsagoulis, N.; Panagiotou, M.; Cokkinos, D.V. Long term thyroid hormone administration reshapes left ventricular chamber and improves cardiac function after myocardial infarction in rats. *Basic Res. Cardiol.* **2008**, *103*, 308–318.

25. Chen, Y.F.; Kobayashi, S.; Chen, J.; Redetzke, R.A.; Said, S.; Liang, Q.; Gerdes, A.M. Short term triiodo-lthyronine treatment inhibits cardiac myocyte apoptosis in border area after myocardial infarction in rats. *J. Mol. Cell. Cardiol.* **2008**, *44*, 180–187.

26. Forini, F.; Lionetti, V.; Ardehali, H.; Pucci, A.; Cecchetti, F.; Ghanefar, M.; Nicolini, G.; Ichikawa, Y.; Nannipieri, M.; Recchia, F.A.; *et al.* Early long-term L-T3 replacement rescues mitochondria and prevents ischemic cardiac remodelling in rats. *J. Cell. Mol. Med.* **2011**, *5*, 514–524.

27. Forini, F.; Kusmic, C.; Nicolini, G.; Mariani, L.; Zucchi, R.; Matteucci, M.; Iervasi, G.; Pitto, L. Triiodothyronine prevents cardiac ischemia/reperfusion mitochondrial impairment and cell loss by regulating miR30a/p53 axis. *Endocrinology* **2014**, *155*, 4581–4590.

28. Saba, A.; Donzelli, R.; Colligiani, D.; Raffaelli, A.; Nannipieri, M.; Kusmic, C.; dos Remedios, C.G.; Simonides, W.S.; Iervasi, G.; Zucchi, R. Quantification of thyroxine and 3,5,3′-triiodo-thyronine in human and animal hearts by a novel liquid chromatography-tandem mass spectrometry method. *Horm. Metab. Res.* **2014**, *46*, 628–634.

29. Forini, F.; Nicolini, G.; Iervasi, G. Mitochondria as Key Targets of cardioprotection in cardiac ischemic disease: Role of thyroid hormone triiodothyronine. *Int. J. Mol. Sci.* **2015**, *16*, 6312–6336.

30. Fischer, F.; Hamann, A.; Osiewacz, H.D. Mitochondrial quality control: An integrated network of pathways. *Trends Biochem. Sci.* **2012**, *37*, 284–292.

31. Williams, R.S.; Benjamin, I.J. Protective responses in the ischemic myocardium. HSPs play an important role in the defense mechanism against IR injury. *J. Clin. Investig.* **2000**, *106*, 813–818.

32. Martin, J.L.; Mestril, R.; HilalDandan, R.; Brunton, L.L.; Dillmann, W.H. Small heat shock proteins and protection against ischemic injury in cardiac myocytes. *Circulation* **1997**, *96*, 4343–4348.

33. Budas, G.R.; Churchill, E.N.; Disatnik, M.H.; Sun, L.; Mochly-Rosen, D. Mitochondrial import of PKCepsilon is mediated byHSO90: A role in cardioprotection from ischaemia andreperfusion injury. *Cardiovasc. Res.* **2010**, *88*, 83–92.

34. Dillmann, W.H. Heat shock proteins and protection against ischemic injury. *Infect. Dis. Obstet. Gynecol.* **1999**, *7*, 55–57.

35. Asea, A.; Kraeft, S.K.; KurtJones, E.A.; Stevenson, M.A.; Chen, L.B.; Finberg, R.W.; Koo, G.C.; Calderwood, S.K. HSP70 stimulates cytokine production through a CD14 dependant pathway, demonstrating its dual role as a chaperone and cytokine. *Nat. Med.* **2000**, *6*, 435–442.

36. Xiao, X.; Benjamin, I.J. Stress response proteins in cardiovascular disease. *Am. J. Hum. Genet.* **1999**, *64*, 685–690.

37. Vander Heide, R.S. Increased expression of HSP27 protects canine myocytes from simulated ischemia reperfusion injury. *Am. J. Physiol. Heart Circ. Physiol.* **2000**, *282*, H935–H941.

38. Ide, T.; Tsutsui, H.; Hayashidani, S.; Kang, D.; Suematsu, N.; Nakamura, K.; Utsumi, H.; Hamasaki, N.; Takeshita, A. Mitochondrial DNA damage and dysfunction associated with oxidative stress in failing hearts after myocardial infarction. *Circ. Res.* **2001**, *88*, 529–535.

39. Kim, Y.; Kim, H.D.; Kim, J. Cytoplasmic ribosomal protein S3 (rpS3) plays a pivotal role in mitochondrial DNA damage surveillance. *Biochim. Biophys. Acta* **2013**, *1833*, 2943–2952.

40. Jankevicius, G.; Hassler, M.; Golia, B.; Rybin, V.; Zacharias, M.; Timinszky, G.; Ladurner, A.G. A family of macrodomain proteins reverses cellular mono-ADP-ribosylation. *Nat. Struct. Mol. Biol.* **2013**, *20*, 508–514.

41. Rosenthal, F.; Feijs, K.L.; Frugier, E.; Bonalli, M.; Forst, A.H.; Imhof, R.; Winkler, H.C.; Fischer, D.; Caflisch, A.; Hassa, P.O.; *et al.* Macrodomain-containing proteins are new mono-ADP-ribosylhydrolases. *Nat. Struct. Mol. Biol.* **2013**, *20*, 502–507.

42. Chen, C.H.; Budas, G.R.; Churchill, E.N.; Disatnik, M.H.; Hurley, T.D.; Mochly-Rosen, D. Activation of aldehyde dehydrogenase-2 reduces ischemic damage to the heart. *Science* **2008**, *321*, 1493–1495.

43. Ma, H.; Guo, R.; Yu, L.; Zhang, Y.; Ren, J. Aldehyde dehydrogenase 2 (ALDH2) rescues myocardial ischaemia/reperfusion injury: Role of autophagy paradox and toxic aldehyde. *Eur. Heart J.* **2011**, *32*, 1025–1038.

44. Zhao, W.; Fan, G.C.; Zhang, Z.G.; Bandyopadhyay, A.; Zhou, X.; Kranias, E.G. Protection of peroxiredoxine II on oxidative stress-induced cardiomyocyte death and apoptosis. *Basic Res. Cardiol.* **2009**, *104*, 377–389.

45. Zhou, Y.; Kok, K.H.; Chun, A.C.; Wong, C.M.; Wu, H.W.; Lin, M.C.; Fung, P.C.; Kung, H.; Jin, D.Y. Mouse peroxiredoxin V is a thioredoxin peroxidase that inhibits p53-induced apoptosis. *Biochem. Biophys. Res. Commun.* **2000**, *268*, 921–927.

46. Oba, D.; Dai, S.; Keith, R.; Dimova, N.; Kingery, J.; Zheng, Y.T.; Zweier, J.; Velayutham, M.; Prabhu, S.D.; Li, Q.; *et al.* Cardiomyocyte restricted overexpression of extracellular superoxide dismutase increases nitric oxide bioavailability and reduces infarct size after ischemia/reperfusion. *Basic Res. Cardiol.* **2012**, *107*, 305.

47. Suzuki, K.; Sawa, Y.; Ichikawa, H.; Kaneda, Y.; Matsuda, H. Myocardial protection with endogenous overexpression of manganese superoxide dismutase. *Ann. Thorac. Surg.* **1999**, *68*, 1266–1271.

48. Shanmuganathan, S.; Hausenloy, D.J.; Duchen, M.R.; Yellon, D.M. Mitochondrial permeability transition pore as a target for cardioprotection in the human heart. *Am. J. Physiol. Heart Circ. Physiol.* **2005**, *289*, H237–H242.

49. Bernardi, P.; di Lisa, F. The mitochondrial permeability transition pore: Molecular nature and role as a target in cardioprotection. *J. Mol. Cell. Cardiol.* **2015**, *78*, 100–106.

50. Szondy, Z.; Mastroberardino, P.G.; Váradi, J.; Farrace, M.G.; Nagy, N.; Bak, I.; Viti, I.; Wieckowski, M.R.; Melino, G.; Rizzuto, R.; *et al.* Tissue transglutaminase (TG2) protects cardiomyocytes against ischemia/reperfusion injury by regulating ATP. *Cell Death Differ.* **2006**, *13*, 1827–1829.

51. Rossin, F.; D'Eletto, M.; Falasca, L.; Sepe, S.; Cocco, S.; Fimia, G.M.; Campanella, M.; Mastroberardino, P.G.; Farrace, M.G.; Piacentini, M. Transglutaminase 2 ablation leads to mitophagy impairment associated with a metabolic shift towards aerobic. *Cell Death Differ.* **2015**, *22*, 408–418.

52. Lapuente-Brun, E.; Moreno-Loshuertos, R.; Acín-Pérez, R.; Latorre-Pellicer, A.; Colás, C.; Balsa, E.; Perales-Clemente, E.; Quirós, P.M.; Calvo, E.; Rodríguez-Hernández, M.A.; *et al.* Supercomplex assembly determines electron flux in the mitochondrial electron transport chain. *Science* **2013**, *28*, 1567–1570.

53. Acin-Perez, R.; Salazar, E.; Brosel, S.; Yang, H.; Schon, E.A.; Manfredi, G. Modulation of mitochondrial protein phosphorylation by soluble adenylyl cyclase ameliorates cytochrome oxidase defects. *EMBO Mol. Med.* **2009**, *1*, 392–406.

54. Salvi, M.; Brunati, A.M.; Toninello, A. Tyrosine phosphorylation in mitochondria: A new frontier in mitochondrial signaling. *Free Radic. Biol. Med.* **2005**, *38*, 1267–1277.

55. Mahapatra, G.; Varughese, A.; Qinqin, J.; Lee, I.; Salomon, A.; Huttemann, M. Role of cytochrome *c* phosphorylation in regulation of respiration and apoptosis. *FASEB J.* **2015**, *29*, 725.

56. Chinopoulos, C. Which way does the citric acid cycle turn during hypoxia? The critical role of α ketoglutarate dehydrogenase complex. *J. Neurosci. Res.* **2013**, *91*, 1030–1043.

231

57. Phillips, D.; Aponte, A.M.; French, S.A.; Chess, D.J.; Balaban, R.S. Succinyl-CoA synthetase is a phosphate target for the activation of mitochondrial metabolism. *Biochemistry* **2009**, *48*, 7140–7149.

58. Taegtmeyer, H.; King, L.M.; Jones, B.E. Energy substrate metabolism, myocardial ischemia, and targets for pharmacotherapy. *Am. J. Cardiol.* **1998**, *82*, 54K–60K.

59. Taegtmeyer, H.; Goodwin, G.W.; Doenst, T.; Frazier, O.H. Substrate metabolism as a determinant for postischemic functional recovery of the heart. *Am. J. Cardiol.* **1997**, *80*, 3A–10A.

60. Olson, A.K.; Bouchard, B.; Ning, X.-H.; Isern, N.; Rosiers, C.D.; Portman, M.A. Triiodothyronine increases myocardial function and pyruvate entry into the citric acid cycle after reperfusion in a model of infant cardiopulmonary bypass. *Am. J. Physiol. Heart Circ. Physiol.* **2012**, *302*, H1086–H1093.

61. Olson, A.K.; Hyti, O.M.; Cohen, G.A.; Ning, X.-H.; Sadilek, M.; Isern, N.; Portman, M.A. Superior cardiac function via anaplerotic pyruvate in the immature swine heart after cardiopulmonary bypass and reperfusion. *Am. J. Physiol. Heart Circ. Physiol.* **2008**, *295*, H2315–H2320.

62. Kjellman, U.; Bjork, K.; Ekroth, R.; Karlsson, H.; Jagenburg, R.; Nilsson, F.; Svensson, G.; Wernerman, J. α-Ketoglutarate for myocardial protection in heart surgery. *Lancet* **1995**, *345*, 552–553.

63. Kjellman, U.W.; Bjork, K.; Ekroth, R.; Karlsson, H.; Jagenburg, R.; Nilsson, F.N.; Svensson, G.; Wernerman, J. Addition of alpha-ketoglutarate to blood cardioplegia improves cardioprotection. *Ann. Thor. Surg.* **1997**, *63*, 1625–1633.

64. Mullis-Jansson, S.L.; Argenziano, M.; Corwin, S.; Homma, S.; Weinberg, A.D.; Williams, M.; Rose, E.A.; Smith, C.R. A randomized double-blind study of the effect of triiodothyronine on cardiac function and morbidity after coronary bypass surgery. *J. Thorac. Cardiovasc. Surg.* **1999**, *117*, 1128–1134.

65. Bilsen, M.; van Nieuwenhoven, F.A.; van der Vusse, G.J. Metabolic remodelling of the failing heart: Beneficial or detrimental? *Cardiovasc. Res.* **2009**, *81*, 420–428.

66. Tanno, M.; Kuno, A. Reversal of metabolic shift in post-infarct-remodelled hearts: Possible novel therapeutic approach. *Cardiovasc. Res.* **2013**, *1*, 195–196.

67. Lou, P.H.; Zhang, L.; Lucchinetti, E.; Heck, M.; Affolter, A.; Gandhi, M.; Kienesberger, P.C.; Hersberger, M.; Clanachan, A.S.; Zaugg, M. Infarct-remodelled hearts with limited oxidative capacity boost fatty acid oxidation after conditioning against ischaemia/reperfusion injury. *Cardiovasc. Res.* **2013**, *97*, 251–261.

68. Goldenthal, M.J.; Weiss, H.R.; Marín-García, J. Bioenergetic remodeling of heart mitochondria by thyroid hormone. *Mol. Cell. Biochem.* **2004**, *265*, 97–106.

69. Kusmic, C.; Barsanti, C.; Matteucci, M.; Vesentini, N.; Pelosi, G.; Abraham, N.G.; L'Abbate, A. Up-regulation of heme oxygenase-1 after infarct initiation reduces mortality, infarct size and left ventricular remodeling: Experimental evidence and proof of concept. *J. Transl. Med.* **2014**.

70. Kristián, T.; Hopkins, I.B.; McKenna, M.C.; Fiskum, G. Isolation of mitochondria with high respiratory control from primary cultures of neurons and astrocytes using nitrogen cavitation. *J. Neurosci. Methods* **2006**, *152*, 136–143.

71. Gillet, L.C.; Navarro, P.; Tate, S.; Röst, H.; Selevsek, N.; Reiter, L.; Bonner, R.; Aebersold, R. Targeted data extraction of the MS/MS spectra generated by data-independent acquisition: A new concept for consistent and accurate proteome analysis. *Mol. Cell. Proteom.* **2012**.

72. Liu, Y.; Hüttenhain, R.; Collins, B.; Aebersold, R. Mass spectrometric protein maps for biomarker discovery and clinical research. *Expert Rev. Mol. Diagn.* **2013**, *13*, 811–825.

Mitochondria as Key Targets of Cardioprotection in Cardiac Ischemic Disease: Role of Thyroid Hormone Triiodothyronine

Francesca Forini, Giuseppina Nicolini and Giorgio Iervasi

Abstract: Ischemic heart disease is the major cause of mortality and morbidity worldwide. Early reperfusion after acute myocardial ischemia has reduced short-term mortality, but it is also responsible for additional myocardial damage, which in the long run favors adverse cardiac remodeling and heart failure evolution. A growing body of experimental and clinical evidence show that the mitochondrion is an essential end effector of ischemia/reperfusion injury and a major trigger of cell death in the acute ischemic phase (up to 48–72 h after the insult), the subacute phase (from 72 h to 7–10 days) and chronic stage (from 10–14 days to one month after the insult). As such, in recent years scientific efforts have focused on mitochondria as a target for cardioprotective strategies in ischemic heart disease and cardiomyopathy. The present review discusses recent advances in this field, with special emphasis on the emerging role of the biologically active thyroid hormone triiodothyronine (T3).

Reprinted from *Int. J. Mol. Sci.* Cite as: Forini, F.; Nicolini, G.; Iervasi, G. Mitochondria as Key Targets of Cardioprotection in Cardiac Ischemic Disease: Role of Thyroid Hormone Triiodothyronine. *Int. J. Mol. Sci.* **2015**, *16*, 6312–6331.

1. Introduction

Acute myocardial infarction (AMI) leading to ischemic heart disease is a major debilitating disease and important cause of death worldwide [1]. Deprivation of oxygen and nutrients following coronary occlusion is the primary cause of damage to the myocardium and its severity depends on the extent and duration of artery obstruction. Although timely, reperfusion effectively reduces short-term mortality, the reperfusion process itself yields additional injury, including cardiomyocyte dysfunction and death, which in the long run prompts adverse cardiac remodeling [1–3]. As a consequence, prevention or limitation of cardiac damage in the early stages of reperfusion is a crucial step in ameliorating patient prognosis.

Multiple lines of evidence show that mitochondrial functional impairments are critical determinants for myocyte loss during the acute ischemic stage, as well as for the progressive decline of surviving myocytes during the subacute and chronic stages [3–6]. Therefore, mitochondrial dysfunction is considered to be one of the major mechanisms in the pathogenesis of ischemia/reperfusion injury (IRI) and cardiomyopathy.

In spite of promising mitochondria-targeted therapeutic strategies emerging from experimental studies, very few have successfully completed clinical trials. As such, the mitochondrion is a potential untapped target for new therapies. Although ischemic pre-conditioning is a potent protective strategy first reported many decades ago [7], its utility in myocardial ischemia (MI) patients with an abrupt onset of disease undermines implementation of preconditioning in the clinical settings. Therefore, most modern approaches focus on the application of pharmacological or ischemic post-conditioning maneuvers to combat reperfusion injury and adverse cardiac remodeling [8].

Along this line, a growing body of clinical and experimental evidence shows that thyroid hormone (TH) supplementation may offer a novel option for cardiac diseases [9–12]. Indeed, 3,5,3'-triiodothyronine (T3), the biologically active form of TH, significantly declines after AMI both in animal models and in patients [13–15], with "low-T3 Syndrome" (low-T3S) being a strong independent prognostic predictor of death and major adverse cardiac events [16]. Consistently, treatment for low-T3S exerts cardioprotective effects in both humans and animal models [17–20].

Since the mitochondrion is a common effector of cardioprotective strategy and a main target of TH action [21–23], this review has a dual purpose: (1) to summarize the mitochondria-targeted noxious pathways and protective signaling that could be exploited to improve post-ischemic cardiac recovery and (2) to integrate classic and novel TH actions in a unified, mitochondria-centered picture that highlights how the crosstalk of TH with those molecular networks favors post-ischemic cardiomyocytes' survival.

2. Triggers of Mitochondrial-Dependent Cardiomyocyte Death in Ischemia/Reperfusion

2.1. Mitochondrial Dysfunction in Ischemia/Reperfusion

A wide spectrum of metabolic and ionic derangements occur in ischemia/reperfusion (I/R), culminating in mitochondrial impairment. Oxygen deprivation during ischemia arrests oxidative phosphorylation, decreasing intracellular ATP and favoring anaerobic glycolysis. The accumulation of lactic acid decreases the intracellular pH. As a consequence, the Na^+/H^+ antiporter is activated in an attempt to restore the pH. The resulting accumulation of cytosolic sodium reverses the direction of the Na^+/Ca^{2+} exchanger, leading to an increase in intracellular Ca^{2+} levels. The mitochondria act as a buffer for intracellular calcium, which ultimately causes calcium overload in the mitochondria [24,25]. This leads to an increase in ROS production from mitochondrial electron transfer complexes

I and III, which consequently causes a decrease in anti-oxidant defenses [26–28]. The increased oxygen tension at the onset of reperfusion results in a greater burst of oxidative stress [29], which worsens mitochondrial dysfunction and alters membrane properties [5,30–32]. Damage to the mitochondrial outer membrane along with activation of the proapoptotic BCL-2 proteins leads to mitochondrial outer membrane permeabilization, release of cytochrome *c*, caspase activation, and apoptosis [33]. Massive oxidative stress can lead to a sudden increase in inner mitochondrial membrane permeability that is attributable to the opening of the so-called permeability transition pore (PTP). Opening of the PTP (PTPO) is accompanied by release of ROS and calcium [34,35]; this can propagate the damage to neighboring mitochondria and culminate in activation of calcium-dependent proteases (calpains) and lipases (cPLA2), inducing necrotic cell death [30,36]. The molecular nature of the PTP remains controversial, but current evidence implicates a matrix protein, Cyclophilin-D (Cyp-D), and two inner membrane proteins, adenine nucleotide translocase (ANT) and the phosphate carrier (PiC) [4,37–39].

An array of stress-responsive signaling pathways activated during early reoxygenation or in post-ischemic wound healing has been implicated in the regulation of these mitochondrial changes, and thus represents potential targets for therapeutic intervention.

2.2. The p38 Mitogen-Activated Protein Kinase Intracellular Signaling

A highly conserved component of myocyte stress-responsiveness in I/R involves signaling through a family of serine-threonine kinase effectors known as p38 mitogen-activated protein kinase (p38MAPK). Four separate p38MAPK isoforms, including p38α, p38β, p38γ, and p38δ, have been identified. Each p38 isoform phosphorylates a diverse array of intracellular proteins including stress-responsive transcription factors [40].

This signaling cascade ultimately converges in mitochondria to enhance oxidative stress and mitochondrial-dependent cardiomyocyte death [41–43]. Among the pro-apoptotic targets activated by p38Mapk in I/R, the tumor suppressor protein p53 and Bax play key roles in determining both acute cell injury and post-ischemic adverse remodeling [44–46]. Also, it has been demonstrated that p38 MAPK plays a causative role in the inhibition of the anti-apoptotic Bcl-2 protein [47]. Accordingly, inhibitors of p38 signaling have been shown to confer protection from IRI [48].

TH Inhibits p38MAPK under Stress Conditions

TH exhibits a prominent role in the regulation of p38MAPK. In the post-ischemic rat brain, thyroxine, T4, treatment was protective through its p38-targeted anti-apoptotic and anti-inflammatory mechanism [49]. In Langendorff-perfused rat heart models of I/R, long-term T4 pretreatment or acute T3 administration markedly improved post-ischemic recovery of left ventricular performance while reducing cardiomyocyte death markers and blunting the activation of p38MAPK [50,51]. As suggested by a subsequent study, this effect was mediated at least in part by the thyroid hormone receptor α1 (TRα1) [52]. Indeed, in a mouse model of AMI, pharmacological inhibition of TRα1 further depressed post-ischemic cardiac function and was accompanied by marked activation of p38MAPK [52].

2.3. Tumor Suppressor Protein p53

2.3.1. p53 and Cardiomyocyte Death: Direct Action

Tumor suppressor protein p53 accumulates in the myocardium after myocardial infarction, and plays an important role in the progression to heart failure. It is well established that p53 can trigger apoptosis through the mitochondrial pathway [53]. For example, it can trans-activate Bax, the pro-apoptotic member of the BCL-2 family that translocates from the cytosol to mitochondria, causing the release of apoptotic proteins [54,55]. Besides its classic role, a broader role in organ homeostasis is just beginning to be understood. It has recently been reported that in response to oxidative stress, p530 accumulates in the mitochondrial matrix and triggers mitochondrial PTPO and necrosis by physical interaction with the PTP regulator Cyclophilin D (Cyp-D) [56]. p53 also plays a critical role in other important processes that regulate mitochondrial integrity but are impaired in I/R, such as mitochondrial morphology and mitophagy [57,58].

2.3.2. p53 Regulation of Mitochondrial Morphology

Mitochondria change their morphology by undergoing either fusion or fission, resulting in either elongated, tubular, interconnected mitochondrial networks or fragmented, discontinuous mitochondria, respectively [59,60]. These two opposing processes are regulated by the mitochondrial fusion proteins: mitofusin (Mfn) 1, Mfn2, and optic atrophy protein 1(OPA1); and the mitochondrial fission proteins: dynamin-related protein 1 (Drp1) and human mitochondrial fission protein 1 (hFis1). The fine balance between mitochondrial fusion and fission within a cell may be upset by a variety of factors, including oxidative stress [61] and ischemia [34,62], which can predispose the cell to apoptosis and mitochondrial PTPO [63], critical mediators of IRI.

p53 affects the mitochondrial dynamic by two opposite mechanisms that disrupt the equilibrium between fission and fusion, promoting cell death. In one way, p53 may upregulate Drp1 with consequent activation of excessive mitochondrial fission [64]. Drp1, in turn, stabilizes p53 in the mitochondria to trigger necrosis [65]. On the other hand, p53 may promote indirect, Bax-mediated, excessive mitochondrial fusion leading to cell necrosis as well [66].

2.3.3. p53 Effect on Mitophagy

In response to stress, cells have developed mitophagy, a defense mechanism that involves selective sequestration and subsequent degradation of the dysfunctional mitochondrion [67]. In I/R, mitophagy functions as an early cardioprotective response, favoring the removal of damaged mitochondria before they can cause activation of cell death [58]. The E3 ubiquitin ligase Parkin was recently discovered to play an important role in targeting damaged mitochondria for removal via autophagy in cardiomyocytes [58]. The proposed mechanism involves Mfn2 activation and Parkin recruitment from the cytosol to depolarized mitochondria [68]. Interestingly, another report showed Parkin localization to depolarized mitochondria even in the absence of Mfn2, which could indicate the presence of alternative mechanisms for Parkin translocation [69].

In the mouse heart, p53 cytosolic accumulation induces mitochondrial dysfunction by binding to Parkin and disturbing its translocation to damaged mitochondria and their subsequent clearance by mitophagy [70]. On the contrary, p53 knock-down preserved mitophagic flux under ischemia without a change in cardiac tissue ATP content [71]. Analysis of autophagic mediators acting downstream of p53 revealed that the TP53-induced glycolysis and apoptosis regulator (TIGAR) mediated the inhibition of myocyte mitophagy responsible for impairment of mitochondrial integrity and subsequent apoptosis, and this process is closely involved in p53-dependent ventricular remodeling after myocardial infarction [71].

3. Promoters of Mitochondria-Mediated Cardioprotection in Ischemia/Reperfusion

3.1. The Reperfusion Injury Salvage Pathway

A central biochemical pathway involved in cytoprotection is the phosphoinositide 3-kinase (PI3K) pathway, also known as the reperfusion injury salvage kinase (RISK) pathway. This pathway consists of a tyrosine kinase receptor (RTK) whose activation results in the recruitment of PI3K. Next, PI3K activates Akt, which in turn phosphorylates downstream kinases [72]. Regarding the heart, the literature on Akt is extensive [73] and has largely established Akt as a key pro-survival kinase in normal cardiac homeostasis and in response to injury. Classically, Akt activation promotes survival via inhibition of pro-apoptotic Bcl-2 family proteins Bax and Bad, limiting mitochondrial outer membrane (OMM) permeabilization and thereby blocking release of cytochrome c and caspase-mediated apoptosis. The RISK pathway is also implicated in PTP regulation and preservation of mitochondrial membrane potential ($\Delta\Psi$m) [74]. The glycogen synthase kinase 3-β (GSK-3β) is a key downstream target of Akt and is inactive when phosphorylated. Thus, GSK-3β phosphorylation by Akt or other upstream mediators results in inhibition of GSK-3β-activated targets. For example, inactivation of GSK-3β by Akt reduces mitochondrial Bax recruitment [75] as well as PTPO [76,77]. Enhancement of p53 activity by GSK-3β and GSK-3β interaction with CypD may have a role in mPTP opening [78,79]. The use of GSK-3β inhibitors in the post-ischemic setting is hampered by the side effect of inhibiting the physiological function of GSK-3β [80]. To overcome this limitation, selective inhibition of GSK-3β mitochondria uptake has been reported as a promising and novel approach to cardioprotection from lethal reperfusion injury [81].

The recruitment of the RISK pathway also induces phosphorylation-dependent activation of the endothelial nitric oxide synthase (eNOS), which is expected to block PTPO through its release of nitric oxide (NO) [82]. In turn, NO triggers the opening of the mitochondrial ATP-dependent potassium channels (mitoKATP) [83], a cardioprotective process that has been causally related to post-conditioning [84]. Furthermore, an increase in NO availability may enhance mitochondrial protein S-nitrosylation (SNO) and promote cardioprotection [85,86]. Finally, the RISK pathway has also been shown to confer cardioprotection against IRI by modulating Mfn1-dependent mitochondrial morphology [86,87].

Role of Thyroid Hormone in the Activation of Reperfusion Injury Salvage Pathway

THs are critically involved in the activation of the RISK pathways in both physiological and stress conditions. Rapid T3-mediated activation of PI3K by cytosolic TRα1, and subsequent activation of the Akt-mTOR signaling pathway, has been proposed as one of the mechanisms by which TH regulates physiological cardiac growth [88]. T3 administration can prevent serum starvation-induced neonatal cardiomyocyte apoptosis via Akt [89]. *In vivo* T4 treatment has been shown to cause phosphorylation of Akt and downstream signaling targets such as GSK-3β and mTOR in rat heart ventricles [90]. The Akt-mediated cardioprotective action of TH was confirmed in an experimental model of rat myocardial ischemia, where early short-term treatment of T3 reduced myocytes apoptosis through activation of Akt [19]. In a recent study, TH was found to have a dose-dependent effect on Akt phosphorylation, which may be of physiological relevance [91]. Mild activation of Akt caused by the replacement dose of TH resulted in favorable effects, while further induction of Akt signaling by higher doses of TH was accompanied by increased mortality and activation of extracellular signal-regulated kinases (ERK), some of the most well-studied kinases in relation to pathological remodeling [92]. This study may be of important therapeutic relevance because it shows that TH replacement therapy may be sufficient to restore cardiac function, while excessive TH doses may be detrimental rather than beneficial.

3.2. Inhibition of p53 Signaling

Given the detrimental effects of p53 in the myocardial IRI, this molecule may be proposed as a central hub in stress-induced apoptosis and necrosis instigated in mitochondria and may act as a novel therapeutic target [93].

Role of Thyroid Hormone in the Inhibition of p53 Signaling

Thyroid hormone is a critical regulator of p53 activity. A p53-centered anti-apoptotic action of TH has been well characterized in tumor cells [94,95]. In a rat model of post-ischemic acute stroke, TH treatment reduced cerebral infarction while limiting cell death through modulation of the p53 targets Bax and BCl2 [49]. On the other hand, p53 is able to hamper TH signaling. Early studies showed that the physical interaction of thyroid hormone receptors (TRs) with p53 inhibited the binding of TRs to the TH-responsive elements (TREs) in a concentration-dependent manner and that this interaction negatively regulated the TRs' signaling pathways [96,97]. Although these data collectively suggest that the cross-talk between p53 and TH may play an important role in physiological and pathological conditions, its role in cardiac disease evolution is only beginning to be explored. A recent paper reported a critical role for TH in inhibiting the p53-dependent activation of mitochondrial-mediated cell death in a model of cardiac I/R [98]. In this study, the low-T3S following the ischemic insult was accompanied by an up-regulation of p53 and activation of its downstream events, such as Bax induction and mitochondrial impairment. Early T3 administration at near-physiological dose improved the recovery of post-ischemic cardiac performance. At the molecular level, T3 blunted p53 and Bax up-regulation in the area at risk (AAR), thus preserving mitochondrial function and decreasing apoptosis and necrosis extent in the AAR [98]. Similarly, in cardiomyocytes exposed to oxidative stress, T3 treatment reduced cell death, preserved mitochondrial biogenesis and membrane potential, and limited p53 upregulation [98,99].

3.3. Targeting Mitochondrial Oxidative Stress

In accordance with a role for mitochondrial dysfunction and ROS production in the pathogenesis of heart disease, it has been shown that targeted mitochondrial ROS scavenging reduces remodeling whereas non-targeted ROS treatment has no effect [32]. This is in agreement with the lack of benefit provided by non-targeted anti-oxidants in the clinical arena [100,101], and supports the general concept of compartmentalized signaling. This concept implies close vicinity of signaling molecules to provide local control over second messengers; in this regard, targeted ROS scavenging, which accumulates manifold at the microdomains of ROS formation in mitochondria, might be more efficient in the specific targeting of cellular ROS signaling. One relevant paper suggests that mitochondria-targeted Bendavia may be extremely effective in preventing reperfusion-induced damage to cardiac mitochondria [102]. Mitoquinone (mitoQ), a coenzyme Q analog, easily crosses phospholipid bilayers and is driven to concentrate within mitochondria by the large electrochemical membrane potential. The respiratory chain reduces mitoQ to its active ubiquinol antioxidant form to limit myocardial I/R injury [103]. The SS-31 (Szeto-Schiller) peptide is also of interest since it is cell-permeable and specifically targeted to inner mitochondrial membranes based on its residue sequence, with an anti-oxidant dimethyltyrosine moiety. SS-31 has been shown to be taken up by the heart in an *ex vivo* reperfusion system and was protective against I/R injury [104]. The peptides SS-02 and SS-31 were also protective against cardiac I/R injury when added during reperfusion [105].

Role of Thyroid Hormone in the Inhibition of Mitochondrial Oxidative Stress

It has recently been shown that TH has a mitochondria-targeted antioxidant protective effect under *in vitro* stress conditions and after myocardial infarction *in vivo* [98,99,106]. In cultured cardiomyocytes, T3 treatment decreased oxidative stress-induced cell death while maintaining mitochondrial function [99]. These effects were prevented by inhibitors of mitoKATP channel opening, suggesting that activation of the mitoKATP channel in rescued mitochondria is an important protective mechanism elicited by T3 against oxidative stress-mediated cell death [99]. In a post-ischemic HF model, TH administration during the post-infarction period leads to normalization of the myocardial performance index, reduction of ROS level, and stimulation of cytosolic and mitochondrial anti-oxidant defenses [106]. In an experimental AMI model, T3 supplementation reduced mitochondrial superoxide production and limited inner mitochondrial membrane depolarization, thus improving mitochondrial function and cell viability [99].

3.4. Inhibition of Mitochondrial Permeability Transition Pore Opening

Since the initial reports on the existence of mitochondrial cyclophilin in the late 1980s, a vast majority of studies have recognized the crucial role of Cyp-D in PTP regulation. In animal models, inhibition of Cyp-D by either pharmacologic targeting [107,108], genetic ablation [109,110], or RNA interference [111], provides strong protection from both reperfusion injury and post ischemic HF. Although chronic pharmacological inhibition of CypD has been shown to cause metabolic reprogramming and worsening of pressure-induced HF [112], its acute inhibition to attenuate lethal IR injury holds great promise for reducing myocardial infarct size in humans [8,113]. Besides Cyp-D, several physiological regulators of mPTP function may be exploited to confer cardioprotection from I/R injury.

241

One important class of endogenous transducers of cell stress signals are the signal transducer and activator of transcription (STAT) proteins. Several STAT isoforms are expressed in the heart; among them, STAT3 is involved in the reduction of post-ischemic myocardial injury [114]. The infarct size reduction by ischemic post-conditioning is also attenuated in STAT3-KO mice [115]. STAT3 has been localized in mitochondria, where it contributes to cardioprotection by stimulating respiration and inhibiting the Ca^{2+}-induced mitochondrial PTPO [116].

Nitric oxide (NO) is another important signaling molecule that has been shown to reduce myocardial injury in a number of ischemia/reperfusion models. For example, brief periods of NO breathing reduced myocardial injury from ischemia/reperfusion in mice and pigs [117–119]. A critical process during NO-induced cardioprotection is to prevent mitochondrial PTPO potentially via targeting of the ANT component of the pore-forming complex [120].

Hypoxia inducible factor 1 α (HIF-1α) is an oxygen-sensitive transcription factor that enables aerobic organisms to adapt to hypoxia. This is achieved through the transcriptional activation of up to 200 genes, many of which are critical to cell survival [121]. Under normoxic conditions, the hydroxylation of HIF-1α by prolyl hydroxylase domain-containing (PHD) enzymes targets it for proteosomal degradation. However, under hypoxic conditions, PHD activity is inhibited, thereby allowing HIF-1α to accumulate and translocate to the nucleus, where it binds to the hypoxia-responsive element sequences of target gene promoters. Experimental studies suggest that stabilization of HIF-1α may protect the heart against the detrimental effects of acute I/R injury [121].

Role of TH in the Inhibition of Permeability Transition Pore Opening

The mechanisms underlying the myocyte-directed protective effect of HIF are not completely clear. A recent paper showed that HIF-1α stabilization, by either a pharmacological or genetic approach, protected the heart against acute IRI by inhibiting mitochondrial PTPO and decreasing mitochondrial oxidative stress [122]. Accordingly, T3 replacement has been shown to induce HIF-1α stabilization in a post-ischemic HF model, which was related to better preserved mitochondrial activity and cardiac performance [99].

3.5. Mitochondrial Biogenesis

Mitochondrial biogenesis has emerged as an important point in the multi-site control of mitochondrial function and a putative target for therapeutic intervention against cardiac IRI [123,124]. Mitochondrial biogenesis includes regulation of mitochondrial protein expression, their assembly within the mitochondrial network, and replication of mitochondrial DNA (mtDNA). Of the 1500 proteins representing the mitochondrial proteome, mtDNA provides 13 subunits of the oxidative phosphorylation system together with ribosomal and transfer RNAs; whereas more than 98% of the mitochondrial protein requirement is encoded by the nuclear genome [125]. Hence, a spatial and temporal coordination of nuclear and mitochondrial genomes is necessary to ensure that all mitochondrial components are available for correct assembly. The master regulator of the process is the nuclear-encoded peroxisome proliferator-activated receptor-γ coactivator-1α (PGC1-α). PGC-1α lacks DNA-binding activity but interacts and coactivates numerous transcription factors driving mitochondrial biogenesis, energy metabolism, fatty acid oxidation, and antioxidant activity [126]. In particular, PGC-1α activates nuclear transcription factors (NTFs) leading to upregulation of nuclear-encoded proteins. Nuclear-encoded proteins are imported into mitochondria through the outer-membrane (TOM) or inner-membrane (TIM) translocase transport machinery. Finally, nuclear- and mitochondrial-encoded subunits of the respiratory chain are assembled [127]. A downregulation of the entire pathway of mitochondrial biogenesis was reported both in AMI and in HF evolution [128,129]. Reduced PGC-1α activity and gene expression have been observed in several experimental models of pathologic cardiac hypertrophy, and HF [130,131] and has been involved in the pathogenesis of human heart disease [132]. It has been shown that in the heart, pathological stressors such as ischemia are associated with a downregulation of mitochondrial biogenesis via PGC-1α activity [133], and that impairment of the PGC-1α-mediated mitochondrial biogenesis increased heart vulnerability to IRI [134]. Accordingly, upregulation of the PGC-1α pathway confers protection against simulated I/R in cardiomyoblast cells [135]. Moreover, the induction of PGC-1α protein upregulates a broad spectrum of ROS detoxification systems, such as superoxide dismutase 2 (SOD2) and glutathione peroxidase-1 [136]. Hence, a putative mechanism whereby the mitochondrial biogenesis program may additionally augment tolerance to cardiac ischemia is via ROS detoxification.

During mitochondrial biogenesis, the coordinated transcription and replication of the mitochondrial genome is carried out via the nuclear-encoded mitochondrial transcription factor A (mtTFA), a downstream effector of PGC1-α signaling. It has long been recognized that post-ischemic adverse remodeling is frequently associated with qualitative and quantitative defects in mtDNA [137–139], and that a decline in mitochondrial function and mtDNA copy number play a major role in the development of post-ischemic heart disease [31,140]. In accordance, targeted disruption of mtTFA specifically within cardiac tissue resulted in a significant decrease in electron transport capacity, spontaneous cardiomyopathy, and cardiac disease [141,142]. Conversely, increasing the expression of mtTFA within cardiac tissue offered protection from adverse remodeling induced by myocardial infarction [143].

Given the causative role of mitochondrial dis-homeostasis in adverse cardiac remodeling, understanding the stimuli, signals, and transducers that govern mitochondrial biogenesis pathways may have critical significance in the treatment of ischemic cardiovascular disorders. In the last few years, the NAD^+-dependent protein deacetylase sirtuin 1 (SIRT1) has emerged as an important regulator of mitochondrial biogenesis [144,145]. Besides its epigenetic role in silencing of transcription by heterochromatin formation through histones modification, SIRT1 influences the activity of PGC-1α through a functional protein–protein interaction [146]. SIRT1 activation of PGC-1α enhances mitochondrial biogenesis, optimizes mitochondrial surface/volume ratio to reduce ROS production, and mounts an antioxidant defense [146,147]. Several lines of evidence show that SIRT1 has pivotal roles in cardiovascular function. Transgenic mice that overexpress SIRT1 in the heart are resistant to oxidative stress-related cardiac hypertrophy and ischemia/reperfusion injury [148,149]. In addition, the putative SIRT1 activator, resveratrol, a recognized mediator of mitochondrial biogenesis, can ameliorate heart ischemia or reperfusion injury, improve vascular functions, and ameliorate Ang II-induced cardiac remodeling [150,151].

Thyroid Hormone Is Key Regulator of Mitochondrial Biogenesis

Thyroid hormone plays a crucial role in regulation of mitochondrial biogenesis in both physiological and pathological conditions [24]. PGC-1α is rapidly and strongly induced by TH. PGC-1α expression and protein levels are increased 6 h after administration of T3, and this action is mediated by a TH responsive element (TRE) in the promoter [152–154]. In a rat model of post-ischemic HF, a low-T3S correlated with PGC-1α and mtTFA downregulation, which corresponded to decreased mitochondrial function in the border zone; T3 replacement rescued myocardial contractility and hemodynamic parameters, while maintaining the expression of PGC-1α and mtTFA and mitochondrial function [99]. Since the PGC-1α pathway is downregulated by p53 activation under oxidative stress conditions, the inhibitory role of T3 on p53 expression may be part of an additional and indirect mechanism by which TH controls PGC1-α levels in the post-cardiac ischemia setting [155].

The reduced PGC1-α level in the post-ischemic low-T3S is consistent with the activation of a fetal metabolic pathway observed in cardiomyopathy that is characterized by a preference for glucose over fat as a substrate for oxidative phosphorylation. Although such changes lower the oxygen consumed per ATP produced, the yield of ATP per substrate also decreases. Such inefficient metabolism lowers ATP and phosphocreatine levels and decreases metabolic reserve and flexibility, leading to pump dysfunction [156,157]. Therefore, T3 supplementation in low T3 post-ischemic cardiomyopathy may favor the normal mitochondrial homeostasis and metabolic flexibility of the heart, preventing adverse cardiac remodeling and HF evolution.

Thyroid-stimulated mitochondrial biogenesis appears to be mediated via specific TRs located in both the nuclear and mitochondrial compartments [21,22]. Wrutniak-Cabello et al. [158] reported the discovery in mitochondria of two N-terminally truncated forms of the T3 nuclear receptor, TRα1, with molecular weights of 43 and 28 kDa, respectively. While the function of p28 remains unknown, p43 is a T3-dependent transcription factor of the mitochondrial genome, acting through dimeric complexes involving at least two other truncated forms of nuclear receptors, mitochondria retinoid X receptor (mtRXR) and mitochondrial peroxisome proliferator-activated receptor (mtPPAR); p43 activation by T3 stimulates mitochondrial protein synthesis, respiratory chain activity, and mitochondriogenesis [23]. Similarly, Saelim et al. [159] reported that T3 bound TH truncated receptor isoforms (TRs) target mitochondria where they modulate inositol 1,4,5 trisphosphate (IP3)-mediated Ca^{2+} signaling [160] to inhibit apoptotic potency.

3.6. MicroRNAs

MicroRNAs (miRNAs) are a subset of regulatory molecules involved in several cellular processes of cardiac remodeling and heart failure (HF) [161–164], and have become an intriguing target for therapeutic interventions [165]. In response to diverse cardiac stresses such as myocardial I/R, miRNAs are reported to be up- or downregulated [166–169]. Some of them have recently attracted attention as regulators of mitochondrial function and mitochondrial cell death signaling in both myocardial I/R and in vitro models of oxidative stress [170–174].

Role of Thyroid Hormone in the Regulation of Cardioprotective miRNA

The miR-30 family members are abundantly expressed in the mature heart, but they are significantly downregulated in experimental I/R and in vitro after oxidative stress [174–176]. Li et al. [173] report that miR-30 family members are able to regulate apoptosis by targeting the mitochondrial fission machinery. In exploring the underlying molecular mechanism, they identified that miR-30 family members inhibited mitochondrial fission by suppressing the expression of p53 and its downstream target, dynamin-related protein 1 (Drp1) [174]. Therefore, maintenance of miR-30 levels in I/R may be regarded as cardioprotective. In a rat model of I/R, early short-term T3 supplementation at near-physiological dose maintained the post-ischemic level of miR-30a, leading to a depression of p53 and inhibition of p53 detrimental effects on mitochondria [98]. In turn, p53 is responsible for the post-ischemic inhibition of miR-499, another highly expressed cardiac miRNA with a key role in the regulation of mitochondrial dynamic [64]. MiR-499 levels are reduced in experimental ischemia as well as in anoxic cardiomyocytes, and this reduction is causally linked to apoptosis and the severity of myocardial infarction and cardiac dysfunction induced by I/R [64]. MiR-499 inhibits cardiomyocyte apoptosis through its suppression of calcineurin-mediated dephosphorylation of Drp1, thereby decreasing Drp1 accumulation in mitochondria and Drp1-mediated activation of the mitochondrial fission program [64].

4. Closing Remarks and Conclusions

In the last three decades, several biochemical pathways conveying I/R deleterious effects to the mitochondria have been characterized. In parallel, several endogenous protective molecules that enhance mitochondrial survival have been identified and, consequently, prevent progression to HF. As depicted in Figure 1, TH acts on both noxious and beneficial pathways to induce cardioprotection from IRI. Therefore, treatment of post-ischemic low-T3S appears to be a promising modality for reducing mitochondrial-driven IRI and preventing progression to HF. However, from a translational point of view, there are still some unsolved issues regarding the dose and timing of TH administration after AMI. To complicate the picture, a low-T3S in the very first hours after AMI is considered protective since it lowers the energetic demand and predisposes the heart to regenerative repair [177]. On the other hand, previous studies have demonstrated that post-MI LV remodeling, a major determinant of morbidity and mortality in overt HF [178,179], is an early process. As a consequence, it is expected that an efficacious intervention aimed at preventing the initial stages of remodeling would better contrast the progression towards HF [176]. In accordance, Henderson *et al.* [18] showed that L-T3 replacement, initiated one week after MI, improved ventricular performance without reversing cardiac remodeling. On the other hand, early T3 replacement limited post-ischemic cardiomyocyte loss and blunted adverse cardiac remodeling [19,98,99]. With TH administration, it is also critical to choose the right dose in order to limit cardiac remodeling and avoid the potentially adverse systemic effects (*i.e.*, thyrotoxicosis). In a previous study, an immediate long-term, but not controlled, supplementation of TH at a high dose in post-MI improved LV function and prevented cardiac remodeling, but also induced a thyrotoxic state [20], which in the long run may lead to heart dysfunction. These results were confirmed in a successive study demonstrating the dose-dependent bimodal effects of TH administration [91].

In conclusion, if we exclude the hyperacute post-MI period, the available evidence suggests that TH should be administered at a physiological or near-physiological dose in the early phase of the post-ischemic wound healing following reactivation of the endogenous regenerative process in order to obtain the maximal protective effect.

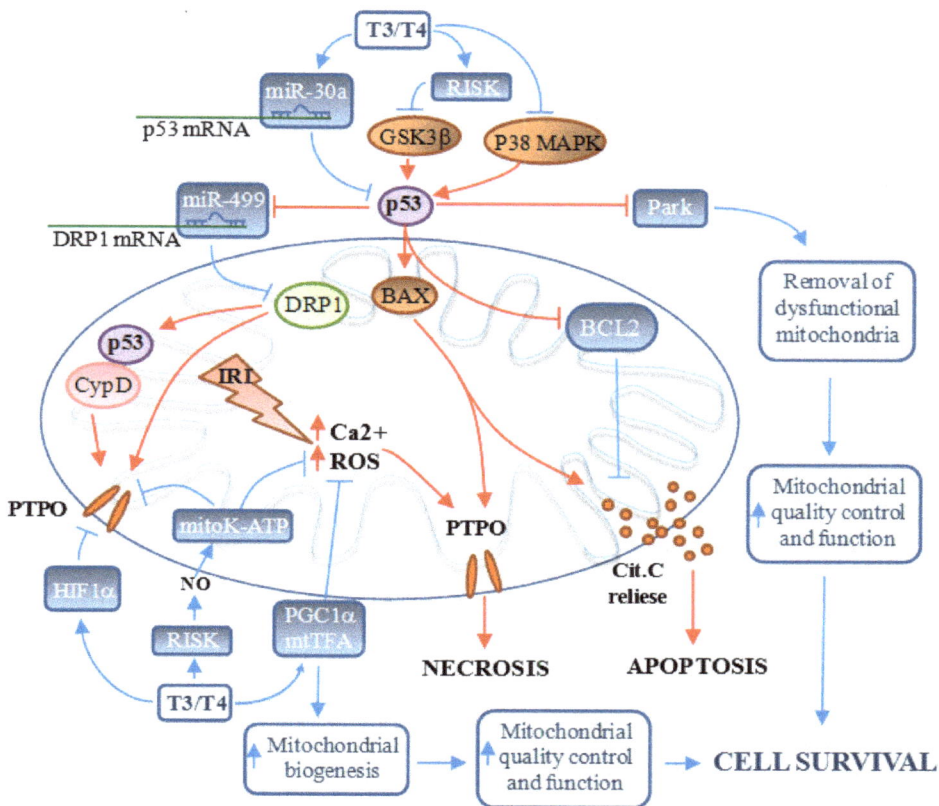

Figure 1. Schematic overview of the role of TH in the modulation of the mitochondrial pro-survival (blue connectors) or pro-death (red connectors) signaling networks that control cardiomyocyte fate in the I/R heart. CypD = Cyclophilin D; DRP1 = dynamin-related protein 1; GSK3β = glycogen synthase kinase 3-β; HIF1α = Hypoxia inducible factor 1 α, IRI = ischemia/reperfusion injuries; mitoK-ATP = mitochondrial ATP-dependent potassium channel; mtTFA = mitochondrial transcription factor A; Park = parkin; PTPO = permeability transition pore opening; PGC1-α = peroxisome proliferator-activated receptor-γ coactivator-1α; RISK = reperfusion injury salvage kinase.

Acknowledgments: This work was funded by the Tuscany Region Research Grant (DGR 1157/2011) "Study of the molecular, biochemical and metabolic mechanisms involved in the cardioprotective effect of T3".

Conflicts of Interest: The authors declare no conflict of interest.

Abbreviations

AMI	acute myocardial infarction
CypD	Cyclophilin D
DRP1	dynamin-related protein
GSK3β	glycogen synthase kinase 3-β
HIF1α	Hypoxia inducible factor 1 α
IRI	ischemia/reperfusion injuries
Low T3 syndrome	Low-T3S
mitoK-ATP	mitochondrial ATP-dependent potassium channel
mtTFA	mitochondrial transcription factor A
Park	parkin
PGC1-α	peroxisome proliferator-activated receptor-γ coactivator-1α
PTPO	permeability transition pore opening
P38MAPK	P38 mitogen activated protein kinase
RISK	reperfusion injury salvage kinase

References

1. Yellon, D.M.; Hausenloy, D.J. Myocardial reperfusion injury. *N. Engl. J. Med.* **2007**, *357*, 1121–1135.
2. Guzy, R.D.; Hoyos, B.; Robin, E.; Chen, H.; Liu, L.; Kyle, D.; Mansfield, K.D.; Simon, M.C.; Hammerling, U.; Schumacker, P.T. Mitochondrial complex III is required for hypoxia-induced ROS production and cellular oxygen sensing. *Cell Metab.* **2005**, *1*, 401–408.
3. Whelan, R.S.; Kaplinskiy, V.; Kitsis, R.N. Cell death in the pathogenesis of heart disease mechanisms and significance. *Annu. Rev. Physiol.* **2010**, *72*, 19–44.
4. Baines, C.P. The mitochondrial permeability transition pore and ischemia-reperfusion injury. *Basic Res. Cardiol.* **2009**, *104*, 181–188.
5. Marín-García, J.; Goldenthal, M.J. Mitochondrial centrality in heart failure. *Heart Fail. Rev.* **2008**, *13*, 137–150.
6. Galluzzi, L.; Kepp, O.; Kroemer, G. Mitochondria: Master regulators of danger signaling. *Nat. Rev. Mol. Cell Biol.* **2012**, *13*, 780–788.
7. Murry, C.E.; Jennings, R.B.; Reimer, K.A. Preconditioning with ischemia: A delay of lethal cell injury in ischemic myocardium. *Circulation* **1986**, *74*, 1124–1136.
8. Ovize, M.; Thibault, H.; Przyklenk, K. Myocardial conditioning: Opportunities for clinical translation. *Circ. Res.* **2013**, *113*, 439–450.
9. Gerdes, A.M.; Iervasi, G. Thyroid replacement therapy and heart failure. *Circulation* **2010**, *122*, 385–393.

10. Pingitore, A.; Chen, Y.; Gerdes, A.M.; Iervasi, G. Acute myocardial infarction and thyroid function: New pathophysiological and therapeutic perspectives. *Ann. Med.* **2012**, *44*, 745–757.

11. Mourouzis, I.; Forini, F.; Pantos, C.; Iervasi, G. Thyroid hormone and cardiac disease: From basic concepts to clinical application. *J. Thyroid Res.* **2011**, *2011*.

12. Nicolini, G.; Pitto, L.; Kusmic, C.; Balzan, S.; Sabatino, L.; Iervasi, G.; Forini, F. New insights into mechanisms of cardioprotection mediated by thyroid hormones. *J. Thyroid Res.* **2013**, *2013*.

13. Hamilton, M.A.; Stevenson, L.W.; Luu, M.; Walden, J.A. Altered thyroid hormone metabolism in advanced heart failure. *J. Am. Coll. Cardiol.* **1990**, *16*, 91–95.

14. Wiersinga, W.M.; Lie, K.I.; Touber, J.L. Thyroid hormones in acute myocardial infarction. *Clin. Endocrinol. (Oxf.)* **1981**, *14*, 367–374.

15. Friberg, L.; Drvota, V.; Bjelak, A.H.; Eggertsen, G.; Ahnve, S. Association between increased levels of reverse triiodothyronine and mortality after acute myocardial infarction. *Am. J. Med.* **2001**, *111*, 699–703.

16. Iervasi, G.; Pingitore, A.; Landi, P.; Raciti, M.; Ripoli, A.; Scarlattini, M.; L'Abbate, A.; Donato, L. Low-T3 syndrome: A strong prognostic predictor of death in patients with heart disease. *Circulation* **2003**, *107*, 708–713.

17. Pingitore, A.; Galli, E.; Barison, A.; Iervasi, A.; Scarlattini, M.; Nucci, D.; L'abbate, A.; Mariotti, R.; Iervasi, G. Acute effects of triiodothyronine (T3) replacement therapy in patients with chronic heart failure and low-T3 syndrome: A randomized, placebo-controlled study. *J. Clin. Endocrinol. Metab.* **2008**, *93*, 1351–1358.

18. Henderson, K.K.; Danzi, S.; Paul, J.T.; Leya, G.; Klein, I.; Samarel, A.M. Physiological replacement of t3 improves left ventricular function in an animal model of myocardial infarction-induced congestive heart failure. *Circ. Heart Fail.* **2009**, *2*, 243–252.

19. Chen, Y.F.; Kobayashi, S.; Chen, J.; Redetzke, R.A.; Said, S.; Liang, Q.; Gerdes, A.M. Short term triiodo-Lthyronine treatment inhibits cardiac myocyte apoptosis in border area after myocardial infarction in rats. *J. Mol. Cell. Cardiol.* **2008**, *44*, 180–187.

20. Pantos, C.; Mourouzis, I.; Markakis, K.; Tsagoulis, N.; Panagiotou, M.; Cokkinos, D.V. Long term thyroid hormone administration reshapes left ventricular chamber and improves cardiac function after myocardial infarction in rats. *Basic Res. Cardiol.* **2008**, *103*, 308–318.

21. Wrutniak-Cabello, C.; Casas, F.; Cabello, G. Thyroid hormone action in mitochondria. *J. Mol. Endocrinol.* **2001**, *26*, 67–77.

22. Goldenthal, M.J.; Ananthakrishnan, R.; Marín-García, J. Nuclear-mitochondrial cross-talk in cardiomyocyte T3 signaling: A time-course analysis. *J. Mol. Cell. Cardiol.* **2005**, *39*, 319–326.

23. Marín-García, J. Thyroid hormone and myocardial mitochondrial biogenesis. *Vascul. Pharmacol.* **2010**, *52*, 120–130.

24. Shintani-Ishida, K.; Inui, M.; Yoshida, K. Ischemia-reperfusion induces myocardial infarction through mitochondrial Ca^{2+} overload. *J. Mol. Cell. Cardiol.* **2012**, *53*, 233–239.

25. Herzig, S.; Maundrell, K.; Martinou, J.C. Life without the mitochondrial calcium uniporter. *Nat. Cell Biol.* **2013**, *15*, 1398–1400.

26. Tompkinsa, A.J.; Burwellb, L.; Digernessc, S.B.; Zaragozac, C.; Holmanc, W.L.; Brookesa, P.S. Mitochondrial dysfunction in cardiac ischemia-reperfusion injury: ROS from complex I, without inhibition. *BBA Mol. Basis Dis.* **2006**, *1762*, 223–231.

27. Dorn, G.W., II. Apoptotic and non-apoptotic programmed cardiomyocyte death in ventricular remodeling. *Cardiovasc. Res.* **2009**, *81*, 465–473.

28. Dröse, S.1.; Brandt, U. Molecular mechanisms of superoxide production by the mitochondrial respiratory chain. *Adv. Exp. Med. Biol.* **2012**, *748*, 145–169.

29. Becker, L.B. New concepts in reactive oxygen species and cardiovascular reperfusion physiology. *Cardiovasc. Res.* **2004**, *61*, 461–470.

30. Assaly, R.; d'Anglemont de Tassigny, A.; Paradis, S.; Jacquin, S.; Berdeaux, A.; Morin, D. Oxidative stress, mitochondrial permeability transition pore opening and cell death during hypoxia-reoxygenation in adult Cardiomyocytes. *Eur. J. Pharmacol.* **2012**, *675*, 6–14.

31. Ide, T.; Tsutsui, H.; Hayashidani, S.; Kang, D.; Suematsu, N.; Nakamura, K.; Utsumi, H.; Hamasaki, N.; Takeshita, A. Mitochondrial DNA damage and dysfunction associated with oxidative stress in failing hearts after myocardial infarction. *Circ. Res.* **2001**, *88*, 529–535.

32. Dai, D.F.; Chen, T.; Szeto, H.; Nieves-Cintrón, M.; Kutyavin, V.; Santana, L.F.; Rabinovitch, P.S. Mitochondrial targeted antioxidant Peptide ameliorates hypertensive cardiomyopathy. *J. Am. Coll. Cardiol.* **2011**, *58*, 73–82.

33. Baines, C.P. The cardiac mitochondrion: Nexus of stress. *Annu. Rev. Physiol.* **2010**, *72*, 61–80.

34. Brady, N.R.; Hamacher-Brady, A.; Gottlieb, R.A. Proapoptotic BCL-2 family members and mitochondrial dysfunction during ischemia/reperfusion injury, a study employing cardiac HL-1 cells and GFP biosensors. *Biochim. Biophys. Acta* **2006**, *1757*, 667–678.

35. Zorov, D.B.; Filburn, C.R.; Klotz, L.O.; Zweier, J.L.; Sollott, S.J. Reactive oxygen species (ROS)-induced ROS release: A new phenomenon accompanying induction of the mitochondrial permeability transition in cardiac myocytes. *J. Exp. Med.* **2000**, *192*, 1001–1014.

36. Garcia-Dorado, D.; Ruiz-Meana, M.; Inserte, J.; Rodriguez-Sinovas, A.; Piper, H.M. Calcium-mediated cell death during myocardial reperfusion. *Cardiovasc. Res.* **2012**, *94*, 168–180.

37. McStay, G.P.; Clarke, S.J.; Halestrap, A.P. Role of critical thiol groups on the matrix surface of the adenine nucleotide translocase in the mechanism of the mitochondrial permeability transition pore. *Biochem. J.* **2002**, *367*, 541–548.

38. Leung, A.W.C.; Varanyuwatana, P.; Halestrap, A.P. The mitochondrial phosphate carrier interacts with cyclophilin D and may play a key role in the permeability transition *J. Biol. Chem.* **2008**, *283*, 26312–26323.

39. Kwong, J.; Davis, C.P.; Baines, M.A.; Sargent, J.; Karch, X.; Wang, X.; Huang, T.; Molkentin, J.D. Genetic deletion of the mitochondrial phosphate carrier desensitizes the mitochondrial permeability transition pore and causes cardiomyopathy. *Cell Death Differ.* **2014**, *21*, 1209–1217.

40. Paul, A.; Wilson, S.; Belham, C.M.; Robinson, C.J.; Scott, P.H.; Gould, G.W.; Plevin, R. Stress-activated protein kinases: Activation, regulation and function. *Cell Signal.* **1997**, *9*, 403–410.

41. Ma, X.L.; Kumar, S.; Gao, F.; Louden, C.S.; Lopez, B.L.; Christopher, T.A.; Wang, C.; Lee, J.C.; Feuerstein, G.Z.; Yue, T.L. Inhibition of p38 mitogen-activated protein kinase decreases cardiomyocyte apoptosis and improves cardiac function after myocardial ischemia and reperfusion. *Circulation* **1999**, *99*, 1685–1689.

42. Dhingra, S.; Sharma, A.K.; Singla, D.K.; Singal, P.K. p38 and ERK1/2 MAPKs mediate the interplay of TNF-α and IL-10 in regulating oxidative stress and cardiac myocyte apoptosis. *Am. J. Physiol. Heart Circ. Physiol.* **2007**, *293*, H3524–H3531.

43. Ashraf, M.I.; Ebner, M.; Wallner, C.; Haller, M.; Khalid, S.; Schwelberger, H.; Koziel, K.; Enthammer, M.; Hermann, M.; Sickinger, S.; *et al.* A p38MAPK/MK2 signaling pathway leading to redox stress, cell death and ischemia/reperfusion injury. *Cell Commun. Signal.* **2014**, *12*.

44. Ren, J.; Zhang, S.; Kovacs, A.; Wang, Y.; Muslin, A.J. Role of p38α MAPK in cardiac apoptosis and remodeling after myocardial infarction. *J. Mol. Cell. Cardiol.* **2005**, *38*, 617–623.

45. Capano, M.; Crompton, M. Bax translocates to mitochondria of heart cells during simulated ischaemia: Involvement of AMP-activated and p38 mitogen-activated protein kinases. *Biochem. J.* **2006**, *395*, 57–64.

46. Martin, E.D.; de Nicola, G.F.; Marber, M.S. New therapeutic targets in cardiology: p38 α mitogen-activated protein kinase for ischemic heart disease. *Circulation* **2012**, *126*, 357–368.

47. Kaiser, R.A.; Bueno, O.F.; Lips, D.J.; Doevendans, P.A.; Jones, F.; Kimball, T.F.; Molkentin, J.D. Targeted inhibition of p38 mitogen-activated protein kinase antagonizes cardiac injury and cell death following ischemia-reperfusion *in vivo. J. Biol. Chem.* **2004**, *279*, 15524–15530.

48. Jeong, C.W.; Yoo, K.Y.; Lee, S.H.; Jeong, H.J.; Lee, C.S.; Kim, S. Curcumin protects against regional myocardial ischemia/reperfusion injury through activation of RISK/GSK-3β and inhibition of p38 MAPK and JNK. *J. Cardiovasc. Pharmacol. Ther.* **2012**, *17*, 387–394.

49. Genovese, T.; Impellizzeri, D.; Ahmad, A.; Cornelius, C.; Campolo, M.; Cuzzocrea, S.; Esposito, E. Post ischaemic thyroid hormone treatment in a rat model of acute stroke. *Brain Res.* **2013**, *1513*, 92–102.

50. Pantos, C.; Malliopoulou, V.; Mourouzis, I.; Karamanoli, E.; Tzeis, S.M.; Carageorgiou, H.; Varonos, D.; Cokkinos, D.V. Long-term thyroxine administration increases HSP70 mRNA expression and attenuates p38 MAP kinase activity in response to ischaemia. *J. Endocrinol.* **2001**, *70*, 207–215.

51. Pantos, C.; Mourouzis, I.; Saranteas, T.; Clavé, G.; Ligeret, H.; Noack-Fraissignes, P.; Renard, P.Y.; Massonneau, M.; Perimenis, P.; Spanou, D.; *et al.* Thyroid hormone improves postischaemic recovery of function while limiting apoptosis: A new therapeutic approach to support hemodynamics in the setting of ischaemia-reperfusion? *Basic Res. Cardiol.* **2009**, *104*, 69–77.

52. Mourouzis, I.; Kostakou, E.; Galanopoulos, G.; Mantzouratou, P.; Pantos, C. Inhibition of thyroid hormone receptor α1 impairs post ischemic cardiac performance after myocardial infarction in mice. *Mol. Cell. Biochem.* **2013**, *379*, 97–105.

53. Vousden, K.H. p53, death star. *Cell* **2000**, *103*, 691–694.

54. Miyashita, T.; Reed, J. Tumor suppressor p53 is a direct transcriptional activator of the human bax gene. *Cell* **1995**, *80*, 293–299.

55. Long, X.; Boluyt, M.O.; Hipolito, M.; Lundberg, M.S.; Zheng, J.S.; O'Neill, L.; Cirielli, C.; Lakatta, E.G.; Crow, M.T. p53 and the hypoxia-induced apoptosis of cultured neonatal rat cardiac myocytes. *J. Clin. Investig.* **1997**, *99*, 2635–2643.

56. Vaseva, A.V.; Marchenko, N.D.; Ji, K.; Tsirka, S.E.; Holzmann, S.; Moll, U.M. p53 opens the mitochondrial permeability transition pore to trigger necrosis. *Cell* **2012**, *149*, 1536–1548.

57. Ong, S.B.; Hausenloy, D.J. Mitochondrial morphology and cardiovascular disease. *Cardiovasc. Res.* **2010**, *88*, 16–29.

58. Dieter, A.; Kubli, Å.B. Gustafsson mitochondria and mitophagy: The yin and yang of cell death control. *Circ. Res.* **2012**, *111*, 1208–1221.

59. Dimmer, K.S.; Scorrano, L. (De)constructing mitochondria: What for? *Physiology* **2006**, *21*, 233–241.

60. Hausenloy, D.J.; Scorrano, L. Targeting cell death. *Clin. Pharmacol. Ther.* **2007**, *82*, 370–373.

61. Shen, T.; Zheng, M.; Cao, C.; Chen, C.; Tang, J.; Zhang, W.; Cheng, H.; Chen, K.H.; Xiao, R.P. Mitofusin-2 is a major determinant of oxidative stress-mediated heart muscle cell apoptosis. *J. Biol. Chem.* **2007**, *282*, 23354–23361.

62. Ong, S.B.; Subrayan, S.; Lim, S.Y.; Yellon, D.M.; Davidson, S.M.; Hausenloy, D.J. Inhibiting mitochondrial fission protects the heart against ischemia/reperfusion injury. *Circulation* **2010**, *121*, 2012–2202.

63. Kong, D.; Xu, L.; Yu, Y.; Zhu, W.; Andrews, D.W.; Yoon, Y.; Kuo, T.H. Regulation of Ca^{2+}-induced permeability transition by Bcl-2 is antagonized by Drpl and hFis1. *Mol. Cell Biochem.* **2005**, *272*, 187–199.

64. Wang, J.; Jiao, J.; Li, Q.; Long, B.; Wang, K.; Liu, J.; Li, Y.; Li, P. miR-499 regulates mitochondrial dynamics by targeting calcineurin and dynamin-related protein-1. *Nat. Med.* **2011**, *17*, 71–78.

65. Guo, X.; Sesaki, H.H.; Qi, X. Drp1 stabilizes p53 on the mitochondria to trigger necrosis under oxidative stress conditions *in vitro* and *in vivo*. *Biochem. J.* **2014**, *461*, 137–146.

66. Whelan, R.S.; Konstantinidis, K.; Wei, A.C.; Chen, Y.; Reyna, D.E.; Jha, S.; Yang, Y.; Calvert, J.W.; Lindsten, T.; Thompson, C.B.; *et al.* Bax regulates primary necrosis through mitochondrial dynamics. *Proc. Natl. Acad. Sci. USA* **2012**, *109*, 6566–6571.

67. Wohlgemuth, S.E.; Calvani, R.; Marzetti, E. The interplay between autophagy and mitochondrial dysfunction in oxidative stress-induced cardiac aging and pathology. *J. Mol. Cell. Cardiol.* **2014**, *71*, 62–70.

68. Chen, Y.; Dorn, G.W., II. PINK1-phosphorylated mitofusin 2 is a Parkin receptor for culling damaged mitochondria. *Science* **2013**, *340*, 471–475.

69. Narendra, D.; Tanaka, A.; Suen, D.F.; Youle, R.J. Parkin is recruited selectively to impaired mitochondria and promotes their autophagy. *J. Cell Biol.* **2008**, *183*, 795–803.

70. Hoshino, A.; Mita, Y.; Okawa, Y.; Ariyoshi, M.; Iwai-Kanai, E.; Ueyama, T.; Ikeda, K.; Ogata, T.; Matoba, S. Cytosolic p53 inhibits Parkin-mediated mitophagy and promotes mitochondrial dysfunction in the mouse heart. *Nat. Commun.* **2013**, *4*.

71. Hoshino, A.; Matoba, S.; Iwai-Kanai, E.; Nakamura, H.; Kimata, M.; Nakaoka, M.; Katamura, M.; Okawa, Y.; Ariyoshi, M.; Mita, Y.; *et al.* p53-TIGAR axis attenuates mitophagy to exacerbate cardiac damage after ischemia. *J. Mol. Cell. Cardiol.* **2012**, *52*, 1175–1184.

72. Cohen, P.; Frame, S. The renaissance of GSK3B. *Nat. Rev. Mol. Cell Biol.* **2001**, *2*, 769–776.

73. Sussman, M.A.; Völkers, M.; Fischer, K.; Bailey, B.; Cottage, C.T.; Din, S.; Gude, N.; Avitabile, D.; Alvarez, R.; Sundararaman, B.; *et al.* Myocardial AKT: The omnipresent nexus. *Physiol. Rev.* **2011**, *91*, 1023–1070.

74. Hausenloy, D.J.; Yellon, D.M. New directions for protecting the heart against ischaemia–reperfusion injury: Targeting the Reperfusion Injury Salvage Kinase (RISK)-pathway. *Cardiovasc. Res.* **2004**, *61*, 448–460.

75. Linseman, D.A.; Butts, B.D.; Precht, T.A.; Phelps, R.A.; Le, S.S.; Laessig, T.A.; Bouchard, R.J.; Florez-McClure, M.L.; Heidenreich, K.A. Glycogen synthase kinase-3β phosphorylates Bax and promotes its mitochondrial localization during neuronal apoptosis. *J. Neurosci.* **2004**, *24*, 9993–10002.

76. Juhaszova, M.; Zorov, D.B.; Kim, S.H.; Pepe, S.; Fu, Q.; Fishbein, K.W.; Ziman, B.D.; Wang, S.; Ytrehus, K.; Antos, C.L.; *et al.* Glycogen synthase kinase-3β mediates convergence of protection signaling to inhibit the mitochondrial permeability transition pore. *J. Clin. Investig.* **2004**, *113*, 1535–1549.

77. Gomez, L.; Paillard, M.; Thibault, H.; Derumeaux, G.; Ovize, M. Inhibition of GSK3β by postconditioning is required to prevent opening of the mitochondrial permeability transition pore during reperfusion. *Circulation* **2008**, *117*, 2761–2768.

78. Watcharasit, P.; Bijur, G.N.; Song, L.; Zhu, J.; Chen, X.; Jope, R.S. Glycogen synthase kinase-3b(GSK3b) binds to and promotes the actions of p53. *J. Biol. Chem.* **2003**, *278*, 48872–48879.

79. Rasola, A.; Sciacovelli, M.; Chiara, F.; Pantic, B.; Brusilow, W.S.; Bernardi, P. Activation of mitochondrial ERK protects cancer cells from death through inhibition of the permeability transition. *Proc. Natl. Acad. Sci. USA* **2010**, *107*, 726–731.

80. Phukan, S.; Babu, V.S.; Kannoji, A.; Hariharan, R.; Balaji, V.N. GSK3β: Role in therapeutic landscape and development of modulators. *Br. J. Pharmacol.* **2010**, *160*, 1–19.

81. Tanno, M.; Kuno, A.; Ishikawa, S.; Miki, T.; Kouzu, H.; Yano, T.; Murase, H.; Tobisawa, T.; Ogasawara, M.; Horio, Y.; *et al.* Translocation of Glycogen Synthase Kinase-3β (GSK-3β), a trigger of permeability transition, is kinase activity dependent and mediated by interaction with voltage-dependent anion channel 2 (VDAC2). *J. Biol. Chem.* **2014**, *289*, 29285–29296.

82. Gross, G.J.; Hsu, A.; Pfeiffer, A.W.; Nithipatikom, K. Roles of endothelial nitric oxide synthase (eNOS) and mitochondrial permeability transition pore (MPTP) in epoxyeicosatrienoic acid (EET)-induced cardioprotection against infarction in intact rat hearts. *J. Mol. Cell. Cardiol.* **2013**, *59*, 20–29.

83. Sasaki, N.; Sato, T.; Ohler, A.; O'Rourke, B.; Marbán, E. Activation of mitochondrial ATP-dependent potassium channels by nitric oxide. *Circulation* **2000**, *101*, 439–445.

84. Penna, C.; Rastaldo, R.; Mancardi, D.; Raimondo, S.; Cappello, S.; Gattullo, D.; Losano, G.; Pagliario, P. Post-conditioning induced cardioprotection requires signaling through a redox-sensitive mechanism, mitochondrial ATP-sensitive K^+ channel and protein kinase C activation. *Basic Res. Cardiol.* **2006**, *101*, 180–189.

85. Gucek, M.; Murphy, E. What can we learn about cardioprotection from the cardiac mitochondrial proteome? *Cardiovasc. Res.* **2010**, *88*, 211–218.

86. Burwell, L.S.; Brookes, P.S. Mitochondria as a target for the cardioprotective effects of nitric oxide in ischemia-reperfusion injury. *Antioxid. Redox Signal.* **2008**, *10*, 579–599.

87. Ong, S.B.; Hall, A.R.; Dongworth, R.K.; Kalkhoran, S.; Pyakurel, A.; Scorrano, L.; Hausenloy, D.J. Akt protects the heart against ischaemia/reperfusion injury by modulating mitochondrial morphology. *Thromb. Haemost.* **2015**, *113*, 513–516.

88. Kenessey, A.; Ojamaa, K. Thyroid hormone stimulates protein synthesis in the cardiomyocyte by activating the Akt-mTOR and p70S6K pathways. *J. Biol. Chem.* **2006**, *28*, 20666–20672.

89. Kuzman, J.A.; Gerdes, A.M.; Kobayashi, S.; Liang, Q. Thyroid hormone activates Akt and prevents serum starvation-induced cell death in neonatal rat cardiomyocytes. *J. Mol. Cell. Cardiol.* **2005**, *39*, 841–844.

90. Kuzman, J.A.; Vogelsang, K.A.; Thomas, T.A.; Gerdes, A.M. L-Thyroxine activates Akt signaling in the heart. *J. Mol. Cell. Cardiol.* **2005**, *39*, 251–258.

91. Mourouzis, I.; Mantzouratou, P.; Galanopoulos, G.; Kostakou, E.; Roukounakis, N.; Kokkinos, A.D.; Cokkinos, D.V.; Pantos, C. Dose-dependent effects of thyroid hormone on post-ischemic cardiac performance: Potential involvement of Akt and ERK signalings. *Mol. Cell. Biochem.* **2012**, *363*, 235–243.

92. Kehat, I.; Davis, J.; Tiburcy, M.; Accornero, F.; Saba-El-Leil, M.K.; Maillet, M.; York, A.J.; Lorenz, J.N.; Zimmermann, W.H.; Meloche, S.; *et al.* Extracellular signal-regulated kinases 1 and 2 regulate the balance between eccentric and concentric cardiac growth. *Circ. Res.* **2010**, *108*, 176–183.

93. Naito, A.T.; Okada, S.; Minamino, T.; Iwanaga, K.; Liu, M.L.; Sumida, T.; Nomura, S.; Sahara, N.; Mizoroki, T.; Takashima, A.; *et al.* Promotion of chip-mediated p53 degradation protects the heart from ischemic injury. *Circ. Res.* **2010**, *106*, 1692–1702.

94. Lin, H.Y.; Davis, P.J.; Tang, H.Y.; Mousa, S.A.; Luidens, M.K.; Hercbergs, A.H.; Davis, F.B. The pro-apoptotic action of stilbene-induced COX-2 in cancer cells: Convergence with the anti-apoptotic effect of thyroid hormone. *Cell Cycle* **2009**, *8*, 1877–1882.

95. Lin, H.Y.; Tang, H.Y.; Keating, T.; Wu, Y.H.; Shih, A.; Hammond, D.; Sun, M.; Hercbergs, A.; Davis, F.B.; Davis, P.J. Resveratrol is pro-apoptotic and thyroid hormone is anti-apoptotic in glioma cells: Both actions are integrin and ERK mediated. *Carcinogenesis* **2008**, *29*, 62–69.

96. Yap, N.; Yu, C.L.; Cheng, S.Y. Modulation of thetrascriptional activity ofthyroid hormone receptor by the tumor suppressor p53. *Proc. Natl. Acad. Sci. USA* **1996**, *93*, 4273–4277.

97. Bhat, M.K.; Yu, C.l.; Yap, N.; Zhan, Q.; Hayashi, Y.; Seth, P. Tumor suppressor p53 is a negative regulator in thyroid hormone receptor signaling pathways. *J. Biol. Chem.* **1997**, *272*, 28989–28993.

98. Forini, F.; Kusmic, C.; Nicolini, G.; Mariani, L.; Zucchi, R.; Matteucci, M.; Iervasi, G.; Pitto, L. Triiodothyronine prevents cardiac ischemia/reperfusion mitochondrial impairment and cell loss by regulating miR30a/p53 axis. *Endocrinology* **2014**, *155*, 4581–4590.

99. Forini, F.; Lionetti, V.; Ardehali, H.; Pucci, A.; Cecchetti, F.; Ghanefar, M.; Nicolini, G.; Ichikawa, Y.; Nannipieri, M.; Recchia, F.A.; *et al.* Early long-term L-T3 replacement rescues mitochondria and prevents ischemic cardiac remodelling in rats. *J. Cell. Mol. Med.* **2011**, *15*, 514–524.

100. Yusuf, S.; Dagenais, G.; Pogue, J.; Bosch, J.; Sleight, P. Vitamin E supplementation and cardiovascular events in high-risk patients. The heart outcomes prevention evaluation study investigators. *N. Engl. J. Med.* **2000**, *342*, 154–160.

101. Hare, J.M.; Mangal, B.; Brown, J.; Fisher, C., Jr.; Freudenberger, R.; Colucci, W.S.; Mann, D.L.; Liu, P.; Givertz, M.M.; Schwarz, R.P.; OPT-CHF Investigators. Impact of oxypurinol in patients with symptomatic heart failure: Results of the OPT-CHF study. *J. Am. Coll. Cardiol.* **2008**, *5*, 2301–2309.

102. Brown, D.A.; Hale, S.L.; Baines, C.P.; del Rio, C.L.; Hamlin, R.L.; Yueyama, Y.; Kijtawornrat, A.; Yeh, S.T.; Frasier, C.R.; Stewart, L.M.; *et al.* Reduction of early reperfusion injury with the mitochondria-targeting peptide bendavia. *J. Cardiovasc. Pharmacol. Ther.* **2014**, *19*, 121–132.

103. Adlam, V.J.; Harrison, J.C.; Porteous, C.M.; James, A.M.; Smith, R.A.; Murphy, M.P.; Sammut, I.A. Targeting an antioxidant to mitochondria decreases cardiac ischemia-reperfusion injury. *FASEB J.* **2005**, *19*, 1088–1095.

104. Szeto, H.H. Mitochondria-targeted cytoprotective peptides for ischemia-reperfusion injury. *Antiox. Redox Signal.* **2008**, *10*, 601–619.

105. Zhao, K.; Zhao, G.M.; Wu, D.; Soong, Y.; Birk, A.V.; Schiller, P.W.; Szeto, H.H. Cell-permeable peptide antioxidants targeted to inner mitochondrial membrane inhibit mitochondrial swelling, oxidative cell death, and reperfusion injury. *J. Biol. Chem.* **2004**, *279*, 34682–34690.

106. De Castro, A.L.; Tavares, A.V.; Campos, C.; Fernandes, R.O.; Siqueira, R.; Conzatti, A.; Bicca, A.M.; Fernandes, T.R.; Sartório, C.L.; Schenkel, P.C.; *et al.* Cardioprotective effects of thyroid hormones ina rat model of myocardial infarction are associated with oxidative stress reduction. *Mol. Cell. Endocrinol.* **2014**, *391*, 22–29.

107. Di Lisa, F.; Menabò, R.; Canton, M.; Barile, M.; Bernardi, P. Opening of the mitochondrial permeability transition pore causes depletion of mitochondrial and cytosolic NAD^+ and is a causative event in the death of myocytes in postischemic reperfusion of the heart. *J. Biol. Chem.* **2001**, *276*, 2571–2575.

108. Clarke, S.J.; McStay, G.P.; Halestrap, A.P. Sanglifehrin A acts as a potent inhibitor of the mitochondrial permeability transition and reperfusion injury of the heart by binding to cyclophilin-D at a different site from cyclosporin A. *J. Biol. Chem.* **2002**, *277*, 34793–34799.

109. Nakayama, H.; Chen, X.; Baines, C.P.; Klevitsky, R.; Zhang, X.; Zhang, H.; Jaleel, N.; Chua, B.H.; Hewett, T.E.; Robbins, J.; *et al.* Ca^{2+}- and mitochondrial-dependent cardiomyocyte necrosis as a primary mediator of heart failure. *J. Clin. Investig.* **2007**, *117*, 2431–2444.

110. Nakagawa, T.; Shimizu, S.; Watanabe, T.; Yamaguchi, O.; Otsu, K.; Yamagata, H.; Inohara, H.; Kubo, T.; Tsujimoto, Y. Cyclophilin D-dependent mitochondrial permeability transition regulates some necrotic but not apoptotic cell death. *Nature* **2005**, *434*, 652–658.

111. Kato, M.; Akao, M.; Matsumoto-Ida, M.; Makiyama, T.; Iguchi, M.; Takeda, T.; Shimizu, S.; Kita, T. The targeting of cyclophilin D by RNAi as a novel cardioprotective therapy: Evidence from two-photon imaging. *Cardiovasc. Res.* **2009**, *83*, 335–344.

112. Elrod, J.W.; Wong, R.; Mishra, S.; Vagnozzi, R.J.; Sakthievel, B.; Goonasekera, S.A.; Karch, J.; Gabel, S.; Farber, J.; Force, T.; *et al.* Cyclophilin D controls mitochondrial pore-dependent Ca^{2+} exchange, metabolic flexibility, and propensity for heart failure in mice. *J. Clin. Investig.* **2010**, *120*, 3680–3687.

113. Piot, C.; Croisille, P.; Staat, P.; Thibault, H.; Rioufol, G.; Mewton, N.; Elbelghiti, R.; Cung, T.T.; Bonnefoy, E.; Angoulvant, D.; *et al.* Effect of cyclosporine on reperfusion injury in acute myocardial infarction. *N. Engl. J. Med.* **2008**, *359*, 473–448.

114. Fuglesteg, B.N.; Suleman, N.; Tiron, C.; Kanhema, T.; Lacerda, L.; Andreasen, T.V.; Sack, M.N.; Janassen, A.K.; Mjos, O.D.; Opie, L.H.; *et al.* Signal transducer and activator of transcription 3 is involved in the cardioprotective signaling pathway activated by insulin therapy at reperfusion. *Bas. Res. Cardiol.* **2008**, *103*, 444–453.

115. Lacerda, L.; Somers, S.; Opie, L.H.; Lecour, S. Ischemic postconditioning protect against reperfusion injury via SAFE pathway. *Cardiovasc. Res.* **2009**, *84*, 201–208.

116. Boengler, K.; Hilfiker-Kleiner, D.; Heusch, G.; Schulz, R. Inhibition of permeability transition pore opening by mitochondrial STAT3 and its role in myocardial ischemia/reperfusion. *Basic Res. Cardiol.* **2010**, *105*, 771–785.

117. Hataishi, R.; Rodrigues, A.C.; Neilan, T.G.; Morgan, J.G.; Buys, E.; Shiva, S.; Tambouret, R.; Jassal, D.S.; Raher, M.J.; Furutani, E.; *et al.* Inhaled nitric oxide decreases infarction size and improves left ventricular function in a murine model of myocardial ischemia-reperfusion injury. *Am. J. Physiol. Heart Circ. Physiol.* **2006**, *291*, H379–H384.

118. Nagasaka, Y.; Fernandez, B.O.; Garcia-Saura, M.F.; Petersen, B.; Ichinose, F.; Bloch, K.D.; Feelisch, M.; Zapol, W.M. Brief periods of nitric oxide inhalation protect against myocardial ischemia-reperfusion injury. *Anesthesiology* **2008**, *109*, 675–682.

119. Liu, X.; Huang, Y.; Pokreisz, P.; Vermeersch, P.; Marsboom, G.; Swinnen, M.; Verbeken, E.; Santos, J.; Pellens, M.; Gillijns, H.; *et al.* Nitric oxide inhalation improves microvascular flow and decreases infarction size after myocardial ischemia and reperfusion. *J. Am. Coll. Cardiol.* **2007**, *50*, 808–817.

120. Wang, G.; Liem, D.A.; Vondriska, T.M.; Honda, H.M.; Korge, P.; Pantaleon, D.M.; Qiao, X.; Wang, Y.; Weiss, J.N.; Ping, P. Nitric oxide donors protect murine myocardium against infarction via modulation of mitochondrial permeability transition. *Am. J. Physiol. Heart Circ. Physiol.* **2005**, *288*, H1290–H1295.

121. Ong, S.G.; Hausenloy, D.J. Hypoxia-inducible factor as a therapeutic target for cardioprotection. *Pharmacol. Ther.* **2012**, *136*, 69–81.

122. Ong, S.G.; Lee, W.H.; Theodorou, L.; Kodo, K.; Lim, S.Y.; Shukla, D.H.; Briston, T.; Kiriakidis, S.; Ashcroft, M.; Davidson, S.M.; *et al.* HIF-1 reduces ischaemia-reperfusion injury in the heart by targeting the mitochondrial permeability transition pore. *Cardiovasc. Res.* **2014**, *104*, 24–36.

123. Benard, G.; Bellance, N.; Jose, C.; Melser, S.; Nouette-Gaulain, K.; Rossignol, R. Multi-site control and regulation of mitochondrial energy production. *Biochim. Biophys. Acta* **2010**, *1797*, 698–709.

124. McLeod, C.J.; Pagel, I.; Sack, M.N. The mitochondrial biogenesis regulatory program adaptation to ischemia-A putative target for therapeutic intervention. *Trends Cardiovasc. Med.* **2005**, *15*, 118–123.

125. Hock, M.B.; Kralli, A. Transcriptional control of mitochondrial biogenesis and function. *Annu. Rev. Physiol.* **2009**, *71*, 177–203.

126. Ventura-Clapier, R.; Garnier, A.; Veksler, V. Transcriptional control of mitochondrial biogenesis: The central role of PGC-1α. *Cardiovasc. Res.* **2008**, *79*, 208–217.

127. Ventura-Clapier, R.; Garnier, A.; Veksler, V.; Joubert, F. Bioenergetics of the failing heart. *Biochim. Biophys. Acta* **2011**, *1813*, 1360–1372.

128. Garnier, A.; Fortin, D.; Deloménie, C.; Momken, I.; Veksler, V.; Ventura-Clapier, R. Depressed mitochondrial transcription factors and oxidative capacity in rat failing cardiac and skeletal muscles. *J. Physiol.* **2003**, *551*, 491–501.

129. Watson, P.A.; Reusch, J.E.; McCune, S.A.; Leinwand, L.A.; Luckey, S.W.; Konhilas, J.P. Restoration of CREB function is linked to completion and stabilization of adaptive cardiac hypertrophy in response to exercise. *Am. J. Physiol.* **2007**, *293*, H246–H259.

130. Lehman, J.J.; Kelly, D.P. Transcriptional activation of energy metabolic switches in the developing and hypertrophied heart. *Clin. Exp. Pharmacol. Physiol.* **2002**, *29*, 339–345.

131. Arany, Z.; Novikov, M.; Chin, S.; Ma, Y.; Rosenzweig, A.; Spiegelman, B.M. Transverse aortic constriction leads to accelerated heart failure in mice lacking PPARγ coactivator 1α. *Proc. Natl. Acad. Sci. USA* **2006**, *103*, 10086–10091.

132. Sihag, S.; Li, A.Y.; Cresci, S.; Sucharov, C.C.; Lehman, J.J. PGC-1α and ERRα target gene down-regulation is a signature of the failing human heart. *J. Mol. Cell. Cardiol.* **2009**, *46*, 201–212.

133. Ahuja, P.; Zhao, P.; Angelis, E.; Ruan, H.; Korge, P.; Olson, A.; Wang, Y.; Jin, E.S.; Jeffrey, F.M.; Portman, M.; *et al.* Myc controls transcriptional regulation of cardiac metabolism and mitochondrial biogenesis in response to pathological stress in mice. *J. Clin. Investig.* **2010**, *120*, 1494–1505.

134. Yan, W.; Zhang, H.; Liu, P.; Wang, H.; Liu, J.; Gao, C.; Liu, Y.; Lian, K.; Yang, L.; Sun, L.; *et al.* Impaired mitochondrial biogenesis due to dysfunctional adiponectin-AMPK-PGC-1α signaling contributing to increased vulnerability in diabetic heart. *Basic Res. Cardiol.* **2013**, *108*.

135. Sun, L.; Zhao, M.; Yu, X.J.; Wang, H.; He, X.; Liu, J.K.; Zang, W.J. Cardioprotection by acetylcholine: A novel mechanism via mitochondrial biogenesis and function involving the PGC-1α pathway. *J. Cell. Physiol.* **2013**, *228*, 1238–1248.

136. St-Pierre, J.; Lin, J.; Krauss, S.; Tarr, P.T.; Yang, R.; Newgard, C.B.; Spiegelman, B.M. Bioenergetic analysis of peroxisome proliferator-activated receptor gamma coactivators 1α and 1β (PGC-1α and PGC-1β) in muscle cells. *J. Biol. Chem.* **2003**, *278*, 26597–26603.

137. Kajander, O.A.; Karhunen, P.J.; Jacobs, H.T. The relationship between somatic mtDNA rearrangements, human heart disease and aging. *Hum. Mol. Genet.* **2002**, *11*, 317–324.

138. Naya, F.J.; Black, B.L.; Wu, H.; Bassel-Duby, R.; Richardson, J.A.; Hill, J.A.; Olson, E.N. Mitochondrial deficiency and cardiac sudden death in mice lacking the MEF2A transcription factor. *Nat. Med.* **2002**, *8*, 1303–1309.

139. Lebrecht, D.; Setzer, B.; Ketelsen, U.P.; Haberstroh, J.; Walker, U.A. Timedependent and tissue-specific accumulation of mtDNA and respiratory chain defects in chronic doxorubicin cardiomyopathy. *Circulation* **2003**, *108*, 2423–2429.

140. Ide, T.; Tsutsui, H.; Kinugawa, S.; Utsumi, H.; Kang, D.; Hattori, N.; Uchida, K.; Arimura, K.; Egashira, K.; Takeshita, A. Mitochondrial electron transport complex I is a potential source of oxygen free radicals in the failing myocardium. *Circ. Res.* **1999**, *85*, 357–363.

141. Li, H.; Wang, J.; Wilhelmsson, H.; Hansson, A.; Thoren, P.; Duffy, J.; Rustin, P.; Larsson, N.G. Genetic modification of survival in tissue-specific knockout mice with mitochondrial cardiomyopathy. *Proc. Natl. Acad. Sci. USA* **2000**, *97*, 3467–3472.

142. Wang, J.; Wilhelmsson, H.; Graff, C.; Li, H.; Oldfors, A.; Rustin, P.; Bruning, J.C.; Kahn, C.R.; Clayton, D.A.; Barsh, G.S.; *et al.* Dilated cardiomyopathy and atrioventricular conduction blocks induced by heart-specific inactivation of mitochondrial DNA gene expression. *Nat. Genet.* **1999**, *21*, 133–137.

143. Ikeuchi, M.; Matsusaka, H.; Kang, D.; Matsushima, S.; Ide, T.; Kubota, T.; Fujiwara, T.; Hamasaki, N.; Takeshita, A.; Sunagawa, K.; *et al.* Overexpression of mitochondrial transcription factor a ameliorates mitochondrial deficiencies and cardiac failure after myocardial infarction. *Circulation* **2005**, *112*, 683–669.

144. Guarente, L. Sirtuins in aging and disease. *Cold Spring Harb. Symp. Quant. Biol.* **2007**, *72*, 483–488.

145. Nakagawa, T.; Guarente, L. Sirtuins at a glance. *J. Cell Sci.* **2011**, *124*, 833–838.
146. Aquilano, K.; Vigilanza, P.; Baldelli, S.; Pagliei, B.; Rotilio, G.; Ciriolo, M.R. Peroxisome proliferator-activated receptor γ co-activator 1α (PGC-1α) and sirtuin 1 (SIRT1) reside in mitochondria. Possible direct function ion mitochondrila biogenesis. *J. Biol. Chem.* **2010**, *285*, 21590–21599.
147. Nemoto, S.; Fergusson, M.M.; Finkel, T. SIRT1 functionally interacts with the metabolic regulator and transcriptional coactivator PGC-1α. *J. Biol. Chem.* **2005**, *280*, 16456–16460.
148. Alcendor, R.R.; Gao, S.; Zhai, P.; Zablocki, D.; Holle, E.; Yu, X.; Tian, B.; Wagner, T.; Vatner, S.F.; Sadoshima, J. Sirt1 regulates aging and resistance to oxidative stress in the heart. *Circ. Res.* **2007**, *100*, 1512–1521.
149. Hsu, C.P.; Zhai, P.; Yamamoto, T.; Maejima, Y.; Matsushima, S.; Hariharan, N.; Shao, D.; Takagi, H.; Oka, S.; Sadoshima, J. Silent information regulator 1 protects the heart from ischemia/reperfusion. *Circulation* **2010**, *122*, 2170–2182.
150. Orallo, F.; Alvarez, E.; Camina, M.; Leiro, J.M.; Gomez, E.; Fernandez, P. The possible implication of trans-Resveratrol in the cardioprotective effects of long-term moderate wine consumption. *Mol. Pharmacol.* **2002**, *61*, 294–302.
151. Biala, A.; Tauriainen, E.; Siltanen, A.; Shi, J.; Merasto, S.; Louhelainen, M.; Martonen, E.; Finckenberg, P.; Muller, D.N.; Mervaala, E. Resveratrol induces mitochondrial biogenesis and ameliorates Ang II-induced cardiac remodeling in transgenic rats harboring human renin and angiotensinogen genes. *Blood Press.* **2010**, *19*, 196–205.
152. Weitzel, J.M.; Iwen, K.A. Coordination of mitochondrial biogenesis by thyroid hormone. *Mol. Cell. Endocrinol.* **2011**, *342*, 1–7.
153. Venditti, P.; Bari, A.; di Stefano, L.; Cardone, A.; della Ragione, F.; D'Esposito, M.; di Meo, S. Involvement of PGC-1, NRF-1, and NRF-2 in metabolic response by rat liver to hormonal and environmental signals. *Mol. Cell. Endocrinol.* **2009**, *305*, 22–29.
154. Wulf, A.; Harneit, A.; Kröger, M.; Kebenko, M.; Wetzel, M.G.; Weitzel, J.M. T3-mediated expression of PGC-1α via a far upstream located thyroid hormone response element. *Mol. Cell. Endocrinol.* **2008**, *287*, 90–95.
155. Villeneuve, C.L.; Guilbeau-Frugier, C.; Sicard, P.; Lairez, O.; Ordener, C.; Duparc, T.; de Paulis, D.; Couderc, B.; Spreux-Varoquaux, O.; Tortosa, F.; *et al.* p53-PGC-1α pathway mediates oxidative mitochondrial damage and cardiomyocyte necrosis induced by monoamine oxidase-A up-regulation: Role in chronic left ventricular dysfunction in mice. *Antioxid. Redox Signal.* **2013**, *18*, 5–18.
156. Ingwall, J.S. Energy metabolism in heart failure and remodelling. *Cardiovasc. Res.* **2009**, *81*, 412–419.
157. Ardehali, H.; Sabbah, H.N.; Burke, M.A.; Sarma, S.; Liu, P.P.; Cleland, J.G.; Maggioni, A.; Fonarow, G.C.; Abel, E.D.; Campia, U.; *et al.* Targeting myocardial substrate metabolism in heart failure: Potential for new therapies. *Eur. J. Heart Fail.* **2012**, *14*, 120–129.
158. Wrutniak-Cabello, C.; Carazo, A.; Casas, F.; Cabello, G. Triiodothyronine mitochondrial receptors: Import and molecular mechanisms. *J. Soc. Biol.* **2008**, *202*, 83–92.

159. Saelim, N.; Holstein, D.; Chocron, E.S.; Camacho, P.; Lechleiter, J.D. Inhibition of apoptotic potency by ligand stimulated thyroid hormone receptors located in mitochondria. *Apoptosis* **2007**, *12*, 1781–1794.

160. Saelim, N.; John, L.M.; Wu, J.; Park, J.S.; Bai, Y.; Camacho, P.; Lechleiter, J.D. Nontranscriptional modulation of intracellular Ca^{2+} signaling by ligand stimulated thyroid hormone receptor. *J. Cell Biol.* **2004**, *167*, 915–924.

161. Van Rooij, E.; Sutherland, L.B.; Qi, X.; Richardson, J.A.; Hill, J.; Olson, E.N. Control of stress-dependent cardiac growth and gene expression by a microRNA. *Science* **2007**, *316*, 575–579.

162. Barringhaus, K.G.; Zamore, P.D. MicroRNAs: Regulating a change of heart. *Circulation* **2009**, *119*, 2217–2224.

163. Divakaran, V.; Mann, D.L. The emerging role of microRNAs in cardiac remodeling and heart failure. *Circ. Res.* **2008**, *103*, 1072–1083.

164. Thum, T.; Galuppo, P.; Wolf, C.; Fiedler, J.; Kneitz, S.; van Laake, L.W.; Doevendans, P.A.; Mummery, C.L.; Borlak, J.; Haverich, A.; *et al.* MicroRNAs in the human heart: A clue to fetal gene reprogramming in heart failure. *Circulation* **2007**, *116*, 258–267.

165. Van Rooij, E.; Marshall, W.; Olson, E. Toward microRNA-based therapeutics for heart disease the sense in antisense. *Circ. Res.* **2008**, *103*, 919–928.

166. Van Rooij, E.; Sutherland, L.B.; Liu, N.; Williams, A.H.; McAnally, J.; Gerard, R.D.; Richardson, J.A.; Olson, E.N. A signature pattern of stress-responsive micrornas that can evoke cardiac hypertrophy and heart failure. *Proc. Natl. Acad. Sci. USA* **2006**, *103*, 18255–18260.

167. Van Rooij, E.; Sutherland, L.B.; Thatcher, J.E.; diMaio, J.M.; Naseem, R.H.; Marshall, W.S.; Hill, J.A.; Olson, E.N. Dysregulation of microRNAs after myocardial infarction reveals a role of miR-29 in cardiac fibrosis. *Proc. Natl. Acad. Sci. USA* **2008**, *105*, 13027–13032.

168. Roy, S.; Khanna, S.; Hussain, S.R.; Biswas, S.; Azad, A.; Rink, C.; Gnyawali, S.; Shilo, S.; Nuovo, G.J.; Sen, C.K. MicroRNA expression in response to murine myocardial infarction: miR-21 regulates fibroblast metalloprotease-2 via phosphatase and tensin homologue. *Cardiovasc. Res.* **2009**, *82*, 21–29.

169. Ren, X.P.; Wu, J.; Wang, X.; Sartor, M.A.; Qian, J.; Jones, K.; Nicolaou, P.; Pritchard, T.J.; Fan, G.C. MicroRNA-320 is involved in the regulation of cardiac ischemia/reperfusion injury by targeting heat-shock protein 20. *Circulation* **2009**, *119*, 2357–2366.

170. Sripada, L.; Tomar, D.; Singh, R. Mitochondria: One of the destinations of miRNAs. *Mitochondrion* **2012**, *12*, 593–599.

171. Ye, Y.; Perez-Polo, J.R.; Qian, J.; Birnbaum, Y. The role of microRNA in modulating myocardial ischemia-reperfusion injury. *Physiol. Genomics* **2011**, *43*, 534–542.

172. Aurora, A.B.; Mahmoud, A.I.; Luo, X.; Johnson, B.A.; van Rooij, E.; Matsuzaki, S.; Humphries, K.M.; Hill, J.A.; Bassel-Duby, R.; Sadek, H.A.; *et al.* MicroRNA-214 protects the mouse heart from ischemic injury by controlling Ca^{2+} overload and cell death. *J. Clin. Investig.* **2012**, *122*, 1222–1232.

173. Wang, X.; Zhang, X.; Ren, X.P.; Chen, J.; Liu, H.; Yang, J.; Medvedovic, M.; Hu, Z.; Fan, G.C. MicroRNA494 targeting both proapoptotic and antiapoptotic proteins protects against ischemia/reperfusion-induced cardiac injury. *Circulation* **2010**, *122*, 1308–1318.

174. Li, J.; Donath, S.; Li, Y.; Qin, D.; Prabhakar, B.; Li, P. miR-30 regulates mitochondrial fission through targeting p53 and the dynamin-related protein-1 pathway. *PLoS Genet.* **2010**, *6*, e1000795.

175. Duisters, R.F.; Tijsen, A.J.; Schroen, B.; Leenders, J.J.; Lentink, V.; van der Made, I.; Herias, V.; van Leeuwen, R.E.; Schellings, M.W.; Barenbrug, P.; *et al.* miR-133 and miR-30 regulate connective tissue growth factor: Implications for a role of microRNAs in myocardial matrix remodeling. *Circ. Res.* **2009**, *104*, 170–178.

176. Gambacciani, C.; Kusmic, C.; Chiavacci, E.; Meghini, F.; Rizzo, M.; Mariani, L.; Pitto, L. miR-29a and miR-30c negatively regulate DNMT3a in cardiac ischemic tissues: Implications for cardiac remodelling. *microRNA Diagn. Ther.* **2013**, *2013*, 34–44.

177. Pantos, C.; Mourouzis, I.; Cokkinos, D.V. Thyroid hormone and cardiac repair/regeneration: From Prometheus myth to reality? *Can. J. Physiol. Pharmacol.* **2012**, *90*, 977–987.

178. Pfeffer, M.A.; Braunwald, E. Ventricular remodeling after myocardial infarction. Experimental observations and clinical implications. *Circulation* **1990**, *81*, 1161–1172.

179. Sigurdsson, A.; Eriksson, S.V.; Hall, C.; Kahan, T.; Swedberg, K. Early neurohormonal effects of trandolapril in patients with left ventricular dysfunction and a recent acute myocardial infarction: A double-blind, randomized, placebo-controlled multicentre study. *Eur. J. Heart Fail.* **2001**, *3*, 69–78.

Mitochondrial Mechanisms in Septic Cardiomyopathy

María Cecilia Cimolai, Silvia Alvarez, Christoph Bode and Heiko Bugger

Abstract: Sepsis is the manifestation of the immune and inflammatory response to infection that may ultimately result in multi organ failure. Despite the therapeutic strategies that have been used up to now, sepsis and septic shock remain a leading cause of death in critically ill patients. Myocardial dysfunction is a well-described complication of severe sepsis, also referred to as septic cardiomyopathy, which may progress to right and left ventricular pump failure. Many substances and mechanisms seem to be involved in myocardial dysfunction in sepsis, including toxins, cytokines, nitric oxide, complement activation, apoptosis and energy metabolic derangements. Nevertheless, the precise underlying molecular mechanisms as well as their significance in the pathogenesis of septic cardiomyopathy remain incompletely understood. A well-investigated abnormality in septic cardiomyopathy is mitochondrial dysfunction, which likely contributes to cardiac dysfunction by causing myocardial energy depletion. A number of mechanisms have been proposed to cause mitochondrial dysfunction in septic cardiomyopathy, although it remains controversially discussed whether some mechanisms impair mitochondrial function or serve to restore mitochondrial function. The purpose of this review is to discuss mitochondrial mechanisms that may causally contribute to mitochondrial dysfunction and/or may represent adaptive responses to mitochondrial dysfunction in septic cardiomyopathy.

Reprinted from *Int. J. Mol. Sci.* Cite as: Cimolai, M.C.; Alvarez, S.; Bode, C.; Bugger, H. Mitochondrial Mechanisms in Septic Cardiomyopathy. *Int. J. Mol. Sci.* **2015**, *16*, 17763–17776.

1. Introduction

Sepsis is responsible for millions of deaths worldwide each year and is a frequent cause of death in people who have been hospitalized [1]. In the United States, 3 in 1000 people suffer from sepsis, and severe sepsis contributes to more than 200,000 deaths per year [2]. Sepsis is the manifestation of the immune and inflammatory response to infection that may ultimately result in multi organ failure. It is believed that the release of pro-inflammatory mediators is not answered by an appropriate anti-inflammatory response and overwhelms the immune system, resulting in an uncontrolled excessive inflammatory state and an inability to neutralize pathogens [3]. Both effects of bacterial endotoxins and exotoxins, as well as an excessive release of a large number of cytokines contribute to organ dysfunction and failure during sepsis and septic shock. Guidelines recommend early treatment with antibiotics, to restore fluid deficits, and vasopressor treatment if necessary, actions that are critical for patient outcomes and overall survival. Besides specifically neutralizing the pathogen with antibiotics, the currently recommended therapeutic strategies are symptomatic, and no specific treatment is available to treat the consequences of specific organ damage in sepsis.

End-organ damage and organ failure in sepsis affects the most significant organs of the body, including the heart. Myocardial dysfunction is a well-described complication of severe sepsis, also referred to as septic cardiomyopathy, which includes both systolic and diastolic dysfunction [4]. Adequate O_2 supply in sepsis suggests that myocardial depression is not related to tissue hypoperfusion but to the presence of circulating depressant factors, or other mechanisms [5]. A number of mechanisms have been proposed to be involved in myocardial dysfunction in sepsis, including toxins, cytokines, nitric oxide, complement activation, apoptosis and energy metabolic derangements [6,7]. The latter two mechanisms imply that mitochondrial function may be compromised in septic cardiomyopathy. Already a century ago, it had been recognized that low arterial oxygen tension, low circulating hemoglobin levels and/or tissue hypoperfusion may lead to tissue hypoxia during sepsis, thereby contributing to impaired ATP generation [8–10]. Others demonstrated that these parameters were not impaired or even improved during sepsis [11,12]. Instead, more recent studies strongly suggest that organ dysfunction in sepsis, including myocardial depression, is rather related to the presence of circulating depressant factors or other mechanisms, leading to cellular energy depletion [5].

Indeed, a number of studies identified myocardial mitochondrial dysfunction in septic conditions both in animal models and humans [13–18]. Since the heart is highly dependent on continuous delivery of ATP to maintain contractile function, impairment in mitochondrial function is energetically detrimental for the heart. For a number of cardiac pathologies, mitochondrial dysfunction and energy depletion have been proposed to significantly contribute to cardiac dysfunction, including prevalent cardiac diseases such as ischemia reperfusion injury, diabetic cardiomyopathy and heart failure [19–21]. While a number of mechanisms have been proposed to causally contribute to myocardial mitochondrial dysfunction in sepsis, a number of other mitochondria-related alterations occur in septic cardiomyopathy the purpose of which may be to restore mitochondrial function. In the following sections, we will review the evidence of mitochondrial dysfunction in septic cardiomyopathy and discuss mitochondrial mechanisms that may causally contribute to mitochondrial dysfunction and/or may represent adaptive responses to mitochondrial dysfunction in septic cardiomyopathy.

2. Myocardial Mitochondrial Dysfunction in Sepsis

In the heart, high-energy phosphates are mainly generated from the oxidation of fatty acids, while glucose and other substrates contribute to myocardial ATP regeneration to a lesser extent. Substrate oxidation yields mainly NADH (reduced nicotinamide adenine dinucleotide) and $FADH_2$ (reduced flavin adenine dinucleotide), which feed electrons into the electron transport chain of oxidative phosphorylation (OXPHOS) via complex I and II. Electron transport results in the reduction of O_2 to H_2O, which is coupled to ATP regeneration by the F_0F_1-ATPase (ATP synthase). ATP exits the mitochondrion and is mainly utilized by myosin ATPases to maintain contractile function, whereas a minor part of ATP is used to maintain cellular ion homeostasis. Of note, the entire pool of myocardial ATP is turned over completely every few seconds, emphasizing that mitochondrial energy substrate metabolism is critically important to maintain ATP availability and thus cardiac function.

Mitochondrial dysfunction is widely discussed as a crucial mechanism of organ dysfunction in sepsis, including the heart, although only few data on mitochondrial function are available from human septic hearts. Takasu and colleagues reported edema of the mitochondrial matrix, associated with cystic alterations of the cristae and collapse into small myelin-like clusters in hearts of septic patients [22]. Soriano *et al.* observed that patients that did not survive sepsis presented a more severe degree of cardiac dysfunction compared to survivors, and non-survivors also showed derangements in mitochondrial cristae [13]. In contrast, numerous animal studies reported myocardial mitochondrial dysfunction in sepsis. The most commonly used animal models of sepsis are the induction of sepsis by injection of an exogenous bacterial toxin, mostly lipopolysaccharide (LPS), or the induction of intestinal leakage by cecal ligation and puncture (CLP) or colon ascendant stent peritonitis (CASP) [23]. Using these models, it was well documented by Smeding and colleagues [24] that, in the majority of the cases, the impairment in cardiac mitochondrial function correlated with decreased cardiac contractility, both measured *in vivo* by echochardiography or *ex vivo* in the isolated heart perfusion. Mitochondrial dysfunction was characterized by decreased rates of State 3 respiration and ATP synthesis, decreased respiratory control ratios and membrane potential, decreased activities of mitochondrial OXPHOS Complexes, increased rates of State 4 respiration, and increased mitochondrial size and fragility [14,25–29]. Several authors reported a decrease in State 3 respiration in the heart using glutamate as a Complex I substrate, whereas an impairment in Complex II-mediated respiration was less consistent [29–31]. Decreased activity or expression of individual OXPHOS Complexes was also widely reported, in particular Complex I. Decreased State 3 respiration was frequently associated with a decreased respiratory control ratio (RCR), although the lower RCR was sometimes related to increased State 4 respiration [26], which could be indicative of mitochondrial uncoupling. In addition, a decrease of the mitochondrial transmembrane potential (Δψ) or increased mitochondrial permeability transition was often observed in animal models of sepsis. Functional impairment of cardiac mitochondria was often associated with structural defects of mitochondria, such as cristae derangement, cleared matrix and mitochondrial swelling in animal models, thereby confirming the findings in human septic hearts as described above. Finally, increased mitochondrial reactive oxygen species (ROS) generation also confirms dysfunction of mitochondria in septic cardiomyopathy [32].

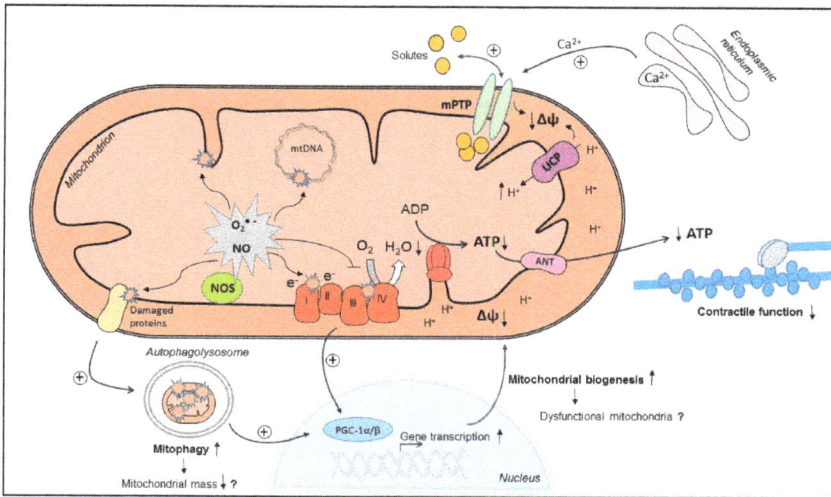

Figure 1. Mitochondrial mechanisms in septic cardiomyopathy: Increased superoxide ($O_2 \cdot ^-$) and nitric oxide (NO) production can cause direct oxidative or nitrosative damage and inhibition of oxidative phosphorylation (OXPHOS) complexes, leading to decreased O_2 consumption and decreased mitochondrial membrane potential (Δψ). In addition, Δψ may drop due to increased uncoupling protein (UCP)-mediated proton leak, increased Ca^{2+}-induced mitochondrial permeability transition pore (mPTP) opening and direct oxidative damage of the inner mitochondrial membrane. As a consequence, mitochondrial ATP regeneration is compromised, and energy depletion may contribute to cardiac contractile dysfunction. Increased mitophagy may eliminate dysfunctional mitochondria, which may be replaced by increased mitochondrial biogenesis, mediated by activation of peroxisome proliferator-activated receptor γ coactivator 1α/β (PGC-1α/β). However, if uncoordinatedly activated, mitophagy and mitochondrial biogenesis may lead to decreased mitochondrial mass and dysfunctional mitochondria (upward pointing arrow: increase; downward pointing arrow: decrease; dashed arrow: possible consequence).

265

Although mitochondrial dysfunction in septic cardiomyopathy is associated with cardiac dysfunction, the underlying mechanisms of mitochondrial dysfunction are still incompletely elucidated. In addition, a number of mitochondrial mechanisms are triggered in septic cardiomyopathy, which may rather serve to restore than impair mitochondrial function, although controversially discussed. In the following sections, we will discuss mitochondrial mechanisms that may causally contribute to mitochondrial dysfunction and/or may represent adaptive responses to mitochondrial dysfunction in septic cardiomyopathy (illustrated in Figure 1).

3. Mechanisms of Myocardial Mitochondrial Dysfunction in Sepsis

3.1. Mitochondrial NO Production and Oxidative Stress

During respiration, mitochondria may reduce O_2 univalently thereby producing $O_2 \cdot^-$ (superoxide anion) as a normal metabolite. Under physiological conditions, only a small percentage (1%–2%) of the O_2 utilized is driven to $O_2 \cdot^-$ formation, which is dismutated to H_2O_2 by Mn-superoxide dismutase. However, situations in which steady-state concentrations of these reactive oxygen species (ROS) are increased can lead to a diverse array of reversible and irreversible toxic modifications on biomolecules, such as protein carbonylation or lipid peroxidation. The continuous mitochondrial production of free radicals and the significant amount of mitochondria within cells makes mitochondria the most important source of these harmful species within the cell. At the same time, mitochondrial enzymes, function and mtDNA are particularly sensitive to ROS-induced damage [33,34]. Mitochondria also produce nitric oxide (NO) through the activity of mitochondrial NO synthase (mtNOS), which physiologically regulates mitochondrial respiration by inhibition of cytochrome c oxidase [35].

Sepsis is an excessive, systemic inflammatory response characterized by massive increases in ROS, NO and inflammatory cytokines [36]. A large body of evidence strongly suggests that excessive mitochondrial production of NO and ROS contribute to mitochondrial dysfunction in sepsis in various tissues, including the heart. Studies demonstrated that NO production, production of $O_2 \cdot^-$ and H_2O_2, global protein nitration, nitrotyrosine content, protein carbonylation and lipid peroxidation are increased in cardiac mitochondria of animals subjected to endotoxemia [14,26,37–39]. Moreover, the antioxidant systems, aimed to scavenge these reactive species, seem to be inhibited as shown by decreased activity of Mn-superoxide dismutase and glutathione peroxidase, and depletion of gutathione [40,41]. Peroxynitrite ($ONOO^-$) is a particularly powerful oxidant, which results from a diffusion-controlled reaction of NO with $O_2 \cdot^-$. In a model of CLP, Escames *et al.* showed increased activity of the inducible mitochondrial nitric oxide synthase (i-mtNOS) in septic mice, which leads to increased mitochondrial $ONOO^-$ levels [42,43]. Increased oxidative

266

stress, impairment in OXPHOS function and a decrease in ATP production were restored by genetic deletion of iNOS (iNOS$^{-/-}$ mice), which also includes the mitochondrial isoform, suggesting a significant role of ONOO$^-$ in myocardial mitochondrial dysfunction in sepsis [43]. This argument is further supported by the fact that treatment with melatonin, an inhibitor of iNOS, prevented the impairment of mitochondrial homeostasis after sepsis, restored ATP production and improved survival. Some studies also showed an improvement of cardiac function by pharmacological inhibition or genetic deletion of NOS in septic animal models [37,44]. One of these studies showed that recovery of NO production and content of NO metabolites in mitochondria was associated with partial improvement of left ventricular (LV) developed pressure. A causal relationship between mitochondrial NO production and cardiac dysfunction needs to be considered carefully though, since NO is produced by several NOS isoforms (not only mitochondrial) at different intracellular locations and in different cell types, and since the role of NO in cardiovascular function and regulation of vascular tone is indisputably contributing to cardiac dysfunction in sepsis.

While increased ROS levels in sepsis may origin from ROS production at different cellular sites, the importance of mitochondrial oxidative stress in animal models of sepsis has been underlined by studies demonstrating beneficial effects by treatment with mitochondria-targeted antioxidants. Treatment with α-lipoic acid, an antioxidant that is reduced in the mitochondria to a more active form (dihyrolipoic acid), led to full recovery of cardiac mitochondrial function during endotoxemia [45]. Furthermore, administration of Coenzyme Q10 that was conjugated to triphenylphosphonium (MitoQ), a powerful mitochondria-targeted antioxidant, not only restored mitochondrial function but also ameliorated myocardial dysfunction in mice and rats treated with LPS [46]. Finally, treatment with vitamin E conjugated to triphenylphosphonium (Mito-VitE) in a rat pneumonia-related sepsis model prevented mitochondrial ROS production and ROS-related damage in the heart, with recovery of mitochondrial structure and function [32].

3.2. Ca^{2+} Handling and Mitochondrial Permeability Transition

On the basis of the chemiosmotic model of energy transduction, the impermeability of the inner mitochondrial membrane is essential to allow the electron transport chain to pump protons across the inner mitochondrial membrane (IMM) and to build up a proton gradient that can be used by the F_0F_1-ATPase to regenerate ATP. Mitochondrial permeability transition is a sudden increase of IMM permeability for solutes with a molecular mass up to 1500 Da, and it is believed that this permeability transition is mediated by a voltage- and Ca^{2+}-dependent, cyclosporin A (CsA)-sensitive, high-conductance channel located in the inner mitochondrial membrane termed mitochondrial permeability transition pore (mPTP). The consequence of mPTP opening is loss of $\Delta\psi$, a decrease in State 3 respiration, mitochondrial swelling and rupture of the outer mitochondrial membrane, ultimately leading to activation of pro-apoptotic pathways and necrotic cell death. Ca^{2+} overload is the primary trigger of mPTP opening, but the sensitivity of the pore is dependent on prevailing conditions, such as oxidative stress, adenine nucleotide depletion, increased inorganic phosphate concentrations and mitochondrial depolarization [47]. Using the LPS-induced endotoxemia model in rats, Hassoun et al. observed decreased Ca^{2+} uptake into the sarcoplasmic reticulum and increased Ca^{2+} leak from this organelle, which was associated with increased mitochondrial Ca^{2+} content. Administration of the sarcoplasmic reticulum Ca^{2+} leak inhibitor dantrolene prevented mitochondrial Ca^{2+} overload, and improved mitochondrial dysfunction and cardiac contractility defects, suggesting that mitochondrial Ca^{2+} overload could be an important mechanism contributing to myocardial mitochondrial dysfunction in sepsis [30]. Larche and colleagues reported decreased survival, myocardial dysfunction, decreased State 3 and increased State 4 respiration, and increased opening of the mPTP in CLP-induced sepsis. Inhibition of the mPTP by CsA treatment improved cardiac function, survival, and also attenuated mitochondrial dysfunction in this study. In addition, the amount of Ca^{2+} required to increase mPTP opening in CLP-treated mice was less than 25% of the Ca^{2+} needed in control animals, indicative of increased sensitivity for Ca^{2+}-induced mPTP opening of the mitochondria in CLP-treated mice [48]. In a different study by Fauvel et al., endotoxemic mice showed decreased LV developed pressure in isolated perfused hearts, and this contractile defect was rescued by in vivo treatment with CsA. Treatment with CsA also decreased caspase activation and cytochrome c release, suggestive of prevention of mPTP opening [49]. Chopra et al. observed severe morphological deformations that could be indicative of pore formation in rats treated with cecal inoculum, accompanied by a mitochondrial $\Delta\psi$ collapse. Restorative effects were observed in both parameters by treatment with 5-hydroxydecanoate (5-HD), a blocker of mitoK$_{ATP}$ (mitochondrial ATP-sensitive K^+ channel). Moreover, sepsis-related decreases in cardiac output and ejection fraction were recovered after inhibition of the mitoK$_{ATP}$ [28]. The primary function of mitoK$_{ATP}$ channels is thought to be the regulation of mitochondrial volume; mitoK$_{ATP}$ channel opening results in mitochondrial uptake of potassium and associated inorganic phosphates, anions, and water [50], and is believed to sensitize outer membrane permeabilization and rupture in this sepsis model. Thus, it is tempting to speculate that in some models of sepsis, an impairment in cytosolic calcium handling may lead to mitochondrial Ca^{2+} overload, which triggers opening of a hypersensitive mPTP, thereby contributing to mitochondrial and contractile dysfunction. Further evidence is required to test whether activation of mitoK$_{ATP}$ contributes to increased sensitivity for mPTP opening. It remains to be mentioned though that some studies reported no change of mPTP opening using the mitochondrial swelling assay, although Ca^{2+} handling was found to be impaired in isolated cardiomyocytes [51].

3.3. Mitochondrial Uncoupling

Mitchell's chemiosmotic theory of energy transduction states that the free energy generated by electron transport through the OXPHOS chain and ultimately transfer on O_2 is used to set up an electrochemical H^+ gradient across the inner mitochondrial membrane. Protons return into the matrix via the F_0 subunit of the F_0F_1-ATPase, thereby regenerating ATP from ADP. Under physiological conditions, some H^+ return into the matrix but bypass the F_0F_1-ATPase (e.g., via uncoupling proteins (UCPs)), resulting in "mitochondrial uncoupling" of ATP synthesis from O_2 consumption. Physiologic uncoupling is used for heat generation in brown adipose tissue or to decrease mitochondrial ROS production [52]. Pathologic conditions with increased uncoupling include diabetic cardiomyopathy, where increased utilization of long-chain fatty acids results in ROS-induced activation of UCPs, resulting in decreased amounts of ATP per oxygen consumed and therefore impaired cardiac efficiency (cardiac work/myocardial O_2 consumption) [53–55]. Similarly, an increase in UCP3 expression may increase mitochondrial uncoupling and decrease cardiac efficiency in chronically infarcted failing rat hearts [56].

The role of UCPs in sepsis remains controversial. Twelve hours following CLP, Roshon and colleagues found a 35% reduction in cardiac efficiency in isolated perfused hearts, probably due to a decrease in cardiac work and associated with increased UCP2 mRNA levels [57]. Similarly, myocardial mRNA content of both UCP2 and UCP3 were increased in LPS-induced endotoxemia in rats, and increased myocardial mRNA and protein expression of UCP2 was observed in a canine model of endotoxin-induced shock associated with decreased phosphocreatine/ATP ratios [41,58]. Thus, UCP-mediated uncoupling may impair cardiac efficiency under septic conditions. In contrast, a very recent study reported that treatment of rat embryonic cardiomyoblasts (H9C2) with LPS plus peptidoglycan G led to increased mRNA expression of UCP2, associated with decreased membrane potential, decreased ATP content, increased ROS and depletion of mtDNA. These effects were aggravated by additional silencing of UCP2, suggesting that increased UCP2 expression may actually exert protective effects [59]. Less and also controversial evidence is available on UCP3. Aguirre et al. showed increased myocardial UCP3 levels in LPS-induced endotoxemia [60]. These authors report decreased State 3 respiration and Complex IV activity in wildtype and UCP3$^{-/-}$ mice treated with LPS. Proton conductance was found unchanged after 24 h of LPS treatment, regardless of the higher levels of UCP3. In contrast, UCP3 levels were increased with a concomitant loss of membrane potential and ATP content in neonatal cardiomyocytes treated with LPS [61].

3.4. Mitochondrial Biogenesis

Mitochondrial biogenesis can be described as growth and division of mitochondria that requires the proper functioning of a number of cellular processes including the coordinated synthesis and import of proteins encoded by the nuclear genome, mtDNA replication, synthesis of proteins encoded by the mitochondrial genome and fusion–fission processes [62]. Mitochondrial biogenesis is triggered by different cellular stressors, such as increased levels of NO, oxidative stress, and an elevated AMP/ATP ratio, among others, resulting in the activation of the peroxisome proliferator-activated receptor y (PPARγ) coactivator (PGC) family of transcriptional coactivators, most importantly PGC-1α and PGC-1β [63,64]. PGC-1α coactivates and/or increases the expression of transcription factors, including ERRα/γ, NRF1/2 and Tfam, that mediate the transcription of nuclear and mitochondrially encoded OXPHOS subunits, nuclear proteins necessary for mtDNA transcription and replication, OXPHOS assembly factors, and components of mitochondrial protein import [65]. By regulating ERRα and PPARα, PGC-1α/β also stimulate the expression of genes encoding for proteins and enzymes of fatty acid oxidation, the main metabolic pathway of fuel oxidation in cardiac tissue.

Several studies have illustrated that mitochondrial biogenesis or the process of generating new mitochondria improves survival in sepsis, and inhibition of mitochondrial biogenesis worsens outcome [25,66]. It is generally thought that loss and/or removal of damaged mitochondria in septic organs is compensated by generation of new mitochondria by increased mitochondrial biogenesis, which should be responsible for increasing survival and reversing organ damage. In the heart, Vanasco *et al.* recently reported increased myocardial levels of PGC1-α and Tfam, accompanied by increased mitochondrial mass and recovery of initially decreased amounts of mtDNA 24 h following LPS treatment [14]. Electron microscopy revealed an increased amount of structurally damaged mitochondria, including swelling, loss and/or disruption of cristae, cleared matrix, and internal vesicles. Despite evidence of mitochondrial biogenesis, mitochondrial State 3 respiration related to Complex I substrates and activity of Complexes I and II were impaired. Others also reported increased expression of mediators of mitochondrial biogenesis, including PGC-1α, NRF-1, and NRF-2, as well as Tfam and the mitochondrial DNA polymerase, associated with mitochondrial structural changes such as swelling and loss of cristae [41]. In neonatal cardiomyocytes, LPS treatment stimulated an increase in Tfam, nuclear accumulation of NRF-1, and expression of PGC-1α, and the changes correlated with signs of mitochondrial dysfunction (decreased mitochondrial membrane potential and ATP content) and markers of autophagy [61]. Taken together, it appears that mitochondrial biogenesis signaling is activated in endotoxemia, which however may not necessarily result in improvement or recovery of mitochondrial function. Russell and colleagues demonstrated that cardiomyocyte-specific overexpression of PGC-1α resulted in a markedly increased mitochondrial biogenesis but also heart failure [67]. Possibly, intensive stimulation of mitochondrial biogenesis might have perturbed the complex process of gene transcription and mitochondrial dynamics in this study, resulting in mitochondria with altered biochemical composition and stoichiometry and thus functional deficits. Thus, while being a compensatory mechanism to overcome loss of dysfunctional mitochondria, uncontrolled triggering of mitochondrial biogenesis in endotoxemia may also result in dysfunctional mitochondria *per se*. Alternatively, mitochondrial biogenesis may simply not be sufficient to compensate for mitochondrial defects in septic conditions.

In contrast to increased mitochondrial biogenesis signaling, Schilling *et al.* showed that LPS injection led to a rapid downregulation of PGC-1α and PGC-1β mRNA levels, accompanied by decreased levels of PPARα, ERRα, and NRF1. Cardiac function was depressed 6 h after the onset of endotoxemia, associated with a decrease in palmitate oxidation in the isolated working heart and increased myocyte lipid accumulation [68]. The total mtDNA copy number and mRNA levels of several enzymes were decreased, and the content of respiratory Complexes I and IV, assessed by 2D-gel electrophoresis and MALDI-TOF mass spectrometry, was reduced in treated rats [41]. Thus, time points and models investigated may affect initiation of mitochondrial biogenesis, and it remains controversial whether activation of mitochondrial biogenesis is actually adaptive or maladaptive under septic and endotoxemic conditions. Of interest, cells overexpressing PGC-1β and mice overexpressing PGC-1β selectively in cardiomyocytes were resistant to the LPS-induced decrease of the expression of PPARα, ERRα, and enzymes of fatty acid oxidation. PGC-1β transgenic mice showed improvement in palmitate oxidation rates in isolated working hearts after treatment with LPS, and echocardiography revealed only a modest decrease in cardiac function [68]. These data suggests that impairment in mitochondrial energetics may importantly contribute to cardiac dysfunction, and that activation of mitochondrial biogenesis may have positive effects in endotoxemia.

When assessing the role of PPARα in endotoxemia, Drosatos *et al.* found that LPS administration reduced the mRNA expression of PGC-1α/β, PPARα, and metabolic PPARα downstream targets. Treatment with a PPARα agonist, however, did not improve cardiac function or alter the mRNA levels of these enzymes after endotoxemia. Interestingly, treatment with rosiglitazone, a PPARγ agonist, led to an improvement of cardiac dysfunction together with improved expression of PGC-1 molecules and Tfam. Moreover, αMHC-PPARγ mice that express PPARγ

constitutively in cardiomyocytes showed no disturbances in cardiac function and lipid accumulation after LPS treatment [16]. The exact mechanism by which PPARγ agonism improves or maintains cardiac function in endotoxemia, and in how far these mechanisms are related to mitochondrial biogenesis, remains to be determined.

3.5. Mitophagy

Recovery of organ function in sepsis is dependent on removal of dysfunctional mitochondria and generation of new, functionally competent mitochondria, as may be achieved by stimulation of mitochondrial biogenesis. Removal of dysfunctional mitochondria occurs via autophagy of mitochondria, or simply "mitophagy", which is a quality control mechanism by which cells eliminate dysfunctional mitochondria [69]. Damaged mitochondria are selectively sequestered in autophagosomes and ultimately degraded following fusion with lysosomes. Different pathways have been described to recognize and eliminate the dysfunctional mitochondria. Activation of the PINK1/Parkin (PARK2) pathway leads to ubiquitinylation of special substrates of mitochondria that can be recognized by the adaptor protein p62, which in turn binds to microtubule-associated protein 1A/1B-light chain 3 (LC3). An ubiquitin-independent pathway involves direct binding of autophagy-related protein 8 (ATG8) family proteins to autophagy receptors, such as NIX (Nip3-like protein X) [63,69]. Mitophagy may serve to eliminate dysfunctional mitochondria before they induce cellular damage or cell death, but may also serve as a cell survival pathway by suppressing apoptosis or as a back-up mechanism when the process of apoptosis is defective [70].

Several studies have shown that autophagy and mitophagy are increased in various organs in sepsis [71,72]. In the heart, several studies observed increased myocardial mRNA or protein levels of ATG3, ATG5, ATG7, LC3 II and p62 in LPS-induced endotoxemia, suggestive of increased rates of autophagy [14,16,73]. In some cases, a reduced number of cardiomyocyte mitochondria was reported in LPS-treated animals [16,25,41] potentially suggesting increased mitochondrial degradation by mitophagy, although unchanged mitochondrial mass was observed in other studies [41,74]. An interesting study was published by Piquereau *et al.* who addressed the role of Parkin/PARK2 in mitochondrial dysfunction in endotoxemia. LPS treatment led to a 50% decrease in cardiac output and stroke volume in wildtype and PARK2$^{-/-}$ mice. While wildtype mice completely recovered cardiac function after 48 h, cardiac function remained impaired in PARK2$^{-/-}$ mice. Similarly, a decrease in OXPHOS complex activities, mitochondrial O_2 consumption rates, and increased susceptibility to mPTP opening were normalized 48 h after endotoxemia induction in wildtype but not PARK2$^{-/-}$ mice. Thus, PARK2-dependent autophagy/mitophagy is required to recover mitochondrial and cardiac function following LPS treatment. In mitochondrial fractions of LPS-treated wildtype mice, levels of polyubiquitinated proteins and PARK2 were increased, and proteins involved in mitochondrial quality control and autophagy (SQTSTM1, LC3 II, BNIP3L/NIX, and BNIP3) were recruited to mitochondria both in wildtype and PARK2$^{-/-}$ mice, showing that septic mitochondria can be targeted for degradation by mitophagy, even in the absence of PARK2. Using electron microscopy, the presence of autophagosome-like structures was confirmed as well [31]. Thus, mitophagy may be increased in endotoxemia and may serve a compensatory role in the recovery of mitochondrial function by eliminating defective mitochondria. The occurrence of mitochondrial biogenesis after the early activation of mitophagy during the course of endotoxemia suggests that an increase in mitochondrial biogenesis may serve to recover an appropriate volume density of mitochondria to restore the full mitochondrial energy-producing capacity of the heart [14,75,76]. Recent studies suggest that the degradation of mitochondria via mitophagy promotes mitochondrial biogenesis directly via a Toll-like receptor 9 (TLR9)-dependent signaling, although the mechanistic links between autophagy and mitochondrial biogenesis are not well characterized yet [72]. In contrast to a beneficial role of autophagy, the controlled elimination of organelles may also be part of apoptosis. Interactions among signaling components of the two pathways (autophagy and apoptosis) have been reported and indicate a complex cross-talk [77]. Thus, it remains unclear whether autophagy is actually an attempt to remove damaged mitochondria or part of a cellular program leading to apoptosis during sepsis. More studies are needed to further define the role of mitophagy in septic cardiomyopathy and to define how mitophagy may directly regulate mitochondrial biogenesis.

4. Conclusions

Cardiac dysfunction is common in patients suffering from severe sepsis and septic shock. At present, treatment mainly includes management of the infectious focus and hemodynamic support. However, recent progress in sepsis research increased our understanding of underlying molecular alterations contributing to organ dysfunction, including mitochondrial dysfunction. In the heart, a number of mechanisms have been proposed to underlie myocardial mitochondrial dysfunction, including excessive production of mitochondrial NO and ROS, increased mPTP opening, and increased mitochondrial uncoupling. Several reports showed recovery of mitochondrial function which was associated with improvement or normalization of cardiac pump function following treatment with drugs that target different mechanisms within mitochondria but that had no influence on the inflammatory response, supporting the idea that impaired myocardial energetics rather than cardiac inflammation *per se* are critical for cardiac dysfunction in sepsis [11,23,63]. We now need more studies using pharmacological modulators to prevent or reverse specific mitochondrial mechanisms to further evaluate the significance of mitochondrial mechanisms in septic cardiomyopathy, and to test whether such treatments may be a promising and feasible therapeutic approach that could potentially be transferred from bench to bedside. Such approaches may, for example, include inhibition of mitochondrial ROS and NO production and inhibition of mPTP opening. In addition, it would be very informative to perform such studies in the different models of sepsis and endotoxemia, which may not only reveal differences in the underlying disease mechanisms but may also allow evaluating whether the treatment may be broadly applicable and effective irrespective of the sepsis model. The increasing interest in sepsis research and the availability of novel research tools promises that effective specific treatments may be available in the future to complement our current treatment options of severe sepsis and septic shock to improve outcome for this highly jeopardized patient cohort.

Acknowledgments: This work was supported by research grants from the German Research Foundation (DFG) and the argentinean National Scientific and Technical Research Council (CONICET).

Author Contributions: María Cecilia Cimolai, Silvia Alvarez, Christoph Bode and Heiko Bugger contributed to the writing of this review.

Conflicts of Interest: The authors declare no conflict of interest.

References

1. Deutschman, C.S.; Tracey, K.J. Sepsis: Current dogma and new perspectives. *Immunity* **2014**, *40*, 463–475.
2. Soong, J.; Soni, N. Sepsis: Recognition and treatment. *Clin. Med.* **2012**, *12*, 276–280.
3. Sagy, M.; Al-Qaqaa, Y.; Kim, P. Definitions and pathophysiology of sepsis. *Curr. Prob. Pediatr. Adolesc. Health Care* **2013**, *43*, 260–263.
4. Court, O.; Kumar, A.; Parrillo, J.E.; Kumar, A. Clinical review: Myocardial depression in sepsis and septic shock. *Crit. Care* **2002**, *6*, 500–508.
5. Cunnion, R.E.; Schaer, G.L.; Parker, M.M.; Natanson, C.; Parrillo, J.E. The coronary circulation in human septic shock. *Circulation* **1986**, *73*, 637–644.
6. Romero-Bermejo, F.J.; Ruiz-Bailen, M.; Gil-Cebrian, J.; Huertos-Ranchal, M.J. Sepsis-induced cardiomyopathy. *Curr. Cardiol. Rev.* **2011**, *7*, 163–183.
7. Rudiger, A.; Singer, M. Mechanisms of sepsis-induced cardiac dysfunction. *Crit. Care Med.* **2007**, *35*, 1599–1608.
8. Vallet, B.; Lund, N.; Curtis, S.E.; Kelly, D.; Cain, S.M. Gut and muscle tissue PO2 in endotoxemic dogs during shock and resuscitation. *J. Appl. Physiol.* **1994**, *76*, 793–800.
9. Joseph, B. On anoxaemia. *Lancet* **1920**, 485–489.
10. Sair, M.; Etherington, P.J.; Curzen, N.P.; Winlove, C.P.; Evans, T.W. Tissue oxygenation and perfusion in endotoxemia. *Am. J. Physiol.* **1996**, *271*, H1620–H1625.
11. Hotchkiss, R.S.; Rust, R.S.; Dence, C.S.; Wasserman, T.H.; Song, S.K.; Hwang, D.R.; Karl, I.E.; Welch, M.J. Evaluation of the role of cellular hypoxia in sepsis by the hypoxic marker [18F] fluoromisonidazole. *Am. J. Physiol.* **1991**, *261*, R965–R972.

12. Fink, M.P. Cytopathic hypoxia. Mitochondrial dysfunction as mechanism contributing to organ dysfunction in sepsis. *Crit. Care Clin.* **2001**, *17*, 219–237.

13. Soriano, F.G.; Nogueira, A.C.; Caldini, E.G.; Lins, M.H.; Teixeira, A.C.; Cappi, S.B.; Lotufo, P.A.; Bernik, M.M.; Zsengeller, Z.; Chen, M.; *et al.* Potential role of poly(adenosine 5'-diphosphate-ribose) polymerase activation in the pathogenesis of myocardial contractile dysfunction associated with human septic shock. *Crit. Care Med.* **2006**, *34*, 1073–1079.

14. Vanasco, V.; Saez, T.; Magnani, N.D.; Pereyra, L.; Marchini, T.; Corach, A.; Vaccaro, M.I.; Corach, D.; Evelson, P.; Alvarez, S. Cardiac mitochondrial biogenesis in endotoxemia is not accompanied by mitochondrial function recovery. *Free Radic. Biol. Med.* **2014**, *77*, 1–9.

15. Tavener, S.A.; Long, E.M.; Robbins, S.M.; McRae, K.M.; van Remmen, H.; Kubes, P. Immune cell toll-like receptor 4 is required for cardiac myocyte impairment during endotoxemia. *Circ. Res.* **2004**, *95*, 700–707.

16. Drosatos, K.; Khan, R.S.; Trent, C.M.; Jiang, H.; Son, N.H.; Blaner, W.S.; Homma, S.; Schulze, P.C.; Goldberg, I.J. Peroxisome proliferator-activated receptor-γ activation prevents sepsis-related cardiac dysfunction and mortality in mice. *Circ. Heart Fail.* **2013**, *6*, 550–562.

17. Escames, G.; Leon, J.; Macias, M.; Khaldy, H.; Acuna-Castroviejo, D. Melatonin counteracts lipopolysaccharide-induced expression and activity of mitochondrial nitric oxide synthase in rats. *FASEB J.* **2003**, *17*, 932–934.

18. Alvarez, S.; Boveris, A. Mitochondrial nitric oxide metabolism in rat muscle during endotoxemia. *Free Radic. Biol. Med.* **2004**, *37*, 1472–1478.

19. Bugger, H.; Abel, E.D. Mitochondria in the diabetic heart. *Cardiovasc. Res.* **2010**, *88*, 229–240.

20. Bugger, H.; Boudina, S.; Hu, X.X.; Tuinei, J.; Zaha, V.G.; Theobald, H.A.; Yun, U.J.; McQueen, A.P.; Wayment, B.; Litwin, S.E.; *et al.* Type 1 diabetic akita mouse hearts are insulin sensitive but manifest structurally abnormal mitochondria that remain coupled despite increased uncoupling protein 3. *Diabetes* **2008**, *57*, 2924–2932.

21. Lesnefsky, E.J.; Chen, Q.; Slabe, T.J.; Stoll, M.S.; Minkler, P.E.; Hassan, M.O.; Tandler, B.; Hoppel, C.L. Ischemia, rather than reperfusion, inhibits respiration through cytochrome oxidase in the isolated, perfused rabbit heart: Role of cardiolipin. *Am. J. Physiol. Heart Circ. Physiol.* **2004**, *287*, H258–H267.

22. Takasu, O.; Gaut, J.P.; Watanabe, E.; To, K.; Fagley, R.E.; Sato, B.; Jarman, S.; Efimov, I.R.; Janks, D.L.; Srivastava, A.; *et al.* Mechanisms of cardiac and renal dysfunction in patients dying of sepsis. *Am. J. Respir. Crit. Care Med.* **2013**, *187*, 509–517.

23. Doi, K.; Leelahavanichkul, A.; Yuen, P.S.; Star, R.A. Animal models of sepsis and sepsis-induced kidney injury. *J. Clin. Investig.* **2009**, *119*, 2868–2878.

24. Smeding, L.; Plotz, F.B.; Groeneveld, A.B.; Kneyber, M.C. Structural changes of the heart during severe sepsis or septic shock. *Shock* **2012**, *37*, 449–456.

25. Reynolds, C.M.; Suliman, H.B.; Hollingsworth, J.W.; Welty-Wolf, K.E.; Carraway, M.S.; Piantadosi, C.A. Nitric oxide synthase-2 induction optimizes cardiac mitochondrial biogenesis after endotoxemia. *Free Radic. Biol. Med.* **2009**, *46*, 564–572.

26. Joshi, M.S.; Julian, M.W.; Huff, J.E.; Bauer, J.A.; Xia, Y.; Crouser, E.D. Calcineurin regulates myocardial function during acute endotoxemia. *Am. J. Respir. Crit. Care Med.* **2006**, *173*, 999–1007.

27. An, J.; Du, J.; Wei, N.; Guan, T.; Camara, A.K.; Shi, Y. Differential sensitivity to LPS-induced myocardial dysfunction in the isolated brown norway and Dahl S rat hearts: Roles of mitochondrial function, NF-κB activation, and TNF-α production. *Shock* **2012**, *37*, 325–332.

28. Chopra, M.; Golden, H.B.; Mullapudi, S.; Dowhan, W.; Dostal, D.E.; Sharma, A.C. Modulation of myocardial mitochondrial mechanisms during severe polymicrobial sepsis in the rat. *PLoS ONE* **2011**, *6*, e21285.

29. Kozlov, A.V.; Staniek, K.; Haindl, S.; Piskernik, C.; Ohlinger, W.; Gille, L.; Nohl, H.; Bahrami, S.; Redl, H. Different effects of endotoxic shock on the respiratory function of liver and heart mitochondria in rats. *Am. J. Physiol. Gastrointest. Liver Physiol.* **2006**, *290*, G543–G549.

30. Hassoun, S.M.; Marechal, X.; Montaigne, D.; Bouazza, Y.; Decoster, B.; Lancel, S.; Neviere, R. Prevention of endotoxin-induced sarcoplasmic reticulum calcium leak improves mitochondrial and myocardial dysfunction. *Crit. Care Med.* **2008**, *36*, 2590–2596.

31. Piquereau, J.; Godin, R.; Deschenes, S.; Bessi, V.L.; Mofarrahi, M.; Hussain, S.N.; Burelle, Y. Protective role of PARK2/Parkin in sepsis-induced cardiac contractile and mitochondrial dysfunction. *Autophagy* **2013**, *9*, 1837–1851.

32. Zang, Q.S.; Sadek, H.; Maass, D.L.; Martinez, B.; Ma, L.; Kilgore, J.A.; Williams, N.S.; Frantz, D.E.; Wigginton, J.G.; Nwariaku, F.E.; *et al.* Specific inhibition of mitochondrial oxidative stress suppresses inflammation and improves cardiac function in a rat pneumonia-related sepsis model. *Am. J. Physiol. Heart Circ. Physiol.* **2012**, *302*, H1847–H1859.

33. Alvarez, S.; Evelson, P.; Cimolai, M.C. Oxygen and nitric oxide metabolism in sepsi. In *Free Radical Pathophysiology*; Alvarez, S., Ed.; Transworld Research Network: Kerala, India, 2008; pp. 223–236.

34. Wallace, D.C.; Shoffner, J.M.; Watts, R.L.; Juncos, J.L.; Torroni, A. Mitochondrial oxidative phosphorylation defects in Parkinson's disease. *Ann. Neurol.* **1992**, *32*, 113–114.

35. Kanai, A.J.; Pearce, L.L.; Clemens, P.R.; Birder, L.A.; VanBibber, M.M.; Choi, S.Y.; de Groat, W.C.; Peterson, J. Identification of a neuronal nitric oxide synthase in isolated cardiac mitochondria using electrochemical detection. *Proc. Natl. Acad. Sci. USA* **2001**, *98*, 14126–14131.

36. Galley, H.F. Oxidative stress and mitochondrial dysfunction in sepsis. *Br. J. Anaesth.* **2011**, *107*, 57–64.

37. Van de Sandt, A.M.; Windler, R.; Godecke, A.; Ohlig, J.; Zander, S.; Reinartz, M.; Graf, J.; van Faassen, E.E.; Rassaf, T.; Schrader, J.; *et al.* Endothelial nos (NOS3) impairs myocardial function in developing sepsis. *Basic Res. Cardiol.* **2013**, *108*, 330.

38. Vanasco, V.; Magnani, N.D.; Cimolai, M.C.; Valdez, L.B.; Evelson, P.; Boveris, A.; Alvarez, S. Endotoxemia impairs heart mitochondrial function by decreasing electron transfer, ATP synthesis and atp content without affecting membrane potential. *J. Bioenerg. Biomembr.* **2012**, *44*, 243–252.

39. Burgoyne, J.R.; Rudyk, O.; Mayr, M.; Eaton, P. Nitrosative protein oxidation is modulated during early endotoxemia. *Nitric Oxide* **2011**, *25*, 118–124.

40. Zang, Q.; Maass, D.L.; Tsai, S.J.; Horton, J.W. Cardiac mitochondrial damage and inflammation responses in sepsis. *Surg. Infect.* **2007**, *8*, 41–54.

41. Suliman, H.B.; Welty-Wolf, K.E.; Carraway, M.; Tatro, L.; Piantadosi, C.A. Lipopolysaccharide induces oxidative cardiac mitochondrial damage and biogenesis. *Cardiovasc. Res.* **2004**, *64*, 279–288.

42. Boveris, A.; Alvarez, S.; Navarro, A. The role of mitochondrial nitric oxide synthase in inflammation and septic shock. *Free Radic. Biol. Med.* **2002**, *33*, 1186–1193.

43. Escames, G.; Lopez, L.C.; Ortiz, F.; Lopez, A.; Garcia, J.A.; Ros, E.; Acuna-Castroviejo, D. Attenuation of cardiac mitochondrial dysfunction by melatonin in septic mice. *FEBS J.* **2007**, *274*, 2135–2147.

44. Xu, C.; Yi, C.; Wang, H.; Bruce, I.C.; Xia, Q. Mitochondrial nitric oxide synthase participates in septic shock myocardial depression by nitric oxide overproduction and mitochondrial permeability transition pore opening. *Shock* **2012**, *37*, 110–115.

45. Vanasco, V.; Cimolai, M.C.; Evelson, P.; Alvarez, S. The oxidative stress and the mitochondrial dysfunction caused by endotoxemia are prevented by α-lipoic acid. *Free Radic. Res.* **2008**, *42*, 815–823.

46. Supinski, G.S.; Murphy, M.P.; Callahan, L.A. MitoQ administration prevents endotoxin-induced cardiac dysfunction. *Am. J. Physiol. Regul. Integr. Comp. Physiol.* **2009**, *297*, R1095–R1102.

47. Halestrap, A.P.; McStay, G.P.; Clarke, S.J. The permeability transition pore complex: Another view. *Biochimie* **2002**, *84*, 153–166.

48. Larche, J.; Lancel, S.; Hassoun, S.M.; Favory, R.; Decoster, B.; Marchetti, P.; Chopin, C.; Neviere, R. Inhibition of mitochondrial permeability transition prevents sepsis-induced myocardial dysfunction and mortality. *J. Am. Coll. Cardiol.* **2006**, *48*, 377–385.

49. Fauvel, H.; Marchetti, P.; Obert, G.; Joulain, O.; Chopin, C.; Formstecher, P.; Neviere, R. Protective effects of cyclosporin a from endotoxin-induced myocardial dysfunction and apoptosis in rats. *Am. J. Respir. Crit. Rare Red.* **2002**, *165*, 449–455.

50. Costa, A.D.; Quinlan, C.L.; Andrukhiv, A.; West, I.C.; Jaburek, M.; Garlid, K.D. The direct physiological effects of mitoKATP opening on heart mitochondria. *Am. J. Physiol. Heart Circ. Physiol.* **2006**, *290*, H406–H415.

51. Bougaki, M.; Searles, R.J.; Kida, K.; Yu, J.; Buys, E.S.; Ichinose, F. Nos3 protects against systemic inflammation and myocardial dysfunction in murine polymicrobial sepsis. *Shock* **2010**, *34*, 281–290.

52. Ricquier, D.; Bouillaud, F. Mitochondrial uncoupling proteins: From mitochondria to the regulation of energy balance. *J. Physiol.* **2000**, *1*, 3–10.

53. Boudina, S.; Sena, S.; Theobald, H.; Sheng, X.; Wright, J.J.; Hu, X.X.; Aziz, S.; Johnson, J.I.; Bugger, H.; Zaha, V.G.; *et al.* Mitochondrial energetics in the heart in obesity-related diabetes: Direct evidence for increased uncoupled respiration and activation of uncoupling proteins. *Diabetes* **2007**, *56*, 2457–2466.

54. Bugger, H.; Abel, E.D. Molecular mechanisms of diabetic cardiomyopathy. *Diabetologia* **2014**, *57*, 660–671.

55. Konig, A.; Bode, C.; Bugger, H. Diabetes mellitus and myocardial mitochondrial dysfunction: Bench to bedside. *Heart Fail. Clin.* **2012**, *8*, 551–561.

56. Murray, A.J.; Cole, M.A.; Lygate, C.A.; Carr, C.A.; Stuckey, D.J.; Little, S.E.; Neubauer, S.; Clarke, K. Increased mitochondrial uncoupling proteins, respiratory uncoupling and decreased efficiency in the chronically infarcted rat heart. *J. Mol. Cell. Cardiol.* **2008**, *44*, 694–700.

57. Roshon, M.J.; Kline, J.A.; Thornton, L.R.; Watts, J.A. Cardiac UCP2 expression and myocardial oxidative metabolism during acute septic shock in the rat. *Shock* **2003**, *19*, 570–576.

58. Wang, X.; Liu, D.; Chai, W.; Long, Y.; Su, L.; Yang, R. The role of uncoupling protein-2 (UCP2) during myocardial dysfunction in a canine model of endotoxin shock. *Shock* **2014**, *3*, 292–297.

59. Zheng, G.; Lyu, J.; Liu, S.; Huang, J.; Liu, C.; Xiang, D.; Xie, M.; Zeng, Q. Silencing of uncoupling protein 2 by small interfering rna aggravates mitochondrial dysfunction in cardiomyocytes under septic conditions. *Int. J. Mol. Med.* **2015**, *35*, 1525–1536.

60. Aguirre, E.; Cadenas, S. GDP and carboxyatractylate inhibit 4-hydroxynonenal-activated proton conductance to differing degrees in mitochondria from skeletal muscle and heart. *Biochim. Biophys. Acta* **2010**, *1797*, 1716–1726.

61. Hickson-Bick, D.L.; Jones, C.; Buja, L.M. Stimulation of mitochondrial biogenesis and autophagy by lipopolysaccharide in the neonatal rat cardiomyocyte protects against programmed cell death. *J. Mol. Cell. Cardiol.* **2008**, *44*, 411–418.

62. Jornayvaz, F.R.; Shulman, G.I. Regulation of mitochondrial biogenesis. *Essays Biochem.* **2010**, *47*, 69–84.

63. Gottlieb, R.A.; Gustafsson, A.B. Mitochondrial turnover in the heart. *Biochim. Biophys. Acta* **2011**, *1813*, 1295–1301.

64. Wenz, T. Regulation of mitochondrial biogenesis and PGC-1α under cellular stress. *Mitochondrion* **2013**, *13*, 134–142.

65. Kelly, D.P.; Scarpulla, R.C. Transcriptional regulatory circuits controlling mitochondrial biogenesis and function. *Genes Dev.* **2004**, *18*, 357–368.

66. Lancel, S.; Hassoun, S.M.; Favory, R.; Decoster, B.; Motterlini, R.; Neviere, R. Carbon monoxide rescues mice from lethal sepsis by supporting mitochondrial energetic metabolism and activating mitochondrial biogenesis. *J. Pharmacol. Exp. Ther.* **2009**, *329*, 641–648.

67. Russell, L.K.; Mansfield, C.M.; Lehman, J.J.; Kovacs, A.; Courtois, M.; Saffitz, J.E.; Medeiros, D.M.; Valencik, M.L.; McDonald, J.A.; Kelly, D.P. Cardiac-specific induction of the transcriptional coactivator peroxisome proliferator-activated receptor γ coactivator-1α promotes mitochondrial biogenesis and reversible cardiomyopathy in a developmental stage-dependent manner. *Circ. Res.* **2004**, *94*, 525–533.

68. Schilling, J.; Lai, L.; Sambandam, N.; Dey, C.E.; Leone, T.C.; Kelly, D.P. Toll-like receptor-mediated inflammatory signaling reprograms cardiac energy metabolism by repressing peroxisome proliferator-activated receptor gamma coactivator-1 signaling. *Circ. Heart Fail.* **2011**, *4*, 474–482.

69. Kubli, D.A.; Gustafsson, A.B. Mitochondria and mitophagy: The yin and yang of cell death control. *Circ. Res.* **2012**, *111*, 1208–1221.

70. Eisenberg-Lerner, A.; Bialik, S.; Simon, H.U.; Kimchi, A. Life and death partners: Apoptosis, autophagy and the cross-talk between them. *Cell Death Differ.* **2009**, *16*, 966–975.

71. Carchman, E.H.; Rao, J.; Loughran, P.A.; Rosengart, M.R.; Zuckerbraun, B.S. Heme oxygenase-1-mediated autophagy protects against hepatocyte cell death and hepatic injury from infection/sepsis in mice. *Hepatology* **2011**, *53*, 2053–2062.

72. Carchman, E.H.; Whelan, S.; Loughran, P.; Mollen, K.; Stratamirovic, S.; Shiva, S.; Rosengart, M.R.; Zuckerbraun, B.S. Experimental sepsis-induced mitochondrial biogenesis is dependent on autophagy, TLR4, and TLR9 signaling in liver. *FASEB J.* **2013**, *27*, 4703–4711.

73. Ceylan-Isik, A.F.; Zhao, P.; Zhang, B.; Xiao, X.; Su, G.; Ren, J. Cardiac overexpression of metallothionein rescues cardiac contractile dysfunction and endoplasmic reticulum stress but not autophagy in sepsis. *J. Mol. Cell. Cardiol.* **2010**, *48*, 367–378.

74. Suliman, H.B.; Welty-Wolf, K.E.; Carraway, M.S.; Schwartz, D.A.; Hollingsworth, J.W.; Piantadosi, C.A. Toll-like receptor 4 mediates mitochondrial DNA damage and biogenic responses after heat-inactivated. *E. coli. FASEB J.* **2005**, *19*, 1531–1533.

75. Crouser, E.D.; Julian, M.W.; Huff, J.E.; Struck, J.; Cook, C.H. Carbamoyl phosphate synthase-1: A marker of mitochondrial damage and depletion in the liver during sepsis. *Crit. Care Med.* **2006**, *34*, 2439–2446.

76. Yuan, H.; Perry, C.N.; Huang, C.; Iwai-Kanai, E.; Carreira, R.S.; Glembotski, C.C.; Gottlieb, R.A. LPS-induced autophagy is mediated by oxidative signaling in cardiomyocytes and is associated with cytoprotection. *Am. J. Physiol. Heart Circ. Physiol.* **2009**, *296*, H470–H479.

77. Nikoletopoulou, V.; Markaki, M.; Palikaras, K.; Tavernarakis, N. Crosstalk between apoptosis, necrosis and autophagy. *Biochim. Biophys. Acta* **2013**, *1833*, 3448–3459.

Mitochondrial Optic Atrophy (OPA) 1 Processing Is Altered in Response to Neonatal Hypoxic-Ischemic Brain Injury

Ana A. Baburamani, Chloe Hurling, Helen Stolp, Kristina Sobotka,
Pierre Gressens, Henrik Hagberg and Claire Thornton

Abstract: Perturbation of mitochondrial function and subsequent induction of cell death pathways are key hallmarks in neonatal hypoxic-ischemic (HI) injury, both in animal models and in term infants. Mitoprotective therapies therefore offer a new avenue for intervention for the babies who suffer life-long disabilities as a result of birth asphyxia. Here we show that after oxygen-glucose deprivation in primary neurons or in a mouse model of HI, mitochondrial protein homeostasis is altered, manifesting as a change in mitochondrial morphology and functional impairment. Furthermore we find that the mitochondrial fusion and cristae regulatory protein, OPA1, is aberrantly cleaved to shorter forms. OPA1 cleavage is normally regulated by a balanced action of the proteases Yme1L and Oma1. However, in primary neurons or after HI *in vivo*, protein expression of YmeIL is also reduced, whereas no change is observed in Oma1 expression. Our data strongly suggest that alterations in mitochondria-shaping proteins are an early event in the pathogenesis of neonatal HI injury.

Reprinted from *Int. J. Mol. Sci.* Cite as: Baburamani, A.A.; Hurling, C.; Stolp, H.; Sobotka, K.; Gressens, P.; Hagberg, H.; Thornton, C. Mitochondrial Optic Atrophy (OPA) 1 Processing Is Altered in Response to Neonatal Hypoxic-Ischemic Brain Injury. *Int. J. Mol. Sci.* **2015**, *16*, 22509–22524.

1. Introduction

Moderate to severe hypoxic-ischemic encephalopathy (HIE), caused by a lack of oxygen or blood flow to the brain around the time of birth, affects 1.5 in every 1000 live births in the UK and far more in the developing world [1–3]. The consequences for babies and parents affected by HIE are devastating; 15%–20% of infants will die in the postnatal period and a further 25% will develop severe and long-lasting neurological impairments [4].

In infants and in animal models of hypoxic-ischemic injury (HI) there is an initial depletion of ATP, phosphocreatine and glucose within the brain followed by a transient recovery to almost physiological levels [5]. However, a second, rapid energy failure facilitates the majority of cell death [6–8]. We and others have shown that HI triggers multiple signaling events such as NMDA/AMPA receptor activation, release of reactive oxygen species and increase in intracellular calcium [9–12]. In addition, cell death after neonatal brain injury is characterized morphologically by a mixed necrotic–necroptotic–apoptotic phenotype depending on time post injury and brain region [10,13,14]. However, data from our lab and others strongly suggest that the common thread linking these diverse mechanisms is mitochondrial dysfunction [15,16].

It is well established that in animal models of neonatal HI, mitochondrial respiration and calcium homeostasis are impaired [17–19]. More recently, it was determined that mitochondrial outer membrane permeabilization (MOMP) mediated mitochondrial dysfunction in rodent neonatal HI models [20]. In response to HI, the Bcl-2 family member Bax is activated and translocates to the mitochondria where it complexes and forms a pore with Bak allowing passage of cytochrome c and apoptosis inducing factor (AIF) into the cytosol. Once released AIF and cytochrome c initiate a cascade resulting in activation of caspases, degradation of DNA and ultimately cell death [21]. As such, genetic and pharmacological inhibition of Bax is protective from HI in immature brain [20,22,23]. In addition to Bax-mediated MOMP, mitochondrial ultrastructure is also altered in response to neonatal HI insult [10] and a wide range of mitochondrial morphologies are observed [13]. We therefore hypothesize that such environmental stress may alter mitochondrial dynamics, particularly in the proteins which regulate fission and fusion. Optic Atrophy 1 (OPA1), a dynamin-related guanosine triphosphatase protein, plays a pivotal role in conducting inner-membrane mitochondrial fusion and therefore regulates both mitochondrial cristae junction formation and fusion of distinct mitochondria [24–26]. Here we present data analyzing the effect of *in vitro* oxygen-glucose deprivation (OGD) and *in vivo* HI on the processing of OPA1.

2. Results

2.1. OGD in C17.2 Cells Alters Mitochondrial Function and Morphology

OGD is a widely used *in vitro* technique to mimic aspects of cell death observed in *in vivo* HI injury. We performed OGD on mouse primary cortical neurons and examined mitochondrial morphology and membrane potential in live cells throughout the insult using JC-1 dye. Aggregates of JC-1 accumulate in mitochondria in which the membrane potential is maintained, exhibiting red fluorescence, whereas the appearance of diffuse green JC-1 monomers throughout the cell indicates dissipation of membrane potential. Neurons were preloaded with JC-1 before exposure to OGD. Mitochondria were clearly visible in the processes of control cells and generally of uniform size (Figure 1a,b, Con). However after 90 min OGD, we observed an increase in green monomeric JC-1 suggesting impaired membrane potential and altered morphology with both mitochondrial aggregates and rounded puncta (Figure 1a,b, OGD). Similar findings were observed in a recent study of rat cortical neurons exposed to OGD [27], where control mitochondria were found to be tubular and OGD-exposed mitochondria rounded or poorly labelled. In order to quantify these changes, we performed time-lapse imaging on isolated neurons and calculated the changes in mitochondrial length over time. We found a significant decrease in the average mitochondrial length after 30 min of OGD (Figure 1c). After 90 min OGD we returned the cultures to growth medium and analyzed them at subsequent time points for the effect of the insult on mitochondrial health. We found that 24 h post insult, citrate synthase activity was significantly reduced indicating impaired TCA cycle function (Figure 1d). This suggests that neurons which survive the initial insult may subsequently exhibit impaired mitochondrial function.

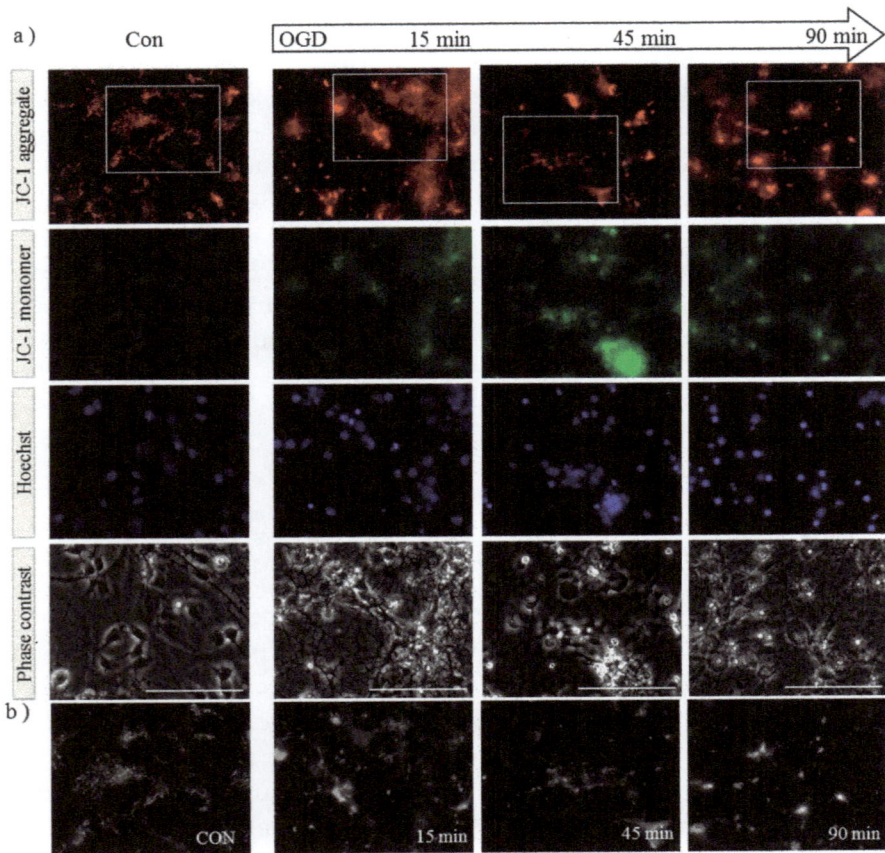

Figure 1. *Cont.*

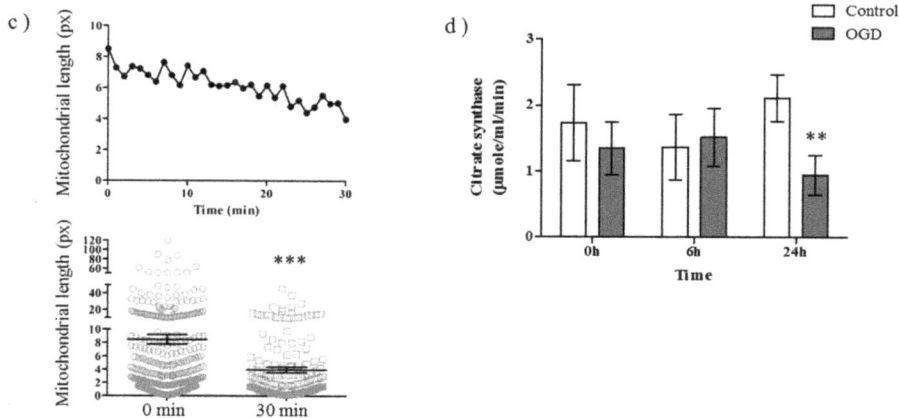

Figure 1. Oxygen-glucose deprivation (OGD) alters mitochondrial membrane potential, morphology and function in primary cortical neurons. (**a**) Primary mouse cortical neurons were loaded with JC-1 dye and Hoechst before exposure to OGD. Cells were imaged live before (Con) and at 15, 45 and 90 min during OGD. Both mitochondrial morphology (as observed in red, top row) and membrane potential (increased green signal, second row) are altered during exposure to OGD. Scale bar represents 100 μm; Figures are representative of three individual experiments: (**b**) Enlargement (3×) of regions defined by white boxes in (**a**). As the experiment progresses, mitochondria morphology appears to alter from tubular structures to round punctate or larger aggregations; (**c**) Primary neurons loaded with JC-1 were imaged every minute during OGD followed by analyses of mitochondrial length. Data shown are an average of 360 mitochondria per time point, and mitochondria from the first and last time points analyzed by student's *t*-test in the panel below, *** $p < 0.001$; (**d**) Primary neurons were subjected to 90 min of OGD followed by up to 24 h incubation in normal growth medium. Lysates were assayed for citrate synthase activity at time points shown following the insult. Data is shown $\pm$ SD, n = 4–6 independent litters, determined by two-way ANOVA followed by a Bonferroni *post-hoc* test, ** $p < 0.01$ for interaction and treatment.

2.2. OPA1 Processing Is Altered after OGD

As there was a distinct alteration in mitochondrial morphology in response to OGD, we examined the expression of key genes involved in mitochondrial fission and fusion. Primary neurons were either untreated or exposed to OGD and RNA extracted at 0, 6 and 24 h post insult. Expression of fission genes (*Drp-1*, *Fis-1*) and fusion genes (*Mitofusin 1*, *Mitofusin 2* and *OPA1*) after OGD were compared with expression in control untreated neurons. We found that there was a small but significant decrease in the expression of *OPA1* mRNA comparing treatment groups (Figure 2a, a = 0.0296, two-way ANOVA for treatment). To further analyze changes in OGD-mediated OPA1 expression, we generated whole cell lysates from control and OGD-treated neurons and determined OPA1 protein expression by western blot at 0, 6 and 24 h post insult. There was a small decrease in the expression of OPA1 apparent at the 6 h timepoint (Figure 2b). Interestingly, OGD appeared to induce the generation of a smaller band and alter the distribution of remaining OPA1 immunoreactivity. There was a proportional shift towards expression of smaller OPA1 moieties most pronounced at 6 h after OGD, compared with control OPA1 expression. (Figure 2b, arrowheads).

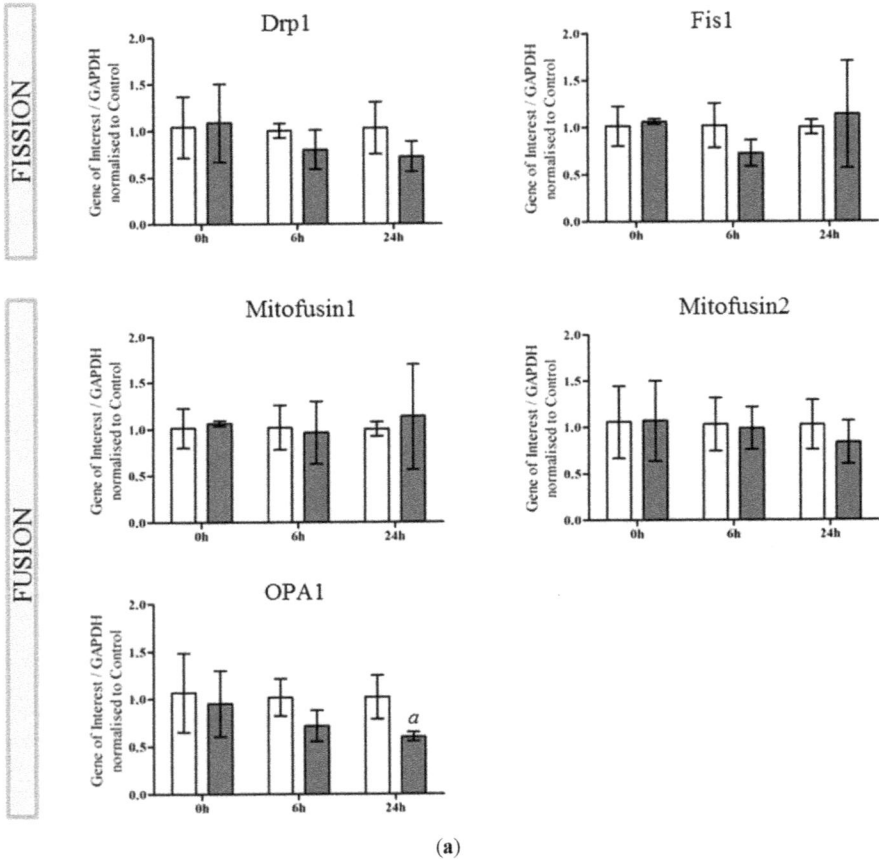

(a)

Figure 2. *Cont.*

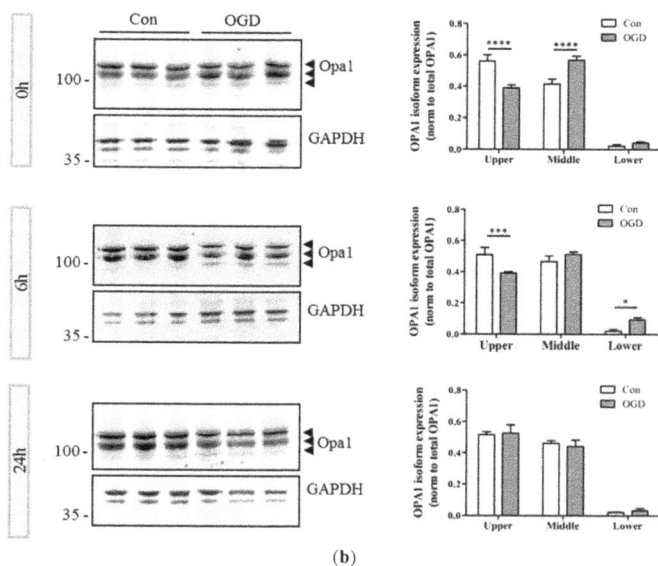

(b)

Figure 2. OPA1 processing is altered after OGD (**a**) mRNA generated from control (white bars) and OGD-treated (grey bars) primary neurons was analyzed by qRT-PCR for changes in expression of fission (*Drp1*, *Fis1*) and fusion genes (*Mitofusin 1* and *2*, *OPA1*). There was a small but significant decrease in OPA1 expression in response to OGD exposure ($a = 0.0296$). Mean data shown $\pm$ SD ($n = 4$–5 independent litters), significance determined by two way ANOVA; (**b**) Protein lysates were generated from primary neurons either immediately after 90 min OGD (0 h) or following 6 or 24 h recovery. Proteins were resolved by SDS-PAGE and OPA1 analyzed by western blot. Equal protein loading was determined by GAPDH expression. Size distribution of the OPA1 immunoreactivity is expressed as a proportion of total OPA1 (arrowheads). There was a significant increase in the expression of the OPA1 lower band at 0 and 6 h, significance determined by two-way ANOVA for treatment and band. If there was a significant interaction, a Bonferroni *post-hoc* was performed. Data are mean $\pm$ SD ($n = 3$ independent litters), * $p < 0.05$, *** $p < 0.001$, **** $p < 0.0001$.

2.3. OGD Reduces Yme1L Protein Expression in Primary Neurons

Alternative splicing of OPA1 generates eight isoforms which depending on variant, will contain S1 and S2 cleavage sites, or an S1 site alone [28]. Previous studies have demonstrated that cleavage at S2 by the intermembrane space AAA-protease Yme1L produces a balance of long and short OPA1 products, optimal for OPA1 function [24,29]. When mitochondrial membrane potential is lost, Oma1, a zinc-metalloprotease which resides on the inner membrane, cleaves OPA1 at the S1 site [30,31]. We therefore examined the expression of Yme1L and Oma1 in primary

285

neurons exposed to OGD. Although gene expression of *Yme1L* was not altered significantly in response to OGD (Yme1L p = 0.0506; Figure 3a), there was a discernible decrease in Yme1L protein expression at the end of OGD (Figure 3b). Conversely, no changes were apparent for either the gene (Figure 3c) or protein expression (Figure 3d) of Oma1.

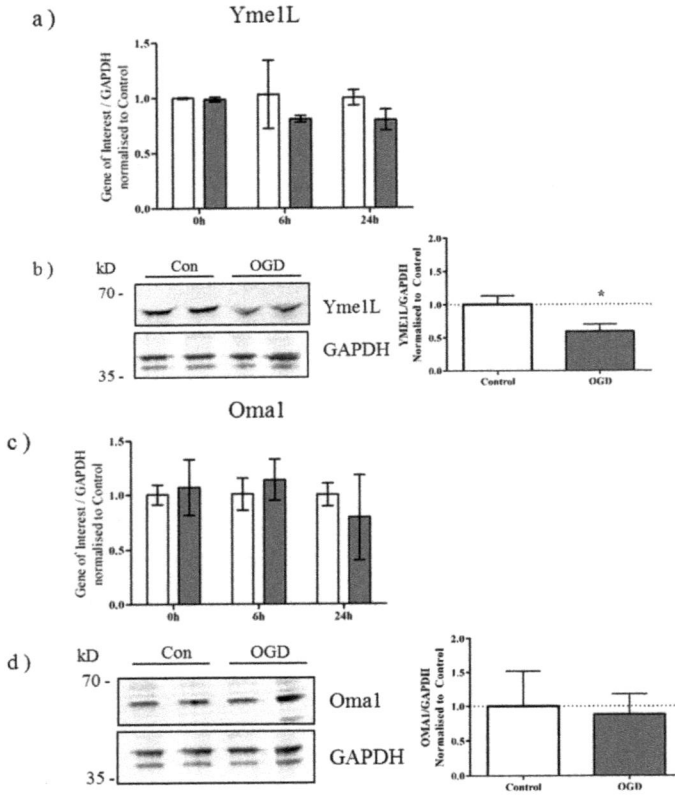

Figure 3. Yme1L protein expression is reduced after OGD (**a**) mRNA generated from control and OGD-treated primary neurons was analyzed by qRT-PCR for changes in expression of Yme1L. Data shown ± SD, n = 3–4 independent litters; (**b**) Protein lysates were generated from primary neurons immediately following OGD and analyzed by western blot for expression of Yme1L. Equal protein loading was determined by GAPDH expression which was used for the quantification (right hand panel). Figure is representative of three individual litters, * p < 0.05 determined by students *t*-test; (**c**) Oma1 gene expression or (**d**) Oma1 protein expression was determined as above with GAPDH for equal loading. Data was analyzed as above. Figure is representative of three individual litters.

2.4. Alterations in OPA1 Processing Are Apparent in Vivo after HI

Finally we determined if these effects occurred *in vivo* in an animal model of term HI. We used the well-characterized Vannucci HI model in mouse P9 pups, which recapitulates aspects of delayed cell death in human perinatal HI [32,33]. Following unilateral carotid artery ligation, pups are exposed to hypoxia for 75 min before returning to normoxia. This allows both hypoxic (contralateral hemisphere) and hypoxic-ischemic (ipsilateral hemisphere) brain tissue to be sampled from the same animal. Brain tissue was harvested at 0, 24 and 48 h post injury, mitochondrial fractions isolated and OPA1, Yme1L and Oma1 analyzed by western blot (Figure 4a). We found that the bias towards cleaved OPA1 was clearly visible in the hypoxic-ischemic samples from the earliest time point, with a significant decrease of upper band intensity correlating with a significant increase in middle band intensity (Figure 4a, 0 h). Furthermore an additional band of a lower molecular weight was clearly visible in the 24 (Figure 4b) and 48 h (Figure 4c) HI samples. We quantified the upper, middle and lower molecular weight OPA1 bands as a proportion of total OPA1 and observed a significant decrease in the upper band after HI but not hypoxia alone, which was accompanied by a significant increase in the expression of the lower form (Figure 4b). In addition to changes in OPA1, there was a distinct trend towards a decrease in Yme1L expression at 0 (Figure 4a) and 24 h (Figure 4b) which appeared to be resolved by 48 h. Throughout the time course of the experiment, Oma1 expression did not appear to vary (Figure 4a–c). In summary, our data suggest that OGD *in vitro* and HI injury *in vivo* result in cleavage of OPA1 to lower molecular weight forms. This observation correlates with an OGD- or HI-mediated decrease in Yme1L expression.

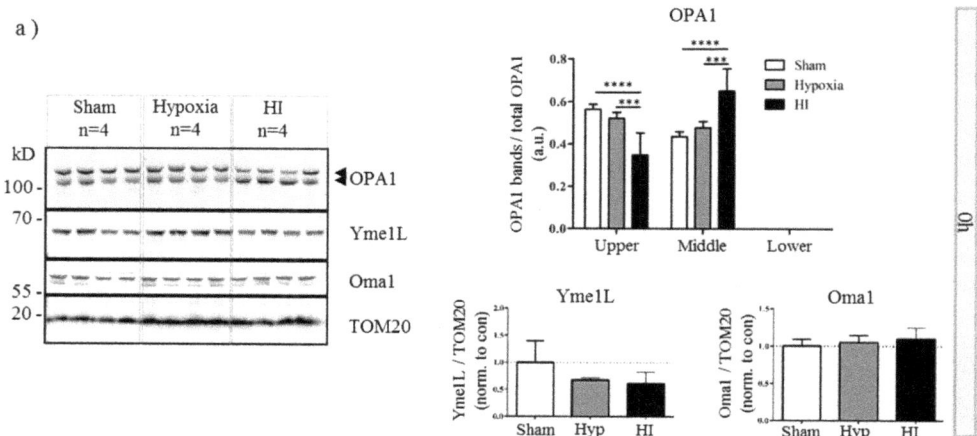

Figure 4. *Cont.*

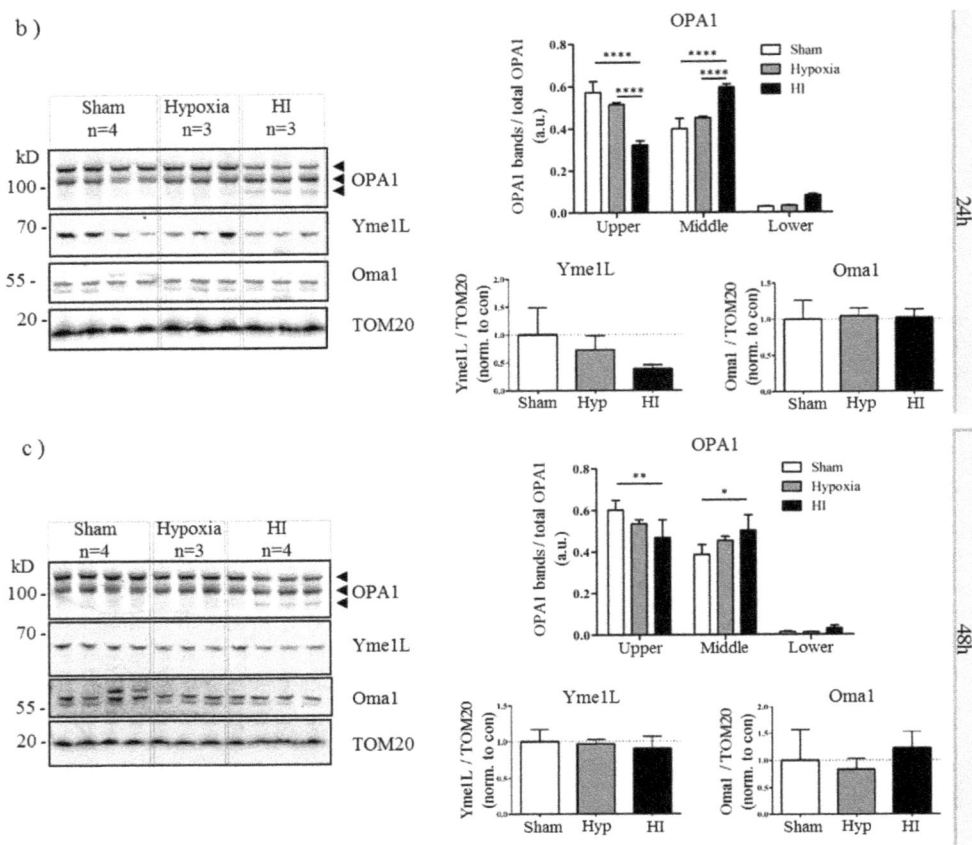

Figure 4. Aberrant processing of OPA1 *in vivo*. Mitochondrial fractions were generated from hypoxic-ischemic (HI), hypoxia alone or sham control mice. Proteins were resolved by SDS-PAGE and analyzed by western blot for OPA1, Yme1L and Oma1 at 0 (**a**), 24 (**b**) and 48 h (**c**) following HI. Tom20 was used as a loading control for mitochondria (bottom panel). Total OPA1 expression was determined by densitometry and OPA1 isoforms or cleavage products expressed as a proportion of that total. Arrowheads indicate alternate forms of OPA1. Mean data shown ± SD. OPA1 significance was determined by two-way ANOVA for band size and treatment followed by a Bonferroni *post-hoc* test * $p < 0.05$, ** $p < 0.01$, *** $p < 0.001$, **** $p < 0.0001$. YME1L and Oma1 data were analyzed with a one-way ANOVA.

3. Discussion

Dysregulation of energy metabolism is a common feature in a number of human diseases including diabetes, cardiovascular and neurodegenerative disorders [34–36].

288

It is therefore unsurprising the mitochondria are center stage in the development of neonatal brain injury, due to additional high energy demands of the immature brain as it develops [15], and protecting mitochondrial function represents a valid target for future therapeutic intervention. Here we present data suggesting that the homeostasis of inner mitochondrial membrane proteins is disrupted in response to OGD *in vitro* and HI *in vivo* and that alterations in protein function are due largely to post-translational modification.

Although we only identified small changes in OPA1 gene expression, our major finding is that OPA1 is rapidly cleaved to shorter forms in response to hypoxia-ischemia. Shortened forms of OPA1 are reported to occur as a result of environmental stress, resulting in fragmented mitochondria [37,38]. Furthermore, actively altering the balance of long (L-OPA1) and short (S-OPA1) forms may favor mitochondrial fission [39], agreeing with the appearance of small punctate mitochondria apparent during the OGD insult in primary neurons (Figure 1).

OPA1 cleavage occurs due to the actions of the ATP-dependent protease Yme1L and the ATP-independent protease Oma1 [30,31]. Yme1L cleaves OPA1 at the S2 site and subsequent products remain fusion-competent [24,29]. However, loss of mitochondrial membrane potential results in cleavage at OPA1 S1 site by Oma1 and generation of fusion-incompetent mitochondria [31]. Our results (Figure 1) and those of others [27,40] suggest that exposure of primary neurons to OGD induces a decrease in mitochondrial membrane potential coupled with changes in mitochondrial morphology (Figure 1, Con *vs.* OGD).

Concomitantly, post insult, we observed a decrease in Yme1L protein expression *in vitro* (Figure 3) and *in vivo* (Figure 4). It is well established that in animal models of HI and *in vitro*, the insult induces a rapid depletion of cellular ATP [8,33,41]. Yme1L activity is ATP-dependent and more recently, its expression was found to be reduced following oxidative stress [42]. These authors also identified Oma1 as an ATP-independent protease which regulated Yme1L expression; resistance to Oma1-mediated Yme1L degradation was conferred by ATP binding. Both our *in vitro* and our *in vivo* results are in line with these findings suggesting that loss of Yme1L may sensitize cells to oxidative stress through dysfunctional mitochondrial dynamics [43]. However, further work is required to determine whether disrupting the balance of Yme1L and Oma1 is critical in neonatal hypoxic-ischemic brain injury.

In the development of neonatal brain injury in response to HI insult, we and others have documented the induction of apoptosis resulting in cytochrome c release from permeabilized mitochondria [13,44–46]. However, the majority of cytochrome c is held within the cristae, relying on cristae reconfiguration to allow it access to move into the intermembrane space once apoptosis is induced. In addition to its role in fusion, OPA1 controls the integrity of the cristae [25] and inhibition of OPA1 leads to cristae disorganization [26]. Our data suggest the possibility that

OPA1 processing may be an early but key step in the propagation of neuronal cell death induced by neonatal hypoxic-ischemic injury. Interestingly, during revision of our manuscript, Sanderson and colleagues identified release of cytochrome c and appearance of degraded OPA1 in the cytosol in response to OGD/reperfusion in primary neurons [47] providing further evidence for our hypothesis. Our future studies will therefore center on the roles of Yme1L and Oma1 in the regulation of OPA1 in order to highlight whether prevention of such cleavage is neuroprotective in neonatal hypoxic-ischemic injury.

4. Experimental Section

4.1. Research Ethics Statement

All animal use was in accordance with local rules (King's College London, Animal Welfare and Ethical Review Board, London, UK) and with the regulations and guidance issued under the Animals (Scientific Procedures) Act (1986) covered by Home Office personal and project licenses.

4.2. Primary Cortical Neuron Preparation

C57/Bl6 pregnant mice (Charles Rivers, Margate, UK) at embryonic day 13.5–15.5 were killed by schedule 1 methods. Embryonic cerebral cortices were dissected and tissue from the same litter pooled. Primary cortical neurons were prepared as described previously [48], plated at a density of 2×10^6 cells/6 cm plate and maintained in neurobasal medium (Life Technologies, Paisley, UK) containing B27® (Life Technologies), L-Glutamine (Sigma, Gillingham, UK) and Streptomycin/AmphotericinB (Life Technologies,). Cultures were maintained at 37 °C, 5% CO_2.

4.3. Neonatal Hypoxia-Ischemia

C57/Bl6 mice (Charles River) at postnatal day 9 (P9) were subjected to unilateral hypoxia-ischemia, essentially according to the Rice–Vannucci model that results in a focal ischemic injury allowing for comparison between an HI (ipsilateral) and hypoxia alone (contralateral) hemisphere [32,33]. Briefly, mice were anesthetized with isoflurane (4.5% for induction and 2% for maintenance) in a mixture of nitrous oxide and oxygen (1:1), with the duration of anesthesia being <5 min per pup. The left common carotid artery was isolated and ligated. Pups recovered for 1–2 h in the parent cage. Litters were placed in a chamber with a humidified hypoxic gas mixture (10% oxygen in nitrogen, 36 °C) for 75 min. Sham control mice were not subjected to surgery or hypoxic chamber. After hypoxic exposure, pups were returned to their biological dams until the conclusion of the experiment. Both females and males pups were used and treatment groups contained pups from at least 3 independent litters.

4.4. Oxygen-Glucose Deprivation (OGD) of Primary Neurons

Primary neurons were cultured for a minimum of 6 days (DIV6) prior to treatment then loaded with JC-1 (5 μM; Life Technologies) and Hoechst (10 μg/mL; Sigma) for 30 min, 37 °C, 5% CO_2. Growth medium was replaced with de-gassed, glucose-free, Neurobasal-A medium (Life Technologies) and culture plates mounted on the EVOS/Chamlide microscope (Life Technologies) where they were maintained at 37 °C in a 95% N_2/5% CO_2 environment for the duration of the OGD. Following OGD, medium was replaced with standard Neurobasal medium (containing additions described above) and cultures returned to 5% CO_2 incubation. For control plates, medium was replaced with standard Neurobasal media at the start of the OGD plates and replaced again with media after 90 min. Cells were collected at 0, 6 or 24 h following treatment.

4.5. MTT Assay

Primary cortical neurons were assayed for mitochondrial reductase activity using the MTT (3-(4,5-Dimethylthiazol-2-yl-)-2,5-diphenyl-2H-tetrazolium bromide; Sigma) assay as described previously [49].

4.6. Citrate Synthase Assay

Primary neurons were lysed in CelLytic MT Cell Lysis Reagent (Sigma), protein concentration determined by BCA assay (Thermo Scientific, Loughborough, UK) and were assayed for enzymatic activity using Citrate Synthase Assay Kit (Sigma) according to manufacturer's instructions. Eight micrograms of total protein was used per reaction on a 96-well plate and independent samples were measured in duplicate. Absorbance was read at 412 nm on a SPECTROstar Nano plate reader using MARS analysis software (BMG Labtech, Aylesbury, UK). Baseline absorbance was measured every 30 s for 5 min, and following addition of oxaloacetic acid, total activity was measured every 10 min for 60 min.

4.7. qRT-PCR

RNA was harvested using the Direct-zol RNA MiniPrep kit (Zymo Research, through Cambridge Biosciences, Cambridge UK) as per manufacturer's instructions. Total RNA (100 ng) was analysed using Taqman gene expression assays and RNA-to-CT kit (Life Technologies) on a StepOneplus Cycler (Life Technologies). Data were normalized to the expression of GAPDH and to controls for each time point using the $2^{-\Delta\Delta Ct}$ method [50]. The following primer pairs were used: *GAPDH* (Mm99999915), *DRP1* (Dmn1; Mm01342903), *Fis1* (Mm00481580), *Mitofusin1* (Mm00612599), *Mitofusin2* (Mm00500120), *OPA1* (Mm01349707), *Oma1* (Mm01260328) and *Yme1L* (Mm00496843) from Life Technologies

4.8. Subcellular Fractionation

For preparation of mitochondrial fractions, mice were sacrificed at 0, 24 and 48 h post HI and brains rapidly dissected into ice-cold subcellular fractionation buffer (250 mM sucrose, 20 mM HEPES pH 7.4, 10 mM KCL, 1.5 mM MgCl$_2$, 1 mM EDTA, 1 mM EGTA, 1 mM DTT, 1× protease inhibitor cocktail). Samples were homogenized in a 2 mL dounce homogenizer on ice, passed through a 27 G needle and incubated on ice for 20 min. The suspension was centrifuged (720× g, 5 min) to remove the nuclear fraction and any unbroken cells. The resulting supernatant was centrifuged (15,000× g, 15 min) to obtain a mitochondrial pellet and a cytosolic supernatant.

4.9. Western Blot

For protein analysis primary neurons were lysed in HEPES buffer A (50 mM HEPES (pH 7.5), 50 mM sodium fluoride, 5 mM sodium pyrophosphate, 1 mM EDTA, +1× protease inhibitors (Sigma)) containing 1% (v/v) Triton X-100 (Sigma). Protein from cell lysates (50 µg) and mitochondrial fractions from brain lysates (30 µg) were resolved on 4%–12% gel (w/v) NuPAGE BisTris gels in MOPs buffer (ThermoFisher Scientific, Hemel Hempstead, UK) and transferred to polyvinylidene fluoride membrane (PVDF, Millipore, Beeston, UK) in NuPAGE transfer buffer (Life Technologies). Membranes were blocked in 5% skim-milk in Tris-buffered saline with 0.1% Tween 20 (TBS-T) and incubated overnight with the following antibodies: anti-mouse OPA1 (1:1000, Cat# 612606, BD Biosciences, Oxford, UK), anti-rabbit YME1 (1:1000, Cat# 11510-1-AP, ProteinTech, Manchester, UK), anti-rabbit OMA1 (for primary neurons; 1:1000, Cat# NBP1-56970, Novusbio, UK and for mitochondrial fractions from brain lysates; 1:750, Cat# 17116-1-AP, ProteinTech), anti-mouse GAPDH (1:2000, Cat# G8795 Sigma) and for mitochondrial fractions Tom20 (1:750, Cat# sc-11415, Santa Cruz Biotechnology, Wembley, UK). Secondary antibodies: Li-cor IRDye Goat anti-rabbit 800 or Goat anti-mouse 680 secondary antibodies were used at 1:10,000 and incubated for 2 h at room temperature. Membranes were imaged on a Li-Cor Odyssey Infrared Imaging System (Li-Cor Imaging Biotechnology UK Ltd., Cambridge, UK) using the manufacturer's Image Studio software for band quantification.

4.10. Data Analysis

Data are expressed as mean ± SD. All primary neuronal culture experiments were performed on 3–6 independent litters as stated in the figure legends. Mitochondrial length was analyzed using Squassh (segmentation and quantification of subcellular shapes) software, part of the Mosaic toolkit in Image J [51]. All statistical analyses were performed using GraphPad Prism 6 Software (GraphPad Software, San Diego, CA, USA). All data were assessed for normality. Data was assessed either by a Student's *t*-test or for multiple conditions, with a two-way ANOVA. If significant, a Bonferroni *post-hoc* analysis was conducted. Analyses used are detailed in table and figure captions. * $p < 0.05$, ** $p < 0.01$, *** $p < 0.001$, **** $p < 0.0001$.

5. Conclusions

We have determined that in primary neurons, mitochondrial membrane potential, morphology and function are impaired in response to OGD. Furthermore, we present the first data suggesting that in a mouse model of neonatal HI, the expression of mitochondrial shaping proteins, such as OPA1 and Yme1L, are altered; *in vitro* and *in vivo*, OPA1 is cleaved to shorter forms and Yme1L expression is reduced. Further studies are required to determine the molecular pathways regulating these events.

Acknowledgments: We gratefully acknowledge the support of the Department of Perinatal Imaging and Health and financial support from Wellcome Trust (WT094823), the Medical Research Council, Action Medical Research, Swedish Medical Research Council (VR 2012-3500), Brain Foundation (HH), Ahlen Foundation (HH), ALF-GBG (426401), ERA-net (EU;VR 529-2014-7551) and the Leducq Foundation (DSRRP34404) to enable this study to be completed. CH was supported by the King's Bioscience Institute and the Guy's and St Thomas' Charity Prize PhD Programme in Biomedical and Translational Science. In addition, the authors acknowledge financial support from the Department of Health via the National Institute for Health Research (NIHR) comprehensive Biomedical Research Centre Award to Guy's & St Thomas' NHS Foundation Trust in partnership with King's College London and King's College Hospital NHS Foundation Trust.

Author Contributions: Claire Thornton conceived and designed the experiments; Ana A. Baburamani, Chloe Hurling and Claire Thornton performed the experiments; Ana A. Baburamani, Chloe Hurling, Helen Stolp, Kristina Sobotka, Pierre Gressens and Henrik Hagberg analyzed the data; Helen Stolp contributed reagents/materials/analysis tools; Ana A. Baburamani, Chloe Hurling, Helen Stolp, Kristina Sobotka, Pierre Gressens and Henrik Hagbergcontributed to the preparation of the manuscript.

Conflicts of Interest: The authors declare no conflict of interest.

References

1. Evans, K.; Rigby, A.S.; Hamilton, P.; Titchiner, N.; Hall, D.M. The relationships between neonatal encephalopathy and cerebral palsy: A cohort study. *J. Obstet. Gynaecol.* **2001**, *21*, 114–120.
2. Smith, J.; Wells, L.; Dodd, K. The continuing fall in incidence of hypoxic-ischaemic encephalopathy in term infants. *BJOG* **2000**, *107*, 461–466.
3. Lawn, J.E.; Bahl, R.; Bergstrom, S.; Bhutta, Z.A.; Darmstadt, G.L.; Ellis, M.; English, M.; Kurinczuk, J.J.; Lee, A.C.; Merialdi, M.; *et al.* Setting research priorities to reduce almost one million deaths from birth asphyxia by 2015. *PLoS Med.* **2011**, *8*, e1000389.
4. Vannucci, R.C.; Perlman, J.M. Interventions for perinatal hypoxic-ischemic encephalopathy. *Pediatrics* **1997**, *100*, 1004–1014.

5. Azzopardi, D.; Wyatt, J.S.; Cady, E.B.; Delpy, D.T.; Baudin, J.; Stewart, A.L.; Hope, P.L.; Hamilton, P.A.; Reynolds, E.O. Prognosis of newborn infants with hypoxic-ischemic brain injury assessed by phosphorus magnetic resonance spectroscopy. *Pediatr. Res.* **1989**, *25*, 445–451.

6. Blumberg, R.M.; Cady, E.B.; Wigglesworth, J.S.; McKenzie, J.E.; Edwards, A.D. Relation between delayed impairment of cerebral energy metabolism and infarction following transient focal hypoxia-ischaemia in the developing brain. *Exp. Brain Res.* **1997**, *113*, 130–137.

7. Gilland, E.; Bona, E.; Hagberg, H. Temporal changes of regional glucose use, blood flow, and microtubule-associated protein 2 immunostaining after hypoxia-ischemia in the immature rat brain. *J. Cereb. Blood Flow Metab.* **1998**, *18*, 222–228.

8. Lorek, A.; Takei, Y.; Cady, E.B.; Wyatt, J.S.; Penrice, J.; Edwards, A.D.; Peebles, D.; Wylezinska, M.; Owen-Reece, H.; Kirkbride, V.; *et al.* Delayed ("secondary") cerebral energy failure after acute hypoxia-ischemia in the newborn piglet: Continuous 48-hour studies by phosphorus magnetic resonance spectroscopy. *Pediatr. Res.* **1994**, *36*, 699–706.

9. Hagberg, H.; Thornberg, E.; Blennow, M.; Kjellmer, I.; Lagercrantz, H.; Thiringer, K.; Hamberger, A.; Sandberg, M. Excitatory amino acids in the cerebrospinal fluid of asphyxiated infants: Relationship to hypoxic-ischemic encephalopathy. *Acta Paediatr.* **1993**, *82*, 925–929.

10. Puka-Sundvall, M.; Gajkowska, B.; Cholewinski, M.; Blomgren, K.; Lazarewicz, J.W.; Hagberg, H. Subcellular distribution of calcium and ultrastructural changes after cerebral hypoxia-ischemia in immature rats. *Brain Res. Dev. Brain Res.* **2000**, *125*, 31–41.

11. Van den Tweel, E.R.; Nijboer, C.; Kavelaars, A.; Heijnen, C.J.; Groenendaal, F.; van Bel, F. Expression of nitric oxide synthase isoforms and nitrotyrosine formation after hypoxia-ischemia in the neonatal rat brain. *J. Neuroimmunol.* **2005**, *167*, 64–71.

12. Wallin, C.; Puka-Sundvall, M.; Hagberg, H.; Weber, S.G.; Sandberg, M. Alterations in glutathione and amino acid concentrations after hypoxia-ischemia in the immature rat brain. *Brain Res. Dev. Brain Res.* **2000**, *125*, 51–60.

13. Northington, F.J.; Zelaya, M.E.; O'Riordan, D.P.; Blomgren, K.; Flock, D.L.; Hagberg, H.; Ferriero, D.M.; Martin, L.J. Failure to complete apoptosis following neonatal hypoxia-ischemia manifests as "continuum" phenotype of cell death and occurs with multiple manifestations of mitochondrial dysfunction in rodent forebrain. *Neuroscience* **2007**, *149*, 822–833.

14. Portera-Cailliau, C.; Price, D.L.; Martin, L.J. Excitotoxic neuronal death in the immature brain is an apoptosis-necrosis morphological continuum. *J. Comp. Neurol.* **1997**, *378*, 70–87.

15. Hagberg, H.; Mallard, C.; Rousset, C.I.; Thornton, C. Mitochondria: Hub of injury responses in the developing brain. *Lancet Neurol.* **2014**, *13*, 217–232.

16. Thornton, C.; Rousset, C.I.; Kichev, A.; Miyakuni, Y.; Vontell, R.; Baburamani, A.A.; Fleiss, B.; Gressens, P.; Hagberg, H. Molecular mechanisms of neonatal brain injury. *Neurol. Res. Int.* **2012**, *2012*.

17. Gilland, E.; Puka-Sundvall, M.; Hillered, L.; Hagberg, H. Mitochondrial function and energy metabolism after hypoxia-ischemia in the immature rat brain: Involvement of NMDA-receptors. *J. Cereb. Blood Flow Metab.* **1998**, *18*, 297–304.

18. Puka-Sundvall, M.; Wallin, C.; Gilland, E.; Hallin, U.; Wang, X.; Sandberg, M.; Karlsson, J.; Blomgren, K.; Hagberg, H. Impairment of mitochondrial respiration after cerebral hypoxia-ischemia in immature rats: Relationship to activation of caspase-3 and neuronal injury. *Brain Res. Dev. Brain Res.* **2000**, *125*, 43–50.

19. Rosenberg, A.A.; Parks, J.K.; Murdaugh, E.; Parker, W.D., Jr. Mitochondrial function after asphyxia in newborn lambs. *Stroke* **1989**, *20*, 674–679.

20. Wang, X.; Carlsson, Y.; Basso, E.; Zhu, C.; Rousset, C.I.; Rasola, A.; Johansson, B.R.; Blomgren, K.; Mallard, C.; Bernardi, P.; *et al.* Developmental shift of cyclophilin D contribution to hypoxic-ischemic brain injury. *J. Neurosci.* **2009**, *29*, 2588–2596.

21. Zhu, C.; Wang, X.; Huang, Z.; Qiu, L.; Xu, F.; Vahsen, N.; Nilsson, M.; Eriksson, P.S.; Hagberg, H.; Culmsee, C.; *et al.* Apoptosis-inducing factor is a major contributor to neuronal loss induced by neonatal cerebral hypoxia-ischemia. *Cell Death Differ.* **2007**, *14*, 775–784.

22. Gibson, M.E.; Han, B.H.; Choi, J.; Knudson, C.M.; Korsmeyer, S.J.; Parsadanian, M.; Holtzman, D.M. BAX contributes to apoptotic-like death following neonatal hypoxia-ischemia: Evidence for distinct apoptosis pathways. *Mol. Med.* **2001**, *7*, 644–655.

23. Wang, X.; Han, W.; Du, X.; Zhu, C.; Carlsson, Y.; Mallard, C.; Jacotot, E.; Hagberg, H. Neuroprotective effect of Bax-inhibiting peptide on neonatal brain injury. *Stroke* **2010**, *41*, 2050–2055.

24. Song, Z.; Chen, H.; Fiket, M.; Alexander, C.; Chan, D.C. OPA1 processing controls mitochondrial fusion and is regulated by mRNA splicing, membrane potential, and Yme1L. *J. Cell Biol.* **2007**, *178*, 749–755.

25. Frezza, C.; Cipolat, S.; Martins de Brito, O.; Micaroni, M.; Beznoussenko, G.V.; Rudka, T.; Bartoli, D.; Polishuck, R.S.; Danial, N.N.; De Strooper, B.; *et al.* OPA1 controls apoptotic cristae remodeling independently from mitochondrial fusion. *Cell* **2006**, *126*, 177–189.

26. Olichon, A.; Baricault, L.; Gas, N.; Guillou, E.; Valette, A.; Belenguer, P.; Lenaers, G. Loss of OPA1 perturbates the mitochondrial inner membrane structure and integrity, leading to cytochrome c release and apoptosis. *J. Biol. Chem.* **2003**, *278*, 7743–7746.

27. Wappler, E.A.; Institoris, A.; Dutta, S.; Katakam, P.V.; Busija, D.W. Mitochondrial dynamics associated with oxygen-glucose deprivation in rat primary neuronal cultures. *PLoS ONE* **2013**, *8*, e63206.

28. Delettre, C.; Griffoin, J.M.; Kaplan, J.; Dollfus, H.; Lorenz, B.; Faivre, L.; Lenaers, G.; Belenguer, P.; Hamel, C.P. Mutation spectrum and splicing variants in the *OPA1* gene. *Hum. Genet.* **2001**, *109*, 584–591.

29. Griparic, L.; Kanazawa, T.; van der Bliek, A.M. Regulation of the mitochondrial dynamin-like protein Opa1 by proteolytic cleavage. *J. Cell Biol.* **2007**, *178*, 757–764.

30. Ehses, S.; Raschke, I.; Mancuso, G.; Bernacchia, A.; Geimer, S.; Tondera, D.; Martinou, J.C.; Westermann, B.; Rugarli, E.I.; Langer, T. Regulation of OPA1 processing and mitochondrial fusion by m-AAA protease isoenzymes and OMA1. *J. Cell Biol.* **2009**, *187*, 1023–1036.

31. Head, B.; Griparic, L.; Amiri, M.; Gandre-Babbe, S.; van der Bliek, A.M. Inducible proteolytic inactivation of OPA1 mediated by the OMA1 protease in mammalian cells. *J. Cell Biol.* **2009**, *187*, 959–966.

32. Rice, J.E., 3rd; Vannucci, R.C.; Brierley, J.B. The influence of immaturity on hypoxic-ischemic brain damage in the rat. *Ann. Neurol.* **1981**, *9*, 131–141.

33. Vannucci, R.C.; Vannucci, S.J. A model of perinatal hypoxic-ischemic brain damage. *Ann. N. Y. Acad. Sci.* **1997**, *835*, 234–249.

34. Biala, A.K.; Dhingra, R.; Kirshenbaum, L.A. Mitochondrial dynamics: Orchestrating the journey to advanced age. *J. Mol. Cell. Cardiol.* **2015**, *83*, 37–43.

35. Lin, M.T.; Beal, M.F. Mitochondrial dysfunction and oxidative stress in neurodegenerative diseases. *Nature* **2006**, *443*, 787–795.

36. Supale, S.; Li, N.; Brun, T.; Maechler, P. Mitochondrial dysfunction in pancreatic β cells. *Trends Endocrinol. Metab.* **2012**, *23*, 477–487.

37. Duvezin-Caubet, S.; Jagasia, R.; Wagener, J.; Hofmann, S.; Trifunovic, A.; Hansson, A.; Chomyn, A.; Bauer, M.F.; Attardi, G.; Larsson, N.G.; *et al.* Proteolytic processing of OPA1 links mitochondrial dysfunction to alterations in mitochondrial morphology. *J. Biol. Chem.* **2006**, *281*, 37972–37979.

38. Ishihara, N.; Fujita, Y.; Oka, T.; Mihara, K. Regulation of mitochondrial morphology through proteolytic cleavage of OPA1. *EMBO J.* **2006**, *25*, 2966–2977.

39. Anand, R.; Wai, T.; Baker, M.J.; Kladt, N.; Schauss, A.C.; Rugarli, E.; Langer, T. The i-AAA protease YME1L and OMA1 cleave OPA1 to balance mitochondrial fusion and fission. *J. Cell Biol.* **2014**, *204*, 919–929.

40. Iijima, T. Mitochondrial membrane potential and ischemic neuronal death. *Neurosci. Res.* **2006**, *55*, 234–243.

41. Azzopardi, D.; Wyatt, J.S.; Hamilton, P.A.; Cady, E.B.; Delpy, D.T.; Hope, P.L.; Reynolds, E.O. Phosphorus metabolites and intracellular pH in the brains of normal and small for gestational age infants investigated by magnetic resonance spectroscopy. *Pediatr. Res.* **1989**, *25*, 440–444.

42. Rainbolt, T.K.; Saunders, J.M.; Wiseman, R.L. YME1L degradation reduces mitochondrial proteolytic capacity during oxidative stress. *EMBO Rep.* **2015**, *16*, 97–106.

43. Stiburek, L.; Cesnekova, J.; Kostkova, O.; Fornuskova, D.; Vinsova, K.; Wenchich, L.; Houstek, J.; Zeman, J. YME1L controls the accumulation of respiratory chain subunits and is required for apoptotic resistance, cristae morphogenesis, and cell proliferation. *Mol. Biol. Cell* **2012**, *23*, 1010–1023.

44. Nijboer, C.H.; Heijnen, C.J.; van der Kooij, M.A.; Zijlstra, J.; van Velthoven, C.T.; Culmsee, C.; van Bel, F.; Hagberg, H.; Kavelaars, A. Targeting the p53 pathway to protect the neonatal ischemic brain. *Ann. Neurol.* **2011**, *70*, 255–264.

45. Zhu, C.; Qiu, L.; Wang, X.; Hallin, U.; Cande, C.; Kroemer, G.; Hagberg, H.; Blomgren, K. Involvement of apoptosis-inducing factor in neuronal death after hypoxia-ischemia in the neonatal rat brain. *J. Neurochem.* **2003**, *86*, 306–317.

46. Zhu, C.; Xu, F.; Fukuda, A.; Wang, X.; Fukuda, H.; Korhonen, L.; Hagberg, H.; Lannering, B.; Nilsson, M.; Eriksson, P.S.; *et al.* X chromosome-linked inhibitor of apoptosis protein reduces oxidative stress after cerebral irradiation or hypoxia-ischemia through up-regulation of mitochondrial antioxidants. *Eur. J. Neurosci.* **2007**, *26*, 3402–3410.

47. Sanderson, T.H.; Raghunayakula, S.; Kumar, R. Neuronal hypoxia disrupts mitochondrial fusion. *Neuroscience* **2015**, *301*, 71–78.

48. Thornton, C.; Bright, N.J.; Sastre, M.; Muckett, P.J.; Carling, D. AMP-activated protein kinase (AMPK) is a tau kinase, activated in response to amyloid beta-peptide exposure. *Biochem. J.* **2011**, *434*, 503–512.

49. Fleiss, B.; Chhor, V.; Rajudin, N.; Lebon, S.; Hagberg, H.; Gressens, P.; Thornton, C. The anti-inflammatory effects of the small molecule Pifithrin-μ on BV2 microglia. *Dev. Neurosci.* **2015**, *37*, 363–375.

50. Livak, K.J.; Schmittgen, T.D. Analysis of relative gene expression data using real-time quantitative PCR and the $2^{-\Delta\Delta Ct}$ method. *Methods* **2001**, *25*, 402–408.

51. Rizk, A.; Paul, G.; Incardona, P.; Bugarski, M.; Mansouri, M.; Niemann, A.; Ziegler, U.; Berger, P.; Sbalzarini, I.F. Segmentation and quantification of subcellular structures in fluorescence microscopy images using Squassh. *Nat. Protoc.* **2014**, *9*, 586–596.

Mitochondria: A Therapeutic Target for Parkinson's Disease?

Yu Luo, Alan Hoffer, Barry Hoffer and Xin Qi

Abstract: Parkinson's disease (PD) is one of the most common neurodegenerative disorders. The exact causes of neuronal damage are unknown, but mounting evidence indicates that mitochondrial-mediated pathways contribute to the underlying mechanisms of dopaminergic neuronal cell death both in PD patients and in PD animal models. Mitochondria are organized in a highly dynamic tubular network that is continuously reshaped by opposing processes of fusion and fission. Defects in either fusion or fission, leading to mitochondrial fragmentation, limit mitochondrial motility, decrease energy production and increase oxidative stress, thereby promoting cell dysfunction and death. Thus, the regulation of mitochondrial dynamics processes, such as fusion, fission and mitophagy, represents important mechanisms controlling neuronal cell fate. In this review, we summarize some of the recent evidence supporting that impairment of mitochondrial dynamics, mitophagy and mitochondrial import occurs in cellular and animal PD models and disruption of these processes is a contributing mechanism to cell death in dopaminergic neurons. We also summarize mitochondria-targeting therapeutics in models of PD, proposing that modulation of mitochondrial impairment might be beneficial for drug development toward treatment of PD.

Reprinted from *Int. J. Mol. Sci.* Cite as: Luo, Y.; Hoffer, A.; Hoffer, B.; Qi, X. Mitochondria: A Therapeutic Target for Parkinson's Disease? *Int. J. Mol. Sci.* **2015**, *16*, 20704–20726.

1. Introduction

Parkinson's disease (PD) is the second most common neurodegenerative disorder after Alzheimer's disease, affecting over 1% of the population older than 60 years of age. Clinically, it is diagnosed primarily based on motor abnormalities including bradykinesia, resting tremor, and cogwheel rigidity [1]. A key characteristic of pathology in PD is the degeneration of the nigrostriatal (NS) dopaminergic pathway which is one of the most important dopamine (DA) pathways in the brain and contains about 80% of the total brain DA. Despite a large number of studies on the pathogenesis of PD, there is still inconclusive evidence about why dopaminergic neurons are selectively degenerated. Currently, there is no effective restorative treatment available for PD, only symptomatic treatment is available.

Among a number of proposed mechanisms involved in PD pathogenesis, mitochondrial dysfunction has been repeatedly implicated as the cause of the death of DA neurons in PD [2–5]. Mitochondria are critical for many cellular functions, such as intermediary metabolism [6,7], redox signaling [8], calcium homeostasis [9–11], cell proliferation [12,13], development [14,15] and cell death [16–18]. Mitochondrial dysfunction is mainly characterized by the generation of reactive oxygen species (ROS), a defect in mitochondrial electron transport complex enzyme activities, ATP depletion, caspase 3 release and depletion of mitochondrial DNA. In this review, we summarize evidence on the critical involvement of mitochondria in both genetic mutation and environmental toxin-induced PD. We propose a causal role for mitochondrial dysfunction in the development of PD, because (1) neurotoxins causing parkinsonism, such as 1-methyl-4-phenyl-1,2,3,6-tetrahydropyridine (MPTP), rotenone, paraquat, induce dopaminergic neuronal death through direct inhibition of mitochondrial complex I activity; (2) mutant proteins from PD-related genes associate with mitochondria where they elicit diverse mitochondrial dysregulation and subsequently cause neuronal degeneration; (3) therapeutic agents that target mitochondrial protein or inhibit mitochondrial damage can reduce neuropathological phenotypes of PD in animal models and cells from PD patients.

2. Mitochondrial Dysfunction in Parkinson's Disease

Aberrant mitochondrial function is one of the major cytopathologies in PD and has been widely accepted as a central pathogenic mechanism underlying PD pathogenesis. Chronic systemic administration of rotenone, a specific complex I inhibitor and a pesticide, results in neuropathologic and behavioral changes in rats that are similar to human PD [19,20]. 1-methyl-4-phenyl-1,2,3,6-tetrahydropyridine (MPTP), a meperidine analog found to cause parkinsonism in humans, exerts its toxic effects through metabolism to 1-methyl-4-phenylpyridinium (MPP$^+$), another complex I inhibitor [19,21]. These compounds have long been used in animal models of PD. Furthermore a number of familial forms of PD are associated with mutations in genes encoding both mitochondrially targeted proteins and proteins involved in mitochondrial function and/or oxidative stress responses, including mutations in PINK-1, DJ-1, parkin, and leucine-rich repeat kinase 2 (LRRK2) [22]. Interestingly, one study reported that the size of mitochondria in dopaminergic neurons in the substantia nigra (SN; susceptible to PD degeneration) is smaller than in neighboring non-dopaminergic neurons or in dopaminergic neurons of the ventral tegmental area which is more resistant in PD, suggesting a basis for the increased vulnerability of SN neurons to subtle changes in mitochondrial maintenance and function [23]. In addition, increased oxidative stress due to mitochondrial compromise in PD model animals has been proposed to contribute to the degeneration of dopaminergic neurons [23].

Consistent with the evidence from basic science, clinical studies also showed that mitochondrial damage plays a predominant role in the development of PD in patients. Mild deficiency in mitochondrial respiratory electron transport chain NADH dehydrogenase (Complex I) activity has been reported in the substantia nigra [24] as well as platelets [25,26] and lymphocytes [27,28] in PD patients, suggesting a systemic inhibition of complex I activity in PD patients. Mitochondrial dysfunction could lead to increased oxidative stress. Indeed oxidative damage to lipids, proteins and DNA has been detected in brain tissue from PD patients [29,30]. A recent proteomic analysis of mitochondria-enriched fractions from post-mortem PD substantia nigra revealed differential expression of multiple mitochondrial proteins in PD brains as compared with control brains [23]. In further support for a "mitochondrial genetics" hypothesis for PD pathophysiology, Bender *et al.* [31] reported higher levels of mitochondrial DNA deletions in nigral neurons from PD patients. Moreover, both Bender *et al.* [31] and Kraytsberg *et al.* [32] reported higher levels of mitochondrial DNA deletions in nigral neurons of aged humans with sharp elevations starting shortly before age 70. This correlates with the known risk factor of age in PD. It is possible that there is an accumulation of mitochondrial dysfunction and of reactive oxygen species (ROS) damage during aging which needs to reach a critical threshold for cellular dysfunction and degeneration to be observed. How a systemic dysregulation in mitochondrial function or oxidative damage leads to tissue or cell type specific vulnerability still remains to be elucidated.

Taken together, although studies over many years on PD indicate an important role of mitochondria in PD-associated pathology, the process by which the mitochondria become dysfunctional in PD and whether correction of mitochondrial defects could provide neuroprotection in PD remain to be determined.

3. Environmental Toxins that Influence Mitochondrial Function

Some toxins used to model DA loss in PD, such as MPTP and rotenone, impair respiratory chain function by inhibiting complex I [33–36]. These complex I inhibitors replicate some of the key motor features of PD and lead to DA neuronal loss. Intravenous injection of the compound MPTP by drug addicts caused a condition that closely resembles the anatomic and clinical features of PD [37,38]. Multiple models have been developed in the laboratory in which the chronic infusion of the pesticide rotenone or combination of herbicides and pesticides [39] lead to a pattern of cell death and DA loss similar to that of PD. The precise role of environmental toxins in the cause of PD remains to be defined, but these data support the hypothesis that environmental toxins could introduce mitochondrial dysfunction and lead to parkinsonism in human. Indeed, epidemiological studies showed that the prevalence of sporadic PD is higher among farming communities [40]. Exposure to pesticides or herbicides elicited a three-to fourfold increased risk of developing PD [41]. All of these data suggest an environmental contribution to the etiology of sporadic PD.

4. Genetic Factors Associated with PD

In a twin study, Tanner *et al.* showed that for early onset cases, monozygotic concordance was twice that of dizygotic concordance, suggesting that genetic factors are important in the early onset PD [42]. Many examples of familial parkinsonism have also been reported (See review [43]). Mutations of several genes have been linked to familial PD and parkinsonian syndromes [44–46].

4.1. Mitochondrial DNA Mutations and Deletions in PD

Mitochondrial DNA (mtDNA) encodes 13 subunits of respiratory chain proteins, including seven complex I, one complex III, three complex IV, and two complex V subunits. Until now, there has been no PD-associated genetic mutations in mtDNA reported [47]. However, mtDNA deletions have been observed in individual dopaminergic neurons dissected from postmortem human substantia nigra tissue [31]. In addition, mutations in the gene encoding human mtDNA polymerase subunit γ (POLG) leads to clinical parkinsonism associated with multiple mtDNA deletions. [48,49]. Furthermore, proliferator-activated coactivator-1 α (PGC-1 α), one of the key regulators of mitochondrial biogenesis, was found to be decreased in PD patients through a genome-wide association study (GWAS) [50]. Directed deletion of transcriptional factor A (TFAM) in mouse DA neurons using the Cre-LoxP system, termed as MitoPark mice, causes marked deletion of mtDNA, severe impairment of oxidative phosphorylation and slowly progressive motor deficits in the DA system that mimic human parkinsonism as well as altered response to L-3,4-dihydroxyphenylalanine (L-DOPA) treatment [51–53]. The MitoPark mice of PD provided direct evidence that mitochondrial dysfunction in DA neurons can causes PD-related phenotypes. Consistently increased level of mtDNA deletions in the striatum of PD patients have been reported [54] and mtDNA deletions were significantly higher in neurons with impaired cytochrome oxidase activity [31,32]. These findings support a mitochondrial genetic contribution in PD.

4.2. Nuclear Gene Mutations Affecting Mitochondrial Function

Many of the PD susceptible genes identified are related to mitochondrial function (Table 1). These genes and their potential contribution to PD have previously been extensively reviewed [43]. PD linked genes that affect mitochondrial function include, but are not limited to, α-synuclein, Parkin, Phosphatase and tensin homologue (PTEN)-induced putative kinase 1 (PINK1), DJ-1 and LRRK2. In this review we will focus on their involvement and effects on dysregulation of mitochondrial dynamics, mitophagy and mitochondrial redox, and mitochondrial protein import.

Alpha-synuclein: α-synuclein is aggregation-prone protein which attains an increased propensity to aggregate because of the presence of its hydrophobic non-amyloid β component domain. Missense mutations in the alpha-synuclein gene are associated with autosomal dominant PD [55]. The fibrillar form of α-synucleinis a major component of Lewy bodies and has been demonstrated to trigger neurotoxicity in PD [56,57].

Parkin: Parkin is a RING finger containing ubiquitin E3 ligase. It is known to mediate poly-ubiquitination of its substrates for proteasomal degradation. Mutations in the parkin gene cause early onset juvenile autosomal recessive PD, and Parkin mutations are the most common cause of young onset PD. The loss of parkin E3 ligase activity results in accumulation of its substrates leading to neurotoxicity in autosomal recessive PD [58,59].

PINK1: Phosphatase and tensin homologue (PTEN)-induced putative kinase 1 (PINK1) is a serine/threonine kinase localized in mitochondria. Mutations in PINK1 are associated with a rare form of autosomal recessive PD. PINK mutations result in the loss of PINK1 function which leads to aberrant phosphorylation of its substrates to cause PD [60,61].

DJ-1: DJ-1 belongs to the peptidase C56 family of proteins. Wild-type DJ-1 can serve as a chaperone, protease, regulator of transcription and autophagy through redox regulation. DJ-1 seems to be cytoprotective specifically under conditions related to oxidative stress. Its protective action is the result of the modification of cysteine residues on DJ-1 to cysteine-sulfinic and cysteine-sulfonic acids under oxidative stress [78]. DJ-1 mutations associated with PD are rare and account for 1%–2% of autosomal recessive early-onset PD [79].

LRRK2: Leucine-rich repeat kinase 2 (LRRK2) is a protein encoded by the PARK8 locus. It has a conserved serine-threonine kinase mitogen-activated protein kinase kinase kinase (MAPKKK) domain, a member of the Roc (Ras of complex) GTPase family [80–82]. To date, there are over 50 variants identified in PD patients. The mutation G2019S (Gly2019 to Ser) that takes place in the MAPKKK domain has been recognized as the most common cause of dominant familial PD and accounts for up to 2% of sporadic PD cases [83]. The G2019S mutant augments the kinase activity of LRRK2, which is associated with increased toxicity in dopaminergic neurons [84].

Table 1. Animal models of Parkinson's disease (PD)-related genes affect mitochondrial function.

Animal Models	Genetic Manipulation in Animals	Motor Phenotypes	PD Pathology and Mitochondrial Function	References
Alpha-Synuclein transgenic mice	hA53T alpha-Synuclein in mice; mPrP promoter	Severe leading to paralysis and premature death	Lewy body-like inclusion in older mice; mitochondrial dysfunction; no dopaminergic neuronal loss	[62–64]
	hA30P alpha-synuclein in mice; mThy-1 promoter	Severe leading to paralysis	Lewy body-like inclusion; sensorimotor neuronal loss in brain stem	[65]
	Alpha-synuclein overexpression in mice (Thy1 promoter)	Progressive declines in spontaneous and motor activity	No DA neuronal degeneration, mitochondrial dysregulation	[66,67]
	hA53T alpha-synuclein expressing in SN DA neurons of mice	Body weight loss; normal locomotion activity	Progressive DA neuronal loss; aberrant mitochondrial inclusion	[68]
LRRK2 transgenic or knock-in mice	LRRK2 R1441G mice; BAC promoter	Rearing activity decrease in older mice	DA neurite degeneration; Tau phosphorylation increase; no DA neuronal degeneration; mitochondrial dysfunction	[69,70]
	LRRK2 G2019S knock-in mice	Absent	Abnormal mitochondrial morphology; mitochondrial dysfunction; no DA neuronal loss	[71]
Parkin$^{-/-}$ mice	Parkin germline inactivation	Conflicting: either absent or subtle motor movement disturbance	Absent	[72,73]
DJ1$^{-/-}$ mice	DJ1 germline inactivation	Age-dependent declines in locomotor activity	Impaired DA update; no DA neuronal degeneration	[74]
PINK1$^{-/-}$ mice	PINK1 germline inactivation	Age-dependent declines in spontaneous activity	Impaired DA update; no DA neuronal degeneration; mitochondrial abnormalities	[75,76]
Mito-Park mice	DAT riven cre; loxed p TFAM	Begins at 3–4 months; declines in spontaneous and rearing activity	Abnormal mitochondrial aggregates; DA reduction in the striatum; progressive DA neuronal degeneration	[51]
Double-mutant mice	A53T alpha-synuclein overexpression in Parkin$^{-/-}$ mice	Absent	Altered mitochondrial structure and morphology; no DA neuronal loss	[77]

4.3. Mitochondrial Dynamics Impairment in Parkinson's Disease

Mitochondria are organized in a highly dynamic tubular network that is continuously reshaped by opposing processes of fusion and fission [85]. Mitochondrial fission and fusion were first observed in yeast, and since have been observed in all mammalian cells [86]. A delicate balance is maintained between fusion and fission to ensure the normal function of mitochondria. Specifically the fusion process is important for mitochondrial interactions and communication, and fission facilitates the segregation of mitochondria into daughter cells and enhances mitochondrial renewal and distribution along cytoskeletal tracks. Fusion and fission events enable the proper exchanging and mixing of mitochondrial membranes and contents. This dynamic process controls not only mitochondrial morphology, but also the subcellular location and function of mitochondria. Defects in either fusion or fission limit mitochondrial motility, decrease energy production and increase oxidative stress, thereby promoting cell dysfunction and death [87,88]. The two opposing processes, fusion and fission, are controlled by evolutionarily conserved large GTPases that belong to the dynamin family of proteins. In mammalian cells, mitochondrial fusion is regulated by mitofusin-1 and -2 (MFN1/2) and optic atrophy 1 (OPA1), whereas mitochondrial fission is controlled by the dynamin-1-related protein, (Drp1) [89,90] and its mitochondrial adaptors such as Fis1, Mff and MiD49/51 [91–93]. Drp1 is primarily found in the cytosol, but it translocates from the cytosol to the mitochondrial surface in response to various cellular stimuli to regulate mitochondrial morphology [1]. At the mitochondrial surface, Drp1 is thought to wrap around the mitochondria to induce fission using its GTPase activity [94].

In terms of PD, no mutations in typical mitochondria fission and fusion genes have yet been identified in PD patients. However, increasing evidence from both toxin models and genetic mutations in PD animal models supports the hypothesis that mitochondrial dynamic regulation and dysfunction are involved in PD. Evidence from toxin-induced PD models support a role for mitochondrial fission/fusion in the pathogenesis of PD. The Parkinsonian neurotoxins, 6-hydroxy dopamine (6-OHDA), rotenone, and MPP$^+$, all induce mitochondrial fragmentation, leading to dopaminergic cell death in neuronal cultures [91,95,96]. Inhibition of pro-fission Drp1 or overexpression of pro-fusion protein mitofusin-1 (Mfn1) using genetic techniques prevents both neurotoxin-induced mitochondrial fission and neuronal cell death [97–99]. Loss of Mfn2 or conditional knockout of neuronal Mfn2 in mice has recently been reported to result in age-dependent motor deficits, followed by the loss of dopaminergic terminals in the striatum [100,101], suggesting a role for Mfn2 in parkinsonism. Also, Drp1 is required for synaptic formation [102] and lack of Drp1 leads to an impairment of brain development in mice [103,104]. Recently it has been reported that the Drp1 is critical for targeting mitochondria to the terminal synapses of dopaminergic neurons and deletion of Drp1 gene in dopaminergic neurons rapidly eliminates DA terminals in the caudate-putamen and causes cell bodies in the midbrain to degenerate and lose α-synuclein [105]. Taken together, all these support that molecular machinery which maintains the balance of fusion and fission dynamics in the cells might contribute to the pathogenesis of PD.

In addition, several genes, whose mutated forms are associated with familial PD, affect mitochondrial dynamics: these include PINK1, Parkin, LRRK2 and DJ-1 [45,94,96,106–110]. Among these genes, the role of PINK1/Parkin pathway in regulation of mitochondrial dynamics seems to be opposite from LRRK2 and DJ-1; mutations in PINK1/Parkin lead to mitochondrial fusion [44,111] whereas mutations in LRRK2 or DJ-1 promote mitochondrial fission [44,112]. Fibroblasts from PD patients carrying PINK1 or Parkin mutations exhibited a more fragmented mitochondrial network, showing mitochondrial dysfunction [113,114]. The mitochondrial network could also be reduced by the depletion of Drp1, or overexpression of OPA1 or Mfn2 [115–118]. Deficiency in DJ-1 in cell lines, cultured neurons and lymphoblasts derived from DJ-1-deficient patients displayed aberrant mitochondrial morphology [119]. Further, in double-mutant mice in which alpha-synuclein mutant is overexpressed and parkin is ablated, severe genotype-, age- and region-dependent mitochondrial morphological alterations were found in neuronal somata. The number of structurally altered mitochondria was significantly increased in the SN of these double-mutants mice [77]. These studies further support the involvement of vulnerable genes in mitochondrial dynamics regulation in PD.

Wild-type LRRK2 interacts and colocalizes with several key regulators of mitochondrial fusion/fission, suggesting that it might have multiple regulatory roles [120]. Furthermore, mutant *LRRK2* G2019S, the most common mutation in the population of familial PD patients, has been recently demonstrated to interact with Drp1 and to promote mitochondrial fragmentation, leading to mitochondrial dysfunction and neuronal abnormalities [121,122]. This fragmentation can be reduced by expression of the dominant-negative Drp1K38A or overexpression of the fusion protein Mfn2 [121,122]. iPS cells derived from PD patients carrying the G2019S mutation show excessive mitochondrial fission, aberrant autophagy and neuronal damage in DA neurons differentiated *in vitro* [123]. More importantly, treatment with P110, a selective peptide inhibitor of Drp1 recently developed in Qi's group [124,125], reduced mitochondrial fragmentation and damage, and corrected excessive autophagy. In this study it was also shown that G2019S mutated LRRK2 protein primarily phosphorylates Drp1 at T595 resulting in aberrant mitochondrial fragmentation [123].

Together, these findings suggest that impairment of mitochondrial dynamics might contribute to the pathogenesis and progression of PD.

5. Mitophagy/Autophagy Impairment in Parkinson's Disease

Autophagy is a process of cellular degradation in which cargos are degraded by autophagosomes fused with lysosomes [126]. In addition to non-selective cargos, autophagosomes can degrade protein aggregates or damaged mitochondria. Mitochondria-associated autophagy (mitophagy) is regulated by autophagy receptors that preferentially bind ubiquitylated mitochondria and subsequently recruit the autophagosome protein light chain 3 (LC3) through their LC3-interaction region (LIR) motif [127]. LC3 in mammals, also known as Atg8 in yeast, plays crucial roles in both autophagosome membrane biogenesis and cargo recognition [128,129]. In yeast, Atg32 functions as a receptor on mitochondria to initiate mitophagy through its interaction with Atg8 [130,131]. In mammals, FUNDC1 [132], p62 [133], BNIP3 [134], and AMBRA1 [135], have been recognized as receptors for mitophagy, all of which bind to LC3 via the LIR motif [136]. The implication of these receptors in mitophagy regulation, however, seems to be dependent of experimental conditions. Whether and how these receptors cooperatively regulate mitophagy remains to be determined.

A number of lines of evidence suggest a critical role for defective autophagy/mitophagy in neurodegeneration in PD. Cultured cells exposed to parkinsonian neurotoxins such as MPP^+, rotenone, or 6-OHDA showed an increased number of autophagosomes and associated neuronal cell death [125,137,138]. In recent years, PINK1-depenent activation of parkin has been recognized as a major pathway of mitophagy [139]. In mammalian cells, parkin is recruited to depolarized mitochondria, which are subsequently eliminated by autophagy. Such parkin recruitment to mitochondria depends on PINK1 accumulation on mitochondria and therefore PINK1 is a key molecule in the signal transduction of mitophagy [140]. The failure of PINK1/parkin-mediated mitophagic process leads to accumulation of damaged mitochondria, which results in an increase in ROS and cell death [141,142]. Because these studies were all conducted in cells with overexpression of PINK1 and Parkin and because Parkin at endogenous levels fail to mediate mitophagy in PD patient cells [143], the matter of whether these proteins, at the endogenous levels, cooperatively regulate mitophagy remains to be validated. Knock-down of DJ-1, another PD-related gene, resulted in decreased mitochondrial membrane potential, increased reactive oxygen species, excessive mitochondrial fragmentation and impaired autophagy [74,119,144]. Interestingly, overexpression of PINK1 and parkin can rescue mitochondrial fragmentation and dysfunction induced by the depletion of DJ-1 [145,146]. Given that wild-type and mutant DJ-1 can interact with PINK1 [147] and that Parkin, as an E3 ligase on mitochondria, catalyzes the ubiquitination of DJ-1 [148], PINK1, Parkin and DJ-1 may be operative in the pathway of PINK1/parkin-mediated mitophagy.

Pathogenic LRRK2-mediated autophagy has been observed in a variety of cell cultures [149,150], in neurons derived from patient-induced pluripotent stem cells [151] and in animal models in which a mutant form of LRRK2 is expressed [152]. Either knock-down of LRRK2 by siRNA or treatment with a LRRK2 kinase inhibitor caused an increase in autophagic fluxin-cultured cells [153,154]. Overexpression of LRRK2, especially mutant forms, seems to suppress autophagy [155]. In contrast, studies from other two groups show that mutant forms of LRRK2 may induce autophagy via an ERK1/2-dependent pathway [149,156]. Although the detailed mechanisms by which LRRK2 and its mutants mediate autophagy are not clear, LRRK2 may disrupt the balance of autophagy through damaging lysosome-related calcium storage and cargo degradation [157,158]. Further, our group recently reported excessive mitophagy in a variety of cells expressing the LRRK2 G2019S mutant, which was accompanied by mitochondrial depolarization, recruitment of p62 to the mitochondria, increased LC3II levels and lysosomal activity as well as death of dopaminergic neurons [123,159]. We showed that the LRRK2 G2019S mutant caused excessive mitophagyby phosphorylating its substrate including fission protein Drp1 and mitochondrial outer membrane protein Bcl_2 [123,159]. Thus, the pathway that LRRK2 mediation of mitophagy might be different from those occurred during autophagy.

Taken together, a large body of studies indicates that the different PD-related genes contribute to the pathogenesis of PD at the intersection of mitochondrial dysfunction and autophagy. The loss of mitochondrial membrane potential, which is associated with mitochondrial dysfunction, seems to be a common signal for mitochondria to be degraded via mitophagy. Moreover, occurrence of mitochondrial fragmentation due to impairment of mitochondrial fusion and fission often precedes the induction of mitophagy. Thus, it is possible that mitophagy may be a multistep process starting with the degradation of profusion/fission proteins, resulting in an imbalance of mitochondrial dynamics and the subsequent clearance of mitochondria [86]. However, the factors that regulate these processes involved in mitochondrial "quality control", including mitochondrial dynamics, mitochondria-associated degradation and mitophagy in PD, remain to be determined.

6. Mitochondrial Redox Signaling in Parkinson's Disease

Decreased Complex I activity in the SN of PD patients and animal models has been repeatedly observed. The defect in complex I results in impairment of electron transport and causes ROS accumulation in mitochondria which lead to neuronal degeneration. Neurotoxins causing parkinsonism, MPP^+ are selectively taken up into dopaminergic neurons in which it inhibits Complex I activity [160]. Rotenone also inhibits Complex I by impairing oxidative phosphorylation [161]. These studies demonstrate a contribution from ROS to the pathogenesis of dopamine neuronal loss in PD. The genetic PD-linked proteins play a significant role in this process. Alpha-synuclein mutant A53T can enter mitochondria where it binds to the Complex I subunit to inhibit Complex I activity, producing ROS [162,163]. Functional studies showed that alpha-synuclein associated with mitochondria induces cytochrome c release, increased calcium and ROS levels resulting in dopaminergic neuronal death [164]. In addition, expression of mutants of PD-related genes Parkin, PINK1, DJ-1 and LRRK2 in cultured cells all increased ROS. This evidence has been well-summarized in other reviews [29,165]. However, it is uncertain if such elevated ROS are directly caused by these mutants or through indirect cellular effects.

Besides ROS, reactive nitrogen species (RNS) mediating nitrosative stress is also implicated in SN neuronal loss in PD [166]. RNS are generated by the reaction of superoxide with nitric oxide (NO), which results in the production of peroxynitrite.

NO inhibits several enzymes including complexes I and IV of the mitochondrial electron transport chain, which in turn lead to ROS generation [167,168]. Increased expression of iNOS and nNOS were observed in basal ganglia of postmortem brain of PD patients [169]. In the mouse MPTP model, there was a significant upregulation of iNOS associated with the gliosis in the SN [170]. Inhibition of nNOS has been reported to protect against neurotoxicity in MPTP-induced PD animal model [171,172]. These observations suggest that NO and its metabolite peroxynitriteare implicated in the pathogenesis of PD.

7. Mitochondrial Protein Import in Parkinson's Disease

Mitochondria possess their own DNA and translational machinery; however, there are only a small number of mitochondrial proteins encoded by mtDNA that are synthesized within the organelle. The majority of mitochondrial proteins are nuclear-encoded and have to be imported into the organelle. The translocase of the outer mitochondrial membrane (TOM complex) plays central roles in controlling protein entry to mitochondria [173]. The TOM complex is the main entry portal for most mitochondrial proteins that are synthesized in cytoplasm. The TOM complex contains seven subunits including TOM40, TOM22, two proprotein receptors, TOM20, and TOM70, and three smaller proteins, TOM5, TOM6 and TOM7. In general, imported proteins bind to one of these receptors. With the assistance of TOM22 and TOM5, the imported proteins pass the TOM 40 channel [174]. TOM complexes also need to cooperate with TIM (Translocase of the Inner Membrane) 23 complexes to import matrix-targeted proteins [175]. Thus, the TOM and the TIM23 complex direct the translocation of oxidative phosphorylation and metabolite transporter proteins to the inner membrane.

A number of studies have reported that PD-associated genes, alpha-synuclein, PINK1 and Parkin can interact with the TOM complex, disrupting the mitochondrial protein import. Alpha-synuclein can enter mitochondria where it mainly localizes at the inner membrane. The import of alpha-synuclein to mitochondria is through the TOM complex [176,177]. In postmortem brain samples of PD and in brain tissues of mice overexpressing alpha-synuclein, TOM 40 levels changed together with the levels of alpha-synuclein [177]. Expression with either wild-type alpha-synuclein or mutant A53T in cultured cells resulted in the loss of TOM40, whereas over-expression of TOM40 prevented the cellular damage caused by expression of either alpha-synuclein wild-type or its mutant A53T [177]. Thus, alpha-synuclein mutants may impair mitochondrial function by suppression of TOM40-dependent mitochondrial protein import pathways. PINK1 is also imported into the mitochondria via the TOM complex channel [178,179] and then is degraded in a membrane potential-dependent manner. Under mitochondrial depolarization PINK1, interacting with the TOM complex, recruits and activates Parkin, leading to degradation of mitochondrial outer membrane proteins and ultimately mitophagy [178]. Thus, pathogenic mutations in PINK1 and parkin may disrupt this pathway, resulting in the accumulation of dysfunctional mitochondria.

8. Potential Therapeutics Targeting Mitochondria for Treatment of PD

Given the importance of mitochondrial dysfunction in the pathogenesis of PD, therapeutics targeting mitochondria have been studied to prevent or treat PD. Although the cellular and pathological phenotypes from neurotoxin-induced and genetic mutant-associated PD models are different, the outcomes of mitochondrial dysfunction, including mitochondrial dynamic impairments, increased ROS, and impaired bioenergetics, seem to be common pathways. Thus, the mechanistic information on mitochondrial dysfunction in PD models provides potential targets for the development of therapeutic approaches for treatment of PD.

Here, we summarize reported therapeutic agents that reduce PD pathology in models of PD (see Table 2). We categorize these agents into (1) modulating PD-related genetic mutants; (2) modulating mitochondrial proteins; and (3) modulating the consequences of mitochondrial dysfunction. In Table 2, we list the neuroprotective effects of these agents especially in either animal models of PD or neuron derived patient-iPS cells.

Modulating PD-related genetic mutants: LRRK2 mutants are the most common genetic mutants in both familial and sporadic PD. Modulation of the LRRK2 kinase domain has been attractive for the development of therapeutics of PD. LRRK2 inhibitors, PF-06447475 and GW5074, have been shown to increase dopaminergic neuronal survival in primary neuronal cultures and neurons- derived from patient iPS cells [180,181]. Increasing glyoxalase activity of DJ-1 by supplying D-lactate and glycolate rescues the requirement for DJ-1 in maintenance of mitochondrial potential, increases cytocatalytic rate of DJ-1, and reduces neuronal death caused by paraquat and down-regulation of PINK1 [182]. However, thus far, these reagents are only tested in cultured cells so far. Whether treatment with these pharmacological agents protect neurons from PD in animal models remains to be determined.

Table 2. Therapeutic agents that target mitochondria for treatment of Parkinson's disease.

Category	Agent	Molecular Action	PD Mode	Therapeutic Effects
Modulating PD-related genes	PF-0644747	LRRK2 kinase inhibitor	Transgenic rat with LRRK2 G2019S	Reduce behavioral and neuropathological phenotypes Reduce inflammation [183]
	GW5074	LRRK2 kinase inhibitor	DA neurons derived from PD patient iPS cells	Suppression of ROS Improve mitochondrial respiration Increase DA neuronal survival [180]
	FX2149	LRRK2 GTPase inhibitor	mouse inflammation model	Reduce neuroinflammation Inhibit microglial activity [184]
	Alda-1	ALDH2 activator	Rotenone- and MPTP-induced animal models	Improve mitochondrial membrane potential Inhibit mitochondrial ROS Reduce dopaminergic cell death [185]
	TRO40303	inhibitor of mitochondrial transition pore	Mice expressing alha-synuclein	Upregulate mitochondrial proteins Increase TH expression [186]
Modulating mitochondrial protein	P110	Peptide inhibitor of Drp1	DA neurons from LRRK2 G2019S PD patient iPS cells	Improve mitochondrial membrane potential Inhibit mitochondrial ROS Increase mitochondrial integrity Reduce autophagy Improve DA neuronal morphology and survival [123]
	Mdivi-1	Inhibitor of mitochondrial fragmentation	MPTP-induced mouse PD model PINK1−/− mouse model	Improve mitochondrial morphology Improve mouse behavioral outcome Reduce DA neuronal loss in SN Restore dopamine level [187]
	Q1	8-OH-quinoline-based iron chelator	MPTP-induced mouse PD model	Reduce DA neuronal degeneration in SN Decrease mitochondrial iron pool [188]
	Rapamycin	mTOR inhibitor	6-OHDA-induced rat PD model	Inhibit oxidative stress Inhibit mitochondrial apoptosis [189]
Modulating mitochondrial dysfunction	Edaravone	ROS scavenger	Rotenone-induced rat PD model	Inhibit mitochondrial apoptosis Reduce ROS [190]
	Melatonin	Antioxidant	Rotenone-induced rat PD model 6-OHDA-induced rat PD model	Suppress calcium level Inhibit mitochondrial ROS Enhance complex I activity [191,192]
	Quercetin	Bioflavonoid	Rotenone-induced rat PD model	Inhibit mitochondrial ROS generation Inhibit p53 level Inhibit nuclear translocation of NF-kappaB Inhibit mitochondrial apoptosis [193]

Table 2. *Cont.*

Category	Agent	Molecular Action	PD Mode	Therapeutic Effects
	CNB-001	Curcumin derivative	MPTP-induced mouse PD model	Improve mitochondrial morphology Inhibit mitochondrial apoptotic pathway Improve mitochondrial membrane potential [194]
	Alpha-Lipoic acid	Antioxidant	Rotenone-induced rat PD model	Increase mitochondrial complex I activity Inhibit ROS generation Increase mitochondrial biogenesis Increase glutathione [195]
	Lycopene	Chemical carotene	Rotenone-induced rat PD model	Inhibit mitochondrial apoptotic pathway Increase SOD activity Increase glutathione Inhibit lipid peroxidation [196]

Modulating mitochondrial proteins: Aldehyde dehydrogenase 2 (ALDH2), located in mitochondrial matrix, functions as a cellular protector against oxidative stress by detoxification of cytotoxic aldehydes. Alda-1 is a small molecule that enhances ALDH enzyme activity and protects against oxidative toxicity [197]. Treatment with Alda-1 can reduce rotenone-induced apoptosis in both SH-SY5Y cells and primary dopaminergic neurons. Moreover, intraperitoneal administration of Alda-1 can improve mitochondrial membrane potential, inhibit mitochondrial ROS and reduce death of tyrosine hydroxylase (TH)-positive dopaminergic neurons in rotenone- or MPTP-induced PD animal models [185]. Cholesterol oximes such as olesoxime and TRO40303 are small molecules that interact with the mitochondrial outer membrane protein VDAC and limit opening of the mitochondrial transition pore in response to oxidative stress [198]. Olesoxime can protect differentiated SHSY-5Y cells from cell death, and reduce neurite retraction and cytoplasmic shrinkage induced by alpha-synuclein overexpression [199]. Low dose TRO40303 upregulates a number of mitochondrial proteins including Drp1 and VDAC and enhances expression of tyrosine hydroxylase in mice overexpressing alpha-synuclein [186]. As mentioned above, our group has developed a peptide inhibitor P110 that selectively blocks the protein-protein interactions between Drp1 and its mitochondrial adaptor Fis1 [125]. Treatment with P110 significantly reduced mitochondrial fragmentation, decreased mitochondrial ROS and improved mitochondrial integrity in dopaminergic neurons exposed to MPP^+ in neurons expressing the LRRK2 G2019S mutation, and in dopaminergic neurons derived from LRRK2 G2019S patient-iPS cells [123,125]. Importantly, we showed that treatment with P110 had minor effects on mitochondrial dynamics and neuronal survival under physiological conditions [123,125]. In addition, Mdivi-1, an inhibitor of mitochondrial fragmentation, was reported to reduce behavioral and neuropathological phenotypes in an MPTP-induced PD mouse model, in addition to its protective effects in dopaminergic neurons exposed to neurotoxins [187].

Modulating the consequences of mitochondrial dysfunction: Modulation of downstream mitochondrial dysfunction may also provide therapeutic opportunities, in both sporadic and familial PD. In the past decades, a number of natural products and small molecules have been reported to protect against neuropathology associated with PD, at least in part through protecting mitochondria. Treatment with these agents has been shown to reduce mitochondrial ROS, inhibit mitochondrial apoptotic pathways, and increase mitochondrial complex I activity. As a consequence, treatment with these agents can reduce neuronal degeneration in PD in culture and in animals. These agents have been extensively reviewed [200–203]. Here, we only list those that have significant protection in animal models of PD (Table 2).

9. Concluding Remarks

Accumulating evidence supports the hypothesis that mitochondrial abnormalities and dysfunction could critically influence neuronal degeneration in both sporadic and faimilial PD. Dysregulation of mitochondrial dynamics and mitophagy have been centrally implicated in the neuropathology of PD. This evidence supports the idea that mitochondrial damage might be a primary cause initiating the progression of PD. Thus, targeting mitochondria may offer the opportunities for drug development to treat neurodegenerative diseases such as parkinsonism.

Conflicts of Interest: The authors declare no conflict of interest.

References

1. Duvoisin, R.C. Overview of Parkinson's disease. *Ann. N. Y. Acad. Sci.* **1992**, *648*, 187–193.
2. Dauer, W.; Przedborski, S. Parkinson's disease: Mechanisms and models. *Neuron* **2003**, *39*, 889–909.
3. Dawson, T.M.; Dawson, V.L. Molecular pathways of neurodegeneration in Parkinson's disease. *Science* **2003**, *302*, 819–822.
4. Yao, Z.; Wood, N.W. Cell death pathways in Parkinson's disease: Role of mitochondria. *Antioxid. Redox Signal.* **2009**, *11*, 2135–2149.
5. Ellis, C.E.; Murphy, E.J.; Mitchell, D.C.; Golovko, M.Y.; Scaglia, F.; Barcelo-Coblijn, G.C.; Nussbaum, R.L. Mitochondrial lipid abnormality and electron transport chain impairment in mice lacking alpha-synuclein. *Mol. Cell. Biol.* **2005**, *25*, 10190–10201.
6. Herrmann, J.M.; Longen, S.; Weckbecker, D.; Depuydt, M. Biogenesis of mitochondrial proteins. *Adv. Exp. Med. Biol.* **2012**, *748*, 41–64.
7. Sas, K.; Robotka, H.; Toldi, J.; Vecsei, L. Mitochondria, metabolic disturbances, oxidative stress and the kynurenine system, with focus on neurodegenerative disorders. *J. Neurol. Sci.* **2007**, *257*, 221–239.
8. Daiber, A. Redox signaling (cross-talk) from and to mitochondria involves mitochondrial pores and reactive oxygen species. *Biochim. Biophys. Acta* **2010**, *1797*, 897–906.
9. Bononi, A.; Missiroli, S.; Poletti, F.; Suski, J.M.; Agnoletto, C.; Bonora, M.; de Marchi, E.; Giorgi, C.; Marchi, S.; Patergnani, S.; *et al.* Mitochondria-associated membranes (MAMs) as hotspot Ca^{2+} signaling units. *Adv. Exp. Med. Biol.* **2012**, *740*, 411–437.
10. Rizzuto, R.; de Stefani, D.; Raffaello, A.; Mammucari, C. Mitochondria as sensors and regulators of calcium signalling. *Nat. Rev. Mol. Cell Biol.* **2012**, *13*, 566–578.
11. Cali, T.; Ottolini, D.; Brini, M. Mitochondrial Ca^{2+} as a key regulator of mitochondrial activities. *Adv. Exp. Med. Biol.* **2012**, *942*, 53–73.
12. Osteryoung, K.W.; Nunnari, J. The division of endosymbiotic organelles. *Science* **2003**, *302*, 1698–704.
13. Antico Arciuch, V.G.; Elguero, M.E.; Poderoso, J.J.; Carreras, M.C. Mitochondrial regulation of cell cycle and proliferation. *Antioxid. Redox Signal.* **2012**, *16*, 1150–1180.

14. Moyes, C.D.; Hood, D.A. Origins and consequences of mitochondrial variation in vertebrate muscle. *Annu. Rev. Physiol.* **2003**, *65*, 177–201.

15. Schatten, H.; Prather, R.S.; Sun, Q.Y. The significance of mitochondria for embryo development in cloned farm animals. *Mitochondrion* **2005**, *5*, 303–321.

16. Giorgi, C.; Baldassari, F.; Bononi, A.; Bonora, M.; de Marchi, E.; Marchi, S.; Missiroli, S.; Patergnani, S.; Rimessi, A.; Suski, J.M.; *et al.* Mitochondrial Ca^{2+} and apoptosis. *Cell Calcium* **2012**, *52*, 36–43.

17. Rasola, A.; Bernardi, P. Mitochondrial permeability transition in Ca^{2+}-dependent apoptosis and necrosis. *Cell Calcium* **2011**, *50*, 222–233.

18. Wang, C.; Youle, R.J. The role of mitochondria in apoptosis*. *Annu. Rev. Genet.* **2009**, *43*, 95–118.

19. Burbach, J.P.; Smits, S.; Smidt, M.P. Transcription factors in the development of midbrain dopamine neurons. *Ann. N. Y. Acad. Sci.* **2003**, *991*, 61–68.

20. Wallen, A.; Perlmann, T. Transcriptional control of dopamine neuron development. *Ann. N. Y. Acad. Sci.* **2003**, *991*, 48–60.

21. Dreyer, S.D.; Zhou, G.; Baldini, A.; Winterpacht, A.; Zabel, B.; Cole, W.; Johnson, R.L.; Lee, B. Mutations in LMX1B cause abnormal skeletal patterning and renal dysplasia in nail patella syndrome. *Nat. Genet.* **1998**, *19*, 47–50.

22. Smidt, M.P.; van Schaick, H.S.; Lanctot, C.; Tremblay, J.J.; Cox, J.J.; van der Kleij, A.A.; Wolterink, G.; Drouin, J.; Burbach, J.P. A homeodomain gene Ptx3 has highly restricted brain expression in mesencephalic dopaminergic neurons. *Proc. Natl Acad. Sci. USA* **1997**, *94*, 13305–13310.

23. Smidt, M.P.; Asbreuk, C.H.; Cox, J.J.; Chen, H.; Johnson, R.L.; Burbach, J.P. A second independent pathway for development of mesencephalic dopaminergic neurons requires Lmx1b. *Nat. Neurosci.* **2000**, *3*, 337–341.

24. Mann, V.M.; Cooper, J.M.; Daniel, S.E.; Srai, K.; Jenner, P.; Marsden, C.D.; Schapira, A.H. Complex I, iron, and ferritin in Parkinson's disease substantia nigra. *Ann. Neurol.* **1994**, *36*, 876–881.

25. Blandini, F.; Nappi, G.; Greenamyre, J.T. Quantitative study of mitochondrial complex I in platelets of parkinsonian patients. *Mov. Disord.* **1998**, *13*, 11–15.

26. Haas, R.H.; Nasirian, F.; Nakano, K.; Ward, D.; Pay, M.; Hill, R.; Shults, C.W. Low platelet mitochondrial complex I and complex II/III activity in early untreated Parkinson's disease. *Ann. Neurol.* **1995**, *37*, 714–722.

27. Barroso, N.; Campos, Y.; Huertas, R.; Esteban, J.; Molina, J.A.; Alonso, A.; Gutierrez-Rivas, E.; Arenas, J. Respiratory chain enzyme activities in lymphocytes from untreated patients with Parkinson disease. *Clin. Chem.* **1993**, *39*, 667–69.

28. Yoshino, H.; Nakagawa-Hattori, Y.; Kondo, T.; Mizuno, Y. Mitochondrial complex I and II activities of lymphocytes and platelets in Parkinson's disease. *J. Neural Transm Parkinsons Dis. Dement. Sect.* **1992**, *4*, 27–34.

29. Dias, V.; Junn, E.; Mouradian, M.M. The role of oxidative stress in Parkinson's disease. *J. Parkinsons Dis.* **2013**, *3*, 461–491.

30. Hwang, O. Role of oxidative stress in Parkinson's disease. *Exp. Neurobiol.* **2013**, *22*, 11–17.

31. Bender, A.; Krishnan, K.J.; Morris, C.M.; Taylor, G.A.; Reeve, A.K.; Perry, R.H.; Jaros, E.; Hersheson, J.S.; Betts, J.; Klopstock, T.; *et al.* High levels of mitochondrial DNA deletions in substantia nigra neurons in aging and Parkinson disease. *Nat. Genet.* **2006**, *38*, 515–517.

32. Kraytsberg, Y.; Kudryavtseva, E.; McKee, A.C.; Geula, C.; Kowall, N.W.; Khrapko, K. Mitochondrial DNA deletions are abundant and cause functional impairment in aged human substantia nigra neurons. *Nat. Genet.* **2006**, *38*, 518–520.

33. Langston, J.W.; Ballard, P.; Tetrud, J.W.; Irwin, I. Chronic Parkinsonism in humans due to a product of meperidine-analog synthesis. *Science* **1983**, *219*, 979–980.

34. Betarbet, R.; Sherer, T.B.; MacKenzie, G.; Garcia-Osuna, M.; Panov, A.V.; Greenamyre, J.T. Chronic systemic pesticide exposure reproduces features of Parkinson's disease. *Nat. Neurosci.* **2000**, *3*, 1301–1306.

35. Testa, C.M.; Sherer, T.B.; Greenamyre, J.T. Rotenone induces oxidative stress and dopaminergic neuron damage in organotypic substantia nigra cultures. *Brain Res. Mol. Brain Res.* **2005**, *134*, 109–118.

36. Smeyne, R.J.; Jackson-Lewis, V. The MPTP model of Parkinson's disease. *Brain Res. Mol. Brain Res.* **2005**, *134*, 57–66.

37. Langston, J.W.; Ballard, P.A., Jr. Parkinson's disease in a chemist working with 1-methyl-4-phenyl-1,2,5,6-tetrahydropyridine. *N. Engl. J. Med.* **1983**, *309*, 310.

38. Kopin, I.J. Toxins and Parkinson's disease: MPTP parkinsonism in humans and animals. *Adv. Neurol.* **1987**, *45*, 137–144.

39. Barbeau, A. Parkinson's disease: Clinical features and etiopahthology. In *Amsterdam Handbook of Clinical Neurology*; Viken, P.J., Bruyn, G.W., Klawans, H.L., Eds.; Elsevier Science Publishers: Amsterdam, The Netherlands, 1986; pp. 87–108.

40. Jenner, P.; Olanow, C.W. Understanding cell death in Parkinson's disease. *Ann. Neurol.* **1998**, *44* (Suppl. S1), S72–S84.

41. Nunes, I.; Tovmasian, L.T.; Silva, R.M.; Burke, R.E.; Goff, S.P. Pitx3 is required for development of substantia nigra dopaminergic neurons. *Proc. Natl. Acad. Sci. USA* **2003**, *100*, 4245–4250.

42. Van den Munckhof, P.; Luk, K.C.; Ste-Marie, L.; Montgomery, J.; Blanchet, P.J.; Sadikot, A.F.; Drouin, J. Pitx3 is required for motor activity and for survival of a subset of midbrain dopaminergic neurons. *Development* **2003**, *130*, 2535–2542.

43. Klein, C.; Westenberger, A. Genetics of Parkinson's disease. *Cold Spring Harb. Perspect. Med.* **2012**, *2*, a008888.

44. Zetterstrom, R.H.; Solomin, L.; Jansson, L.; Hoffer, B.J.; Olson, L.; Perlmann, T. Dopamine neuron agenesis in Nurr1-deficient mice. *Science* **1997**, *276*, 248–250.

45. Wallen, A.; Zetterstrom, R.H.; Solomin, L.; Arvidsson, M.; Olson, L.; Perlmann, T. Fate of mesencephalic AHD2-expressing dopamine progenitor cells in NURR1 mutant mice. *Exp. Cell Res.* **1999**, *253*, 737–46.

46. Smidt, M.P.; Smits, S.M.; Burbach, J.P. Molecular mechanisms underlying midbrain dopamine neuron development and function. *Eur. J. Pharmacol.* **2003**, *480*, 75–88.

47. Franco-Iborra, S.; Vila, M.; Perier, C. The Parkinson Disease Mitochondrial Hypothesis: Where Are We at? *Neuroscientist* **2015**.

48. Luoma, P.; Melberg, A.; Rinne, J.O.; Kaukonen, J.A.; Nupponen, N.N.; Chalmers, R.M.; Oldfors, A.; Rautakorpi, I.; Peltonen, L.; Majamaa, K.; *et al.* Parkinsonism, premature menopause, and mitochondrial DNA polymerase gamma mutations: Clinical and molecular genetic study. *Lancet* **2004**, *364*, 875–882.

49. Henchcliffe, C.; Beal, M.F. Mitochondrial biology and oxidative stress in Parkinson disease pathogenesis. *Nat. Clin. Pract. Neurol.* **2008**, *4*, 600–609.

50. Zheng, B.; Liao, Z.; Locascio, J.J.; Lesniak, K.A.; Roderick, S.S.; Watt, M.L.; Eklund, A.C.; Zhang-James, Y.; Kim, P.D.; Hauser, M.A.; *et al.* PGC-1α, a potential therapeutic target for early intervention in Parkinson's disease. *Sci. Transl Med.* **2010**, *2*, 52ra73.

51. Ekstrand, M.I.; Terzioglu, M.; Galter, D.; Zhu, S.; Hofstetter, C.; Lindqvist, E.; Thams, S.; Bergstrand, A.; Hansson, F.S.; Trifunovic, A.; *et al.* Progressive parkinsonism in mice with respiratory-chain-deficient dopamine neurons. *Proc. Natl. Acad. Sci. USA* **2007**, *104*, 1325–1330.

52. Ekstrand, M.I.; Galter, D. The MitoPark Mouse—An animal model of Parkinson's disease with impaired respiratory chain function in dopamine neurons. *Parkinsonism Relat. Disord.* **2009**, *15* (Suppl. S3), S185–S188.

53. Gellhaar, S.; Marcellino, D.; Abrams, M.B.; Galter, D. Chronic L-DOPA induces hyperactivity, normalization of gait and dyskinetic behavior in MitoPark mice. *Genes Brain Behav.* **2015**, *14*, 260–270.

54. Ikebe, S.; Tanaka, M.; Ohno, K.; Sato, W.; Hattori, K.; Kondo, T.; Mizuno, Y.; Ozawa, T. Increase of deleted mitochondrial DNA in the striatum in Parkinson's disease and senescence. *Biochem. Biophys. Res. Commun.* **1990**, *170*, 1044–1048.

55. Lee, V.M.; Trojanowski, J.Q. Mechanisms of Parkinson's disease linked to pathological alpha-synuclein: New targets for drug discovery. *Neuron* **2006**, *52*, 33–38.

56. Peelaerts, W.; Bousset, L.; van der Perren, A.; Moskalyuk, A.; Pulizzi, R.; Giugliano, M.; van den Haute, C.; Melki, R.; Baekelandt, V. Synuclein strains cause distinct synucleinopathies after local and systemic administration. *Nature* **2015**, *522*, 340–344.

57. Selkoe, D.; Dettmer, U.; Luth, E.; Kim, N.; Newman, A.; Bartels, T. Defining the native state of alpha-synuclein. *Neurodegener. Dis.* **2014**, *13*, 114–117.

58. Riess, O.; Jakes, R.; Kruger, R. Genetic dissection of familial Parkinson's disease. *Mol. Med. Today* **1998**, *4*, 438–444.

59. De Silva, H.R.; Khan, N.L.; Wood, N.W. The genetics of Parkinson's disease. *Curr. Opin. Genet. Dev.* **2000**, *10*, 292–298.

60. Tan, J.M.; Dawson, T.M. Parkin blushed by PINK1. *Neuron* **2006**, *50*, 527–529.

61. Kubo, S.; Hattori, N.; Mizuno, Y. Recessive Parkinson's disease. *Mov. Disord.* **2006**, *21*, 885–893.

62. Giasson, B.I.; Duda, J.E.; Quinn, S.M.; Zhang, B.; Trojanowski, J.Q.; Lee, V.M. Neuronal alpha-synucleinopathy with severe movement disorder in mice expressing A53T human alpha-synuclein. *Neuron* **2002**, *34*, 521–533.

63. Lee, M.K.; Stirling, W.; Xu, Y.; Xu, X.; Qui, D.; Mandir, A.S.; Dawson, T.M.; Copeland, N.G.; Jenkins, N.A.; Price, D.L. Human alpha-synuclein-harboring familial Parkinson's disease-linked Ala-53 —> Thr mutation causes neurodegenerative disease with alpha-synuclein aggregation in transgenic mice. *Proc. Natl. Acad. Sci. USA* **2002**, *99*, 8968–8973.

64. Gispert, S.; Del Turco, D.; Garrett, L.; Chen, A.; Bernard, D.J.; Hamm-Clement, J.; Korf, H.W.; Deller, T.; Braak, H.; Auburger, G.; *et al.* Transgenic mice expressing mutant A53T human alpha-synuclein show neuronal dysfunction in the absence of aggregate formation. *Mol. Cell. Neurosci.* **2003**, *24*, 419–429.

65. Neumann, M.; Kahle, P.J.; Giasson, B.I.; Ozmen, L.; Borroni, E.; Spooren, W.; Muller, V.; Odoy, S.; Fujiwara, H.; Hasegawa, M.; *et al.* Misfolded proteinase K-resistant hyperphosphorylated alpha-synuclein in aged transgenic mice with locomotor deterioration and in human alpha-synucleinopathies. *J. Clin. Investig.* **2002**, *110*, 1429–11439.

66. Rockenstein, E.; Mallory, M.; Hashimoto, M.; Song, D.; Shults, C.W.; Lang, I.; Masliah, E. Differential neuropathological alterations in transgenic mice expressing alpha-synuclein from the platelet-derived growth factor and Thy-1 promoters. *J. Neurosci. Res.* **2002**, *68*, 568–578.

67. Fleming, S.M.; Salcedo, J.; Fernagut, P.O.; Rockenstein, E.; Masliah, E.; Levine, M.S.; Chesselet, M.F. Early and progressive sensorimotor anomalies in mice overexpressing wild-type human alpha-synuclein. *J. Neurosci.* **2004**, *24*, 9434–9440.

68. Chen, L.; Xie, Z.; Turkson, S.; Zhuang, X. A53T human alpha-synuclein overexpression in transgenic mice induces pervasive mitochondria macroautophagy defects preceding dopamine neuron degeneration. *J. Neurosci.* **2015**, *35*, 890–905.

69. Li, Y.; Liu, W.; Oo, T.F.; Wang, L.; Tang, Y.; Jackson-Lewis, V.; Zhou, C.; Geghman, K.; Bogdanov, M.; Przedborski, S.; *et al.* Mutant LRRK2(R1441G) BAC transgenic mice recapitulate cardinal features of Parkinson's disease. *Nat. Neurosci.* **2009**, *12*, 826–828.

70. Dranka, B.P.; Gifford, A.; McAllister, D.; Zielonka, J.; Joseph, J.; O'Hara, C.L.; Stucky, C.L.; Kanthasamy, A.G.; Kalyanaraman, B. A novel mitochondrially-targeted apocynin derivative prevents hyposmia and loss of motor function in the leucine-rich repeat kinase 2 (LRRK2(R1441G)) transgenic mouse model of Parkinson's disease. *Neurosci. Lett.* **2014**, *583*, 159–164.

71. Yue, M.; Hinkle, K.M.; Davies, P.; Trushina, E.; Fiesel, F.C.; Christenson, T.A.; Schroeder, A.S.; Zhang, L.; Bowles, E.; Behrouz, B.; *et al.* Progressive dopaminergic alterations and mitochondrial abnormalities in LRRK2 G2019S knock-in mice. *Neurobiol. Dis.* **2015**, *78*, 172–195.

72. Goldberg, M.S.; Fleming, S.M.; Palacino, J.J.; Cepeda, C.; Lam, H.A.; Bhatnagar, A.; Meloni, E.G.; Wu, N.; Ackerson, L.C.; Klapstein, G.J.; *et al.* Parkin-deficient mice exhibit nigrostriatal deficits but not loss of dopaminergic neurons. *J. Biol. Chem.* **2003**, *278*, 43628–43635.

73. Perez, F.A.; Palmiter, R.D. Parkin-deficient mice are not a robust model of parkinsonism. *Proc. Natl. Acad. Sci. USA* **2005**, *102*, 2174–2179.

74. Lopert, P.; Patel, M. Brain mitochondria from DJ-1 knockout mice show increased respiration-dependent hydrogen peroxide consumption. *Redox Biol.* **2014**, *2*, 667–672.

75. Akundi, R.S.; Huang, Z.; Eason, J.; Pandya, J.D.; Zhi, L.; Cass, W.A.; Sullivan, P.G.; Bueler, H. Increased mitochondrial calcium sensitivity and abnormal expression of innate immunity genes precede dopaminergic defects in Pink1-deficient mice. *PLoS ONE* **2011**, *6*, e16038.

76. Gispert, S.; Ricciardi, F.; Kurz, A.; Azizov, M.; Hoepken, H.H.; Becker, D.; Voos, W.; Leuner, K.; Muller, W.E.; Kudin, A.P.; *et al.* Parkinson phenotype in aged PINK1-deficient mice is accompanied by progressive mitochondrial dysfunction in absence of neurodegeneration. *PLoS ONE* **2009**, *4*, e5777.

77. Stichel, C.C.; Zhu, X.R.; Bader, V.; Linnartz, B.; Schmidt, S.; Lubbert, H. Mono- and double-mutant mouse models of Parkinson's disease display severe mitochondrial damage. *Hum. Mol. Genet.* **2007**, *16*, 2377–2393.

78. Bonifati, V.; Oostra, B.A.; Heutink, P. Linking DJ-1 to neurodegeneration offers novel insights for understanding the pathogenesis of Parkinson's disease. *J. Mol. Med. (Berl.)* **2004**, *82*, 163–174.

79. Abou-Sleiman, P.M.; Healy, D.G.; Wood, N.W. Causes of Parkinson's disease: Genetics of DJ-1. *Cell Tissue Res.* **2004**, *318*, 185–188.

80. Zimprich, A.; Biskup, S.; Leitner, P.; Lichtner, P.; Farrer, M.; Lincoln, S.; Kachergus, J.; Hulihan, M.; Uitti, R.J.; Calne, D.B.; *et al.* Mutations in LRRK2 cause autosomal-dominant parkinsonism with pleomorphic pathology. *Neuron* **2004**, *44*, 601–607.

81. Paisan-Ruiz, C.; Jain, S.; Evans, E.W.; Gilks, W.P.; Simon, J.; van der Brug, M.; de Munain, A.L.; Aparicio, S.; Gil, A.M.; Khan, N.; *et al.* Cloning of the gene containing mutations that cause PARK8-linked Parkinson's disease. *Neuron* **2004**, *44*, 595–600.

82. West, A.B.; Moore, D.J.; Choi, C.; Andrabi, S.A.; Li, X.; Dikeman, D.; Biskup, S.; Zhang, Z.; Lim, K.L.; Dawson, V.L.; *et al.* Parkinson's disease-associated mutations in LRRK2 link enhanced GTP-binding and kinase activities to neuronal toxicity. *Hum. Mol. Genet.* **2007**, *16*, 223–232.

83. Cookson, M.R. The role of leucine-rich repeat kinase 2 (LRRK2) in Parkinson's disease. *Nat. Rev. Neurosci.* **2010**, *11*, 791–797.

84. West, A.B.; Moore, D.J.; Biskup, S.; Bugayenko, A.; Smith, W.W.; Ross, C.A.; Dawson, V.L.; Dawson, T.M. Parkinson's disease-associated mutations in leucine-rich repeat kinase 2 augment kinase activity. *Proc. Natl. Acad. Sci. USA* **2005**, *102*, 16842–16847.

85. Simon, H.H.; Saueressig, H.; Wurst, W.; Goulding, M.D.; O'Leary, D.D. Fate of midbrain dopaminergic neurons controlled by the engrailed genes. *J. Neurosci.* **2001**, *21*, 3126–3134.

86. Yu-chin Su, X.Q. Impairment of mitochondrial dynamics: A target for the treatment of neurological disorders? *Future Med.* **2013**, *8*, 333–346.

87. Tanner, C.M.; Goldman, S.M. Epidemiology of Parkinson's disease. *Neurol. Clin.* **1996**, *14*, 317–35.

88. Caradoc-Davies, T.H.; Weatherall, M.; Dixon, G.S.; Caradoc-Davies, G.; Hantz, P. Is the prevalence of Parkinson's disease in New Zealand really changing? *Acta Neurol. Scand.* **1992**, *86*, 40–44.

89. Morgante, L.; Rocca, W.A.; di Rosa, A.E.; de Domenico, P.; Grigoletto, F.; Meneghini, F.; Reggio, A.; Savettieri, G.; Castiglione, M.G.; Patti, F.; *et al.* Prevalence of Parkinson's disease and other types of parkinsonism: A door-to-door survey in three Sicilian municipalities. The Sicilian Neuro-Epidemiologic Study (SNES) Group. *Neurology* **1992**, *42*, 1901–1907.

90. Mutch, W.J.; Dingwall-Fordyce, I.; Downie, A.W.; Paterson, J.G.; Roy, S.K. Parkinson's disease in a Scottish city. *Br. Med. J. (Clin. Res. Ed.)* **1986**, *292*, 534–536.

91. Polymeropoulos, M.H.; Higgins, J.J.; Golbe, L.I.; Johnson, W.G.; Ide, S.E.; di Iorio, G.; Sanges, G.; Stenroos, E.S.; Pho, L.T.; Schaffer, A.A.; *et al.* Mapping of a gene for Parkinson's disease to chromosome 4q21-q23. *Science* **1996**, *274*, 1197–1199.

92. Dekker, M.C.; Bonifati, V.; van Duijn, C.M. Parkinson's disease: Piecing together a genetic jigsaw. *Brain* **2003**, *126*, 1722–1733.

93. Langston, J.W. Epidemiology *versus* genetics in Parkinson's disease: Progress in resolving an age-old debate. *Ann. Neurol.* **1998**, *44* (Suppl. S1), S45–S52.

94. Bowers, W.J.; Howard, D.F.; Federoff, H.J. Gene therapeutic strategies for neuroprotection: Implications for Parkinson's disease. *Exp. Neurol.* **1997**, *144*, 58–68.

95. Kitada, T.; Asakawa, S.; Hattori, N.; Matsumine, H.; Yamamura, Y.; Minoshima, S.; Yokochi, M.; Mizuno, Y.; Shimizu, N. Mutations in the parkin gene cause autosomal recessive juvenile parkinsonism. *Nature* **1998**, *392*, 605–608.

96. Lincoln, S.; Vaughan, J.; Wood, N.; Baker, M.; Adamson, J.; Gwinn-Hardy, K.; Lynch, T.; Hardy, J.; Farrer, M. Low frequency of pathogenic mutations in the ubiquitin carboxy-terminal hydrolase gene in familial Parkinson's disease. *Neuroreport* **1999**, *10*, 427–429.

97. Golbe, L.I.; Di Iorio, G.; Bonavita, V.; Miller, D.C.; Duvoisin, R.C. A large kindred with autosomal dominant Parkinson's disease. *Ann. Neurol.* **1990**, *27*, 276–282.

98. Kruger, R.; Kuhn, W.; Muller, T.; Woitalla, D.; Graeber, M.; Kosel, S.; Przuntek, H.; Epplen, J.T.; Schols, L.; Riess, O. Ala30Pro mutation in the gene encoding alpha-synuclein in Parkinson's disease. *Nat. Genet.* **1998**, *18*, 106–108.

99. Polymeropoulos, M.H.; Lavedan, C.; Leroy, E.; Ide, S.E.; Dehejia, A.; Dutra, A.; Pike, B.; Root, H.; Rubenstein, J.; Boyer, R.; *et al.* Mutation in the alpha-synuclein gene identified in families with Parkinson's disease. *Science* **1997**, *276*, 2045–2047.

100. Pham, A.H.; Meng, S.; Chu, Q.N.; Chan, D.C. Loss of Mfn2 results in progressive, retrograde degeneration of dopaminergic neurons in the nigrostriatal circuit. *Hum. Mol. Genet.* **2012**, *21*, 4817–4826.

101. Lee, S.; Sterky, F.H.; Mourier, A.; Terzioglu, M.; Cullheim, S.; Olson, L.; Larsson, N.G. Mitofusin 2 is necessary for striatal axonal projections of midbrain dopamine neurons. *Hum. Mol. Genet.* **2012**, *21*, 4827–35.

102. Li, H.; Chen, Y.; Jones, A.F.; Sanger, R.H.; Collis, L.P.; Flannery, R.; McNay, E.C.; Yu, T.; Schwarzenbacher, R.; Bossy, B.; *et al.* Bcl-xL induces Drp1-dependent synapse formation in cultured hippocampal neurons. *Proc. Natl. Acad. Sci. USA* **2008**, *105*, 2169–2174.

103. Ishihara, N.; Nomura, M.; Jofuku, A.; Kato, H.; Suzuki, S.O.; Masuda, K.; Otera, H.; Nakanishi, Y.; Nonaka, I.; Goto, Y.; *et al.* Mitochondrial fission factor Drp1 is essential for embryonic development and synapse formation in mice. *Nat. Cell. Biol.* **2009**, *11*, 958–966.

104. Wakabayashi, J.; Zhang, Z.; Wakabayashi, N.; Tamura, Y.; Fukaya, M.; Kensler, T.W.; Iijima, M.; Sesaki, H. The dynamin-related GTPase Drp1 is required for embryonic and brain development in mice. *J. Cell Biol.* **2009**, *186*, 805–816.

105. Berthet, A.; Margolis, E.B.; Zhang, J.; Hsieh, I.; Zhang, J.; Hnasko, T.S.; Ahmad, J.; Edwards, R.H.; Sesaki, H.; Huang, E.J.; *et al.* Loss of mitochondrial fission depletes axonal mitochondria in midbrain dopamine neurons. *J. Neurosci.* **2014**, *34*, 14304–14317.

106. Zarranz, J.J.; Alegre, J.; Gomez-Esteban, J.C.; Lezcano, E.; Ros, R.; Ampuero, I.; Vidal, L.; Hoenicka, J.; Rodriguez, O.; Atares, B.; *et al.* The new mutation, E46K, of alpha-synuclein causes Parkinson and Lewy body dementia. *Ann. Neurol.* **2004**, *55*, 164–173.

107. Bonifati, V.; Rizzu, P.; Squitieri, F.; Krieger, E.; Vanacore, N.; van Swieten, J.C.; Brice, A.; van Duijn, C.M.; Oostra, B.; Meco, G.; *et al.* DJ-1(PARK7), a novel gene for autosomal recessive, early onset parkinsonism. *Neurol. Sci.* **2003**, *24*, 159–160.

108. Iwatsubo, T.; Ito, G.; Takatori, S.; Hannno, Y.; Kuwahara, T. Pathogenesis of Parkinson's disease: Implications from familial Parkinson's disease. *Rinsho Shinkeigaku* **2005**, *45*, 899–901.

109. Bialecka, M.; Hui, S.; Klodowska-Duda, G.; Opala, G.; Tan, E.K.; Drozdzik, M. Analysis of LRRK 2 G 2019 S and I 2020 T mutations in Parkinson's disease. *Neurosci. Lett.* **2005**, *390*, 1–3.

110. Smith, R.G. The aging process: Where are the drug opportunities? *Curr. Opin. Chem. Biol.* **2000**, *4*, 371–376.

111. Le, W.; Conneely, O.M.; He, Y.; Jankovic, J.; Appel, S.H. Reduced Nurr1 expression increases the vulnerability of mesencephalic dopamine neurons to MPTP-induced injury. *J. Neurochem.* **1999**, *73*, 2218–2221.

112. Xiao, Q.; Castillo, S.O.; Nikodem, V.M. Distribution of messenger RNAs for the orphan nuclear receptors Nurr1 and Nur77 (NGFI-B) in adult rat brain using in situ hybridization. *Neuroscience* **1996**, *75*, 221–230.

113. Rakovic, A.; Grunewald, A.; Seibler, P.; Ramirez, A.; Kock, N.; Orolicki, S.; Lohmann, K.; Klein, C. Effect of endogenous mutant and wild-type PINK1 on Parkin in fibroblasts from Parkinson disease patients. *Hum. Mol. Genet.* **2010**, *19*, 3124–3137.

114. Rakovic, A.; Grunewald, A.; Kottwitz, J.; Bruggemann, N.; Pramstaller, P.P.; Lohmann, K.; Klein, C. Mutations in PINK1 and Parkin Impair Ubiquitination of Mitofusins in Human Fibroblasts. *PLoS ONE* **2011**, *6*, e16746.

115. Jin, H.J.; Li, C.G. Tanshinone IIA and Cryptotanshinone Prevent Mitochondrial Dysfunction in Hypoxia-Induced H9c2 Cells: Association to Mitochondrial ROS, Intracellular Nitric Oxide, and Calcium Levels. *Evid. Based Complement. Alternat. Med.* **2013**, *2013*, 610694.

116. Poole, A.C.; Thomas, R.E.; Andrews, L.A.; McBride, H.M.; Whitworth, A.J.; Pallanck, L.J. The PINK1/Parkin pathway regulates mitochondrial morphology. *Proc. Natl. Acad. Sci. USA* **2008**, *105*, 1638–1643.

117. Yang, Y.; Ouyang, Y.; Yang, L.; Beal, M.F.; McQuibban, A.; Vogel, H.; Lu, B. Pink1 regulates mitochondrial dynamics through interaction with the fission/fusion machinery. *Proc. Natl. Acad. Sci. USA* **2008**, *105*, 7070–7075.

118. Yu, W.; Sun, Y.; Guo, S.; Lu, B. The PINK1/Parkin pathway regulates mitochondrial dynamics and function in mammalian hippocampal and dopaminergic neurons. *Hum. Mol. Genet.* **2011**, *20*, 3227–3240.

119. Wang, X.; Petrie, T.G.; Liu, Y.; Liu, J.; Fujioka, H.; Zhu, X. Parkinson's disease-associated DJ-1 mutations impair mitochondrial dynamics and cause mitochondrial dysfunction. *J. Neurochem.* **2012**, *121*, 830–839.

120. Ryan, B.J.; Hoek, S.; Fon, E.A.; Wade-Martins, R. Mitochondrial dysfunction and mitophagy in Parkinson's: From familial to sporadic disease. *Trends Biochem. Sci.* **2015**, *40*, 200–210.

121. Niu, J.; Yu, M.; Wang, C.; Xu, Z. Leucine-rich repeat kinase 2 disturbs mitochondrial dynamics via Dynamin-like protein. *J. Neurochem.* **2012**, *122*, 650–658.

122. Wang, X.; Yan, M.H.; Fujioka, H.; Liu, J.; Wilson-Delfosse, A.; Chen, S.G.; Perry, G.; Casadesus, G.; Zhu, X. LRRK2 regulates mitochondrial dynamics and function through direct interaction with DLP1. *Hum. Mol. Genet.* **2012**, *21*, 1931–1944.

123. Su, Y.C.; Qi, X. Inhibition of excessive mitochondrial fission reduced aberrant autophagy and neuronal damage caused by LRRK2 G2019S mutation. *Hum. Mol. Genet.* **2013**, *22*, 4545–4561.

124. Guo, X.; Disatnik, M.H.; Monbureau, M.; Shamloo, M.; Mochly-Rosen, D.; Qi, X. Inhibition of mitochondrial fragmentation diminishes Huntington's disease-associated neurodegeneration. *J. Clin. Investig.* **2013**, *123*, 5371–5388.

125. Qi, X.; Qvit, N.; Su, Y.C.; Mochly-Rosen, D. A novel Drp1 inhibitor diminishes aberrant mitochondrial fission and neurotoxicity. *J. Cell Sci.* **2013**, *126*, 789–802.

126. Yamamoto, A.; Yue, Z. Autophagy and its normal and pathogenic states in the brain. *Annu. Rev. Neurosci.* **2014**, *37*, 55–78.

127. Redmann, M.; Dodson, M.; Boyer-Guittaut, M.; Darley-Usmar, V.; Zhang, J. Mitophagy mechanisms and role in human diseases. *Int. J. Biochem. Cell Biol.* **2014**, *53*, 127–33.

128. Fass, E.; Amar, N.; Elazar, Z. Identification of essential residues for the C-terminal cleavage of the mammalian LC3: A lesson from yeast Atg8. *Autophagy* **2007**, *3*, 48–50.

129. Kabeya, Y.; Mizushima, N.; Ueno, T.; Yamamoto, A.; Kirisako, T.; Noda, T.; Kominami, E.; Ohsumi, Y.; Yoshimori, T. LC3, a mammalian homologue of yeast Apg8p, is localized in autophagosome membranes after processing. *EMBO J.* **2000**, *19*, 5720–5728.

130. Kanki, T.; Wang, K.; Cao, Y.; Baba, M.; Klionsky, D.J. Atg32 is a mitochondrial protein that confers selectivity during mitophagy. *Dev. Cell* **2009**, *17*, 98–109.

131. Okamoto, K.; Kondo-Okamoto, N.; Ohsumi, Y. Mitochondria-anchored receptor Atg32 mediates degradation of mitochondria via selective autophagy. *Dev. Cell* **2009**, *17*, 87–97.

132. Liu, L.; Feng, D.; Chen, G.; Chen, M.; Zheng, Q.; Song, P.; Ma, Q.; Zhu, C.; Wang, R.; Qi, W.; *et al.* Mitochondrial outer-membrane protein FUNDC1 mediates hypoxia-induced mitophagy in mammalian cells. *Nat. Cell Biol.* **2012**, *14*, 177–185.

133. Narendra, D.; Kane, L.A.; Hauser, D.N.; Fearnley, I.M.; Youle, R.J. p62/SQSTM1 is required for Parkin-induced mitochondrial clustering but not mitophagy; VDAC1 is dispensable for both. *Autophagy* **2010**, *6*, 1090–1106.

134. Shi, R.Y.; Zhu, S.H.; Li, V.; Gibson, S.B.; Xu, X.S.; Kong, J.M. BNIP3 interacting with LC3 triggers excessive mitophagy in delayed neuronal death in stroke. *CNS Neurosci. Ther.* **2014**, *20*, 1045–1055.

135. Strappazzon, F.; Nazio, F.; Corrado, M.; Cianfanelli, V.; Romagnoli, A.; Fimia, G.M.; Campello, S.; Nardacci, R.; Piacentini, M.; Campanella, M.; *et al.* AMBRA1 is able to induce mitophagy via LC3 binding, regardless of PARKIN and p62/SQSTM1. *Cell. Death Differ.* **2015**, *22*, 419–432.

136. Johansen, T.; Lamark, T. Selective autophagy mediated by autophagic adapter proteins. *Autophagy* **2011**, *7*, 279–296.

137. Wu, F.; Xu, H.D.; Guan, J.J.; Hou, Y.S.; Gu, J.H.; Zhen, X.C.; Qin, Z.H. Rotenone impairs autophagic flux and lysosomal functions in Parkinson's disease. *Neuroscience* **2015**, *284*, 900–911.

138. Arsikin, K.; Kravic-Stevovic, T.; Jovanovic, M.; Ristic, B.; Tovilovic, G.; Zogovic, N.; Bumbasirevic, V.; Trajkovic, V.; Harhaji-Trajkovic, L. Autophagy-dependent and -independent involvement of AMP-activated protein kinase in 6-hydroxydopamine toxicity to SH-SY5Y neuroblastoma cells. *Biochim. Biophys. Acta* **2012**, *1822*, 1826–1836.

139. Youle, R.J.; Narendra, D.P. Mechanisms of mitophagy. *Nat. Rev. Mol. Cell Biol.* **2011**, *12*, 9–14.

140. Narendra, D.P.; Youle, R.J. Targeting mitochondrial dysfunction: Role for PINK1 and Parkin in mitochondrial quality control. *Antioxid. Redox Signal.* **2011**, *14*, 1929–1938.

141. Ashrafi, G.; Schlehe, J.S.; LaVoie, M.J.; Schwarz, T.L. Mitophagy of damaged mitochondria occurs locally in distal neuronal axons and requires PINK1 and Parkin. *J. Cell Biol.* **2014**, *206*, 655–670.

142. Geisler, S.; Holmstrom, K.M.; Skujat, D.; Fiesel, F.C.; Rothfuss, O.C.; Kahle, P.J.; Springer, W. PINK1/Parkin-mediated mitophagy is dependent on VDAC1 and p62/SQSTM1. *Nat. Cell Biol.* **2010**, *12*, 119–131.

143. Rakovic, A.; Shurkewitsch, K.; Seibler, P.; Grunewald, A.; Zanon, A.; Hagenah, J.; Krainc, D.; Klein, C. Phosphatase and tensin homolog (PTEN)-induced putative kinase 1 (PINK1)-dependent ubiquitination of endogenous Parkin attenuates mitophagy: Study in human primary fibroblasts and induced pluripotent stem cell-derived neurons. *J. Biol. Chem.* **2013**, *288*, 2223–2237.

144. McCoy, M.K.; Cookson, M.R. DJ-1 regulation of mitochondrial function and autophagy through oxidative stress. *Autophagy* **2011**, *7*, 531–532.

145. Thomas, K.J.; McCoy, M.K.; Blackinton, J.; Beilina, A.; van der Brug, M.; Sandebring, A.; Miller, D.; Maric, D.; Cedazo-Minguez, A.; Cookson, M.R. DJ-1 acts in parallel to the PINK1/parkin pathway to control mitochondrial function and autophagy. *Hum. Mol. Genet.* **2011**, *20*, 40–50.

146. Irrcher, I.; Aleyasin, H.; Seifert, E.L.; Hewitt, S.J.; Chhabra, S.; Phillips, M.; Lutz, A.K.; Rousseaux, M.W.; Bevilacqua, L.; Jahani-Asl, A.; *et al.* Loss of the Parkinson's disease-linked gene DJ-1 perturbs mitochondrial dynamics. *Hum. Mol. Genet.* **2010**, *19*, 3734–46.

147. Tang, B.; Xiong, H.; Sun, P.; Zhang, Y.; Wang, D.; Hu, Z.; Zhu, Z.; Ma, H.; Pan, Q.; Xia, J.H.; *et al.* Association of PINK1 and DJ-1 confers digenic inheritance of early-onset Parkinson's disease. *Hum. Mol. Genet.* **2006**, *15*, 1816–1825.

148. Xiong, H.; Wang, D.; Chen, L.; Choo, Y.S.; Ma, H.; Tang, C.; Xia, K.; Jiang, W.; Ronai, Z.; Zhuang, X.; *et al.* Parkin, PINK1, and DJ-1 form a ubiquitin E3 ligase complex promoting unfolded protein degradation. *J. Clin. Investig.* **2009**, *119*, 650–660.

149. Plowey, E.D.; Cherra, S.J., 3rd; Liu, Y.J.; Chu, C.T. Role of autophagy in G2019S-LRRK2-associated neurite shortening in differentiated SH-SY5Y cells. *J. Neurochem.* **2008**, *105*, 1048–1056.

150. Yakhine-Diop, S.M.; Bravo-San Pedro, J.M.; Gomez-Sanchez, R.; Pizarro-Estrella, E.; Rodriguez-Arribas, M.; Climent, V.; Aiastui, A.; Lopez de Munain, A.; Fuentes, J.M.; Gonzalez-Polo, R.A. G2019S LRRK2 mutant fibroblasts from Parkinson's disease patients show increased sensitivity to neurotoxin 1-methyl-4-phenylpyridinium dependent of autophagy. *Toxicology* **2014**, *324*, 1–9.

151. Sanchez-Danes, A.; Richaud-Patin, Y.; Carballo-Carbajal, I.; Jimenez-Delgado, S.; Caig, C.; Mora, S.; di Guglielmo, C.; Ezquerra, M.; Patel, B.; Giralt, A.; *et al.* Disease-specific phenotypes in dopamine neurons from human iPS-based models of genetic and sporadic Parkinson's disease. *EMBO Mol. Med.* **2012**, *4*, 380–395.

152. Lachenmayer, M.L.; Yue, Z. Genetic animal models for evaluating the role of autophagy in etiopathogenesis of Parkinson disease. *Autophagy* **2012**, *8*, 1837–1838.

153. Alegre-Abarrategui, J.; Christian, H.; Lufino, M.M.; Mutihac, R.; Venda, L.L.; Ansorge, O.; Wade-Martins, R. LRRK2 regulates autophagic activity and localizes to specific membrane microdomains in a novel human genomic reporter cellular model. *Hum. Mol. Genet.* **2009**, *18*, 4022–4034.

154. Saez-Atienzar, S.; Bonet-Ponce, L.; Blesa, J.R.; Romero, F.J.; Murphy, M.P.; Jordan, J.; Galindo, M.F. The LRRK2 inhibitor GSK2578215A induces protective autophagy in SH-SY5Y cells: Involvement of Drp-1-mediated mitochondrial fission and mitochondrial-derived ROS signaling. *Cell Death Dis.* **2014**, *5*, e1368.

155. Gomez-Suaga, P.; Luzon-Toro, B.; Churamani, D.; Zhang, L.; Bloor-Young, D.; Patel, S.; Woodman, P.G.; Churchill, G.C.; Hilfiker, S. Leucine-rich repeat kinase 2 regulates autophagy through a calcium-dependent pathway involving NAADP. *Hum. Mol. Genet.* **2012**, *21*, 511–525.

156. Bravo-San Pedro, J.M.; Niso-Santano, M.; Gomez-Sanchez, R.; Pizarro-Estrella, E.; Aiastui-Pujana, A.; Gorostidi, A.; Climent, V.; Lopez de Maturana, R.; Sanchez-Pernaute, R.; Lopez de Munain, A.; *et al.* The LRRK2 G2019S mutant exacerbates basal autophagy through activation of the MEK/ERK pathway. *Cell. Mol. Life Sci.* **2013**, *70*, 121–136.

157. Gomez-Suaga, P.; Hilfiker, S. LRRK2 as a modulator of lysosomal calcium homeostasis with downstream effects on autophagy. *Autophagy* **2012**, *8*, 692–693.

158. Orenstein, S.J.; Kuo, S.H.; Tasset, I.; Arias, E.; Koga, H.; Fernandez-Carasa, I.; Cortes, E.; Honig, L.S.; Dauer, W.; Consiglio, A.; *et al.* Interplay of LRRK2 with chaperone-mediated autophagy. *Nat. Neurosci.* **2013**, *16*, 394–406.

159. Su, Y.C.; Guo, X.; Qi, X. Threonine 56 phosphorylation of Bcl-2 is required for LRRK2 G2019S-induced mitochondrial depolarization and autophagy. *Biochim. Biophys. Acta* **2015**, *1852*, 12–21.

160. Ramsay, R.R.; Salach, J.I.; Singer, T.P. Uptake of the neurotoxin 1-methyl-4-phenylpyridine (MPP+) by mitochondria and its relation to the inhibition of the mitochondrial oxidation of NAD$^+$-linked substrates by MPP+. *Biochem. Biophys. Res. Commun.* **1986**, *134*, 743–748.

161. Marey-Semper, I.; Gelman, M.; Levi-Strauss, M. The high sensitivity to rotenone of striatal dopamine uptake suggests the existence of a constitutive metabolic deficiency in dopaminergic neurons from the substantia nigra. *Eur. J. Neurosci.* **1993**, *5*, 1029–1034.

162. Chinta, S.J.; Mallajosyula, J.K.; Rane, A.; Andersen, J.K. Mitochondrial alpha-synuclein accumulation impairs complex I function in dopaminergic neurons and results in increased mitophagy *in vivo*. *Neurosci. Lett.* **2010**, *486*, 235–239.

163. Devi, L.; Raghavendran, V.; Prabhu, B.M.; Avadhani, N.G.; Anandatheerthavarada, H.K. Mitochondrial import and accumulation of alpha-synuclein impair complex I in human dopaminergic neuronal cultures and Parkinson disease brain. *J. Biol. Chem.* **2008**, *283*, 9089–9100.

164. Martin, L.J.; Pan, Y.; Price, A.C.; Sterling, W.; Copeland, N.G.; Jenkins, N.A.; Price, D.L.; Lee, M.K. Parkinson's disease alpha-synuclein transgenic mice develop neuronal mitochondrial degeneration and cell death. *J. Neurosci.* **2006**, *26*, 41–50.

165. Zuo, L.; Motherwell, M.S. The impact of reactive oxygen species and genetic mitochondrial mutations in Parkinson's disease. *Gene* **2013**, *532*, 18–23.

166. Nakamura, T.; Prikhodko, O.A.; Pirie, E.; Nagar, S.; Akhtar, M.W.; Oh, C.K.; McKercher, S.R.; Ambasudhan, R.; Okamoto, S.I.; Lipton, S.A. Aberrant protein S-nitrosylation contributes to the pathophysiology of neurodegenerative diseases. *Neurobiol. Dis.* **2015**.

167. Van Muiswinkel, F.L.; Steinbusch, H.W.; Drukarch, B.; de Vente, J. Identification of NO-producing and -receptive cells in mesencephalic transplants in a rat model of Parkinson's disease: A study using NADPH-d enzyme- and NOSc/cGMP immunocytochemistry. *Ann. N. Y. Acad. Sci.* **1994**, *738*, 289–304.

168. Gu, Z.; Nakamura, T.; Lipton, S.A. Redox reactions induced by nitrosative stress mediate protein misfolding and mitochondrial dysfunction in neurodegenerative diseases. *Mol. Neurobiol.* **2010**, *41*, 55–72.

169. Levecque, C.; Elbaz, A.; Clavel, J.; Richard, F.; Vidal, J.S.; Amouyel, P.; Tzourio, C.; Alperovitch, A.; Chartier-Harlin, M.C. Association between Parkinson's disease and polymorphisms in the nNOS and iNOS genes in a community-based case-control study. *Hum. Mol. Genet.* **2003**, *12*, 79–86.

170. Joniec, I.; Ciesielska, A.; Kurkowska-Jastrzebska, I.; Przybylkowski, A.; Czlonkowska, A.; Czlonkowski, A. Age- and sex-differences in the nitric oxide synthase expression and dopamine concentration in the murine model of Parkinson's disease induced by 1-methyl-4-phenyl-1,2,3,6-tetrahydropyridine. *Brain Res.* **2009**, *1261*, 7–19.

171. Watanabe, Y.; Kato, H.; Araki, T. Protective action of neuronal nitric oxide synthase inhibitor in the MPTP mouse model of Parkinson's disease. *Metab. Brain Dis.* **2008**, *23*, 51–69.

172. Castagnoli, K.; Palmer, S.; Castagnoli, N., Jr. Neuroprotection by (R)-deprenyl and 7-nitroindazole in the MPTP C57BL/6 mouse model of neurotoxicity. *Neurobiology (Bp)* **1999**, *7*, 135–149.

173. Dolezal, P.; Likic, V.; Tachezy, J.; Lithgow, T. Evolution of the molecular machines for protein import into mitochondria. *Science* **2006**, *313*, 314–318.

174. De Marcos-Lousa, C.; Sideris, D.P.; Tokatlidis, K. Translocation of mitochondrial inner-membrane proteins: Conformation matters. *Trends Biochem. Sci.* **2006**, *31*, 259–267.

175. Baker, M.J.; Frazier, A.E.; Gulbis, J.M.; Ryan, M.T. Mitochondrial protein-import machinery: Correlating structure with function. *Trends Cell. Biol.* **2007**, *17*, 456–464.

176. Gottschalk, W.K.; Lutz, M.W.; He, Y.T.; Saunders, A.M.; Burns, D.K.; Roses, A.D.; Chiba-Falek, O. The Broad Impact of TOM40 on Neurodegenerative Diseases in Aging. *J. Parkinsons Dis. Alzheimers Dis.* **2014**, *1*, 12.

177. Bender, A.; Desplats, P.; Spencer, B.; Rockenstein, E.; Adame, A.; Elstner, M.; Laub, C.; Mueller, S.; Koob, A.O.; Mante, M.; *et al.* TOM40 mediates mitochondrial dysfunction induced by alpha-synuclein accumulation in Parkinson's disease. *PLoS ONE* **2013**, *8*, e62277.

178. Okatsu, K.; Kimura, M.; Oka, T.; Tanaka, K.; Matsuda, N. Unconventional PINK1 localization to the outer membrane of depolarized mitochondria drives Parkin recruitment. *J. Cell Sci.* **2015**, *128*, 964–978.

179. Kato, H.; Lu, Q.; Rapaport, D.; Kozjak-Pavlovic, V. Tom70 is essential for PINK1 import into mitochondria. *PLoS ONE* **2013**, *8*, e58435.

180. Cooper, O.; Seo, H.; Andrabi, S.; Guardia-Laguarta, C.; Graziotto, J.; Sundberg, M.; McLean, J.R.; Carrillo-Reid, L.; Xie, Z.; Osborn, T.; *et al.* Pharmacological rescue of mitochondrial deficits in iPSC-derived neural cells from patients with familial Parkinson's disease. *Sci. Transl. Med.* **2012**, *4*, 141ra90.

181. Reinhardt, P.; Schmid, B.; Burbulla, L.F.; Schondorf, D.C.; Wagner, L.; Glatza, M.; Hoing, S.; Hargus, G.; Heck, S.A.; Dhingra, A.; *et al.* Genetic correction of a LRRK2 mutation in human iPSCs links parkinsonian neurodegeneration to ERK-dependent changes in gene expression. *Cell. Stem Cell.* **2013**, *12*, 354–367.

182. Toyoda, Y.; Erkut, C.; Pan-Montojo, F.; Boland, S.; Stewart, M.P.; Muller, D.J.; Wurst, W.; Hyman, A.A.; Kurzchalia, T.V. Products of the Parkinson's disease-related glyoxalase DJ-1, D-lactate and glycolate, support mitochondrial membrane potential and neuronal survival. *Biol. Open* **2014**, *3*, 777–784.

183. Daher, J.P.; Abdelmotilib, H.A.; Hu, X.; Volpicelli-Daley, L.A.; Moehle, M.S.; Fraser, K.B.; Needle, E.; Chen, Y.; Steyn, S.J.; Galatsis, P.; *et al.* LRRK2 Pharmacological Inhibition Abates alpha-Synuclein Induced Neurodegeneration. *J. Biol. Chem.* **2015**, *290*, 19433–19444.

184. Li, T.; He, X.; Thomas, J.M.; Yang, D.; Zhong, S.; Xue, F.; Smith, W.W. A novel GTP-binding inhibitor, FX2149, attenuates LRRK2 toxicity in Parkinson's disease models. *PLoS ONE* **2015**, *10*, e0122461.

185. Chiu, C.C.; Yeh, T.H.; Lai, S.C.; Wu-Chou, Y.H.; Chen, C.H.; Mochly-Rosen, D.; Huang, Y.C.; Chen, Y.J.; Chen, C.L.; Chang, Y.M.; *et al.* Neuroprotective effects of aldehyde dehydrogenase 2 activation in rotenone-induced cellular and animal models of parkinsonism. *Exp. Neurol.* **2015**, *263*, 244–253.

186. Richter, F.; Gao, F.; Medvedeva, V.; Lee, P.; Bove, N.; Fleming, S.M.; Michaud, M.; Lemesre, V.; Patassini, S.; de La Rosa, K.; *et al.* Chronic administration of cholesterol oximes in mice increases transcription of cytoprotective genes and improves transcriptome alterations induced by alpha-synuclein overexpression in nigrostriatal dopaminergic neurons. *Neurobiol. Dis.* **2014**, *69*, 263–275.

187. Rappold, P.M.; Cui, M.; Grima, J.C.; Fan, R.Z.; de Mesy-Bentley, K.L.; Chen, L.; Zhuang, X.; Bowers, W.J.; Tieu, K. Drp1 inhibition attenuates neurotoxicity and dopamine release deficits *in vivo*. *Nat. Commun.* **2014**, *5*, 5244.

188. Mena, N.P.; Garcia-Beltran, O.; Lourido, F.; Urrutia, P.J.; Mena, R.; Castro-Castillo, V.; Cassels, B.K.; Nunez, M.T. The novel mitochondrial iron chelator 5-((methylamino)methyl)-8-hydroxyquinoline protects against mitochondrial-induced oxidative damage and neuronal death. *Biochem. Biophys. Res. Commun.* **2015**, *463*, 787–792.

189. Jiang, J.; Zuo, Y.; Gu, Z. Rapamycin protects the mitochondria against oxidative stress and apoptosis in a rat model of Parkinson's disease. *Int. J. Mol. Med.* **2013**, *31*, 825–832.

190. Xiong, N.; Xiong, J.; Khare, G.; Chen, C.; Huang, J.; Zhao, Y.; Zhang, Z.; Qiao, X.; Feng, Y.; Reesaul, H.; *et al.* Edaravone guards dopamine neurons in a rotenone model for Parkinson's disease. *PLoS ONE* **2011**, *6*, e20677.

191. Dabbeni-Sala, F.; di Santo, S.; Franceschini, D.; Skaper, S.D.; Giusti, P. Melatonin protects against 6-OHDA-induced neurotoxicity in rats: A role for mitochondrial complex I activity. *FASEB J.* **2001**, *15*, 164–170.

192. Zaitone, S.A.; Hammad, L.N.; Farag, N.E. Antioxidant potential of melatonin enhances the response to L-dopa in 1-methyl 4-phenyl 1,2,3,6-tetrahydropyridine-parkinsonian mice. *Pharmacol. Rep.* **2013**, *65*, 1213–1226.

193. Karuppagounder, S.S.; Madathil, S.K.; Pandey, M.; Haobam, R.; Rajamma, U.; Mohanakumar, K.P. Quercetin up-regulates mitochondrial complex-I activity to protect against programmed cell death in rotenone model of Parkinson's disease in rats. *Neuroscience* **2013**, *236*, 136–148.

194. Jayaraj, R.L.; Elangovan, N.; Dhanalakshmi, C.; Manivasagam, T.; Essa, M.M. CNB-001, a novel pyrazole derivative mitigates motor impairments associated with neurodegeneration via suppression of neuroinflammatory and apoptotic response in experimental Parkinson's disease mice. *Chem. Biol. Interact.* **2014**, *220*, 149–157.

195. Abdin, A.A.; Sarhan, N.I. Intervention of mitochondrial dysfunction-oxidative stress-dependent apoptosis as a possible neuroprotective mechanism of alpha-lipoic acid against rotenone-induced parkinsonism and L-dopa toxicity. *Neurosci. Res.* **2011**, *71*, 387–395.

196. Kaur, H.; Chauhan, S.; Sandhir, R. Protective effect of lycopene on oxidative stress and cognitive decline in rotenone induced model of Parkinson's disease. *Neurochem. Res.* **2011**, *36*, 1435–1443.

197. Chen, C.H.; Budas, G.R.; Churchill, E.N.; Disatnik, M.H.; Hurley, T.D.; Mochly-Rosen, D. Activation of aldehyde dehydrogenase-2 reduces ischemic damage to the heart. *Science* **2008**, *321*, 1493–1145.

198. Bordet, T.; Buisson, B.; Michaud, M.; Drouot, C.; Galea, P.; Delaage, P.; Akentieva, N.P.; Evers, A.S.; Covey, D.F.; Ostuni, M.A.; *et al.* Identification and characterization of cholest-4-en-3-one, oxime (TRO19622), a novel drug candidate for amyotrophic lateral sclerosis. *J. Pharmacol. Exp. Ther.* **2007**, *322*, 709–720.

199. Gouarne, C.; Tracz, J.; Paoli, M.G.; Deluca, V.; Seimandi, M.; Tardif, G.; Xilouri, M.; Stefanis, L.; Bordet, T.; Pruss, R.M. Protective role of olesoxime against wild-type alpha-synuclein-induced toxicity in human neuronally differentiated SHSY-5Y cells. *Br. J. Pharmacol.* **2015**, *172*, 235–245.

200. Fernandez-Moriano, C.; Gonzalez-Burgos, E.; Gomez-Serranillos, M.P. Mitochondria-Targeted Protective Compounds in Parkinson's and Alzheimer's Diseases. *Oxid. Med. Cell. Longev.* **2015**, *2015*, 408927.

201. Valadas, J.S.; Vos, M.; Verstreken, P. Therapeutic strategies in Parkinson's disease: What we have learned from animal models. *Ann. N. Y. Acad. Sci.* **2015**, *1338*, 16–37.

202. Procaccio, V.; Bris, C.; de la Barca, J.M.C.; Oca, F.; Chevrollier, A.; Amati-Bonneau, P.; Bonneau, D.; Reynier, P. Perspectives of drug-based neuroprotection targeting mitochondria. *Rev. Neurol. (Paris)* **2014**, *170*, 390–400.

203. Yadav, A.; Agarwal, S.; Tiwari, S.K.; Chaturvedi, R.K. Mitochondria: Prospective targets for neuroprotection in Parkinson's disease. *Curr. Pharm. Des.* **2014**, *20*, 5558–5573.

The Role of Mitochondrial DNA in Mediating Alveolar Epithelial Cell Apoptosis and Pulmonary Fibrosis

Seok-Jo Kim, Paul Cheresh, Renea P. Jablonski, David B. Williams and David W. Kamp

Abstract: Convincing evidence has emerged demonstrating that impairment of mitochondrial function is critically important in regulating alveolar epithelial cell (AEC) programmed cell death (apoptosis) that may contribute to aging-related lung diseases, such as idiopathic pulmonary fibrosis (IPF) and asbestosis (pulmonary fibrosis following asbestos exposure). The mammalian mitochondrial DNA (mtDNA) encodes for 13 proteins, including several essential for oxidative phosphorylation. We review the evidence implicating that oxidative stress-induced mtDNA damage promotes AEC apoptosis and pulmonary fibrosis. We focus on the emerging role for AEC mtDNA damage repair by 8-oxoguanine DNA glycosylase (OGG1) and mitochondrial aconitase (ACO-2) in maintaining mtDNA integrity which is important in preventing AEC apoptosis and asbestos-induced pulmonary fibrosis in a murine model. We then review recent studies linking the sirtuin (SIRT) family members, especially SIRT3, to mitochondrial integrity and mtDNA damage repair and aging. We present a conceptual model of how SIRTs modulate reactive oxygen species (ROS)-driven mitochondrial metabolism that may be important for their tumor suppressor function. The emerging insights into the pathobiology underlying AEC mtDNA damage and apoptosis is suggesting novel therapeutic targets that may prove useful for the management of age-related diseases, including pulmonary fibrosis and lung cancer.

Reprinted from *Int. J. Mol. Sci.* Cite as: Kim, S.-J.; Cheresh, P.; Jablonski, R.P.; Williams, D.B.; Kamp, D.W. The Role of Mitochondrial DNA in Mediating Alveolar Epithelial Cell Apoptosis and Pulmonary Fibrosis. *Int. J. Mol. Sci.* **2015**, *16*, 21486–21513.

1. Introduction

Pulmonary fibrosis is characterized by an over abundant accumulation of extracellular matrix (ECM) collagen deposition in the distal lung interstitial tissue in association with an injured overlying epithelium and activated myofibroblasts. Idiopathic pulmonary fibrosis (IPF) is the most common variety of lung fibrosis and carries a sobering mortality approaching 50% at 3–4 years [1]. Although many of the cellular and molecular mechanisms underlying the pathophysiology of lung fibrosis have emerged from numerous studies over the past several decades, the precise pathways involved, their regulation, and the role of crosstalk between cells are not fully understood. With the exception of two FDA-approved drug therapies (pirfenidone and nintenanib) emerging in the fall of 2014, there are no effective therapies for patients with IPF. Furthermore, these two drugs primarily slow disease progression rather than improve lung function or symptoms. A better understanding of the pathobiology of pulmonary fibrosis is critically important in the design of more useful therapies.

As will be reviewed herein, the extent of alveolar epithelial cell (AEC) injury, repair, and aging are emerging as critical determinants underlying pulmonary fibrosis. The purpose of this review is to highlight our current understanding of the causal role of AEC mitochondrial DNA (mtDNA) damage following oxidative stress in promoting AEC apoptosis and pulmonary fibrosis. Although oxidative mtDNA damage in other cell types (*i.e.*, vascular endothelial cells, macrophages, fibroblasts, *etc.*) are likely important, we concentrate on the lung epithelium given its prominent role in the pathophysiology of lung fibrosis. In particular, we focus on asbestosis (pulmonary fibrosis arising following asbestos exposure) as it shares radiographic and pathologic features with IPF though IPF is more common and carries a worse prognosis. Our group is using the asbestos paradigm to better understand the pathophysiologic mechanisms underlying pulmonary fibrosis. We begin with a brief overview of the evidence implicating that oxidative stress induces mtDNA damage and thereby promotes AEC apoptosis and pulmonary fibrosis. We explore the evidence that mitochondrial-derived reactive oxygen species (ROS) trigger an AEC mtDNA damage response and apoptosis that can promote lung fibrosis and other degenerative lung diseases (*i.e.*, lung cancer and chronic obstructive pulmonary disease [COPD]). We discuss the emerging role for AEC mtDNA damage repair by 8-oxoguanine DNA glycosylase (OGG1) and mitochondrial aconitase (ACO-2) in maintaining mtDNA integrity, which is important in preventing AEC apoptosis and asbestos-induced pulmonary fibrosis. We review the emerging evidence on the important crosstalk between mitochondrial ROS production, mtDNA damage, p53 activation, OGG1, and ACO-2 acting as a mitochondrial redox-sensor involved in mtDNA maintenance in animal models of lung fibrosis. We summarize recent studies linking the sirtuin (SIRT) family members, especially SIRT3, to mitochondrial integrity and mtDNA damage repair and aging. SIRT3 is considered the 'guardian of the mitochondria' because it is the major mitochondrial deacetylase controlling mitochondrial metabolism playing important roles in mtDNA integrity and the prevention of aging. Finally, we present a conceptual model of how SIRTs modulate ROS-driven mitochondrial metabolism that may be important for cell survival as well as their tumor suppressor function. A general hypothetical model linking mtDNA damage and mitochondrial dysfunction to diverse degenerative diseases, including pulmonary fibrosis, aging, and tumorigenesis is shown in Figure 1. Specifically, herein we focus on a proposed model of AEC mtDNA damage in mediating AEC-intrinsic apoptosis and pulmonary fibrosis as illustrated in Figure 2. Collectively, these studies are revealing novel insights into the pathobiology underlying AEC mtDNA damage and apoptosis that should provide the rationale for developing novel therapeutic targets for managing age-related diseases such as pulmonary fibrosis, COPD, and lung cancer.

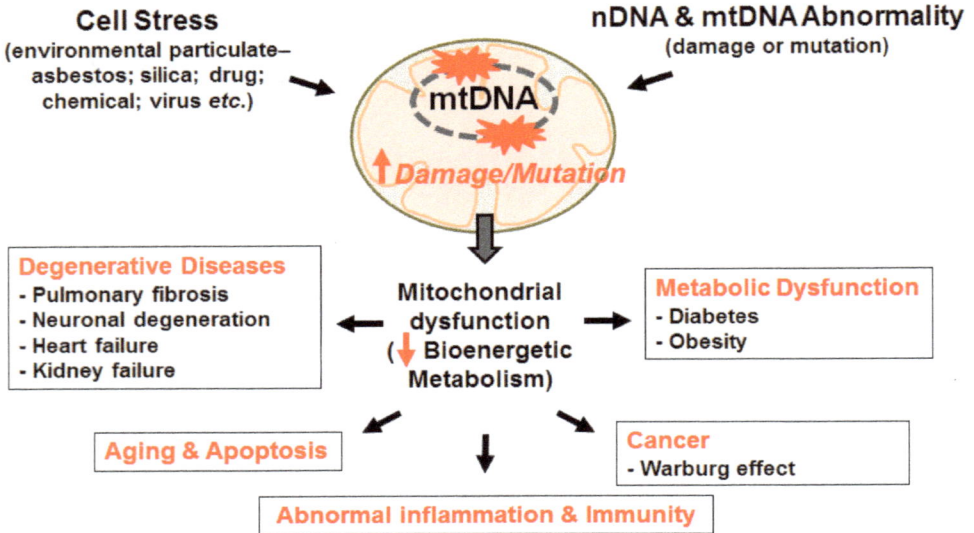

Figure 1. Hypothetical model whereby mtDNA damage induces diverse degenerative diseases. MtDNA damage and mutation can be induced by cell stress from environmental particulates and/or DNA abnormality. MtDNA damage/mutation cause mitochondrial dysfunction reducing bioenergeneric metabolism that can promote degenerative diseases, metabolic dysfunction, aging, apoptosis and cancer. Red-up arrow, increase; red-down arrow, decrease.

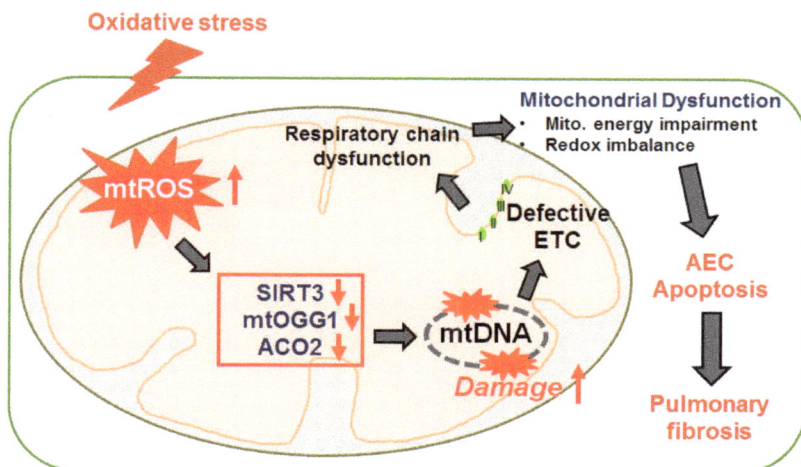

Figure 2. Proposed model of mtDNA damage in mediating AEC intrinsic apoptosis and pulmonary fibrosis. Oxidative stress-induced mtROS induces mtDNA damage by decreasing SIRT3, ACo-2 and mtOGG1 in AEC. MtDNA damage causes a defective ETC that can promote mitochondrial dysfunction, AEC apoptosis, and pulmonary fibrosis. I, II, III, IV, four different complex of ETC in mitochondria. Red-up arrow, increase; red-down arrow, decrease.

2. The Mitochondria, mtDNA, and ROS—The Basics

Mitochondria are maternally inherited and have an essential cellular function of generating energy in the form of ATP via respiration, hence are the "powerhouse" of the cell. Mitochondria also are critically important in regulating complex survival signals that determine whether cells live or die and are closely involved in additional functions, such as cellular differentiation, growth, and cell cycle control [2]. The number of mitochondria in a cell ranges from one to several thousand, a range largely determined by the tissue type and organism [3,4]. Human mtDNA is composed of 16,569 nucleotide bases and encodes 13 polypeptides of the electron transport chain (ETC), 22 transfer RNAs, and two ribosomal RNAs located in the inner mitochondrial membrane (IMM) matrix [4,5]. MtDNA, which encodes approximately 3% of all mitochondrial proteins, is present in multiple copies (~100) per cell, whereas nearly 1200 nuclear DNA (nDNA)-encoded mitochondrial proteins are translated in the cytosol and imported into the mitochondria [3–5]. Interestingly, the extent of mtDNA alterations that occurs during development affects the presence and emergence of mtDNA genetic variations and mutations primarily in a region termed the D-loop [6,7]. This D-loop on the mtDNA forms the basis of forensic medicine in human identification and has been a useful tool in molecular anthropological studies on human origins [7]. Some proteins encoded by nDNA have been shown to be essential for maintaining mtDNA integrity including OGG1, ACO-2, mitochondrial transcription factor A (Tfam), and others [8–13]. However, the mitochondrial proteins involved in mtDNA repair (mostly base excision repair [BER]) are all nuclear encoded and highly dependent on the nDNA repair machinery systems [14,15]. Notably, compared to nDNA, mtDNA is ~50-fold more sensitive to oxidative damage, in part due to its proximity to the ETC and ROS production, lack of a histone protective shield overlying the mtDNA, and relatively limited DNA repair mechanisms [12,13,16,17]. As compared to nDNA, oxidative stress-induced mtDNA damage has a mutation rate that is 10-fold greater [18–23]. Not surprisingly, mtDNA damage and subsequent mutations can lead to mitochondrial dysfunction, including the collapse in the mitochondrial membrane potential ($\Delta\Psi_m$) and release of pro-apoptogenic agents which drives disease formation, aging, and tumorigenesis (Figure 1) [12,19].

ROS, such as hydroxyl radicals (•HO), superoxide anions ($O_2^{\bullet-}$), hydrogen peroxide (H_2O_2), and others are primarily generated under physiologic conditions from the mitochondrial ETC but also from other intracellular sources [12,18,24–29]. Although low levels of ROS are important for promoting cell survival signaling pathways and antioxidant defenses, higher ROS levels, as occurs in the setting of disease or aging, cause oxidative damage to biomolecules, such as DNA, protein, and lipids. Oxidative DNA damage can result in apoptosis and senescence, tumorigenesis, and degenerative diseases ([7,12,18,24–26,28], Figure 1). Furthermore, mtDNA variants within cells can affect both energy and non-energy pathways (complement, inflammatory, and apoptotic) supporting a paradigm shift in thinking about the role of mitochondria beyond simply energy production [7,12,30].

Biological tissues, especially the lungs, are exposed to both extrinsic sources of ROS (e.g., tobacco, asbestos, silica, radiation, bleomycin, and other drugs) and intrinsic sources (such as those from inflammatory, mesenchymal, epithelial, and endothelial cells) primarily via the mitochondrial ETC as well as numerous enzyme systems including Nicotinamide adenine dinucleotide phosphate (NADPH) oxidases (NOXs), xanthine oxidase, and nitric oxide synthase (NOS) [12,19,31]. Intracellular ROS can also be generated from redox-active ferrous (Fe^{2+}) iron within or on the surface of asbestos fibers via the Haber-Weiss reaction [31,32], resulting in the accumulation of •HO and other free radicals. As reviewed in detail elsewhere, several lines of evidence demonstrate that ROS play a role in pulmonary fibrotic disease: (1) oxidized lipids and proteins have been identified from the exhaled air, BAL fluid, and lung tissue of patients with fibrotic lung disease [33,34]; and (2) Bleomycin-induced pulmonary fibrosis (the most common animal fibrosis model) is associated with increased levels of ROS, oxidized proteins, DNA, and lipids [35,36]; (3) Increased oxidative DNA damage is seen in IPF, silicosis, and asbestosis patients, as well in experimental animal models of silicosis or asbestos-induced lung fibrosis [37]; (4) Antioxidants and iron chelators can attenuate fibrosis induced by bleomycin or asbestos in rodent models [35,38]. Additionally, there is some evidence implying that mitochondria-generated ROS of lung parenchymal cells mediate pulmonary fibrosis [36]. Exogenous toxins, such as asbestos fibers, can also induce mitochondrial ROS in lung epithelial cells and macrophages; both of which are important target cells implicated in pulmonary fibrosis (see for reviews: [31,35,37]). However, more work is needed to determine the precise molecular mechanisms involved as well as any cross-talk between cell types. A key role of alveolar macrophage (AM) mitochondrial ROS in mediating asbestosis has been suggested in the studies by Carter and colleagues [39–42]. These investigators showed that mitochondrial Rac-1 levels are elevated in AM from patients with asbestosis, that Rac-1 augments asbestos-induced AM H_2O_2 production, and that ROS production is reduced by knockdown of the iron-sulfur complex III in the mitochondrial ETC. These data implicate H_2O_2 production via electron transfer from Rac-1 to complex III may activate cellular injury pathways that promote asbestosis.

Although the precise role of H_2O_2-induced AEC mtDNA damage in mediating pulmonary fibrosis is unclear, a possible causal role for H_2O_2 in promoting lung fibrosis is supported by several lines of evidence that we recently reviewed in detail elsewhere [35], and briefly summarize herein; some of the key points are: (1) catalase, a H_2O_2 scavenger, blocks H_2O_2-induced human IPF fibroblast activation [34]; and prevents asbestos-induced fibrosis in rats [43,44]; (2) glutathione (GSH), an antioxidant, is diminished in IPF lungs and epithelial lining fluid [34,45,46]; and (3) although n-acetyl cysteine (NAC), a GSH precursor, attenuates bleomycin-induced fibrosis in rodents and increases lung GSH levels, NAC administration to patients with IPF was recently proven no better than placebo [47,48]. Similar to H_2O_2, considerable evidence reviewed elsewhere, implicates NOXs, especially NOX1, 2, and 4 isoforms, in the pathogenesis of pulmonary fibrosis, including AEC apoptosis and apoptosis-resistant myofibroblasts [49–53].

3. AEC Apoptosis and Lung Fibrosis—Role of the Mitochondria

3.1. AEC Aging, Apoptosis and Pulmonary Fibrosis

Accumulating evidence firmly implicate that "exaggerated" aging lung has a crucial role in the pathogenesis of lung fibrosis, although the detailed molecular mechanisms involved are not fully established (see for reviews: [1,54]). The nine proposed pivotal hallmarks mediating the "aging phenotype" include: genomic instability, telomere shortening, epigenetic alterations, deficient proteostasis, dysregulated nutrient sensing, mitochondrial dysfunction, cellular senescence/apoptosis, stem cell depletion, and distorted intercellular communication [55]. However, many of the nine aging pathways are implicated in humans with IPF, including AEC DNA damage, activation of epigenetic signaling, shortened alveolar type II (AT2—the distal lung epithelial stem cell) cell telomeres, AEC mitochondria-mediated (intrinsic) apoptosis, and activated endoplasmic reticulum (ER) stress response in apoptotic AECs (see for reviews: [1,37,54,56–60]). Herein we focus on mtDNA damage since it is an early event in oxidant-exposed cells that may contribute to the inflammatory, fibrogenic and malignant potential of asbestos [37,57,61,62]. Notably, genome-wide association studies (GWAS) have established an important role for aberrant DNA repair pathways in patients with IPF [55,63–65]. As reviewed in detail elsewhere [7,66], mutations in maternally-inherited mtDNA encoding for key genes of mitochondrial energy-generating oxidative phosphorylation, rather than Mendelian nuclear genetic principles, better accounts for the complex clinical-pathological features of many common degenerative and metabolic disease whose tissue stem cells are bioenergetically abnormal (Figure 1).

In contrast to catastrophic lytic/necrotic cell death that can trigger an inflammatory response, apoptosis is a regulated, ATP-dependent process of cell death that results in the elimination of cells with extensive DNA damage without eliciting an inflammatory response. Apoptotic cellular responses occur by two mechanisms: (1) the extrinsic (death receptor) pathway and (2) the intrinsic (mitochondria-regulated) pathway. Others [19,67] as well as ourselves [35,37] have extensively reviewed these pathways recently we, therefore, confine our comments to some of the more recent updates centered on mitochondrial dysfunction and mtDNA damage in driving AEC apoptosis.

Considerable evidence reviewed elsewhere convincingly demonstrates that AEC apoptosis is one of the key pathophysiologic events hindering normal lung repair and thus promoting pulmonary fibrosis [1,35,37,52,56–59,68–71]. Briefly summarizing some of the key findings includes the following: (1) patients with IPF and animal models of asbestos- and silica-induced pulmonary fibrosis show significant lung epithelial cell injury, ER stress and apoptosis; (2) fibrogenic dusts, such as asbestos and silica, can induce both lytic and apoptotic AEC death in part by generating ROS derived from the mitochondria or NOXs; (3) DNA damage, which is a strong activator of intrinsic apoptosis, occurs in the AEC of human patients with IPF and murine models of asbestos-induced lung fibrosis and can activate p53, an important DNA damage response molecule; (4) protein S-glutathionylation, in part through effects occurring in the ER, mediates redox-based alterations in the FAS death receptor important for triggering extrinsic lung epithelial cell apoptosis and pulmonary fibrosis; and (5) blocking $\alpha v \beta 6$ integrin release from injured lung epithelial cells, prevents latent TGF-β activation and pulmonary fibrosis following radiation or bleomycin exposure.

Perhaps the most convincing evidence implicating AEC apoptosis in the pathophysiology of pulmonary fibrosis is that genetic approaches targeting apoptosis

334

of Alveolar epithelial type 2 cell (AT2 cells) in mice and humans demonstrate an important role for AECs (see for reviews: [56,58]). A primary requirement for AEC death and inadequate epithelial cell repair in causing pulmonary fibrosis as first pointed out by Haschek and Witschi 35 years ago has now been elegantly verified by various transgenic murine models of pulmonary fibrosis and genetic mutations in 10 different surfactant protein C (SPC) BRICHOS domain mutations that are only evident in AT2 cells in humans with interstitial pulmonary fibrosis [56]. A clear genetic predisposition to developing IPF is evident in 5%–20% of patients [1,56,58,59,63–65]. Notably, many mutations associated with the development of pulmonary fibrosis are only expressed in epithelial cells (*i.e.*, surfactant C and A2 genes, MUC5b), while others are more ubiquitously expressed. Although a detailed discussion of these mutations is beyond the scope of this article, SPC mutations can induce AEC ER stress response that promotes AEC apoptosis and pulmonary fibrosis [59]. Among the most common gene mutations seen in patients with IPF and familial pulmonary fibrosis involve telomerase (TERT and TERTC). Unlike surfactant mutations, telomerase mutations are ubiquitously expressed in cells, especially in stem cells [72]. Shortened telomeres, which are associated with aging-related diseases due to oxidative stress, are evident in the majority of AT2 cells (the distal lung epithelial stem cells) in patients with lung fibrosis [72,73]. One study showed that AT2 cells from 97% of 62 IPF patients (both sporadic and familial) had shortened telomeres [73]. Although mitochondrial dysfunction and ROS production appear important in driving telomerase-dependent cell senescence, the role of mtDNA damage is uncertain [74]. The finding that AEC telomerase shortening, alone, does not trigger lung fibrosis, nor augment bleomycin-induced lung fibrosis in mice, strongly implicates other genetic and/or environmental factors are likely crucial [75].

3.2. Mitochondria-Regulated AEC Apoptosis

The intrinsic apoptotic death pathway is activated by various fibrotic stimuli (*i.e.*, ROS, DNA damage, asbestos, *etc.*) that stimulate pro-apoptotic Bcl-2 family members action at the mitochondria that results in increased permeability of the outer mitochondrial membrane, reduced $\Delta\Psi_m$, and the release of numerous apoptotic proteins (*i.e.*, cytochrome c, *etc.*) that subsequently activate pro-apoptotic caspase-9 and caspase-3 (see for reviews: [19,35,37,58,67,68]). Notably, a crucial role for mtDNA damage in driving intrinsic apoptosis was established in cell-sorting studies demonstrating that persistent mtDNA damage results in the collapse in the $\Delta\Psi_m$ and intrinsic apoptosis [76]. Bleomycin-induced fibrosis in mice is blocked in the pro-apoptotic Bid deficient mice [77]. Bid is activated by the death receptor pathway and triggers intrinsic apoptosis by blocking anti-apoptotic Bcl-2 molecules and thereby enabling Bax/Bak-mediated apoptosis [77]. However, additional studies are warranted to better understand how Bcl-2 family members modulate mtDNA integrity to impact AEC survival and prevent lung fibrosis. More recently, two groups have established that PTEN-induced putative kinase 1 (PINK1) deficiency impairs AEC mitochondrial function in patients with IPF and PINK1-deficient mice have increased AEC intrinsic apoptosis and lung fibrosis following viral-induced ER stress or bleomycin exposure [78,79]. Interestingly, the pro-fibrotic cytokine TGF-β may be protective to lung epithelial by promoting PINK1 expression and attenuating AEC apoptosis that drives lung fibrosis [79]. Collectively, these studies firmly support an important role of AEC mitochondria-regulated apoptosis in the pathophysiology of pulmonary fibrosis. Additional studies are required to further characterize the precise molecular mechanisms involved, the role of mtDNA damage, and crosstalk between AEC and macrophages. Furthermore, the translational significance of any identified targets in animals exposed to fibrogenic agents will need to be investigated in humans with IPF.

Others as well as our group have been using the asbestos paradigm to better inform our understanding of how oxidative stress resulting from fiber exposure promotes lung epithelial cell intrinsic apoptosis important in the development of pulmonary fibrosis. As we have extensively reviewed these studies from our group elsewhere [35,37], we summarize herein only some of the salient supporting findings including: (1) asbestos fibers are internalized by AECs soon after exposure, resulting in the production of iron-derived ROS, mtDNA damage, and intrinsic apoptosis as evidenced by decreased $\Delta\Psi_m$, mitochondrial cytochrome c release into the cytosol, and activation of caspase-9 and 3 (but not caspase-8); (2) these deleterious actions by asbestos on AECs are blocked by phytic acid (an iron chelator), benzoic acid (a free-radical scavenger), and overexpression of Bcl-XL; and (3) an important role for mitochondrial ROS is suggested by the findings that asbestos-induced AEC intrinsic apoptosis and p53 activation are blocked in cells unable to produce mitochondrial ROS and that asbestos preferentially induces mitochondrial ROS production as assessed using a highly sensitive rho-GFP probes targeted to the mitochondria or cytosol. Furthermore, as reviewed below (see Section 3.3), we find a direct relationship between asbestos-induced mtDNA damage and intrinsic AEC apoptosis. Mossman and colleagues showed that activated protein kinase delta (PKCδ) migrates to the mitochondria of lung epithelial cells *in vitro* and *in vivo* following asbestos exposure and is crucial for promoting asbestos-induced mitochondria-regulated apoptosis and fibrosis via mechanisms dependent upon pro-apoptotic Bim activation [80,81]. Taken together, mitochondrial ROS production and PKCδ activation following asbestos exposure appear important for inducing p53 activation and intrinsic lung epithelial cell apoptosis. However, the role of mtDNA integrity in modulating p53 and PKCδ activation requires additional study.

3.3. mtDNA Damage and Repair—Role in Cancer and Lung Fibrosis

Prompt repair of damaged mtDNA is important given the accumulating evidence convincingly showing that mtDNA damage and mutations are linked to various pathologic conditions, including lung fibrosis (see for reviews: [7,12,18,19,82]). MtDNA mutations accumulate in tissues with aging in part via disruptions in mitochondrial quality control pathways. MtDNA deletions that occur early in development can become widely disseminated throughout the body and cause spontaneous mitochondrial dysfunction [7]. Further, mtDNA deletions [83] and mutations [84] can arise in cells of various tissues throughout life and their accumulation modulates aging and longevity [7,84]. These findings suggest that the accumulation of mtDNA mutations arising from mtDNA damage caused by aging, environmental exposure, and other forms of oxidative stress support the "mitochondrial theory of aging" that may be crucial in depleting the longevity of important stem cells (*i.e.*, AT2 cells in the distal lung) and promoting the pathobiology of degenerative diseases and tumorigenesis [3,7,12].

Cancer is characterized by altered energy metabolism (Warburg effect) arising from mtDNA mutations and changes in mtDNA copy number [85,86]. Of 41 human lung, bladder, and head and neck tumors examined, mutated mtDNA occurred 19–220 times more frequently than nDNA [87]. The importance of mtDNA mutations in lung cancer is supported by the observation that over 40% of patients with lung cancer demonstrate mutations in their mtDNA [88]. MtDNA mutations can compromise ETC function and contribute to altered metabolism driving accelerated aerobic glycolysis in the setting of metastatic progression [89]. Additionally, severe mtDNA damage promotes mitochondrial genome deletion [90]. Tumor cells lacking mtDNA can acquire mtDNA of host origin, resulting in sequential recovery of respiration from primary to metastatic tumor cells [91]. There is also evidence that mtDNA mutations can preferentially accumulate in non-small cell lung cancer (NSCLC) tissues as compared to matched blood samples [88]. Taken together, these studies demonstrate that mtDNA damage and mutations occur in malignant cells, including lung cancers, and that preservation of mtDNA integrity may be an innovative preventative therapeutic target.

Base excision repair (BER), which is the major mtDNA repair mechanism, has been reviewed in detail elsewhere [18]. All mtDNA repair proteins are nuclear-encoded and imported into mitochondria. 8OHdG, the most common of ~50 DNA base changes that occur with oxidative stress, is highly mutagenic in replicating cells by causing G:C→A:T transversions that can contribute to tumorigenesis and aging [12,18]. Mutations in the hOGG1 gene occur in patients with lung cancer and other malignancies [92]. OGG1 is over three-fold more active in the mitochondria as compared to the nucleus and $Ogg1^{-/-}$ mice have a 20-fold increase in liver mitochondrial 8OHdG levels [18,92]. Mitochondria-targeted OGG1 (mt-OGG1)

337

over-expression prevents mitochondria-regulated apoptosis caused by oxidative stress, including AEC apoptosis following asbestos exposure [8,9,12,13,62,93,94]. OGG1 has two isoforms (α and β), yet curiously the βOGG1 isoform has negligible DNA repair activity despite being in 50-fold excess as compared to the αOGG1 isoform in the mitochondria [95]. This suggests that βOGG1 isoform plays a role in mitigating mitochondrial oxidative stress independent of its BER activity. We recently showed that overexpression of mt-OGG1 or mt-OGG1 mutants incapable of DNA repair promote AEC survival despite high levels of asbestos-induced mitochondrial ROS stress [9]. Although one mechanism by which mt-OGG1 preserves mitochondrial function is by increasing mtDNA repair, we identified a novel function of OGG1 in chaperoning mitochondrial aconitase (ACO-2) from oxidative degradation and, thereby, preserving mtDNA integrity [9]. ACO-2, a mitochondrial tricarboxylic acid cycle (TCA) enzyme, is a sensitive marker of oxidative stress and, notably, preserves mtDNA in yeast independent of ACO-2 activity [96,97]. We also reported that oxidative stress (asbestos or H_2O_2) preferentially induces mtDNA than nuclear DNA damage in AEC both *in vitro* as well as *in vivo* and that OGG1 preservation of ACO-2 is crucial for preventing asbestos-induced AEC mtDNA damage, intrinsic apoptosis, and pulmonary fibrosis [8,98]. Additional support for a protective role of ACO-2 is that ACO-2 inactivation has been linked to decreased lifespan in yeast and progressive neurodegenerative diseases in humans [99]. Thus, ACO-2 appears to have a dual function in the TCA cycle for mitochondrial bioenergy production as well as for preserving mtDNA. The precise molecular mechanisms by which mt-OGG1 and ACO-2 interact to preserve AEC mtDNA integrity are not fully understood. Furthermore, additional studies are necessary to assess the translational significance of AEC OGG1 and ACO-2 in preserving mtDNA integrity and preventing pulmonary fibrosis.

3.4. Animals Models of Pulmonary Fibrosis—Role of Mitochondrial ROS, mtDNA Damage, and Mitochondrial Dysfunction

Accumulating evidence reviewed above strongly implicates mitochondrial ROS production, mtDNA damage, and mitochondrial dysfunction (in part due to PINK1 deficiency) in the pathophysiology of AEC apoptosis and pulmonary fibrosis. Herein we review some of the other more recent animal lung fibrosis models that have better informed our understanding of the field. Gadzhar *et al.* [36] examined adult Wistar rat lungs at various time points after a single intratracheal dose of bleomycin and observed lung fibrosis, as measured by Ashcroft scores, collagen, and TGF-β levels at day 14. Evidence of ROS, as assessed by malondialdehyde (MDA) production, was noted as early as 24 h after bleomycin treatment and continued to increase over 14 days. By day seven, mtDNA deletions were significantly elevated and disruption of the mitochondrial architecture as assessed by electron microscopy was noted in lung tissue. Notably, these mitochondrial abnormalities resulted in dysfunction of ETC subunits encoded by mtDNA, but not nDNA, and mtDNA deletions and mtDNA-encoded ETC dysfunction were directly associated with pulmonary TGF-β levels that were predictive of developing lung fibrosis in a multivariate model.

As pulmonary fibrosis is a disease of aging, Hecker *et al.* [53] compared the capacity of young (two months) and aged (18 months) mice to repair bleomycin-induced lung injury and fibrosis. Although the severity of lung fibrosis in each group was similar at three weeks following bleomycin exposure, the aged mice were unable to resolve fibrotic lung injury at two months whereas young mice were largely free of fibrotic injury. Persistent lung fibrosis in the aged mice was characterized by the accumulation of senescent and apoptosis-resistant myofibroblasts, as well as sustained alterations in redox balance resulting from the elevated expression of NOX4 and an impaired capacity to induce the nuclear factor erythroid 2-related factor 2 (Nrf2)-mediated antioxidant response. Human IPF lung tissues also exhibited the imbalance between NOX4 and Nrf2, as well as NOX4 mediated senescence and apoptosis resistance in IPF fibroblasts. Genetic and pharmacological inhibition of NOX4 (with NOX4 siRNA and GKT137831) in older mice with established fibrosis attenuated the senescent and apoptosis-resistant myofibroblast phenotype and led to a reversal of persistent fibrosis. These findings implicating Nrf2 are in accord with prior studies showing that Nrf2 knockout mice are more sensitive to bleomycin and paraquat-induced lung injury than their wild-type counterparts, that primary lung fibroblasts isolated from IPF patients, as compared to healthy controls, have decreased Nrf2 expression and a myofibroblast (pro-fibrotic) differentiated phenotype, and that treatment with sulfaphane, an Nrf2 activator, increases antioxidant levels that results in decreased ROS levels, myofibroblastic de-differentiation, and TGF-β profibrotic effects. Although the role of the NOX4/Nrf2 pathway in AEC and mtDNA damage is unknown, these studies firmly suggest that the restoration of NOX4-Nrf2 redox imbalance in myofibroblasts may be an important therapeutic target.

Our group recently reported that mice globally deficient in $Ogg1^{-/-}$ are more prone to pulmonary fibrosis following asbestos exposure than their wild-type counterparts due in part to increased AEC mtDNA damage and apoptosis [98]. Interestingly, compared to AT2 cells isolated from WT mice, AT2 cells from $Ogg1^{-/-}$ mice have increased mtDNA damage, reduced ACO-2 expression, and increased p53 expression at baseline and these changes were augmented following crocidolite asbestos exposure for three weeks. Collectively, these data support a key role for AEC OGG1 and ACO-2 in the maintenance of mtDNA necessary for preventing AEC apoptosis and pulmonary fibrosis. These findings implicating AEC mtDNA damage signaling in mediating pulmonary fibrosis parallels work by other groups implicating mtDNA damage in the pathophysiology of diverse conditions, such as atherosclerosis, cardiac fibrosis/heart failure, diaphragmatic dysfunction from mechanical ventilation, and cancer [100–106]. Interestingly, mitochondria-targeted OGG1 diminishes ventilator-induced lung injury in mice by reducing the levels of mtDNA damage in the lungs [107]. Accumulating evidences also support an important association between p53, OGG1, and ACO-2, including (1) p53 regulates *OGG1* gene transcription in colon and renal epithelial cells [93]; (2) p53 deficient cells have reduced OGG1 protein expression and activity [93]; (3) p53 can reduce *Aco-2* gene expression [108]; (4) p53 activation is required for oxidant-induced apoptosis in *OGG1*-deficient human fibroblasts [93]; and (5) p53 sensitizes HepG2 cells to oxidative stress by reducing mtDNA [109]. Collectively, these data support a key role for p53 in modulating AEC mtDNA damage in the pro-fibrotic lung response following asbestos exposure that also has important implications for our understanding of the malignant potential of asbestos fibers. However, the precise molecular mechanisms by which OGG1, ACO-2, and p53 coordinately regulate mtDNA integrity in AEC as well as the translational significance of these findings in humans await further study.

A number of animal models of pulmonary fibrosis beyond the scope of this review have implicated excess plasminogen activator inhibitor (PAI-1) in augmenting AEC apoptosis. Shetty and colleagues have recently published a number of elegant studies showing an important dichotomous role of PAI-1 in promoting AEC apoptosis but reducing fibroblast proliferation and collagen production in the pathobiology of lung fibrosis [110–112]. Similar to older studies, these investigators showed that lung injury and pulmonary fibrosis are more evident in mice deficient in urokinase-type plasminogen activator (uPA), whereas mice deficient in PAI-1 are protected. To explore whether changes in AT2 cell uPA and PAI-1 contribute to epithelial-mesenchymal transition (EMT), AT2 cells from patients with IPF and COPD, and mice with bleomycin-, transforming growth factor β-, or passive cigarette smoke-induced lung injury all had reduced expression of E-cadherin and zona occludens-1, whereas collagen-I and α-smooth muscle actin (markers of EMT) were increased along with a parallel increase in PAI-1 and reduced uPA expression [110]. These studies suggest that induction of PAI-1 and inhibition of uPA during fibrotic lung injury promotes EMT in AT2 cells. These same investigators showed that fibroblasts isolated from human IPF lungs and from mice with bleomycin-induced lung fibrosis had an increased rate of proliferation compared with normal lung fibroblasts [111]. Basal expression of plasminogen activator inhibitor-1 (PAI-1) in human and murine fibroblasts was reduced, whereas collagen-I and α-smooth muscle actin were markedly elevated. In contrast, AT2 cells surrounding the fibrotic foci, as well as those isolated from IPF lungs, showed increased caspase-3 and PAI-1 activation with a parallel reduction in uPA expression. PAI-1 depletion and enforced expression studies in cultured fibroblast confirmed the inverse relationship between PAI-1 activation and collagen production. The authors suggested that depletion of PAI-1 in fibroblasts promotes an activated collagen producing cell that is resistant to senescence/apoptosis whereas activated PAI-1 augments AT2 cell apoptosis important for the propagation of lung fibrosis. Using a silica-induced model of lung fibrosis, these investigators showed p53-mediated changes in the uPA system promote lung fibrosis in part by reducing caveolin-1 scaffolding domain peptide (CSP), which is necessary for inhibiting p53 expression and silica-induced lung injury [112]. Notably, as compared to untreated WT mice, silica-exposed WT mice treated with CSP inhibited PAI-1, augmented uPA expression and prevented AEC apoptosis by suppressing p53. The authors suggested that silica-induced lung fibrosis is driven by important crosstalk between the p53-uPA fibrinolytic system in AT2 cells and provide support for a novel pharmacologic target (*i.e.*, CSP), in modulating this pathway. In contrast, another group working with murine fibroblasts recently showed that increased PAI-1 may drive age-related and bleomycin-induced pulmonary fibrosis at least in part by blocking fibroblast apoptosis [113]. Additional studies are required to better understand how precisely PAI-1 affects AEC and fibroblast mtDNA damage response, mitochondrial function and apoptosis as well as how this impacts lung fibrosis.

Mitochondrial ROS and mtDNA can also trigger NACHT, LRR, and PYD domains-containing protein 3 (NALP3) inflammasome signaling important in driving lung fibrosis following asbestos or silica exposure [37,57,114,115]. The mtDNA released into the circulation can act as a sentinel molecule triggering a DNA damage-associated molecular pattern (DAMP) that activates innate immune responses, especially toll like receptor (TLR)-9 signaling, leading to change a phenotype in lung fibroblasts and tissue injury, including lung fibrosis [12,116]. Gu and associates recently showed that intratracheal instillation of mtDNA into murine lungs triggers infiltration of inflammatory cells and production of inflammatory cytokines (*i.e.*, IL-1β, IL-6, and TNF-β) and that these effects were blocked when the lungs were pretreated with TLR-9 siRNA [117]. This suggests that the mtDNA DAMPs can activate innate immune signaling in the lungs via the TLR-9 pathway. Interestingly, Kuck and colleagues showed that mitochondria-targeted OGG1 protein infusion mitigates mtDNA DAMP formation and TLR-9-dependent vascular injury induced by intratracheally-instilled bacteria [118]. Using mice lacking vimentin, a type III intermediate filament, Dos Santos and colleagues demonstrated an important role for vimentin in regulating NLRP3 inflammasome signaling that promotes acute lung injury, IL-1β expression, alveolar epithelial barrier permeability, and lung fibrosis following exposure to lipopolysaccharide (LPS), crocidolite asbestos, and bleomycin [119]. Notably, a direct interaction between vimentin and NLRP3 was demonstrated as well as an important role for macrophages based upon the finding that bone marrow chimeric mice lacking vimentin have decreased lung fibrosis as well as levels of caspase-1 and IL-1β. Collectively, these recent studies provide insight into the role of mtDNA and vimentin in regulating the NLRP3 inflammasome signaling important in promoting lung inflammation and fibrosis.

4. The Role of Sirtuins in Mitochondrial Integrity, mtDNA Damage Repair, and Aging

4.1. Overview of the Sirtuin (SIRT) Family Members

The yeast silent information regulator protein (SIR2) is a highly-conserved protein that has been linked to increased longevity via maintenance of genomic stability in a variety of organisms, including *Drosophila melanogaster* and *Caenorhabditis elegans* [120–122]. The homologous mammalian sirtuin family (SIRTs) consists of seven identified members to date, which are localized to the nucleus (SIRT1, SIRT6, SIRT7), cytoplasm (SIRT2), and mitochondria (SIRT3-5), respectively. All sirtuins contain a conserved core domain with NAD$^+$ binding activity, while most sirtuins catalyze NAD$^+$-dependent deacetylation of lysine residues though SIRT4 is known to have ADP-ribosyltransferase activity and SIRT5 both desuccinylase and demalonyase activity. Members of the sirtuin family play a role in maintenance of genomic stability at multiple levels by participating in DNA repair, altering chromatin structure and function via histone deacetylation [120], and via adaptation of cellular metabolic flow and energy demand [121]. As such, sirtuins have been considered by many to be the guardians of the genome [122].

4.2. The Role of SIRTs and Mitochondria; Normal and General Diseases

Changes in mitochondrial number and function are implicated in the pathogenesis of aging and many of its associated diseases, such as Parkinson's disease, presbycusis, diabetes and the metabolic syndrome, malignancy, and fibrosis. Though changes in mitochondria have been linked to many disease processes, the underlying detailed molecular mechanisms have yet to be fully elucidated as noted above. Major changes in the mitochondria occur with aging, including an increase in the $\Delta\psi_m$ with subsequent increases in ROS production and oxidative damage to mtDNA and other cellular macromolecules. Maintenance of intracellular redox balance is critical to cellular homeostasis and survival. Certain mtDNA haplogroups are associated with an increase in ROS production; these may be associated with a protective effect in early life due to protection against infection, though with age may in fact be maladaptive due to the presence of chronic oxidative stress [7].

Crosstalk between the mitochondria and nucleus has emerged as a potentially powerful regulator underlying age-related diseases. Using a mitochondrial cybrid model of age-related macular degeneration, Kenney *et al.* [30] showed varying bioenergetic profiles among mtDNA haplogroups and went further to demonstrate different gene expression profiles for both mitochondrial-encoded genes involved in cellular respiration and nuclear-encoded genes involved in inflammation and the alternative complement and apoptosis pathways. Based on their findings, they propose a constant interaction between the mitochondrial and nuclear DNA whereby the mtDNA haplotype sets a baseline bioenergetics profile for the cell which interacts with environmental factors to contribute to oxidative stress, mitochondrial dysfunction, cell death and disease. As extensively reviewed in detail elsewhere [18,123], considerable evidence demonstrate that preservation of the mitochondrial genome by various mechanisms beyond the scope of this review is critical for ensuring a functional mitochondria and, thereby, contributing to nuclear DNA stability and cell survival.

As emerging evidence reviewed above implicates the mitochondria as the sentinel organelle governing the cell's cytotoxic response to oxidative stress in the lung, important questions emerge about how the redox balance in the mitochondria is maintained [12]. A crucial role of the mitochondria is to balance energy generation via oxidative phosphorylation with cellular nutrient supply; failure to do so may result in damaging levels or ROS, with resultant mtDNA damage and decreased mitochondrial biogenesis, or decreased cellular availability of ATP. SIRT3 has emerged as a key regulator in cellular ROS balance due to its role modulating mitochondrial homeostasis via deacetylation of multiple mediators of energy metabolism and cellular ROS, including members of the ETC, TCA cycle, and mitochondrial enzymes which detoxify ROS. As such, SIRT3 is considered the "gatekeeper" of mitochondrial integrity [124]. Figure 3 highlights some of the key mitochondrial SIRT3 proteins whose functions are post-translationally regulated by acetylation. Given the hypothesized relation between chronic oxidative stress, SIRTs, and aging [122,124], it is interesting that SIRT3 is the only sirtuin that has been closely associated with longevity in humans based upon the presence of a variable nucleotide tandem repeat (VNTR) enhancer within the SIRT3 gene that is associated with increased survival in the elderly [125].

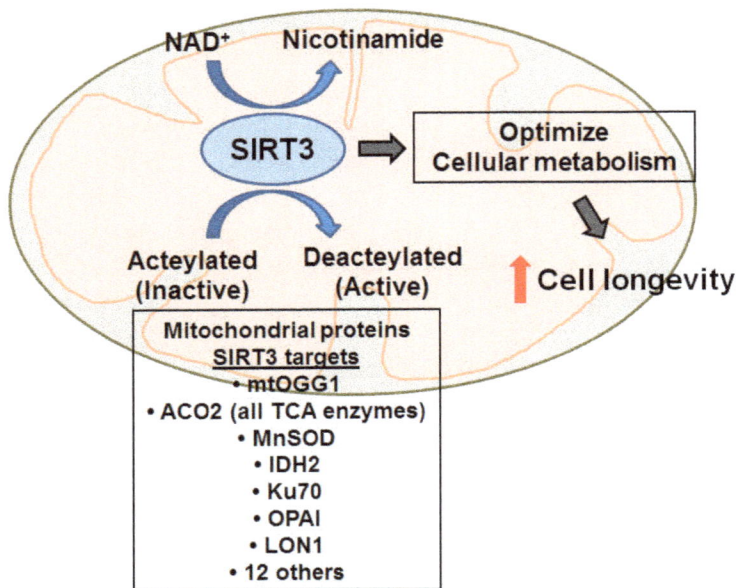

Figure 3. SIRT3 mitochondrial protein deacetylation targets modulate cell metabolism and longevity. Red-up arrow, increase.

4.3. The Role of SIRT3 in mtDNA Damage, Aging and Diseases Including Lung

As reviewed briefly above, repair of mtDNA damage following oxidative injury is a complex process that remains incompletely understood. Emerging evidence suggest an important role for SIRT3 in mtDNA repair. SIRT3 deaceylates and activates many mitochondrial proteins involved in energy metabolism (including all the enzymes involved in the TCA cycle such as ACO-2, isocitrate dehydrogenase 2 [IDH2], *etc.*), ETC members, antioxidant defenses, and mtDNA repair (OGG1) [121,124,126]. Murine SIRT3 depletion studies have shown that SIRT3 is necessary for cellular resistance against genotoxic and oxidative stress by preserving mitochondrial function and genomic stability. SIRT3 deficiency promotes intrinsic apoptosis by several mechanisms identified to date, including (1) augmenting mitochondrial ROS production due to acetylation and inactivation of manganese superoxide dismutase (MnSOD) and IDH2 [124,127]; (2) increasing cyclophilin D activity and down-stream mitochondrial permeability transition (MPT) [128,129]; (3) promoting Bax-mediated apoptosis by acetylating Ku70 [129]; and (4) attenuating p53-mediated growth arrest [130,131]. Notably, a recent study using glioma and renal epithelial tumor cells showed that mitochondrial OGG1 is a direct SIRT3 deacetylation target and that SIRT3 deficiency reduces the oxygen consumption rate (OCR) and 8OHdG mtDNA BER that increases mtDNA damage and intrinsic apoptosis [126]. Further, mtDNA mutations/deletions can reduce SIRT3 expression [132,133]. Although unclear in AT2 cells, loss of SIRT3 in keratinocytes augments ROS levels, reduces NAD$^+$ levels, and promotes epidermal terminal differentiation [134]. These findings suggest SIRT3 may play an important role in maintaining AT2 cell mtDNA and cell survival in the setting of oxidative stress.

A decade ago, a novel function of ACO-2 was discovered, whereby this TCA cycle enzyme was also found to associate with mtDNA nucleoids and stabilize mtDNA in the setting of mtDNA instability [98]. ACO-2 activity is sensitive to the redox state of the cell, suggesting that in the setting of oxidative stress ACO-2 can be relocated from the TCA cycle to the nucleosome to aid in stabilization of the mtDNA with subsequent removal of oxidized Aco-2 by Lon protease [135]. As noted earlier, in an asbestos-induced model of pulmonary fibrosis, oxidant-induced mtDNA damage was blocked by overexpression of ACO-2, a mitochondria-targeted OGG1, and an OGG1 mutant incapable of DNA repair that chaperones ACO-2 [8,9]. The precise mechanism of mtDNA protection by mutant mt-OGG1 comparable to WT mt-OGG1 is not established but may involve blockade of oxidative modification sites on ACO-2, which are necessary for subsequent degradation by the Lon protease. Support for this possibility is that MG132, an inhibitor of Lon protease activity, prevents oxidant-induced reductions in ACO-2 activity in AECs [9]. Another possibility is via effects of SIRT3 on Lon protease since both proteins co-precipitate in breast cancer cells exposed to oxidative and hypoxic stress and SIRT3 silencing results in hyperacetylation and inactivation of Lon [136]. As activity of Lon can favor the transition from aerobic respiration to anaerobic glycolysis, regulate proteins of the TCA cycle and degrade proteins following oxidative damage, these findings suggest a unique coupling between Lon, SIRT3, OGG1, ACO-2, mtDNA repair and the metabolic state of the cell.

Chronic oxidative stress with resultant depletion of SIRT3 and perturbations in mitochondrial function and biogenesis has been increasingly identified as a vital element in the pathogenesis of age-related diseases. Sirt3-deficient (Sirt3$^{-/-}$)mice have no obvious phenotype but are susceptible to age-linked diseases including metabolic syndrome, cancer, cardiac hypertrophy-fibrosis/CHF, hearing loss, acute renal injury, neurodegeneration, and radiation-induced fibrosis [124,137–143]. Notably, Sirt3$^{-/-}$ mice exposed to irradiation have reduced liver ACO-2 activity and increased mitochondrial p53 expression [140]. Aged mouse hematopoietic stem cells (mHSCs) have a diminished capacity to respond to oxidative injury induced by H$_2$O$_2$ resulting in decreased cell survival and increased apoptosis [144]. SIRT3 expression is reduced in aged mHSCs and SIRT3 overexpression in turn restored the proliferative ability of these cells. Whether SIRT3 affords similar changes in AT2 cells, the stem cell of the distal alveolar epithelium, is unknown. Interestingly, Huang et al. [145] showed that rat bone marrow mesenchymal stem cells can differentiate into AT2

345

cells and alleviate bleomycin-induced lung fibrosis when injected at the same time as bleomycin exposure. In skeletal muscle, aging is associated with oxidative damage to cellular macromolecules; exercise training has been associated with an increase in SIRT3 expression, in both young and old subjects, as well as reduced acetylation of IDH-2, a SIRT3 deacetylase target involved in cellular ROS detoxification [146]. In a murine model of acute kidney injury (AKI), cisplatin-induced oxidative injury decreased renal SIRT3 expression and increased mitochondrial fragmentation [141]. $Sirt3^{-/-}$ mice were more susceptible to cisplatin-induced AKI as compared to their wild-type counterparts and treatment with antioxidants restored SIRT3 expression and activity and protected against renal dysfunction in the wild-type mice but not in $Sirt3^{-/-}$ animals. In an *in vitro* cortical neuron model of oxidative stress, overexpression of SIRT3 increases mtDNA and mitochondrial biogenesis that has pro-survival effects [147]. In a rodent model of pre-diabetes, oxidative stress induced by a high-fat diet was associated with disruption of the SIRT3/PGC-1α axis with impairment in mitochondrial bioenergetics and decreased testicular mtDNA and adenylate energy charge [148]. A potential role of targeting SIRT3 in diabetes was supported by a recent study showing that overexpression of SIRT3 can mitigate palmitate-mediated pancreatic β cell dysfunction [149].

Murine models of SIRT3 deficiency confirm its protective role in multiple models of age-associated diseases related to oxidative stress. Lysine deacetylation has emerged as a key post-translational protein modification employed to activate mitochondrial signaling [150,151]. Mice deficient in SIRT3 exhibit striking increases in global protein acetylation; effects that were notably not seen in SIRT4- and SIRT5-deficient mice, suggesting that SIRT3 serves as the key mitochondrial deacetylase [152]. In a model of cardiac fibrosis, $Sirt3^{-/-}$ mice develop severe interstitial cardiac fibrosis and hypertrophy following application of a hypertrophic stimulus, whereas mice engineered to overexpress SIRT3 maintain normal cardiac structure and function [137]. These investigators showed that the mechanism of SIRT3 protection was by activation of the FOXO3a-dependent antioxidant genes catalase and MnSOD which reduced cellular ROS [137]. Another group recently showed that $Sirt3^{-/-}$ mice developed age-related cardiac dysfunction likely due to increased acetylation of various mitochondrial energy producing proteins resulting in myocardial energy depletion [153]. SIRT3 also seems important in pulmonary arterial hypertension (PAH) since pulmonary artery smooth muscle cells (PASMC) from patients with PAH display down-regulation of SIRT3; $Sirt3^{-/-}$ mice spontaneously develop PAH and SIRT3 restoration in a rodent model and $Sirt3^{-/-}$ mice PASMC attenuates the disease phenotype [154]. In a murine model of presbycusis, which is postulated to be due to chronic oxidative stress, decreased SIRT3 expression in the central auditory cortex was noted with concurrent increases in ROS, mtDNA damage, and SOD2 acetylation [155]. Taken together, these data strongly support the protective role of SIRT3 against oxidative stress in a variety of cell types and murine models of various degenerative diseases. Additional studies are necessary to better define the role of mtDNA integrity in mediating the beneficial effects of SIRT3 as well as the translational significance in humans, including pulmonary fibrosis.

4.4. Therapeutic Approach: Resveratrol/Viniferin/Honokiol

The first techniques used to enhance sirtuin expression involved caloric restriction due to the putative link between SIRT3 deacetylase activity and the metabolic state of the cell [124]. The role of SIRT3 in modulating mitochondrial energy metabolism will be discussed in detail the final section below. Exercise training has been associated with increased SIRT3 expression and recently has been linked to a decrease in mitochondrial protein acetylation, increased ACO-2 and MnSOD activity, and improved mitochondrial biogenesis in a cardiac model of doxorubicin-induced oxidative injury [156].

Recently, small molecule sirtuin inducers have been under investigation. Resveratrol (RSV), a naturally occurring polyphenol, extends the lifespan of diverse model organisms in a Sir2 (a silent information regulator)-dependent fashion [157]. RSV induces both SIRT1 and SIRT3 expression via stimulation of NADH dehydrogenases and mitochondrial complex I resulting in increases in the mitochondrial NAD^+/NADH ratio [158]. Treatment with RSV has been shown to abrogate hyperoxia-mediated acute liver injury [158], airway remodeling and hyperreactivity in a model of allergic asthma [159], insulin resistance [160], and pulmonary fibrosis in a bleomycin mouse model [161]. Activation of SIRT3 via RSV in a murine model of cardiac fibrosis attenuated collagen deposition and cardiac hypertrophy via regulation of the TGF-β/Smad pathway; SIRT3 is required for the cardio-protective effects revealing a potential link between mitigation of oxidative stress and the signaling pathways responsible for fibrosis [142]. Viniferin, a natural dehydrodimer of RSV, may have increased antioxidant potential compared to RSV and has been shown to protect vascular endothelial cells from oxidative stress by suppressing intracellular ROS production [162]. Further, SIRT3 was shown to mediate the neuroprotective effect of viniferin in models of Huntington disease via an increase in the number of mtDNA copies, attenuation of the loss of mitochondrial membrane potential, and enhanced activity of MnSOD [163]. Pillai and colleagues showed that honokiol, a natural occurring compound from the bark of magnolia trees with anti-inflammatory, anti-oxidative, anti-tumor, and neuroprotective properties, reverses murine cardiac hypertrophy and fibrosis by a SIRT3-dependent mechanism [164]. Although it is unknown whether these small molecule sirtuin inducers are protective to AEC mtDNA and cell survival following exposure to oxidative stress, they represent an exciting therapeutic strategy for future studies addressing age-related diseases attributed to chronic oxidative stress, such as pulmonary fibrosis.

5. MtDNA and Metabolism in Mitochondria

5.1. Mitochondrial Metabolism—The Basics

As noted earlier, the mitochondria, long championed as the "powerhouse of the cell", are now well-recognized as being critically important in overall cellular health, intracellular signaling, and disease pathobiology [12]. However, that does not diminish the importance of the organelle's ATP-producing function, as mitochondrial metabolism is the main mechanism by which a cell synthesizes ATP. Mitochondrial metabolism is governed primarily by one aspect: the presence of oxygen. Under normal conditions, in the presence of oxygen, mitochondria utilize both oxygen and electrochemical gradients for the production of ATP via the ETC, known as oxidative phosphorylation (OXPHOS), and the TCA (or Krebs) cycle, while in the absence of oxygen the mitochondria rely on glycolysis and fatty acid oxidation for ATP production. Glycolysis produces pyruvate as an end product, which is then used by the TCA to produce the reducing equivalents (e.g., NADH and $FADH_2$) necessary for the ETC to convert ADP to ATP. Fatty acid oxidation, which occurs in the mitochondrial matrix, breaks fatty acids down from triacytlglycerols by β-oxidation and forms both acetyl-CoA (which is converted to citrate in the TCA cycle) and ATP [165]. The interaction between cytosolic liquid triacytlglycerol droplets and mitochondria in fatty acid oxidation must maintain a fine balance, as lipids can either inhibit or uncouple OXPHOS [166]. These anaerobic cycles produce less ATP per cycle, but can be much faster making them indispensable for cells that are not always under "ideal" conditions (i.e., oxidative stress). Cells typically utilize these anaerobic pathways under low-oxygen conditions or when the cell requires energy faster than can be produced by O_2-dependent means. However, mitochondrial damage, be it due to mtDNA damage, ROS, or ETC complex inhibition, can lead the cell to favor these alternate ATP producing pathways. The Warburg effect, or aerobic glycolysis, is where the cell favors the glycolytic pathway over the ETC even in the presence of O_2, a phenomenon nearly synonymous with a cancerous phenotype as it can rapidly produce copious amounts of ATP, which are useful for the rapid growth of cancer cells [167]. However, excessive glycolysis, beyond that which is utilized by cancerous cells, can promote cell stasis, excessive mtDNA damage, and cytotoxicity [168].

5.2. Role of Mitochondria-Derived ROS, mtDNA Damage, and Mitochondrial Metabolism

ROS, especially $O_2^{\bullet-}$, is a natural byproduct of the ETC due to inefficiencies in electron transfer between the four complexes (I, II, III, and IV). ROS are not inherently detrimental, but increased mitochondrial metabolism or damage within the OXPHOS pathway (*i.e.*, buildup of acetylated metabolic intermediates, ETC complex damage) increases cellular ROS, and, as detailed above, can cause mtDNA damage and harm components of the ETC, which further promotes ROS production. NAD^+, a natural byproduct of metabolism, is produced when NADH is reduced by electrons from the ETC; with an impaired ETC, cells instead use glycolytic pathways or fatty acid oxidation and, as such, metabolic reducing equivalents (*i.e.*, NADH) build up in the mitochondria, reducing NAD^+ formation and down-stream NAD^+-dependent processes (e.g., SIRT3 deacetylation—see below) [165]. Decreased NAD^+ levels promote OXPHOS dysfunction leading to metabolic failure in cells, which is a common phenotype in the aged [169]. However, strategies to augment NAD^+ levels can prevent apoptosis-induced complex I inhibition, the depletion of intracellular ATP, and preserve mtDNA integrity, possibly by preventing an increase in BAX expression (6). Additionally, ACO-2, a TCA cycle intermediary metabolite, can stabilize the mtDNA and, thereby, linking mtDNA integrity with metabolic efficiency [135]. However, the mechanism by which ACO-2 toggles between these two ACO-2 functions is not well understood. Notably, mutations in *ACO-2* gene can cause syndromic optic neuropathy with encephalopathy and cerebellar atrophy in humans [170].

Disruption of mitochondrial dynamics is an early event in ROS-induced intrinsic apoptotic cell death. Excessive ROS production can induce ER Ca^{2+} release, reduce the cell's abilty to produce autophagosomes, and contribute to degenerative diseases via effects on mtDNA [12,171]. OXPHOS dysregulation arising from mtDNA mutations/damage results in an increase in the invasive cell phenotype via effects on matrix metalloproteinase (MMP) family members. It also mimics lack-of-oxygen by causing increased lactate production and acidification of the extracellular environment, and can act parallel to (and independent of) hypoxia inducible factor (HIF)-1α activation to increase angiogenesis and glucose uptake [172]. Inflammatory signaling typically increases oxygen consumption (OCR), ROS production, and $\Delta\psi_m$, but does not significantly affect spare respiratory capacity (SCR—the difference between a cell's maximal, uncoupled respiration, and its normal, basal respiration) when normalized to mitochondria content [173]. With progressive cellular stress, mitochondrial rupture and release of mtDNA can occur, which can lead to TLR-9/NLRP3 inflammasome activation and a vicious inflammatory cycle [115,117].

Mitophagy is the cellular mechanism in place to discard damaged and ineffective mitochondria with extensive mtDNA damage [3,174]. Healthy mitochondria can fuse, and such fusion acts as a buffer against mitophagy, ensuring that, optimally, fully-functional mitochondria will be spared, while those mitochondria that cannot fuse will be more easily mitophagized [3,174]. Full mitochondrial fusion can also allow for mtDNA and nucleoid transfer, thereby diminishing/alleviating mtDNA damage. However, fusion can still occur in mitochondria with extensive mtDNA damage if initiated by a healthy mitochondrion, which can revive the metabolic functioning of the damaged mitochondrion. Metabolically inefficient mitochondria that overexpress mitofusion genes can escape mitophagy [174]. Some inherited mitochondrial haplotypes display significant differences in ATP and ROS production and ETC complex expression; those that exhibit relatively lower productions of ATP and ROS appear to be more inclined toward dysfunction and resultant disease independent of nuclear DNA mutations [30]. As reviewed in detail elsewhere [7], there has been extensive study of mitochondrial lineage and how the inheritance of mitochondrial mutations can affect disease progression and severity. Notably, altered proteins resulting from damaged nDNA can allow mtDNA damage to accumulate [7,12]. For example, PINK1 and PARK2 mutations prevent the proper functioning of autophagosomes thereby preventing damaged mitochondria from being destroyed, leading to even greater mtDNA damage.

Although the role of mitochondrial metabolic pathways in regulating AEC function and maintaining mtDNA integrity are not well established, there is some information in other cell types. For example, in Huntington's disease there are increased numbers of large, unhealthy mitochondria with decreased $\Delta\psi_m$, PGC-1α, ATP production, ETC function, ADP-uptake, glucose metabolism, and SRC [175]. PGC-1α, a regulator of both metabolism and mitochondrial biogenesis, is upregulated by RIP1 (receptor-interacting protein 1); a decrease in RIP1 leads to an increase of dsDNA breakage and oxidative glycolysis, a reduction in the NAD$^+$ pools, and suppression of cell proliferation [168]. RIP1 maintains cancer cell glycolytic metabolism such that enhanced cell proliferation can occur, but its loss results in excessive glycolysis and p53-mediated cell stasis. Hyperglycemic damage in diabetic heart cells can cause alterations to mtDNA as a result of alterations in the supply of metabolic substrates (i.e., NAD(P)$^+$/NAD(P)H, GSH/GSSG, and TrxSH$_2$/TrxSS) that can affect mitochondrial health [166]. Using a diabetic heart mouse model, these same investigators showed that cardiac muscle contraction and Ca^{2+} signaling are reduced largely because of alterations in complex I resulting from mitochondrial metabolic damage. Impaired pulmonary artery endothelial cell regeneration is linked to mitochondrial DNA deletion, and hypoxia-reoxygenation reduces p53, PGC-1α, ATP, $\Delta\psi_m$, caspase-induced apoptosis, the levels of mitofusin 1 and 2, and mitophagy [176]. Hypoxic conditions also render significant damage to mtDNA, through increased complex III O$_2^{\bullet-}$ production [12]. Lung fibrosis in bleomycin-treated rats is associated with significant mtDNA damage, dysfunction of mtDNA-encoded ETC subunits, and increased ROS production and TGFβ levels [36]. Although the primary cell type mediating these effects was not identified, future studies focusing on mitochondrial metabolism in AECs will be of considerable interest. We reason that AEC mtDNA damage that accumulates over time can promote "aged" AT2 cells that are more prone to triggering pulmonary fibrosis following an environmental or oxidative stress insult.

5.3. Role of Sirtuins and Energy Metabolism in Mitochondria

Convincing evidences implicate SIRT-induced mitochondrial dysfunction in aging as well as diseases associated with aging (*i.e.*, cancer, neurodegeneration, CHF, and metabolic syndrome/diabetes) and will likely prove important in the pathophysiology of lung disease, such as IPF, asbestosis, and lung cancer [177–179]. The paradigm that is emerging by which SIRT3 affects cell metabolism and longevity is illustrated in Figure 3. As noted above (see Section 4), SIRT3 deaceylates and activates numerous mitochondrial proteins involved in diverse functions involving mitochondrial energy metabolism (including all the enzymes involved in the TCA cycle such as ACO-2, *etc.*), ETC members, antioxidant defenses, and mtDNA repair. Global loss of the *Sirt3* gene augments mitochondrial ROS levels, reduces OCR, NAD^+ levels, and ATP production, and promotes radiation-induced genotoxic stress response and a tumor permissive phenotype [124,126,127,180,181]. There is a negative correlation between ROS generated by complex 1 in the ETC and lifespan, and bypassing complex 1 increases lifespan in the fruit fly by mechanisms that are uncertain, but may involve altered sirtuins and cellular NAD^+ levels [178,182]. Given the "mitochondrial gatekeeper" role of SIRT3 for post-translational modification of mitochondrial proteins, not surprisingly SIRT3 is implicated in modulating mitochondrial metabolism. For example, Bass and colleagues demonstrated that circadian control of SIRT3 activity resulted in rhythms in the acetylation and activity of oxidative enzymes and respiration in isolated mitochondria, and that NAD^+ supplementation restored mitochondrial protein deacetylation and augmented OCR in circadian mutant mice [183]. Mitochondrial NAD^+ levels are also regulated by SIRT3 activity [182]. Lower levels of NAD^+ due to aging have also been found to produce a hypoxic-like state that inhibits intracellular signaling [184]. There is some evidence that simply replacing the cellular NAD^+ pool promotes recovery of cells from the aging metabolic phenotype [169]. PGC-1α-dependent SIRT3 signaling prevents complex 1 damage and preserves ETC function [185]. Increased PGC-1α levels have also been implicated in mitochondrial biogenesis, via a serine protease Omi-dependent mechanism [186]. Loss of SIRT3 can exacerbate the Parkinson's phenotype in mice through inhibition of complex 1, leading to excess ROS production and mtDNA damage, resulting in neuronal death [143].

Other SIRTs, especially SIRT1, may also have important roles in mitochondrial metabolism. Cells and tissues from old animals exhibit an increase of SIRT1 expression that is associated with decreased SIRT1 activity, NAD^+ levels, and ETC activity and augmented levels of nuclear p53, PARP, FOXO1, and oxidation of proteins, DNA, and lipids [178,187,188]. Despite SIRT3's key mitochondrial function, there is some evidence suggesting that its activity is highly dependent upon the activity of the nuclear-localized SIRT1 effects on RELB [173]. Additional studies are necessary to better understand the roles of SIRT3 and other sirtuins on AEC mitochondrial metabolism in the setting of oxidative stress as well as the translational significance in human lung diseases.

6. Conclusions

The available evidence reviewed herein convincingly shows that AEC mtDNA plays a key role in modulating mitochondrial function and apoptosis. Furthermore, AEC mtDNA damage following oxidative or environmental stress may be important in promoting pro-fibrotic signaling as can be seen in IPF, asbestosis, and other fibrotic lung diseases. A hypothetical model of AEC mtDNA damage in mediating intrinsic apoptosis that we reviewed is shown in Figure 2. Furthermore, we explored the data implicating that ongoing mtDNA damage may also be important in promoting other degenerative conditions, aging, and cancer (Figure 1). Emerging findings also suggest that SIRT3 and mitochondrial metabolism may be important in attenuating the deleterious effects of oxidative stress in a variety of cell types and murine models of degenerative diseases. Thus, the mitochondria clearly function not only as the "powerhouse" of the cell, but also regulate important cellular functions and signaling that determine cell life and death decisions. The role of OGG1, ACO-2, and small molecule SIRT3 inducers in modulating mtDNA damage in AEC as well as other cells and how this translates into disease pathophysiology await further study. We reason that the OGG1/ACO-2/SIRT3/mtDNA axis is important in regulating complex cell signaling that promotes pulmonary fibrosis as well as other degenerative disease of the lungs and tumor development.

Acknowledgments: This work was supported by VA Merit and NIH grant RO1 ES020357 to David W. Kamp, and NIH/NHLBI training grant 2T32HL076139-11A1 to Renea P. Jablonski.

Author Contributions: Seok-Jo Kim: Wrote sections of the manuscript, prepared figures and references, and edited the document. Paul Cheresh: Wrote sections of the manuscript. Renea P. Jablonski: Wrote sections of the manuscript. David B. Williams: Wrote sections of the manuscript. David W. Kamp: Wrote sections of the manuscript and edited the document.

Conflicts of Interest: The authors declare no conflict of interest.

References

1. Selman, M.; Pardo, A. Revealing the pathogenic and aging-related mechanisms of the enigmatic idiopathic pulmonary fibrosis. An integral model. *Am. J. Respir. Crit. Care Med.* **2014**, *189*, 1161–1172.
2. Trifunovic, A.; Wredenberg, A.; Falkenberg, M.; Spelbrink, J.N.; Rovio, A.T.; Bruder, C.E.; Bohlooly, Y.M.; Gidlof, S.; Oldfors, A.; Wibom, R.; *et al.* Premature ageing in mice expressing defective mitochondrial DNA polymerase. *Nature* **2004**, *429*, 417–423.

3.   Held, N.M.; Houtkooper, R.H. Mitochondrial quality control pathways as determinants of metabolic health. *Bioessays* **2015**, *37*, 867–876.

4.   Miller, F.J.; Rosenfeldt, F.L.; Zhang, C.; Linnane, A.W.; Nagley, P. Precise determination of mitochondrial DNA copy number in human skeletal and cardiac muscle by a pcr-based assay: Lack of change of copy number with age. *Nucleic Acids Res.* **2003**, *31*, e61.

5.   Bell, O.; Tiwari, V.K.; Thoma, N.H.; Schubeler, D. Determinants and dynamics of genome accessibility. *Nat. Rev. Genet.* **2011**, *12*, 554–564.

6.   Reeve, A.K.; Krishnan, K.J.; Turnbull, D. Mitochondrial DNA mutations in disease, aging, and neurodegeneration. *Ann. N. Y. Acad. Sci.* **2008**, *1147*, 21–29.

7.   Wallace, D.C. A mitochondrial bioenergetic etiology of disease. *J. Clin. Investig.* **2013**, *123*, 1405–1412.

8.   Kim, S.J.; Cheresh, P.; Williams, D.; Cheng, Y.; Ridge, K.; Schumacker, P.T.; Weitzman, S.; Bohr, V.A.; Kamp, D.W. Mitochondria-targeted Ogg1 and aconitase-2 prevent oxidant-induced mitochondrial DNA damage in alveolar epithelial cells. *J. Biol. Chem.* **2014**, *289*, 6165–6176.

9.   Panduri, V.; Liu, G.; Surapureddi, S.; Kondapalli, J.; Soberanes, S.; de Souza-Pinto, N.C.; Bohr, V.A.; Budinger, G.R.; Schumacker, P.T.; Weitzman, S.A.; *et al.* Role of mitochondrial hOGG1 and aconitase in oxidant-induced lung epithelial cell apoptosis. *Free Radic. Biol. Med.* **2009**, *47*, 750–759.

10. Nakabeppu, Y. Regulation of intracellular localization of human MTH1, OGG1, and MYH proteins for repair of oxidative DNA damage. *Prog. Nucleic Acid. Res. Mol. Biol.* **2001**, *68*, 75–94.

11. Vartanian, V.; Lowell, B.; Minko, I.G.; Wood, T.G.; Ceci, J.D.; George, S.; Ballinger, S.W.; Corless, C.L.; McCullough, A.K.; Lloyd, R.S. The metabolic syndrome resulting from a knockout of the NEIL1 DNA glycosylase. *Proc. Natl. Acad. Sci. USA* **2006**, *103*, 1864–1869.

12. Schumacker, P.T.; Gillespie, M.N.; Nakahira, K.; Choi, A.M.; Crouser, E.D.; Piantadosi, C.A.; Bhattacharya, J. Mitochondria in lung biology and pathology: More than just a powerhouse. *Am. J. Physiol. Lung. Cell Mol. Physiol.* **2014**, *306*, L962–L974.

13. Chouteau, J.M.; Obiako, B.; Gorodnya, O.M.; Pastukh, V.M.; Ruchko, M.V.; Wright, A.J.; Wilson, G.L.; Gillespie, M.N. Mitochondrial DNA integrity may be a determinant of endothelial barrier properties in oxidant-challenged rat lungs. *Am. J. Physiol. Lung Cell Mol. Physiol.* **2011**, *301*, L892–L898.

14. Cerritelli, S.M.; Frolova, E.G.; Feng, C.; Grinberg, A.; Love, P.E.; Crouch, R.J. Failure to produce mitochondrial DNA results in embryonic lethality in Rnaseh1 null mice. *Mol. Cell* **2003**, *11*, 807–815.

15. Simsek, D.; Furda, A.; Gao, Y.; Artus, J.; Brunet, E.; Hadjantonakis, A.K.; van Houten, B.; Shuman, S.; McKinnon, P.J.; Jasin, M. Crucial role for DNA ligase III in mitochondria but not in Xrcc1-dependent repair. *Nature* **2011**, *471*, 245–248.

16. Ballinger, S.W.; Patterson, C.; Yan, C.N.; Doan, R.; Burow, D.L.; Young, C.G.; Yakes, F.M.; van Houten, B.; Ballinger, C.A.; Freeman, B.A.; *et al.* Hydrogen peroxide- and peroxynitrite-induced mitochondrial DNA damage and dysfunction in vascular endothelial and smooth muscle cells. *Circ. Res.* **2000**, *86*, 960–966.

17. Yakes, F.M.; van Houten, B. Mitochondrial DNA damage is more extensive and persists longer than nuclear DNA damage in human cells following oxidative stress. *Proc. Natl. Acad. Sci. USA* **1997**, *94*, 514–519.

18. Bohr, V.A.; Stevnsner, T.; de Souza-Pinto, N.C. Mitochondrial DNA repair of oxidative damage in mammalian cells. *Gene* **2002**, *286*, 127–134.

19. Kroemer, G.; Galluzzi, L.; Brenner, C. Mitochondrial membrane permeabilization in cell death. *Physiol. Rev.* **2007**, *87*, 99–163.

20. Enns, G.M. The contribution of mitochondria to common disorders. *Mol. Genet. Metab.* **2003**, *80*, 11–26.

21. Brandon, M.; Baldi, P.; Wallace, D.C. Mitochondrial mutations in cancer. *Oncogene* **2006**, *25*, 4647–4662.

22. Cline, S.D. Mitochondrial DNA damage and its consequences for mitochondrial gene expression. *Biochim. Biophys. Acta* **2012**, *1819*, 979–991.

23. Tuppen, H.A.; Blakely, E.L.; Turnbull, D.M.; Taylor, R.W. Mitochondrial DNA mutations and human disease. *Biochim. Biophys. Acta* **2010**, *1797*, 113–128.

24. Van Houten, B.; Woshner, V.; Santos, J.H. Role of mitochondrial DNA in toxic responses to oxidative stress. *DNA Repair* **2006**, *5*, 145–152.

25. Figueira, T.R.; Barros, M.H.; Camargo, A.A.; Castilho, R.F.; Ferreira, J.C.; Kowaltowski, A.J.; Sluse, F.E.; Souza-Pinto, N.C.; Vercesi, A.E. Mitochondria as a source of reactive oxygen and nitrogen species: From molecular mechanisms to human health. *Antioxid. Redox Signal.* **2013**, *18*, 2029–2074.

26. Balaban, R.S.; Nemoto, S.; Finkel, T. Mitochondria, oxidants, and aging. *Cell* **2005**, *120*, 483–495.

27. Lin, M.T.; Beal, M.F. Mitochondrial dysfunction and oxidative stress in neurodegenerative diseases. *Nature* **2006**, *443*, 787–795.

28. Turrens, J.F. Mitochondrial formation of reactive oxygen species. *J. Physiol.* **2003**, *552*, 335–344.

29. Aaij, R.; Abellan Beteta, C.; Adeva, B.; Adinolfi, M.; Adrover, C.; Affolder, A.; Ajaltouni, Z.; Albrecht, J.; Alessio, F.; Alexander, M.; *et al.* First observation of *CP* violation in the decays of B_s^0 mesons. *Phys. Rev. Lett.* **2013**, *110*, 221601.

30. Kenney, M.C.; Chwa, M.; Atilano, S.R.; Falatoonzadeh, P.; Ramirez, C.; Malik, D.; Tarek, M.; Caceres-del-Carpio, J.; Nesburn, A.B.; Boyer, D.S.; *et al.* Inherited mitochondrial DNA variants can affect complement, inflammation and apoptosis pathways: Insights into mitochondrial-nuclear interactions. *Hum. Mol. Genet.* **2014**, *23*, 3537–3551.

31. Kamp, D.W.; Graceffa, P.; Pryor, W.A.; Weitzman, S.A. The role of free radicals in asbestos-induced diseases. *Free Radic. Biol. Med.* **1992**, *12*, 293–315.

32. Turci, F.; Tomatis, M.; Lesci, I.G.; Roveri, N.; Fubini, B. The iron-related molecular toxicity mechanism of synthetic asbestos nanofibres: A model study for high-aspect-ratio nanoparticles. *Chemistry* **2011**, *17*, 350–358.

33. Faner, R.; Rojas, M.; Macnee, W.; Agusti, A. Abnormal lung aging in chronic obstructive pulmonary disease and idiopathic pulmonary fibrosis. *Am. J. Respir. Crit. Care Med.* **2012**, *186*, 306–313.

34. Kliment, C.R.; Oury, T.D. Oxidative stress, extracellular matrix targets, and idiopathic pulmonary fibrosis. *Free Radic. Biol. Med.* **2010**, *49*, 707–717.

35. Cheresh, P.; Kim, S.J.; Tulasiram, S.; Kamp, D.W. Oxidative stress and pulmonary fibrosis. *Biochim. Biophys. Acta* **2013**, *1832*, 1028–1040.

36. Gazdhar, A.; Lebrecht, D.; Roth, M.; Tamm, M.; Venhoff, N.; Foocharoen, C.; Geiser, T.; Walker, U.A. Time-dependent and somatically acquired mitochondrial DNA mutagenesis and respiratory chain dysfunction in a scleroderma model of lung fibrosis. *Sci. Rep.* **2014**, *4*, 5336.

37. Liu, G.; Cheresh, P.; Kamp, D.W. Molecular basis of asbestos-induced lung disease. *Annu. Rev. Pathol.* **2013**, *8*, 161–187.

38. Oury, T.D.; Thakker, K.; Menache, M.; Chang, L.Y.; Crapo, J.D.; Day, B.J. Attenuation of bleomycin-induced pulmonary fibrosis by a catalytic antioxidant metalloporphyrin. *Am. J. Respir. Cell Mol. Biol.* **2001**, *25*, 164–169.

39. Osborn-Heaford, H.L.; Ryan, A.J.; Murthy, S.; Racila, A.M.; He, C.; Sieren, J.C.; Spitz, D.R.; Carter, A.B. Mitochondrial Rac1 GTPase import and electron transfer from cytochrome c are required for pulmonary fibrosis. *J. Biol. Chem.* **2012**, *287*, 3301–3312.

40. Murthy, S.; Ryan, A.; He, C.; Mallampalli, R.K.; Carter, A.B. Rac1-mediated mitochondrial H_2O_2 generation regulates *MMP-9* gene expression in macrophages via inhibition of SP-1 and AP-1. *J. Biol. Chem.* **2010**, *285*, 25062–25073.

41. Murthy, S.; Adamcakova-Dodd, A.; Perry, S.S.; Tephly, L.A.; Keller, R.M.; Metwali, N.; Meyerholz, D.K.; Wang, Y.; Glogauer, M.; Thorne, P.S.; *et al.* Modulation of reactive oxygen species by rac1 or catalase prevents asbestos-induced pulmonary fibrosis. *Am. J. Physiol. Lung Cell Mol. Physiol.* **2009**, *297*, L846–L855.

42. He, C.; Murthy, S.; McCormick, M.L.; Spitz, D.R.; Ryan, A.J.; Carter, A.B. Mitochondrial Cu, Zn-superoxide dismutase mediates pulmonary fibrosis by augmenting H_2O_2 generation. *J. Biol. Chem.* **2011**, *286*, 15597–15607.

43. Mossman, B.T.; Marsh, J.P.; Sesko, A.; Hill, S.; Shatos, M.A.; Doherty, J.; Petruska, J.; Adler, K.B.; Hemenway, D.; Mickey, R.; *et al.* Inhibition of lung injury, inflammation, and interstitial pulmonary fibrosis by polyethylene glycol-conjugated catalase in a rapid inhalation model of asbestosis. *Am. Rev. Respir. Dis.* **1990**, *141*, 1266–1271.

44. Gao, F.; Kinnula, V.L.; Myllarniemi, M.; Oury, T.D. Extracellular superoxide dismutase in pulmonary fibrosis. *Antioxid. Redox Signal.* **2008**, *10*, 343–354.

45. Cantin, A.M.; Hubbard, R.C.; Crystal, R.G. Glutathione deficiency in the epithelial lining fluid of the lower respiratory tract in idiopathic pulmonary fibrosis. *Am. Rev. Respir. Dis.* **1989**, *139*, 370–372.

46. Waghray, M.; Cui, Z.; Horowitz, J.C.; Subramanian, I.M.; Martinez, F.J.; Toews, G.B.; Thannickal, V.J. Hydrogen peroxide is a diffusible paracrine signal for the induction of epithelial cell death by activated myofibroblasts. *FASEB J.* **2005**, *19*, 854–856.

47. Hagiwara, S.I.; Ishii, Y.; Kitamura, S. Aerosolized administration of *N*-acetylcysteine attenuates lung fibrosis induced by bleomycin in mice. *Am. J. Respir. Crit. Care Med.* **2000**, *162*, 225–231.

48. Martinez, F.J.; de Andrade, J.A.; Anstrom, K.J.; King, T.E., Jr.; Raghu, G. Randomized trial of acetylcysteine in idiopathic pulmonary fibrosis. *N. Engl. J. Med.* **2014**, *370*, 2093–2101.

49. Crestani, B.; Besnard, V.; Boczkowski, J. Signalling pathways from NADPH oxidase-4 to idiopathic pulmonary fibrosis. *Int. J. Biochem. Cell Biol.* **2011**, *43*, 1086–1089.

50. Hecker, L.; Cheng, J.; Thannickal, V.J. Targeting nox enzymes in pulmonary fibrosis. *Cell Mol. Life Sci.* **2012**, *69*, 2365–2371.

51. Hecker, L.; Vittal, R.; Jones, T.; Jagirdar, R.; Luckhardt, T.R.; Horowitz, J.C.; Pennathur, S.; Martinez, F.J.; Thannickal, V.J. NADPH oxidase-4 mediates myofibroblast activation and fibrogenic responses to lung injury. *Nat. Med.* **2009**, *15*, 1077–1081.

52. Carnesecchi, S.; Deffert, C.; Donati, Y.; Basset, O.; Hinz, B.; Preynat-Seauve, O.; Guichard, C.; Arbiser, J.L.; Banfi, B.; Pache, J.C.; *et al.* A key role for NOX4 in epithelial cell death during development of lung fibrosis. *Antioxid. Redox Signal.* **2011**, *15*, 607–619.

53. Hecker, L.; Logsdon, N.J.; Kurundkar, D.; Kurundkar, A.; Bernard, K.; Hock, T.; Meldrum, E.; Sanders, Y.Y.; Thannickal, V.J. Reversal of persistent fibrosis in aging by targeting Nox4-Nrf2 redox imbalance. *Sci. Transl. Med.* **2014**, *6*, 231ra247.

54. Thannickal, V.J. Mechanistic links between aging and lung fibrosis. *Biogerontology* **2013**, *14*, 609–615.

55. Lopez-Otin, C.; Blasco, M.A.; Partridge, L.; Serrano, M.; Kroemer, G. The hallmarks of aging. *Cell* **2013**, *153*, 1194–1217.

56. Uhal, B.D.; Nguyen, H. The Witschi Hypothesis revisited after 35 years: Genetic proof from SP-C BRICHOS domain mutations. *Am. J. Physiol. Lung Cell Mol. Physiol.* **2013**, *305*, L906–L911.

57. Mossman, B.T.; Lippmann, M.; Hesterberg, T.W.; Kelsey, K.T.; Barchowsky, A.; Bonner, J.C. Pulmonary endpoints (lung carcinomas and asbestosis) following inhalation exposure to asbestos. *J. Toxicol. Environ. Health B Crit. Rev.* **2011**, *14*, 76–121.

58. Noble, P.W.; Barkauskas, C.E.; Jiang, D. Pulmonary fibrosis: Patterns and perpetrators. *J. Clin. Investig.* **2012**, *122*, 2756–2762.

59. Tanjore, H.; Blackwell, T.S.; Lawson, W.E. Emerging evidence for endoplasmic reticulum stress in the pathogenesis of idiopathic pulmonary fibrosis. *Am. J. Physiol. Lung Cell Mol. Physiol.* **2012**, *302*, L721–L729.

60. Weiss, C.H.; Budinger, G.R.; Mutlu, G.M.; Jain, M. Proteasomal regulation of pulmonary fibrosis. *Proc. Am. Thorac. Soc.* **2010**, *7*, 77–83.

61. Brody, A.R.; Overby, L.H. Incorporation of tritiated thymidine by epithelial and interstitial cells in bronchiolar-alveolar regions of asbestos-exposed rats. *Am. J. Pathol.* **1989**, *134*, 133–140.

62. Shukla, A.; Jung, M.; Stern, M.; Fukagawa, N.K.; Taatjes, D.J.; Sawyer, D.; van Houten, B.; Mossman, B.T. Asbestos induces mitochondrial DNA damage and dysfunction linked to the development of apoptosis. *Am. J. Physiol. Lung Cell Mol. Physiol.* **2003**, *285*, L1018–L1025.

63. Fingerlin, T.E.; Murphy, E.; Zhang, W.; Peljto, A.L.; Brown, K.K.; Steele, M.P.; Loyd, J.E.; Cosgrove, G.P.; Lynch, D.; Groshong, S.; *et al.* Genome-wide association study identifies multiple susceptibility loci for pulmonary fibrosis. *Nat. Genet.* **2013**, *45*, 613–620.

64. Kropski, J.A.; Pritchett, J.M.; Zoz, D.F.; Crossno, P.F.; Markin, C.; Garnett, E.T.; Degryse, A.L.; Mitchell, D.B.; Polosukhin, V.V.; Rickman, O.B.; *et al.* Extensive phenotyping of individuals at risk for familial interstitial pneumonia reveals clues to the pathogenesis of interstitial lung disease. *Am. J. Respir. Crit. Care Med.* **2015**, *191*, 417–426.

65. Cogan, J.D.; Kropski, J.A.; Zhao, M.; Mitchell, D.B.; Rives, L.; Markin, C.; Garnett, E.T.; Montgomery, K.H.; Mason, W.R.; McKean, D.F.; *et al.* Rare variants in RTEL1 are associated with familial interstitial pneumonia. *Am. J. Respir. Crit. Care Med.* **2015**, *191*, 646–655.

66. Wallace, D.C.; Chalkia, D. Mitochondrial DNA genetics and the heteroplasmy conundrum in evolution and disease. *Cold Spring Harb. Perspect. Med.* **2013**, *5*, a021220.

67. Galluzzi, L.; Vitale, I.; Abrams, J.M.; Alnemri, E.S.; Baehrecke, E.H.; Blagosklonny, M.V.; Dawson, T.M.; Dawson, V.L.; El-Deiry, W.S.; Fulda, S.; *et al.* Molecular definitions of cell death subroutines: Recommendations of the nomenclature committee on cell death 2012. *Cell Death Differ.* **2012**, *19*, 107–120.

68. Huang, S.X.; Jaurand, M.C.; Kamp, D.W.; Whysner, J.; Hei, T.K. Role of mutagenicity in asbestos fiber-induced carcinogenicity and other diseases. *J. Toxicol. Environ. Health B Crit. Rev.* **2011**, *14*, 179–245.

69. Anathy, V.; Roberson, E.C.; Guala, A.S.; Godburn, K.E.; Budd, R.C.; Janssen-Heininger, Y.M. Redox-based regulation of apoptosis: S-glutathionylation as a regulatory mechanism to control cell death. *Antioxid. Redox Signal.* **2012**, *16*, 496–505.

70. Anathy, V.; Roberson, E.; Cunniff, B.; Nolin, J.D.; Hoffman, S.; Spiess, P.; Guala, A.S.; Lahue, K.G.; Goldman, D.; Flemer, S.; *et al.* Oxidative processing of latent fas in the endoplasmic reticulum controls the strength of apoptosis. *Mol. Cell. Biol.* **2012**, *32*, 3464–3478.

71. Horan, G.S.; Wood, S.; Ona, V.; Li, D.J.; Lukashev, M.E.; Weinreb, P.H.; Simon, K.J.; Hahm, K.; Allaire, N.E.; Rinaldi, N.J.; *et al.* Partial inhibition of integrin αvβ6 prevents pulmonary fibrosis without exacerbating inflammation. *Am. J. Respir. Crit. Care Med.* **2008**, *177*, 56–65.

72. Armanios, M. Telomerase and idiopathic pulmonary fibrosis. *Mutat. Res.* **2012**, *730*, 52–58.

73. Alder, J.K.; Chen, J.J.; Lancaster, L.; Danoff, S.; Su, S.C.; Cogan, J.D.; Vulto, I.; Xie, M.; Qi, X.; Tuder, R.M.; *et al.* Short telomeres are a risk factor for idiopathic pulmonary fibrosis. *Proc. Natl. Acad. Sci. USA* **2008**, *105*, 13051–13056.

74. Passos, J.F.; Saretzki, G.; Ahmed, S.; Nelson, G.; Richter, T.; Peters, H.; Wappler, I.; Birket, M.J.; Harold, G.; Schaeuble, K.; *et al.* Mitochondrial dysfunction accounts for the stochastic heterogeneity in telomere-dependent senescence. *PLoS Biol.* **2007**, *5*, e110.

75. Degryse, A.L.; Xu, X.C.; Newman, J.L.; Mitchell, D.B.; Tanjore, H.; Polosukhin, V.V.; Jones, B.R.; McMahon, F.B.; Gleaves, L.A.; Phillips, J.A., 3rd; *et al.* Telomerase deficiency does not alter bleomycin-induced fibrosis in mice. *Exp. Lung Res.* **2012**, *38*, 124–134.

76. Santos, J.H.; Hunakova, L.; Chen, Y.; Bortner, C.; van Houten, B. Cell sorting experiments link persistent mitochondrial DNA damage with loss of mitochondrial membrane potential and apoptotic cell death. *J. Biol. Chem.* **2003**, *278*, 1728–1734.

77. Budinger, G.R.; Mutlu, G.M.; Eisenbart, J.; Fuller, A.C.; Bellmeyer, A.A.; Baker, C.M.; Wilson, M.; Ridge, K.; Barrett, T.A.; Lee, V.Y.; *et al.* Proapoptotic bid is required for pulmonary fibrosis. *Proc. Natl. Acad. Sci. USA.* **2006**, *103*, 4604–4609.

78. Bueno, M.; Lai, Y.C.; Romero, Y.; Brands, J.; St Croix, C.M.; Kamga, C.; Corey, C.; Herazo-Maya, J.D.; Sembrat, J.; Lee, J.S.; *et al.* Pink1 deficiency impairs mitochondrial homeostasis and promotes lung fibrosis. *J. Clin. Investig.* **2015**, *125*, 521–538.

79. Patel, A.S.; Song, J.W.; Chu, S.G.; Mizumura, K.; Osorio, J.C.; Shi, Y.; El-Chemaly, S.; Lee, C.G.; Rosas, I.O.; Elias, J.A.; *et al.* Epithelial cell mitochondrial dysfunction and pink1 are induced by transforming growth factor-beta1 in pulmonary fibrosis. *PLoS ONE* **2015**, *10*, e0121246.

80. Lounsbury, K.M.; Stern, M.; Taatjes, D.; Jaken, S.; Mossman, B.T. Increased localization and substrate activation of protein kinase Cδ in lung epithelial cells following exposure to asbestos. *Am. J. Pathol.* **2002**, *160*, 1991–2000.

81. Buder-Hoffmann, S.A.; Shukla, A.; Barrett, T.F.; MacPherson, M.B.; Lounsbury, K.M.; Mossman, B.T. A protein kinase Cδ-dependent protein kinase D pathway modulates ERK1/2 and JNK1/2 phosphorylation and Bim-associated apoptosis by asbestos. *Am. J. Pathol.* **2009**, *174*, 449–459.

82. Corral-Debrinski, M.; Horton, T.; Lott, M.T.; Shoffner, J.M.; Beal, M.F.; Wallace, D.C. Mitochondrial DNA deletions in human brain: Regional variability and increase with advanced age. *Nat. Genet.* **1992**, *2*, 324–329.

83. Michikawa, Y.; Mazzucchelli, F.; Bresolin, N.; Scarlato, G.; Attardi, G. Aging-dependent large accumulation of point mutations in the human mtDNA control region for replication. *Science* **1999**, *286*, 774–779.

84. Kujoth, G.C.; Hiona, A.; Pugh, T.D.; Someya, S.; Panzer, K.; Wohlgemuth, S.E.; Hofer, T.; Seo, A.Y.; Sullivan, R.; Jobling, W.A.; *et al.* Mitochondrial DNA mutations, oxidative stress, and apoptosis in mammalian aging. *Science* **2005**, *309*, 481–484.

85. DeBerardinis, R.J.; Mancuso, A.; Daikhin, E.; Nissim, I.; Yudkoff, M.; Wehrli, S.; Thompson, C.B. Beyond aerobic glycolysis: Transformed cells can engage in glutamine metabolism that exceeds the requirement for protein and nucleotide synthesis. *Proc. Natl. Acad. Sci. USA* **2007**, *104*, 19345–19350.

86. Vander Heiden, M.G.; Cantley, L.C.; Thompson, C.B. Understanding the warburg effect: The metabolic requirements of cell proliferation. *Science* **2009**, *324*, 1029–1033.

87. Fliss, M.S.; Usadel, H.; Caballero, O.L.; Wu, L.; Buta, M.R.; Eleff, S.M.; Jen, J.; Sidransky, D. Facile detection of mitochondrial DNA mutations in tumors and bodily fluids. *Science* **2000**, *287*, 2017–2019.

88. Wang, Z.; Choi, S.; Lee, J.; Huang, Y.T.; Chen, F.; Zhao, Y.; Lin, X.; Neuberg, D.; Kim, J.; Christiani, D.C. Mitochondrial variations in non-small cell lung cancer (NSCLC) survival. *Cancer Inform.* **2015**, *14*, 1–9.

89. Ishikawa, K.; Takenaga, K.; Akimoto, M.; Koshikawa, N.; Yamaguchi, A.; Imanishi, H.; Nakada, K.; Honma, Y.; Hayashi, J. Ros-generating mitochondrial DNA mutations can regulate tumor cell metastasis. *Science* **2008**, *320*, 661–664.

90. King, M.P.; Attardi, G. Injection of mitochondria into human cells leads to a rapid replacement of the endogenous mitochondrial DNA. *Cell* **1988**, *52*, 811–819.

91. Tan, A.S.; Baty, J.W.; Dong, L.F.; Bezawork-Geleta, A.; Endaya, B.; Goodwin, J.; Bajzikova, M.; Kovarova, J.; Peterka, M.; Yan, B.; *et al.* Mitochondrial genome acquisition restores respiratory function and tumorigenic potential of cancer cells without mitochondrial DNA. *Cell Metab.* **2015**, *21*, 81–94.

92. Chevillard, S.; Radicella, J.P.; Levalois, C.; Lebeau, J.; Poupon, M.F.; Oudard, S.; Dutrillaux, B.; Boiteux, S. Mutations in *OGG1*, a gene involved in the repair of oxidative DNA damage, are found in human lung and kidney tumours. *Oncogene* **1998**, *16*, 3083–3086.

93. Youn, C.K.; Song, P.I.; Kim, M.H.; Kim, J.S.; Hyun, J.W.; Choi, S.J.; Yoon, S.P.; Chung, M.H.; Chang, I.Y.; You, H.J. Human 8-oxoguanine DNA glycosylase suppresses the oxidative stress induced apoptosis through a p53-mediated signaling pathway in human fibroblasts. *Mol. Cancer Res.* **2007**, *5*, 1083–1098.

94. Ruchko, M.; Gorodnya, O.; LeDoux, S.P.; Alexeyev, M.F.; Al-Mehdi, A.B.; Gillespie, M.N. Mitochondrial DNA damage triggers mitochondrial dysfunction and apoptosis in oxidant-challenged lung endothelial cells. *Am. J. Physiol. Lung Cell Mol. Physiol.* **2005**, *288*, L530–L535. ·

95. Hashiguchi, K.; Stuart, J.A.; de Souza-Pinto, N.C.; Bohr, V.A. The C-terminal aO helix of human Ogg1 is essential for 8-oxoguanine DNA glycosylase activity: The mitochondrial β-Ogg1 lacks this domain and does not have glycosylase activity. *Nucleic Acids Res.* **2004**, *32*, 5596–5608.

96. Bulteau, A.L.; Ikeda-Saito, M.; Szweda, L.I. Redox-dependent modulation of aconitase activity in intact mitochondria. *Biochemistry* **2003**, *42*, 14846–14855.

97. Chen, X.J.; Wang, X.; Kaufman, B.A.; Butow, R.A. Aconitase couples metabolic regulation to mitochondrial DNA maintenance. *Science* **2005**, *307*, 714–717.

98. Cheresh, P.; Morales-Nebreda, L.; Kim, S.J.; Yeldandi, A.; Williams, D.B.; Cheng, Y.; Mutlu, G.M.; Budinger, G.R.; Ridge, K.; Schumacker, P.T.; *et al.* Asbestos-induced pulmonary fibrosis is augmented in 8-oxoguanine DNA glycosylase knockout mice. *Am. J. Respir. Cell Mol. Biol.* **2015**, *52*, 25–36.

99. Park, L.C.; Albers, D.S.; Xu, H.; Lindsay, J.G.; Beal, M.F.; Gibson, G.E. Mitochondrial impairment in the cerebellum of the patients with progressive supranuclear palsy. *J. Neurosci. Res.* **2001**, *66*, 1028–1034.

100. Przybylowska, K.; Kabzinski, J.; Sygut, A.; Dziki, L.; Dziki, A.; Majsterek, I. An association selected polymorphisms of *XRCC1*, *OGG1* and *MUTYH* gene and the level of efficiency oxidative DNA damage repair with a risk of colorectal cancer. *Mutat. Res.* **2013**, *745–746*, 6–15.

101. Duan, W.X.; Hua, R.X.; Yi, W.; Shen, L.J.; Jin, Z.X.; Zhao, Y.H.; Yi, D.H.; Chen, W.S.; Yu, S.Q. The association between *OGG1* Ser326Cys polymorphism and lung cancer susceptibility: A meta-analysis of 27 studies. *PLoS ONE* **2012**, *7*, e35970.

102. Elahi, A.; Zheng, Z.; Park, J.; Eyring, K.; McCaffrey, T.; Lazarus, P. The human OGG1 DNA repair enzyme and its association with orolaryngeal cancer risk. *Carcinogenesis* **2002**, *23*, 1229–1234.

103. Wang, J.; Wang, Q.; Watson, L.J.; Jones, S.P.; Epstein, P.N. Cardiac overexpression of 8-oxoguanine DNA glycosylase 1 protects mitochondrial DNA and reduces cardiac fibrosis following transaortic constriction. *Am. J. Physiol. Heart Circ. Physiol.* **2011**, *301*, H2073–H2080.

104. Tsutsui, H. Oxidative stress in heart failure: The role of mitochondria. *Intern. Med.* **2001**, *40*, 1177–1182.

105. Ding, Z.; Liu, S.; Wang, X.; Khaidakov, M.; Dai, Y.; Mehta, J.L. Oxidant stress in mitochondrial DNA damage, autophagy and inflammation in atherosclerosis. *Sci. Rep.* **2013**, *3*, 1077.

106. Picard, M.; Jung, B.; Liang, F.; Azuelos, I.; Hussain, S.; Goldberg, P.; Godin, R.; Danialou, G.; Chaturvedi, R.; Rygiel, K.; *et al.* Mitochondrial dysfunction and lipid accumulation in the human diaphragm during mechanical ventilation. *Am. J. Respir. Crit. Care Med.* **2012**, *186*, 1140–1149.

107. Hashizume, M.; Mouner, M.; Chouteau, J.M.; Gorodnya, O.M.; Ruchko, M.V.; Potter, B.J.; Wilson, G.L.; Gillespie, M.N.; Parker, J.C. Mitochondrial-targeted DNA repair enzyme 8-oxoguanine DNA glycosylase 1 protects against ventilator-induced lung injury in intact mice. *Am. J. Physiol. Lung Cell Mol. Physiol.* **2013**, *304*, L287–L297.

108. Tsui, K.H.; Feng, T.H.; Lin, Y.F.; Chang, P.L.; Juang, H.H. P53 downregulates the gene expression of mitochondrial aconitase in human prostate carcinoma cells. *Prostate* **2011**, *71*, 62–70.

109. Koczor, C.A.; Torres, R.A.; Fields, E.J.; Boyd, A.; Lewis, W. Mitochondrial matrix P53 sensitizes cells to oxidative stress. *Mitochondrion* **2013**, *13*, 277–281.

110. Marudamuthu, A.S.; Bhandary, Y.P.; Shetty, S.K.; Fu, J.; Sathish, V.; Prakash, Y.; Shetty, S. Role of the urokinase-fibrinolytic system in epithelial-mesenchymal transition during lung injury. *Am. J. Pathol.* **2015**, *185*, 55–68.

111. Marudamuthu, A.S.; Shetty, S.K.; Bhandary, Y.P.; Karandashova, S.; Thompson, M.; Sathish, V.; Florova, G.; Hogan, T.B.; Pabelick, C.M.; Prakash, Y.S.; *et al.* Plasminogen activator inhibitor-1 suppresses profibrotic responses in fibroblasts from fibrotic lungs. *J. Biol. Chem.* **2015**, *290*, 9428–9441.

112. Bhandary, Y.P.; Shetty, S.K.; Marudamuthu, A.S.; Fu, J.; Pinson, B.M.; Levin, J.; Shetty, S. Role of p53-fibrinolytic system cross-talk in the regulation of quartz-induced lung injury. *Toxicol. Appl. Pharmacol.* **2015**, *283*, 92–98.

113. Huang, W.T.; Akhter, H.; Jiang, C.; MacEwen, M.; Ding, Q.; Antony, V.; Thannickal, V.J.; Liu, R.M. Plasminogen activator inhibitor 1, fibroblast apoptosis resistance, and aging-related susceptibility to lung fibrosis. *Exp. Gerontol.* **2015**, *61*, 62–75.

114. Cassel, S.L.; Eisenbarth, S.C.; Iyer, S.S.; Sadler, J.J.; Colegio, O.R.; Tephly, L.A.; Carter, A.B.; Rothman, P.B.; Flavell, R.A.; Sutterwala, F.S. The Nalp3 inflammasome is essential for the development of silicosis. *Proc. Natl. Acad. Sci. USA* **2008**, *105*, 9035–9040.

115. Zhou, R.; Yazdi, A.S.; Menu, P.; Tschopp, J. A role for mitochondria in Nlrp3 inflammasome activation. *Nature* **2011**, *469*, 221–225.

116. Kirillov, V.; Siler, J.T.; Ramadass, M.; Ge, L.; Davis, J.; Grant, G.; Nathan, S.D.; Jarai, G.; Trujillo, G. Sustained activation of toll-like receptor 9 induces an invasive phenotype in lung fibroblasts: Possible implications in idiopathic pulmonary fibrosis. *Am. J. Pathol.* **2015**, *185*, 943–957.

117. Gu, X.; Wu, G.; Yao, Y.; Zeng, J.; Shi, D.; Lv, T.; Luo, L.; Song, Y. Intratracheal administration of mitochondrial DNA directly provokes lung inflammation through the TLR9-p38 mapk pathway. *Free Radic. Biol. Med.* **2015**, *83*, 149–158.

118. Kuck, J.L.; Obiako, B.O.; Gorodnya, O.M.; Pastukh, V.M.; Kua, J.; Simmons, J.D.; Gillespie, M.N. Mitochondrial DNA damage-associated molecular patterns mediate a feed-forward cycle of bacteria-induced vascular injury in perfused rat lungs. *Am. J. Physiol. Lung Cell Mol. Physiol.* **2015**, *308*, L1078–L1085.

119. Dos Santos, G.; Rogel, M.R.; Baker, M.A.; Troken, J.R.; Urich, D.; Morales-Nebreda, L.; Sennello, J.A.; Kutuzov, M.A.; Sitikov, A.; Davis, J.M.; *et al.* Vimentin regulates activation of the Nlrp3 inflammasome. *Nat. Commun.* **2015**, *6*, 6574.

120. Vaquero, A. The conserved role of sirtuins in chromatin regulation. *Int. J. Dev. Biol.* **2009**, *53*, 303–322.

121. Verdin, E.; Hirschey, M.D.; Finley, L.W.; Haigis, M.C. Sirtuin regulation of mitochondria: Energy production, apoptosis, and signaling. *Trends Biochem. Sci.* **2010**, *35*, 669–675.

122. Bosch-Presegue, L.; Vaquero, A. Sirtuins in stress response: Guardians of the genome. *Oncogene* **2014**, *33*, 3764–3775.

123. Kaniak-Golik, A.; Skoneczna, A. Mitochondria-nucleus network for genome stability. *Free Radic. Biol. Med.* **2015**, *82*, 73–104.

124. Kincaid, B.; Bossy-Wetzel, E. Forever young: SIRT3 a shield against mitochondrial meltdown, aging, and neurodegeneration. *Front. Aging Neurosci.* **2013**, *5*, 48.

125. Bellizzi, D.; Rose, G.; Cavalcante, P.; Covello, G.; Dato, S.; de Rango, F.; Greco, V.; Maggiolini, M.; Feraco, E.; Mari, V.; *et al.* A novel vntr enhancer within the *SIRT3* gene, a human homologue of *SIR2*, is associated with survival at oldest ages. *Genomics* **2005**, *85*, 258–263.

126. Cheng, Y.; Ren, X.; Gowda, A.S.; Shan, Y.; Zhang, L.; Yuan, Y.S.; Patel, R.; Wu, H.; Huber-Keener, K.; Yang, J.W.; *et al.* Interaction of Sirt3 with OGG1 contributes to repair of mitochondrial DNA and protects from apoptotic cell death under oxidative stress. *Cell Death Dis.* **2013**, *4*, e731.

127. Chen, Y.; Fu, L.L.; Wen, X.; Wang, X.Y.; Liu, J.; Cheng, Y.; Huang, J. Sirtuin-3 (sirt3), a therapeutic target with oncogenic and tumor-suppressive function in cancer. *Cell Death Dis.* **2014**, *5*, e1047.

128. Shulga, N.; Pastorino, J.G. Ethanol sensitizes mitochondria to the permeability transition by inhibiting deacetylation of cyclophilin-d mediated by sirtuin-3. *J. Cell Sci.* **2010**, *123*, 4117–4127.

129. Sundaresan, N.R.; Samant, S.A.; Pillai, V.B.; Rajamohan, S.B.; Gupta, M.P. SIRT3 is a stress-responsive deacetylase in cardiomyocytes that protects cells from stress-mediated cell death by deacetylation of Ku70. *Mol. Cell. Biol.* **2008**, *28*, 6384–6401.

130. Li, S.; Banck, M.; Mujtaba, S.; Zhou, M.M.; Sugrue, M.M.; Walsh, M.J. P53-induced growth arrest is regulated by the mitochondrial sirt3 deacetylase. *PLoS ONE* **2010**, *5*, e10486.

131. Kawamura, Y.; Uchijima, Y.; Horike, N.; Tonami, K.; Nishiyama, K.; Amano, T.; Asano, T.; Kurihara, Y.; Kurihara, H. Sirt3 protects *in vitro*-fertilized mouse preimplantation embryos against oxidative stress-induced p53-mediated developmental arrest. *J. Clin. Investig.* **2010**, *120*, 2817–2828.

132. D'Aquila, P.; Rose, G.; Panno, M.L.; Passarino, G.; Bellizzi, D. *SIRT3* gene expression: A link between inherited mitochondrial DNA variants and oxidative stress. *Gene* **2012**, *497*, 323–329.

133. Wu, Y.T.; Lee, H.C.; Liao, C.C.; Wei, Y.H. Regulation of mitochondrial F_0F_1ATPase activity by Sirt3-catalyzed deacetylation and its deficiency in human cells harboring 4977bp deletion of mitochondrial DNA. *Biochim. Biophys. Acta* **2013**, *1832*, 216–227.

134. Bause, A.S.; Matsui, M.S.; Haigis, M.C. The protein deacetylase SIRT3 prevents oxidative stress-induced keratinocyte differentiation. *J. Biol. Chem.* **2013**, *288*, 36484–36491.

135. Shadel, G.S. Mitochondrial DNA, aconitase "wraps" it up. *Trends Biochem. Sci.* **2005**, *30*, 294–296.

136. Gibellini, L.; Pinti, M.; Beretti, F.; Pierri, C.L.; Onofrio, A.; Riccio, M.; Carnevale, G.; de Biasi, S.; Nasi, M.; Torelli, F.; *et al.* Sirtuin 3 interacts with lon protease and regulates its acetylation status. *Mitochondrion* **2014**, *18*, 76–81.

137. Sundaresan, N.R.; Gupta, M.; Kim, G.; Rajamohan, S.B.; Isbatan, A.; Gupta, M.P. Sirt3 blocks the cardiac hypertrophic response by augmenting Foxo3a-dependent antioxidant defense mechanisms in mice. *J. Clin. Investig.* **2009**, *119*, 2758–2771.

138. Someya, S.; Xu, J.; Kondo, K.; Ding, D.; Salvi, R.J.; Yamasoba, T.; Rabinovitch, P.S.; Weindruch, R.; Leeuwenburgh, C.; Tanokura, M.; *et al.* Age-related hearing loss in C57BL/6J mice is mediated by Bak-dependent mitochondrial apoptosis. *Proc. Natl. Acad. Sci. USA* **2009**, *106*, 19432–19437.

139. Someya, S.; Yu, W.; Hallows, W.C.; Xu, J.; Vann, J.M.; Leeuwenburgh, C.; Tanokura, M.; Denu, J.M.; Prolla, T.A. Sirt3 mediates reduction of oxidative damage and prevention of age-related hearing loss under caloric restriction. *Cell* **2010**, *143*, 802–812.

140. Coleman, M.C.; Olivier, A.K.; Jacobus, J.A.; Mapuskar, K.A.; Mao, G.; Martin, S.M.; Riley, D.P.; Gius, D.; Spitz, D.R. Superoxide mediates acute liver injury in irradiated mice lacking sirtuin 3. *Antioxid. Redox Signal.* **2014**, *20*, 1423–1435.

141. Morigi, M.; Perico, L.; Rota, C.; Longaretti, L.; Conti, S.; Rottoli, D.; Novelli, R.; Remuzzi, G.; Benigni, A. Sirtuin 3-dependent mitochondrial dynamic improvements protect against acute kidney injury. *J. Clin. Investig.* **2015**, *125*, 715–726.

142. Chen, T.; Li, J.; Liu, J.; Li, N.; Wang, S.; Liu, H.; Zeng, M.; Zhang, Y.; Bu, P. Activation of SIRT3 by resveratrol ameliorates cardiac fibrosis and improves cardiac function via the TGF-β/Smad3 pathway. *Am. J. Physiol. Heart Circ. Physiol.* **2015**, *308*, H424–H434.

143. Liu, L.; Peritore, C.; Ginsberg, J.; Kayhan, M.; Donmez, G. SIRT3 attenuates MPTP-induced nigrostriatal degeneration via enhancing mitochondrial antioxidant capacity. *Neurochem. Res.* **2015**, *40*, 600–608.

144. Wang, X.Q.; Shao, Y.; Ma, C.Y.; Chen, W.; Sun, L.; Liu, W.; Zhang, D.Y.; Fu, B.C.; Liu, K.Y.; Jia, Z.B.; *et al.* Decreased SIRT3 in aged human mesenchymal stromal/stem cells increases cellular susceptibility to oxidative stress. *J. Cell Mol. Med.* **2014**, *18*, 2298–2310.

145. Huang, K.; Kang, X.; Wang, X.; Wu, S.; Xiao, J.; Li, Z.; Wu, X.; Zhang, W. Conversion of bone marrow mesenchymal stem cells into type II alveolar epithelial cells reduces pulmonary fibrosis by decreasing oxidative stress in rats. *Mol. Med. Rep.* **2015**, *11*, 1685–1692.

146. Johnson, M.L.; Irving, B.A.; Lanza, I.R.; Vendelbo, M.H.; Konopka, A.R.; Robinson, M.M.; Henderson, G.C.; Klaus, K.A.; Morse, D.M.; Heppelmann, C.; *et al.* Differential effect of endurance training on mitochondrial protein damage, degradation, and acetylation in the context of aging. *J. Gerontol. A Biol. Sci. Med. Sci.* **2014**.

147. Dai, S.H.; Chen, T.; Wang, Y.H.; Zhu, J.; Luo, P.; Rao, W.; Yang, Y.F.; Fei, Z.; Jiang, X.F. Sirt3 protects cortical neurons against oxidative stress via regulating mitochondrial Ca^{2+} and mitochondrial biogenesis. *Int. J. Mol. Sci.* **2014**, *15*, 14591–14609.

148. Rato, L.; Duarte, A.I.; Tomas, G.D.; Santos, M.S.; Moreira, P.I.; Socorro, S.; Cavaco, J.E.; Alves, M.G.; Oliveira, P.F. Pre-diabetes alters testicular PGC1-α/SIRT3 axis modulating mitochondrial bioenergetics and oxidative stress. *Biochim. Biophys. Acta* **2014**, *1837*, 335–344.

149. Kim, M.; Lee, J.S.; Oh, J.E.; Nan, J.; Lee, H.; Jung, H.S.; Chung, S.S.; Park, K.S. SIRT3 overexpression attenuates palmitate-induced pancreatic β-cell dysfunction. *PLoS ONE* **2015**, *10*, e0124744.

150. Choudhary, C.; Kumar, C.; Gnad, F.; Nielsen, M.L.; Rehman, M.; Walther, T.C.; Olsen, J.V.; Mann, M. Lysine acetylation targets protein complexes and co-regulates major cellular functions. *Science* **2009**, *325*, 834–840.

151. Haigis, M.C.; Deng, C.X.; Finley, L.W.; Kim, H.S.; Gius, D. SIRT3 is a mitochondrial tumor suppressor: A scientific tale that connects aberrant cellular ROS, the warburg effect, and carcinogenesis. *Cancer Res.* **2012**, *72*, 2468–2472.

152. Lombard, D.B.; Alt, F.W.; Cheng, H.L.; Bunkenborg, J.; Streeper, R.S.; Mostoslavsky, R.; Kim, J.; Yancopoulos, G.; Valenzuela, D.; Murphy, A.; *et al.* Mammalian Sir2 homolog SIRT3 regulates global mitochondrial lysine acetylation. *Mol. Cell. Biol.* **2007**, *27*, 8807–8814.

363

153. Koentges, C.; Pfeil, K.; Schnick, T.; Wiese, S.; Dahlbock, R.; Cimolai, M.C.; Meyer-Steenbuck, M.; Cenkerova, K.; Hoffmann, M.M.; Jaeger, C.; *et al.* SIRT3 deficiency impairs mitochondrial and contractile function in the heart. *Basic Res. Cardiol.* **2015**, *110*, 493.

154. Paulin, R.; Dromparis, P.; Sutendra, G.; Gurtu, V.; Zervopoulos, S.; Bowers, L.; Haromy, A.; Webster, L.; Provencher, S.; Bonnet, S.; *et al.* Sirtuin 3 deficiency is associated with inhibited mitochondrial function and pulmonary arterial hypertension in rodents and humans. *Cell Metab.* **2014**, *20*, 827–839.

155. Zeng, L.; Yang, Y.; Hu, Y.; Sun, Y.; Du, Z.; Xie, Z.; Zhou, T.; Kong, W. Age-related decrease in the mitochondrial sirtuin deacetylase Sirt3 expression associated with ROS accumulation in the auditory cortex of the mimetic aging rat model. *PLoS ONE* **2014**, *9*, e88019.

156. Marques-Aleixo, I.; Santos-Alves, E.; Mariani, D.; Rizo-Roca, D.; Padrao, A.I.; Rocha-Rodrigues, S.; Viscor, G.; Torrella, J.R.; Ferreira, R.; Oliveira, P.J.; *et al.* Physical exercise prior and during treatment reduces sub-chronic doxorubicin-induced mitochondrial toxicity and oxidative stress. *Mitochondrion* **2015**, *20*, 22–33.

157. Wood, J.G.; Rogina, B.; Lavu, S.; Howitz, K.; Helfand, S.L.; Tatar, M.; Sinclair, D. Sirtuin activators mimic caloric restriction and delay ageing in metazoans. *Nature* **2004**, *430*, 686–689.

158. Desquiret-Dumas, V.; Gueguen, N.; Leman, G.; Baron, S.; Nivet-Antoine, V.; Chupin, S.; Chevrollier, A.; Vessieres, E.; Ayer, A.; Ferre, M.; *et al.* Resveratrol induces a mitochondrial complex I-dependent increase in NADH oxidation responsible for sirtuin activation in liver cells. *J. Biol. Chem.* **2013**, *288*, 36662–36675.

159. Royce, S.G.; Dang, W.; Yuan, G.; Tran, J.; El Osta, A.; Karagiannis, T.C.; Tang, M.L. Resveratrol has protective effects against airway remodeling and airway hyperreactivity in a murine model of allergic airways disease. *Pathobiol. Aging Age Relat. Dis.* **2011**, *1*.

160. Haohao, Z.; Guijun, Q.; Juan, Z.; Wen, K.; Lulu, C. Resveratrol improves high-fat diet induced insulin resistance by rebalancing subsarcolemmal mitochondrial oxidation and antioxidantion. *J. Physiol. Biochem.* **2015**, *71*, 121–131.

161. Sener, G.; Topaloglu, N.; Sehirli, A.O.; Ercan, F.; Gedik, N. Resveratrol alleviates bleomycin-induced lung injury in rats. *Pulm. Pharmacol. Ther.* **2007**, *20*, 642–649.

162. Zghonda, N.; Yoshida, S.; Ezaki, S.; Otake, Y.; Murakami, C.; Mliki, A.; Ghorbel, A.; Miyazaki, H. Epsilon-viniferin is more effective than its monomer resveratrol in improving the functions of vascular endothelial cells and the heart. *Biosci. Biotechnol. Biochem.* **2012**, *76*, 954–960.

163. Fu, J.; Jin, J.; Cichewicz, R.H.; Hageman, S.A.; Ellis, T.K.; Xiang, L.; Peng, Q.; Jiang, M.; Arbez, N.; Hotaling, K.; *et al.* Trans-(–)-ε-viniferin increases mitochondrial sirtuin 3 (SIRT3), activates AMP-activated protein kinase (AMPK), and protects cells in models of huntington disease. *J. Biol. Chem.* **2012**, *287*, 24460–24472.

164. Pillai, V.B.; Samant, S.; Sundaresan, N.R.; Raghuraman, H.; Kim, G.; Bonner, M.Y.; Arbiser, J.L.; Walker, D.I.; Jones, D.P.; Gius, D.; *et al.* Honokiol blocks and reverses cardiac hypertrophy in mice by activating mitochondrial Sirt3. *Nat. Commun.* **2015**, *6*, 6656.

165. Nunnari, J.; Suomalainen, A. Mitochondria: In sickness and in health. *Cell* **2012**, *148*, 1145–1159.

166. Aon, M.A.; Tocchetti, C.G.; Bhatt, N.; Paolocci, N.; Cortassa, S. Protective mechanisms of mitochondria and heart function in diabetes. *Antioxid. Redox Signal.* **2015**, *22*, 1563–1586.

167. Gatenby, R.A.; Gillies, R.J. Why do cancers have high aerobic glycolysis? *Nat. Rev. Cancer* **2004**, *4*, 891–899.

168. Chen, W.; Wang, Q.; Bai, L.; Chen, W.; Wang, X.; Tellez, C.S.; Leng, S.; Padilla, M.T.; Nyunoya, T.; Belinsky, S.A.; *et al.* RIP1 maintains DNA integrity and cell proliferation by regulating PGC-1α-mediated mitochondrial oxidative phosphorylation and glycolysis. *Cell Death Differ.* **2014**, *21*, 1061–1070.

169. Mendelsohn, A.R.; Larrick, J.W. Partial reversal of skeletal muscle aging by restoration of normal NAD$^+$ levels. *Rejuv. Res.* **2014**, *17*, 62–69.

170. Metodiev, M.D.; Gerber, S.; Hubert, L.; Delahodde, A.; Chretien, D.; Gerard, X.; Amati-Bonneau, P.; Giacomotto, M.C.; Boddaert, N.; Kaminska, A.; *et al.* Mutations in the tricarboxylic acid cycle enzyme, aconitase 2, cause either isolated or syndromic optic neuropathy with encephalopathy and cerebellar atrophy. *J. Med. Genet.* **2014**, *51*, 834–838.

171. Cha, M.Y.; Kim, D.K.; Mook-Jung, I. The role of mitochondrial DNA mutation on neurodegenerative diseases. *Exp. Mol. Med.* **2015**, *47*, e150.

172. Van Waveren, C.; Sun, Y.; Cheung, H.S.; Moraes, C.T. Oxidative phosphorylation dysfunction modulates expression of extracellular matrix—Remodeling genes and invasion. *Carcinogenesis* **2006**, *27*, 409–418.

173. Liu, T.F.; Vachharajani, V.; Millet, P.; Bharadwaj, M.S.; Molina, A.J.; McCall, C.E. Sequential actions of SIRT1-RELB-SIRT3 coordinate nuclear-mitochondrial communication during immunometabolic adaptation to acute inflammation and sepsis. *J. Biol. Chem.* **2015**, *290*, 396–408.

174. Ashrafi, G.; Schwarz, T.L. The pathways of mitophagy for quality control and clearance of mitochondria. *Cell Death Differ.* **2013**, *20*, 31–42.

175. Ayala-Pena, S. Role of oxidative DNA damage in mitochondrial dysfunction and huntington's disease pathogenesis. *Free Radic. Biol. Med.* **2013**, *62*, 102–110.

176. Diebold, I.; Hennigs, J.K.; Miyagawa, K.; Li, C.G.; Nickel, N.P.; Kaschwich, M.; Cao, A.; Wang, L.; Reddy, S.; Chen, P.I.; *et al.* BMPR2 preserves mitochondrial function and DNA during reoxygenation to promote endothelial cell survival and reverse pulmonary hypertension. *Cell Metab.* **2015**, *21*, 596–608.

177. De Cavanagh, E.M.; Inserra, F.; Ferder, L. Angiotensin II blockade: A strategy to slow ageing by protecting mitochondria? *Cardiovasc Res.* **2011**, *89*, 31–40.

178. Radak, Z.; Koltai, E.; Taylor, A.W.; Higuchi, M.; Kumagai, S.; Ohno, H.; Goto, S.; Boldogh, I. Redox-regulating sirtuins in aging, caloric restriction, and exercise. *Free Radic. Biol. Med.* **2013**, *58*, 87–97.

179. Boyette, L.B.; Tuan, R.S. Adult stem cells and diseases of aging. *J. Clin. Med.* **2014**, *3*, 88–134.

180. Tao, R.; Coleman, M.C.; Pennington, J.D.; Ozden, O.; Park, S.H.; Jiang, H.; Kim, H.S.; Flynn, C.R.; Hill, S.; Hayes McDonald, W.; *et al.* Sirt3-mediated deacetylation of evolutionarily conserved lysine 122 regulates MNSoD activity in response to stress. *Mol. Cell* **2010**, *40*, 893–904.

181. Kim, H.S.; Patel, K.; Muldoon-Jacobs, K.; Bisht, K.S.; Aykin-Burns, N.; Pennington, J.D.; van der Meer, R.; Nguyen, P.; Savage, J.; Owens, K.M.; *et al.* SIRT3 is a mitochondria-localized tumor suppressor required for maintenance of mitochondrial integrity and metabolism during stress. *Cancer Cell* **2010**, *17*, 41–52.

182. Stefanatos, R.; Sanz, A. Mitochondrial complex I: A central regulator of the aging process. *Cell Cycle* **2011**, *10*, 1528–1532.

183. Peek, C.B.; Affinati, A.H.; Ramsey, K.M.; Kuo, H.Y.; Yu, W.; Sena, L.A.; Ilkayeva, O.; Marcheva, B.; Kobayashi, Y.; Omura, C.; *et al.* Circadian clock NAD+ cycle drives mitochondrial oxidative metabolism in mice. *Science* **2013**, *342*, 1243417.

184. Gomes, A.P.; Price, N.L.; Ling, A.J.; Moslehi, J.J.; Montgomery, M.K.; Rajman, L.; White, J.P.; Teodoro, J.S.; Wrann, C.D.; Hubbard, B.P.; *et al.* Declining NAD+ induces a pseudohypoxic state disrupting nuclear-mitochondrial communication during aging. *Cell* **2013**, *155*, 1624–1638.

185. Zhou, X.; Chen, M.; Zeng, X.; Yang, J.; Deng, H.; Yi, L.; Mi, M.T. Resveratrol regulates mitochondrial reactive oxygen species homeostasis through Sirt3 signaling pathway in human vascular endothelial cells. *Cell Death Dis.* **2014**, *5*, e1576.

186. Xu, R.; Hu, Q.; Ma, Q.; Liu, C.; Wang, G. The protease Omi regulates mitochondrial biogenesis through the GSK3β/PGC-1α pathway. *Cell Death Dis.* **2014**, *5*, e1373.

187. Massudi, H.; Grant, R.; Braidy, N.; Guest, J.; Farnsworth, B.; Guillemin, G.J. Age-associated changes in oxidative stress and NAD+ metabolism in human tissue. *PLoS ONE* **2012**, *7*, e42357.

188. Braidy, N.; Guillemin, G.J.; Mansour, H.; Chan-Ling, T.; Poljak, A.; Grant, R. Age related changes in NAD+ metabolism oxidative stress and sirt1 activity in wistar rats. *PLoS ONE* **2011**, *6*, e19194.

Current Experience in Testing Mitochondrial Nutrients in Disorders Featuring Oxidative Stress and Mitochondrial Dysfunction: Rational Design of Chemoprevention Trials

Giovanni Pagano, Annarita Aiello Talamanca, Giuseppe Castello,
Mario D. Cordero, Marco d'Ischia, Maria Nicola Gadaleta, Federico V. Pallardó,
Sandra Petrović, Luca Tiano and Adriana Zatterale

Abstract: An extensive number of pathologies are associated with mitochondrial dysfunction (MDF) and oxidative stress (OS). Thus, mitochondrial cofactors termed "mitochondrial nutrients" (MN), such as α-lipoic acid (ALA), Coenzyme Q10 (CoQ10), and L-carnitine (CARN) (or its derivatives) have been tested in a number of clinical trials, and this review is focused on the use of MN-based clinical trials. The papers reporting on MN-based clinical trials were retrieved in MedLine up to July 2014, and evaluated for the following endpoints: (a) treated diseases; (b) dosages, number of enrolled patients and duration of treatment; (c) trial success for each MN or MN combinations as reported by authors. The reports satisfying the above endpoints included total numbers of trials and frequencies of randomized, controlled studies, *i.e.*, 81 trials testing ALA, 107 reports testing CoQ10, and 74 reports testing CARN, while only 7 reports were retrieved testing double MN associations, while no report was found testing a triple MN combination. A total of 28 reports tested MN associations with "classical" antioxidants, such as antioxidant nutrients or drugs. Combinations of MN showed better outcomes than individual MN, suggesting forthcoming clinical studies. The criteria in study design and monitoring MN-based clinical trials are discussed.

Reprinted from *Int. J. Mol. Sci.* Cite as: Pagano, G.; Talamanca, A.A.; Castello, G.; Cordero, M.D.; d'Ischia, M.; Gadaleta, M.N.; Pallardó, F.V.; Petrović, S.; Tiano, L.; Zatterale, A. Current Experience in Testing Mitochondrial Nutrients in Disorders Featuring Oxidative Stress and Mitochondrial Dysfunction: Rational Design of Chemoprevention Trials. *Int. J. Mol. Sci.* **2014**, *15*, 20169–20201.

1. Introduction

The functions of mitochondria, unconfined to bioenergetic pathways, were linked to OS by studies in early 1990's reporting on reactive oxygen species (ROS) formation as by-products of oxygen metabolism [1,2]. The implications for OS in Parkinson's disease were discovered by Di Monte *et al.* [3] and associated with MDF by a correlation of complex I–III deficiency with lower-than-normal levels of CoQ10, a cofactor in the oxidative phosphorylation (OXPHOS) pathway [4,5].

The association of OS and MDF has been reported in in mitochondrial diseases [6,7], and our recent reviews evaluated the literature showing that OS/MDF is involved in broad-ranging pathologies, including some genetic diseases [8], aging and age-associated disorders, neurologic and psychiatric diseases, malignancies and autoimmune diseases [9]. This state-of-art has prompted a growing body of literature from clinical studies, aimed at compensating OS/MDF by means of MN administration [10–14]. These endogenous cofactors, such as ALA, CoQ10 and CARN (or CARN derivatives) are essential in mitochondrial functions.

ALA exists in two redox states, and the reduced thiol form is a potent mitochondrial antioxidant, a metal chelator and a glutathione (GSH) repletor. It is involved as an essential cofactor in Krebs cycle for mitochondrial α-ketoacid dehydrogenases. CoQ10 is a benzoquinone derivative playing a central role in the mitochondrial respiratory chain through shuttling between three redox forms, the quinone, the semiquinone and the hydroquinone. It acts as a carrier accepting electrons from mitochondrial complex I and complex II and transferring them to complex III. CARN (bioactive L-form) is biosynthesized from lysine and methionine and is concentrated in tissues that use fatty acids as their primary source of energy. It is involved in the transport of long-chain fatty acids from the intermembraneous space to the matrix in the mitochondria for generation of metabolic energy.

A number of studies have demonstrated MN deficiencies in several diseases [9], either related to deficiencies in OXPHOS activities [15–18] or caused by their degradation due to OS-related by-products as, e.g., by ALA oxidative modification induced by 4-hydroxynonenal in Alzheimer disease brain [19]. The present review is to provide an overall survey of the currently published clinical trials having tested each of the above MN, or their combinations, and/or their associations with "typical" antioxidants, including antioxidant nutrients and/or herbal preparations and/or disease-specific drugs. For each disease, or disease group, the total numbers of clinical trials were recorded, along with the relative frequencies of controlled, randomized trials. No further attempt was made to evaluate the clinical and/or laboratory outcomes of the evaluated reports. The major objective of the present survey was providing an overall state-of-art, and suggesting some prospects in study design and in monitoring the outcomes of forthcoming clinical trials. These should be aimed at verifying any specific MN deficiency(ies) associated to a given disorder, and at prompting adequate follow-up in monitoring the effects, if any, of mitochondria-targeted interventions.

2. Methods

A MedLine retrieval up to July 2014 was carried out for each MN. The papers reporting on clinical trials for each MN were evaluated according to: (a) treated diseases (or disease groups); (b) dosages, number of enrolled patients and duration of treatment; (c) numbers of trials and frequencies of randomized, controlled studies (opposed to open-label and pilot studies); and (d) an empyrical "success ratio" (SR) of trials testing each of the individual MN or their combinations, as reported by authors; SR was calculated by dividing the number of successful results by the total of trials for each agent. No clinical trials involving healthy volunteers were included for evaluation. The reports failing to provide clear-cut data for dosages, the numbers of patients and/or treatment duration were not included in survey, nor were included self-repeating reports of previous or contemporary studies.

3. α-Lipoic Acid

As shown in Table 1, ALA testing was reported in 81 evaluated clinical trials on a total of 2980 patients, since the pioneering report by Marshall *et al.* on alcohol-related liver disease [10], and since early studies in 1990's in counteracting diabetes-associated neuropathies [20–22]. Both Type 1 and Type 2 diabetes mellitus (DM) have been the major focus for ALA administration, encompassing 42 trials (of which 30 controlled trials) on a total of 2980 patients and with a SR = 0.93 [20–61]. A major outcome of ALA treatment in diabetic patients consisted of the amelioration of neurologic damage [22–24,26–31,33–35,37,39,42,46,49,51,52,60,61]. Moreover, ALA-treated diabetic patients displayed decreased serum concentrations of thiobarbituric acid reactive substances (TBARS) [20,43], increased insulin sensitivity [21,25,36,58], an improvement in endothelium function [38,43,55], decreased erectile dysfunction [59] and decreased urinary PGF2α-isoprostanes, a marker of oxidative damage [50].

Beyond the literature of ALA-centered clinical trials in DM, a broad clinical use of ALA is recognized as a prescription generic drug for DM in Germany (German Drug Index), thus the use of ALA in diabetic patients may be seen as an established practice in the specialist community.

Other clinical trials successfully tested ALA in neurological diseases [62–70], in liver and metabolic diseases [10,71–77], and in heart and vessel diseases [78–81] (Table 1).

Lesser, if any, positive effects were reported for three ALA trials in kidney diseases [82–85]. Other clinical trials tested the effects of ALA administration in genetic and mitochondrial diseases [86–88], burning mouth syndrome [89–93], cancer cachexia [94,95], mitochondrial function in HIV-1-related lipoatrophy [96,97], in vitiligo [98], and in osteoporosis [99].

Table 1. Clinical studies utilizing α-lipoic acid aimed at compensating oxidative stress (OS)/mitochondrial dysfunction (MDF)-related pathogenetic mechanisms.

Diseases/Conditions	No. Studies (Controlled Studies)	No. Treated Patients	Success Ratio	References
Type 1 and Type 2 diabetes	42 (30)	2980	0.93	[20–61]
Neurological diseases	9 (5)	509	0.89	[62–70]
Liver and metabolic diseases	8 (5)	417	0.86	[10,71–77]
Heart and vessel diseases	4 (4)	137	1.00	[78–81]
Kidney diseases	4 (3)	288	0.25	[82–85]
Genetic and mitochondrial diseases	3 (2)	129	0.33	[86–88]
Burning mouth syndrome	5 (5)	293	0.60	[89–93]
Other diseases §	6 (4)	525	1.00	[94–99]
Total	81	5278		

§ Malignancies [94,95]; HIV [96,97]; vitiligo [98]; osteoporosis [99].

Among these studies, some reports investigated the comparative effects of different MN formulations, and/or provided evidence for combined effects in clinical and laboratry endpoints.

Li *et al.* [79] showed that ALA administration to patients with acute coronary syndrome caused a decrease in an OS marker, 8-*iso*-prostaglandin F2α and an increase in aldehyde dehydrogenase-2, responsible for acetaldehyde oxidation in ethanol metabolism that also provides protection against OS. Protective effects were reported by Martins *et al.* [87] in sickle cell trait subjects and sickle cell patients following ALA administration, with a significant increase in blood catalase (CAT) and a significant decrease in malondialdehyde (MDA) and in carbonyl levels.

A clinical trial reported by Galasko *et al.* [64] tested the effects of combined ALA with Vit E and Vit C (E/C/ALA), or CoQ10, or placebo in three groups of patients with Alzheimer's disease, finding a decrease in CSF F2-isoprostane levels in the E/C/ALA group that suggested a reduction of OS in the brain. However, this treatment raised the caution of faster cognitive decline.

Altogether, the current body of evidence may suggest further interventions with ALA in several disorders, beyond DM and the other disorders where ALA treatment led both to clinical improvements and to compensation of OS-related endpoints.

4. Coenzyme Q10

A recent review has focused on the multiple implications of CoQ10 deficiency and of CoQ10 administration encompassing an extensive number of disorders [100]. Out of 101 reports on clinical trials testing CoQ10, 39 studies were focused on support to patients with heart and vessel disorders or undergoing heart surgery [11,12,81,101–136], including 32 controlled studies, as shown in Table 2. The pioneering studies by the groups of Tanaka [11] and of Langsjoen [12] since 1982 provided the avenue for a number of clinical trials assessing the successful outcomes of CoQ10 administration to patients with coronary artery disease, cardiomyopathy, heart failure, or heart surgery, with an overall SR = 0.89. Among these studies, Dai *et al.* [105] evaluated mitochondrial function in patients with ischemic left ventricular systolic dysfunction receiving CoQ10 *vs.* placebo, in terms of plasma lactate/pyruvate (LP) ratio. After an 8-week treatment, CoQ10-treated patients had significant increases in plasma CoQ10 concentration, brachial flow-mediated dilation, and decreased LP ratio, showing positive effects both on heart function and in balancing mitochondrial activities [105]. Lee *et al.* [106] tested CoQ10 administration in patients with coronary artery disease and found a decrease in the inflammatory markers C-reactive protein (hs-CRP) and interleukin 6 (IL-6), and decreased MDA and superoxide dismutase (SOD) activities, along with increased CoQ10 levels. Direct evidence for a CoQ10-induced compensation of mitochondrial function was reported by Rosenfeldt *et al.* [133] in patients undergoing heart surgery, which resulted in increased CoQ10 levels in serum, atrial trabeculae, and isolated mitochondria compared with patients receiving placebo, with an improvement in mitochondrial respiration, as adenosine diphosphate/oxygen ratio, and a decrease in mitochondrial MDA content.

Table 2. Clinical studies utilizing coenzyme Q10 aimed at compensating OS/MDF-related pathogenetic mechanisms.

Diseases/Conditions	No. Studies (Controlled Studies)	No. Treated Patients	Success Ratio	References
Heart and vessel diseases	39 (32)	3386	0.89	[11,12,81,101–136]
Genetic and mitochondrial diseases	18 (8)	680	0.75	[13,14,88,137–151]
Neurological diseases	16 (10)	1185	0.87	[65,152–166]
Type 1 and Type 2 Diabetes	9 (7)	370	0.89	[41,166–174]
Malignancies	6 (2)	301	0.67	[175–180]
Kidney diseases	4 (2)	171	0.50	[181–184]
Other diseases §	15 (13)	555	0.59	[76,185–198]
Total	107	6648		

§ Psoriasis [76,196]; metabolic syndrome [185–187]; statin-induced myalgias [188–192]; bronchial asthma [193]; pre-eclampsia [194]; mucocutaneous infections [195]; cataract surgery [197]; idiopathic infertility [198].

Eighteen studies (of which 8 controlled studies) have focused on clinical trials testing CoQ10 in a number of genetic and mitochondrial diseases [13,14,88,137–151] on 680 patients (Table 2) with different success across the different diseases investigated (SR = 0.75).

Among these studies, Friedreich ataxia (FRDA) is relevant for a deficiency in the gene encoding frataxin, a mitochondrial protein implicated in iron metabolism and glutathione balance [139], and lower-than-normal CoQ10 levels [139,140]. FRDA was investigated by Cooper's group [140,141] for CoQ10 and Vit E administration finding a significant improvement in cardiac and skeletal muscle bioenergetics and in International Co-operative Ataxia Ratings Scale. Rodriguez *et al.* [88], as mentioned above, tested the effect of a combination therapy with CoQ10 with ALA and creatine monohydrate on several outcome variables in patients with mitochondrial diseases, both achieving amelioration of clinical and of OS/MDF endpoints [88]. Limited evidence for a CoQ10-induced improvement in brain and muscle bioenergetics in patients with mitochondrial diseases was reported by Barbiroli et al. [142] and by Glover et al. [145].

A set of studies was focused on 16 clinical trials (of which 10 controlled trials) testing CoQ10 supplementation in 1185 patients with eight neurologic disorders [65,152–166].

The above-mentioned report by Galasko et al. [65] tested CoQ10 as an alternative treatment to E/C/ALA in patients with Alzheimer's disease, failing to report any effects of CoQ10, unlike E/C/ALA. A study by Shults et al. [152] in patients with Parkinson's disease (PD) tested the effects of three CoQ10 dosages (300, 600, or 1200 mg/d) that were evaluated with the Unified Parkinson Disease Rating Scale at the baseline and up to 16-month visits. Less disability developed in subjects assigned to CoQ10 than in those assigned to placebo, with a significant benefit in patients receiving the highest dosage [152]. Moderate beneficial effects of CoQ10 in PD patients were also reported by Müller et al. [153], though at the low dosage (360 mg/d) reported by Shults et al. [152]. A study by Storch et al. [154] found that nanoparticular CoQ10 at a dosage of 300 mg/d was well tolerated, though failing to display symptomatic effects in midstage PD.

Sanoobar et al. [155,156] tested CoQ10 in multiple sclerosis (MS) patients, and found reduced OS and increased antioxidant enzyme activity [155], with a significant decrease of tumor necrosis factor alpha (TNF-α) and IL-6 levels [156] in the CoQ10 group compared to placebo group.

A series of studies focused on the role of CoQ10 in fibromyalgia (FM) in long-term treatment of fibromyalgic patients, resulting in significant pain reduction, fatigue, and morning tiredness [157–160]. CoQ10-treated patients underwent significant reduction in the pain visual scale and in tender points, recovery of inflammation, antioxidant enzymes, mitochondrial biogenesis, and AMPK gene expression levels [157,158]. A recent report by Miyamae et al. [159] measured plasma levels of ubiquinone-10, ubiquinol-10, free cholesterol, cholesterol esters, and free fatty acids in patients with juvenile FM vs. healthy control subjects. Plasma level of ubiquinol-10 was significantly decreased and the ratio of ubiquinone-10 to total CoQ10 was significantly increased in juvenile FM relative to healthy controls, with compensated plasma levels of lipid endpoints [159]. A previous uncontrolled trial by Lister [160] tested the effects of combined CoQ10 and and Ginkgo biloba extract reporting improved quality-of-life in FM patients. Altogether, independent studies support the usefulness of CoQ10 in FM pathogenesis and point to CoQ10 utilization in FM both in inducing clinical improvements and in compensating FM-associated OS/MDF.

No evidence of CoQ10-induced beneficial effects were found in a Phase II clinical trial in patients with amyotrophic lateral sclerosis [161].

Migraine has been investigated for deficiency of CoQ10, which showed lower-than-normal levels in patients with pediatric and adolescent migraine [162]. Clinical trials testing CoQ10 in migraine patients provided overall positive, though apparently transient outcomes [162–164].

Forester *et al.* [165] tested high-dose CoQ10 in patients with geriatric bipolar depression suggesting a reduction in depression symptom severity.

Stamelou *et al.* [166] tested CoQ10 in patients with progressive supranuclear palsy (PSP), and found that patients receiving CoQ10 displayed decreased concentration of low-energy phosphates, with increased ratio of high-energy phosphates to low-energy phosphates. Clinically, the PSP rating scale and the Frontal Assessment Battery improved upon CoQ10 treatment compared to placebo [166].

Nine clinical trials (seven controlled trials) were conducted by testing CoQ10 in 370 patients with Type 1 and Type 2 DM, particularly focusing on DM-associated vascular complications and on statin-induced CoQ10 depletion [167–174]. The outcomes of these studies showed beneficial effects of CoQ10, better if combined with fenofibrate [169]. By considering the established advantages of ALA treatment in DM patients [20–61], one may suggest further clinical trials toward the clinical use of ALA and CoQ10 combinations in DM. This trial design was performed by Palacka *et al.* with successful outcomes both in improving heart left ventricular function and in decreasing lactate dehydrogenase activity in DM patients treated with ALA and CoQ10 [41].

Six clinical trials have been reported testing CoQ10 in association with other agents (e.g., riboflavin and niacin) in patients with advanced malignancies under antineoplastic therapy as supportive or palliative treatments [175–180]. Some of these trials reported survival prolongation [177], and an early study by Lockwood *et al.* [176] reported partial and complete regression of breast cancer in patients treated with CoQ10. A clinical trial by Rusciani *et al.* [180] tested the effects of CoQ10 and interferon α-2b in melanoma patients, and found significantly decreased recurrence rates *vs.* patients receiving interferon only.

Four clinical trials tested the CoQ10 in patients with chronic kidney disease or undergoing hemodialysis, or submitted to statin treatment [181–184]. Sakata *et al.* [182] reported CoQ10 administration in hemodialysis patients as partially effective for suppressing OS. Shojaei *et al.* [184] tested CoQ10 and CARN, separately or in association, in hemodialysis patients who were on statin treatment. This study showed that supplementation with CoQ10 and CARN reduced serum levels of lipoprotein(a) in maintenance hemodialysis patients treated with statins. Altogether, the clinical trials testing CoQ10 in patients with kidney diseases failed to report clear-cut beneficial effects (SR = 0.50).

Controversial results were reported from clinical trials aimed at testing CoQ10 in counteracting various dysmetabolic conditions [185–187], statin-induced myalgias [188–192], bronchial asthma [193], pre-eclampsia [194], cutaneous infections [195], psoriasis [76,196], cataract surgery [197], and idiopathic infertility [198].

5. L-Carnitine and Acetyl- or Propionyl-Carnitine

Multi-decade long investigations showed the crucial roles of CARN in mitochondrial physiology, along with CARN deficiency in several disorders. As reviewed by Gilbert since 1985 [199], CARN deficiency results in accumulation of neutral lipid deposits within skeletal muscle, myocardium and liver, with mitochondria aggregates in skeletal muscle and myocardium. The efficacy of CARN *vs.* acetyl-CARN (ALC) was compared both *in vitro* and in aging rat brain since early studies [200,201]. CARN and ALC were similar in elevating carnitine levels in plasma and brain as well as in increasing ambulatory activity of old rats. However, ALC but not CARN was able to decrease the level of oxidative stress biomarkers in the brain of old rats [201]. Subsequent investigations pointed to a role of ALC in the reactivation of mitochondrial biogenesis in aging through the increased expression of PGC-1α signaling pathway [202,203].

As shown in Table 3, out of 74 evaluated studies, 18 clinical trials (of which 16 controlled trials) tested the effects of CARN in kidney diseases, treating a total of 427 patients with end-stage renal disease under hemodialysis, based on the rationale that CARN levels sharply decrease during hemodialysis, thus impairing response to recombinant human erythropoietin (rHuEPO) [181,184,204–220]. One study utilized both CARN and CoQ10—separately or in association—finding positive effects of both supplements, yet without evidence for improvement effects following combined supplementation [184]. Some reports failed to find significant improvements following CARN administration [205–208,213,220] (overall SR = 0.58). One might note that out of four trials adopting *i.v.* CARN (or PLC) administration, three resulted in significant improvements in the tested endpoints, such as Medical Outcomes Short Form-36, rHuEPO requirement, hemodynamic flow, endothelial profile and homocysteine levels, or inflammatory status as significant decrease in C-reactive protein [210,211,217].

Table 3. Clinical studies utilizing L-carnitine (CARN) (or acetyl-CARN (ALC)) aimed at compensating OS/MDF-related pathogenetic mechanisms.

Diseases/Conditions	No. Studies (Controlled Studies)	No. Treated Patients	Success Ratio	References
Kidney diseases	18 (16)	427	0.58	[184,204–220]
Type 1 and Type 2 Diabetes	13 (9)	1894	1.00	[221–233]
Heart and vessel diseases	9 (6)	359	1.00	[78,114,234–240]
Liver diseases	6 (2)	275	1.00	[241–246]
Neurological diseases	9 (9)	384	1.00	[68,247–254]
Malignancies	7 (6)	699	0.70	[255–261]
Genetic diseases	6 (4)	202	1.00	[262–267]
HIV	4 (1)	93	1.00	[97,268–270]
Other diseases [§]	2 (2)	99	1.00	[271,272]
Total	74	4432		

[§] Major surgery [271]; ulcerative colitis [272].

Unlike renal diseases, all of 13 clinical trials (9 controlled trials) supplementing CARN (or ALC, or PLC) to patients with Type 1 and Type 2 DM to a total of 1894 patients succeeded achieving improvements in disease status [221–233]. Long-term infusion of CARN or ALC in DM patients was tested by the group of Giancaterini [221,222], resulting in improved insulin sensitivity, and in decreased lactate levels, suggesting activation of pyruvate dehydrogenase, whose activity is depressed in the insulin resistant status. Three independent studies found a long-term treatment with ALC or with PLC effective and well tolerated in improving neurophysiological parameters and in reducing pain [223,226,229]. Derosa *et al.* reported that CARN significantly lowered the plasma lipoprotein(a) *vs.* placebo in hypercholesterolemic patients with DM [224]. A clinical trial with PLC in patients with DM and peripheral arterial disease was reported by Ragozzino *et al.* [225], who found significant increases in maximal walking distance, initial claudication distance, and a decrease in dosage of oral antihyperglycemic agents. Two studies investigated the effects of CARN or ALC treatment on lipidemic profile of DM patients [223,224], failing to detect a decrease of lipidemic profile of CARN alone to this endpoint, yet Solfrizzi *et al.* [228] found significant effects from combined treatment with CARN and simvastatin in lowering Lp(a) serum levels in patients with DM than with simvastatin alone. A study by Malaguarnera *et al.* [230] testing CARN in DM patients found significant improvements compared with the placebo group in a number of OS endpoints, proinflammatory markers and lipidemic profile.

A clinical trial was conducted by Ruggenenti *et al.* [231] by testing the effects of ALC on glucose disposal rate in patients with metabolic syndrome and hypertension, reporting that ALC ameliorated arterial hypertension, insulin resistance, glucose tolerance, and hypoadiponectinemia. Patients with Type 1 DM were treated by Uzun *et al.* [233] with CARN to evaluate changes in their neuropathy frequency and nerve conduction velocity, and found that CARN treatment in early stages improved neuropathy in Type 1 DM patients.

Altogether, the body of literature of clinical trials testing CARN—or CARN derivatives—in patients with DM provides substantial evidence for the amelioration of multiple DM-related endpoints that encourage the clinical use of CARN—or CARN derivatives—in DM patients. One may note that no report was found of combined treatments of DM patients using CARN with either ALA or CoQ10, despite the body of positive evidence accumulated in clinical trials testing ALA, or CoQ10 in DM patients (as shown Tables 1 and 2).

A series of 9 clinical trials (6 controlled trials) were reported testing CARN (or ALC, or PLC) in 359 patients with heart and vessel diseases [78,114,234–240]. A study by McMackin et al. [78] reported on the effects of combined ALC and ALA treatment in patients with coronary artery disease, and found significant improvement as increased brachial artery diameter, a trend to decreased systolic blood pressure for the whole patient group, with a significant effect in the subgroup with blood pressure above the median and in the subgroup with metabolic syndrome. An analogous investigation was reported by Kumar et al. [114], who tested the effects of carni Q-gel (CARN and ubiquinol) in patients with heart failure. Serum concentration of IL-6 was significantly decreased in the intervention group without such changes in the control group. TNF-α, which was comparable at baseline, also showed a greater decline in the carni Q-gel group compared to the placebo group. Serum CoQ10 showed a significant increase in the carni Q-gel group as compared to the control group. The symptom scale indicated that the majority of treated patients showed a significant improvement compared to the placebo group [114]. Altogether, both of these clinical trials using associations of ALC with ALA [78] or CARN with CoQ10 (as ubiquinol) [114] showed definite improvements both in heart and vessel performance and in related laboratory endpoints.

Relevant mechanistic data on PLC treatment in patients with peripheral arterial disease (PAD) were obtained by Loffredo et al. [234], who measured serum levels of nitrite and nitrate (NO_x) and 8-hydroxy-2'-deoxyguanosine (8-OHdG), and maximal walking distance (MWD) in PAD patients. Serum levels of 8-OHdG were significantly increased in PAD patients, and serum levels of NO_x were significantly decreased. Patients treated with PLC showed a significant increase of MWD and in NO_x, while 8-OHdG levels underwent a significant decrease, unlike the patients given placebo [234].

Improved short-term exercise capacity, diastolic function and symptoms, WHO heart functional class were reported in patients receiving CARN [237,238], with improvements in diastolic function and in myocardial hypoxanthine concentration [239,240].

A total of 275 patients affected by various liver disorders were treated with CARN or ALC in 6 clinical trials (of which 2 controlled studies) [241–246]. Łapiński and Grzeszczuk [241] tested CARN (or L-ornitine L-aspartate) in patients with liver cirrhosis, and evaluated the outcomes as changes in serum concentrations of ammonia, cholesterol and triglycerides. A significant improvement was observed both in the patients treated with CARN and in those treated with L-ornitine L-aspartate. However, the individuals treated with CARN showed a significant increase of serum cholesterol and triglyceride concentrations.

Patients with non-alcoholic fatty liver disease were treated with CARN by Lim *et al.* [242]; by measuring a liver function test, peripheral blood mitochondrial DNA and 8-oxo-dG analysis. The results showed a decrease in ALT, AST, and total bilirubin in treated patients, not in controls. In the CARN treated patients, mitochondrial DNA copy number was significantly increased, unlike the control group. 8-oxo-dG levels showed a non-significant decrease in CARN group while it tended to increase in the control group. The group of Malaguarnera *et al.* has conducted a series of clinical trials testing the effects of CARN in patients with hepatic encephalopathy or with HCV-induced chronic hepatitis [243–246]. Beneficial effects in CARN-treated patients included significant differences in AST, ALT, viremia, Hb, RBC, WBC and platelets. Also significant improvements were detected in neurologic scores and fatigue symptoms in hepatic encephalopathy following CARN administration [244,245].

Nine controlled clinical trials for CARN—mostly ALC—were conducted in a total of 384 patients with some neurological disorders, including Alzheimer's disease, multiple sclerosis, migraine, sciatica, fibromyalgia, chronic fatigue syndrome, and amyotrophic lateral sclerosis [68,247–254]. A report by Memeo and Loiero [68] compared the effects of treating sciatica patients with either ALA or ALC, and found significant protective effects from both of these agents detected as improvements from baseline in neuropathy on electromyography.

Tomassini *et al.* [247] tested the efficacy of ALC *vs.* amantadine, a widely used drug in treating patients with multiple sclerosis (MS)-related fatigue. The results showed significant effects of ALC compared with amantadine for the Fatigue Severity Scale, suggesting that ALC is better tolerated and more effective than amantadine for the treatment of MS-related fatigue.

Patients with Alzheimer's disease and vascular dementia were treated with ALC by Gavrilova *et al.* [248], and the outcomes were assessed with MMSE and CGI scales, and a battery of neuropsychological tests. The treatment effect of ALC was significantly higher than in the placebo group.

Tarighat Esfanjani *et al.* [249] evaluated the effects of magnesium, CARN, and concurrent magnesium-CARN supplementation in migraine prophylaxis. Migrainous patients were randomly assigned into three intervention groups:

magnesium oxide *vs.* CARN *vs.* Mg-CARN, and a control group. The results showed a significant reduction in all migraine indicators in all studied groups, yet without a clear-cut difference in-between the three supplementation regimens.

Four independent clinical trials tested CARN effects [250–253] in patients with chronic fatigue syndrome (CFS), or fibromyalgia (FM), or narcolepsy (NL). Pistone *et al.* [250] treated elderly subjects with onset of fatigue following slight physical activity. Wessely and Powell fatigue scores decreased significantly in subjects taking CARN *vs.* the placebo group, with significant improvements in total fat mass, total muscle mass, total cholesterol, LDL-C, HDL-C, triglycerides, apoA1, and apoB [250]. Miyagawa *et al.* [251] tested CARN administration in NL patients and found a significant reduction of excessive daytime sleepiness in CARN-treated NL patients *vs.* those receiving placebo. A clinical trial testing the effects of ALC in FM patients was reported by Rossini *et al.* [252]. The "total myalgic score" and the number of positive tender points declined significantly in the ALC-treated patients *vs.* the placebo group, and a significant difference was observed for depression and musculo-skeletal pain [252]. The effects of ALC and PLC were tested in patients with CFS by Vermeulen and Scholte [253]. This clinical trial compared ALC, PLC, and their combination. Clinical global impression of change after treatment showed considerable improvement in 59% of the patients in the ALC group and 63% in the PLC group, but less in the ALC plus PLC group (37%). In the ALC group the changes in plasma CARN levels correlated with clinical improvement [253].

A clinical trial by Beghi *et al.* tested the effects of ALC plus riluzole *vs.* riluzole and placebo on disability and mortality of patients with amyotrophic lateral clerosis [254]. Patients receiving ALC became non-self-sufficient to significantly less extent compared to those receiving placebo, and median survival was significantly extended in ALC-treated patients *vs.* placebo group [254].

Based on CARN depletion in advanced cancer, palliative treatments by CARN supplementation have been proposed by a series of seven clinical trials, of which six controlled trials [255–261]. Unfortunately, the outcomes showed a treatment-related increase of CARN serum levels, yet CARN or ALC supplementation did not improve fatigue in patients, or even increased chemotherapy-induced peripheral neuropathy [261]; otherwise, positive results were associated with treatments with multiple agents [258,259], thus casting doubts as to the specific efficacy of CARN supplementation.

A few genetic disorders were the focus of six clinical trials, of which 4 controlled trials, using CARN or ALC [262–267]. A clinical trial was carried out by Schöls *et al.* [262] by testing the administration of CARN *vs.* creatine in patients with Friedreich's ataxia. After CARN treatment, patients had a significantly improved phosphocreatine recovery compared to baseline, while creatine effects did not reach significance.

Two independent clinical trials tested the effects of CARN in thalassemic patients, concurring to report improved cardiac function [263,264]. Moreover, Karimi *et al.* [264] tested the effect of combination therapy of hydroxyurea with CARN and magnesium chloride on hematologic parameters and cardiac function of patients with β-thalassemia intermedia. Patients were randomly divided into four groups: group A (hydroxyurea alone); group B (hydroxyurea and CARN); group C (hydroxyurea and magnesium chloride); and group D (hydroxyurea, CARN and magnesium chloride). In groups B, C, and D, mean Hb and hematocrit significantly increased during 6-month treatment [264].

Three independent studies tested the effects of CARN or ALC in patients with autistic spectrum disorders (ASD) [265–267]. Ellaway *et al.* [265] tested CARN in a group of Rett syndrome females, and found significantly improved sleep efficiency and expressive speech, compared to control Rett syndrome patients. A multi-center clinical trial testing ALC on the attention deficit hyperactivity disorder in fragile X syndrome boys was reported by Torrioli *et al.* [266]. Patients treated with ALC, compared with the placebo group, showed a stronger reduction of hyperactivity and improvement of social behavior [266]. Geier *et al.* [267] tested CARN in patients with ASD that were randomly assigned to receive a controlled regimen liquid CARN or placebo for 3 months. Treated patients showed significant improvements in Childhood Autism Rating Scale, modified clinical global impression, and Autism Treatment Evaluation Checklist scores.

Four clinical trials (of which one controlled trial) have tested the CARN, or ALC, aimed at improving different complications of HIV infection [97,268–270]. The above-cited report by Milazzo *et al.* [97] tested ALC *vs.* a combination of ALA and *N*-acetylcysteine (NAC) for a number of OS/MDF and other endpoints in HIV-1-infected patients with lipoatrophy, showing that either ALC, or ALA/NAC supplementation exerts a protective role on mitochondrial function in HIV-1-infected patients [97]. Two closely related studies [268,269] tested the effects of ALC in mitigating the neuropathy associated with antiretroviral toxicity in HIV-infected patients. Mean pain intensity score was significantly reduced, while electrophysiological parameters did not significantly change during treatment [269].

A controlled clinical trial was reported by Pignatelli *et al.* [271] on the effect of CARN on OS and platelet activation after major surgery, whose trauma is associated with an increased ROS production. At baseline and after treatment, OS was evaluated by detection of circulating levels of soluble NOX2-derived peptide, and by analyzing platelet ROS formation. The results showed an increase of OS endpoints in the placebo group compared with the baseline after the surgical intervention, while the CARN-treated group did not significantly differ from the baseline.

Patients with ulcerative colitis (UC) were treated by Mikhailova *et al.* [272] by PLC testing a clinical/endoscopic response. Patients with mild-to-moderate UC receiving stable oral aminosalicylate or thiopurine therapy were randomised to receive PLC or placebo. A significant response was found in patients receiving PLC as clinical/endoscopic response *vs.* those receiving placebo [272].

6. Treatments with Mitochondrial Nutrient Combinations

Out of 262 clinical trials reviewed here (Tables 1–3), only a tiny minority of seven reports referred to the combined use of two MN, ALA and/or CoQ10 and/or CARN [41,68,76,78,81,88,184], as shown in Table 4. Only four clinical trials tested the ALA/CoQ10 combination [41,76,81,88], while three clinical trials tested CARN combined with either ALA or CoQ10 [68,78,184]. Surprisingly, no retrieved report mentioned the combination of the three cofactors in any clinical trial; to the best of our knowledge; only Mattiazzi *et al.* [273] reported on the combined *in vitro* use of ALA, CoQ10 and CARN in improving a mitochondrial (mtDNA T8993G (NARP)) mutation on T8993G mutated cells.

Table 4. Clinical studies utilizing multiple mitochondrial nutrients (α-lipoic acid and/or coenzyme Q10 and/or L-carnitine) aimed at compensating OS/MDF-related pathogenetic mechanisms.

Diseases/Conditions	α-Lipoic Acid	Coenzyme Q10 (Daily Dose, mg/d)	L-Carnitine	References
Heart and vessel diseases	100	100	–	[81]
	400	–	1000	[78]
Mitochondrial diseases	600	240	–	[88]
Type 2 diabetes	100	60	–	[41]
Psoriasis	150	50	–	[76]
Sciatalgia	600	–	1180	[68]
Kidney diseases	–	100	500	[184]

The report by Rodriguez *et al.* [88] tested the combined use of ALA, CoQ10 and creatine in a group of patients with different mitochondrial diseases; these may be viewed as the "ideal" target for administration of MN, in view of supporting known MDF [6,7,9,14,274–276] that involve OS, decreased ATP production, and involvement of anaerobic energetic pathways. The combination therapy resulted in lower resting plasma lactate and OS markers (urinary 8-isoprostanes), and improved energetic performance [88].

Beyond mitochondrial diseases, a substantial body of evidence has been accumulated pointing to the involvement of OS/MDF in a number of heterogeneous diseases [9].

The combined use of ALA and CoQ10 was tested in patients with Type 2 DM [41] and in cardiac surgery patients [81]. One may note that both the report by

Palacka *et al.* [41] and that by Leong *et al.* [81] fulfilled the expected clinical outcomes, along with the finding of decreased plasma lipid peroxides [41].

A number of clinical trials have been conducted by testing MN in combinations with classical antioxidants and/or herbal compounds and/or disease-specific drugs.

A total of 27 reports were evaluated concerning the therapies utilizing ALA or CoQ10 combinations with various antioxidants and/or herbal compounds and/or drugs (Tables 5 and 6). As shown in Table 5, a study utilizing ALA and Vit E alone or in combination failed to find changes in the lipid profile or insulin sensitivity of Type 2 DM patients [44]. Supplementation of quercetin plus Vit C or ALA alone did not change the blood biomarkers of inflammation and disease severity of rheumatoid arthritis patients under conventional treatments [64].

A combined administration with ALA, CoQ10, magnesium orotate, ω-3 polyunsaturated fatty acids (ω-3 PUFA) and selenium was associated by Leong *et al.* [81] with improved redox status and reduced myocardial damage in patients undergoing heart surgery.

A clinical trial in Alzheimer's disease patients showed that supplementation with ALA, Vit E and Vit C reduced OS in the brain [65]. The beneficial effects of combination therapy were also observed in Parkinson's disease (PD) patients receiving homocysteine-lowering therapy (folate and Vit B) and ALA supplementation, suggesting that combination therapy may prevent bone loss in PD patients [277]. A combination therapy of ALA and transdermal testosterone provided amelioration of on erectile dysfunction and quality of life in patients with Type 2 DM [59]. Treatment with ALA and γ-linolenic acid was proposed for controlling symptoms and improving the evolution of carpal tunnel syndrome [67]. Antioxidant supplementation with Vit C, Vit E and ALA was reported as a safe, though ineffective, treatment for dementia in individuals with Down syndrome [86].

Combined supplementations were tested in liver disorders, such as chronic hepatitis virus C infection. Melhem *et al.* [72] reported on a combination of ALA with antioxidative preparations (glycyrrhizin, *Schisandra*, silymarin, Vit C, L-glutathione, and Vit E) in patients with chronic hepatitis C, with favorable response rate.

Twelve clinical trials have reported on combination regimens with CoQ10 and various antioxidants (Table 6). Witte *et al.* [112] tested a combination of CoQ10 and high-dose micronutrients that was found to improve left ventricular function and quality of life in elderly patients with chronic heart failure. Shargorodsky *et al.* [128] tested a combination therapy with CoQ10, Vit C, Vit E and selenium in patients with multiple cardiovascular risk factors, finding an improvement in glucose and lipid metabolism, and decreased blood pressure.

Table 5. Clinical studies utilizing mitochondrial nutrients and antioxidants and/or herbal compounds aimed at compensating OS/MDF-related pathogenetic mechanisms.

Diseases/Conditions	α-Lipoic Acid + Other Agent(s) (Daily Dose)	References
Type 1 and Type 2 diabetes	ALA 100 mg, CoQ10 60 mg, Vit E 200 mg	[41]
	ALA 600 mg, Vit E 800 mg	[44]
	ALA 800 mg, pyridoxine 80 mg	[56]
	ALA 600 mg, transdermal testosterone 50 mg	[59]
	ALA 2 × 600 mg, allopurinol 300 mg, nicotinamide 2 × 750 mg	[60]
Heart surgery	ALA 100 mg, CoQ10 100 mg, Mg orotate 400 mg, ω-3 PUFA 300 mg, Se 200 μg	[81]
Metabolic syndrome	ALA 600 mg, Vit E 100 IU	[77]
Cancer-related anorexia/cachexia	ALA 300 mg, polyphenols 400 mg, carbocysteine 2.7 g, Vit E 400 mg, Vit A 30,000 IU, Vit C 500 mg, (n-3)-PUFA 2 cans	[94]
Alzheimer disease	ALA 900 mg, CoQ10 400 mg, Vit C 500 mg, Vit E 800 IU	[65]
Parkinson's disease	ALA 1200 mg, folate 5 mg, Vit B12 1500 μg	[277]
Carpal tunnel syndrome	ALA 600 mg, γ-linolenic acid 360 mg, Vit B6 150 mg, Vit B1 100 mg, Vit B12 500 μg	[67]
Down syndrome	ALA 600 mg, Vit E 900 IU, Vit C 200 mg	[86]
HCV infection	ALA 300 mg, glycyrrhizin 1 g, *Schisandra* 1.5 g, silymarin 750 mg, Vit C 6 g, L-glutathione 300 mg, Vit E 800 IU	[72]
Psoriasis	ALA 600 mg, CoQ10 50 mg, resveratrol 20 mg, Vit E 36 mg, Krill oil 300 mg, Vitis vinifera seed oil 30 mg, Se 27 mg	[76]
Osteoporosis	ALA 2 × 300 mg, Vit C 30 mg, Vit E 5 mg, Se 2.75 mg	[99]
Vitiligo	ALA 100 mg, Vit C 100 mg, Vit E 40 mg, PUFA 12%	[98]

Table 6. Clinical studies utilizing mitochondrial nutrients and antioxidants and/or herbal compounds aimed at compensating OS/MDF-related pathogenetic mechanisms.

Diseases/Conditions	Coenzyme Q10 + Other Agent(s) (Daily Dose)	References
Heart surgery	CoQ10 100 mg, ALA 100 mg, Mg orotate 400 mg, ω-3 PUFA 300 mg, Se 200 μg	[81]
Chronic heart failure	CoQ10 150 mg, Ca 250 mg, Mg 150 mg, Zn 15 mg, Cu 1.2 mg, Se 50 μg, Vit A 800 μg, thiamine 200 mg, riboflavin 2 mg, Vit B_6 200 mg, folate 5 mg, Vit B_{12} 200 μg, Vit C 500 mg, Vit E 400 mg, Vit D 10 μg	[112,124]
Cardiovascular diseases	CoQ10 120 mg, Vit C 1 g, Vit E 400 IU, Se 200 μg	[128]
Cardiovascular mortality	CoQ10 200 mg, organic Se 200 μg	[136]
Breast cancer	CoQ10 100 mg, riboflavin 10 mg, niacin 50 mg	[175]
	CoQ10 90 mg, Vit C 2850 mg, Vit E 2500 IU, β-carotene 32.5 IU, Se 387 mg, γ-linolenic acid 1.2 g, n-3 fatty acids 3.5 g	[176]
	CoQ10 300 mg, Vit E 300 IU	[177]
Prostate cancer	CoQ10 200 mg, Vit C 750 mg, Vit E 350 mg, Se 200 μg	[179]
Friedreich ataxia	CoQ10 400 mg, Vit E 2100 IU	[140]
	CoQ10 600 mg, Vit E 2100 IU	[141]
Fibromyalgia	CoQ10 200 mg, Gingko biloba extract 200 mg	[160]

A clinical trial utilizing CoQ10 and a mixture of antioxidants was reported by Hertz and Lister [175] to improve survival of patients with end-stage cancer. An early report by Lockwood *et al.* [176] found that supplementation with CoQ10 and antioxidants, γ-linolenic and *n*-3 fatty acids significantly improved clinical conditions in breast cancer patients. A trial by Premkumar *et al.* [177], also conducted on breast cancer patients, reported that a supplementation with CoQ10, riboflavin and niacin decreased the levels of pro-angiogenic factors and increased the levels of anti-angiogenic factors, and reduced tumor burden. On the other hand, a combination therapy with CoQ10 and Vit E did not result in improvement of fatigue symptoms in breast cancer patients, according to Lesser *et al.* [178]. Negative results were obtained by Hoenjet *et al.* [179] by testing a combined supplement containing CoQ10, Vit E, Vit C and selenium in patients with hormonally untreated carcinoma of the prostate, which did not affect serum level of PSA or hormone levels.

Two studies by Hart and co-workers [140,141] reported on a combined treatment with CoQ10 and Vit E in patients with Friedreich ataxia finding sustained improvement in mitochondrial energy production and a significant improvement in cardiac function. A combination therapy with CoQ10 and *Ginkgo biloba* extract significantly improved quality-of-life in patients with clinically diagnosed fibromyalgia [160].

Taken together, the literature on clinical trials testing MN and "classical" antioxidants may provide crucial data on the therapeutic efficacy of compounds aimed at improving mitochondrial health and several OS/MDF-related endpoints, suggesting the design of improved treatments of OS/MDF-related diseases.

7. State-of-Art: Critical Remarks

The present survey of published clinical trials testing MN may reflect the current state-of-art across broad-ranging disorders that are usually focused in separate medical disciplines. On the other hand, the clinical trials testing MN published to date have been conducted on some selected diseases, far below the potential scope of investigation, by considering the extensive range of OS/MDF-related disorders [9]. Trying to interpret this delay, one might find one or more of the following explanations: (a) a disease is commonly recognized to rely on other etiologic grounds than OS/MDF: this is the case, e.g., of Fanconi anemia and of other genetic diseases [8,278]; (b) very rare diseases discourage patient recruitment in view of clinical trials as, e.g., progerias [8,279]; (c) the specialist community may be committed in established therapeutic strategies, thus disregarding potential adjuvant interventions.

Other two remarks may be addressed to the state-of-art in the literature of clinical trials using MN. First, the reviewed studies were mostly oriented to clinical endpoints of the investigated disorders, by underscoring any relevant mechanistic aspects enabling to elucidate trial's results. Second, and most relevant, the vast majority of the reviewed trials tested one MN only, by disregarding: (a) the complementary roles of these cofactors in mitochondrial functions; (b) their different physico-chemical behaviors in water- *vs.* lipid-phase; and (c) the scarce, if any, knowledge as to specific MDF in a given disorder and the appropriate requirements of ALA, or CoQ10, or CARN (or its derivatives) supplementation. Thus, the as yet prevailing design of trials using only one MN may fail to achieve the desired outcomes of OS/MDF compensation and hence of improving clinical endpoints. This limitation, along with some possible shortage in dosages and/or in trial duration might have contributed to some unsuccessful trial outcomes. One may note that supplement combinations of more than one MN, or the use of ALA or CoQ10 with classical antioxidants may have contributed to the overall success of these trials (see Tables 4–6). Thus, one may anticipate the optimization of future clinical trials in OS/MDF-related disorders by means of combined MN treatments, possibly associated with "classical" antioxidants.

The choice of MN formulations may have critical implications on the expected success of MN-based clinical trials. An outstanding aspect in the design of a MN-based clinical trial is represented by the type of formulation chosen. The oxidative state of molecules, chirality, and the vehiculants used in the formulation may have dramatic implications in terms of bioavailability and interaction with the recipient biological system. In the case of CoQ10, bioavailability is known to be influenced by modality of administration, with higher plasma levels reached by administration of the same amount divided in multiple doses, and by vehiculant type with soft gels containing triglyceride dispersing medium showing a better bioavailability in comparison with the crystalline form [280]. Another issue, also linked with bioavailability, is the oxidative state of the active compound in the formulation. Most of the studies cited used ubiquinone, however recently also the reduced form of CoQ10 (ubiquinol) has been used in several clinical trials [113,114,198]. Direct use of ubiquinol displays the advantages of improved absorption and consequently enhanced bioavailability, and direct exposure of the reduced (and active) antioxidant form to digestive epithelial tissue; finally, ubiquinol may influence the ubiquinone/total CoQ10 ratio in conditions where reductive systems are sub-optimal as it has been shown in the aging process [281].

α-Lipoic acid is characterized by an enhanced bioavailability compared to CoQ10, with absorption rates around 30%–40% of an oral dose of ALA [282]. Moreover, while ALA being synthethized by biological systems is exclusively in the R-lipoic acid chiral form, formulations are often (R/S) or (+/−) ALA, composed by a 50/50 mixture of R- and S-ALA. It has been shown that bioavailability of the two isomers is different, R-lipoic acid showing peak plasma concentrations in pharmacokinetic studies 40%–50% higher than S-LA, suggesting that R-LA is better absorbed than S-LA [282].

Regarding CARN and its derivatives, especially ALC, the state-of-art provides extensive information as to the efficacy of CARN, ALC and PLC. The advantages of using ALC *vs.* CARN have been provided [200–203], along with successful use of ALC [68,78,222,223,226,231,246–248,252–254,261,266–268] or PLC [210,225,235,239,253,272] in a number of clinical trials. Acetyl-L-carnitine probably acts as an acetylating agent and its involvement in epigenetic modifications of proteins might explain its success in aging and in OS/MDF-related disorders [283–285]. Future research on CARN/ALC/PLC supplementation should address the correlation of supplement dosage, changes and maintenance of tissue concentrations, and metabolic and functional changes and outcomes [286,287]. An analogous research strategy was highlighted by Camp *et al.* [288] on nutritional interventions for inborn errors of metabolism, suggesting combined strategies in improving the clinical management of an extensive number of disorders.

8. Prospects of Clinical Trials in OS/MDF-Related Disorders

The scheme depicted in Figure 1 points to the prerequisite of identifying the OS and MDF endpoints that are altered in a given disorder, often incompletely elucidated. A choice of OS endpoints should include some established parameters including oxidative damage to DNA, lipids, carbohydrates and proteins, and changes in the glutathione system. A few selected MDF endpoints should be addressed to verify the deficiency, if any, of ALA (to date almost unexplored), CoQ10, and CARN. Other relevant endpoints in assessing an *in vivo* MDF should include mitochondrial DNA content, ATP production, OXPHOS activities, lactate/pyruvate ratio, and expression of mitochondrial antioxidant activities (MnSOD and peroxiredoxin 3). This set of measurements should clarify the quantitative aspects of the disorder to be focused in a clinical trial design. Once the characterization of OS/MDF endpoints is accomplished, the target outcomes of the clinical trials may be defined.

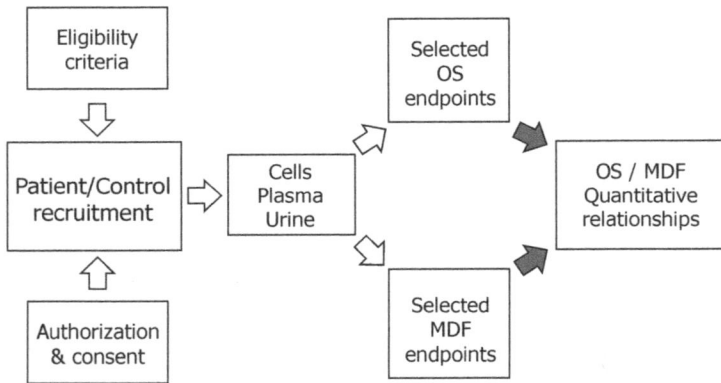

Figure 1. Preliminary database to be achieved in design of mitochondrial nutrients (MN)-based clinical trials.

As shown in Figure 2, the OS/MDF endpoints for a given disease, found as abnormal in the preliminary step, shall be the grounds for choosing the best appropriate supplements to be administered to patients. These supplements may include one or—more reasonably—more MN, as discussed above. The follow-up period will be of paramount relevance; indeed, short admnistration regimens may offer hints for efficacy and safety, yet they may conceal long-run effects. Moreover, in diseases with multi-year-long progression, expecting the desired outcomes may be a quite elusive task. Thus, follow-up ought to be designed for a reasonably long duration with, e.g., six-month intervals in results analysis, allowing us to monitor the outcomes of multi-year-long administration. The ultimate goal will be achieved by compensating OS/MDF endpoints, along with the amelioration of disease progression.

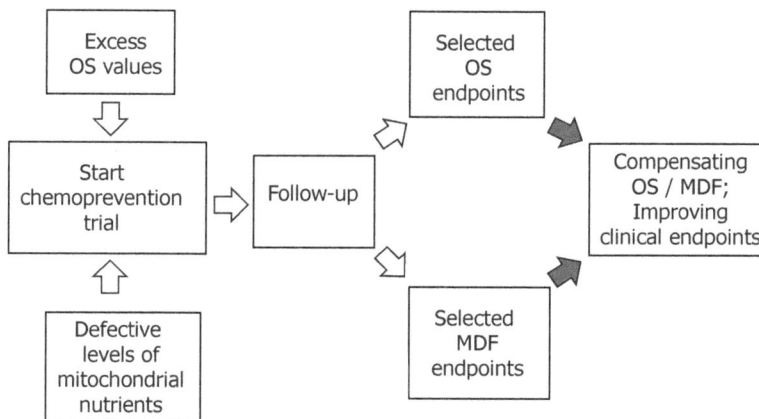

Figure 2. Follow-up procedures combining evaluation of MDF and OS endpoints, along with disease specific clinical endpoints.

Materializing the above prospect may have dramatic consequences in the success of to-be clinical trials for OS/MDF-related disorders, trespassing from "hopes", or from single-parameter information, toward a rational approach that shall foresee properly targeted interventions, then with rational expectations, rather than attempts.

Acknowledgments: Sandra Petrović was supported by the Ministry of Education, Science and Technological Development of the Republic of Serbia (Grant No. 173046).

Conflicts of Interest: The authors declare no conflict of interest.

References

1. Richter, C.C.; Kass, G.E. Oxidative stress in mitochondria: its relationship to cellular Ca^{2+} homeostasis, cell death, proliferation, and differentiation. *Chem. Biol. Interact.* **1991**, *77*, 1–23.
2. Sohal, R.S.; Brunk, U.T. Mitochondrial production of pro-oxidants and cellular senescence. *Mutat. Res./DNAging* **1992**, *275*, 295–304.
3. Di Monte, D.A.; Chan, P.; Sandy, M.S. Glutathione in Parkinson's disease: A link between oxidative stress and mitochondrial damage? *Ann. Neurol.* **1992**, *32*, S111–S115.
4. Beal, M.F. Therapeutic approaches to mitochondrial dysfunction in Parkinson's disease. *Parkinsonism Relat. Disord.* **2009**, *15*, S189–S194.
5. Mischley, L.K.; Allen, J.; Bradley, R. Coenzyme Q10 deficiency in patients with Parkinson's disease. *J. Neurol. Sci.* **2012**, *318*, 72–75.

6. Parikh, S.; Goldstein, A.; Koenig, M.K.; Scaglia, F.; Enns, G.M.; Saneto, R.; Mitochondrial Medicine Society Clinical Directors Working Group; Clinical Director's Work Group. Practice patterns of mitochondrial disease physicians in North America. Part 2: Treatment, care and management. *Mitochondrion* **2013**, *13*, 681–687.

7. Tarnopolsky, M.A. The mitochondrial cocktail: Rationale for combined nutraceutical therapy in mitochondrial cytopathies. *Adv. Drug Deliv. Rev.* **2008**, *60*, 1561–1567.

8. Pallardó, F.V.; Lloret, A.; Lebel, M.; d'Ischia, M.; Cogger, V.C.; le Couteur, D.G.; Gadaleta, M.N.; Castello, G.; Pagano, G. Mitochondrial dysfunction in some oxidative stress-related genetic diseases: Ataxia-Telangiectasia, Down Syndrome, Fanconi Anaemia and Werner Syndrome. *Biogerontology* **2010**, *11*, 401–419.

9. Pagano, G.; Aiello Talamanca, A.; Castello, G.; Cordero, M.D.; d'Ischia, M.; Gadaleta, M.N.; Pallardó, F.V.; Petrović, S.; Tiano, L.; Zatterale, A. Oxidative stress and mitochondrial dysfunction across broad-ranging pathologies: Toward a rational design of chemoprevention strategies by means of mitochondrial nutrients. *Oxid. Med. Cell. Longev.* **2014**, *2014*, 541230.

10. Marshall, A.W.; Graul, R.S.; Morgan, M.Y.; Sherlock, S. Treatment of alcohol-related liver disease with thioctic acid: A six month randomised double-blind trial. *Gut* **1982**, *23*, 1088–1093.

11. Tanaka, J.; Tominaga, R.; Yoshitoshi, M.; Matsui, K.; Komori, M.; Sese, A.; Yasui, H.; Tokunaga, K. Coenzyme Q10: The prophylactic effect on low cardiac output following cardiac valve replacement. *Ann. Thorac. Surg.* **1982**, *33*, 145–151.

12. Langsjoen, P.H.; Vadhanavikit, S.; Folkers, K. Response of patients in classes III and IV of cardiomyopathy to therapy in a blind and crossover trial with coenzyme Q10. *Proc. Natl. Acad. Sci. USA* **1985**, *82*, 4240–4244.

13. Ogasahara, S.; Nishikawa, Y.; Yorifuji, S.; Soga, F.; Nakamura, Y.; Takahashi, M.; Hashimoto, S.; Kono, N.; Tarui, S. Treatment of Kearns–Sayre syndrome with coenzyme Q10. *Neurology* **1986**, *36*, 45–53.

14. Scarlato, G.N.; Bresolin, N.I.; Moroni, I.; Doriguzzi, C.; Castelli, E.; Comi, G.; Angelini, C.; Carenzi, A. Multicenter trial with ubidecarenone: Treatment of 44 patients with mitochondrial myopathies. *Rev. Neurol.* **1991**, *147*, 542–548.

15. Mayr, J.A.; Zimmermann, F.A.; Fauth, C.; Bergheim, C.; Meierhofer, D.; Radmayr, D.; Zschocke, J.; Koch, J.; Sperl, W. Lipoic acid synthetase deficiency causes neonatal-onset epilepsy, defective mitochondrial energy metabolism, and glycine elevation. *Am. J. Hum. Genet.* **2011**, *89*, 792–797.

16. Liu, J. The effects and mechanisms of mitochondrial nutrient alpha-lipoic acid on improving age-associated mitochondrial and cognitive dysfunction: An overview. *Neurochem. Res.* **2008**, *33*, 194–203.

17. Maes, M.; Mihaylova, I.; Kubera, M.; Uytterhoeven, M.; Vrydags, N.; Bosmans, E. Coenzyme Q10 deficiency in myalgic encephalomyelitis/chronic fatigue syndrome (ME/CFS) is related to fatigue, autonomic and neurocognitive symptoms and is another risk factor explaining the early mortality in ME/CFS due to cardiovascular disorder. *Neuro Endocrinol. Lett.* **2009**, *30*, 470–476.

18. Tiano, L.; Busciglio, J. Mitochondrial dysfunction and Down's syndrome: Is there a role for coenzyme Q10? *Biofactors* **2011**, *37*, 386–392.

19. Hardas, S.S.; Sultana, R.; Clark, A.M.; Beckett, T.L.; Szweda, L.I.; Murphy, M.P.; Butterfield, D.A. Oxidative modification of lipoic acid by HNE in Alzheimer disease brain. *Redox Biol.* **2013**, *1*, 80–85.

20. Kähler, W.; Kuklinski, B.; Rühlmann, C.; Plötz, C. Diabetes mellitus—A free radical-associated disease. Results of adjuvant antioxidant supplementation. *Z. Gesamte Inn. Med.* **1993**, *48*, 223–232.

21. Jacob, S.; Henriksen, E.J.; Schiemann, A.L.; Simon, I.; Clancy, D.E.; Tritschler, H.J.; Jung, W.I.; Augustin, H.J.; Dietze, G.J. Enhancement of glucose disposal in patients with type 2 diabetes by alpha-lipoic acid. *Arzneimittelforschung* **1995**, *45*, 872–874.

22. Ziegler, D.; Hanefeld, M.; Ruhnau, K.J.; Meissner, H.P.; Lobisch, M.; Schütte, K.; Gries, F.A. Treatment of symptomatic diabetic peripheral neuropathy with the anti-oxidant alpha-lipoic acid. A 3-week multicentre randomized controlled trial (ALADIN Study). *Diabetologia* **1995**, *38*, 1425–1433.

23. Ziegler, D.; Schatz, H.; Conrad, F.; Gries, F.A.; Ulrich, H.; Reichel, G. Effects of treatment with the antioxidant alpha-lipoic acid on cardiac autonomic neuropathy in NIDDM patients. A 4-month randomized controlled multicenter trial (DEKAN Study). Deutsche Kardiale Autonome Neuropathie. *Diabetes Care* **1997**, *20*, 369–373.

24. Strokov, I.A.; Kozlova, N.A.; Mozolevskiĭ, IuV; Miasoedov, S.P.; Iakhno, N.N. The efficacy of the intravenous administration of the trometamol salt of thioctic (alpha-lipoic) acid in diabetic neuropathy. *Zh. Nevrol. Psikhiatr. Im. S. S. Korsakova* **1999**, *99*, 18–22.

25. Jacob, S.P.; Ruus, P.R.; Hermann, R.; Tritschler, H.J.; Maerker, E.; Renn, W.; Augustin, H.J.; Dietze, G.J.; Rett, K. Oral administration of RAC-alpha-lipoic acid modulates insulin sensitivity in patients with type-2 diabetes mellitus: A placebo-controlled pilot trial. *Free Radic. Biol. Med.* **1999**, *27*, 309–314.

26. Ziegler, D.; Hanefeld, M.; Ruhnau, K.J.; Hasche, H.; Lobisch, M.; Schütte, K.; Kerum, G.; Malessa, R. Treatment of symptomatic diabetic polyneuropathy with the antioxidant alpha-lipoic acid: A 7-month multicenter randomized controlled trial (ALADIN III study). ALADIN III Study Group. Alpha-lipoic acid in diabetic neuropathy. *Diabetes Care* **1999**, *22*, 1296–1301.

27. Reljanovic, M.G.; Reichel, G.K.; Rett, K.; Lobisch, M.; Schuette, K.; Möller, W.; Tritschler, H.J.; Mehnert, H. Treatment of diabetic polyneuropathy with the antioxidant thioctic acid (alpha-lipoic acid): A two year multicenter randomized double-blind placebo-controlled trial (ALADIN II). Alpha lipoic acid in diabetic neuropathy. *Free Radic. Res.* **1999**, *31*, 171–179.

28. Ruhnau, K.J.; Meissner, H.P.; Finn, J.R.; Reljanovic, M.; Lobisch, M.; Schütte, K.; Nehrdich, D.; Tritschler, H.J.; Mehnert, H.; Ziegler, D. Effects of 3-week oral treatment with the antioxidant thioctic acid (alpha-lipoic acid) in symptomatic diabetic polyneuropathy. *Diabet. Med.* **1999**, *16*, 1040–1043.

29. Negrişanu, G.; Roşu, M.; Bolte, B.; Lefter, D.; Dabelea, D. Effects of 3-month treatment with the antioxidant alpha-lipoic acid in diabetic peripheral neuropathy. *Rom. J. Intern. Med.* **1999**, *37*, 297–306.

30. Androne, L.; Gavan, N.A.; Veresiu, I.A.; Orasan, R. In vivo effect of lipoic acid on lipid peroxidation in patients with diabetic neuropathy. *In Vivo* **2000**, *14*, 327–330.

31. Haak, E.; Usadel, K.H.; Kusterer, K.; Amini, P.; Frommeyer, R.; Tritschler, H.J.; Haak, T. Effects of alpha-lipoic acid on microcirculation in patients with peripheral diabetic neuropathy. *Exp. Clin. Endocrinol. Diabetes* **2000**, *108*, 168–174.

32. Evans, J.L.; Heymann, C.J.; Goldfine, I.D.; Gavin, L.A. Pharmacokinetics, tolerability, and fructosamine-lowering effect of a novel, controlled-release formulation of alpha-lipoic acid. *Endocr. Pract.* **2002**, *8*, 29–35.

33. Ametov, A.S.; Barinov, A.; Dyck, P.J.; Hermann, R.; Kozlova, N.; Litchy, W.J.; Low, P.A.; Nehrdich, D.; Novosadova, M.; O'Brien, P.C.; et al. The sensory symptoms of diabetic polyneuropathy are improved with alpha-lipoic acid: The SYDNEY trial. *Diabetes Care* **2003**, *26*, 770–776.

34. Hahm, J.R.; Kim, B.J.; Kim, K.W. Clinical experience with thioctacid (thioctic acid) in the treatment of distal symmetric polyneuropathy in Korean diabetic patients. *J. Diabetes Complicat.* **2004**, *18*, 79–85.

35. Ziegler, D.; Ametov, A.; Barinov, A.; Dyck, P.J.; Gurieva, I.; Low, P.A.; Munzel, U.; Yakhno, N.; Raz, I.; Novosadova, M.; et al. Oral treatment with alpha-lipoic acid improves symptomatic diabetic polyneuropathy: The SYDNEY 2 trial. *Diabetes Care* **2006**, *29*, 2365–2370.

36. Kamenova, P. Improvement of insulin sensitivity in patients with type 2 diabetes mellitus after oral administration of alpha-lipoic acid. *Hormones* **2006**, *5*, 251–258.

37. Liu, F.; Zhang, Y.; Yang, M.; Liu, B.; Shen, Y.D.; Jia, W.P.; Xiang, K.S. Curative effect of alpha-lipoic acid on peripheral neuropathy in type 2 diabetes: A clinical study. *Zhonghua Yi Xue Za Zhi* **2007**, *87*, 2706–2709.

38. Xiang, G.D.; Sun, H.L.; Zhao, L.S.; Hou, J.; Yue, L.; Xu, L. The antioxidant alpha-lipoic acid improves endothelial dysfunction induced by acute hyperglycaemia during OGTT in impaired glucose tolerance. *Clin. Endocrinol.* **2008**, *68*, 716–723.

39. Bureković, A.; Terzić, M.; Alajbegović, S.; Vukojević, Z.; Hadzić, N. The role of alpha-lipoic acid in diabetic polyneuropathy treatment. *Bosn. J. Basic Med. Sci.* **2008**, *8*, 341–245.

40. Heinisch, B.B.; Francesconi, M.; Mittermayer, F.; Schaller, G.; Gouya, G.; Wolzt, M.; Pleiner, J. Alpha-lipoic acid improves vascular endothelial function in patients with type 2 diabetes: A placebo-controlled randomized trial. *Eur. J. Clin. Investig.* **2010**, *40*, 148–154.

41. Palacka, P.; Kucharska, J.; Murin, J.; Dostalova, K.; Okkelova, A.; Cizova, M.; Waczulikova, I.; Moricova, S.; Gvozdjakova, A. Complementary therapy in diabetic patients with chronic complications: A pilot study. *Bratisl. Lek. Listy* **2010**, *111*, 205–211.

42. Gu, X.M.; Zhang, S.S.; Wu, J.C.; Tang, Z.Y.; Lu, Z.Q.; Li, H.; Liu, C.; Chen, L.; Ning, G. Efficacy and safety of high-dose α-lipoic acid in the treatment of diabetic polyneuropathy. *Zhonghua Yi Xue Za Zhi* **2010**, *90*, 2473–2476.

43. Xiang, G.; Pu, J.; Yue, L.; Hou, J.; Sun, H. α-Lipoic acid can improve endothelial dysfunction in subjects with impaired fasting glucose. *Metabolism* **2011**, *60*, 480–485.

44. De Oliveira, A.M.; Rondó, P.H.; Luzia, L.A.; D'Abronzo, F.H.; Illison, V.K. The effects of lipoic acid and α-tocopherol supplementation on the lipid profile and insulin sensitivity of patients with type 2 diabetes mellitus: A randomized, double-blind, placebo-controlled trial. *Diabetes Res. Clin. Pract.* **2011**, *92*, 253–260.

45. Ansar, H.; Mazloom, Z.; Kazemi, F.; Hejazi, N. Effect of alpha-lipoic acid on blood glucose, insulin resistance and glutathione peroxidase of type 2 diabetic patients. *Saudi Med. J.* **2011**, *32*, 584–588.

46. Ziegler, D.; Low, P.A.; Litchy, W.J.; Boulton, A.J.; Vinik, A.I.; Freeman, R.; Samigull, R.; Tritschler, H.; Munzel, U.; Maus, J.; *et al.* Efficacy and safety of antioxidant treatment with α-lipoic acid over 4 years in diabetic polyneuropathy: The NATHAN 1 trial. *Diabetes Care* **2011**, *34*, 2054–2060.

47. Haritoglou, C.; Gerss, J.; Hammes, H.P.; Kampik, A.; Ulbig, M.W.; RETIPON Study Group. Alpha-lipoic acid for the prevention of diabetic macular edema. *Ophthalmologica* **2011**, *226*, 127–137.

48. Rahman, S.T.; Merchant, N.; Haque, T.; Wahi, J.; Bhaheetharan, S.; Ferdinand, K.C.; Khan, B.V. The impact of lipoic acid on endothelial function and proteinuria in quinapril-treated diabetic patients with stage I hypertension: Results from the QUALITY study. *J. Cardiovasc. Pharmacol. Ther.* **2012**, *17*, 139–145.

49. Bertolotto, F.; Massone, A. Combination of alpha lipoic acid and superoxide dismutase leads to physiological and symptomatic improvements in diabetic neuropathy. *Drugs R&D* **2012**, *12*, 29–34.

50. Porasuphatana, S.; Suddee, S.; Nartnampong, A.; Konsil, J.; Harnwong, B.; Santaweesuk, A. Glycemic and oxidative status of patients with type 2 diabetes mellitus following oral administration of alpha-lipoic acid: A randomized double-blinded placebo-controlled study. *Asia Pac. J. Clin. Nutr.* **2012**, *21*, 12–21.

51. Ibrahimpasic, K. Alpha lipoic acid and glycaemic control in diabetic neuropathies at type 2 diabetes treatment. *Med. Arch.* **2013**, *67*, 7–9.

52. Tankova, T.; Koev, D.; Dakovska, L. Alpha-lipoic acid in the treatment of autonomic diabetic neuropathy (controlled, randomized, open-label study). *Rom. J. Intern. Med.* **2004**, *42*, 457–464.

53. Huang, E.A.; Gitelman, S.E. The effect of oral alpha-lipoic acid on oxidative stress in adolescents with type 1 diabetes mellitus. *Pediat. Diabetes* **2008**, *9*, 69–73.

54. Mollo, R.; Zaccardi, F.; Scalone, G.; Scavone, G.; Rizzo, P.; Navarese, E.P.; Manto, A.; Pitocco, D.; Lanza, G.A.; Ghirlanda, G.; *et al.* Effect of α-lipoic acid on platelet reactivity in type 1 diabetic patients. *Diabetes Care* **2012**, *35*, 196–197.

55. Morcos, M.; Borcea, V.; Isermann, B.; Gehrke, S.; Ehret, T.; Henkels, M.; Schiekofer, S.; Hofmann, M.; Amiral, J.; Tritschler, H.; *et al.* Effect of alpha-lipoic acid on the progression of endothelial cell damage and albuminuria in patients with diabetes mellitus: An exploratory study. *Diabetes Res. Clin. Pract.* **2001**, *5*, 175–183.

56. Noori, N.; Tabibi, H.; Hosseinpanah, F.; Hedayati, M.; Nafar, M. Effects of combined lipoic acid and pyridoxine on albuminuria, advanced glycation end-products, and blood pressure in diabetic nephropathy. *Int. J. Vitam. Nutr. Res.* **2013**, *83*, 77–85.

57. Hegazy, S.K.; Tolba, O.A.; Mostafa, T.M.; Eid, M.A.; el-Afify, D.R. Alpha-lipoic acid improves subclinical left ventricular dysfunction in asymptomatic patients with type 1 diabetes. *Rev. Diabet. Stud.* **2013**, *10*, 58–67.

58. Huang, Z.; Wan, X.; Liu, J.; Deng, W.; Chen, A.; Liu, L.; Liu, J.; Wei, G.; Li, H.; Fang, D.; *et al.* Short-term continuous subcutaneous insulin infusion combined with insulin sensitizers rosiglitazone, metformin, or antioxidant α-lipoic acid in patients with newly diagnosed type 2 diabetes mellitus. *Diabetes Technol. Ther.* **2013**, *15*, 859–869.

59. Mitkov, M.D.; Aleksandrova, I.Y.; Orbetzova, M.M. Effect of transdermal testosterone or alpha-lipoic acid on erectile dysfunction and quality of life in patients with type 2 diabetes mellitus. *Folia Med.* **2013**, *55*, 55–63.

60. Pop-Busui, R.; Stevens, M.J.; Raffel, D.M.; White, E.A.; Mehta, M.; Plunkett, C.D.; Brown, M.B.; Feldman, E.L. Effects of triple antioxidant therapy on measures of cardiovascular autonomic neuropathy and on myocardial blood flow in type 1 diabetes: A randomized controlled trial. *Diabetologia* **2013**, *56*, 1835–1844.

61. Zhang, X.; Zhang, Y.; Gao, X.; Wu, J.; Jiao, X.; Zhao, J.; Lv, X. Investigating the role of backward walking therapy in alleviating plantar pressure of patients with diabetic peripheral neuropathy. *Arch. Phys. Med. Rehabil.* **2014**, *95*, 832–839.

62. Yadav, V.; Marracci, G.; Lovera, J.; Woodward, W.; Bogardus, K.; Marquardt, W.; Shinto, L.; Morris, C.; Bourdette, D. Lipoic acid in multiple sclerosis: A pilot study. *Mult. Sclerosis* **2005**, *11*, 159–165.

63. Khalili, M.; Azimi, A.; Izadi, V.; Eghtesadi, S.; Mirshafiey, A.; Sahraian, M.A.; Motevalian, A.; Norouzi, A.; Sanoobar, M.; Eskandari, G.; *et al.* Does lipoic acid consumption affect the cytokine profile in multiple sclerosis patients: A double-blind, placebo-controlled, randomized clinical trial. *Neuroimmunomodulation* **2014**, *21*, 291–296.

64. Bae, S.C.; Jung, W.J.; Lee, E.J.; Yu, R.; Sung, M.K. Effects of antioxidant supplements intervention on the level of plasma inflammatory molecules and disease severity of rheumatoid arthritis patients. *J. Am. Coll. Nutr.* **2009**, *28*, 56–62.

65. Galasko, D.R.; Peskind, E.; Clark, C.M.; Quinn, J.F.; Ringman, J.M.; Jicha, G.A.; Cotman, C.; Cottrell, B.; Montine, T.J.; Thomas, R.G.; *et al.* Alzheimer's disease cooperative study. Antioxidants for Alzheimer disease: A randomized clinical trial with cerebrospinal fluid biomarker measures. *Arch. Neurol.* **2012**, *69*, 836–841.

66. Magis, D.; Ambrosini, A.; Sándor, P.; Jacquy, J.; Laloux, P.; Schoenen, J. A randomized double-blind placebo-controlled trial of thioctic acid in migraine prophylaxis. *Headache* **2007**, *47*, 52–57.

67. Di Geronimo, G.; Caccese, A.F.; Caruso, L.; Soldati, A.; Passaretti, U. Treatment of carpal tunnel syndrome with alpha-lipoic acid. *Eur. Rev. Med. Pharmacol. Sci.* **2009**, *13*, 133–139.

68. Memeo, A.; Loiero, M. Thioctic acid and acetyl-L-carnitine in the treatment of sciatic pain caused by a herniated disc: A randomized, double-blind, comparative study. *Clin. Drug Investig.* **2008**, *28*, 495–500.

69. Ranieri, M.; Sciuscio, M.; Cortese, A.M.; Santamato, A.; di Teo, L.; Ianieri, G.; Bellomo, R.G.; Stasi, M.; Megna, M. The use of alpha-lipoic acid (ALA), gamma linolenic acid (GLA) and rehabilitation in the treatment of back pain: Effect on health-related quality of life. *Int. J. Immunopathol. Pharmacol.* **2009**, *22*, 45–50.

70. Shinto, L.; Quinn, J.; Montine, T.; Dodge, H.H.; Woodward, W.; Baldauf-Wagner, S.; Waichunas, D.; Bumgarner, L.; Bourdette, D.; Silbert, L.; *et al.* A randomized placebo-controlled pilot trial of omega-3 fatty acids and alpha lipoic acid in Alzheimer's disease. *J. Alzheimers Dis.* **2014**, *38*, 111–120.

71. Dünschede, F.; Erbes, K.; Kircher, A.; Westermann, S.; Seifert, J.; Schad, A.; Oliver, K.; Kiemer, A.K.; Theodor, J. Reduction of ischemia reperfusion injury after liver resection and hepatic inflow occlusion by alpha-lipoic acid in humans. *World J. Gastroenterol.* **2006**, *12*, 6812–6817.

72. Melhem, A.; Stern, M.; Shibolet, O.; Israeli, E.; Ackerman, Z.; Pappo, O.; Hemed, N.; Rowe, M.; Ohana, H.; Zabrecky, G.; *et al.* Treatment of chronic hepatitis C virus infection via antioxidants: Results of a phase I clinical trial. *J. Clin. Gastroenterol.* **2005**, *39*, 737–742.

73. Koh, E.H.; Lee, W.J.; Lee, S.A.; Kim, E.H.; Cho, E.H.; Jeong, E.; Kim, D.W.; Kim, M.S.; Park, J.Y.; Park, K.G.; *et al.* Effects of alpha-lipoic acid on body weight in obese subjects. *Am. J. Med.* **2011**, *124*, 85.e1-8.

74. McNeilly, A.M.; Davison, G.W.; Murphy, M.H.; Nadeem, N.; Trinick, T.; Duly, E.; Novials, A.; McEneny, J. Effect of α-lipoic acid and exercise training on cardiovascular disease risk in obesity with impaired glucose tolerance. *Lipids Health Dis.* **2011**, *10*, 217.

75. Zhang, Y.; Han, P.; Wu, N.; He, B.; Lu, Y.; Li, S.; Liu. Y.; Zhao, S.; Liu, L.; Li, Y. Amelioration of lipid abnormalities by α-lipoic acid through antioxidative and anti-inflammatory effects. *Obesity* **2011**, *19*, 1647–1653.

76. Skroza, N.; Proietti, I.; Bernardini, N.; la Viola, G.; Nicolucci, F.; Pampena, R.; Tolino, E.; Zuber, S.; Mancini, M.T.; Soccodato, V.; *et al.* Efficacy of food supplement to improve metabolic syndrome parameters in patients affected by moderate to severe psoriasis during anti-TNFα treatment. *G. Ital. Dermatol. Venereol.* **2013**, *148*, 661–665.

77. Manning, P.J.; Sutherland, W.H.; Williams, S.M.; Walker, R.J.; Berry, E.A.; de Jong, S.A.; Ryalls, A.R. The effect of lipoic acid and vitamin E therapies in individuals with the metabolic syndrome. *Nutr. Metab. Cardiovasc. Dis.* **2013**, *23*, 543–549.

78. McMackin, C.J.; Widlansky, M.E.; Hamburg, N.M. Effect of combined treatment with alpha-lipoic acid and acetyl-L-carnitine on vascular function and blood pressure in patients with coronary artery disease. *J. Clin. Hypertens.* **2007**, *9*, 249–255.

79. Li, R.J.; Ji, W.Q.; Pang, J.J.; Wang, J.L.; Chen, Y.G.; Zhang, Y. Alpha-lipoic acid ameliorates oxidative stress by increasing aldehyde dehydrogenase-2 activity in patients with acute coronary syndrome. *Tohoku J. Exp. Med.* **2013**, *229*, 45–51.

80. Vincent, H.K.; Bourguignon, C.M.; Vincent, K.R.; Taylor, A.G. Effects of alpha-lipoic acid supplementation in peripheral arterial disease: A pilot study. *J. Altern. Complement. Med.* **2007**, *13*, 577–584.

81. Leong, J.Y.; van der Merwe, J.; Pepe, S.; Bailey, M.; Perkins, A.; Lymbury, R.; Esmore, D.; Marasco, S.; Rosenfeldt, F. Perioperative metabolic therapy improves redox status and outcomes in cardiac surgery patients: A randomised trial. *Heart Lung Circ.* **2010**, *19*, 584–591.

82. Chang, J.W.; Lee, E.K.; Kim, T.H.; Min, W.K.; Chun, S.; Lee, K.U.; Kim, S.B.; Park, J.S. Effects of alpha-lipoic acid on the plasma levels of asymmetric dimethylarginine in diabetic end-stage renal disease patients on hemodialysis: A pilot study. *Am. J. Nephrol.* **2007**, *27*, 70–74.

83. Ramos, L.F.; Kane, J.; McMonagle, E.; Le, P.; Wu, P.; Shintani, A.; Ikizler, T.A.; Himmelfarb, J. Effects of combination tocopherols and alpha lipoic acid therapy on oxidative stress and inflammatory biomarkers in chronic kidney disease. *J. Ren. Nutr.* **2011**, *21*, 211–218.

84. Khabbazi, T.; Mahdavi, R.; Safa, J.; Pour-Abdollahi, P. Effects of alpha-lipoic acid supplementation on inflammation, oxidative stress, and serum lipid profile levels in patients with end-stage renal disease on hemodialysis. *J. Ren. Nutr.* **2012**, *22*, 244–250.

85. Himmelfarb, J.; Ikizler, T.A.; Ellis, C.; Wu, P.; Shintani, A.; Dalal, S.; Kaplan, M.; Chonchol, M.; Hakim, R.M. Provision of antioxidant therapy in hemodialysis (PATH): A randomized clinical trial. *J. Am. Soc. Nephrol.* **2014**, *25*, 623–633.

86. Lott, I.T.; Doran, E.; Nguyen, V.Q.; Tournay, A.; Head, E.; Gillen, D.L. Down syndrome and dementia: A randomized, controlled trial of antioxidant supplementation. *Am. J. Med. Genet. A* **2011**, *155A*, 1939–1948.

87. Martins, V.D.; Manfredini, V.; Peralba, M.C.; Benfato, M.S. Alpha-lipoic acid modifies oxidative stress parameters in sickle cell trait subjects and sickle cell patients. *Clin. Nutr.* **2009**, *28*, 192–197.

88. Rodriguez, M.C.; MacDonald, J.R.; Mahoney, D.J.; Parise, G.; Beal, M.F.; Tarnopolsky, M.A. Beneficial effects of creatine, CoQ10, and lipoic acid in mitochondrial disorders. *Muscle Nerve* **2007**, *35*, 235–242.

89. Femiano, F.; Gombos, F.; Scully, C.; Busciolano, M.; de Luca, P. Burning mouth syndrome (BMS): Controlled open trial of the efficacy of alpha-lipoic acid (thioctic acid) on symptomatology. *Oral Dis.* **2000**, *6*, 274–277.

90. Carbone, M.; Pentenero, M.; Carrozzo, M.; Ippolito, A.; Gandolfo, S. Lack of efficacy of alpha-lipoic acid in burning mouth syndrome: A double-blind, randomized, placebo-controlled study. *Eur. J. Pain* **2009**, *13*, 492–496.

91. López-Jornet, P.; Camacho-Alonso, F.; Leon-Espinosa, S. Efficacy of alpha lipoic acid in burning mouth syndrome: A randomized, placebo-treatment study. *J. Oral. Rehabil.* **2009**, *36*, 52–57.

92. Marino, R.; Torretta, S.; Capaccio, P.; Pignataro, L.; Spadari, F. Different therapeutic strategies for burning mouth syndrome: Preliminary data. *J. Oral Pathol. Med.* **2010**, *39*, 611–616.

93. López-D'Alessandro, E.; Escovich, L. Combination of alpha lipoic acid and gabapentin, its efficacy in the treatment of Burning Mouth Syndrome: A randomized, double-blind, placebo controlled trial. *Med. Oral Patol. Oral Cir. Bucal.* **2011**, *16*, e635–e640.

94. Mantovani, G.; Macciò, A.; Madeddu, C.; Gramignano, G.; Lusso, M.R.; Serpe, R.; Massa, E.; Astara, G.; Deiana, L. A phase II study with antioxidants, both in the diet and supplemented, pharmaconutritional support, progestagen, and anti-cyclooxygenase-2 showing efficacy and safety in patients with cancer-related anorexia/cachexia and oxidative stress. *Cancer Epidemiol. Biomark. Prev.* **2006**, *15*, 1030–1034.

95. Mantovani, G.; Macciò, A.; Madeddu, C.; Gramignano, G.; Serpe, R.; Massa, E.; Dessì, M.; Tanca, F.M.; Sanna, E.; Deiana, L.; *et al.* Randomized phase III clinical trial of five different arms of treatment for patients with cancer cachexia: Interim results. *Nutrition* **2008**, *24*, 305–313.

96. Jariwalla, R.J.; Lalezari, J.; Cenko, D.; Mansour, S.E.; Kumar, A.; Gangapurkar, B.; Nakamura, D. Restoration of blood total glutathione status and lymphocyte function following alpha-lipoic acid supplementation in patients with HIV infection. *J. Altern. Complement. Med.* **2008**, *14*, 139–146.

97. Milazzo, L.; Menzaghi, B.; Caramma, I.; Nasi, M.; Sangaletti, O.; Cesari, M.; Zanone Poma, B.; Cossarizza, A.; Antinori, S.; Galli, M. Effect of antioxidants on mitochondrial function in HIV-1-related lipoatrophy: A pilot study. *AIDS Res. Hum. Retroviruses* **2010**, *26*, 1207–1214.

98. Dell'Anna, M.L.; Mastrofrancesco, A.; Sala, R.; Venturini, M.; Ottaviani, M.; Vidolin, A.P.; Leone, G.; Calzavara, P.G.; Westerhof, W.; Picardo, M. Antioxidants and narrow band-UVB in the treatment of vitiligo: A double-blind placebo controlled trial. *Clin. Exp. Dermatol.* **2007**, *32*, 631–636.

99. Mainini, G.; Rotondi, M.; di Nola, K.; Pezzella, M.T.; Iervolino, S.A.; Seguino, E.; D'Eufemia, D.; Iannicelli, I.; Torella, M. Oral supplementation with antioxidant agents containing alpha lipoic acid: Effects on postmenopausal bone mass. *Clin. Exp. Obstet. Gynecol.* **2012**, *39*, 489–493.

100. Garrido-Maraver, J.; Cordero, M.D.; Oropesa-Avila, M.; Oropesa-Avila, M.; Vega, A.F.; de la Mata, M.; Pavon, A.D.; Alcocer-Gomez, E.; Calero, C.P.; Paz, M.V.; *et al.* Clinical applications of coenzyme Q10. *Front. Biosci.* **2014**, *19*, 619–633.

101. Singh, R.B.; Niaz, M.A. Serum concentration of lipoprotein(a) decreases on treatment with hydrosoluble coenzyme Q10 in patients with coronary artery disease: Discovery of a new role. *Int. J. Cardiol.* **1999**, *68*, 23–29.

102. Singh, R.B.; Niaz, M.A.; Rastogi, S.S.; Shukla, P.K.; Thakur, A.S. Effect of hydrosoluble coenzyme Q10 on blood pressures and insulin resistance in hypertensive patients with coronary artery disease. *J. Hum. Hypertens.* **1999**, *13*, 203–208.

103. Tiano, L.; Belardinelli, R.; Carnevali, P.; Principi, F.; Seddaiu, G.; Littarru, G.P. Effect of coenzyme Q10 administration on endothelial function and extracellular superoxide dismutase in patients with ischaemic heart disease: A double-blind, randomized controlled study. *Eur. Heart J.* **2007**, *28*, 2249–2255.

104. Belardinelli, R.; Muçaj, A.; Lacalaprice, F.; Solenghi, M.; Principi, F.; Tiano, L.; Littarru, G.P. Coenzyme Q10 improves contractility of dysfunctional myocardium in chronic heart failure. *Biofactors* **2005**, *25*, 137–145.

105. Dai, Y.L.; Luk, T.H.; Yiu, K.H.; Wang, M.; Yip, P.M.; Lee, S.W.; Li, S.W.; Tam, S.; Fong, B.; Lau, C.P.; *et al.* Reversal of mitochondrial dysfunction by coenzyme Q10 supplement improves endothelial function in patients with ischaemic left ventricular systolic dysfunction: A randomized controlled trial. *Atherosclerosis* **2011**, *216*, 395–401.

106. Lee, B.J.; Huang, Y.C.; Chen, S.J.; Lin, P.T. Coenzyme Q10 supplementation reduces oxidative stress and increases antioxidant enzyme activity in patients with coronary artery disease. *Nutrition* **2012**, *28*, 250–255.

107. Hofman-Bang, C.; Rehnqvist, N.; Swedberg, K.; Wiklund, I.; Aström, H. Coenzyme Q10 as an adjunctive in the treatment of chronic congestive heart failure. The Q10 Study Group. *J. Card. Fail.* **1995**, *1*, 101–107.

108. Munkholm, H.; Hansen, H.H.; Rasmussen, K. Coenzyme Q10 treatment in serious heart failure. *Biofactors* **1999**, *9*, 285–289.

109. Khatta, M.; Alexander, B.S.; Krichten, C.M.; Fisher, M.L.; Freudenberger, R.; Robinson, S.W.; Gottlieb, S.S. The effect of coenzyme Q10 in patients with congestive heart failure. *Ann. Intern. Med.* **2000**, *132*, 636–640.

110. Berman, M.; Erman, A.; Ben-Gal, T.; Dvir, D.; Georghiou, G.P.; Stamler, A.; Vered, Y.; Vidne, B.A.; Aravot, D. Coenzyme Q10 in patients with end-stage heart failure awaiting cardiac transplantation: A randomized, placebo-controlled study. *Clin. Cardiol.* **2004**, *27*, 295–299.

111. Damian, M.S.; Ellenberg, D.; Gildemeister, R.; Lauermann, J.; Simonis, G.; Sauter, W.; Georgi, C. Coenzyme Q10 combined with mild hypothermia after cardiac arrest: A preliminary study. *Circulation* **2004**, *110*, 3011–3016.

112. Witte, K.K.; Nikitin, N.P.; Parker, A.C.; von Haehling, S.; Volk, H.D.; Anker, S.D.; Clark, A.L.; Cleland, J.G. The effect of micronutrient supplementation on quality-of-life and left ventricular function in elderly patients with chronic heart failure. *Eur. Heart J.* **2005**, *26*, 2238–2244.

113. Langsjoen, P.H.; Langsjoen, A.M. Supplemental ubiquinol in patients with advanced congestive heart failure. *Biofactors* **2008**, *32*, 119–128.

114. Kumar, A.; Singh, R.B.; Saxena, M.; Niaz, M.A.; Josh, S.R.; Chattopadhyay, P.; Mechirova, V.; Pella, D.; Fedacko, J. Effect of carni Q-gel (ubiquinol and carnitine) on cytokines in patients with heart failure in the Tishcon study. *Acta Cardiol.* **2007**, *62*, 349–354.

115. Belcaro, G.; Cesarone, M.R.; Dugall, M.; Hosoi, M.; Ippolito, E.; Bavera, P.; Grossi, M.G. Investigation of pycnogenol in combination with coenzyme Q10 in heart failure patients (NYHA II/III). *Panminerva Med.* **2010**, *52*, 21–25.

116. Fumagalli, S.; Fattirolli, F.; Guarducci, L.; Cellai, T.; Baldasseroni, S.; Tarantini, F.; di Bari, M.; Masotti, G.; Marchionni, N. Coenzyme Q10 terclatrate and creatine in chronic heart failure: A randomized, placebo-controlled, double-blind study. *Clin. Cardiol.* **2011**, *34*, 211–217.

117. Kuklinski, B.; Weissenbacher, E.; Fähnrich, A. Coenzyme Q10 and antioxidants in acute myocardial infarction. *Mol. Aspects Med.* **1994**, *15*, s143–s147.

118. Singh, R.B.; Wander, G.S.; Rastogi, A.; Shukla, P.K.; Mittal, A.; Sharma, J.P.; Mehrotra, S.K.; Kapoor, R.; Chopra, R.K. Randomized, double-blind placebo-controlled trial of coenzyme Q10 in patients with acute myocardial infarction. *Cardiovasc. Drugs Ther.* **1998**, *12*, 347–353.

119. Singh, R.B.; Neki, N.S.; Kartikey, K.; Pella, D.; Kumar, A.; Niaz, M.A.; Thakur, A.S. Effect of coenzyme Q10 on risk of atherosclerosis in patients with recent myocardial infarction. *Mol. Cell. Biochem.* **2003**, *246*, 75–82.

120. Permanetter, B.; Rössy, W.; Klein, G.; Weingartner, F.; Seidl, K.F.; Blömer, H. Ubiquinone (coenzyme Q10) in the long-term treatment of idiopathic dilated cardiomyopathy. *Eur. Heart J.* **1992**, *13*, 1528–1533.

121. Langsjoen, P.H.; Folkers, K.; Lyson, K.; Muratsu, K.; Lyson, T.; Langsjoen, P. Pronounced increase of survival of patients with cardiomyopathy when treated with coenzyme Q10 and conventional therapy. *Int. J. Tissue React.* **1990**, *12*, 163–168.

122. Morisco, C.; Trimarco, B.; Condorelli, M. Effect of coenzyme Q10 therapy in patients with congestive heart failure: A long-term multicenter randomized study. *Clin. Investig.* **1993**, *71*, S134–S136.

123. Ma, A.W.; Zhang, W.; Liu, Z. Effect of protection and repair of injury of mitochondrial membrane-phospholipid on prognosis in patients with dilated cardiomyopathy. *Blood Press. Suppl. 1* **1996**, *3*, 53–55.

124. Sacher, H.L.; Sacher, M.L.; Landau, S.W.; Kersten, R.; Dooley, F.; Sacher, A.; Sacher, M.; Dietrick, K.; Ichkhan, K. The clinical and hemodynamic effects of coenzyme Q10 in congestive cardiomyopathy. *Am. J. Ther.* **1997**, *4*, 66–72.

125. Soongswang, J.; Sangtawesin, C.; Durongpisitkul, K.; Laohaprasitiporn, D.; Nana, A.; Punlee, K.; Kangkagate, C. The effect of coenzyme Q10 on idiopathic chronic dilated cardiomyopathy in children. *Pediatr. Cardiol.* **2005**, *26*, 361–366.

126. Caso, G.; Kelly, P.; McNurlan, M.A.; Lawson, W.E. Effect of coenzyme Q10 on myopathic symptoms in patients treated with statins. *Am. J. Cardiol.* **2007**, *99*, 1409–1412.

127. Kocharian, A.; Shabanian, R.; Rafiei-Khorgami, M.; Kiani, A.; Heidari-Bateni, G. Coenzyme Q10 improves diastolic function in children with idiopathic dilated cardiomyopathy. *Cardiol. Young* **2009**, *19*, 501–506.

128. Shargorodsky, M.; Debby, O.; Matas, Z.; Zimlichman, R. Effect of long-term treatment with antioxidants (vitamin C, vitamin E, coenzyme Q10 and selenium) on arterial compliance, humoral factors and inflammatory markers in patients with multiple cardiovascular risk factors. *Nutr. Metab.* **2010**, *7*, 55.

129. Judy, W.V.; Stogsdill, W.W.; Folkers, K. Myocardial preservation by therapy with coenzyme Q10 during heart surgery. *Clin. Investig.* **1993**, *71*, S155–S161.

130. Taggart, D.P.; Jenkins, M.; Hooper, J.; Hadjinikolas, L.; Kemp, M.; Hue, D.; Bennett, G. Effects of short-term supplementation with coenzyme Q10 on myocardial protection during cardiac operations. *Ann. Thorac. Surg.* **1996**, *61*, 829–833.

131. Chello, M.; Mastroroberto, P.; Romano, R.; Castaldo, P.; Bevacqua, E.; Marchese, A.R. Protection by coenzyme Q10 of tissue reperfusion injury during abdominal aortic cross-clamping. *J. Cardiovasc. Surg.* **1996**, *37*, 229–235.

132. Zhou, M.; Zhi, Q.; Tang, Y.; Yu, D.; Han, J. Effects of coenzyme Q10 on myocardial protection during cardiac valve replacement and scavenging free radical activity *in vitro*. *J. Cardiovasc. Surg.* **1999**, *40*, 355–361.

133. Rosenfeldt, F.; Marasco, S.; Lyon, W.; Wowk, M.; Sheeran, F.; Bailey, M.; Esmore, D.; Davis, B.; Pick, A.; Rabinov, M.; *et al.* Coenzyme Q10 therapy before cardiac surgery improves mitochondrial function and *in vitro* contractility of myocardial tissue. *J. Thorac. Cardiovasc. Surg.* **2005**, *129*, 25–32.

134. Keith, M.; Mazer, C.D.; Mikhail, P.; Jeejeebhoy, F.; Briet, F.; Errett, L. Coenzyme Q10 in patients undergoing (coronary artery bypass graft) CABG: Effect of statins and nutritional supplementation. *Nutr. Metab. Cardiovasc. Dis.* **2008**, *18*, 105–111.

135. Makhija, N.; Sendasgupta, C.; Kiran, U.; Lakshmy, R.; Hote, M.P.; Choudhary, S.K.; Airan, B.; Abraham, R. The role of oral coenzyme Q10 in patients undergoing coronary artery bypass graft surgery. *J. Cardiothorac. Vasc. Anesth.* **2008**, *22*, 832–839.

136. Alehagen, U.; Johansson, P.; Björnstedt, M.; Rosén, A.; Dahlström, U. Cardiovascular mortality and *N*-terminal-proBNP reduced after combined selenium and coenzyme Q10 supplementation: A 5-year prospective randomized double-blind placebo-controlled trial among elderly Swedish citizens. *Int. J. Cardiol.* **2013**, *167*, 1860–1866.

137. Miles, M.V.; Patterson, B.J.; Chalfonte-Evans, M.L.; Horn, P.S.; Hickey, F.J.; Schapiro, M.B.; Steele, P.E.; Tang, P.H.; Hotze, S.L. Coenzyme Q10 (ubiquinol-10) supplementation improves oxidative imbalance in children with trisomy 21. *Pediatr. Neurol.* **2007**, *37*, 398–403.

138. Tiano, L.; Padella, L.; Santoro, L.; Carnevali, P.; Principi, F.; Brugè, F.; Gabrielli, O.; Littarru, G.P. Prolonged coenzyme Q10 treatment in Down syndrome patients: Effect on DNA oxidation. *Neurobiol. Aging* **2012**, *33*, 626.e1-8.

139. Sparaco, M.; Gaeta, L.M.; Santorelli, F.M.; Passarelli, C.; Tozzi, G.; Bertini, E.; Simonati, A.; Scaravilli, F.; Taroni, F.; Duyckaerts, C.; *et al.* Friedreich's ataxia: Oxidative stress and cytoskeletal abnormalities. *J. Neurol. Sci.* **2009**, *287*, 111–118.

140. Cooper, J.M.; Korlipara, L.V.; Hart, P.E.; Bradley, J.L.; Schapira, A.H. Coenzyme Q10 and vitamin E deficiency in Friedreich's ataxia: Predictor of efficacy of vitamin E and coenzyme Q10 therapy. *Eur. J. Neurol.* **2008**, *15*, 1371–1379.

141. Hart, P.E.; Lodi, R.; Rajagopalan, B.; Bradley, J.L.; Crilley, J.G.; Turner, C.; Blamire, A.M.; Manners, D.; Styles, P.; Schapira, A.H.; *et al.* Antioxidant treatment of patients with Friedreich ataxia: Four-year follow-up. *Arch. Neurol.* **2005**, *62*, 621–626.

142. Barbiroli, B.; Frassineti, C.; Martinelli, P.; Iotti, S.; Lodi, R.; Cortelli, P.; Montagna, P. Coenzyme Q10 improves mitochondrial respiration in patients with mitochondrial cytopathies. An *in vivo* study on brain and skeletal muscle by phosphorous magnetic resonance spectroscopy. *Cell. Mol. Biol.* **1997**, *43*, 741–749.

143. Abe, K.; Matsuo, Y.; Kadekawa, J.; Inoue, S.; Yanagihara, T. Effect of coenzyme Q10 in patients with mitochondrial myopathy, encephalopathy, lactic acidosis, and stroke-like episodes (MELAS): Evaluation by noninvasive tissue oximetry. *J. Neurol. Sci.* **1999**, *162*, 65–68.

144. Hanisch, F.; Zierz, S. Only transient increase of serum CoQ subset 10 during long-term CoQ10 therapy in mitochondrial ophthalmoplegia. *Eur. J. Med. Res.* **2003**, *8*, 485–491.

145. Glover, E.I.; Martin, J.; Maher, A.; Thornhill, R.E.; Moran, G.R.; Tarnopolsky, M.A. A randomized trial of coenzyme Q10 in mitochondrial disorders. *Muscle Nerve* **2010**, *42*, 739–748.

146. Huntington Study Group. A randomized, placebo-controlled trial of coenzyme Q10 and remacemide in Huntington's disease. *Neurology* **2001**, *57*, 397–404.

147. Huntington Study Group Pre2CARE Investigators; Hyson, H.C.; Kieburtz, K.; Shoulson, I.; McDermott, M.; Ravina, B.; de Blieck, E.A.; Cudkowicz, M.E.; Ferrante, R.J.; Como, P.; *et al.* Safety and tolerability of high-dosage coenzyme Q10 in Huntington's disease and healthy subjects. *Mov. Disord.* **2010**, *25*, 1924–1928.

148. Spurney, C.F.; Rocha, C.T.; Henricson, E.; Florence, J.; Mayhew, J.; Gorni, K.; Pasquali, L.; Pestronk, A.; Martin, G.R.; Hu, F.; *et al.* CINRG pilot trial of coenzyme Q10 in steroid-treated Duchenne muscular dystrophy. *Muscle Nerve* **2011**, *44*, 174–178.

149. Folkers, K.; Wolaniuk, J.; Simonsen, R.; Morishita, M.; Vadhanavikit, S. Biochemical rationale and the cardiac response of patients with muscle disease to therapy with coenzyme Q10. *Proc. Natl. Acad. Sci. USA* **1985**, *82*, 4513–4516.

150. Folkers, K.; Simonsen, R. Two successful double-blind trials with coenzyme Q10 (vitamin Q10) on muscular dystrophies and neurogenic atrophies. *Biochim. Biophys. Acta* **1995**, *1271*, 281–286.

151. Linnane, A.W.; Kopsidas, G.; Zhang, C.; Yarovaya, N.; Kovalenko, S.; Papakostopoulos, P.; Eastwood, H.; Graves, S.; Richardson, M. Cellular redox activity of coenzyme Q10: Effect of CoQ10 supplementation on human skeletal muscle. *Free Radic. Res.* **2002**, *36*, 445–453.

152. Shults, C.W.; Oakes, D.; Kieburtz, K.; Beal, M.F.; Haas, R.; Plumb, S.; Juncos, J.L.; Nutt, J.; Shoulson, I.; Carter, J.; *et al.* Effects of coenzyme Q10 in early Parkinson disease: Evidence of slowing of the functional decline. *Arch. Neurol.* **2002**, *59*, 1541–1550.

153. Müller, T.; Büttner, T.; Gholipour, A.F.; Kuhn, W. Coenzyme Q10 supplementation provides mild symptomatic benefit in patients with Parkinson's disease. *Neurosci. Lett.* **2003**, *341*, 201–204.

154. Storch, A.; Jost, W.H.; Vieregge, P.; Spiegel, J.; Greulich, W.; Durner, J.; Müller, T.; Kupsch, A.; Henningsen, H.; Oertel, W.H.; *et al.* Randomized, double-blind, placebo-controlled trial on symptomatic effects of coenzyme Q10 in Parkinson disease. *Arch. Neurol.* **2007**, *64*, 938–944.

155. Sanoobar, M.; Eghtesadi, S.; Azimi, A.; Khalili, M.; Jazayeri, S.; Reza Gohari, M. Coenzyme Q10 supplementation reduces oxidative stress and increases antioxidant enzyme activity in patients with relapsing-remitting multiple sclerosis. *Int. J. Neurosci.* **2013**, *123*, 776–782.

156. Sanoobar, M.; Eghtesadi, S.; Azimi, A.; Khalili, M.; Khodadadi, B.; Jazayeri, S.; Gohari, M.R.; Aryaeian, N. Coenzyme Q10 supplementation ameliorates inflammatory markers in patients with multiple sclerosis: A double blind, placebo, controlled randomized clinical trial. *Nutr. Neurosci.* **2014**.

157. Cordero, M.D.; Cano-García, F.J.; Alcocer-Gómez, E.; de Miguel, M.; Sánchez-Alcázar, J.A. Oxidative stress correlates with headache symptoms in fibromyalgia: Coenzyme Q10 effect on clinical improvement. *PLoS One* **2012**, *7*, e35677.

158. Cordero, M.D.; Alcocer-Gómez, E.; de Miguel, M.; Culic, O.; Carrión, A.M.; Alvarez-Suarez, J.M.; Bullón, P.; Battino, M.; Fernández-Rodríguez, A.; Sánchez-Alcazar, J.A. Can coenzyme q10 improve clinical and molecular parameters in fibromyalgia? *Antioxid. Redox Signal.* **2013**, *19*, 1356–1361.

159. Miyamae, T.; Seki, M.; Naga, T.; Uchino, S.; Asazuma, H.; Yoshida, T.; Iizuka, Y.; Kikuchi, M.; Imagawa, T.; Natsumeda, Y.; *et al.* Increased oxidative stress and coenzyme Q10 deficiency in juvenile fibromyalgia: Amelioration of hypercholesterolemia and fatigue by ubiquinol-10 supplementation. *Redox Rep.* **2013**, *18*, 12–19.

160. Lister, R.E. An open, pilot study to evaluate the potential benefits of coenzyme Q10 combined with *Ginkgo biloba* extract in fibromyalgia syndrome. *J. Int. Med. Res.* **2002**, *30*, 195–199.

161. Kaufmann, P.; Thompson, J.L.; Levy, G.; Buchsbaum, R.; Shefner, J.; Krivickas, L.S.; Katz, J.; Rollins, Y.; Barohn, R.J.; Jackson, C.E.; *et al.* Phase II trial of CoQ10 for ALS finds insufficient evidence to justify phase III. *Ann. Neurol.* **2009**, *66*, 235–244.

162. Rozen, T.D.; Oshinsky, M.L.; Gebeline, C.A.; Bradley, K.C.; Young, W.B.; Shechter, A.L.; Silberstein, S.D. Open label trial of coenzyme Q10 as a migraine preventive. *Cephalalgia* **2002**, *22*, 137–141.

163. Hershey, A.D.; Powers, S.W.; Vockell, A.L.; Lecates, S.L.; Ellinor, P.L.; Segers, A.; Burdine, D.; Manning, P.; Kabbouche, M.A. Coenzyme Q10 deficiency and response to supplementation in pediatric and adolescent migraine. *Headache* **2007**, *47*, 73–80.

164. Slater, S.K.; Nelson, T.D.; Kabbouche, M.A.; LeCates, S.L.; Horn, P.; Segers, A.; Manning, P.; Powers, S.W.; Hershey, A.D. A randomized, double-blinded, placebo-controlled, crossover, add-on study of CoEnzyme Q10 in the prevention of pediatric and adolescent migraine. *Cephalalgia* **2011**, *31*, 897–905.

165. Forester, B.P.; Zuo, C.S.; Ravichandran, C.; Harper, D.G.; Du, F.; Kim, S.; Cohen, B.M.; Renshaw, P.F. Coenzyme Q10 effects on creatine kinase activity and mood in geriatric bipolar depression. *J. Geriatr. Psychiatry Neurol.* **2012**, *25*, 43–50.

166. Stamelou, M.; Reuss, A.; Pilatus, U.; Magerkurth, J.; Niklowitz, P.; Eggert, K.M.; Krisp, A.; Menke, T.; Schade-Brittinger, C.; Oertel, W.H.; *et al.* Short-term effects of coenzyme Q10 in progressive supranuclear palsy: A randomized, placebo-controlled trial. *Mov. Disord.* **2008**, *23*, 942–949.

167. Watts, G.F.; Playford, D.A.; Croft, K.D.; Ward, N.C.; Mori, T.A.; Burke, V. Coenzyme Q10 improves endothelial dysfunction of the brachial artery in type II diabetes mellitus. *Diabetologia* **2002**, *45*, 420–426.

168. Hodgson, J.M.; Watts, G.F.; Playford, D.A.; Burke, V.; Croft, K.D. Coenzyme Q10 improves blood pressure and glycaemic control: A controlled trial in subjects with type 2 diabetes. *Eur. J. Clin. Nutr.* **2002**, *56*, 1137–1142.

169. Playford, D.A.; Watts, G.F.; Croft, K.D.; Burke, V. Combined effect of coenzyme Q10 and fenofibrate on forearm microcirculatory function in type 2 diabetes. *Atherosclerosis* **2003**, *168*, 169–179.

170. Chew, G.T.; Watts, G.F.; Davis, T.M.; Stuckey, B.G.; Beilin, L.J.; Thompson, P.L.; Burke, V.; Currie, P.J. Hemodynamic effects of fenofibrate and coenzyme Q10 in type 2 diabetic subjects with left ventricular diastolic dysfunction. *Diabetes Care* **2008**, *31*, 1502–1509.

171. Hamilton, S.J.; Chew, G.T.; Watts, G.F. Coenzyme Q10 improves endothelial dysfunction in statin-treated type 2 diabetic patients. *Diabetes Care* **2009**, *32*, 810–812.

172. Kolahdouz Mohammadi, R.; Hosseinzadeh-Attar, M.J.; Eshraghian, M.R.; Nakhjavani, M.; Khorami, E.; Esteghamati, A. The effect of coenzyme Q10 supplementation on metabolic status of type 2 diabetic patients. *Minerva Gastroenterol. Dietol.* **2013**, *59*, 231–236.

173. Suzuki, S.; Hinokio, Y.; Ohtomo, M.; Hirai, M.; Hirai, A.; Chiba, M.; Kasuga, S.; Satoh, Y.; Akai, H.; Toyota, T. The effects of coenzyme Q10 treatment on maternally inherited diabetes mellitus and deafness, and mitochondrial DNA 3243 (A to G) mutation. *Diabetologia* **1998**, *41*, 584–588.

174. Henriksen, J.E.; Andersen, C.B.; Hother-Nielsen, O.; Vaag, A.; Mortensen, S.A.; Beck-Nielsen, H. Impact of ubiquinone (coenzyme Q10) treatment on glycaemic control, insulin requirement and well-being in patients with type 1 diabetes mellitus. *Diabet. Med.* **1999**, *16*, 312–318.

175. Hertz, N.; Lister, R.E. Improved survival in patients with end-stage cancer treated with coenzyme Q10 and other antioxidants: A pilot study. *J. Int. Med. Res.* **2009**, *37*, 1961–1971.

176. Lockwood, K.; Moesgaard, S.; Folkers, K. Partial and complete regression of breast cancer in patients in relation to dosage of coenzyme Q10. *Biochem. Biophys. Res. Commun.* **1994**, *199*, 1504–1508.

177. Premkumar, V.G.; Yuvaraj, S.; Sathish, S.; Shanthi, P.; Sachdanandam, P. Anti-angiogenic potential of Coenzyme Q10, riboflavin and niacin in breast cancer patients undergoing tamoxifen therapy. *Vascul. Pharmacol.* **2008**, *48*, 191–201.

178. Lesser, G.J.; Case, D.; Stark, N.; Williford, S.; Giguere, J.; Garino, L.A.; Naughton, M.J.; Vitolins, M.Z.; Lively, M.O.; Shaw, E.G.; *et al.* A randomized, double-blind, placebo-controlled study of oral coenzyme Q10 to relieve self-reported treatment-related fatigue in newly diagnosed patients with breast cancer. *J. Support Oncol.* **2013**, *11*, 31–42.

179. Hoenjet, K.M.; Dagnelie, P.C.; Delaere, K.P.; Wijckmans, N.E.; Zambon, J.V.; Oosterhof, G.O. Effect of a nutritional supplement containing vitamin E, selenium, vitamin c and coenzyme Q10 on serum PSA in patients with hormonally untreated carcinoma of the prostate: A randomised placebo-controlled study. *Eur. Urol.* **2005**, *47*, 433–439.

180. Rusciani, L.; Proietti, I.; Paradisi, A.; Rusciani, A.; Guerriero, G.; Mammone, A.; de Gaetano, A.; Lippa, S. Recombinant interferon alpha-2b and coenzyme Q10 as a postsurgical adjuvant therapy for melanoma: A 3-year trial with recombinant interferon-alpha and 5-year follow-up. *Melanoma Res.* **2007**, *17*, 177–183.

181. Gazdíková, K.; Gvozdjáková, A.; Kucharská, J.; Spustová, V.; Braunová, Z.; Dzúrik, R. Effect of coenzyme Q10 in patients with kidney diseases. *Cas. Lek. Cesk.* **2001**, *140*, 307–310.

182. Sakata, T.; Furuya, R.; Shimazu, T.; Odamaki, M.; Ohkawa, S.; Kumagai, H. Coenzyme Q10 administration suppresses both oxidative and antioxidative markers in hemodialysis patients. *Blood Purif.* **2008**, *26*, 371–378.

183. Mori, T.A.; Burke, V.; Puddey, I.; Irish, A.; Cowpland, C.A.; Beilin, L.; Dogra, G.; Watts, G.F. The effects of omega-3 fatty acids and coenzyme Q10 on blood pressure and heart rate in chronic kidney disease: A randomized controlled trial. *J. Hypertens.* **2009**, *27*, 1863–1872.

184. Shojaei, M.; Djalali, M.; Khatami, M.; Siassi, F.; Eshraghian, M. Effects of carnitine and coenzyme Q10 on lipid profile and serum levels of lipoprotein(a) in maintenance hemodialysis patients on statin therapy. *Iran J. Kidney Dis.* **2011**, *5*, 114–118.

185. Kaikkonen, J.; Nyyssönen, K.; Tomasi, A.; Iannone, A.; Tuomainen, T.P.; Porkkala-Sarataho, E.; Salonen, J.T. Antioxidative efficacy of parallel and combined supplementation with coenzyme Q10 and d-alpha-tocopherol in mildly hypercholesterolemic subjects: A randomized placebo-controlled clinical study. *Free Radic. Res.* **2000**, *33*, 329–340.

186. Lee, Y.J.; Cho, W.J.; Kim, J.K.; Lee, D.C. Effects of coenzyme Q10 on arterial stiffness, metabolic parameters, and fatigue in obese subjects: A double-blind randomized controlled study. *J. Med. Food* **2011**, *14*, 386–390.

187. Young, J.M.; Florkowski, C.M.; Molyneux, S.L.; McEwan, R.G.; Frampton, C.M.; Nicholls, M.G.; Scott, R.S.; George, P.M. A randomized, double-blind, placebo-controlled crossover study of coenzyme Q10 therapy in hypertensive patients with the metabolic syndrome. *Am. J. Hypertens.* **2012**, *25*, 261–270.

188. Mabuchi, H.; Nohara, A.; Kobayashi, J.; Kawashiri, M.A.; Katsuda, S.; Inazu, A.; Koizumi, J.; Hokuriku Lipid Research Group. Effects of CoQ10 supplementation on plasma lipoprotein lipid, CoQ10 and liver and muscle enzyme levels in hypercholesterolemic patients treated with atorvastatin: A randomized double-blind study. *Atherosclerosis* **2007**, *195*, e182–e189.

189. Young, J.M.; Florkowski, C.M.; Molyneux, S.L.; McEwan, R.G.; Frampton, C.M.; George, P.M.; Scott, R.S. Effect of coenzyme Q10 supplementation on simvastatin-induced myalgia. *Am. J. Cardiol.* **2007**, *100*, 1400–1403.

190. Bookstaver, D.A.; Burkhalter, N.A.; Hatzigeorgiou, C. Effect of coenzyme Q10 supplementation on statin-induced myalgias. *Am. J. Cardiol.* **2012**, *110*, 526–529.

191. Zlatohlavek, L.; Vrablik, M.; Grauova, B.; Motykova, E.; Ceska, R. The effect of coenzyme Q10 in statin myopathy. *Neuro Endocrinol. Lett.* **2012**, *33*, 98–101.

192. Fedacko, J.; Pella, D.; Fedackova, P.; Hänninen, O.; Tuomainen, P.; Jarcuska, P.; Lopuchovsky, T.; Jedlickova, L.; Merkovska, L.; Littarru, G.P. Coenzyme Q10 and selenium in statin-associated myopathy treatment. *Can. J. Physiol. Pharmacol.* **2013**, *91*, 165–170.

193. Gvozdjáková, A.; Kucharská, J.; Bartkovjaková, M.; Gazdíková, K.; Gazdík, F.E. Coenzyme Q10 supplementation reduces corticosteroids dosage in patients with bronchial asthma. *Biofactors* **2005**, *25*, 235–240.

194. Teran, E.; Hernandez, I.; Nieto, B.; Tavara, R.; Ocampo, J.E.; Calle, A. Coenzyme Q10 supplementation during pregnancy reduces the risk of pre-eclampsia. *Int. J. Gynaecol. Obstet.* **2009**, *105*, 43–45.

195. De Luca, C.; Kharaeva, Z.; Raskovic, D.; Pastore, P.; Luci, A.; Korkina, L. Coenzyme Q10, vitamin E, selenium, and methionine in the treatment of chronic recurrent viral mucocutaneous infections. *Nutrition* **2012**, *28*, 509–514.

196. Kharaeva, Z.; Gostova, E.; de Luca, C.; Raskovic, D.; Korkina, L. Clinical and biochemical effects of coenzyme Q10, vitamin E, and selenium supplementation to psoriasis patients. *Nutrition* **2009**, *25*, 295–302.

197. Fogagnolo, P.; Sacchi, M.; Ceresara, G.; Paderni, R.; Lapadula, P.; Orzalesi, N.; Rossetti, L. The effects of topical coenzyme Q10 and vitamin E D-α-tocopheryl polyethylene glycol 1000 succinate after cataract surgery: A clinical and *in vivo* confocal study. *Ophthalmologica* **2013**, *229*, 26–31.

198. Safarinejad, M.R.; Safarinejad, S.; Shafiei, N.; Safarinejad, S. Effects of the reduced form of coenzyme Q10 (ubiquinol) on semen parameters in men with idiopathic infertility: A double-blind, placebo controlled, randomized study. *J. Urol.* **2012**, *188*, 526–531.

199. Gilbert, E.F. Carnitine deficiency. *Pathology* **1985**, *17*, 161–171.

200. Siliprandi, N.; Siliprandi, D.; Ciman, M. Stimulation of oxidation of mitochondrial fatty acids and of acetate by acetylcarnitine. *Biochem. J.* **1965**, *96*, 777–780.

201. Liu, J.; Head, E.; Kuratsune, H.; Cotman, C.W.; Ames, B.N. Comparison of the effects of L-carnitine and acetyl-L-carnitine on carnitine levels, ambulatory activity, and oxidative stress biomarkers in the brain of old rats. *Ann. N. Y. Acad. Sci.* **2004**, *1033*, 117–131.

202. Musicco, C.; Capelli, V.; Pesce, V.; Timperio, A.M.; Calvani, M.; Mosconi, L.; Cantatore, P.; Gadaleta, M.N. Rat liver mitochondrial proteome: Changes associated with aging and acetyl-L-carnitine treatment. *J. Proteomics* **2011**, *74*, 2536–2547.

203. Pesce, V.; Nicassio, L.; Fracasso, F.; Musicco, C.; Cantatore, P.; Gadaleta, M.N. Acetyl-L-carnitine activates the peroxisome proliferator-activated receptor-γ coactivators PGC-1α/PGC-1β-dependent signaling cascade of mitochondrial biogenesis and decreases the oxidized peroxiredoxins content in old rat liver. *Rejuvenation Res.* **2012**, *15*, 136–139.

204. Bertoli, M.; Battistella, P.A.; Vergani, L.; Naso, A.; Gasparotto, M.L.; Romagnoli, G.F.; Angelini, C. Carnitine deficiency induced during hemodialysis and hyperlipidemia: Effect of replacement therapy. *Am. J. Clin. Nutr.* **1981**, *34*, 1496–1500.

205. Kletzmayr, J.; Mayer, G.; Legenstein, E.; Heinz-Peer, G.; Leitha, T.; Hörl, W.H.; Kovarik, J. Anemia and carnitine supplementation in hemodialyzed patients. *Kidney Int. Suppl.* **1999**, *69*, S93–S106.

206. Thomas, S.; Fischer, F.P.; Mettang, T.; Pauli-Magnus, C.; Weber, J.; Kuhlmann, U. Effects of L-carnitine on leukocyte function and viability in hemodialysis patients: A double-blind randomized trial. *Am. J. Kidney Dis.* **1999**, *34*, 678–687.

207. Chazot, C.; Blanc, C.; Hurot, J.M.; Charra, B.; Jean, G.; Laurent, G. Nutritional effects of carnitine supplementation in hemodialysis patients. *Clin. Nephrol.* **2003**, *59*, 24–30.

208. Vaux, E.C.; Taylor, D.J.; Altmann, P.; Rajagopalan, B.; Graham, K.; Cooper, R.; Bonomo, Y.; Styles, P. Effects of carnitine supplementation on muscle metabolism by the use of magnetic resonance spectroscopy and near-infrared spectroscopy in end-stage renal disease. *Nephron Clin. Pract.* **2004**, *97*, c41–c48.

209. Steiber, A.L.; Davis, A.T.; Spry, L.; Strong, J.; Buss, M.L.; Ratkiewicz, M.M.; Weatherspoon, L.J. Carnitine treatment improved quality-of-life measure in a sample of Midwestern hemodialysis patients. *J. Parenter. Enteral. Nutr.* **2006**, *30*, 10–15.

210. Signorelli, S.S.; Fatuzzo, P.; Rapisarda, F.; Neri, S.; Ferrante, M.; Oliveri Conti, G.; Fallico, R.; di Pino, L.; Pennisi, G.; Celotta, G.; *et al.* A randomised, controlled clinical trial evaluating changes in therapeutic efficacy and oxidative parameters after treatment with propionyl L-carnitine in patients with peripheral arterial disease requiring haemodialysis. *Drugs Aging* **2006**, *23*, 263–270.

211. Rathod, R.; Baig, M.S.; Khandelwal, P.N.; Kulkarni, S.G.; Gade, P.R.; Siddiqui, S. Results of a single blind, randomized, placebo-controlled clinical trial to study the effect of intravenous L-carnitine supplementation on health-related quality of life in Indian patients on maintenance hemodialysis. *Indian J. Med. Sci.* **2006**, *60*, 143–153.

212. Duranay, M.; Akay, H.; Yilmaz, F.M.; Senes, M.; Tekeli, N.; Yücel, D. Effects of L-carnitine infusions on inflammatory and nutritional markers in haemodialysis patients. *Nephrol. Dial. Transplant.* **2006**, *21*, 3211–3214.

213. Verrina, E.; Caruso, U.; Calevo, M.G.; Emma, F.; Sorino, P.; de Palo, T.; Lavoratti, G.; Turrini Dertenois, L.; Cassanello, M.; Cerone, R.; *et al.* Effect of carnitine supplementation on lipid profile and anemia in children on chronic dialysis. *Pediatr. Nephrol.* **2007**, *22*, 727–733.

214. Fatouros, I.G.; Douroudos, I.; Panagoutsos, S.; Pasadakis, P.; Nikolaidis, M.G.; Chatzinikolaou, A.; Sovatzidis, A.; Michailidis, Y.; Jamurtas, A.Z.; Mandalidis, D.; *et al.* Effects of L-carnitine on oxidative stress responses in patients with renal disease. *Med. Sci. Sports Exerc.* **2010**, *42*, 1809–1818.

215. Hakeshzadeh, F.; Tabibi, H.; Ahmadinejad, M.; Malakoutian, T.; Hedayati, M. Effects of L-carnitine supplement on plasma coagulation and anticoagulation factors in hemodialysis patients. *Renal Failure* **2010**, *32*, 1109–1114.

216. Tabibi, H.; Hakeshzadeh, F.; Hedayati, M.; Malakoutian, T. Effects of L-carnitine supplement on serum amyloid A and vascular inflammation markers in hemodialysis patients: A randomized controlled trial. *J. Ren. Nutr.* **2011**, *21*, 485–491.

217. Suchitra, M.M.; Ashalatha, V.L.; Sailaja, E.; Rao, A.M.; Reddy, V.S.; Bitla, A.R.; Sivakumar, V.; Rao, P.V. The effect of L-carnitine supplementation on lipid parameters, inflammatory and nutritional markers in maintenance hemodialysis patients. *Saudi J. Kidney Dis. Transpl.* **2011**, *22*, 1155–1159.

218. Naini, A.E.; Sadeghi, M.; Mortazavi, M.; Moghadasi, M.; Harandi, A.A. Oral carnitine supplementation for dyslipidemia in chronic hemodialysis patients. *Saudi J. Kidney Dis. Transpl.* **2012**, *23*, 484–488.

219. Sgambat, K.; Frank, L.; Ellini, A.; Sable, C.; Moudgil, A. Carnitine supplementation improves cardiac strain rate in children on chronic hemodialysis. *Pediatr. Nephrol.* **2012**, *27*, 1381–1387.

220. Mercadal, L.; Coudert, M.; Vassault, A.; Pieroni, L.; Debure, A.; Ouziala, M.; Depreneuf, H.; Fumeron, C.; Servais, A.; Bassilios, N.; *et al.* L-Carnitine treatment in incident hemodialysis patients: The multicenter, randomized, double-blinded, placebo-controlled CARNIDIAL trial. *Clin. J. Am. Soc. Nephrol.* **2012**, *7*, 1836–1842.

221. Mingrone, G.; Greco, A.V.; Capristo, E.; Benedetti, G.; Giancaterini, A.; de Gaetano, A.; Gasbarrini, G. L-Carnitine improves glucose disposal in type 2 diabetic patients. *J. Am. Coll. Nutr.* **1999**, *18*, 77–82.

222. Giancaterini, A.; de Gaetano, A.; Mingrone, G.; Gniuli, D.; Liverani, E.; Capristo, E.; Greco, A.V. Acetyl-L-carnitine infusion increases glucose disposal in type 2 diabetic patients. *Metabolism* **2000**, *49*, 704–708.

223. De Grandis, D.; Minardi, C. Acetyl-L-carnitine (levacecarnine) in the treatment of diabetic neuropathy. A long-term, randomised, double-blind, placebo-controlled study. *Drugs R&D* **2002**, *3*, 223–231.

224. Derosa, G.; Cicero, A.F.; Gaddi, A.; Mugellini, A.; Ciccarelli, L.; Fogari, R. The effect of L-carnitine on plasma lipoprotein(a) levels in hypercholesterolemic patients with type 2 diabetes mellitus. *Clin. Ther.* **2003**, *25*, 1429–1439.

225. Ragozzino, G.; Mattera, E.; Madrid, E.; Salomone, P.; Fasano, C.; Gioia, F.; Acerra, G.; del Guercio, R.; Federico, P. Effects of propionyl-carnitine in patients with type 2 diabetes and peripheral vascular disease: Results of a pilot trial. *Drugs R&D* **2004**, *5*, 185–190.

226. Sima, A.A.; Calvani, M.; Mehra, M.; Amato, A. Acetyl-L-Carnitine Study Group. Acetyl-L-carnitine improves pain, nerve regeneration, and vibratory perception in patients with chronic diabetic neuropathy: An analysis of two randomized placebo-controlled trials. *Diabetes Care* **2005**, *28*, 89–94.

227. Rahbar, A.R.; Shakerhosseini, R.; Saadat, N.; Taleban, F.; Pordal, A.; Gollestan, B. Effect of L-carnitine on plasma glycemic and lipidemic profile in patients with type II diabetes mellitus. *Eur. J. Clin. Nutr.* **2005**, *59*, 592–596.

228. Solfrizzi, V.; Capurso, C.; Colacicco, A.M.; D'Introno, A.; Fontana, C.; Capurso, S.A.; Torres, F.; Gadaleta, A.M.; Koverech, A.; Capurso, A.; *et al.* Efficacy and tolerability of combined treatment with L-carnitine and simvastatin in lowering lipoprotein(a) serum levels in patients with type 2 diabetes mellitus. *Atherosclerosis* **2006**, *188*, 455–461.

405

229. Signorelli, S.S.; Neri, S.; di Pino, L.; Marchese, G.; Ferrante, M.; Oliveri Conti, G.; Fallico, R.; Celotta, G.; Pennisi, G.; Anzaldi, M. Effect of PLC on functional parameters and oxidative profile in type 2 diabetes-associated PAD. *Diabetes Res. Clin. Pract.* **2006**, *72*, 231–237.

230. Malaguarnera, M.; Vacante, M.; Avitabile, T.; Malaguarnera, M.; Cammalleri, L.; Motta, M. L-Carnitine supplementation reduces oxidized LDL cholesterol in patients with diabetes. *Am. J. Clin. Nutr.* **2009**, *89*, 71–76.

231. Ruggenenti, P.; Cattaneo, D.; Loriga, G.; Ledda, F.; Motterlini, N.; Gherardi, G.; Orisio, S.; Remuzzi, G. Ameliorating hypertension and insulin resistance in subjects at increased cardiovascular risk: Effects of acetyl-L-carnitine therapy. *Hypertension* **2009**, *54*, 567–574.

232. Molfino, A.; Cascino, A.; Conte, C.; Ramaccini, C.; Rossi Fanelli, F.; Laviano, A. Caloric restriction and L-carnitine administration improves insulin sensitivity in patients with impaired glucose metabolism. *J. Parenter. Enteral. Nutr.* **2010**, *34*, 295–299.

233. Uzun, N.; Sarikaya, S.; Uluduz, D.; Aydin, A. Peripheric and automatic neuropathy in children with type 1 diabetes mellitus: The effect of L-carnitine treatment on the peripheral and autonomic nervous system. *Electromyogr. Clin. Neurophysiol.* **2005**, *45*, 343–351.

234. Loffredo, L.; Pignatelli, P.; Cangemi, R.; Andreozzi, P.; Panico, M.A.; Meloni, V.; Violi, F. Imbalance between nitric oxide generation and oxidative stress in patients with peripheral arterial disease: Effect of an antioxidant treatment. *J. Vasc. Surg.* **2006**, *44*, 525–530.

235. De Marchi, S.; Zecchetto, S.; Rigoni, A.; Prior, M.; Fondrieschi, L.; Scuro, A.; Rulfo, F.; Arosio, E. Propionyl-L-carnitine improves endothelial function, microcirculation and pain management in critical limb ischemia. *Cardiovasc. Drugs Ther.* **2012**, *26*, 401–408.

236. Goldenberg, N.A.; Krantz, M.J.; Hiatt, W.R. L-Carnitine plus cilostazol *versus* cilostazol alone for the treatment of claudication in patients with peripheral artery disease: A multicenter, randomized, double-blind, placebo-controlled trial. *Vasc. Med.* **2012**, *17*, 145–154.

237. Xu, X.Q.; Jing, Z.C.; Jiang, X.; Zhao, Q.H.; He, J.; Dai, L.Z.; Wu, W.H.; Li, Y.; Yao, J. Clinical efficacy of intravenous L-carnitine in patients with right-sided heart failure induced by pulmonary arterial hypertension. *Zhonghua Xin Xue Guan Bing Za Zhi* **2010**, *38*, 152–155.

238. Serati, A.R.; Motamedi, M.R.; Emami, S.; Varedi, P.; Movahed, M.R. L-Carnitine treatment in patients with mild diastolic heart failure is associated with improvement in diastolic function and symptoms. *Cardiology* **2010**, *116*, 178–182.

239. Lango, R.; Smoleński, R.T.; Rogowski, J.; Siebert, J.; Wujtewicz, M.; Słomińska, E.M.; Lysiak-Szydłowska, W.; Yacoub, M.H. Propionyl-L-carnitine improves hemodynamics and metabolic markers of cardiac perfusion during coronary surgery in diabetic patients. *Cardiovasc. Drugs Ther.* **2005**, *19*, 267–275.

240. Xiang, D.; Sun, Z.; Xia, J.; Dong, N.; Du, X.; Chen, X. Effect of L-carnitine on cardiomyocyte apoptosis and cardiac function in patients undergoing heart valve replacement operation. *J. Huazhong Univ. Sci. Technol. Med. Sci.* **2005**, *25*, 501–504.

241. Łapiński, T.W.; Grzeszczuk, A. The impact of carnitine on serum ammonia concentration and lipid metabolism in patients with alcoholic liver cirrhosis. *Pol. Merkur. Lekarski* **2003**, *15*, 38–41.

242. Lim, C.Y.; Jun, D.W.; Jang, S.S.; Cho, W.K.; Chae, J.D.; Jun, J.H. Effects of carnitine on peripheral blood mitochondrial DNA copy number and liver function in non-alcoholic fatty liver disease. *Korean J. Gastroenterol.* **2010**, *55*, 384–389.

243. Malaguarnera, M.; Pistone, G.; Astuto, M.; dell'Arte, S.; Finocchiaro, G.; lo Giudice, E.; Pennisi, G. L-Carnitine in the treatment of mild or moderate hepatic encephalopathy. *Dig. Dis.* **2003**, *21*, 271–275.

244. Malaguarnera, M.; Vacante, M.; Giordano, M.; Motta, M.; Bertino, G.; Pennisi, M.; Neri, S.; Malaguarnera, M.; Li Volti, G.; Galvano, F. L-Carnitine supplementation improves hematological pattern in patients affected by HCV treated with Peg interferon-α 2b plus ribavirin. *World J. Gastroenterol.* **2011**, *17*, 4414–4420.

245. Neri, S.; Pistone, G.; Saraceno, B.; Pennisi, G.; Luca, S.; Malaguarnera, M. L-Carnitine decreases severity and type of fatigue induced by interferon-alpha in the treatment of patients with hepatitis C. *Neuropsychobiology* **2003**, *47*, 94–97.

246. Malaguarnera, M.; Vacante, M.; Giordano, M.; Pennisi, G.; Bella, R.; Rampello, L.; Malaguarnera, M.; Li Volti, G.; Galvano, F. Oral acetyl-L-carnitine therapy reduces fatigue in overt hepatic encephalopathy: A randomized, double-blind, placebo-controlled study. *Am. J. Clin. Nutr.* **2011**, *93*, 799–808.

247. Tomassini, V.; Pozzilli, C.; Onesti, E.; Pasqualetti, P.; Marinelli, F.; Pisani, A.; Fieschi, C. Comparison of the effects of acetyl L-carnitine and amantadine for the treatment of fatigue in multiple sclerosis: Results of a pilot, randomised, double-blind, crossover trial. *J. Neurol. Sci.* **2004**, *218*, 103–108.

248. Gavrilova, S.I.; Kalyn, IaB.; Kolykhalov, I.V.; Roshchina, I.F.; Selezneva, N.D. Acetyl-L-carnitine (carnicetine) in the treatment of early stages of Alzheimer's disease and vascular dementia. *Zh. Nevrol. Psikhiatr. Im. S. S. Korsakova* **2011**, *111*, 16–22.

249. Tarighat Esfanjani, A.; Mahdavi, R.; Ebrahimi Mameghani, M.; Talebi, M.; Nikniaz, Z.; Safaiyan, A. The effects of magnesium, L-carnitine, and concurrent magnesium-L-carnitine supplementation in migraine prophylaxis. *Biol. Trace Elem. Res.* **2012**, *150*, 42–48.

250. Pistone, G.; Marino, A.; Leotta, C.; dell'Arte, S.; Finocchiaro, G.; Malaguarnera, M. Levocarnitine administration in elderly subjects with rapid muscle fatigue: Effect on body composition, lipid profile and fatigue. *Drugs Aging* **2003**, *20*, 761–767.

251. Miyagawa, T.; Kawamura, H.; Obuchi, M.; Ikesaki, A.; Ozaki, A.; Tokunaga, K.; Inoue, Y.; Honda, M. Effects of oral L-carnitine administration in narcolepsy patients: A randomized, double-blind, cross-over and placebo-controlled trial. *PLoS One* **2013**, *8*, e53707.

252. Rossini, M.; di Munno, O.; Valentini, G.; Bianchi, G.; Biasi, G.; Cacace, E.; Malesci, D.; la Montagna, G.; Viapiana, O.; Adami, S. Double-blind, multicenter trial comparing acetyl L-carnitine with placebo in the treatment of fibromyalgia patients. *Clin. Exp. Rheumatol.* **2007**, *25*, 182–188.

253. Vermeulen, R.C.; Scholte, H.R. Exploratory open label, randomized study of acetyl- and propionyl-carnitine in chronic fatigue syndrome. *Psychosom. Med.* **2004**, *66*, 276–282.

254. Beghi, E.; Pupillo, E.; Bonito, V.; Buzzi, P.; Caponnetto, C.; Chiò, A.; Corbo, M.; Giannini, F.; Inghilleri, M.; Bella, V.L.; *et al.* Randomized double-blind placebo-controlled trial of acetyl-L-carnitine for ALS. *Amyotroph. Lateral Scler. Frontotemporal Degener.* **2013**, *14*, 397–405.

255. Cruciani, R.A.; Dvorkin, E.; Homel, P.; Malamud, S.; Culliney, B.; Lapin, J.; Portenoy, R.K.; Esteban-Cruciani, N. Safety, tolerability and symptom outcomes associated with L-carnitine supplementation in patients with cancer, fatigue, and carnitine deficiency: A phase I/II study. *J. Pain Symptom Manag.* **2006**, *32*, 551–559.

256. Cruciani, R.A.; Dvorkin, E.; Homel, P.; Culliney, B.; Malamud, S.; Lapin, J.; Portenoy, R.K.; Esteban-Cruciani, N. L-Carnitine supplementation in patients with advanced cancer and carnitine deficiency: A double-blind, placebo-controlled study. *J. Pain Symptom Manag.* **2009**, *37*, 622–631.

257. Cruciani, R.A.; Zhang, J.J.; Manola, J.; Cella, D.; Ansari, B.; Fisch, M.J. L-Carnitine supplementation for the management of fatigue in patients with cancer: An eastern cooperative oncology group phase III, randomized, double-blind, placebo-controlled trial. *J. Clin. Oncol.* **2012**, *30*, 3864–3869.

258. Mantovani, G. Randomised phase III clinical trial of 5 different arms of treatment on 332 patients with cancer cachexia. *Eur. Rev. Med. Pharmacol. Sci.* **2010**, *14*, 292–301.

259. Madeddu, C.; Dessì, M.; Panzone, F.; Serpe, R.; Antoni, G.; Cau, M.C.; Montaldo, L.; Mela, Q.; Mura, M.; Astara, G.; *et al.* Randomized phase III clinical trial of a combined treatment with carnitine + celecoxib ± megestrol acetate for patients with cancer-related anorexia/cachexia syndrome. *Clin. Nutr.* **2012**, *31*, 176–182.

260. Kraft, M.; Kraft, K.; Gärtner, S.; Mayerle, J.; Simon, P.; Weber, E.; Schütte, K.; Stieler, J.; Koula-Jenik, H.; Holzhauer, P.; *et al.* L-Carnitine-supplementation in advanced pancreatic cancer (CARPAN) a randomized multicentre trial. *Nutr. J.* **2012**, *11*, 52.

261. Hershman, D.L.; Unger, J.M.; Crew, K.D.; Minasian, L.M.; Awad, D.; Moinpour, C.M.; Hansen, L.; Lew, D.L.; Greenlee, H.; Fehrenbacher, L.; *et al.* Randomized double-blind placebo-controlled trial of acetyl-L-carnitine for the prevention of taxane-induced neuropathy in women undergoing adjuvant breast cancer therapy. *J. Clin. Oncol.* **2013**, *31*, 2627–2633.

262. Schöls, L.; Zange, J.; Abele, M.; Schillings, M.; Skipka, G.; Kuntz-Hehner, S.; van Beekvelt, M.C.; Colier, W.N.; Müller, K.; Klockgether, T.; *et al.* L-Carnitine and creatine in Friedreich's ataxia. A randomized, placebo-controlled crossover trial. *J. Neural Transm.* **2005**, *112*, 789–796.

263. El-Beshlawy, A.; Ragab, L.; Fattah, A.A.; Ibrahim, I.Y.; Hamdy, M.; Makhlouf, A.; Aoun, E.; Hoffbrand, V.; Taher, A. Improvement of cardiac function in thalassemia major treated with L-carnitine. *Acta Haematol.* **2004**, *111*, 143–148.

264. Karimi, M.; Mohammadi, F.; Behmanesh, F.; Samani, S.M.; Borzouee, M.; Amoozgar, H.; Haghpanah, S. Effect of combination therapy of hydroxyurea with L-carnitine and magnesium chloride on hematologic parameters and cardiac function of patients with beta-thalassemia intermedia. *Eur. J. Haematol.* **2010**, *84*, 52–58.

265. Ellaway, C.J.; Peat, J.; Williams, K.; Leonard, H.; Christodoulou, J. Medium-term open label trial of L-carnitine in Rett syndrome. *Brain Dev.* **2001**, *23*, S85–S89.

266. Torrioli, M.G.; Vernacotola, S.; Peruzzi, L.; Tabolacci, E.; Mila, M.; Militerni, R.; Musumeci, S.; Ramos, F.J.; Frontera, M.; Sorge, G.; *et al.* A double-blind, parallel, multicenter comparison of L-acetylcarnitine with placebo on the attention deficit hyperactivity disorder in fragile X syndrome boys. *Am. J. Med. Genet. A* **2008**, *146*, 803–812.

267. Geier, D.A.; Kern, J.K.; Davis, G.; King, P.G.; Adams, J.B.; Young, J.L.; Geier, M.R. A prospective double-blind, randomized clinical trial of levocarnitine to treat autism spectrum disorders. *Med. Sci. Monit.* **2011**, *17*, PI15–PI23.

268. Osio, M.; Muscia, F.; Zampini, L.; Nascimbene, C.; Mailland, E.; Cargnel, A.; Mariani, C. Acetyl-L-carnitine in the treatment of painful antiretroviral toxic neuropathy in human immunodeficiency virus patients: An open label study. *J. Peripher. Nerv. Syst.* **2006**, *11*, 72–76.

269. Youle, M.; Osio, M.; ALCAR Study Group. A double-blind, parallel-group, placebo-controlled, multicentre study of acetyl L-carnitine in the symptomatic treatment of antiretroviral toxic neuropathy in patients with HIV-1 infection. *HIV Med.* **2007**, *8*, 241–250.

270. Benedini, S.; Perseghin, G.; Terruzzi, I.; Scifo, P.; Invernizzi, P.L.; del Maschio, A.; Lazzarin, A.; Luzi, L. Effect of L-acetylcarnitine on body composition in HIV-related lipodystrophy. *Horm. Metab. Res.* **2009**, *41*, 840–845.

271. Pignatelli, P.; Tellan, G.; Marandola, M.; Carnevale, R.; Loffredo, L.; Schillizzi, M.; Proietti, M.; Violi, F.; Chirletti, P.; Delogu, G. Effect of L-carnitine on oxidative stress and platelet activation after major surgery. *Acta Anaesthesiol. Scand.* **2011**, *55*, 1022–1028.

272. Mikhailova, T.L.; Sishkova, E.; Poniewierka, E.; Zhidkov, K.P.; Bakulin, I.G.; Kupcinskas, L.; Lesniakowski, K.; Grinevich, V.B.; Malecka-Panas, E.; Ardizzone, S.; *et al.* Randomised clinical trial: The efficacy and safety of propionyl-L-carnitine therapy in patients with ulcerative colitis receiving stable oral treatment. *Aliment. Pharmacol. Ther.* **2011**, *34*, 1088–1097.

273. Mattiazzi, M.; Vijayvergiya, C.; Gajewski, C.D.; deVivo, D.C.; Lenaz, G.; Wiedmann, M.; Manfredi, G. The mtDNA *T8993G* (*NARP*) mutation results in an impairment of oxidative phosphorylation that can be improved by antioxidants. *Hum. Mol. Genet.* **2004**, *13*, 869–879.

274. Chinnery, P.F. Mitochondrial Disorders Overview. Synonyms: Mitochondrial Encephalomyopathies, Mitochondrial Myopathies, Oxidative Phosphorylation Disorders, Respiratory Chain Disorders. Available online: http://www.ncbi.nlm.nih.gov/books/NBK1224 (accessed on 3 November 2014).

275. Di Mauro, S.; Mancuso, M. Mitochondrial diseases: Therapeutic approaches. *Biosci. Rep.* **2007**, *27*, 125–137.

276. Smith, R.A.; Murphey, M.P. Mitochondria-targeted antioxidants as therapies. *Discov. Med.* **2011**, *11*, 106–114.

277. Lee, S.H.; Kim, M.J.; Kim, B.J.; Kim, S.R.; Chun, S.; Ryu, J.S.; Kim, G.S.; Lee, M.C.; Koh, J.M.; Chung, S.J. Homocysteine-lowering therapy or antioxidant therapy for bone loss in Parkinson's disease. *Mov. Disord.* **2010**, *25*, 332–340.

278. Pagano, G.; Aiello Talamanca, A.; Castello, G.; d'Ischia, M.; Pallardó, F.V.; Petrović, S.; Porto, B.; Tiano, L.; Zatterale, A. From clinical description to *in vitro* and animal studies, and backwards to patients: Oxidative stress and mitochondrial dysfunction in Fanconi anaemia. *Free Radic. Biol. Med.* **2013**, *58*, 118–125.

279. Seco-Cervera, M.; Spis, M.; García-Giménez, J.L. Oxidative stress and antioxidant response in fibroblasts from Werner and Atypical Werner Syndromes. *Aging* **2014**, *6*, 231–245.

280. Bhagavan, H.N.; Chopra, R.K. Plasma coenzyme Q10 response to oral ingestion of coenzyme Q10 formulations. *Mitochondrion* **2007**, *7*, S78–S88.

281. Roginsky, V.A.; Tashlitsky, V.N.; Skulachev, V.P. Chain-breaking antioxidant activity of reduced forms of mitochondria-targeted quinones, a novel type of geroprotectors. *Aging* **2009**, *1*, 481–489.

282. Teichert, J.; Hermann, R.; Ruus, P.; Preiss, R. Plasma kinetics, metabolism, and urinary excretion of alpha-lipoic acid following oral administration in healthy volunteers. *J. Clin. Pharmacol.* **2003**, *43*, 1257–1267.

283. Pettegrew, J.W.; Levine, J.; McClure, R.J. Acetyl-L-carnitine physical-chemical, metabolic and therapeutic properties: Relevance for its mode of action in Alzheimer'disease and geriatric depression. *Mol. Psychiatry* **2000**, *5*, 616–632.

284. Madiraju, P.; Pande, S.V.; Prentki, M.; Murthy Madiraju, S.R. Mitochondrial acetylcarnitine provides acetyl groups for nuclear histone acetylation. *Epigenetics* **2009**, *4*, 4–6.

285. Rosca, M.G.; Lemieux, H.; Hoppel, C.L. Mitochondria in the ederly: Is acetylcarnitine a rejuvinator? *Adv. Drug Deliv. Rev.* **2009**, *61*, 1332–1342.

286. Breithaupt-Grögler, K.; Niebch, G.; Schneider, E.; Erb, K.; Hermann, R.; Blume, H.H.; Schug, B.S.; Belz, G.G. Dose-proportionality of oral thioctic acid—Coincidence of assessments via pooled plasma and individual data. *Eur. J. Pharm. Sci.* **1999**, *8*, 57–65.

287. Rebouche, C.J. Kinetics, pharmacokinetics, and regulation of L-carnitine and acetyl-L-carnitine metabolism. *Ann. N. Y. Acad. Sci.* **2004**, *1033*, 30–41.

288. Camp, K.M.; Lloyd-Puryear, M.A.; Yao, L.; Groft, S.C.; Parisi, M.A.; Mulberg, A.; Gopal-Srivastava, R.; Cederbaum, S.; Enns, G.M.; Ershow, A.G.; *et al.* Expanding research to provide an evidence base for nutritional interventions for the management of inborn errors of metabolism. *Mol. Genet. Metab.* **2013**, *109*, 319–328.

Treatment Strategies that Enhance the Efficacy and Selectivity of Mitochondria-Targeted Anticancer Agents

Josephine S. Modica-Napolitano and Volkmar Weissig

Abstract: Nearly a century has passed since Otto Warburg first observed high rates of aerobic glycolysis in a variety of tumor cell types and suggested that this phenomenon might be due to an impaired mitochondrial respiratory capacity in these cells. Subsequently, much has been written about the role of mitochondria in the initiation and/or progression of various forms of cancer, and the possibility of exploiting differences in mitochondrial structure and function between normal and malignant cells as targets for cancer chemotherapy. A number of mitochondria-targeted compounds have shown efficacy in selective cancer cell killing in pre-clinical and early clinical testing, including those that induce mitochondria permeability transition and apoptosis, metabolic inhibitors, and ROS regulators. To date, however, none has exhibited the standards for high selectivity and efficacy and low toxicity necessary to progress beyond phase III clinical trials and be used as a viable, single modality treatment option for human cancers. This review explores alternative treatment strategies that have been shown to enhance the efficacy and selectivity of mitochondria-targeted anticancer agents *in vitro* and *in vivo*, and may yet fulfill the clinical promise of exploiting the mitochondrion as a target for cancer chemotherapy.

Reprinted from *Int. J. Mol. Sci.* Cite as: Modica-Napolitano, J.S.; Weissig, V. Treatment Strategies that Enhance the Efficacy and Selectivity of Mitochondria-Targeted Anticancer Agents. *Int. J. Mol. Sci.* **2015**, *16*, 17394–17416.

1. Introduction

Despite enormous investments in the areas of basic research and medical science during the past few decades, cancer remains a leading health threat worldwide. Today in the United States alone, it is estimated that one in four adult men and one in five adult women are at risk of dying from cancer [1]. A resurgence of interest in the study of mitochondria has led to the discovery of several notable differences in the structure and function of this organelle between normal and cancer cells, and various attempts have been made to exploit these differences as novel and site specific targets for chemotherapy. Although a number of mitochondria-targeted compounds have shown some efficacy in selective cancer cell killing in pre-clinical and early clinical testing, the success of mitochondria-targeted therapeutic agents as a single modality

411

treatment option for human cancers has been quite limited. This article presents an overview of mitochondria structure and function, especially as it relates to those differences found between normal and cancer cells, and highlights the progress made in exploiting this organelle as a target for chemotherapy. In addition, it summarizes three alternative treatment strategies that enhance the efficacy and selectivity of mitochondria-targeted anticancer agents *in vitro* and *in vivo* and offer the promise of therapeutic benefit. These include: mitochondria-targeted drug delivery systems; photodynamic therapy; and combination chemotherapy.

2. Mitochondria Structure and Function

In living cells, mitochondria are dynamic organelles comprising a network of long, filamentous structures that can be seen extending, contracting, fragmenting and fusing with one another as they move in three dimensions throughout the cytoplasm [2,3]. In electron micrographs of fixed tissue specimens, mitochondria appear as oval shaped particles similar in size to the bacterium *Escherichia coli* (1–2 microns long × 0.5–1.0 microns wide) and bound by two membranes. The outer membrane encloses the entire contents of the organelle. The inner membrane, which folds inward to form cristae, encloses the inner space, or matrix. Interestingly, the surface area of the inner mitochondrial membrane correlates with the degree of metabolic activity of the cell, and can vary considerably from cell type to cell type, or within a given cell depending upon its functional state. Mitochondria contain the enzymes and cofactors involved in a number of important metabolic reactions and pathways, including the tricarboxylic acid (TCA) cycle, oxidative phosphorylation, fatty acid degradation, the urea cycle, and gluconeogenesis. In mammalian cells, the matrix also typically contains up to 10,000 copies of a 16.6 kb closed circular double helical molecule of mitochondrial DNA (mtDNA), which is compacted *in vivo* to form a nucleoprotein complex, or nucleoid [4]. Although representing less than 1% of the total cellular DNA, mtDNA encodes two rRNAs, twenty-two tRNAs and thirteen highly hydrophobic polypeptide subunit components of four different respiratory enzyme Complexes (I, III, IV and V) that are localized to the inner mitochondrial membrane.

Mitochondria are considered the "powerhouse" of eukaryotic cells because of their central role in the process of aerobic metabolism. In carbohydrate metabolism, this begins when pyruvate, the end product of glycolysis, is transported from the cytosol into the mitochondrial matrix to undergo oxidative decarboxylation via the pyruvate dehydrogenase complex. In lipid metabolism, this begins when fatty acids are transported into the mitochondrial matrix to undergo sequential rounds of oxidative decarboxylation via the β-oxidation pathway. In either case, the resultant metabolic product is acetyl coA, which is further oxidized in the mitochondrial matrix via the TCA cycle. The net metabolic yield of the TCA cycle includes two molecules of CO_2, one molecule of GTP (the energetic equivalent of ATP), three molecules of reduced nicotinamide adenine dinucleotide (NADH), and one molecule of reduced flavin adenine dinucleotide ($FADH_2$). NADH and $FADH_2$ go on to serve as respiratory substrates for oxidative phosphorylation, which couples the oxidation of these high-energy electron donors to the synthesis of ATP. In this process, electrons are transferred from NADH and $FADH_2$ to oxygen via four multi-subunit electron transfer complexes located on the inner mitochondrial membrane. Complexes I, III and IV of the mitochondrial electron transfer chain assemble into functional supramolecular complexes, called respirasomes [5]. These three respiratory complexes also serve as proton pumps at which the energy derived from the transfer of electrons down the electron transport chain (ETC) is coupled to the translocation of protons from the matrix space outward to the space between the inner and outer mitochondrial membranes (*i.e.*, inter-membrane space). Under normal physiological conditions, the inner mitochondrial membrane is relatively impermeable to the backflow of protons and an electrochemical gradient is established across the membrane. The energy stored in this proton gradient, the proton-motive force, is then used to drive the synthesis of ATP from ADP and P_i via the inner membrane bound enzyme, mitochondrial ATP sythetase (Complex V). Oxidative phosphorylation supplies the vast majority of ATP produced by a cell under aerobic conditions.

Mitochondria are the main intracellular source of reactive oxygen species (ROS) in most tissues. It has been estimated that under physiological conditions, 1%–2% of the molecular oxygen consumed is converted to ROS molecules as a byproduct of oxidative phosphorylation [6]. ROS production can occur when a small fraction of reducing equivalents from Complex I or Complex III of the mitochondrial electron transport chain "leak" electrons directly to molecular oxygen, generating the superoxide anion O_2^-. Mitochondrial superoxide dismutase converts O_2^- to H_2O_2, which can then acquire an additional electron from a reduced transition metal to generate the highly reactive hydroxyl radical ˙OH. There is increasing evidence that Complex II can also be a major regulator of mitochondrial ROS production under physiological and pathophysiological circumstances [7,8]. ROS play an important role as signaling molecules that mediate changes in cell proliferation, differentiation, and gene transcription [9,10]. Uncontrolled ROS activity, or oxidative stress, can damage intracellular protein and lipid components, and affect the integrity of biological membranes. High levels of ROS can also damage both nuclear and mtDNA. The mitochondrial genome is especially susceptible to ROS damage due to its proximity to the site of ROS production (*i.e.*, the ETC), as well as the fact that it has no introns or protective histones and a limited capacity for DNA repair. Thus, oxidative stress can impair mitochondrial function directly at the level of mitochondrial enzyme complexes, or as a consequence of its genotoxicity to mtDNA.

Severe or prolonged oxidative stress can lead to irreversible oxidative damage and cell death [11].

Mitochondria also play a key role in mediating intrinsic apoptosis, an energy dependent cell death pathway regulated by numerous positive and negative signaling factors that exist in dynamic equilibrium [12]. Distally, intrinsic apoptosis can be induced by a variety of physiological or pathological cell stressors, such as toxins, viral infections, hypoxia, hyperthermia, free radicals, and DNA damage. Proximately, the intrinsic pathway is induced by the loss of anti-apoptotic proteins, (e.g., Bcl-2 and Bcl-x) or by activation of pro-apoptotic proteins (e.g., Bax and Bak). Intrinsic apoptosis involves mitochondrial outer membrane permeabilization (MOMP), the critical, irreversible step in the pathway that commits the cell to ultimate destruction. MOMP is followed by the release of cytochrome c and other apoptogenic proteins from the mitochondrial inter-membrane space. Once released into the cytosol, these proteins activate a caspase cascade, which leads to the proteolytic cleavage of intracellular proteins, DNA degradation, formation of apoptotic bodies, and other morphological changes that are considered hallmarks of apoptotic cell death. Both the intrinsic apoptotic pathway and the extrinsic apoptotic pathway, which involves cell membrane receptor-mediated interactions, play significant roles in normal development, tissue remodeling, aging, wound healing, immune response, and maintaining homeostasis in the adult human body.

3. Some Notable Differences between Mitochondria of Cancer Cells and Normal Cells

Nearly a century has passed since Otto Warburg first observed high rates of aerobic glycolysis in a variety of tumor cell types and suggested that this phenomenon might be due to an impaired respiratory capacity in these cells [13]. Warburg's observations prompted many scientists to focus their investigative efforts on the mitochondria of cancer cells in an attempt to understand the underlying basis for the "Warburg Effect", *i.e.*, enhanced glucose uptake, high rate of glycolysis in the presence of sufficient oxygen, and an increase in lactic acid as a byproduct of the glycolytic pathway. It is now known that at least some cancer cells possess a normal capacity for oxidative phosphorylation and can, under certain conditions, generate a majority of their ATP from this process [14–21]. In addition, recent evidence suggests that the enhanced glucose uptake and metabolic shift toward aerobic glycolysis in cancer cells is more likely due to their greater need for glucose metabolites, which serve as precursors for the biosynthesis of nucleic acids, amino acids, and lipids in these rapidly dividing cell populations [22], rather than to any specific impairment in respiratory function. In the years since Warburg's initial observations, however, a number of notable differences between the mitochondria of normal and transformed cells have been identified [23–28]. These include differences in the size, number and shape of the organelle, the rates of protein synthesis and organelle turnover, and the polypeptide and lipid profiles of the inner mitochondrial membrane. Metabolic aberrations specifically associated with mitochondrial bioenergetic function in cancer cells include differences with regard to preference for respiratory substrates, rates of electron and anion transport, calcium uptake and retention, and decreased activities of certain enzymes integral to the process of oxidative phosphorylation, such as cytochrome c oxidase [29,30], adenine nucleotide translocase [31–33], and mitochondrial ATPase [34]. The mitochondrial membrane potential has also been shown to be significantly higher in carcinoma cells than in normal epithelial cells [35–37].

Alterations in mitochondrial genome sequence have also been linked to a variety of cancers [38–40]. Some are germ-line mutations. Among these, a human polymorphic variant in the NADH dehydrogenase 3 (*ND3*) gene at nt 10,398 (nt G10398A) that alters the structure of Complex I in the mitochondrial ETC was associated with an increased risk for invasive breast cancer in African–American women [38,41], the A12308G mutation in tRNA[Leu(CUN)] was associated with increased risk of both renal and prostate cancers [42], and a variant in a non-coding region of mtDNA (16189T>C) was associated with increased susceptibility to endometrial cancer [43]. Somatic mutations in the mitochondrial genome are more common and have been observed in a wide variety of cancers, including ovarian, uterine, liver, lung, colon, gastric, brain, bladder, prostate, and breast cancer, melanoma and leukemia [26]. The displacement loop (or D-loop) region, a triple stranded non-coding sequence of mtDNA (np 16024-516) that houses cis regulatory elements required for replication and transcription of the molecule, has been shown to be a mutational "hot spot" in human cancer. However, mutations in genes encoding the polypeptide subunits of enzymes involved in oxidative phosphorylation also occur and can be of functional significance. Some of these are thought to be adaptive mutations that confer a selective advantage under the harsh growth conditions of the tumor microenvironment [40]. Others have been shown to be involved directly in tumor initiation and/or progression. For example, introduction of the pathogenic mtDNA *ATP6* T8993G mutation into the PC3 prostate cancer cell line through cybrid transfer produced tumors in nude mice that were 7-fold greater in size than those produced by wild-type cybrids [39]. Additionally, mutations in the mtDNA gene encoding NADH dehydrogenase subunit 6 (*ND6*) produced a deficiency in respiratory Complex I activity that was associated with an enhanced metastatic potential of tumor cells [44].

In general, tumor cells also exhibit higher levels of ROS than normal cells [9], and oxidative stress has been suggested to underlie the development and/or maintenance of the malignant phenotype. As noted previously, oxidative stress can cause somatic mutations in mtDNA. Evidence suggests that the converse is also true, *i.e.*, certain mutations in mtDNA, especially those in genes encoding ETC enzyme subunits, can cause ROS overproduction. Oncogene activation is also known to enhance the production of mitochondrial ROS, which has been implicated as a mechanism for K-RAS and MYC-mediated cell transformation [45,46]. In tumor cells, oxidative stress activates signaling pathways that promote cell growth and metastasis. One such pathway involves hypoxia-inducible factor (HIF), which regulates the transcription of a large number of genes that facilitate cell survival at low oxygen pressures [47]. Under the hypoxic conditions of tumor cell growth, mitochondria act as O_2 sensors and further enhance ROS generation as an adaptive response [48]. ROS overproduction stabilizes the HIF-α subunit, facilitating its dimerization with the HIF-β subunit. This activates a number of different genes, including those mediating a metabolic shift toward glycolysis, angiogenesis, and metastasis. ROS have also been shown to activate MAP kinase and phosphoinositide 3-kinase pathways, which are important for cell proliferation and survival [9], and to up-regulate the expression of matrix metalloproteinases (MMPs) and Snail proteins, which are involved in epithelial-to-mesenchymal transition and metastasis, respectively [49].

Inhibition of the intrinsic apoptotic pathway is also observed in a number of hematopoietic malignancies and solid tumors, and has been implicated in cancer initiation, progression and metastasis [50,51]. This is thought to occur as a result of dysregulation of mitochondrial outer membrane proteins of the Bcl-2 family, and may involve overexpression or enhanced function of anti-apoptotic proteins, under-expression or loss of function of pro-apoptotic proteins, or a combination of both. For example, malignant chronic lymphocytic leukemia (CLL) cells express high levels of anti-apoptotic Bcl-2 and low levels of pro-apoptotic proteins such as Bax [52]. Interestingly, the progression of CLL is thought to be due to reduced apoptosis rather than increased proliferation *in vivo* [53]. Overexpression of Bcl-2 has also been shown to inhibit apoptosis in prostate [54], lung, colorectal and gastric cancers [55,56], neuroblastoma, glioblastoma, and breast carcinoma cells [57]. An imbalance in the expression of the anti- and pro-apoptotic Bcl-2 family of proteins is thought to stabilize the outer mitochondrial membrane, prevent MOMP and the release of cytochrome c, and ultimately, inhibit programmed cell death. This failure of normal cell turnover contributes to cell accumulation, transformation, and survival under extreme conditions, such as the hypoxic or acidic environments common in tumors. Interestingly, the inhibition of apoptosis that results from dysregulation of Bcl-2 protein expression has also been shown to underlie the development of drug resistance in cancer cells. For example, the overexpression Bcl-XL protects murine pro-lymphocytic cells from a wide variety of apoptotic stimuli and confers a multidrug resistance phenotype [58], and drug-induced apoptosis in B-CLL cells cultured *in vitro* is inversely related to Bcl-2/Bax ratios [52].

4. Mitochondria-Targeted Drugs That Show Selective Cancer Cell Killing

During the past few decades, scientists have been exploring the possibility that certain structural and functional differences that exist between the mitochondria of normal and transformed cells might serve as targets for selective cell killing by novel and site-specific anticancer agents. Recently, the term "mitocan" (an acronym for mitochondria and cancer) has been proposed to classify mitochondria-targeted anticancer agents, especially those that induce mitochondrial destabilization [59]. A number of these compounds have shown efficacy in selective cancer cell killing in pre-clinical and early clinical testing (see Table 1 for a representative sampling).

Table 1. Representative mitochondria-targeted compounds that exhibit selective cancer cell killing.

Class	Compound	Mode of Action	Demonstrated Efficacy	References
OxPhos Inhibitors	Rhodamine 123	ATP Synthase inhibitor	Preclinical (*in vitro, in vivo*)	[60–62]
	Dequalinium Chloride	Complex I inhibitor	Preclinical (*in vitro, in vivo*)	[63,64]
	AA-1	ATP Synthase inhibitor	Preclinical (*in vitro, in vivo*)	[65]
	MKT-077	General inhibition of ETC enzymes	Preclinical (*in vitro, in vivo*) / Clinical, Phase I	[66–69]
	Metformin	Complex I inhibitor	Preclinical (*in vitro, in vivo*) / Clinical, Phase I	[70–89]
ROS Regulators	Elesclomol	Enhanced ROS production	Preclinical (*in vitro, in vivo*) / Clinical, Phase I	[90–92]
	Bezielle	Enhanced ROS production	Preclinical (*in vitro, in vivo*) / Clinical, Phase I	[93–99]
Intrinsic Apoptosis Inducers	ABT-737	BH3 mimetic	Preclinical (*in vitro, in vivo*)	[100–102]
	ABT-263 (Navitoclax)	BH3 mimetic	Preclinical (*in vitro, in vivo*) / Clinical, Phase I/II	[103–105]
	Gossypol	BH3 mimetic	Preclinical (*in vitro, in vivo*)	[106,107]
	GX15-070 (Obatoclax)	BH3 mimetic	Preclinical (*in vitro, in vivo*)	[108,109]
	HA14-1	BH3 mimetic	Preclinical (*in vitro, in vivo*)	[110,111]

Among the earliest known mitochondria-targeted anticancer agents are the delocalized lipophilic cations (DLCs). Due to their lipophilicity and positive charge, these compounds selectively accumulate in the mitochondria of carcinoma cells in response to a higher, negative inside membrane potential (e.g., approximately 160 mV in carcinoma *vs.* 100 mV in control epithelial cells) [36,37]. Several DLCs have exhibited efficacy in carcinoma cell killing *in vitro* and *in vivo* [60–69,112,113], including the class prototype Rhodamine 123 (Rh123), dequalinium chloride (DECA), and the thiopyrylium AA-1. Although all DLCs are taken up into mitochondria by a common mechanism and display dose dependent mitochondrial toxicity, their

417

specific mechanism of action can be quite varied. For example, Rh123 and AA-1 inhibit mitochondrial ATP synthesis at the level of F0F1-ATPase activity [62,65,113], while DECA and certain DLC thiacarbocyanines interfere with NADH-ubiquinone reductase (ETC Complex I) activity [64,112]. Another DLC, the water-soluble rhodacyanine dye analogue MKT-077, was shown to cause a more generalized deleterious effect on respiratory function through membrane perturbation and consequent inhibition of membrane-bound enzymes [67]. MKT-077 was the first DLC with a favorable pharmacological and toxicological profile and showed great promise as a selective anticancer agent in preclinical studies [66]. Phase I trials were undertaken to evaluate the safety and pharmacokinetics of MKT-077, but were halted due to recurrent but reversible renal toxicity in about half of the patients treated [68]. It was determined, however, that it is feasible to target mitochondria with rhodacyanine analogues if drugs with higher therapeutic indices could be developed [69].

More recently, evidence suggests that the widely prescribed anti-diabetic biguanide derivative, metformin, may also be effective in the prevention and treatment of human cancer via inhibition of mitochondrial respiratory function. Retrospective analyses show an association between the use of metformin and diminished cancer risk, progression and mortality in diabetic patients [70–74]. *In vitro* laboratory studies demonstrate that metformin has a direct and selective inhibitory effect on breast, colon, ovary, pancreas, lung, and prostate cancer cell lines [75–79]. In addition, at doses that had no effect on the viability of non-cancer stem cells, metformin inhibited transformation and selectively killed cancer stem cells resistant to chemotherapeutic agents [80]. *In vivo*, metformin inhibits the growth of spontaneous and carcinogen-induced tumors, and impacts tumor growth in mouse xenograft and syngeneic models [81–85]. Furthermore, prospective studies investigating the therapeutic efficacy of metformin use in non-diabetic cancer patients suggest its promise for the chemoprevention of colorectal cancer and treatment of early breast cancer [86–88]. It has been postulated that the therapeutic effects of metformin may be associated with both direct (insulin-independent) and indirect (insulin-dependent) actions of the drug [74]. However, results of a recent study showed that the direct inhibition of cancer cell mitochondrial Complex I by metformin was required to decrease cell proliferation *in vitro* and tumorigenesis *in vivo* [89]. Interestingly, it has been shown that cancer cell lines harboring mutations in mtDNA encoded Complex I subunits or having impaired glucose utilization exhibit enhanced biguanide sensitivity when grown under the low glucose conditions seen in the tumor microenvironment [114]. Metformin is a very safe and well-tolerated drug that is now prescribed to almost 120 million people in the world for the treatment of type II diabetes. Clinical trials using metformin alone and in combination with conventional anticancer agents in non-diabetic patients are ongoing and should clarify its potential use in cancer therapy.

Mitochondria-targeted ROS regulators have also shown efficacy as anticancer agents. Although the generally higher endogenous levels of ROS in tumor *versus* normal cells contribute to the development and/or maintenance of the malignant phenotype, they also render cancer cells more vulnerable to irreversible oxidative damage and consequent cell death. Therefore, pro-oxidant pharmacological agents that either enhance ROS production or inhibit ROS scavenging activity have the potential to increase ROS level beyond the threshold of lethality in cancer cells while leaving normal cells viable [115]. One such compound that targets mitochondria is elesclomol (STA-4783), an investigational, first-in-class small molecule that has been shown to enhance ROS production and induce a transcriptional gene profile characteristic of an oxidative stress response *in vitro*. Interestingly, the antioxidant N-acetylcysteine blocks elesclomol induced gene expression and apoptosis, indicating that ROS generation is the primary mechanism of cytotoxicity of the drug [115]. Comparative growth assays using the yeast model *S. cerevisiae* demonstrated that elesclomol interacts with the mitochondrial ETC to generate high levels of ROS and induce apoptosis [90]. In the same study, elesclomol was shown to interact similarly with the ETC in human melanoma cells. Elesclomol was granted fast-track designation by the FDA in 2006 for the treatment of metastatic melanoma. A randomized, double-blind, controlled SYMMETRY study evaluating the combination of paclitaxel and elesclomol in patients with advanced melanoma was stopped after all patients were enrolled because the addition of elesclomol to paclitaxel did not significantly improve progression free survival in unselected patients [91]. Studies are ongoing to determine the effect of elesclomol treatment alone and in combination with paclitaxel in patients with acute myeloid leukemia, and ovarian cancer [92].

Bezielle (BZL101), an aqueous extract from the herb *Scutellaria barbata*, is another ROS regulator that displays selective cytotoxicity against a variety of cancers *in vitro* and *in vivo* [93–95]. Early studies showed that in tumor cells, but not in non-transformed cells, Bezielle induces ROS production and causes severe DNA damage followed by hyperactivation of PARP-1, depletion of the cellular ATP and NAD, inhibition of glycolysis, and cell death [96]. It was later shown that treatment of tumor cells with Bezielle induces progressively higher levels of both mitochondrial superoxide and peroxide type ROS, and that Bezielle inhibits oxidative phosphorylation [97]. In addition, tumor cells lacking functional mitochondria did not generate mitochondrial superoxide and were protected from cell death in the presence of Bezielle, supporting the hypothesis that mitochondria are the primary target of the compound [97]. Bezielle has shown promising efficacy and excellent safety in the early phase clinical trials for advanced breast cancer [98,99].

Mitochondria-targeted compounds that induce outer membrane permeabilization and intrinsic apoptosis in cancer cells also show potential as anti-cancer agents. As previously discussed, BCL-2 family proteins, which share one or more of the four BCL-2 homology domains (BH1–BH4), regulate the intrinsic apoptotic pathway. Anti-apoptotic members of the family (such as BCL-2, BCL-X$_L$, BCL-W and MCL-1), which are overexpressed in many cancers, function by sequestering the pro-apoptotic executioners of the MOMP (such as BAX and BAK). Inhibition of programmed cell death is antagonized by BH3-only proteins, a BCL-2 protein subfamily comprised of only the α-helical BH3 domain. These small proteins interact with anti-apoptotic molecules in their BH3-binding groove, causing the release and activation of BAX/BAK and inducing apoptosis [116]. Certain small molecules mimic the effect of BH3-only proteins. Among these BH3 mimetics, the synthetically derived ABT-737 has been shown to induce BAX/BAK-dependent apoptosis in a variety of cancer cell lines *in vitro*, and to display antitumor effects as a single agent *in vivo* [100–102]. Navitoclax (ABT-263), a potent, orally bioavailable analog of ABT-737 with similar biological activity, was shown to elicit complete tumor regression in small cell lung cancer (SCLC) and acute lymphoblastic leukemia xenograft models [103]. A phase I clinical study investigating the single-agent activity of navitoclax in the treatment of recurrent SCLC yielded encouraging preliminary safety and efficacy data [104]. However, in a subsequent phase II study navitoclax treatment induced only a low positive response and was limited by a dose-dependent and clinically significant thrombocytopenia [105]. Since both ABT-737 and navitoclax have been shown to potentiate the efficacy of standard cytotoxic agents against a variety of cancers [103,117–121], combinatorial regimens may ultimately prove a more promising therapeutic strategy for these compounds. Pre-clinical and clinical studies have shown that several other BH3 mimetics, such as the natural polyphenolic compound gossypol, and the synthetic compounds GX15-070 (obatoclax) and HA14-1 (ethyl 2-amino-6-bromo-4-(1-cyano-2-ethoxy-2-oxoethyl)-4H-chromene-3-carboxylate), also demonstrate anti-cancer activity, supporting the therapeutic potential of this class of mitochondria-targeted agents in the treatment of human cancer [106–111].

5. Alternative Treatment Strategies that Enhance the Efficacy and Selectivity of Mitochondria-Targeted Anticancer Agents

The fact that several mitochondria-targeted compounds have exhibited potent cancer cell killing in pre-clinical and early clinical studies is encouraging, and further research and testing of these compounds as viable, single modality treatment options for human cancers is warranted. However, the current limitations of this approach suggest the need also to explore the use of alternative treatment strategies in an effort to improve the efficacy and selectivity of these anticancer agents. Presented below (and summarized in Table 2) are three treatment strategies that have been shown *in vitro* and *in vivo* to enhance the selective cancer cell killing of several compounds known to have direct or indirect effects on mitochondrial function. It is proposed that by expanding the application of these strategies to include additional mitochondria-targeted compounds already known to exhibit significant preclinical and clinical anticancer activity as single agents (e.g., oxidative phosphorylation inhibitors, ROS regulators, and apoptosis inducers), the therapeutic efficacy of these compounds might also be improved.

Table 2. Treatment strategies that have been shown to enhance the efficacy and selectivity of anticancer agents.

Strategy	Carrier/Class	Anticancer Agent	References
Mitochondria-Targeted Drug Delivery Systems	TPP$^+$-conjugated molecules	Vitamin E succinate	[122,123]
		Coenzyme Q	[124]
	DQAsomes	Paclitaxel	[125–127]
		Curcumin	[128]
		Resveratrol	[129]
	STPP$^+$ liposomes	Paclitaxel	[130,131]
		Doxorubicin	[132]
	Mito-targeted nanontubes	Platinum (IV)	[133]
Photodynamic Therapy	Cationic photosensitizers	EDKC	[134]
		Rh123	[135]
		MKT-077	[136]
	Non-cationic photosensitizers	Pba	[137–143]
		BBr2	[144]
Combination Chemotherapy	Inhibitors of glycolysis and oxidative phosphorylation	2-DG plus metformin	[145,146]
	Inhibitors of two or more mitochondrial target sites	AZT plus MKT-077	[147]

5.1. Mitochondria-Targeted Drug Delivery Systems

Over the past several decades, attempts have been made to develop mitochondriotropic drug delivery systems for a variety of therapeutic purposes. One early strategy employed mitochondrial protein-import machinery to deliver macromolecules to mitochondria. For example, a mitochondrial signal sequence was used to direct green fluorescent protein to mitochondria to allow the visualization of mitochondria within living cells [148]. Another strategy employed conjugation with well-established mitochondriotropic cations, such as triphenylphosphonium (TPP$^+$) to successfully target low-molecular weight molecules to mammalian mitochondria. These molecules rapidly permeate lipid bilayers and, in response to the plasma and mitochondrial membrane potentials (negative inside), accumulate several hundredfold inside the organelle. One study demonstrated that significant doses of the TPP-conjugated antioxidants coenzyme Q or vitamin E could be fed safely to mice over long periods, and achieve steady-state distributions within the heart, brain, liver, and muscle [149]. These results showed that mitochondria-targeted bioactive molecules can be administered orally, leading to their accumulation at potentially therapeutic concentrations in those tissues most affected by mitochondrial dysfunction. More recently, mitochondria-targeted, TPP-conjugated vitamin E succinate has been shown to act preferentially on cancer cells, suppressing mitochondrial function and mtDNA transcription and blocking proliferation at low concentrations [122], and inducing apoptosis at higher concentrations [123]. In another study, Mito-Q (coenzyme-Q conjugated to an alkyl triphenylphosphonium cation) and Mito-CP (a 5-membered nitroxide, CP, conjugated to a TPP cation) potently inhibited the proliferation of breast cancer cells (MCF-7 and MDA-MB-231) [124] and human colon cancer cells (HCT-116) [45], further demonstrating the anticancer potential of TPP-conjugated molecules.

A quantitative structure activity relationship (QSAR) model was developed to facilitate guided synthesis and selection of optimal mitochondriotropic structures [150]. In theory, any compound that acts on mitochondria can be chemically modified to become mitochondriotropic. However, there are limitations to this strategy. First, not all potentially therapeutic compounds with molecular targets at or inside mammalian mitochondria find their way to mitochondria once inside a cell. This is because the intracellular distribution of a low-molecular weight compound is strongly affected not only by its own physico-chemical properties, but also by the cytoskeletal network, dissolved macromolecules, and dispersed organelles. Furthermore, any chemical modification that renders a compound mitochondriotropic may adversely affect its inherent pharmacological activity. In contrast, pharmaceutical nanocarriers offer an alternative approach to improve the intracellular disposition of potentially therapeutic compounds. The benefit of this strategy is that all chemistry can be carried out on the components of the nanocarrier, leaving the pharmacological profile of the compound unaltered [151]. Furthermore, nanocarrier delivery can overcome several limitations for the therapeutic use of free compounds, such as lack of water solubility, non-specific biodistribution and targeting, and low therapeutic indices.

The idea that nanocarriers could serve as effective mitochondria-targeted drug delivery systems arose in the late 1990s with the accidental discovery of the vesicle-forming capacity of dequalinium chloride, a cationic bolaamphiphile comprising two quinaldinium rings linked by ten methylene groups [152]. The compound was found to self-assemble into liposome-like vesicles, called DQAsomes (DeQAlinium-based lipoSOMES), and to have a strong affinity for mitochondria [153,154]. Follow-up studies confirmed the suitability of DQAsomes for the delivery of bioactive compounds to mitochondria, and DQAsomes are now considered the prototype for all vesicular mitochondria-specific nanocarriers [155]. *In vitro* and *in vivo* studies have shown that DQAsomal preparations of the anticancer agent paclitaxel increase the solubility of the drug by a factor of 3000, and enhance its efficiency in triggering apoptosis by direct action on mitochondria [125–127]. More recently, DQAsomes have been used for the pulmonary delivery of curcumin [128], a potent antioxidant with anti-inflammatory and potential anticancer properties. Due to its water-insolubility, however, curcumin's bioavailability following oral administration is extremely low. Curcumin encapsulated into DQAsomes displays enhanced antioxidant activity in comparison to the free compound.

Interestingly, a mitochondria-targeting drug delivery system in which dequalinium chloride has been covalently linked to the hydrophilic distal end of polyethylene glycol-distearoylphosphatidylethanolamine (DQA-PEG(2000)-DSPE) has also been prepared [129]. These nanocarriers were used to deliver resveratrol to mitochondria in human lung adenocarcinoma A549 cells, resistant A549/cDDP cells, A549 and A549/cDDP tumor spheroids as well as the xenografted resistant A549/cDDP cancers in nude mice. Results demonstrated that the mitochondrial targeting of resveratrol induced apoptosis in both non-resistant and resistant cancer cells by dissipating the mitochondria membrane potential, releasing cytochrome c and increasing the activities of caspase 9 and 3 [129]. DQAsomes have also been used to deliver an artificial mini-mitochondrial genome construct encoding Green Fluorescence Protein (GFP) to the mitochondrial compartment of a mouse macrophage cell line resulting in the expression of GFP mRNA and protein [156]. Though the transfection efficiency for GFP was very low this work constitutes the very first reported successful transgene expression inside mitochondria within living mammalian cells.

Conventional liposomes are another type of pharmaceutical nanocarrier that can also be rendered mitochondria-specific via the surface attachment of known mitochondriotropic residues, such as the cation TPP [157–160]. Preparation of liposomes in the presence of hydrophilic molecules, which have been artificially hydrophobized via linkage to fatty acid or phospholipid derivatives, results in the covalent "anchoring" of the hydrophilic moiety to the liposomal surface [161,162]. In 2005, TPP cations were conjugated to stearyl residues (yielding stearyl-TPP, or STPP), and STPP-bearing liposomes were first shown to exhibit *in vitro* mitochondriotropism [157]. The same group later demonstrated that surface modification of nanocarriers with mitochondriotropic TPP cations facilitates the efficient subcellular delivery of a model compound, ceramide, to mitochondria of mammalian cells and improves its cytotoxic and pro-apoptotic activities *in vitro* and *in vivo* [158]. More recently, STPP liposomes have been used as nanocarriers to enhance the efficacy of mitochondria-targeted anticancer agents. For example, paclitaxel loaded STPP liposomes were shown to co-localize with mitochondria and to significantly increase cytotoxicity by paclitaxel in a drug resistant ovarian carcinoma cell line [130]. The improvement in cytotoxicity was found to result from the increased accumulation of paclitaxel in mitochondria, as well as from the specific toxicity of STPP towards the resistant cell line. Mechanistic studies revealed that the cytotoxicity of STPP was associated with a decrease in mitochondrial membrane potential and other hallmarks related to caspase-independent cell death. Interestingly, mitochondriotropic STPP liposomes can be made to exhibit even greater cancer cell specificity with the addition of another ligand, folic acid. Cancer cell-specific targeting via surface modification with these dual ligands has been shown to enhance the cellular and mitochondrial delivery of doxorubicin in KB cells, and produce a synergistic effect on ROS production and cytotoxicity in this tumor cell line [132].

The preparation of TPP-surface modified liposomes utilizing an alternative hydrophobic anchor for TPP cations has also been described. For example, a d-alpha-tocopheryl polyethylene glycol 1000 succinate-triphenylphosphine conjugate (TPGS1000-TPP) was synthesized as the mitochondrial targeting molecule and incorporated into the membranes of paclitaxel-loaded liposomes [131]. The paclitaxel loaded TPGS1000-TPP conjugated liposomes were shown to selectively accumulate in the mitochondria. This targeted delivery of paclitaxel caused the release of cytochrome c, initiated a cascade of caspase 9 and 3 reactions, and enhanced apoptosis by activating pro-apoptotic pathways and inhibiting anti-apoptotic pathways. In comparison with taxol and regular paclitaxel liposomes, the mitochondria targeted paclitaxel liposomes exhibited the strongest anticancer efficacy against drug resistant lung cancer cells *in vitro* and in a nude mouse xenograft model *in vivo*, suggesting a potential therapeutic treatment for drug-resistant lung cancer.

A number of other TPP$^+$ modified nanocarriers have shown promise as effective mitochondrial specific drug delivery systems. One novel mitochondriotropic nanocarrier based on an oligolysine scaffold with the addition of two triphenylphosphonium cations per oligomer, and another based on a 5 poly(amidoamine) dendrimer conjugated with TPP$^+$, were shown to be efficiently taken up by cells and display a high degree of mitochondrial specificity [163,164]. A TPP-conjugated, mitochondria-targeted nano delivery system for coenzyme Q10 (CoQ10) has also been shown to reach mitochondria and to deliver CoQ10 in adequate quantities [165]. The multifunctional nanocarrier is composed of poly(ethylene glycol), polycaprolactone and triphenylphosphonium bromide and was synthesized using a combination of click chemistry with ring-opening polymerization followed by self-assembly into nanosized micelles. A potential disadvantage of this system, however, is the localization of the mitochondrial targeting moiety, which is seated between the two polymers, *i.e.*, between the poly(ethylene glycol) and polycaprolactone units.

In a different approach, TPP$^+$ was linked to the PEG side of a PLGA-PEG-based block copolymer, thereby enhancing the availability of the targeting moiety for any potential interaction with mitochondrial membranes [166]. In a follow-up study, Zinc phtalocyanine (ZnPc) was encapsulated inside PLGA-b-PEG-TPP polymer nanoparticles. By targeting ZnPc to the mitochondria, singlet oxygen was locally produced inside the mitochondria to effectively initiate apoptosis [167]. Interestingly, TPP-conjugated poly(ethylene imine) hyperbranched polymer nanoassemblies were also shown to successfully deliver doxorubicin to the mitochondria of human prostate carcinomas cells and cause rapid and severe cytotoxicity within few hours of incubation, even at sub-micromolar incubation concentrations [168].

The mitochondrial cationic dye, rhodamine-110, has also been used for rendering carbon nanotubes (CNTs) mitochondriotropic. In one study, multi-walled carbon nanotubes (MWCNTs) were functionalized with either mitochondrial-targeting fluorescent rhodamine-110 (MWCNT-Rho) or non-targeting fluorescein (MWCNT-Fluo) as a control [133]. Results demonstrated that MWCNT-Rho co-localized well with mitochondria (*ca.* 80% co-localization) in contrast to MWCNT-Fluo, which showed poor association with mitochondria (*ca.* 21% co-localization). In addition, platinum (IV), a prodrug of cis-platin, displayed significantly enhanced cytotoxicity towards several cancer cell lines when incorporated into mitochondria-targeted carbon nanotubes in comparison to non-targeted formulations [133]. MWCNTs have also been functionalized with peptides having a mitochondria-targeted peptide sequence (MTS). The association of such MWCNT-MTS conjugates with mitochondria inside murine macrophages and HeLa cells has been confirmed by wide-field epifluorescence microscopy, confocal laser scanning microscopy and transmission electron microscopy (TEM). The localization of the MTS-MWCNT conjugates with mitochondria was further confirmed by analyzing the isolated organelles using TEM [169]. The use of nanoparticles for the delivery of small molecule anticancer agents has thus shown past success and holds much promise for further development and therapeutic application.

5.2. Photodynamic Therapy

Photodynamic therapy (PDT) involves the use of a photoreactive drug, or photosensitizer, that is selectively taken up or retained by target cells or tissues. Upon administration of light of a specific wavelength, the photosensitizer becomes activated from a ground state to an excited state. As the photosensitizer returns to the ground state, the energy is transferred to molecular oxygen, thus generating ROS and inducing cellular toxicity in the particular areas of tissue that have been exposed to light [170]. There has been considerable interest in PDT as a treatment modality for a variety of cancers [170,171]. Photofrin, which was first used in PDT in 1993 for the prophylactic treatment of bladder cancer, is the most common photosensitizer in clinical use today. However, a number of other photosensitizers have been approved for clinical use or have undergone clinical testing to treat cancers of the head and neck, brain, lung, pancreas, intraperitoneal cavity, breast, prostate and skin. The selectivity of a photosensitizer and its site of action within a cell contribute to the efficacy of PDT. Evidence suggests that subcellular localization is more important than photochemical reactivity in terms of overall cell killing, and that mitochondrial localization represents a highly desirable property for the development of highly specific and efficient photosensitizers for photodynamic therapy applications [172].

Cationic photosensitizers are particularly promising as potential PDT agents. Like other DLCs, these compounds are concentrated by cells and into mitochondria in response to negative-inside transmembrane potentials, and are thus selectively accumulated in the mitochondria of carcinoma cells. In combination with localized photoirradiation, the cationic photosensitizer can be converted to a reactive and highly toxic species, thus enhancing its selectivity for and toxicity to carcinoma cells, and providing a means of highly specific tumor cell killing without injury to normal cells. Several cationic photosensitizers have shown promise for use in PDT. For example, selective photoxicity of carcinomas *in vitro* and *in vivo* has been observed for a series of triarylmethane derivatives [173] and the kryptocyanine EDKC [134]. Both Rh123 and the chalcogenapyrylium dye 8b have been evaluated as photosensitizers for the photochemotherapy of malignant gliomas [135,174]. In another study, photoactivation of the selective anticancer agent MKT-077 was shown to enhance its mitochondrial toxicity [136]. As expected, the mechanisms of mitochondrial toxicity exhibited by these compounds are varied, and range from specific inhibition of mitochondrial enzymes to non-specific perturbation of mitochondrial function due to singlet oxygen production.

Non-cationic photosensitizers that target mitochondria have also shown promise for use in PDT. Pheophorbide a (Pba), is a chlorophyll breakdown product isolated from silkworm excreta and the Chinese medicinal herb, Scutellaria barbarta [137,175]. Because Pba absorbs light at longer wavelengths than the first-generation photosensitizer photofrin, tissue penetration is enhanced. Pba has been shown to accumulate in mitochondria and cause apoptosis in a variety of cancer cells, including leukemia, and uterine, breast, pancreatic, colon and hepatocellular carcinoma [137–143]. *In vivo* animal studies have supported the efficacy of Pba-PDT in preventing tumor cell growth. [139,143]. In addition, the tetra-aryl brominated porphyrin and the corresponding diaryl derivative are also promising sensitizers with good photodynamic properties that have the ability to accumulate in mitochondria and induce cell death in human melanoma and colorectal adenocarcinoma *in vitro* and *in vivo* [144]. These results have positive implications for the use of mitochondria-targeted PDT compounds in cancer therapy.

426

5.3. Combination Chemotherapy

As noted previously, the two major pathways for cellular ATP production are glycolysis and mitochondrial oxidative phosphorylation. The high rate of aerobic glycolysis in cancer cells makes them particularly vulnerable to chemotherapeutic agents that inhibit glycolytic enzymes. For example, 2-deoxy-D-glucose (2DG), 3-bromopyruvate (3-BrPA), and lonidamine, which inhibit the hexokinase (HK) catalyzed first step in glycolysis, each have demonstrated significant anticancer activity against a variety of cell types *in vitro* and *in vivo* [176–181]. Unfortunately, the therapeutic efficacy of these compounds as single agents appears to be quite limited. Perhaps this is due to the fact that many cancer cells have functionally competent mitochondria and can overcome inhibition of the glycolytic pathway by increasing mitochondrial ATP production.

Recent evidence suggests that combination chemotherapy, simultaneously aimed at both glycolytic and mitochondrial pathways for ATP production, can be a more effective chemotherapeutic approach for the selective cytotoxicity of cancer cells. In one study [145], the *in vitro* antitumor activity 2DG alone was found insufficient to promote tumor cell death in human breast cancer and osteosarcoma cell lines, reflecting its limited efficacy in clinical trials. However, the combination of 2DG and metformin led to significant cell death associated with a decrease in cellular ATP. Gene expression analysis and functional assays revealed that metformin compromised OXPHOS. Furthermore, forced energy restoration with methyl pyruvate reversed the cell death induced by 2DG and metformin, suggesting a critical role of energetic deprivation in the underlying mechanism of cell death. The combination of 2DG and metformin also inhibited tumor growth and metastasis in mouse xenograft tumor models [145]. In another study, the combination of 2DG and metformin was shown to inhibit both mitochondrial respiration and glycolysis in prostate cancer cells leading to a severe depletion in cellular ATP. This combination of drugs induced a 96% inhibition of cell viability in LNCaP prostate cancer cells, a cytotoxic effect that was much greater than that induced by treatment with either drug alone. In contrast, only a moderate effect by the combination of 2DG and metformin on cell viability was observed in normal prostate epithelial cells [146].

The selective tumor cell killing by mitochondria-targeted DLCs can also be enhanced by combination with anticancer agents having alternative mitochondria target sites. For example, 3-azido deoxythymidine (AZT) as a single agent was found to induce a dose-dependent inhibition of cell growth of several human carcinoma cells, yet cause no significant effect on the growth of control epithelial cells [147]. Combination treatment employing a constant concentration of a delocalized lipophilic cation (dequalinium chloride or MKT-077) plus varying concentrations of AZT enhanced the AZT-induced cytotoxicity of carcinoma cells up to four-fold. The drug combination of constant DLC and varying AZT had no significant effect on the growth of control cells. Furthermore, clonogenic assays demonstrated up to 20-fold enhancement of selective carcinoma cell killing by combination *vs.* single agent treatment, depending on the specific drug combination and concentrations used. It was hypothesized that the efficacy of the AZT/DLC drug combination in carcinoma cell killing may be based on a dual selectivity involving inhibition of mitochondrial energy metabolism and inhibition of DNA synthesis due to limited deoxythymidine monophosphate availability [147].

Although limited in scope and number, the results of these drug combination studies are encouraging. More importantly, they suggest that additional studies should be undertaken to assess the anticancer activity of novel combinations of metabolic inhibitors targeting both major pathways of ATP production, and of novel combinations of compounds that target different sites in mitochondria.

6. Summary and Concluding Remarks

A persistent challenge in cancer therapy is to find ways to improve the efficacy and selectivity of a therapeutic compound while minimizing its systemic toxicity and treatment-limiting side effects. The central role that mitochondria play in the life and death of a cell, together with the many differences found to exist between the mitochondria of normal and transformed cells, make them prime targets for anticancer agents. However, despite the fact that a number of mitochondria-targeted compounds have exhibited potent and selective cancer cell killing in preclinical and early clinical testing, currently none has achieved the standards for high selectivity and efficacy and low toxicity necessary to progress beyond phase III clinical trials and to be used as a viable, single modality treatment option for human cancers. The limitations of this approach suggest the need to explore the use of alternative treatment strategies to enhance the efficacy and selectivity of mitochondria-targeted anticancer agents. Mitochondria-targeted drug delivery systems, photodynamic therapy, and combination chemotherapy are three strategies that have been shown to enhance the efficacy and selectivity of certain mitochondria-targeted anticancer agents *in vitro* and *in vivo*. These strategies enhance the effects of potential therapeutic agents either by delivering them directly to the site of action (mitochondria-targeted drug delivery systems), or by increasing their potency once they have reached their target site (PDT, combination chemotherapy). It is proposed that by expanding the application of these strategies to include additional mitochondria-targeted compounds that have already demonstrated significant preclinical and clinical anticancer activity as single agents, including but not limited to those summarized in this review, the therapeutic efficacy of these compounds might also be improved. New and ongoing research in this area is warranted, and may yet fulfill the clinical promise of exploiting the mitochondrion as a target for cancer chemotherapy.

Author Contributions: This review was a joint effort between Josephine S. Modica-Napolitano and Volkmar Weissig. Both contributed to the development, research and writing of the article.

Conflicts of Interest: The authors declare no conflict of interest.

References

1. American Cancer Society. Lifetime Risk of Developing or Dying from Cancer. Available online: http://www.cancer.org/cancer/cancerbasics/lifetime-probability-of-developing-or-dying-from-cancer (accessed on 18 June 2015).
2. Chen, H.; Chan, D.C. Emerging functions of mammalian mitochondrial fusion and fission. *Hum. Mol. Genet.* **2005**, *14*, R283–R289.
3. Mishra, P.; Chan, D.C. Mitochondrial dynamics and inheritance during cell division, development and disease. *Nat. Rev. Mol. Cell Biol.* **2014**, *15*, 634–646.
4. Kukat, C.; Larsson, N.G. mtDNA makes a U-turn for the mitochondrial nucleoid. *Trends Cell Biol.* **2013**, *23*, 457–463.
5. Shagger, H.; Pfeiffer, K. Supercomplexes in the respiratory chains of yeast and mammalian mitochondria. *EMBO J.* **2000**, *19*, 1777–1783.
6. Ott, M.; Gogvadze, V.; Orrenius, S.; Zhivotovsky, B. Mitochondria, oxidative stress and cell death. *Apoptosis* **2007**, *12*, 913–922.
7. Quinlan, C.L.; Orr, A.L.; Perevoshchikova, I.V.; Treberg, J.R.; Ackrell, B.A.; Brand, M.D. Mitochondrial Complex II can generate reactive oxygen species at high rates in both the forward and reverse reactions. *J. Biol. Chem.* **2012**, *287*, 27255–27264.
8. Drose, S. Differential effects of Complex II on mitochondrial ROS production and their relation to cardioprotective pre- and post-conditioning. *Biochim. Biophys. Acta* **2013**, *1827*, 578–587.
9. Weinberg, F.; Chandel, N.S. Reactive oxygen species-dependent signaling regulates cancer. *Cell. Mol. Life Sci.* **2009**, *66*, 3663–3673.
10. Kamata, H.; Hirata, H. Redox regulation of cellular signalling. *Cell Signal.* **1999**, *11*, 1–14.
11. Lee, Y.J.; Shacter, E. Oxidative stress inhibits apoptosis in human lymphoma cells. *J. Biol. Chem.* **1999**, *274*, 19792–19798.
12. Elmore, S. Apoptosis: A review of programmed cell death. *Toxicol. Pathol.* **2007**, *35*, 495–516.
13. Warburg, O.; Dickens, F. *The Metabolism of Tumors*; Arnold Constable: London, UK, 1930.
14. Fan, J.; Kamphorst, J.J.; Mathew, R.; Chung, M.K.; White, E.; Shlomi, T.; Rabinowitz, J.D. Glutamine-driven oxidative phosphorylation is a major ATP source in transformed mammalian cells in both normoxia and hypoxia. *Mol. Syst. Biol.* **2013**, *9*, 712.
15. Tan, A.S.; Baty, J.W.; Dong, L.F.; Bezawork-Geleta, A.; Endaya, B.; Goodwin, J.; Bajzikova, M.; Kovarova, J.; Peterka, M.; Yan, B.; *et al.* Mitochondrial genome acquisition restores respiratory function and tumorigenic potential of cancer cells without mitochondrial DNA. *Cell Metab.* **2015**, *21*, 81–94.

16. LeBleu, V.S.; O'Connell, J.T.; Gonzalez Herrera, K.N.; Wikman-Kocher, H.; Pantel, K.; Haigis, M.C.; de Carvalho, F.M.; Damascena, A.; Domingos Chinen, L.T.; Rocha, R.M.; *et al.* PGC-1 mediates mitochondrial biogenesis and oxidative phosphorylation to promote metastasis. *Nat. Cell Biol.* **2014**, *16*, 992–1015.

17. Viale, A.; Pettazzoni, P.; Lyssiotis, C.A.; Ying, H.; Sánchez, N.; Marchesini, M.; Carugo, A.; Green, T.; Seth, S.; Giuliani, V.; *et al.* Oncogene ablation-resistant pancreatic cancer cells depend on mitochondrial function. *Nature* **2014**, *514*, 628–632.

18. Lu, C.L.; Qin, L.; Liu, H.C.; Candas, D.; Fan, M.; Li, J.J. Tumor cells switch to mitochondrial oxidative phosphorylation under radiation via mTOR-mediated hexokinase II inhibition—A Warburg-reversing effect. *PLoS ONE* **2015**, *10*, e0121046.

19. Guppy, M.; Leedman, P.; Zu, X.L.; Russell, V. Contribution by different fuels and metabolic pathways to the total ATP turnover of proliferating MCF-7 breast cancer cells. *Biochem. J.* **2002**, *364*, 309–315.

20. Lagadinou, E.D.; Sach, A.; Callahan, K.; Rossi, R.M.; Neering, S.J.; Minhajuddin, M.; Ashton, J.M.; Pei, S.; Grose, V.; O'Dwyer, K.M.; *et al.* Bcl-2 inhibition targets oxidative phosphorylation and selectively eradicates quiescent human leukemia stem cells. *Cell Stem Cell* **2013**, *12*, 329–341.

21. Vlashi, E.; Lagadec, C.; Vergnes, L.; Matsutani, T.; Masui, K.; Poulou, M.; Popescu, R.; Della Donna, L.; Evers, P.; Dekmezian, C.; *et al.* Metabolic state of glioma stem cells and nontumorigenic cells. *Proc. Natl. Acad. Sci. USA* **2011**, *108*, 16062–16067.

22. Vander Heiden, M.G.; Lunt, S.Y.; Dayton, T.L.; Fiske, B.P.; Israelsen, W.J.; Mattaini, K.R.; Vokes, N.I.; Stephanopoulos, G.; Cantley, L.C.; Metallo, C.M.; *et al.* Metabolic pathway alterations that support cell proliferation. *Cold Spring Harb. Symp. Quant. Biol.* **2011**, *76*, 325–334.

23. Pedersen, P.L. Tumor mitochondria and the bioenergetics of cancer cells. *Prog. Exp. Tumor Res.* **1978**, *22*, 190–274.

24. Modica-Napolitano, J.S.; Singh, K.K. Mitochondria as targets for detection and treatment of cancer. *Expert Rev. Mol. Med.* **2002**, *4*, 1–19.

25. Modica-Napolitano, J.S.; Singh, K.K. Mitochondrial dysfunction in cancer. *Mitochondrion* **2004**, *4*, 755–762.

26. Modica-Napolitano, J.S.; Kulawiec, M.; Singh, K.K. Mitochondria and human cancer. *Curr. Mol. Med.* **2007**, *7*, 121–31.

27. Kroemer, G. Mitochondria in cancer. *Oncogene* **2006**, *25*, 4630–4632.

28. Fogg, V.C.; Lanning, N.J.; MacKeigan, J.P. Mitochondria in cancer: At the crossroads of life and death. *Chin. J. Cancer* **2011**, *30*, 526–539.

29. Modica-Napolitano, J.S.; Touma, S.E. Functional differences in mitochondrial enzymes from normal epithelial and carcinoma cells. In *Mitochondrial Dysfunction in Pathogenesis*; Keystone Symposia: Silverthorne, CO, USA, 2000.

30. Sun, A.S.; Sepkowitz, K.; Geller, S.A. A study of some mitochondrial and peroxisomal enzymes in human colonic adenocarcinoma. *Lab. Investig.* **1981**, *44*, 13–17.

31. Chan, S.H.; Barbour, R.L. Adenine nucleotide transport in hepatoma mitochondria. Characterization of factors influencing the kinetics of ADP and ATP uptake. *Biochim. Biophys. Acta* **1983**, *723*, 104–113.

32. Sul, H.S.; Shrago, E.; Goldfarb, S.; Rose, F. Comparison of the adenine nucleotide translocase in hepatomas and rat liver mitochondria. *Biochim. Biophys. Acta* **1979**, *551*, 148–155.

33. Woldegiorgis, G.; Shrago, E. Adenine nucleotide translocase activity and sensitivity to inhibitors in hepatomas. Comparison of the ADP/ATP carrier in mitochondria and in a purified reconstituted liposome system. *J. Biol. Chem.* **1985**, *260*, 7585–7590.

34. Pedersen, P.L.; Morris, H.P. Uncoupler-stimulated adenosine triphosphatase activity. Deficiency in intact mitochondria from Morris hepatomas and ascites tumor cells. *J. Biol. Chem.* **1974**, *249*, 3327–3334.

35. Johnson, L.V.; Walsh, M.L.; Bockus, B.J.; Chen, L.B. Monitoring of relative mitochondrial membrane potential in living cells by fluorescence microscopy. *J. Cell Biol.* **1981**, *88*, 526–535.

36. Davis, S.; Weiss, M.J.; Wong, J.R.; Lampidis, T.J.; Chen, L.B. Mitochondrial and plasma membrane potentials cause unusual accumulation and retention of rhodamine 123 by human breast adenocarcinoma-derived MCF-7 cells. *J. Biol. Chem.* **1985**, *260*, 13844–13850.

37. Modica-Napolitano, J.S.; Aprille, J.R. Basis for the selective cytotoxicity of rhodamine 123. *Cancer Res.* **1987**, *47*, 4361–4365.

38. Canter, J.A.; Kallianpur, A.R.; Parl, F.F.; Millikan, R.C. Mitochondrial DNA G10398A polymorphism and invasive breast cancer in African–American women. *Cancer Res.* **2005**, *65*, 8028–8033.

39. Petros, J.A.; Baumann, A.K.; Ruiz-Pesini, E.; Amin, M.B.; Sun, C.Q.; Hall, J.; Lim, S.; Issa, M.M.; Flanders, W.D.; Hosseini, S.H.; *et al.* mtDNA mutations increase tumorigenicity in prostate cancer. *Proc. Natl. Acad. Sci. USA* **2005**, *102*, 719–724.

40. Brandon, M.; Baldi, P.; Wallace, D.C. Mitochondrial mutations in cancer. *Oncogene* **2006**, *25*, 4647–4662.

41. Kulawiec, M.; Owens, K.M.; Singh, K.K. mtDNA G10398A variant in African–American women with breast cancer provides resistance to apoptosis and promotes metastasis in mice. *J. Hum. Genet.* **2009**, *54*, 647–654.

42. Booker, L.M.; Habermacher, G.M.; Jessie, B.C.; Sun, Q.C.; Baumann, A.K.; Amin, M.; Lim, S.D.; Fernandez-Golarz, C.; Lyles, R.H.; Brown, M.D.; *et al.* North American white mitochondrial haplogroups in prostate and renal cancer. *J. Urol.* **2006**, *175*, 468–472.

43. Liu, V.W.; Wang, Y.; Yang, H.J.; Tsang, P.C.; Ng, T.Y.; Wong, L.C.; Nagley, P.; Ngan, H.Y. Mitochondrial DNA variant 16189T>C is associated with susceptibility to endometrial cancer. *Hum. Mutat.* **2003**, *22*, 173–174.

44. Ishikawa, K.; Takenaga, K.; Akimoto, M.; Koshikawa, N.; Yamaguchi, A.; Imanishi, H.; Nakada, K.; Honma, Y.; Hayashi, J. ROS-generating mitochondrial DNA mutations can regulate tumor cell metastasis. *Science* **2008**, *320*, 661–664.

45. Weinberg, F.; Hamanaka, R.; Wheaton, W.W.; Weinberg, S.; Joseph, J.; Lopez, M.; Kalyanaraman, B.; Mutlu, G.M.; Budinger, G.R.; Chandel, N.S. Mitochondrial metabolism and ROS generation are essential for Kras-mediated tumorigenicity. *Proc. Natl. Acad. Sci. USA* **2010**, *107*, 8788–8793.

46. Vafa, O.; Wade, M.; Kern, S.; Beeche, M.; Pandita, T.K.; Hampton, G.M.; Wahl, G.M. c-Myc can induce DNA damage, increase reactive oxygen species, and mitigate p53 function: A mechanism for oncogene-induced genetic instability. *Mol. Cell* **2002**, *9*, 1031–1044.

47. Fruehauf, J.P.; Meyskens, F.L., Jr. Reactive oxygen species: A breath of life or death? *Clin. Cancer Res.* **2007**, *13*, 789–794.

48. Guzy, R.D.; Schumacker, P.T. Oxygen sensing by mitochondria at Complex III: The paradox of increased reactive oxygen species during hypoxia. *Exp. Physiol.* **2006**, *91*, 807–819.

49. Cannito, S.; Novo, E.; di Bonzo, L.V.; Busletta, C.; Colombatto, S.; Parola, M. Epithelial-mesenchymal transition: From molecular mechanisms, redox regulation to implications in human health and disease. *Antioxid. Redox Signal.* **2010**, *12*, 1383–1430.

50. Reed, J.C. Dysregulation of apoptosis in cancer. *J. Clin. Oncol.* **1999**, *17*, 2941–2953.

51. Wong, R. Apoptosis in cancer: From pathogenesis to treatment. *J. Exp. Clin. Cancer Res.* **2011**, *30*, 87.

52. Pepper, C.; Hoy, T.; Bentley, D.P. Bcl-2/Bax ratios in chronic lymphocytic leukaemia and their correlation with *in vitro* apoptosis and clinical resistance. *Br. J. Cancer* **1997**, *76*, 935–938.

53. Goolsby, C.; Paniagua, M.; Tallman, M.; Gartenhaus, R.B. Bcl-2 regulatory pathway is functional in chronic lymphocytic leukaemia. *Cytom. Part B Clin. Cytom.* **2005**, *63*, 36–46.

54. Raffo, A.J.; Perlman, H.; Chen, M.W.; Day, M.L.; Streitman, J.S.; Buttyan, R. Overexpression of Bcl-2 protects prostate cancer cells from apoptosis *in vitro* and confers resistance to androgen depletion *in vivo*. *Cancer Res.* **1995**, *55*, 4438–4445.

55. Kitada, S.; Pedersen, I.M.; Schimmer, A.D.; Reed, J.C. Dysregulation of apoptosis genes in hematopoietic malignancies. *Oncogene* **2002**, *21*, 3459–3474.

56. Kirkin, V.; Joos, S.; Zornig, M. The role of Bcl-2 family members in tumorigenesis. *Biochim. Biophys. Acta* **2004**, *1644*, 229–249.

57. Fulda, S.; Meyer, E.; Debatin, K.M. Inhibition of TRAIL-induced apoptosis by Bcl-2 overexpression. *Oncogene* **2000**, *21*, 2283–2294.

58. Minn, A.J.; Rudin, C.M.; Boise, L.H.; Thompson, C.B. Expression of Bcl-XL can confer a multidrug resistance phenotype. *Blood* **1995**, *86*, 1903–1910.

59. Neuzil, J.; Dong, L.F.; Rohlena, J.; Truksa, J.; Ralph, S.J. Classification of mitocans, anti-cancer drugs acting on mitochondria. *Mitochondrion* **2013**, *13*, 199–208.

60. Bernal, S.D.; Lampidis, T.J.; Summerhayes, I.C.; Chen, L.B. Rhodamine-123 selectively reduces clonal growth of carcinoma cells *in vitro*. *Science* **1982**, *218*, 1117–1119.

61. Bernal, S.D.; Lampidis, T.J.; McIsaac, R.M.; Chen, L.B. Anticarcinoma activity *in vivo* of rhodamine 123, a mitochondrial-specific dye. *Science* **1983**, *222*, 169–172.

62. Modica-Napolitano, J.S.; Weiss, M.J.; Chen, L.B.; Aprille, J.R. Rhodamine 123 inhibits bioenergetic function in isolated rat liver mitochondria. *Biochem. Biophys. Res. Commun.* **1984**, *118*, 717–723.

63. Bleday, R.; Weiss, M.J.; Salem, R.R.; Wilson, R.E.; Chen, L.B.; Steele, G., Jr. Inhibition of rat colon tumor isograft growth with dequalinium chloride. *Arch. Surg.* **1986**, *121*, 1272–1275.

64. Weiss, M.J.; Wong, J.R.; Ha, C.S.; Bleday, R.; Salem, R.R.; Steele, G.D., Jr.; Chen, L.B. Dequalinium, a topical antimicrobial agent, displays anticarcinoma activity based on selective mitochondrial accumulation. *Proc. Natl. Acad. Sci. USA* **1987**, *84*, 5444–5448.

65. Sun, X.; Wong, J.R.; Song, K.; Hu, J.; Garlid, K.D.; Chen, L.B. AA1, a newly synthesized monovalent lipophilic cation, expresses potent *in vivo* antitumor activity. *Cancer Res.* **1994**, *54*, 1465–1471.

66. Koya, K.; Li, Y.; Wang, H.; Ukai, T.; Tatsuta, N.; Kawakami, M.; Shishido, T.; Chen, L.B. MKT-077, a novel rhodacyanine dye in clinical trials, exhibits anticarcinoma activity in preclinical studies based on selective mitochondrial accumulation. *Cancer Res.* **1996**, *56*, 538–543.

67. Modica-Napolitano, J.S.; Koya, K.; Weisberg, E.; Brunelli, B.T.; Li, Y.; Chen, L.B. Selective damage to carcinoma mitochondria by the rhodacyanine MKT-077. *Cancer Res.* **1996**, *56*, 544–550.

68. Britten, C.D.; Rowinsky, E.K.; Baker, S.D.; Weiss, G.R.; Smith, L.; Stephenson, J.; Rothenberg, M.; Smetzer, L.; Cramer, J.; Collins, W.; *et al.* A phase I and pharmacokinetic study of the mitochondrial-specific rhodacyanine dye analog MKT 077. *Clin. Cancer Res.* **2000**, *6*, 42–49.

69. Propper, D.J.; Braybrooke, J.P.; Taylor, D.J.; Lodi, R.; Styles, P.; Cramer, J.A.; Collins, W.C.J.; Levitt, N.C.; Talbot, D.C.; Ganesan, T.S.; *et al.* Phase I trial of the selective mitochondrial toxin MKT 077 in chemo-resistant solid tumours. *Ann. Oncol.* **1999**, *10*, 923–927.

70. Evans, J.M.; Donnelly, L.A.; Emslie-Smith, A.M.; Alessi, D.R.; Morris, A.D. Metformin and reduced risk of cancer in diabetic patients. *BMJ* **2005**, *330*, 1304–1305.

71. Libby, G.; Donnelly, L.A.; Donnan, P.T.; Alessi, D.R.; Morris, A.D.; Evans, J.M. New users of metformin are at low risk of incident cancer: A cohort study among people with type 2 diabetes. *Diabetes Care* **2009**, *32*, 1620–1625.

72. Murtola, T.J.; Tammela, T.L.; Lahtela, J.; Auvinen, A. Antidiabetic medication and prostate cancer risk: A population-based case-control study. *Am. J. Epidemiol.* **2008**, *168*, 925–931.

73. Jiralerspong, S.; Angulo, A.M.; Hung, M.C. Expanding the arsenal: Metformin for the treatment of triple-negative breast cancer? *Cell Cycle* **2009**, *8*, 2681.

74. Dowling, R.J.; Niraula, S.; Stambolic, V.; Goodwin, P.J. Metformin in cancer: Translational challenges. *J. Mol. Endocrinol.* **2012**, *48*, R31–R43.

75. Ben Sahra, I.; Laurent, K.; Loubat, A.; Giorgetti-Peraldi, S.; Colosetti, P.; Auberger, P.; Tanti, J.F.; Le Marchand-Brustel, Y.; Bost, F. The antidiabetic drug metformin exerts an antitumoral effect *in vitro* and *in vivo* through a decrease of cyclin D1 level. *Oncogene* **2008**, *27*, 3576–3586.

76. Zakikhani, M.; Dowling, R.; Fantus, I.G.; Sonenberg, N.; Pollak, M. Metformin is an AMP kinase-dependent growth inhibitor for breast cancer cells. *Cancer Res.* **2006**, *66*, 10269–10273.

77. Gotlieb, W.H.; Saumet, J.; Beauchamp, M.C.; Gu, J.; Lau, S.; Pollak, M.N.; Bruchim, I. *In vitro* metformin antineoplastic activity in epithelial ovarian cancer. *Gynecol. Oncol.* **2008**, *110*, 246–250.

78. Wang, L.W.; Li, Z.S.; Zou, D.W.; Jin, Z.D.; Gao, J.; Xu, G.M. Metformin induces apoptosis of pancreatic cancer cells. *World J. Gastroenterol.* **2008**, *14*, 7192–7198.

79. Buzzai, M.; Jones, R.G.; Amaravadi, R.K.; Lum, J.J.; DeBerardinis, R.J.; Zhao, F.; Viollet, B.; Thompson, C.B. Systemic treatment with the antidiabetic drug metformin selectively impairs p53-deficient tumor cell growth. *Cancer Res.* **2007**, *67*, 6745–6752.

80. Hirsch, H.A.; Iliopoulos, D.; Tsichlis, P.N.; Struhl, K. Metformin selectively targets cancer stem cells, and acts together with chemotherapy to block tumor growth and prolong remission. *Cancer Res.* **2009**, *69*, 507–511.

81. Anisimov, V.N.; Berstein, L.M.; Egormin, P.A.; Piskunova, T.S.; Popovich, I.G.; Zabezhinski, M.A.; Kovalenko, I.G.; Poroshina, T.E.; Semenchenko, A.V.; Provinciali, M.; *et al.* 2005 Effect of metformin on life span and on the development of spontaneous mammary tumors in HER-2/neu transgenic mice. *Exp. Gerontol.* **2005**, *40*, 685–693.

82. Huang, X.; Wullschleger, S.; Shpiro, N.; McGuire, V.A.; Sakamoto, K.; Woods, Y.L.; McBurnie, W.; Fleming, S.; Alessi, D.R. Important role of the LKB1–AMPK pathway in suppressing tumorigenesis in PTEN-deficient mice. *Biochem. J.* **2008**, *412*, 211–221.

83. Memmott, R.M.; Mercado, J.R.; Maier, C.R.; Kawabata, S.; Fox, S.D.; Dennis, P.A. Metformin prevents tobacco carcinogen–induced lung tumorigenesis. *Cancer Prev. Res.* **2010**, *3*, 1066–1076.

84. Algire, C.; Zakikhani, M.; Blouin, M.J.; Shuai, J.H.; Pollak, M. 2008 Metformin attenuates the stimulatory effect of a high-energy diet on *in vivo* LLC1 carcinoma growth. *Endocr. Relat. Cancer* **2008**, *15*, 833–839.

85. Phoenix, K.N.; Vumbaca, F.; Fox, M.M.; Evans, R.; Claffey, K.P. Dietary energy availability affects primary and metastatic breast cancer and metformin efficacy. *Breast Cancer Res. Treat.* **2010**, *123*, 333–344.

86. Hosono, K.; Endo, H.; Takahashi, H.; Sugiyama, M.; Sakai, E.; Uchiyama, T.; Suzuki, K.; Iida, H.; Sakamoto, Y.; Yoneda, K.; *et al.* Metformin suppresses colorectal aberrant crypt foci in a short-term clinical trial. *Cancer Prev. Res.* **2010**, *3*, 1077–1083.

87. Niraula, S.; Dowling, R.J.; Ennis, M.; Chang, M.C.; Done, S.J.; Hood, N.; Escallon, J.; Leong, W.L.; McCready, D.R.; Reedijk, M.; *et al.* Metformin in early breast cancer: A prospective window of opportunity neoadjuvant study. *Breast Cancer Res. Treat.* **2012**, *135*, 821–830.

88. Hadad, S.; Iwamoto, T.; Jordan, L.; Purdie, C.; Bray, S.; Baker, L.; Jellema, G.; Deharo, S.; Hardie, D.G.; Pusztai, L.; *et al.* 2011 Evidence for biological effects of metformin in operable breast cancer: A pre-operative, window-of-opportunity, randomized trial. *Breast Cancer Res. Treat.* **2011**, *128*, 783–794.

89. Wheaton, W.W.; Weinberg, S.E.; Hamanaka, R.B.; Soberanes, S.; Sullivan, L.B.; Anso, E.; Glasauer, A.; Dufour, E.; Mutlu, G.M.; Budigner, G.R.S.; *et al.* Metformin inhibits mitochondrial Complex I of cancer cells to reduce tumorigenesis. *eLife* **2014**, *3*, e02242.

90. Blackman, R.K.; Cheung-Ong, K.; Gebbia, M.; Proia, D.A.; He, S.; Kepros, J.; Jonneaux, A.; Marchetti, P.; Kluza, J.; Rao, P.E.; *et al.* Mitochondrial electron transport is the cellular target of the oncology drug elesclomol. *PLoS ONE* **2012**, *7*, e29798.

91. O'Day, S.J.; Eggermont, A.M.; Chiarion-Sileni, V.; Kefford, R.; Grob, J.J.; Mortier, L.; Robert, C.; Schachter, J.; Testori, A.; Mackiewicz, J.; *et al.* Final results of phase III SYMMETRY study: Randomized, double-blind trial of elesclomol plus paclitaxel *versus* paclitaxel alone as treatment for chemotherapy-naive patients with advanced melanoma. *J. Clin. Oncol.* **2013**, *31*, 1211–1218.

92. ClinicalTrials.gov. Available online: https://clinicaltrials.gov/ct2/results?term= elesclomol&Search=Search (accessed on 19 May 2015).

93. Dai, Z.J.; Wang, X.J.; Li, Z.F.; Ji, Z.Z.; Ren, H.T.; Tang, W.; Liu, X.X.; Kang, H.F.; Guan, H.T.; Song, L.Q. *Scutellaria barbate* extract induces apoptosis of hepatoma H22 cells via the mitochondrial pathway involving caspase-3. *World J. Gastroenterol.* **2008**, *14*, 7321–7328.

94. Kim, E.K.; Kwon, K.B.; Han, M.J.; Song, M.Y.; Lee, J.H.; Ko, Y.S.; Shin, B.C.; Yu, J.; Lee, Y.R.; Ryu, D.G.; *et al.* Induction of G1 arrest and apoptosis by *Scutellaria barbata* in the human promyelocytic leukemia HL-60 cell line. *Int. J. Mol. Med.* **2007**, *20*, 123–128.

95. Marconett, C.N.; Morgenstern, T.J.; san Roman, A.K.; Sundar, S.N.; Singhal, A.K.; Firestone, G.L. BZL101, a phytochemical extract from the *Scutellaria barbata* plant, disrupts proliferation of human breast and prostate cancer cells through distinct mechanisms dependent on the cancer cell phenotype. *Cancer Biol. Ther.* **2010**, *10*, 397–405.

96. Fong, S.; Shoemaker, M.; Cadaoas, J.; Lo, A.; Liao, W.; Tagliaferri, M.; Cohen, I.; Shtivelman, E. Molecular mechanisms underlying selective cytotoxic activity of BZL101, an extract of *Scutellaria barbata*, towards breast cancer cells. *Cancer Biol. Ther.* **2008**, *7*, 577–586.

97. Chen, V.; Staub, R.E.; Fong, S.; Tagliaferri, M.; Cohen, I.; Shtivelman, E. Bezielle selectively targets mitochondria of cancer cells to inhibit glycolysis and OXPHOS. *PLoS ONE* **2012**, *7*, e30300.

98. Rugo, H.; Shtivelman, E.; Perez, A.; Vogel, C.; Franco, S.; Chiu, E.T.; Melisko, M.; Tagliaferri, M.; Cohen, I.; Shoemaker, M.; *et al.* Phase I trial and antitumor effects of BZL101 for patients with advanced breast cancer. *Breast Cancer Res. Treat.* **2007**, *105*, 17–28.

99. Perez, A.T.; Arun, B.; Tripathy, D.; Tagliaferri, M.A.; Shaw, H.S.; Kimmick, G.G.; Cohen, I.; Shtivelman, E.; Caygill, K.A.; Grady, D.; *et al.* A phase 1B dose escalation trial of *Scutellaria barbata* (BZL101) for patients with metastatic breast cancer. *Breast Cancer Res. Treat.* **2010**, *120*, 111–118.

100. Oltersdorf, T.; Elmore, S.W.; Shoemaker, A.R.; Armstrong, R.C.; Augeri, D.J.; Belli, B.A.; Bruncko, M.; Deckwerth, T.L.; Dinges, J.; Hajduk, P.J.; *et al.* An inhibitor of Bcl-2 family proteins induces regression of solid tumours. *Nature* **2005**, *435*, 677–681.

101. Konopleva, M.; Contractor, R.; Tsao, T.; Samudio, I.; Ruvolo, P.P.; Kitada, S.; Deng, X.; Zhai, D.; Shi, Y.X.; Sneed, T.; *et al.* Mechanisms of apoptosis sensitivity and resistance to the BH3 mimetic ABT-737 in acute myeloid leukemia. *Cancer Cell* **2006**, *10*, 375–388.

102. Hann, C.L.; Daniel, V.C.; Sugar, E.A.; Dobromilskaya, I.; Murphy, S.C.; Cope, L.; Lin, X.; Hierman, J.S.; Wilburn, D.L.; Neil Watkins, D.; *et al.* Therapeutic efficacy of ABT-737, a selective inhibitor of Bcl-2, in small cell lung cancer. *Cancer Res.* **2008**, *68*, 2321–2328.

103. Tse, C.; Shoemaker, A.R.; Adickes, J.; Anderson, M.G.; Chen, J.; Jin, S.; Johnson, E.F.; Marsh, K.C.; Mitten, M.J.; Nimmer, P.; *et al.* ABT-263: A potent and orally bioavailable Bcl-2 family inhibitor. *Cancer Res.* **2008**, *68*, 3421–3428.

104. Gandhi, L.; Camidge, D.R.; de Oliveira, M.R.; Bonomi, P.; Gandara, D.; Khaira, D.; Hann, C.L.; McKeegan, E.M.; Litvinovich, E.; Hemken, P.M.; *et al.* Phase I study of navitoclax (ABT-263), a novel Bcl-2 family inhibitor, in patients with small-cell lung cancer and other solid tumors. *J. Clin. Oncol.* **2011**, *29*, 909–916.

105. Rudin, C.M.; Hann, C.L.; Garon, E.B.; de Oliveira, M.R.; Bonomi, P.D.; Camidge, D.R.; Chu, Q.; Giaccone, G.; Khaira, D.; Ramalingam, S.S.; *et al.* Phase II study of single-agent navitoclax (ABT-263) and biomarker correlates in patients with relapsed small cell lung cancer. *Clin. Cancer Res.* **2012**, *18*, 3163–3169.

106. Sadahira, K.; Sagawa, M.; Nakazato, T.; Uchida, H.; Ikeda, Y.; Okamoto, S.; Nakajima, H.; Kizaki, M. Gossypol induces apoptosis in multiple myeloma cells by inhibition of interleukin-6 signaling and Bcl-2/Mcl-1 pathway. *Int. J. Oncol.* **2014**, *45*, 2278–2286.

107. Kline, M.P.; Rajkumar, S.V.; Timm, M.M.; Kimlinger, T.K.; Haug, J.L.; Lust, J.A.; Greipp, P.R.; Kumar, S. R-(−)-gossypol (AT-101) activates programmed cell death in multiple myeloma cells. *Exp. Hematol.* **2008**, *36*, 568–576.

108. Konopleva, M.; Watt, J.; Contractor, R.; Tsao, T.; Harris, D.; Estrov, Z.; Bornmann, W.; Kantarjian, H.; Viallet, J.; Samudio, I.; *et al.* Mechanisms of antileukemic activity of the novel BH3 mimetic GX15–070 (obatoclax). *Cancer Res.* **2008**, *68*, 3413–3420.

109. Nguyen, M.; Marcellus, R.C.; Roulston, A.; Watson, M.; Serfass, L.; Madiraju, S.R.M.; Goulet, D.; Viallet, J.; Bélec, L.; Billot, X.; *et al.* Small molecule obatoclax (GX15–070) antagonizes MCL-1 and overcomes MCL-1-mediated resistance to apoptosis. *Proc. Natl. Acad. Sci. USA* **2007**, *104*, 19512–19517.

110. Heikaus, S.; van den Berg, L.; Kempf, T.; Mahotka, C.; Gabbert, H.E.; Ramp, U. HA14–1 is able to reconstitute the impaired mitochondrial pathway of apoptosis in renal cell carcinoma cell lines. *Cell Oncol.* **2008**, *30*, 419–433.

111. Rehman, K.; Tariq, M.; Akash, M.S.; Gillani, Z.; Qazi, M.H. Effect of HA14–1 on apoptosis-regulating proteins in HeLa cells. *Chem. Biol. Drug Des.* **2014**, *83*, 317–323.

112. Anderson, W.M.; Wood, J.M.; Anderson, A.C. Inhibition of mitochondrial and Paracoccus denitrificans NADH-ubiquinone reductase by oxacarbocyanine dyes. A structure-activity study. *Biochem. Pharmacol.* **1993**, *45*, 691–696.

113. Rideout, D.; Bustamante, A.; Patel, J. Mechanism of inhibition of FaDu hypopharyngeal carcinoma cell growth by tetraphenylphosphonium chloride. *Int. J. Cancer* **1994**, *57*, 247–253.

114. Birsoy, K.; Possemato, R.; Lorbeer, F.K.; Bayraktar, E.C.; Thiru, P.; Yucel, B.; Wang, T.; Chen, W.W.; Clish, C.B.; Sabatini, D.M. Metabolic determinants of cancer cell sensitivity to glucose limitation and biguanides. *Nature* **2014**, *508*, 108–112.

115. Kong, Q.; Beel, J.A.; Lillehei, K.O. A threshold concept for cancer therapy. *Med. Hypotheses* **2000**, *55*, 29–35.

116. Billard, C. BH3 mimetics: Status of the field and new developments. *Mol. Cancer Ther.* **2013**, *12*, 1691–1700.

117. Shoemaker, A.R.; Oleksijew, A.; Bauch, J.; Belli, B.A.; Borre, T.; Bruncko, M.; Deckwirth, T.; Frost, D.J.; Jarvis, K.; Joseph, M.K.; *et al.* A small-molecule inhibitor of Bcl-X$_L$ potentiates the activity of cytotoxic drugs *in vitro* and *in vivo*. *Cancer Res.* **2006**, *66*, 8731–8739.

118. Hikita, H.; Takehara, T.; Shimizu, S.; Kodama, T.; Shigekawa, M.; Iwase, K.; Hosui, A.; Miyagi, T.; Tatsumi, T.; Ishida, H.; *et al.* The Bcl-xL inhibitor, ABT-737, efficiently induces apoptosis and suppresses growth of hepatoma cells in combination with sorafenib. *Hepatology* **2010**, *52*, 1310–1321.

119. Jain, H.V.; Meyer-Hermann, M. The molecular basis of synergism between carboplatin and ABT-737 therapy targeting ovarian carcinomas. *Cancer Res.* **2011**, *71*, 705–715.

120. Zall, H.; Weber, A.; Besch, R.; Zantl, N.; Hacker, G. Chemotherapeutic drugs sensitize human renal cell carcinoma cells to ABT-737 by a mechanism involving the Noxa-dependent inactivation of Mcl-1 or A1. *Mol. Cancer* **2010**, *9*, 164.

121. Tan, N.; Malek, M.; Zha, J.; Yue, P.; Kassees, R.; Berry, L.; Fairbrother, W.J.; Sampath, D.; Belmont, L.D. Navitoclax enhances the efficacy of taxanes in non-small cell lung cancer models. *Clin. Cancer Res.* **2011**, *17*, 1394–1404.

122. Truksa, J.; Dong, L.F.; Rohlena, J.; Stursa, J.; Vondrusova, M.; Goodwin, J.; Nguyen, M.; Kluckova, K.; Rychtarcikova, Z.; Lettlova, S.; *et al.* Mitochondrially targeted vitamin E succinate modulates expression of mitochondrial DNA transcripts and mitochondrial biogenesis. *Antiox. Redox Signal.* **2015**, *22*, 883–900.

123. Dong, L.F.; Jameson, V.J.A.; Tilly, D.; Cerny, J.; Mahdavian, E.; Marín-Hernandez, A.; Hernandez-Esquivel, L.; Rodríguez-Enríquez, S.; Stursa, J.; Witting, P.K.; *et al.* Mitochondrial targeting of vitamin E succinate enhances its pro-apoptotic and anti-cancer activity via mitochondrial Complex II. *J. Biol. Chem.* **2011**, *286*, 3717–3728.

124. Rao, V.A.; Klein, S.R.; Bonar, S.J.; Zielonka, J.; Mizuno, N.; Dickey, J.S.; Keller, P.W.; Joseph, J.; Kalyanaraman, B.; Shacter, E. The antioxidant transcription factor Nrf2 negatively regulates autophagy and growth arrest induced by the anticancer redox agent mitoquinone. *J. Biol. Chem.* **2010**, *285*, 34447–34459.

125. D'Souza, G.G.M.; Cheng, S.M.; Boddapati, S.V.; Horobin, R.W.; Weissig, V. Nanocarrier-assisted sub-cellular targeting to the site of mitochondria improves the pro-apoptotic activity of paclitaxel. *J. Drug Target.* **2008**, *16*, 578–585.

126. Paliwal, R.; Rai, S.; Vaidya, B.; Gupta, P.N.; Mahor, S.; Khatri, K.; Goyal, A.K.; Rawat, A.; Vyas, S.P. Cell-selective mitochondrial targeting: Progress in mitochondrial medicine. *Curr. Drug Deliv.* **2007**, *4*, 211–224.

127. Biswas, S.; Dodwadkar, N.S.; Deshpande, P.P.; Torchilin, V.P. Liposomes loaded with paclitaxel and modified with novel triphenylphosphonium-PEG-PE conjugate possess low toxicity, target mitochondria and demonstrate enhanced antitumor effects *in vitro* and *in vivo*. *J. Control. Release* **2012**, *159*, 393–402.

128. Zupancic, S.; Kocbek, P.; Zariwala, M.G.; Renshaw, D.; Gul, M.O.; Elsaid, Z.; Taylor, K.M.; Somavarapu, S. Design and development of novel mitochondrial targeted nanocarriers, DQAsomes for curcumin inhalation. *Mol. Pharm.* **2014**, *11*, 2334–2345.

129. Wang, X.X.; Li, Y.B.; Yao, H.J.; Ju, R.J.; Zhang, Y.; Li, R.J.; Yu, Y.; Zhang, L.; Lu, W.L. The use of mitochondrial targeting resveratrol liposomes modified with a dequalinium polyethylene glycol-distearoylphosphatidyl ethanolamine conjugate to induce apoptosis in resistant lung cancer cells. *Biomaterials* **2011**, *32*, 5673–5687.

130. Solomon, M.A.; Shah, A.A.; D'Souza, G.G. *In vitro* assessment of the utility of stearyl triphenyl phosphonium modified liposomes in overcoming the resistance of ovarian carcinoma Ovcar-3 cells to paclitaxel. *Mitochondrion* **2013**, *13*, 464–472.

131. Zhou, J.; Zhao, W.Y.; Ma, X.; Ju, R.J.; Li, X.Y.; Li, N.; Sun, M.G.; Shi, J.F.; Zhnag, C.X.; Lu, W.L. The anticancer efficacy of paclitaxel liposomes modified with mitochondrial targeting conjugate in resistant lung cancer. *Biomaterials* **2013**, *34*, 3626–3638.

132. Malhi, S.S.; Budhiraja, A.; Arora, S.; Chaudhari, K.R.; Nepali, K.; Kumar, R.; Sohi, H.; Murthy, R.S. Intracellular delivery of redox cycler-doxorubicin to the mitochondria of cancer cell by folate receptor targeted mitocancerotropic liposomes. *Int. J. Pharm.* **2012**, *432*, 63–74.

133. Yoong, S.L.; Wong, B.S.; Zhou, Q.L.; Chin, C.F.; Li, J.; Venkatesan, T.; Ho, H.K.; Yu, V.; Ang, W.H.; Pastorin, G. Enhanced cytotoxicity to cancer cells by mitochondria-targeting MWCNTs containing platinum(IV) prodrug of cisplatin. *Biomaterials* **2014**, *35*, 748–759.

134. Ara, G.; Aprille, J.R.; Malis, C.D.; Kane, S.B.; Cincotta, L.; Foley, J.; Bonventre, J.V.; Oseroff, A.R. Mechanisms of mitochondrial photosensitization by the cationic dye, N,N'-bis(2-ethyl-l,3-dioxylene)kryptocyanine (EDKC): Preferential inactivation of Complex I in the electron transport chain. *Cancer Res.* **1987**, *47*, 6580–6585.

135. Powers, S.K.; Pribil, S.; Gillespie, G.Y.; Watkins, P.J. Laser photochemotherapy of rhodamine-123 sensitized human glioma cells *in vitro*. *J. Neurosurg.* **1986**, *64*, 918–923.

136. Modica-Napolitano, J.S.; Brunelli, B.T.; Koya, K.; Chen, L.B. Photoactivation enhances the mitochondrial toxicity of the cationic rhodacyanine MKT-077. *Cancer Res.* **1998**, *58*, 71–75.

137. Chan, J.Y.; Tang, P.M.; Hon, P.M.; Au, S.W.; Tsui, S.K.; Waye, M.M.; Kong, S.K.; Mak, T.C.; Fung, K.P. Pheophorbide *a*, a major antitumor component purified from *Scutellaria barbata*, induces apoptosis in human hepatocellular carcinoma cells. *Planta Med.* **2006**, *72*, 28–33.

138. Li, W.T.; Tsao, H.W.; Chen, Y.Y.; Cheng, S.W.; Hsu, Y.C. A study on the photodynamic properties of chlorophyll derivatives using human hepatocellular carcinoma cells. *Photochem. Photobiol. Sci.* **2007**, *6*, 1341–1348.

139. Hajri, A.; Coffy, S.; Vallat, F.; Evrard, S.; Marescaux, J.; Aprahamian, M. Human pancreatic carcinoma cells are sensitive to photodynamic therapy *in vitro* and *in vivo*. *Br. J. Surg.* **1999**, *86*, 899–906.

140. Hibasami, H.; Kyohkon, M.; Ohwaki, S.; Katsuzaki, H.; Imai, K.; Nakagawa, M.; Ishi, Y.; Komiya, T. Pheophorbide *a*, a moiety of chlorophyll *a*, induces apoptosis in human lymphoid leukemia molt 4B cells. *Int. J. Mol. Med.* **2000**, *6*, 277–279.

141. Tang, P.M.; Liu, X.Z.; Zhang, D.M.; Fong, W.P.; Fung, K.P. Pheophorbide *a* based photodynamic therapy induces apoptosis via mitochondrial-mediated pathway in human uterine carcinosarcoma. *Cancer Biol. Ther.* **2009**, *8*, 533–539.

142. Hoi, S.W.; Wong, H.M.; Chan, J.Y.; Yue, G.G.; Tse, G.M.; Law, B.K.; Fong, W.P.; Fung, K.P. Photodynamic therapy of Pheophorbide *a* inhibits the proliferation of human breast tumour via both caspase-dependent and -independent apoptotic pathways in *in vitro* and *in vivo* models. *Phytother. Res.* **2012**, *26*, 734–742.

143. Hajri, A.; Wack, S.; Meyer, C.; Smith, M.K.; Leberquier, C.; Kedinger, M.; Aprahamian, M. *In vitro* and *in vivo* efficacy of Photofrin®and pheophorbide *a*, a bacteriochlorin, in photodynamic therapy of colonic cancer cells. *Photochem. Photobiol.* **2002**, *75*, 140–148.

144. Laranjo, M.; Serra, A.C.; Abrantes, M.; Piñeiro, M.; Gonçalves, A.C.; Casalta-Lopes, J.; Carvalho, L.; Sarmento-Ribeiro, A.B.; Rocha-Gonsalves, A.; Botelho, F. 2-Bromo-5-hydroxyphenylporphyrins for photodynamic therapy: Photosensitization efficiency, subcellular localization and *in vivo* studies. *Photodiagn. Photodyn. Ther.* **2013**, *10*, 51–61.

145. Cheong, J.H.; Park, E.S.; Liang, J.; Dennison, J.B.; Tsavachidou, D.; Nguyen-Charles, C.; Cheng, K.W.; Hall, H.; Zhang, D.; Lu, Y.; *et al.* Dual inhibition of tumor energy pathway by 2-deoxyglucose and metformin is effective against a broad spectrum of preclinical cancer models. *Mol. Cancer Ther.* **2011**, *10*, 2350–2362.

146. Ben Sahra, I.; Laurent, K.; Giuliano, S.; Larbret, F.; Ponzio, G.; Gounon, P.; Le Marchand-Brustel, Y.; Giorgetti-Peraldi, S.; Cormont, M.; Bertolotto, C.; *et al.* Targeting cancer cell metabolism: The combination of metformin and 2-deoxyglucose induces p53-dependent apoptosis in prostate cancer cells. *Cancer Res.* **2010**, *70*, 2465–2475.

147. Modica-Napolitano, J.S.; Nalbandian, R.; Kidd, M.E.; Nalbandian, A.; Nguyen, C.C. The selective *in vitro* cytotoxicity of carcinoma cells by AZT is enhanced by concurrent treatment with delocalized lipophilic cations. *Cancer Lett.* **2003**, *19*, 859–868.

148. Westermann, B.; Neupert, W. Mitochondria-targeted green fluorescent proteins: Convenient tools for the study of organelle biogenesis in *Saccharomyces cerevisiae*. *Yeast* **2000**, *16*, 1421–1427.

149. Smith, R.A.; Porteous, C.M.; Gane, A.M.; Murphy, M.P. Delivery of bioactive molecules to mitochondria *in vivo*. *Proc. Natl. Acad. Sci. USA* **2003**, *100*, 5407–5412.

150. Horobin, R.W.; Trapp, S.; Weissig, V. Mitochondriotropics: A review of their mode of action, and their applications for drug and DNA delivery to mammalian mitochondria. *J. Control. Release* **2007**, *121*, 125–136.

151. D'Souza, G.G.M.; Weissig, V. An introduction to subcellular and nanomedicine: Current trends and future developments. In *Organelle-Specific Pharmaceutical Nanotechnology*; John Wiley & Sons, Inc.: Hoboken, NJ, USA, 2010; pp. 1–13.

152. Weissig, V.; Lasch, J.; Erdos, G.; Meyer, H.W.; Rowe, T.C.; Hughes, J. DQAsomes: A novel potential drug and gene delivery system made from dequalinium. *Pharm. Res.* **1998**, *15*, 334–337.

153. Weissig, V.; Torchilin, V.P. Towards mitochondrial gene therapy: DQAsomes as a strategy. *J. Drug Target.* **2001**, *9*, 1–13.

154. Weissig, V.; Torchilin, V.P. Cationic bolasomes with delocalized charge centers as mitochondria-specific DNA delivery systems. *Adv. Drug Deliv. Rev.* **2001**, *49*, 127–149.

155. Weissig, V. DQAsomes as the prototype of mitochondria-targeted pharmaceutical nanocarriers: Preparation, characterization, and use. *Methods Mol. Biol.* **2015**, *1265*, 1–11.

156. Lyrawati, D.; Trounson, A.; Cram, D. Expression of GFP in the mitochondrial compartment using DQAsome-mediated delivery of an artificial mini-mitochondrial genome. *Pharm. Res.* **2011**, *28*, 2848–2862.

157. Boddapati, S.V.; Tongcharoensirikul, P.; Hanson, R.N.; D'Souza, G.G.; Torchilin, V.P.; Weissig, V. Mitochondriotropic liposomes. *J. Liposome Res.* **2005**, *15*, 49–58.

158. Boddapati, S.V.; D'Souza, G.G.; Erdogan, S.; Torchilin, V.P.; Weissig, V. Organelle-targeted nanocarriers: Specific delivery of liposomal ceramide to mitochondria enhances its cytotoxicity *in vitro* and *in vivo*. *Nano Lett.* **2008**, *8*, 2559–2263.

159. Boddapati, S.V.; D'Souza, G.G.; Weissig, V. Liposomes for drug delivery to mitochondria. *Methods Mol. Biol.* **2010**, *605*, 295–303.

160. Weissig, V.; Boddapati, S.V.; Cheng, S.M.; D'Souza, G.G. Liposomes and liposome-like vesicles for drug and DNA delivery to mitochondria. *J. Liposome Res.* **2006**, *16*, 249–264.

161. Weissig, V.; Lasch, J.; Klibanov, A.L.; Torchilin, V.P. A new hydrophobic anchor for the attachment of proteins to liposomal membranes. *FEBS Lett.* **1986**, *202*, 86–90.

162. Weissig, V.; Lasch, J.; Gregoriadis, G. Covalent coupling of sugars to liposomes. *Biochim. Biophys. Acta* **1989**, *1003*, 54–57.

163. Theodossiou, T.A.; Sideratou, Z.; Tsiourvas, D.; Paleos, C.M. A novel mitotropic oligolysine nanocarrier: Targeted delivery of covalently bound D-Luciferin to cell mitochondria. *Mitochondrion* **2011**, *11*, 982–986.

164. Biswas, S.; Dodwadkar, N.S.; Piroyan, A.; Torchilin, V.P. Surface conjugation of triphenylphosphonium to target poly(amidoamine) dendrimers to mitochondria. *Biomaterials* **2012**, *33*, 4773–4782.

165. Sharma, A.; Soliman, G.M.; Al-Hajaj, N.; Sharma, R.; Maysinger, D.; Kakkar, A. Design and evaluation of multifunctional nanocarriers for selective delivery of coenzyme Q10 to mitochondria. *Biomacromolecules* **2012**, *13*, 239–252.

166. Marrache, S.; Dhar, S. Engineering of blended nanoparticle platform for delivery of mitochondria-acting therapeutics. *Proc. Natl. Acad. Sci. USA* **2012**, *109*, 16288–16293.

167. Pathak, R.K.; Kolishetti, N.; Dhar, S. Targeted nanoparticles in mitochondrial medicine. *Wiley Interdiscip. Rev. Nanomed. Nanobiotechnol.* **2015**, *7*, 315–329.

168. Theodossiou, T.A.; Sideratou, Z.; Katsarou, M.E.; Tsiourvas, D. Mitochondrial delivery of doxorubicin by triphenylphosphonium-functionalized hyperbranched nanocarriers results in rapid and severe cytotoxicity. *Pharm. Res.* **2013**, *30*, 2832–2842.

440

169. Battigelli, A.; Russier, J.; Venturelli, E.; Fabbro, C.; Petronilli, V.; Bernardi, P.; Da Ros, T.; Prato, M.; Bianco, A. Peptide-based carbon nanotubes for mitochondrial targeting. *Nanoscale* **2013**, *5*, 9110–9117.

170. Dougherty, T.J.; Gomer, C.J.; Henderson, B.W.; Jori, G.; Kessel, D.; Korbelik, M.; Moan, J.; Peng, Q. Photodynamic therapy. *J. Natl. Cancer Inst.* **1998**, *90*, 889–905.

171. Dolmans, D.E.; Fukumura, D.; Jain, R.K. Photodynamic therapy for cancer. *Nat. Rev. Cancer* **2003**, *3*, 380–387.

172. Oliveira, C.S.; Turchiello, R.; Kowaltowski, A.J.; Indig, G.L.; Baptista, M.S. Major determinants of photoinduced cell death: Subcellular localization versus photosensitization efficiency. *Free Radic. Biol. Med.* **2011**, *51*, 824–833.

173. Modica-Napolitano, J.S.; Joyal, J.L.; Ara, G.; Oseroff, A.R.; Aprille, J.R. Mitochondrial toxicity of cationic photosensitizers for photochemotherapy. *Cancer Res.* **1990**, *50*, 7876–7881.

174. Powers, S.K.; Walstad, D.L.; Brown, J.T.; Detty, M.; Watkins, P.J. Photosensitization of human glioma cells by chalcogenapyrylium dyes. *J. Neurooncol.* **1989**, *7*, 179–188.

175. Park, Y.J.; Lee, W.Y.; Hahn, B.S.; Han, M.J.; Yang, W.I.; Kim, B.S. Chlorophyll derivatives—A new photosensitizer for photodynamic therapy of cancer in mice. *Yonsei Med. J.* **1989**, *30*, 212–218.

176. Zhang, X.D.; Deslandes, E.; Villedieu, M.; Poulain, L.; Duval, M.; Gauduchon, P.; Scwartz, L.; Icard, P. Effect of 2-deoxy-D-glucose on various malignant cell lines *in vitro*. *Anticancer Res.* **2006**, *26*, 3561–3566.

177. Zhang, D.; Li, J.; Wang, F.; Hu, J.; Wang, S.; Sun, Y. 2-Deoxy-D-glucose targeting of glucose metabolism in cancer cells as a potential therapy. *Cancer Lett.* **2014**, *355*, 176–183.

178. Ko, Y.H.; Pedersen, P.L.; Geschwind, J.F. Glucose catabolism in the rabbit VX2 tumor model for liver cancer: Characterization and targeting hexokinase. *Cancer Lett.* **2001**, *173*, 83–91.

179. Pedersen, P.L. 3-Bromopyruvate (3BP) a fast acting, promising, powerful, specific, and effective "small molecule" anti-cancer agent taken from labside to bedside: Introduction to a special issue. *J. Bioenerg. Biomembr.* **2012**, *44*, 1–6.

180. Oudard, S.; Poirson, F.; Miccoli, L.; Bourgeois, Y.; Vassault, A.; Poisson, M.; Magdelénat, H.; Dutrillaux, B.; Poupon, M.F. Mitochondria-bound hexokinase as target for therapy of malignant gliomas. *Int. J. Cancer* **1995**, *62*, 216–222.

181. Pulselli, R.; Amadio, L.; Fanciulli, M.; Floridi, A. Effect of lonidamine on the mitochondrial potential *in situ* in Ehrlich ascites tumor cells. *Anticancer Res.* **1996**, *16*, 419–423.

Targeting Mitochondrial Function to Treat Quiescent Tumor Cells in Solid Tumors

Xiaonan Zhang, Angelo de Milito, Maria Hägg Olofsson, Joachim Gullbo, Padraig D'Arcy and Stig Linder

Abstract: The disorganized nature of tumor vasculature results in the generation of microenvironments characterized by nutrient starvation, hypoxia and accumulation of acidic metabolites. Tumor cell populations in such areas are often slowly proliferating and thus refractory to chemotherapeutical drugs that are dependent on an active cell cycle. There is an urgent need for alternative therapeutic interventions that circumvent growth dependency. The screening of drug libraries using multicellular tumor spheroids (MCTS) or glucose-starved tumor cells has led to the identification of several compounds with promising therapeutic potential and that display activity on quiescent tumor cells. Interestingly, a common theme of these drug screens is the recurrent identification of agents that affect mitochondrial function. Such data suggest that, contrary to the classical Warburg view, tumor cells in nutritionally-compromised microenvironments are dependent on mitochondrial function for energy metabolism and survival. These findings suggest that mitochondria may represent an "Achilles heel" for the survival of slowly-proliferating tumor cells and suggest strategies for the development of therapy to target these cell populations.

Reprinted from *Int. J. Mol. Sci.* Cite as: Zhang, X.; de Milito, A.; Olofsson, M.H.; Gullbo, J.; D'Arcy, P.; Linder, S. Targeting Mitochondrial Function to Treat Quiescent Tumor Cells in Solid Tumors. *Int. J. Mol. Sci.* **2015**, *16*, 27313–27326.

1. Solid Tumors Contain Cell Populations with Limited Sensitivity to Treatment

Cancer drug discovery is focused on the development of agents that selectively inhibit the growth of cancer cells while maintaining a therapeutic window towards normal cells. While a relatively simple idea, this strategy is confounded by the fact that cancer cells within a tumor mass are not uniform entities, but rather a heterogeneous cell population with different growth properties and thus drug sensitivities. Non-proliferative or quiescent cells in particular are notoriously difficult to treat and represent the greatest hurdle to the elimination of cancer in patients [1,2]. While this is due to a number of factors, both intrinsic and extrinsic, a major contributor is the cell cycle dependence of most conventional chemotherapy drugs. This, the cytotoxic effect of drugs, such as topoisomerase inhibitors and microtubule stabilizing agents, is cell cycle dependent, whereas quiescent cells will

442

be relatively unaffected by these agents [3]. There is a growing recognition of the importance in developing therapies that target both the proliferating and quiescent tumor cell population in order to improve patient outcome [2].

Traditional chemotherapy is usually administered at three-week intervals to facilitate recovery of dividing normal cells, such as those of the bone marrow. It is important to realize that tumor cells, as well as normal cells, may survive treatment and recover between treatment cycles. Such surviving cells have the potential to repopulate the tumor between treatment cycles [2,4]. This problem may not have been given the attention it deserves, and it was only recently shown that chemotherapy leads to reoxygenation and increased proliferation of previously hypoxic and quiescent cells in solid tumors [5]. Tumor repopulation is therefore likely to constitute a major problem in clinical oncology and may be a contributing factor for the limited success of the treatment of patients with advanced malignant disease.

The sequence of how chemotherapeutic drugs are administered will influence their therapeutic effects [6]. Sequence dependence may mainly be due to cell cycle perturbations. Drug interactions should be particularly considered when planning combinations of cytotoxic and cytostatic agents in order not to diminish the efficacy of cell cycle-active chemotherapeutic agents [2]. Cytostatic agents should preferably be administered between courses of chemotherapy to inhibit repopulation, but not given immediately prior to the next round of chemotherapy [2]. Inhibitors of the mTOR/PI3K pathway represent examples of cytostatic agents that have the potential to inhibit tumor repopulation [2,7,8]. A rapamycin analogue was demonstrated to increase the *in vivo* effect of the cytotoxic drug docetaxel when administered between docetaxel treatments to xenografted animals [7]. In a recent study, tumor repopulation was reported to be diminished by treatment with the cyclooxygenase-2 (COX2) inhibitor celecoxib [9].

2. Avascular Areas of Solid Tumors Contain Quiescent Cell Populations

Proliferating tumor cells exist in a fine balance between nutrient supply and demand. Growing tumor cells generally outgrow the supporting vasculature, ultimately leading to cell populations distantly situated (>100 µm) from blood vessels [10]. The vasculature at the tumor site is generally poorly organized [11] with a high degree of compression on the supplying blood vessels, leading to the disrupted flow of nutrients and oxygen to the tumor tissue [12]. In addition, the insufficient perfusion between vessels and tumor tissue, combined with the high production of acidic metabolites, contributes to the development of regions of high acidity within solid tumors [13]. As a consequence, the majority of solid tumors will display regions of increased levels of hypoxia coupled with nutrient starvation and low pH [14] (Figure 1).

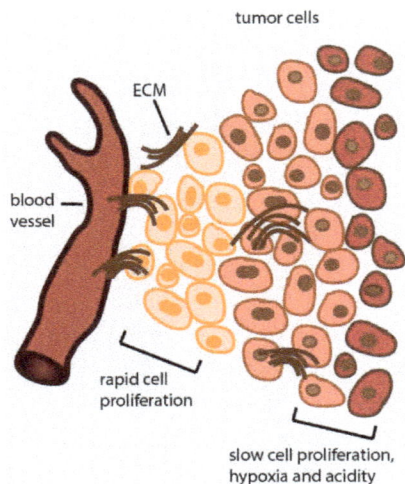

Figure 1. Development of heterogeneity in solid tumors. Continuous outgrowth of the vasculature results in the generation of tumor microenvironments that are characterized by hypoxia and nutrient starvation. Cells in these environments are slowly proliferating and are relatively insensitive to cell cycle-active cancer therapeutics. Extracellular matrix (ECM).

Griffin, Kerbel and coworkers [15] coined the term "multicellular resistance" to describe the combination of pharmacokinetic obstacles that limit drug penetrance, as well as cell–cell interactions that lead to altered expression of proteins that are important for cellular sensitivity to anticancer agents. The issue of efficient drug penetrance into tumor tissue, essential to achieve therapeutically-relevant drug concentrations [16,17], is important to consider when developing drugs for solid tumors [18]. Quiescent cells and hypoxic cells show gene signatures that are distinct from cells in well-vascularized areas. Such signatures have been reported to be associated with prognosis and with drug sensitivity [19–21].

Considering the clinical problem of regrowth resistance, it would be desirable to identify drugs that show activity on quiescent cells in avascular areas of solid tumors. Such drugs must have a therapeutic window that allows targeting non-proliferating tumor cells while sparing non-proliferating cells in healthy tissues. In addition, such drugs must be able to penetrate into the deep tumor parenchyma.

3. Conditional Drug Screening Aimed at Targeting Glucose-Starved Tumor Cells

Cell-based screening for the identification of cytotoxic drugs is usually performed using cell lines grown as monolayers on plastic support. Some culture media contain glucose at concentrations only occurring in severely diabetic individuals, and cells

444

are maintained in atmospheric oxygen levels. The National Cancer Institute has screened >100,000 compounds' antiproliferative activity on a panel of cancer cell lines (the NCI_{60} cell panel) under such conditions [22]. This has resulted in a wealth of information and the identification of a large number of substances. Most of these drugs are, however, expected to show activity on proliferating tumor cells and may be less effective on slowly-proliferating cells in avascular areas. Unphysiological conditions are not only used during screening; almost all studies of apoptosis induction and cell signaling are performed using tumor cells grown on plastic support in nutrient- and oxygen-rich conditions.

The mean glucose concentration in colon carcinoma tissue is ~2% of the plasma glucose concentration (0.12 *vs.* 5.6 mM) [23]. The cells in solid tumors are therefore in a steady state of nutrition depletion. Deprivation of cultured cancer cell lines of glucose has been reported to lead to resistance to many conventional anticancer agents [24], and it has therefore been of interest to identify agents that inhibit cancer cell viability under conditions of nutrient starvation. A number of drugs have been identified by the group of Esumi and coworkers using this approach. Kigamicin D [24], arctigenin [25], efrapeptin F [26] and pyrvinium pamoate [27] are examples of compounds with preferential antiproliferative activity on tumor cells grown under nutrient-deprived conditions. Kigamicin D is one of the compounds found to be cytotoxic to glucose-starved cancer cells, but not to cells grown in nutrient-rich standard media. Kigamicin D suppressed the in vivo tumor growth of pancreas cancer cell lines in nude mice [24]. Arctigenin is an antitumor antibiotic that displays preferential cytotoxicity under conditions of nutrient starvation and also shows strong in vivo activity against pancreas cancer xenografts [25]. Interestingly, arctigenin was reported to inhibit mitochondrial respiration and to induce a bioenergetic catastrophe [28]. Esumi and coworkers also identified the drug efrapeptin F as preferentially toxic to nutrient-deprived cancer cells [26]. Efrapeptin F has been demonstrated to inhibit the activity of the mitochondrial F1F0-ATPase (complex V) [29].

Pyrvinium pamoate (PP) is an anthelminthic drug (effective against parasitic worms) of particular interest. PP was reported to be extremely toxic to a number of cancer cell lines under conditions of glucose starvation [30]. PP also inhibited the growth of human colon cancer multicellular tumor spheroids (MCTS) and showed antitumor activity *in vivo* [30]. PP inhibited the hypoxic electron transfer chain, the NADH-fumarate reductase system, in mitochondria of tumor cells [27]. Fumarate reductase (FRD) activity has been demonstrated in species such as bacteria and helminths, and evidence that human cancer cells have FRD activity has also been presented [31]. The activity level was reported to be low, but to be increased during culture under hypoxic and glucose-deprived conditions. By inhibiting FRD activity, PP may preferentially target the energy production of cells in hypoxic regions in

445

tumors. PP was also identified by direct screening of MCTS and demonstrated to be a strong inhibitor of oxidative phosphorylation (OXPHOS) [32]. The drug screening findings described here led Esumi and coworkers to address the question of whether nutrient-deprived cancer cells are sensitive to mitochondrial inhibition in general, and this was indeed found to be the case [26].

Activation of the Akt signaling pathway was found to be essential for the ability of cancer cells to survive under low glucose conditions [25,33]. Tolerance to nutrient deprivation was also reported to be associated with the expression of AMP-activated protein kinase (AMPK), an enzyme important for protection from metabolic stress [34]. The screening hits identified by Esumi and coworkers were generally found to block the activation of the Akt pathway [24,25,27].

4. Drug Screening Efforts Using Spheroid Models Mimicking the Tumor Microenvironment and Heterogeneity

The 3D multicellular tumor spheroid model (MCTS) was developed in order to provide a more accurate mimic of the conditions of solid tumors [35,36]. MCTS are heterogeneous with a well-defined geometry, containing proliferating cell populations at surface layers and quiescent cells in the core [36]. HCT116 colon cancer MTCS are shown in Figure 2. In this example, the culture medium contained 25 mM glucose, an unphysiological concentration. The advantage of the use of high glucose concentrations is the development of larger areas of hypoxic cores (staining for pimonidazole adducts), useful for screening purposes (see below). HCT116 MTCS core regions are negative for proliferation marker Ki67, but positive for p27^{Kip1}. A possible mechanism for the upregulation of p27^{Kip1} is the downregulation of ERK signaling observed in the MTCS (Figure 2). p27^{Kip1} Protein stability is regulated by a Skp2-dependent mechanism that, in turn, is dependent on MAP (mitogen-activated protein) kinase signaling [37].

Tumor spheroids develop gradients of oxygen, nutrients and catabolites [38]. Oxygen tension values fall steeply in the viable rims of MCTS, suggesting that oxygen consumption rates (OCR) are similar in the inner and outer parts of spheroids [39]. Mitochondrial mass per cell volume has been reported to be constant irrespective of location in MCTS [40,41], whereas mitochondrial function appears reduced in deeper spheroid layers [41]. The core regions of MCTS have lower levels of glucose [42,43] compared to more peripheral layers. The thickness of the viable rim of tumor spheroids is dependent on glucose availability: large necrotic cores are observed in MCTS grown in low-glucose medium [39]. It has been reported that tumor cells grown as monolayers or MCTS show a similar rate of glucose consumption [44,45], suggesting that cells do not increase glucose consumption during hypoxic conditions in MTCS. Lactate levels do, however, increase in large-sized spheroids [46], leading to a more acidic pH in core regions [47]. This finding could be due to limited diffusion of lactate from core regions in MCTS, leading to accumulation.

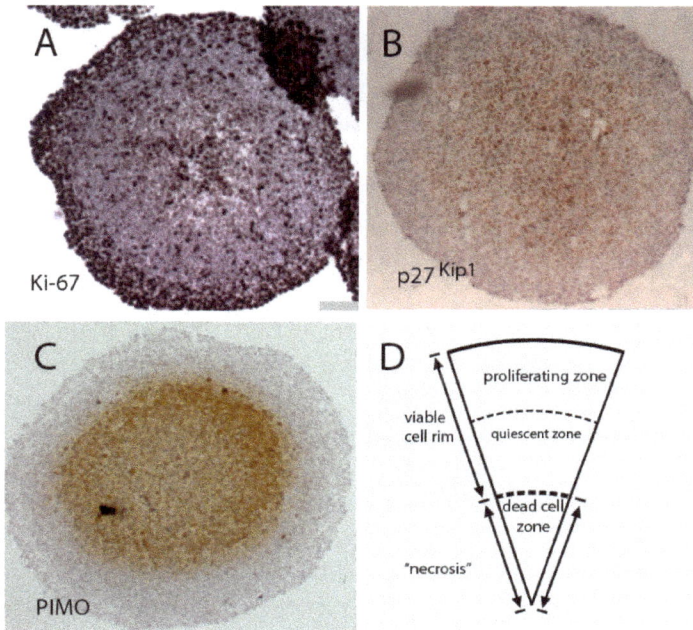

Figure 2. Properties of HCT116 colon cancer spheroids. HCT116 spheroids were grown in 96-well plates for five days, fixed, sectioned and stained with different antibodies. Proliferating cells (Ki-67 positive) are present in peripheral layers (**A**) and quiescent cells (p27^{Kip1} positive) in the core (**B**); Core layers are also severely hypoxic and positive for pimonidazole (PIMO) adducts (**C**); Large spheroids also contain central areas of necrosis (**D**).

MCTS are more resistant to chemotherapy compared to monolayer cultures [48,49]. Resistance is not unexpected, considering the presence of quiescent cells (resistant to cell-cycle-active drugs) and the requirement for drug diffusion to reach the core. High drug hydrophobicity (high log p) was reported to be required for strong antiproliferative activity in spheroid models [49]. The aspect of hydrophobicity and penetration is not necessarily considered during the development of drugs for solid tumors.

The utilization of MCTS for drug screening is an attractive strategy for anticancer drug development, since a number of relevant factors, such as nutrient starvation, hypoxia and drug penetration, are all accounted for. Traditional methods of MCTS production results in spheroids that are quite heterogeneous in size and not useful for screening [35]. Methods have subsequently been developed where spheroids are produced in microtiter plates, resulting in one similarly-sized spheroid per well [50]. Such MCTS can be used for drug screening campaigns [48,51,52]. Methods using microfluidics-generated double-emulsion droplets for spheroid culture have also been described [53], but whether these can be used for screening is unclear. Different read-outs of cell viability and apoptosis can be used for screening, including acid phosphatase [54], green fluorescent protein expression [32], methylene blue [55], caspase-cleaved keratin 18 [51] and high-content analysis using fluorescent dyes [56].

A colon MCTS screen using 10,000 compounds from ChemBridge (San Diego, CA, USA) resulted in the identification of VLX600, a drug that induces preferential cell death of tumor cells in spheroid cores and has *in vivo* activity [57]. The compound induces AMPK phosphorylation in tumor cell lines, but not in immortalized cells. VLX600 was found to reduce mitochondrial OXPHOS, particularly the rate of uncoupled respiration. This effect was associated with a strong decrease in MCTS hypoxia, measured as pimonidazole-stained fraction [57], and also reduced hypoxia in colon carcinoma xenografts *in vivo* (Fryknäs *et al.*, unpublished data). As might be expected, exposure to VLX600 sensitized colon cancer cells to glucose starvation. The precise molecular mechanism of action of VLX600 has been elucidated using gene expression profiling and shown to be iron chelation (Fryknäs *et al.*, unpublished data). The activity of the compound on quiescent tumor cells was hypothesized to be due to these cells showing a decreased metabolic plasticity (*i.e.*, limited ability to switch between different modes of energy production) [57]. Two other screens of MCTS have been described using smaller drug libraries consisting of bioactive molecules. Wentzel *et al.* [56] screened two drug libraries, one consisting of 640 FDA-approved drugs and another containing 480 drugs with known mechanisms of action. In a separate study, 1600 compounds with documented clinical history were screened using MCTS [32]. The use of chemical libraries containing drugs with known modes of action has the advantage of giving an immediate recognition of the mechanisms of cytotoxic activity. Compounds that affect mitochondrial function were identified in both of these screens. All hits from Wentzel *et al.* [56] interfered with the proper function of the respiratory chain, either by acting as inhibitors or uncouplers of the respiratory chain. Thus, the exact target in the respiratory chain seems to be irrelevant, as complex I inhibitors induce similar phenotypes as complex III/V inhibitors or uncouplers of the respiratory chain. The authors interpreted their findings to suggest that low levels of glucose in MCTS core regions do not allow glycolysis to provide sufficient ATP (adenosine triphosphate), resulting in cells with dependence on respiration for survival. Supplementation of glucose was indeed found to decrease the toxicity of the screening hits. These authors also showed that the combination of mitochondrial inhibitors and conventional cytostatic drugs led to increased levels of cell death in MCTS [56]. The study by Senkowski and coworkers [32] led to the identification of five compounds with selective activity on MCTS: closantel, nitazoxanide, niclosamide, pyrvinium pamoate and salinomycin. These compounds have all been described to target mitochondrial function by different mechanisms. Niclosamide, closantel and nitazoxanide share an identical pharmacophore and have been demonstrated to possess uncoupling activity [58–60]. Pyrvinium pamoate also inhibits the mitochondrial OXPHOS and was discussed above. The identification of salinomycin by MCTS screening is interesting considering that the same drug was identified in a screen for agents active on cancer stem cells [61]. Salinomycin has also been described to inhibit OXPHOS [62].

5. Drug Screening Using Cancer Stem Cells

Cancer stem cells (CSC) have attracted an enormous interest from the scientific community during recent years. According to the cancer stem cell model, a limited subset of cancer cells has the capacity to propagate tumors based on their capacity for self-renewal and their ability to generate differentiated progeny [63]. CSCs have been reported to display increased resistance to conventional chemotherapeutic agents and to radiation [64,65] and are therefore likely candidates for therapy-resistant cells that give rise to tumor recurrences. In a recent study, tumor repopulation between chemotherapy cycles was shown to be due to stimulation of CSC proliferation by prostaglandin E2 (PGE-2) release by neighboring cells [9]. Whether CSCs survive chemotherapy and repopulate tumors is, however, somewhat controversial [66]. Similar to quiescent cells in spheroid cores, CSC are thought to reside in a hypoxic niche [67]. It remains possible that the position of tumor cells in relation to the vasculature, rather than stemness, determines the susceptibility to therapy and the potential for recurrence.

A number of groups have used CSC in drug screening campaigns (reviewed in [68]). Gupta et al. [61] performed a drug screen for compounds that would display selective toxicity for breast CSCs. This endeavor resulted in the identification of the antibacterial drug salinomycin [61]. The authors reported that salinomycin inhibits mammary tumor growth *in vivo* and that the drug induced alterations in gene expression, indicating a loss of the CSC phenotype. Salinomycin is an OXPHOS inhibitor [32,62] and has also been identified by MCTS screening [32]. The therapeutic window of salinomycin is limited, and it is presently unclear whether this drug can be developed for clinical applications [69]. Ovarian cancer and breast cancer stem-like cells have also been used for drug screening, resulting in the identification of niclosamide [70,71], a mitochondrial uncoupler [59]. Niclosamide also has activity on colon cancer MCTS [32]. Sztiller-Sikorska utilized a library of natural products to screen for drugs with activity against melanoma cells with self-renewing capability [72]. One of the compounds identified was streptonigrin, shown to decrease the respiration of isolated rat liver mitochondria [73].

Metformin is believed to be the most commonly-prescribed anti-diabetic drug in the world and functions by increasing cellular glucose uptake [74]. Metformin inhibits mitochondrial OXPHOS [75,76], but this effect is controversial, since it occurs at high drug concentrations and may be physiologically irrelevant [77]. Metformin also inhibits 5′-adenosine monophosphate (AMP)-activated kinase (AMPK) [78] and was recently shown to be an inhibitor of mitochondrial glycerophosphate dehydrogenase (mGPD) activity [79]. Metformin was reported to selectively induce apoptosis of pancreatic CSC, by a mechanism involving a bioenergetic catastrophe associated with ROS induction and reduced mitochondrial transmembrane potential [80]. The frequency of tumor incidence in diabetic patients is decreased by metformin use [81], and metformin shows anti-neoplastic activity in mouse animal models [82,83]. For a recent comprehensive review on metformin, see [77].

6. Tumor Cell Metabolism Is Dependent on Oxidative Phosphorylation (OXPHOS)

Increased glycolysis under aerobic conditions ("Warburg effect") is observed in most cancer cells [84]. This effect is currently believed to be associated with the increased rates of proliferation of tumor cells where glycolysis contributes the building blocks required for anabolic processes [85]. Upregulation of glycolysis is also necessary for the synthesis of reducing equivalents (NADPH) via the pentose phosphate pathway [86]. The Warburg effect may be explained by the increased expression of the M2-form of pyruvate kinase observed in rapidly-proliferating cells [87]. The observations of elevated glycolysis in tumor cells have led to efforts aimed at targeting this process. The glucose analogue 2-deoxy-D-glucose (2-DG) is transported into cells and phosphorylated by hexokinase. The phosphorylated form of 2-DG cannot be metabolized further, leading to its accumulation and inhibition of glycolysis. 3-bromopyruvate (3-BP) is a potent inhibitor of glycolysis and an alkylating agent. 3-BP has been demonstrated to inhibit hexokinase-II activity *in vitro* [88]. A number of other targets have also been described, and the effects of 3-BP appear to be mediated mainly by its alkylating properties, notably towards thiols [89]. Dichloroacetate (DCA) is an inhibitor of pyruvate dehydrogenase kinase, a mitochondrial enzyme pyruvate dehydrogenase that converts pyruvate to acetyl CoA (Coenzyme A). Exposure leads to the induction of a switch from aerobic glycolysis to glucose oxidation, subsequently leading to a decreased mitochondrial membrane potential and sensitization to apoptosis [90]. A phase I study for DCA has been initiated [91]. Cancer cell metabolism is emerging as a major arena for the development of cancer therapeutics; this review is not intended to provide a comprehensive overview of this area of research (see [92–94]).

Although Warburg hypothesized that mitochondrial bioenergetics is defective in tumor cells [95], it has become increasingly clear that mitochondria in fact are essential for the proliferation of cancer cells [96,97]. Studies of the energy budgets of various cancer cells under normoxic conditions concluded that most of the ATP in cancer cells in fact emanates from mitochondria. Zu and Guppy reported that in a panel of 31 tumor cell lines, the average contribution of glycolysis was 17%, compared to 20% in normal cells [98]. These results are supported by later studies: mitochondrial respiration indeed continues to operate normally at rates proportional to oxygen supply [86]. Mandujano-Tinoco and coworkers reported that OXPHOS was the predominant source of ATP both in quiescent and proliferating cell layers in MTCS (93% of total ATP was derived from OXPHOS in quiescent cells, compared to 98% in proliferating cells) [99]. Downregulation of OXPHOS in energy-demanding tumors cells is, in fact, counterintuitive considering the inefficiency of glycolysis as a means of ATP production. Mitochondrial metabolism is considered to be important for tumor growth, both by providing energy in the form of ATP, as well as various metabolic intermediates necessary for anabolic reactions [100]. Direct evidence for the dependence of tumor cells on mitochondria comes from the study of ρ^0 tumor cells, where mtDNA has been eliminated through growth in ethidium bromide. ρ^0 Tumor cells display reduced growth rates and a decreased tumorigenicity *in vivo* [101–103]. Furthermore, overexpression of TFAM (mitochondrial transcription factor A) leads to increased mitochondrial biomass and stimulates cell proliferation of cancer cells [104]. Two known regulators of mitochondrial metabolism, PGC-1α (peroxisome proliferator-activated receptor gamma coactivator 1-α) and ERRα (estrogen-related receptor α), have also been reported to stimulate the proliferation various forms of carcinomas and of melanomas [105].

The commonly-used deduction that energy production under hypoxic conditions is met by increased glycolysis does not account for the fact that also glucose levels are limited in hypovascular tumors [23,106,107]. Other sources of energy production are available and have been proposed to be used by cells in hypovascular areas. The reductive metabolism of glutamine-derived α-ketoglutarate to be used for the synthesis of acetyl-CoA is one such pathway [108]. Increased utilization of glutamine for energy generation has also been demonstrated in cells exposed to acidic conditions, occurring in the deep tumor parenchyma. Energy generation will in this instance require oxygen and mitochondrial function, since glutamine supports mitochondrial cell respiration through the TCA (tricarboxylic acid) cycle [109].

Some evidence points to increased OXPHOS activity in human tumors. Histochemical staining of the *in situ* enzymatic activity of cytochrome C oxidase (COX; complex IV) showed abundant activity in tumor cells compared to adjacent stromal cells and normal epithelial cells [110]. Similar results were observed for complex I and II activities [110]. Additional studies have shown that the expression of components involved in mitochondrial biogenesis (nuclear respiratory factor 1 (NRF1), mitochondrial transcription factor A (TFAM) and mitochondrial transcription factor B1 (TFB1M)), mitochondrial translation and mitochondrial lipid biosynthesis (Golgi phosphoprotein 3 (GOLPH3) and GOLPH3L) are upregulated in human breast carcinoma cells and downregulated in adjacent stromal cells [111]. Furthermore, the elevated expression of the mitochondrial markers TIMM17A and TOMM34 is associated with poor clinical outcome [112–114].

7. Mitochondrial OXPHOS as a Therapeutic Opportunity to Target Non-Proliferating Tumor Cells

Mitochondrial inhibitors have been known to possess anti-tumor activity for many years [26,115,116]. Examples of such compounds are the dye Rho123 [115], which inhibits OXPHOS [117], the complex V inhibitor efrapeptin F [26] and mitochondria-targeted lipophilic cations [118]. More recently, screening of a library of FDA-approved drugs for drugs with tumor cell selectivity led to the identification of the antimicrobial drug tigecycline. This drug functions as an inhibitor of mitochondrial translation [119,120].

We conclude that the drug screening efforts using MCTS, glucose-starved cells and CSC reviewed in this article have frequently resulted in the identification of mitochondrial inhibitors. In some reports, all screening hits were found to have effects on OXPHOS. When interpreting results from drug screens, it is important to keep in mind that different targets may show differences in "druggability". Thus, mitochondrial energy metabolism, which is dependent on a large number of components and intact membrane structures, may be more sensitive to drug intervention than other potential targets. It is nevertheless interesting that mitochondrial inhibitors are encountered in screens performed using different cell lines, cancer stem cells, different culture conditions (glucose-starved monolayers, MCTS) and different platforms (screening libraries, read-outs, *etc.*). This has led us and others to hypothesize that energy production in tumor cells situated in the deep tumor parenchyma is vulnerable and cannot tolerate even limited decreases in OXPHOS [56,57].

Acknowledgments: The authors thank Vetenskapsrådet, Cancerfonden, Radiumhemmets forskningsfonder and Barncancerfonden for support.

Author Contributions: All authors contributed to writing this paper.

Conflicts of Interest: Stig Linder and Padraig D'Arcy are shareholders in Vivolux AB (Mölndal, Sweden), which develops VLX600 for cancer therapy.

References

1. Durand, R.E. Distribution and activity of antineoplastic drugs in a tumor model. *J. Natl. Cancer Inst.* **1989**, *81*, 146–152.

2. Kim, J.J.; Tannock, I.F. Repopulation of cancer cells during therapy: An important cause of treatment failure. *Nat. Rev. Cancer* **2005**, *5*, 516–525.

3. Ozawa, S.; Sugiyama, Y.; Mitsuhashi, J.; Inaba, M. Kinetic analysis of cell killing effect induced by cytosine arabinoside and cisplatin in relation to cell cycle phase specificity in human colon cancer and Chinese hamster cells. *Cancer Res.* **1989**, *49*, 3823–3828.

4. Malaise, E.; Tubiana, M. Growth of the cells of an experimental irradiated fibrosarcoma in the C3H mouse. *C. R. Acad. Sci. Hebd. Seances Acad. Sci. D* **1966**, *263*, 292–295.

5. Saggar, J.K.; Tannock, I.F. Chemotherapy rescues hypoxic tumor cells and induces their reoxygenation and repopulation—An effect that is inhibited by the hypoxia-activated prodrug TH-302. *Clin. Cancer Res.* **2015**, *21*, 2107–2114.

6. Shah, M.A.; Schwartz, G.K. The relevance of drug sequence in combination chemotherapy. *Drug Resist. Update* **2000**, *3*, 335–356.

7. Wu, L.; Birle, D.C.; Tannock, I.F. Effects of the mammalian target of rapamycin inhibitor CCI-779 used alone or with chemotherapy on human prostate cancer cells and xenografts. *Cancer Res.* **2005**, *65*, 2825–2831.

8. Hernlund, E.; Olofsson, M.H.; Fayad, W.; Fryknas, M.; Lesiak-Mieczkowska, K.; Zhang, X.; Brnjic, S.; Schmidt, V.; D'Arcy, P.; Sjoblom, T.; *et al*. The phosphoinositide 3-kinase/mammalian target of rapamycin inhibitor NVP-BEZ235 is effective in inhibiting regrowth of tumour cells after cytotoxic therapy. *Eur. J. Cancer* **2012**, *48*, 396–406.

9. Kurtova, A.V.; Xiao, J.; Mo, Q.; Pazhanisamy, S.; Krasnow, R.; Lerner, S.P.; Chen, F.; Roh, T.T.; Lay, E.; Ho, P.L.; *et al*. Blocking PGE2-induced tumour repopulation abrogates bladder cancer chemoresistance. *Nature* **2015**, *517*, 209–213.

10. Thomlinson, R.H.; Gray, L.H. The histological structure of some human lung cancers and the possible implications for radiotherapy. *Br. J. Cancer* **1955**, *9*, 539–549.

11. Less, J.R.; Skalak, T.C.; Sevick, E.M.; Jain, R.K. Microvascular architecture in a mammary carcinoma: Branching patterns and vessel dimensions. *Cancer Res.* **1991**, *51*, 265–273.

12. Padera, T.P.; Stoll, B.R.; Tooredman, J.B.; Capen, D.; di Tomaso, E.; Jain, R.K. Pathology: Cancer cells compress intratumour vessels. *Nature* **2004**, *427*, 695.

13. Fang, J.S.; Gillies, R.D.; Gatenby, R.A. Adaptation to hypoxia and acidosis in carcinogenesis and tumor progression. *Semin. Cancer Biol.* **2008**, *18*, 330–337.

14. Brown, J.M.; Wilson, W.R. Exploiting tumour hypoxia in cancer treatment. *Nat. Rev. Cancer* **2004**, *4*, 437–447.

15. Kobayashi, H.; Man, S.; Graham, C.H.; Kapitain, S.J.; Teicher, B.A.; Kerbel, R.S. Acquired multicellular-mediated resistance to alkylating agents in cancer. *Proc. Natl. Acad. Sci. USA* **1993**, *90*, 3294–3298.

16. Tannock, I.F.; Lee, C.M.; Tunggal, J.K.; Cowan, D.S.; Egorin, M.J. Limited penetration of anticancer drugs through tumor tissue: A potential cause of resistance of solid tumors to chemotherapy. *Clin. Cancer Res.* **2002**, *8*, 878–884.

17. Minchinton, A.I.; Tannock, I.F. Drug penetration in solid tumours. *Nat. Rev. Cancer* **2006**, *6*, 583–592.

18. Tannock, I.F. Tumor physiology and drug resistance. *Cancer Metastasis Rev.* **2001**, *20*, 123–132.

19. Lu, X.; Yan, C.H.; Yuan, M.; Wei, Y.; Hu, G.; Kang, Y. *In vivo* dynamics and distinct functions of hypoxia in primary tumor growth and organotropic metastasis of breast cancer. *Cancer Res.* **2010**, *70*, 3905–3914.

20. Halle, C.; Andersen, E.; Lando, M.; Aarnes, E.K.; Hasvold, G.; Holden, M.; Syljuasen, R.G.; Sundfor, K.; Kristensen, G.B.; Holm, R.; et al. Hypoxia-induced gene expression in chemoradioresistant cervical cancer revealed by dynamic contrast-enhanced MRI. *Cancer Res.* **2012**, *72*, 5285–5295.

21. Kolosenko, I.; Fryknas, M.; Forsberg, S.; Johnsson, P.; Cheon, H.; Holvey-Bates, E.G.; Edsbacker, E.; Pellegrini, P.; Rassoolzadeh, H.; Brnjic, S.; et al. Cell crowding induces interferon regulatory factor 9, which confers resistance to chemotherapeutic drugs. *Int. J. Cancer* **2015**, *136*, E51–E61.

22. Shoemaker, R.H.; Scudiero, D.A.; Melillo, G.; Currens, M.J.; Monks, A.P.; Rabow, A.A.; Covell, D.G.; Sausville, E.A. Application of high-throughput, molecular-targeted screening to anticancer drug discovery. *Curr. Top. Med. Chem.* **2002**, *2*, 229–246.

23. Hirayama, A.; Kami, K.; Sugimoto, M.; Sugawara, M.; Toki, N.; Onozuka, H.; Kinoshita, T.; Saito, N.; Ochiai, A.; Tomita, M.; et al. Quantitative metabolome profiling of colon and stomach cancer microenvironment by capillary electrophoresis time-of-flight mass spectrometry. *Cancer Res.* **2009**, *69*, 4918–4925.

24. Lu, J.; Kunimoto, S.; Yamazaki, Y.; Kaminishi, M.; Esumi, H. Kigamicin D, a novel anticancer agent based on a new anti-austerity strategy targeting cancer cells' tolerance to nutrient starvation. *Cancer Sci.* **2004**, *95*, 547–552.

25. Awale, S.; Lu, J.; Kalauni, S.K.; Kurashima, Y.; Tezuka, Y.; Kadota, S.; Esumi, H. Identification of arctigenin as an antitumor agent having the ability to eliminate the tolerance of cancer cells to nutrient starvation. *Cancer Res.* **2006**, *66*, 1751–1757.

26. Momose, I.; Ohba, S.; Tatsuda, D.; Kawada, M.; Masuda, T.; Tsujiuchi, G.; Yamori, T.; Esumi, H.; Ikeda, D. Mitochondrial inhibitors show preferential cytotoxicity to human pancreatic cancer PANC-1 cells under glucose-deprived conditions. *Biochem. Biophys. Res. Commun.* **2010**, *392*, 460–466.

27. Tomitsuka, E.; Kita, K.; Esumi, H. An anticancer agent, pyrvinium pamoate inhibits the NADH-fumarate reductase system—A unique mitochondrial energy metabolism in tumour microenvironments. *J. Biochem.* **2012**, *152*, 171–183.

28. Gu, Y.; Qi, C.; Sun, X.; Ma, X.; Zhang, H.; Hu, L.; Yuan, J.; Yu, Q. Arctigenin preferentially induces tumor cell death under glucose deprivation by inhibiting cellular energy metabolism. *Biochem. Pharmacol.* **2012**, *84*, 468–476.

29. Cross, R.L.; Kohlbrenner, W.E. The mode of inhibition of oxidative phosphorylation by efrapeptin (A23871). Evidence for an alternating site mechanism for ATP synthesis. *J. Biol. Chem.* **1978**, *253*, 4865–4873.

30. Esumi, H.; Lu, J.; Kurashima, Y.; Hanaoka, T. Antitumor activity of pyrvinium pamoate, 6-(dimethylamino)-2-[2-(2,5-dimethyl-1-phenyl-1*H*-pyrrol-3-yl)ethenyl]-1-methyl-quinolinium pamoate salt, showing preferential cytotoxicity during glucose starvation. *Cancer Sci.* **2004**, *95*, 685–690.

31. Tomitsuka, E.; Kita, K.; Esumi, H. The NADH-fumarate reductase system, a novel mitochondrial energy metabolism, is a new target for anticancer therapy in tumor microenvironments. *Ann. N. Y. Acad. Sci.* **2010**, *1201*, 44–49.

32. Senkowski, W.; Zhang, X.; Olofsson, M.H.; Isacson, R.; Hoglund, U.; Gustafsson, M.; Nygren, P.; Linder, S.; Larsson, R.; Fryknas, M. Three-dimensional cell culture-based screening identifies the anthelmintic drug nitazoxanide as a candidate for treatment of colorectal cancer. *Mol. Cancer Ther.* **2015**.

33. Izuishi, K.; Kato, K.; Ogura, T.; Kinoshita, T.; Esumi, H. Remarkable tolerance of tumor cells to nutrient deprivation: Possible new biochemical target for cancer therapy. *Cancer Res.* **2000**, *60*, 6201–6207.

34. Kato, K.; Ogura, T.; Kishimoto, A.; Minegishi, Y.; Nakajima, N.; Miyazaki, M.; Esumi, H. Critical roles of AMP-activated protein kinase in constitutive tolerance of cancer cells to nutrient deprivation and tumor formation. *Oncogene* **2002**, *21*, 6082–6090.

35. Sutherland, R.M. Cell and environment interactions in tumor microregions: The multicell spheroid model. *Science* **1988**, *240*, 177–184.

36. Hirschhaeuser, F.; Menne, H.; Dittfeld, C.; West, J.; Mueller-Klieser, W.; Kunz-Schughart, L.A. Multicellular tumor spheroids: An underestimated tool is catching up again. *J. Biotechnol.* **2010**, *148*, 3–15.

37. Motti, M.L.; De Marco, C.; Califano, D.; De Gisi, S.; Malanga, D.; Troncone, G.; Persico, A.; Losito, S.; Fabiani, F.; Santoro, M.; *et al.* Loss of p27 expression through RAS→BRAF→MAP kinase-dependent pathway in human thyroid carcinomas. *Cell Cycle* **2007**, *6*, 2817–2825.

38. Mueller-Klieser, W. Multicellular spheroids. A review on cellular aggregates in cancer research. *J. Cancer Res. Clin. Oncol.* **1987**, *113*, 101–122.

39. Mueller-Klieser, W.; Freyer, J.P.; Sutherland, R.M. Influence of glucose and oxygen supply conditions on the oxygenation of multicellular spheroids. *Br. J. Cancer* **1986**, *53*, 345–353.

40. Bredel-Geissler, A.; Karbach, U.; Walenta, S.; Vollrath, L.; Mueller-Klieser, W. Proliferation-associated oxygen consumption and morphology of tumor cells in monolayer and spheroid culture. *J. Cell. Physiol.* **1992**, *153*, 44–52.

41. Kunz-Schughart, L.A.; Habbersett, R.C.; Freyer, J.P. Impact of proliferative activity and tumorigenic conversion on mitochondrial function of fibroblasts in 2D and 3D culture. *Cell Biol. Int.* **2001**, *25*, 919–930.

42. Casciari, J.J.; Sotirchos, S.V.; Sutherland, R.M. Glucose diffusivity in multicellular tumor spheroids. *Cancer Res.* **1988**, *48*, 3905–3909.

43. Teutsch, H.F.; Goellner, A.; Mueller-Klieser, W. Glucose levels and succinate and lactate dehydrogenase activity in EMT6/Ro tumor spheroids. *Eur. J. Cell Biol.* **1995**, *66*, 302–307.

44. Li, C.K. The role of glucose in the growth of 9L multicell tumor spheroids. *Cancer* **1982**, *50*, 2074–2078.

45. Li, C.K. The glucose distribution in 9L rat brain multicell tumor spheroids and its effect on cell necrosis. *Cancer* **1982**, *50*, 2066–2073.

46. Kunz-Schughart, L.A.; Doetsch, J.; Mueller-Klieser, W.; Groebe, K. Proliferative activity and tumorigenic conversion: Impact on cellular metabolism in 3-D culture. *Am. J. Physiol. Cell Physiol.* **2000**, *278*, C765–C780.

47. Acker, H.; Carlsson, J.; Mueller-Klieser, W.; Sutherland, R.M. Comparative pO2 measurements in cell spheroids cultured with different techniques. *Br. J. Cancer* **1987**, *56*, 325–327.

48. Friedrich, J.; Ebner, R.; Kunz-Schughart, L.A. Experimental anti-tumor therapy in 3-D: Spheroids—Old hat or new challenge? *Int. J. Radiat. Biol.* **2007**, *83*, 849–871.

49. Fayad, W.; Rickardson, L.; Haglund, C.; Olofsson, M.H.; D'Arcy, P.; Larsson, R.; Linder, S.; Fryknas, M. Identification of agents that induce apoptosis of multicellular tumour spheroids: Enrichment for mitotic inhibitors with hydrophobic properties. *Chem. Biol. Drug Des.* **2011**, *78*, 547–557.

50. Kelm, J.M.; Timmins, N.E.; Brown, C.J.; Fussenegger, M.; Nielsen, L.K. Method for generation of homogeneous multicellular tumor spheroids applicable to a wide variety of cell types. *Biotechnol. Bioeng.* **2003**, *83*, 173–180.

51. Herrmann, R.; Fayad, W.; Schwarz, S.; Berndtsson, M.; Linder, S. Screening for compounds that induce apoptosis of cancer cells grown as multicellular spheroids. *J. Biomol. Screen.* **2008**, *13*, 1–8.

52. Friedrich, J.; Seidel, C.; Ebner, R.; Kunz-Schughart, L.A. Spheroid-based drug screen: Considerations and practical approach. *Nat. Protoc.* **2009**, *4*, 309–324.

53. Chan, H.F.; Zhang, Y.; Ho, Y.P.; Chiu, Y.L.; Jung, Y.; Leong, K.W. Rapid formation of multicellular spheroids in double-emulsion droplets with controllable microenvironment. *Sci. Rep.* **2013**, *3*, 3462.

54. Friedrich, J.; Eder, W.; Castaneda, J.; Doss, M.; Huber, E.; Ebner, R.; Kunz-Schughart, L.A. A reliable tool to determine cell viability in complex 3-D culture: The acid phosphatase assay. *J. Biomol. Screen.* **2007**, *12*, 925–937.

55. Mellor, H.R.; Ferguson, D.J.; Callaghan, R. A model of quiescent tumour microregions for evaluating multicellular resistance to chemotherapeutic drugs. *Br. J. Cancer* **2005**, *93*, 302–309.

56. Wenzel, C.; Riefke, B.; Grundemann, S.; Krebs, A.; Christian, S.; Prinz, F.; Osterland, M.; Golfier, S.; Rase, S.; Ansari, N.; *et al.* 3D high-content screening for the identification of compounds that target cells in dormant tumor spheroid regions. *Exp. Cell Res.* **2014**, *323*, 131–143.

57. Zhang, X.; Fryknas, M.; Hernlund, E.; Fayad, W.; De Milito, A.; Olofsson, M.H.; Gogvadze, V.; Dang, L.; Pahlman, S.; Schughart, L.A.; *et al.* Induction of mitochondrial dysfunction as a strategy for targeting tumour cells in metabolically compromised microenvironments. *Nat. Commun.* **2014**, *5*, 3295.

58. De Carvalho, L.P.; Darby, C.M.; Rhee, K.Y.; Nathan, C. Nitazoxanide disrupts membrane potential and intrabacterial pH homeostasis of *Mycobacterium tuberculosis*. *ACS Med. Chem. Lett.* **2011**, *2*, 849–854.

59. Jurgeit, A.; McDowell, R.; Moese, S.; Meldrum, E.; Schwendener, R.; Greber, U.F. Niclosamide is a proton carrier and targets acidic endosomes with broad antiviral effects. *PLoS Pathog.* **2012**, *8*, e1002976.

60. Skuce, P.J.; Fairweather, I. The effect of the hydrogen ionophore closantel upon the pharmacology and ultrastructure of the adult liver fluke *Fasciola hepatica*. *Parasitol. Res.* **1990**, *76*, 241–250.

61. Gupta, P.B.; Onder, T.T.; Jiang, G.; Tao, K.; Kuperwasser, C.; Weinberg, R.A.; Lander, E.S. Identification of selective inhibitors of cancer stem cells by high-throughput screening. *Cell* **2009**, *138*, 645–659.

62. Mitani, M.; Yamanishi, T.; Miyazaki, Y.; Otake, N. Salinomycin effects on mitochondrial ion translocation and respiration. *Antimicrob. Agents Chemother.* **1976**, *9*, 655–660.

63. Clarke, M.F.; Dick, J.E.; Dirks, P.B.; Eaves, C.J.; Jamieson, C.H.; Jones, D.L.; Visvader, J.; Weissman, I.L.; Wahl, G.M. Cancer stem cells—Perspectives on current status and future directions: AACR Workshop on cancer stem cells. *Cancer Res.* **2006**, *66*, 9339–9344.

64. Costello, R.T.; Mallet, F.; Gaugler, B.; Sainty, D.; Arnoulet, C.; Gastaut, J.A.; Olive, D. Human acute myeloid leukemia CD34$^+$/CD38$^-$ progenitor cells have decreased sensitivity to chemotherapy and Fas-induced apoptosis, reduced immunogenicity, and impaired dendritic cell transformation capacities. *Cancer Res.* **2000**, *60*, 4403–4411.

65. Dean, M.; Fojo, T.; Bates, S. Tumour stem cells and drug resistance. *Nat. Rev. Cancer* **2005**, *5*, 275–284.

66. Hegde, G.V.; de la Cruz, C.; Eastham-Anderson, J.; Zheng, Y.; Sweet-Cordero, E.A.; Jackson, E.L. Residual tumor cells that drive disease relapse after chemotherapy do not have enhanced tumor initiating capacity. *PLoS ONE* **2012**, *7*, e45647.

67. Ito, K.; Suda, T. Metabolic requirements for the maintenance of self-renewing stem cells. *Nat. Rev. Mol. Cell Biol.* **2014**, *15*, 243–256.

68. Lv, J.; Shim, J.S. Existing drugs and their application in drug discovery targeting cancer stem cells. *Arch. Pharm. Res.* **2015**, *38*, 1617–1626.

69. Story, P.; Doube, A. A case of human poisoning by salinomycin, an agricultural antibiotic. *N. Z. Med. J.* **2004**, *117*, 1190.

70. Yo, Y.T.; Lin, Y.W.; Wang, Y.C.; Balch, C.; Huang, R.L.; Chan, M.W.; Sytwu, H.K.; Chen, C.K.; Chang, C.C.; Nephew, K.P.; *et al.* Growth inhibition of ovarian tumor-initiating cells by niclosamide. *Mol. Cancer Ther.* **2012**, *11*, 1703–1712.

71. Wang, Y.C.; Chao, T.K.; Chang, C.C.; Yo, Y.T.; Yu, M.H.; Lai, H.C. Drug screening identifies niclosamide as an inhibitor of breast cancer stem-like cells. *PLoS ONE* **2013**, *8*, e74538.

72. Sztiller-Sikorska, M.; Koprowska, K.; Majchrzak, K.; Hartman, M.; Czyz, M. Natural compounds' activity against cancer stem-like or fast-cycling melanoma cells. *PLoS ONE* **2014**, *9*, e90783.

73. Inouye, Y.; Okada, H.; Uno, J.; Arai, T.; Nakamura, S. Effects of streptonigrin derivatives and sakyomicin A on the respiration of isolated rat liver mitochondria. *J. Antibiot.* **1986**, *39*, 550–556.

74. Brunmair, B.; Staniek, K.; Gras, F.; Scharf, N.; Althaym, A.; Clara, R.; Roden, M.; Gnaiger, E.; Nohl, H.; Waldhausl, W.; et al. Thiazolidinediones, like metformin, inhibit respiratory complex I: A common mechanism contributing to their antidiabetic actions? *Diabetes* **2004**, *53*, 1052–1059.

75. Owen, M.R.; Doran, E.; Halestrap, A.P. Evidence that metformin exerts its anti-diabetic effects through inhibition of complex 1 of the mitochondrial respiratory chain. *Biochem. J.* **2000**, *348 Pt 3*, 607–614.

76. Andrzejewski, S.; Gravel, S.P.; Pollak, M.; St-Pierre, J. Metformin directly acts on mitochondria to alter cellular bioenergetics. *Cancer Metab.* **2014**, *2*, 12.

77. He, L.; Wondisford, F.E. Metformin action: Concentrations matter. *Cell Metab.* **2015**, *21*, 159–162.

78. Zhou, G.; Myers, R.; Li, Y.; Chen, Y.; Shen, X.; Fenyk-Melody, J.; Wu, M.; Ventre, J.; Doebber, T.; Fujii, N.; et al. Role of AMP-activated protein kinase in mechanism of metformin action. *J. Clin. Investig.* **2001**, *108*, 1167–1174.

79. Madiraju, A.K.; Erion, D.M.; Rahimi, Y.; Zhang, X.M.; Braddock, D.T.; Albright, R.A.; Prigaro, B.J.; Wood, J.L.; Bhanot, S.; MacDonald, M.J.; et al. Metformin suppresses gluconeogenesis by inhibiting mitochondrial glycerophosphate dehydrogenase. *Nature* **2014**, *510*, 542–546.

80. Lonardo, E.; Cioffi, M.; Sancho, P.; Sanchez-Ripoll, Y.; Trabulo, S.M.; Dorado, J.; Balic, A.; Hidalgo, M.; Heeschen, C. Metformin targets the metabolic achilles heel of human pancreatic cancer stem cells. *PLoS ONE* **2013**, *8*, e76518.

81. Evans, J.M.; Donnelly, L.A.; Emslie-Smith, A.M.; Alessi, D.R.; Morris, A.D. Metformin and reduced risk of cancer in diabetic patients. *BMJ* **2005**, *330*, 1304–1305.

82. Hirsch, H.A.; Iliopoulos, D.; Tsichlis, P.N.; Struhl, K. Metformin selectively targets cancer stem cells, and acts together with chemotherapy to block tumor growth and prolong remission. *Cancer Res.* **2009**, *69*, 7507–7511.

83. Rocha, G.Z.; Dias, M.M.; Ropelle, E.R.; Osorio-Costa, F.; Rossato, F.A.; Vercesi, A.E.; Saad, M.J.; Carvalheira, J.B. Metformin amplifies chemotherapy-induced AMPK activation and antitumoral growth. *Clin. Cancer Res.* **2011**, *17*, 3993–4005.

84. Warburg, O.; Wind, F.; Negelein, E. The metabolism of tumors in the body. *J. Gen. Physiol.* **1927**, *8*, 519–530.

85. Vander Heiden, M.G.; Cantley, L.C.; Thompson, C.B. Understanding the Warburg effect: The metabolic requirements of cell proliferation. *Science* **2009**, *324*, 1029–1033.

86. Chen, X.; Qian, Y.; Wu, S. The Warburg effect: Evolving interpretations of an established concept. *Free Radic. Biol. Med.* **2015**, *79*, 253–263.

87. Christofk, H.R.; Vander Heiden, M.G.; Harris, M.H.; Ramanathan, A.; Gerszten, R.E.; Wei, R.; Fleming, M.D.; Schreiber, S.L.; Cantley, L.C. The M2 splice isoform of pyruvate kinase is important for cancer metabolism and tumour growth. *Nature* **2008**, *452*, 230–233.

88. Ko, Y.H.; Pedersen, P.L.; Geschwind, J.F. Glucose catabolism in the rabbit VX2 tumor model for liver cancer: Characterization and targeting hexokinase. *Cancer Lett.* **2001**, *173*, 83–91.

89. Shoshan, M.C. 3-Bromopyruvate: Targets and outcomes. *J. Bioenerg. Biomembr.* **2012**, *44*, 7–15.

90. Bonnet, S.; Archer, S.L.; Allalunis-Turner, J.; Haromy, A.; Beaulieu, C.; Thompson, R.; Lee, C.T.; Lopaschuk, G.D.; Puttagunta, L.; Bonnet, S.; *et al.* A mitochondria-K^+ channel axis is suppressed in cancer and its normalization promotes apoptosis and inhibits cancer growth. *Cancer Cell* **2007**, *11*, 37–51.

91. Dunbar, E.M.; Coats, B.S.; Shroads, A.L.; Langaee, T.; Lew, A.; Forder, J.R.; Shuster, J.J.; Wagner, D.A.; Stacpoole, P.W. Phase 1 trial of dichloroacetate (DCA) in adults with recurrent malignant brain tumors. *Investig. New Drugs* **2014**, *32*, 452–464.

92. Ward, P.S.; Thompson, C.B. Metabolic reprogramming: A cancer hallmark even warburg did not anticipate. *Cancer Cell* **2012**, *21*, 297–308.

93. Parks, S.K.; Chiche, J.; Pouyssegur, J. Disrupting proton dynamics and energy metabolism for cancer therapy. *Nat. Rev. Cancer* **2013**, *13*, 611–623.

94. Galluzzi, L.; Kepp, O.; Vander Heiden, M.G.; Kroemer, G. Metabolic targets for cancer therapy. *Nat. Rev. Drug Discov.* **2013**, *12*, 829–846.

95. Warburg, O. On respiratory impairment in cancer cells. *Science* **1956**, *124*, 269–270.

96. Wallace, D.C. Mitochondria and cancer. *Nat. Rev. Cancer* **2012**, *12*, 685–698.

97. Weinberg, S.E.; Chandel, N.S. Targeting mitochondria metabolism for cancer therapy. *Nat. Chem. Biol.* **2015**, *11*, 9–15.

98. Zu, X.L.; Guppy, M. Cancer metabolism: Facts, fantasy, and fiction. *Biochem. Biophys. Res. Commun.* **2004**, *313*, 459–465.

99. Mandujano-Tinoco, E.A.; Gallardo-Pérez, J.C.; Marín-Hernández, A.; Moreno-Sánchez, R.; Rodríguez-Enríquez, S. Anti-mitochondrial therapy in human breast cancer multi-cellular spheroids. *Biochim. Biophys. Acta* **2013**, *1833*, 541–551.

100. Mullen, A.R.; DeBerardinis, R.J. Genetically-defined metabolic reprogramming in cancer. *Trends Endocrinol. Metab.* **2012**, *23*, 552–559.

101. Zinkewich-Peotti, K.; Parent, M.; Morais, R. On the tumorigenicity of mitochondrial DNA-depleted avian cells. *Cancer Lett.* **1991**, *59*, 119–124.

102. Morais, R.; Zinkewich-Peotti, K.; Parent, M.; Wang, H.; Babai, F.; Zollinger, M. Tumor-forming ability in athymic nude mice of human cell lines devoid of mitochondrial DNA. *Cancer Res.* **1994**, *54*, 3889–3896.

103. Hayashi, J.; Takemitsu, M.; Nonaka, I. Recovery of the missing tumorigenicity in mitochondrial DNA-less HeLa cells by introduction of mitochondrial DNA from normal human cells. *Somat. Cell Mol. Genet.* **1992**, *18*, 123–129.

104. Han, B.; Izumi, H.; Yasuniwa, Y.; Akiyama, M.; Yamaguchi, T.; Fujimoto, N.; Matsumoto, T.; Wu, B.; Tanimoto, A.; Sasaguri, Y.; *et al.* Human mitochondrial transcription factor A functions in both nuclei and mitochondria and regulates cancer cell growth. *Biochem. Biophys. Res. Commun.* **2011**, *408*, 45–51.

105. Bhalla, K.; Hwang, B.J.; Dewi, R.E.; Ou, L.; Twaddel, W.; Fang, H.B.; Vafai, S.B.; Vazquez, F.; Puigserver, P.; Boros, L.; *et al.* PGC1α promotes tumor growth by inducing gene expression programs supporting lipogenesis. *Cancer Res.* **2011**, *71*, 6888–6898.

106. Eskey, C.J.; Koretsky, A.P.; Domach, M.M.; Jain, R.K. Role of oxygen *vs.* glucose in energy metabolism in a mammary carcinoma perfused *ex vivo*: Direct measurement by 31P NMR. *Proc. Natl. Acad. Sci. USA* **1993**, *90*, 2646–2650.

107. Dang, C.V.; Semenza, G.L. Oncogenic alterations of metabolism. *Trends Biochem. Sci.* **1999**, *24*, 68–72.

108. Metallo, C.M.; Gameiro, P.A.; Bell, E.L.; Mattaini, K.R.; Yang, J.; Hiller, K.; Jewell, C.M.; Johnson, Z.R.; Irvine, D.J.; Guarente, L.; *et al.* Reductive glutamine metabolism by IDH1 mediates lipogenesis under hypoxia. *Nature* **2012**, *481*, 380–384.

109. Corbet, C.; Draoui, N.; Polet, F.; Pinto, A.; Drozak, X.; Riant, O.; Feron, O. The SIRT1/HIF2α axis drives reductive glutamine metabolism under chronic acidosis and alters tumor response to therapy. *Cancer Res.* **2014**, *74*, 5507–5519.

110. Whitaker-Menezes, D.; Martinez-Outschoorn, U.E.; Flomenberg, N.; Birbe, R.C.; Witkiewicz, A.K.; Howell, A.; Pavlides, S.; Tsirigos, A.; Ertel, A.; Pestell, R.G.; *et al.* Hyperactivation of oxidative mitochondrial metabolism in epithelial cancer cells *in situ*: Visualizing the therapeutic effects of metformin in tumor tissue. *Cell Cycle* **2011**, *10*, 4047–4064.

111. Sotgia, F.; Whitaker-Menezes, D.; Martinez-Outschoorn, U.E.; Salem, A.F.; Tsirigos, A.; Lamb, R.; Sneddon, S.; Hulit, J.; Howell, A.; Lisanti, M.P. Mitochondria "fuel" breast cancer metabolism: Fifteen markers of mitochondrial biogenesis label epithelial cancer cells, but are excluded from adjacent stromal cells. *Cell Cycle* **2012**, *11*, 4390–4401.

112. Xu, X.; Qiao, M.; Zhang, Y.; Jiang, Y.; Wei, P.; Yao, J.; Gu, B.; Wang, Y.; Lu, J.; Wang, Z.; *et al.* Quantitative proteomics study of breast cancer cell lines isolated from a single patient: Discovery of TIMM17A as a marker for breast cancer. *Proteomics* **2010**, *10*, 1374–1390.

113. Aleskandarany, M.A.; Negm, O.H.; Rakha, E.A.; Ahmed, M.A.; Nolan, C.C.; Ball, G.R.; Caldas, C.; Green, A.R.; Tighe, P.J.; Ellis, I.O. TOMM34 expression in early invasive breast cancer: A biomarker associated with poor outcome. *Breast Cancer Res. Treat.* **2012**, *136*, 419–427.

114. Salhab, M.; Patani, N.; Jiang, W.; Mokbel, K. High TIMM17A expression is associated with adverse pathological and clinical outcomes in human breast cancer. *Breast Cancer* **2012**, *19*, 153–160.

115. Bernal, S.D.; Lampidis, T.J.; McIsaac, R.M.; Chen, L.B. Anticarcinoma activity *in vivo* of rhodamine 123, a mitochondrial-specific dye. *Science* **1983**, *222*, 169–172.

116. Lampidis, T.J.; Bernal, S.D.; Summerhayes, I.C.; Chen, L.B. Selective toxicity of rhodamine 123 in carcinoma cells *in vitro*. *Cancer Res.* **1983**, *43*, 716–720.

117. Abou-Khalil, S.; Abou-Khalil, W.H.; Planas, L.; Tapiero, H.; Lampidis, T.J. Interaction of rhodamine 123 with mitochondria isolated from drug-sensitive and -resistant Friend leukemia cells. *Biochem. Biophys. Res. Commun.* **1985**, *127*, 1039–1044.

118. Cheng, G.; Zielonka, J.; Dranka, B.P.; McAllister, D.; Mackinnon, A.C., Jr.; Joseph, J.; Kalyanaraman, B. Mitochondria-targeted drugs synergize with 2-deoxyglucose to trigger breast cancer cell death. *Cancer Res.* **2012**.

119. Schimmer, A.D.; Skrtic, M. Therapeutic potential of mitochondrial translation inhibition for treatment of acute myeloid leukemia. *Expert Rev. Hematol.* **2012**, *5*, 117–119.

120. Skrtic, M.; Sriskanthadevan, S.; Jhas, B.; Gebbia, M.; Wang, X.; Wang, Z.; Hurren, R.; Jitkova, Y.; Gronda, M.; Maclean, N.; *et al.* Inhibition of mitochondrial translation as a therapeutic strategy for human acute myeloid leukemia. *Cancer Cell* **2011**, *20*, 674–688.

Mitochondrial Transcription Factor A and Mitochondrial Genome as Molecular Targets for Cisplatin-Based Cancer Chemotherapy

Kimitoshi Kohno, Ke-Yong Wang, Mayu Takahashi, Tomoko Kurita, Yoichiro Yoshida, Masakazu Hirakawa, Yoshikazu Harada, Akihiro Kuma, Hiroto Izumi and Shinji Matsumoto

Abstract: Mitochondria are important cellular organelles that function as control centers of the energy supply for highly proliferative cancer cells and regulate apoptosis after cancer chemotherapy. Cisplatin is one of the most important chemotherapeutic agents and a key drug in therapeutic regimens for a broad range of solid tumors. Cisplatin may directly interact with mitochondria, which can induce apoptosis. The direct interactions between cisplatin and mitochondria may account for our understanding of the clinical activity of cisplatin and development of resistance. However, the basis for the roles of mitochondria under treatment with chemotherapy is poorly understood. In this review, we present novel aspects regarding the unique characteristics of the mitochondrial genome in relation to the use of platinum-based chemotherapy and describe our recent work demonstrating the importance of the mitochondrial transcription factor A (mtTFA) expression in cancer cells.

Reprinted from *Int. J. Mol. Sci.* Cite as: Kohno, K.; Wang, K.-Y.; Takahashi, M.; Kurita, T.; Yoshida, Y.; Hirakawa, M.; Harada, Y.; Kuma, A.; Izumi, H.; Matsumoto, S. Mitochondrial Transcription Factor A and Mitochondrial Genome as Molecular Targets for Cisplatin-Based Cancer Chemotherapy. *Int. J. Mol. Sci.* **2015**, *16*, 19836–19847.

1. Introduction

Mitochondria produce the energy required to maintain cellular functions as well as command the process of apoptosis in order to inhibit these functions [1,2]. Cancer cells are immortalized as a result of various genetic and epigenetic changes in the genome. One of the major causes of cancer is the acquisition of the driving force of cellular proliferation induced by a mutation and/or the methylation of oncogenes and tumor suppressor genes. Another factor is resistance to apoptosis, which is also acquired due to genetic or epigenetic changes in apoptosis-related genes. These processes are closely related to the onset of drug resistance and the phenotypic characteristics of cancer stem cells. However, it is unknown whether the direct cytotoxic effects or apoptosis-inducing ability of anti-cancer agents are effective as cancer chemotherapy. Cancer cells are resistant to conventional chemotherapy and radiotherapy regimens, and various molecular mechanisms underlying the onset of resistance to therapy have been proposed [3,4]. For example, cellular survival signal transduction systems protect mitochondrial integrity against drug-induced stress, and apoptosis-related molecules are orchestrated around mitochondria to induce and execute apoptosis in cancer cells. These molecules also play important roles in the development of drug resistance in cancer cells. It is therefore intriguing to explore how the crosstalk of apoptosis-related molecules regulates various mitochondrial functions leading to drug resistance. This review describes some of the putative mechanisms of action of cisplatin at the site of the mitochondrial genome and in terms of the mtTFA expression, which may play an important role in the cellular functions of cancer cells and the prognosis of cancer patients.

2. Cisplatin and DNA

Cisplatin, cisplatinum or *cis*-diamminedichloroplatinum (II) is a well-known chemotherapeutic drug [5,6]. Among various anti-cancer agents, cisplatin is one of the most effective and widely used anti-cancer agents for the treatment of several types of solid tumors [7]. Special attention has been paid to its molecular mechanisms of action [8]. The cytotoxic effects of cisplatin are thought to be mediated primarily by the generation of nuclear DNA adducts, which, if not repaired, cause cell death as a consequence of the inhibition of DNA replication and transcription. However, the ability of cisplatin to induce nuclear DNA (nDNA) damage per se is not sufficient to explain its high degree of effectiveness.

3. Calcium Ion, Endoplasmic Reticulum (ER) Stress, ROS/Redox System and Apoptosis

Bcl-2 family proteins control mitochondrial outer membrane permeability and regulate apoptosis [9,10]. Among Bcl-2 family proteins, anti-apoptotic proteins, Bcl-2 and Bcl-xL localize at the mitochondria outer membrane and thereby inhibit cytochrome c release. On the other hand, apoptosis inducing pro-apoptotic proteins, Bad, Bid and Bax, localize from cytoplasm to mitochondria following various apoptosis signals and promote cytochrome c release [11]. Calcium ion is well-known to play an important role in the mitochondrial control of apoptosis [12].

It has been reported that cisplatin induces calpain-mediated Bid cleavage in association with increased intracellular calcium ion levels and calpain-cleaved Bid induces cytochrome c release from mitochondria [13]. The endoplasmic reticulum (ER) is a specialized organelle for the folding and trafficking of proteins and a cellular target of cisplatin [14]. ER stress signaling is involved in cisplatin-induced apoptosis, as cisplatin has been shown to induce the calpain-dependent activation of ER-specific caspase leading to apoptosis [15]. ER stress is also linked to the production of reactive oxygen species (ROS). ROS/Redox signaling is prominently

463

associated with the progression of human malignancies [16]. Furthermore, ROS/Redox signaling is a critical stress response pathway associated with cancer chemotherapy. Oxidative damage is observed *in vivo* following exposure to cisplatin in several tissues, suggesting the role of oxidative stress in the pathogenesis of cisplatin-induced dose-limiting toxicities [17,18]. However, the mechanisms underlying the cisplatin-induced generation of ROS and their contribution to cisplatin cytotoxicity in normal and cancer cells remain poorly understood.

ROS generation occurs independently of the amount of cisplatin-induced nDNA damage and takes place in mitochondria as a consequence of the impairment of protein synthesis. The contribution of cisplatin-induced mitochondrial dysfunction in determining the cytotoxic effects of this drug varies among cells and depends on the mitochondrial redox status and integrity of mitochondrial DNA (mtDNA). Cisplatin-induced cellular pathways and responses are summarized in Figure 1.

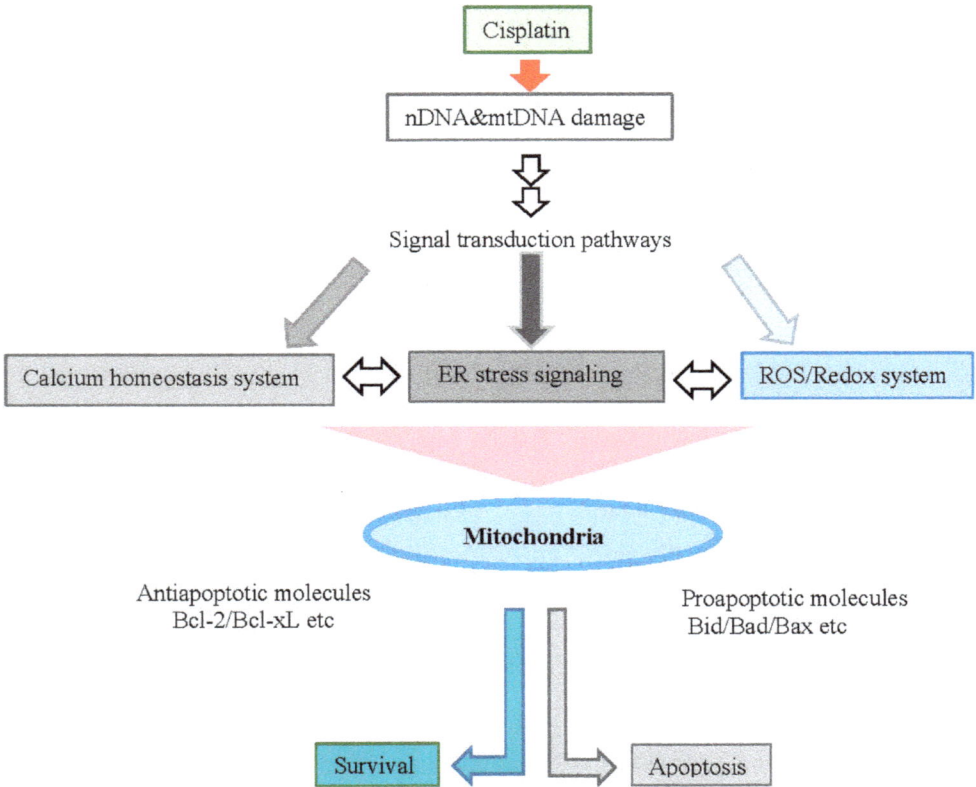

Figure 1. Cellular pathways and apoptosis induced by cisplatin. Cisplatin activates signal transduction pathways and may directly interact with mitochondria, which can induce apoptosis.

4. mtTFA and Cisplatin Resistance

The development of resistance to cisplatin is a major obstacle in terms of the clinical outcomes of cancer patients [19–21]. Several mechanisms are thought to be involved in the onset of cisplatin resistance, including decreased intracellular drug accumulation, increased levels of cellular thiols, an increased nucleotide excision-repair activity and a decreased mismatch-repair activity [22,23]. Two transporters, copper-transporting P-type ATPase 7B (ATP7B) and ATP-binding cassette, sub-family C, member 2 (ABCC2), may be involved in cisplatin efflux and resistance [22,23]. Genome-wide analyses have been shown to be a powerful method for understanding drug resistance. A recent report showed that the Son of sevenless/the mitogen-activated protein kinase/extracellular signal-regulated kinase (SOS/MAPK/ERK) pathway is activated in cisplatin-resistant cells [24]. This pathway mediates the degradation of the proapoptotic molecule Bim resulting in the inhibition of mitochondria-dependent apoptotic pathways [25].

One antioxidative factor, the level of cellular glutathione, has been shown to be involved in the development of cisplatin resistance. Interestingly, upregulation of the ABC transporter and cellular glutathione is a characteristic of cancer stem cells (CSCs) [26–28]. Therefore, CSCs show multidrug resistance and radioresistance. Recently, it has been demonstrated that the expression of CD44, especially variant isoforms (CD44v) among major CSC markers, contributes to ROS defenses via upregulation of the synthesis of reduced glutathione [29]. These observations suggest that CSCs exhibit natural drug-resistant phenotypes and that drug-induced resistant cells display acquired resistant phenotypes. These data also indicate that CSCs mitochondria might differ from those of non-CSCs. However, little is known about the mitochondrial features related to energy production and the ROS/Redox system of CSCs. Therefore, defining these features will be critical for developing mitochondria-targeted therapeutics [30].

In general, the molecules responsible for these phenotypes are upregulated in cisplatin-resistant cells, which indicates that the transcription factors activated in response to cisplatin may play crucial roles in the development of drug resistance [19,20]. Interestingly, we have found that mitochondrial transcription factor A (mtTFA) is upregulated in cisplatin-resistant cells [31]. To date, several nuclear transcription factors involved in the development of resistance against platinum-containing agents have been identified, including Y-box binding protein-1 (YB-1) [32], nuclear factor I/B [33], activating transcription factor 4 (ATF4) [34], zinc-finger factor 143 (ZNF143) [35], and Clock [36]. Several of these transcription factors are regulated by E-box binding transcription factors, which regulate the epithelial-mesenchymal transition [37]. Furthermore, both YB-1 and ZNF143 lack the high-mobility group (HMG) domain and are capable of binding preferentially to cisplatin-modified DNA in addition to HMG domain proteins, such as mtTFA. It has therefore been proposed that various mitochondrial functions may be regulated by the circadian clock system [38], which controls cancer cell proliferation and angio/stromagenesis via WNT signaling [39,40].

Mitochondrial dysfunction is associated not only with cancer progression, but also with chemoresistance [41], and a relationship of mtDNA mutations with cisplatin-induced apoptosis and/or cisplatin resistance has been reported. These mtDNA mutations endow cancer cells with chemoresistance [42]. It has also recently been reported that nuclear co-activators, including peroxisome proliferator-activated receptor gamma co-activator-1 (PGC-1), are upregulated and compensate for respiratory chain defects due to mtDNA mutations in cisplatin-resistant cells [43]. Interestingly, a frameshift mutation of the nuclear mtTFA gene has been reported in colorectal cancer cells with microsatellite instability [44]. These cancer cells express truncated mtTFA and show more resistance to cisplatin-induced apoptosis. These data also show that mitochondrial dysfunction due to genetic changes in both mtDNA and nDNA is closely associated with chemoresistance.

5. Mitochondrial DNA as a Target for Cisplatin

Mitochondria are closely related to carcinogenesis, cancer progression and chemoresistance. The genes encoded by mtDNA are tightly packed together with minimal noncoding regions. Although the number of mitochondria and mtDNA in cancer cells has not been extensively studied, a reduction in the mtDNA copy number has been reported in various human cancers [45]. Furthermore, it is well known that mtDNA is more susceptible than nDNA to damage from reactive oxygen species and chemicals, including anti-cancer agents, due to either a limited capacity for DNA repair or the presence of nucleosome-free structures [46,47].

Cisplatin is a major DNA-targeting agent and the most potent key drug for treating solid tumors among anti-cancer agents. It has been shown that the mitochondrial DNA adduct levels are higher than the nDNA adduct levels and that both a higher degree of initial binding and lack of removal of cisplatin-DNA adducts appear to contribute to the preferential formation of cisplatin-mtDNA adducts. The effects of cisplatin arise from its ability to damage DNA, with the major adducts formed being intrastrand d(GpG) crosslinks. Therefore, the target DNA sequence of cisplatin is G-stretch. Table 1 shows the number of G-stretch sequences in the mtDNA of various species in comparison with that observed in nDNA. As shown in Table 1, the number of G-stretch sequences is higher in primates than in other species. Furthermore, the number of G-stretch sequences in mtDNA is significantly higher than that noted in nDNA. These findings indicate that the number of G-stretch sequences may be related to the higher initial binding capacity of cisplatin to mtDNA.

Table 1. Number of cisplatin-targeted DNA sequences in mitochondrial DNA.

Species	Total Number of mtDNA		GG	GGG	GGGG	GGGGG
Human	16,565	L chain	426	73	15	4
		H chain	1772	624	224	69
		total	2198	697	239	73
Gorilla	16,364	L chain	425	71	16	5
		H chain	1712	596	216	72
		total	2137	667	232	77
Rat	16,300	L chain	397	63	14	4
		H chain	1299	377	99	32
		total	1696	440	113	36
Mouse	16,300	L chain	397	58	11	3
		H chain	1104	288	72	19
		total	1501	346	83	22
Xenopus	17,553	L chain	445	72	12	1
		H chain	1091	259	60	8
		total	1536	331	72	9
Drosophila	16,019	L chain	599	230	16	1
		H chain	770	358	78	30
		total	1369	588	94	31
Expectation		Double strand	2071	518	129	32
		Single strand	1035	259	65	16
Nuclear DNA		Double strand	1919 ± 319	501 ± 77	118 ± 25	35 ± 14

Expectation indicates the calculated number of each G-stretch sequence in the same number of nucleotide sequences of human mitochondria. Nuclear DNA indicates the average number of G-stretch sequences in human *YB-1*, *Sp-1* and *ZNF143* genes by way of example.

It has been demonstrated that the formation of DNA adducts is increased under acidic conditions [48]. In general, perturbation of the intracellular pH of highly proliferative cancer cells is observed [49]. If the pH around mtDNA is higher than that noted around nDNA, this finding may explain the higher initial binding capacity of cisplatin to mtDNA.

6. Cellular Functions of mtTFA

mtTFA is a 25-kDa protein encoded by a nuclear gene and imported to the mitochondria, where it is required for both the transcription and maintenance of mitochondrial DNA. mtTFA preferentially recognizes cisplatin-damaged DNA as well as oxidized DNA. Increased apoptosis is observed in mtTFA knockout animals, suggesting that mtTFA is involved in the process of apoptosis. However, the roles of mtTFA have not been extensively studied in cancer cells [50]. We recently reported the nuclear localization of mtTFA [51]. In addition, the proportion of nuclear-localized mtTFA varies among different cancer cells, and DNA microarray and chromatin immunoprecipitation assays have shown that mtTFA regulates the transcription of nuclear genes. Furthermore, the overexpression of mtTFA enhances the growth of cancer cell lines, whereas the downregulation of mtTFA inhibits the growth of these cells by regulating mtTFA target genes, such as baculoviral IAP repeat-containing 5 (BIRC5; also known as survivin) [52] and BCL2L [53]. Moreover, knockdown of the mtTFA expression induces p21-dependent G1 cell cycle arrest. These results imply that mtTFA functions in both nuclei and mitochondria to promote cell growth.

mtTFA is upregulated by treatment with cisplatin as well as the tumor suppressor p53. p53 is a multifunctional tumor suppressor protein that interacts with a variety of proteins with both positive and negative effects. Furthermore, mutation of the p53 gene is often observed in a wide variety of tumors. Mutated p53 gene products are stabilized and accumulate in the cytoplasm of cancer cells. A fraction of p53 proteins localize in the mitochondria at the onset of p53-dependent apoptosis, although not during p53-independent apoptosis [54]. Using immunochemical coprecipitation, we observed the binding of mtTFA with p53 [55]. The interaction between mtTFA and p53 requires the high mobility group-box B1 or high mobility group-box B2 of mtTFA and amino acids 363–376 of p53. In addition, the binding of mtTFA to cisplatin-modified DNA is significantly enhanced by p53, whereas binding to oxidized DNA is inhibited. Meanwhile, mtTFA preferentially recognizes cisplatin-damaged DNA, as well as oxidized DNA, and the conformational alteration induced by 8-oxo-dG in DNA differs from that induced by cisplatin. Therefore, the different effects of p53 on the binding of mtTFA may depend on structural differences in damaged DNA. Further analyses of the mechanisms by which mtTFA recognizes damaged DNA would be desirable for understanding the differential modulation of mtTFA binding by p53.

The interaction between mtTFA and p53 may be involved in transcriptional regulation, and, similar to transcription factors, both proteins have a DNA-binding domain. Our findings suggest that the interaction of p53 with mtTFA may play an important role in apoptosis. In addition, apoptosis is increased in cells lacking mtTFA [56]. These results suggest a model in which p53 inhibits the ability of mtTFA to execute cell death signaling. Hence, it is possible that manipulating the mtTFA function may be used as a mitochondria-targeted cancer treatment to enhance apoptosis.

It has previously been demonstrated that the expression levels of both mtTFA and mitochondrial antioxidant protein thioredoxin2 (TRX2) are upregulated in cisplatin-resistant cell lines [31]. In addition, TRX2 directly interacts with mtTFA and enhances its damaged DNA binding activity. These results suggest that TRX2 functions as an antioxidant as well as supporting the mtTFA function.

ZNF143 has been shown to be a cisplatin-inducible gene that regulates the mitochondrial ribosomal protein S11 [57]. Interestingly, there is one ZNF143 binding region in the promoter region of the mtTFA gene. The involvement of ZNF143 in cell growth and the protection of cells from oxidative damage, as well as the effects of cisplatin treatment, has recently been reported [35,58,59]. Furthermore, a strong ZNF143 expression was previously found to show a significant correlation with pathologically moderate to poor differentiation and highly invasive characteristics in 183 paraffin-embedded tumor samples from patients with lung adenocarcinoma [60]. Based on these reports, we next discuss the clinical implications of mtTFA. A schematic summary of the mutual relationship of transcription factors, including mtTFA, is shown in Figure 2 and the major factors described in this review are listed with detailed information in Table 2.

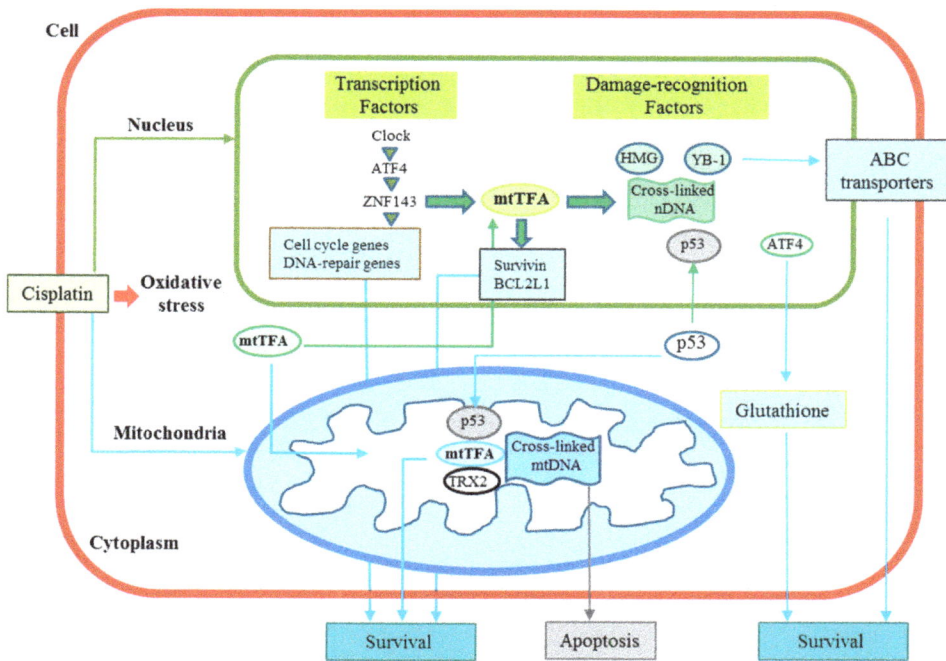

Figure 2. Cisplatin-induced signaling, cisplatin resistance and transcription system in cancer cells discussed in this review. mtTFA functions in both nuclei and mitochondria to not only interact with cisplatin-modified DNA, but also regulate the nuclear and mitochondrial gene expression.

Table 2. Selected transcription factors associated with drug resistance and target genes. "+" means that factors can recognize cisplatin-crosslinks or can render cells resistant.

Factors	Selected Target Genes	Selected Interacting Factors	Damage Recognition	Drug Resistance	Other Functions	References
mtTFA	*Survivin BcLL2*	p53, TRX2	+	+	Cell growth, anti-apoptosis	[54,55]
ZNF143	*mtTFA*	p73	+	+	Cell cycle, DNA repair	[35,57]
ATF4	*ZNF143*			+	Glutathione biosynthesis	[34,36]
Clock	*ATF4, Tip60*			+	Circadian rhythm, DNA repair	[36,38]
YB-1	*MDR1*	p53, PCNA, Topo1	+	+	Endothelial cell growth	[61–63]

7. Clinical Implications of the mtTFA Expression in Tumors

Mitochondrial transcription factor A (mtTFA) is necessary for both the transcription and maintenance of mitochondrial DNA (mtDNA). However, mtTFA is also localized in the nucleus and regulates nuclear genes. To date, the expression of mtTFA has not been thoroughly elucidated in the clinical setting. Studies of clinical specimens have recently investigated the relationships between clinicopathological factors, the prognosis and the immunohistochemical expression of mtTFA, as shown in Table 3.

Table 3. Associations of the mtTFA expression and malignant characteristics in human cancers. B cell lymphoma 2 like 1 (BCL2L1), Folinic acid plus 5-fluorouracil plus oxaliplatin (FOLFOX).

Tumor	No. of Cases	Malignant Characteristics	Reference
Serous ovarian	60	Poor prognosis, up-regulation of BCL2L1 expression	[64]
Pancreatic ductal adenocarcinoma	70	Poor prognosis, up-regulation of survivin expression	[65]
Colorectal	105	Poor prognosis	[66]
Metastatic colorectal	59	Poor clinical outcome with FOLFOX treatment	[67]
Endometrial	245	Invasion and metastasis, p53 mutation, poor prognosis	[68]

An immunohistochemical analysis of the mtTFA expression in 60 tissue samples of serous ovarian cancer showed 56.7% of the serous ovarian cancer patients to be positive for mtTFA, whereas 43.3% were negative [64]. A significant correlation was also reported between the nuclear mtTFA expression and the BCL2L1 expression in seven ovarian cancer cell lines as well as specimens of clinical ovarian cancer. Furthermore, a univariate survival analysis showed that the overall five-year survival rate is significantly worse for patients with mtTFA-positive cancer *versus* mtTFA-negative cancer.

The correlations between the mtTFA expression and the survivin index as well as a poor prognosis were recently assessed using 70 paraffin-embedded tumor samples from patients with surgically-resected pancreatic adenocarcinoma [65]. The results suggested that mtTFA is a prognostic factor for a poor outcome of human cancer and may function as an antiapoptotic factor, regulating target genes, such as BCL2L1 and survivin. In another study, clinical specimens from 105 colorectal patients were immunohistochemically stained using a polyclonal anti-mtTFA antibody [66]. Consequently, a total of 47 (44.8%) of the 105 patients with colorectal cancer were determined to have a positive mtTFA expression, and a positive expression of mtTFA was found to significantly correlate with both lymph node and distant metastasis in addition to an advanced TNM stage. Furthermore, the survival of the patients with a positive mtTFA expression was significantly worse than that of the patients with a negative mtTFA expression. Therefore, a positive mtTFA expression appears to be a useful marker of tumor progression and a poor prognosis in patients with colorectal cancer.

Whether the expression of mtTFA predicts the clinical outcomes of patients with metastatic colorectal cancer treated with modified 5-fluorouracil, leucovorin and oxaliplatin 6 (mFOLFOX6) was recently evaluated [67]. In that study, 59 patients with metastatic lesions of colorectal cancer treated with mFOLFOX6 were analyzed. As a result, a strong expression of mtTFA was detected in eight of 33 cases of a complete response/partial response (24.2%) and 18 of 26 cases of SD/PD (69.2%), indicating that the mtTFA expression significantly correlates with the response to chemotherapy ($p < 0.01$). These results suggest that immunohistochemical studies of mtTFA may be useful for predicting the clinical outcomes of metastatic colorectal cancer patients treated with FOLFOX.

The relationships between the immunohistochemical expression of mtTFA and various clinicopathological variables in 245 cases of endometrioid adenocarcinoma were also recently evaluated [68]. In that report, the mtTFA expression in the endometrioid adenocarcinomas was shown to be significantly associated with the surgical stage, myometrial invasion, lymphovascular space invasion, cervical invasion and lymph node metastasis. In addition, a correlation analysis between the mtTFA and p53 expression levels using the Pearson test showed a significant correlation, and a univariate survival analysis showed that the 10-year overall survival rate of the patients with mtTFA-positive endometrioid adenocarcinoma was significantly worse than that of the patients with mtTFA-negative endometrioid adenocarcinoma. Therefore, a positive mtTFA expression may be a useful marker of tumor progression and a poor prognosis in patients with endometrioid adenocarcinoma.

8. Conclusions and Perspectives

Recent studies have demonstrated the important role of mitochondria in cancer biology. This review focused on the mitochondrial genome and transcription factor A. Since mtTFA proteins have been shown to be highly expressed in cancer and drug-resistant cells compared to normal cells, and the mtTFA expression is upregulated by signals of oxidative and DNA damage stress, this protein may potentially serve as a promising target in cancer chemotherapy. Promising strategies include inhibition of the expression and/or function of mtTFA specifically in cancer cells. Small interference RNA can inhibit specific gene expression levels. mtTFA functions with other transcription factors such as mtTFB1 and B2. Inhibiting the interaction with mtTFBs is expected to be a good strategy using peptides, and both RNA and peptide drugs are good candidates using cancer-specific drug delivery systems [69,70]. Furthermore, the mtTFA expression and mitochondrial genome possess attractive characteristics for platinum-based chemotherapy, and the mtTFA expression reflects a poor prognosis in patients with solid cancers. These findings indicate that further research on mitochondria may provide novel and unique therapeutic interventions for overcoming cancer.

Acknowledgments: This work was partly supported by Grants-in-Aids for Scientific Research from the Ministry of Education, Culture, Sports, Science and Technology of Japan (No. 24501323).

Author Contributions: All authors participated in developing the ideas presented in this manuscript, researching the literature and writing parts of the text. Kimitoshi Kohno and Ke-Yong Wang performed final editing of both text and images.

Conflicts of Interest: The authors declare no conflicts of interest.

References

1.   Wallace, D.C. Mitochondria and cancer. *Nat. Rev. Cancer* **2012**, *12*, 685–698.
2.   Guha, M.; Avadhani, N.G. Mitochondrial retrograde signaling at the crossroads of tumor bioenergetics, genetics and epigenetics. *Mitochondrion* **2013**, *13*, 577–591.
3.   Igney, F.H.; Krammer, P.H. Death and anti-death: Tumour resistance to apoptosis. *Nat. Rev. Cancer* **2002**, *2*, 277–288.
4.   Holohan, C.; van Schaeybroeck, S.; Longley, D.B.; Johnston, P.G. Cancer drug resistance: An evolving paradigm. *Nat. Rev. Cancer* **2013**, *13*, 714–726.
5.   Cohen, S.M.; Lippard, S.J. Cisplatin: From DNA damage to cancer chemotherapy. *Prog. Nucl. Acid Res. Mol. Biol.* **2001**, *67*, 93–130.
6.   Wang, D.; Lippard, S.J. Cellular processing of platinum anticancer drugs. *Nat. Rev. Drug Discov.* **2005**, *4*, 307–320.
7.   Kelland, L. The resurgence of platinum-based cancer chemotherapy. *Nat. Rev. Cancer* **2007**, *7*, 573–584.
8.   Zamble, D.B.; Lippard, S.J. Cisplatin and DNA repair in cancer chemotherapy. *Trends Biochem. Sci.* **1995**, *20*, 435–439.
9.   Correia, C.; Lee, S.H.; Meng, X.W.; Vincelette, N.D.; Knorr, K.L.; Ding, H.; Nowakowski, G.S.; Dai, H.; Kaufmann, S.H. Emerging understanding of Bcl-2 biology: Implications for neoplastic progression and treatment. *Biochim. Biophys. Acta* **2015**, *1853*, 1658–1671.

10. Frenzel, A.; Grespi, F.; Chmelewskij, W.; Villunger, A. Bcl2 family proteins in carcinogenesis and the treatment of cancer. *Apoptosis* **2009**, *14*, 584–596.

11. Lopez, J.; Tait, S.W. Mitochondrial apoptosis: Killing cancer using the enemy within. *Br. J. Cancer* **2015**, *112*, 957–962.

12. Orrenius, S.; Gogvadze, V.; Zhivotovsky, B. Calcium and mitochondria in the regulation of cell death. *Biochem. Biophys. Res. Commun.* **2015**, *460*, 72–81.

13. Mandic, A.; Viktorsson, K.; Strandberg, L.; Heiden, T.; Hansson, J.; Linder, S.; Shoshan, M.C. Calpain-mediated Bid cleavage and calpain-independent Bak modulation: Two separate pathways in cisplatin-induced apoptosis. *Mol. Cell. Biol.* **2002**, *22*, 3003–3013.

14. Linder, S.; Shoshan, M.C. Lysosomes and endoplasmic reticulum: Targets for improved, selective anticancer therapy. *Drug Resist. Updat.* **2005**, *8*, 199–204.

15. Xu, Y.; Wang, C.; Li, Z. A new strategy of promoting cisplatin chemotherapeutic efficiency by targeting endoplasmic reticulum stress. *Mol. Clin. Oncol.* **2014**, *2*, 3–7.

16. Simran, S.S.; Paul, T.S. Mitochondrial ROS in cancer: Initiators, amplifiers or an Achilles' heel? *Nat. Rev. Cancer* **2014**, *14*, 709–721.

17. Chirino, Y.I.; Pedraza, C.J. Role of oxidative and nitrosative stress in cisplatin-induced nephrotoxicity. *Exp. Toxicol. Pathol.* **2009**, *61*, 223–242.

18. Rybak, L.P.; Whitworth, C.A.; Mukherjea, D.; Ramkumar, V. Mechanisms of cisplatin-induced ototoxicity and prevention. *Hear. Res.* **2007**, *226*, 157–167.

19. Torigoe, T.; Izumi, H.; Ishiguchi, H.; Yoshida, Y.; Tanabe, M.; Yoshida, T.; Igarashi, T.; Niina, I.; Wakasugi, T.; Imaizumi, T.; *et al.* Cisplatin resistance and transcription factors. *Curr. Med. Chem. Anticancer Agents* **2005**, *5*, 15–27.

20. Kohno, K.; Uchiumi, T.; Niina, I.; Wakasugi, T.; Igarashi, T.; Momii, Y.; Yoshida, T.; Matsuo, K.; Miyamoto, N.; Izumi, H. Transcription factors and drug resistance. *Eur. J. Cancer* **2005**, *41*, 2577–2586.

21. Galluzzi, L.; Vitale, I.; Michels, J.; Brenner, C.; Szabadkai, G.; Harel-Bellan, A.; Castedo, M.; Kroemer, G. Systems biology of cisplatin resistance: Past, present and future. *Cell Death Dis.* **2014**, *5*, e1257.

22. Shen, D.W.; Pouliot, L.M.; Hall, M.D.; Gottesman, M.M. Cisplatin resistance: A cellular self-defense mechanism resulting from multiple epigenetic and genetic changes. *Pharmacol. Rev.* **2012**, *64*, 706–721.

23. Galluzzi, L.; Senovilla, L.; Vitale, I.; Michels, J.; Martins, I.; Kepp, O.; Castedo, M.; Kroemer, G. Molecular mechanisms of cisplatin resistance. *Oncogene* **2012**, *31*, 1869–1883.

24. Kong, L.R.; Chua, K.N.; Sim, W.J.; Ng, H.C.; Bi, C.; Ho, J.; Nga, M.E.; Pang, Y.H.; Ong, W.R.; Soo, R.A.; *et al.* MEK Inhibition overcomes cisplatin resistance conferred by SOS/MAPK pathway activation in squamous cell carcinoma. *Mol. Cancer Ther.* **2015**, *14*, 1750–1760.

25. Crawford, N.; Chacko, A.D.; Savage, K.I.; McCoy, F.; Redmond, K.; Longley, D.B.; Fennell, D.A. Platinum resistant cancer cells conserve sensitivity to BH3 domains and obatoclax induced mitochondrial apoptosis. *Apotosis* **2011**, *16*, 311–320.

26. Nguyen, L.V.; Vanner, R.; Dirks, P.; Eaves, C.J. Cancer stem cells: An evolving concept. *Nat. Rev. Cancer* **2012**, *12*, 133–143.

27. Visvader, J.E.; Lindeman, G.J. Cancer stem cells in solid tumours: Accumulating evidence and unresolved questions. *Nat. Rev. Cancer* **2008**, *8*, 755–768.

28. Dean, M.; Fojo, T.; Bates, S. Tumour stem cells and drug resistance. *Nat. Rev. Cancer* **2005**, *5*, 275–284.

29. Nagano, O.; Okazaki, S.; Saya, H. Redox regulation in stem-like cancer cells by CD44 variant isoforms. *Oncogene* **2013**, *32*, 5191–5198.

30. Song, I.S.; Jeong, J.Y.; Jeong, S.H.; Kim, H.K.; Ko, K.S.; Rhee, B.D.; Kim, N.; Han, J. Mitochondria as therapeutic targets for cancer stem cells. *World J. Stem Cells* **2015**, *7*, 418–427.

31. Kidani, A.; Izumi, H.; Yoshida, Y.; Kashiwagi, E.; Ohmori, H.; Tanaka, T.; Kuwano, M.; Kohno, K. Thioredoxin2 enhances the damaged DNA binding activity of mtTFA through direct interaction. *Int. J. Oncol.* **2009**, *35*, 1435–1440.

32. Kohno, K.; Izumi, H.; Uchiumi, T.; Ashizuka, M.; Kuwano, M. The pleiotropic functions of the Y-box-binding protein, YB-1. *Bioessays* **2003**, *25*, 691–698.

33. Kashiwagi, E.; Izumi, H.; Yasuniwa, Y.; Baba, R.; Doi, Y.; Kidani, A.; Arao, T.; Nishio, K.; Naito, S.; Kohno, K. Enhanced expression of nuclear factor I/B in oxaliplatin-resistant human cancer cell lines. *Cancer Sci.* **2011**, *102*, 382–386.

34. Tanabe, M.; Izumi, H.; Ise, T.; Higuchi, S.; Yamori, T.; Yasumoto, K.; Kohno, K. Activating transcription factor 4 increases the cisplatin resistance of human cancer cell lines. *Cancer Res.* **2003**, *63*, 8592–8595.

35. Wakasugi, T.; Izumi, H.; Uchiumi, T.; Suzuki, H.; Arao, T.; Nishio, K.; Kohno, K. ZNF143 interacts with p73 and is involved in cisplatin resistance through the transcriptional regulation of DNA repair genes. *Oncogene* **2007**, *26*, 5194–5203.

36. Igarashi, T.; Izumi, H.; Uchiumi, T.; Nishio, K.; Arao, T.; Tanabe, M.; Uramoto, H.; Sugio, K.; Yasumoto, K.; Sasaguri, Y.; *et al.* Clock and ATF4 transcription system regulates drug resistance in human cancer cell lines. *Oncogene* **2007**, *26*, 4749–4760.

37. Lamouille, S.; Xu, J.; Derynck, R. Molecular mechanisms of epithelial-mesenchymal transition. *Nat. Rev. Mol. Cell Biol.* **2014**, *15*, 178–196.

38. Bailey, S.M.; Udoh, U.S.; Young, M.E. Circadian regulation of metabolism. *J. Endocrinol.* **2014**, *222*, R75–R96.

39. Yasuniwa, Y.; Izumi, H.; Wang, K.Y.; Shimajiri, S.; Sasaguri, Y.; Kawai, K.; Kasai, H.; Shimada, T.; Miyake, K.; Kashiwagi, E.; *et al.* Circadian disruption accelerates tumor growth and angio/stromagenesis through a Wnt signaling pathway. *PLoS ONE* **2010**, *5*, e15330.

40. Izumi, H.; Wang, K.Y.; Morimoto, Y.; Sasaguri, Y.; Kohno, K. Circadian disruption and cancer risk: A new concept of stromal niche (review). *Int. J. Oncol.* **2014**, *44*, 364–370.

41. Guaragnella, N.; Giannattasio, S.; Moro, L. Mitochondrial dysfunction in cancer chemoresistance. *Biochem. Pharmacol.* **2014**, *92*, 62–72.

42. Mizutani, S.; Miyato, Y.; Shidara, Y.; Asoh, S.; Tokunaga, A.; Tajiri, T.; Ohta, S. Mutations in the mitochondrial genome confer resistance of cancer cells to anticancer drugs. *Cancer Sci.* **2009**, *100*, 1680–1687.

43. Yao, Z.; Jones, A.W.; Fassone, E.; Sweeney, M.G.; Lebiedzinska, M.; Suski, J.M.; Wieckowski, M.R.; Tajeddine, N.; Hargreaves, I.P.; Yasukawa, T.; *et al.* PGC-1β mediates adaptive chemoresistance associated with mitochondrial DNA mutations. *Oncogene* **2013**, *32*, 2592–2600.

44. Guo, J.; Zheng, L.; Liu, W.; Wang, X.; Wang, Z.; Wang, Z.; French, A.J.; Kang, D.; Chen, L.; Thibodeau, S.N.; *et al.* Frequent truncating mutation of TFAM induces mitochondrial DNA depletion and apoptotic resistance in microsatellite-unstable colorectal cancer. *Cancer Res.* **2011**, *71*, 2978–2987.

45. Campbell, C.T.; Kolesar, J.E.; Kaufman, B.A. Mitochondrial transcription factor A regulates mitochondrial transcription initiation, DNA packaging, and genome copy number. *Biochim. Biophys. Acta* **2012**, *1819*, 921–929.

46. Olivero, O.A.; Semino, C.; Kassim, A.; Lopez-Larraza, D.M.; Poirier, M.C. Preferential binding of cisplatin to mitochondrial DNA of Chinese hamster ovary cells. *Mutat. Res.* **1995**, *346*, 221–230.

47. See comment in PubMed Commons below Olivero, O.A.; Chang, P.K.; Lopez-Larraza, D.M.; Semino-Mora, M.C.; Poirier, M.C. Preferential formation and decreased removal of cisplatin-DNA adducts in Chinese hamster ovary cell mitochondrial DNA as compared to nuclear DNA. *Mutat. Res.* **1997**, *391*, 79–86.

48. Murakami, T.; Shibuya, I.; Ise, T.; Chen, Z.S.; Akiyama, S.; Nakagawa, M.; Izumi, H.; Nakamura, T.; Matsuo, K.; Yamada, Y.; *et al.* Elevated expression of vacuolar proton pump genes and cellular pH in cisplatin resistance. *Int. J. Cancer* **2001**, *93*, 869–874.

49. Reshkin, S.J.; Greco, M.R.; Cardone, R.A. Role of pHi, and proton transporters in oncogene-driven neoplastic transformation. *Philos. Trans. R. Soc. Lond. B* **2014**, *369*, 20130100.

50. Yoshida, Y.; Hoshino, S.; Izumi, H.; Kohno, K.; Yamashita, Y. New roles of mitochondrial transcription factor A in cancer. *J. Phys. Chem. Biophys.* **2011**, *1*, 101.

51. Han, B.; Izumi, H.; Yasuniwa, Y.; Akiyama, M.; Yamaguchi, T.; Fujimoto, N.; Matsumoto, T.; Wu, B.; Tanimoto, A.; Sasaguri, Y.; *et al.* Human mitochondrial transcription factor A functions in both nuclei and mitochondria and regulates cancer cell growth. *Biochem. Biophys. Res. Commun.* **2011**, *408*, 45–51.

52. Altieri, D.C. Survivin, cancer networks and pathway-directed drug discovery. *Nat. Rev. Cancer* **2008**, *8*, 61–70.

53. Cory, S.; Adams, J.M. The Bcl2 family: Regulators of the cellular life-or-death switch. *Nat. Rev. Cancer* **2002**, *2*, 647–656.

54. Vaseva, A.V.; Moll, U.M. The mitochondrial p53 pathway. *Biochim. Biophys. Acta* **2009**, *1787*, 414–420.

55. Yoshida, Y.; Izumi, H.; Torigoe, T.; Ishiguchi, H.; Itoh, H.; Kang, D.; Kohno, K. P53 physically interacts with mitochondrial transcription factor A and differentially regulates binding to damaged DNA. *Cancer Res.* **2003**, *63*, 3729–3734.

56. Wang, J.; Silva, J.P.; Gustafsson, C.M.; Rustin, P.; Larsson, N.G. Increased *in vivo* apoptosis in cells lacking mitochondrial DNA gene expression. *Proc. Natl Acad. Sci. USA* **2001**, *98*, 4038–4043.

57. Ishiguchi, H.; Izumi, H.; Torigoe, T.; Yoshida, Y.; Kubota, H.; Tsuji, S.; Kohno, K. ZNF143 activates gene expression in response to DNA damage and binds to cisplatin-modified DNA. *Int. J. Cancer* **2004**, *111*, 900–909.

58. Izumi, H.; Yasuniwa, Y.; Akiyama, M.; Yamaguchi, T.; Kuma, A.; Kitamura, N.; Kohno, K. Forced expression of ZNF143 restrains cancer cell growth. *Cancers* **2011**, *3*, 3909–3920.

59. Izumi, H.; Wakasugi, T.; Shimajiri, S.; Tanimoto, A.; Sasaguri, Y.; Kashiwagi, E.; Yasuniwa, Y.; Akiyama, M.; Han, B.; Wu, Y.; *et al.* Role of ZNF143 in tumor growth through transcriptional regulation of DNA replication and cell-cycle-associated genes. *Cancer Sci.* **2010**, *101*, 2538–2545.

60. Kawatsu, Y.; Kitada, S.; Uramoto, H.; Li, Z.; Takeda, T.; Kimura, T.; Horie, S.; Tanaka, F.; Sasaguri, Y.; Izumi, H.; *et al.* The combination of strong expression of ZNF143 and high MIB-1 labelling index independently predicts shorter disease-specific survival in lung adenocarcinoma. *Br. J. Cancer.* **2014**, *110*, 2583–2592.

61. Takahashi, M.; Shimajiri, S.; Izumi, H.; Hirano, G.; Kashiwagi, E.; Yasuniwa, Y.; Wu, Y.; Han, B.; Akiyama, M.; Nishizawa, S.; *et al.* Y-box binding protein-1 is a novel molecular target for tumor vessels. *Cancer Sci.* **2010**, *101*, 1367–1373.

62. Wu, Y.; Wang, K.Y.; Li, Z.; Liu, Y.P.; Izumi, H.; Uramoto, H.; Nakayama, Y.; Ito, K.I.; Kohno, K. Y-box binding protein 1 enhances DNA topoisomerase 1 activity and sensitivity to camptothecin via direct interaction. *J. Exp. Clin. Cancer Res.* **2014**, *33*, 112–117.

63. Wu, Y.; Wang, K.Y.; Li, Z.; Liu, Y.P.; Izumi, H.; Yamada, S.; Uramoto, H.; Nakayama, Y.; Ito, K.I.; Kohno, K. Y-box binding protein 1 expression in gastric cancer subtypes and association with cancer neovasculature. *Clin. Transl. Oncol.* **2015**, *17*, 152–159.

64. Kurita, T.; Izumi, H.; Kagami, S.; Kawagoe, T.; Toki, N.; Matsuura, Y.; Hachisuga, T.; Kohno, K. Mitochondrial transcription factor A regulates *BCL2L1* gene expression and is a prognostic factor in serous ovarian cancer. *Cancer Sci.* **2012**, *103*, 239–444.

65. Kimura, T.; Kitada, S.; Uramoto, H.; Li, Z.; Kawatsu, Y.; Takeda, T.; Horie, S.; Nabeshima, A.; Noguchi, H.; Sasaguri, Y.; *et al.* The combination of strong immunohistochemical mtTFA expression and a high survivin index predicts a shorter disease-specific survival in pancreatic ductal adenocarcinoma. *Histol. Histopathol.* **2015**, *30*, 193–204.

66. Nakayama, Y.; Yamauchi, M.; Minagawa, N.; Torigoe, T.; Izumi, H.; Kohno, K.; Yamaguchi, K. Clinical significance of mitochondrial transcription factor A expression in patients with colorectal cancer. *Oncol. Rep.* **2012**, *27*, 1325–1330.

67. Yoshida, Y.; Hasegawa, J.; Nezu, R.; Kim, Y.K.; Hirota, M.; Kawano, K.; Izumi, H.; Kohno, K. Clinical usefulness of mitochondrial transcription factor A expression as a predictive marker in colorectal cancer patients treated with FOLFOX. *Cancer Sci.* **2011**, *102*, 578–582.

68. Toki, N.; Kagami, S.; Kurita, T.; Kawagoe, T.; Matsuura, Y.; Hachisuga, T.; Matsuyama, A.; Hashimoto, H.; Izumi, H.; Kohno, K. Expression of mitochondrial transcription factor A in endometrial carcinomas: Clinicopathologic correlations and prognostic significance. *Virchows Arch.* **2010**, *456*, 387–393.

69. Burnett, J.C.; Rossi, J.J. RNA-based therapeutics: Current progress and future prospects. *Chem. Biol.* **2012**, *19*, 60–71.

70. Pirogova, E.; Istivan, T.; Gan, E.; Cosic, I. Advances in methods for therapeutic peptide discovery, design and development. *Curr. Pharm. Biotechnol.* **2011**, *12*, 1117–1127.

Mitochondria-Derived Reactive Oxygen Species Play an Important Role in Doxorubicin-Induced Platelet Apoptosis

Zhicheng Wang, Jie Wang, Rufeng Xie, Ruilai Liu and Yuan Lu

Abstract: Doxorubicin (DOX) is an effective chemotherapeutic agent; however; its use is limited by some side effects; such as cardiotoxicity and thrombocytopenia. DOX-induced cardiotoxicity has been intensively investigated; however; DOX-induced thrombocytopenia has not been clearly elucidated. Here we show that DOX-induced mitochondria-mediated intrinsic apoptosis and glycoprotein (GP)Ibα shedding in platelets. DOX did not induce platelet activation; whereas; DOX obviously reduced adenosine diphosphate (ADP)- and thrombin-induced platelet aggregation; and impaired platelet adhesion on the von Willebrand factor (vWF) surface. In addition; we also show that DOX induced intracellular reactive oxygen species (ROS) production and mitochondrial ROS generation in a dose-dependent manner. The mitochondria-targeted ROS scavenger Mito-TEMPO blocked intracellular ROS and mitochondrial ROS generation. Furthermore; Mito-TEMPO reduced DOX-induced platelet apoptosis and GPIbα shedding. These data indicate that DOX induces platelet apoptosis; and impairs platelet function. Mitochondrial ROS play a pivotal role in DOX-induced platelet apoptosis and GPIbα shedding. Therefore; DOX-induced platelet apoptosis might contribute to DOX-triggered thrombocytopenia; and mitochondria-targeted ROS scavenger would have potential clinical utility in platelet-associated disorders involving mitochondrial oxidative damage.

Reprinted from *Int. J. Mol. Sci.* Cite as: Wang, Z.; Wang, J.; Xie, R.; Liu, R.; Lu, Y. Mitochondria-Derived Reactive Oxygen Species Play an Important Role in Doxorubicin-Induced Platelet Apoptosis. *Int. J. Mol. Sci.* **2015**, *16*, 11087–11099.

1. Introduction

Doxorubicin (DOX) has been used for the treatment of solid tumors and hematologic malignancy. However, DOX therapy has some side effects, such as thrombocytopenia [1,2]. Up to now, the pathogenesis of DOX-induced thrombocytopenia is not completely understood. Recently, several studies have reported that DOX can induce platelet cytotoxicity and procoagulant activity [2,3]. It has been generally accepted the anticancer effects of DOX via inducing apoptosis of malignant cell [4,5]. Platelet apoptosis induced by either physiological or chemical compounds occurs widely *in vitro* or *in vivo* [6–10], which might play important

478

roles in controlling the number of circulating platelets or in the development of platelet-related diseases. Accumulating evidences indicate that platelet apoptosis might play a key role in chemotherapeutic agents induced-thrombocytopenia [8,9].

DOX localizes to the mitochondria and is highly susceptible to enzymatic reduction to generate ROS, which can cause mitochondrial swelling and ultrastructural changes and alter mitochondrial function [11]. Recently, most studies supported that the major mechanism of DOX-induced apoptosis was related to excessive generation of intracellular ROS [11–16]. Mitochondria are considered the main intracellular source of ROS [17]. ROS are produced at very low levels during mitochondrial respiration under normal physiological conditions. The formation of ROS occurs when unpaired electrons escape the electron transport chain and react with molecular oxygen, generating ROS. Complexes I, II, and III of the electron transport chain are the major potential loci for ROS generation [18,19]. Recently, several studies reported that NADPH oxidase 4 (NOX4) localizes to mitochondria, and NOX4 is a novel source of ROS produced in the mitochondria [20,21]. ROS degradation is performed by endogenous enzymatic antioxidants, such as superoxide dismutase, catalase, and non-enzymatic antioxidants, such as glutathione, ascorbic acid [17]. Under physiological conditions, ROS are maintained at proper levels by a balance between its synthesis and its elimination. An increase in ROS generation, a decrease in antioxidant capacity, or a combination both will lead to oxidative stress [17].

In recent years, mitochondria-targeted ROS antagonists and mitochondrial ROS detection probes have been developed. Thus, with the advent of such tools, the importance of mitochondrial ROS in cell signaling, proliferation and apoptosis gradually attracted much attention. For example, Cheung *et al.* [12] recently reported that SIRT3 prevents DOX-induced mitochondrial ROS production in H9c2 cardiomyocytes. Increased mitochondrial ROS is a significant contributor to the development of DOX-induced myopathy in both cardiac and skeletal muscle fibers [13].

We recently reported that mitochondrial ROS play important roles in hyperthermia-induced platelets [7]. In the present study, using mitochondria-targeted ROS scavenger and mitochondrial ROS detection probe, we explored whether DOX induces mitochondrial ROS production, and whether mitochondria-targeted ROS scavenger has a protective effect on DOX-induced platelet apoptosis.

2. Results

2.1. Doxorubicin (DOX) Dose-Dependently Induces $\Delta\Psi m$ Depolarization and Phosphatidylserine (PS) Exposure in Platelets

In order to investigate whether DOX could induce platelet apoptosis, platelets were incubated with different concentrations of DOX. The effect of DOX on platelet ΔΨm depolarization and PS exposure is analyzed by flow cytometry. We found that DOX dose-dependently induced ΔΨm depolarization and PS exposure (Figure 1A,B). In order to investigate the effect of incubation time on apoptosis, platelets were incubated with DOX for different times. The data indicate that DOX time-dependently induced ΔΨm depolarization and PS exposure (Figure 1C,D). Therefore, in order to obtain obvious apoptotic events, 3 h incubation was selected for the following experiments.

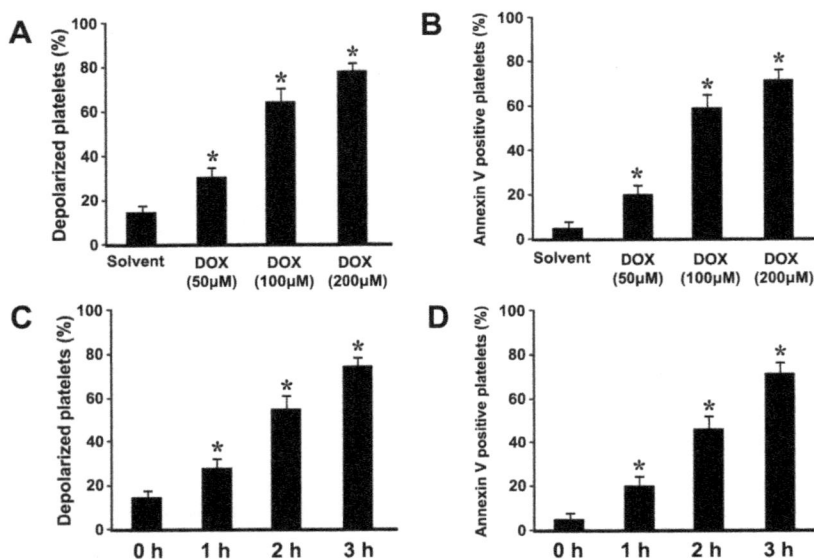

Figure 1. Doxorubicin (DOX) induced mitochondrial inner transmembrane potential (ΔΨm) depolarization and PS exposure. (**A–D**) Platelets were incubated with different concentrations of DOX or solvent control (**A,B**), or incubated with DOX (200 μM) at 37 °C for different times (**C,D**). Treated platelets were incubated with tetramethylrhodamine ethyl ester (TMRE) (**A,C**), or fluorescein isothiocyanate (FITC)-conjugated annexin V (**B,D**), and analyzed by flow cytometry. ΔΨm depolarization was quantified as the percentage of depolarized platelets. Means ± SEM from three independent experiments are shown (**A,C**). PS exposure was quantified as the percentage of PS positive platelets. Means ± SEM of the percentage of PS positive platelets from three independent experiments are shown (**B,D**). * $p < 0.017$ (after a Bonferroni correction) compared with solvent control.

2.2. DOX Dose-Dependently Induces Mitochondrial Translocation of Bax, Cytochrome C Release, and Caspase-3 Activation in Platelets

Pro-apoptotic protein Bax translocation to the mitochondria is a key event that regulates the release of apoptogenic factors like cytochrome C from the mitochondria, which leads to activation of caspases such as executioner caspase-3 [22]. Thus, to further explore whether DOX could induce mitochondrial translocation of Bax and cytochrome C release, platelets were incubated with different concentrations of DOX and subjected to isolation and analysis of cytosolic and mitochondrial fractions. We found that DOX dose-dependently promoted mitochondrial translocation of Bax and cytochrome C release (Figure 2A,B). Meanwhile, caspase-3 activation was examined in DOX-treated platelets. Compared with the control, the 17-kDa caspase-3 fragment, which indicated the activation of caspase-3, dose-dependently increased in platelets treated with DOX (Figure 2C). Taken together, these data suggested that DOX induced apoptotic cascades leading to platelet apoptosis.

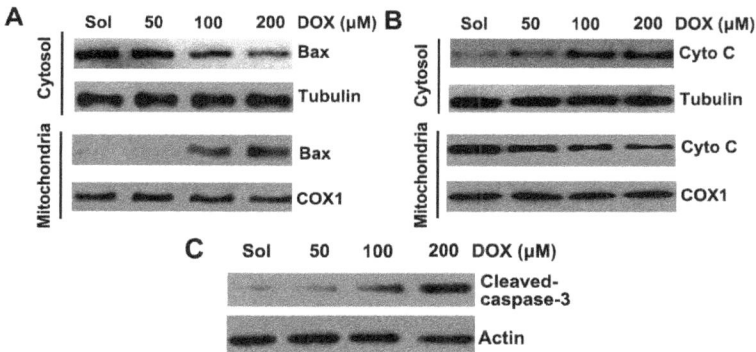

Figure 2. DOX induced mitochondrial translocation of Bax, cytochrome C release, and caspase-3 activation. (**A,B**) Platelets were incubated with different concentrations of DOX or solvent. Treated platelets were lysed, and cytosol and mitochondrial fractions were isolated and analyzed by Western blot with anti-Bax (**A**), and anti-cytochrome C antibodies (**B**), COX1 and tubulin were used as internal controls; (**C**) Platelets were incubated with different concentrations of DOX or solvent. Treated platelets were lysed and analyzed by Western blot with anti-caspase-3 antibody. Actin levels were assayed to demonstrate equal protein loading. Representative data of three independent experiments are presented. Cytochrome C is labeled as Cyto C.

2.3. DOX Impairs Platelet Function

Platelets play a central role in maintaining integrity of endothelium and biological hemostasis. To investigate the effect of DOX on platelet function, platelets were treated with different concentrations of DOX or solvent control, and then examined for platelet aggregation and adhesion. ADP- and thrombin-induced platelet aggregations were reduced in DOX-treated platelets in a dose-dependent manner (Figure 3A,B). Furthermore, compared with solvent control, DOX-treated platelets displayed a significant decrease in adhering on the vWF surface in dose-dependent manner (Figure 3C). Taken together, these data indicate that platelet functions are impaired by DOX.

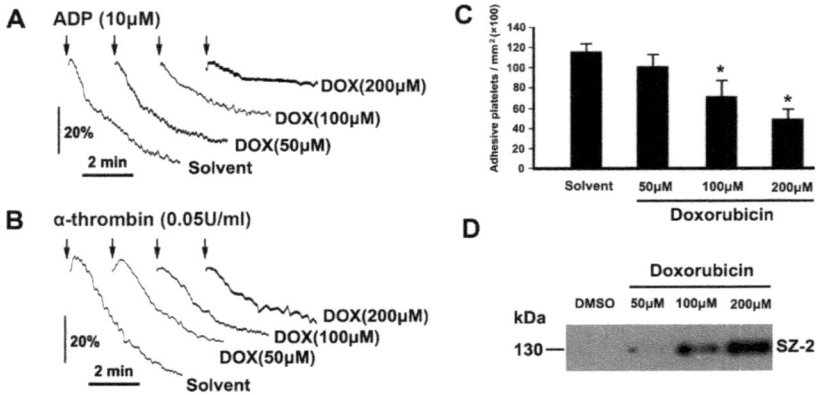

Figure 3. DOX impaired platelet function. (**A,B**) PRP or washed platelets were incubated with different concentrations of DOX or solvent. Platelet aggregation was induced by addition of ADP (**A**) or thrombin (**B**); representative traces from three independent experiments are shown; (**C**) Platelets were incubated with different concentrations of DOX or solvent. Treated platelets were perfused into vWF-coated glass capillary. The results from three independent experiments are shown as the means $\pm$ SEM of cell number/mm^2. * $p < 0.017$ (after Bonferroni correction) as compared with solvent; (**D**) Platelets were incubated with different concentrations of DOX or solvent. Treated platelets were centrifuged, and the supernatants were analyzed by Western blot with SZ-2. Representative data of three independent experiments are presented.

The interaction of GPIbα with vWF at sites of injured blood vessel walls initiates platelet adhesion under flow conditions [23]. GPIbα shedding is a physiological regulatory mechanism leading to platelet dysfunction [23]. In order to investigate whether GPIbα shedding is involved in DOX-induced platelet dysfunction, GPIbα shedding was examined in platelets incubated with DOX. We found that glycocalicin, which is a cleaved production of GPIbα, gradually increased with increasing concentration of DOX (Figure 3D).

2.4. DOX Dose-Dependently Increases Intracellular ROS and Mitochondrial ROS Production in Platelets

In order to investigate whether DOX augments intracellular ROS levels in platelets, we determined platelet ROS levels using DCFDA. As shown in Figure 4A, DOX dose-dependently induced ROS production. As a positive control, A23187

482

significantly induces intracellular ROS production (Figure 4A). Several potential sources of ROS have been suggested, including the mitochondria and NADPH oxidase. Several reports support a role for NADPH oxidase in DOX-induced nuclear cell apoptosis [15,16]. In order to investigate the sources of ROS in DOX-treated anuclear platelets, apocynin, which is an inhibitor of NADPH oxidase, and Mito-TEMPO, which is a mitochondria-targeted ROS antagonist, were used. We found that DOX-induced ROS production was partly inhibited by apocynin, and was obviously inhibited by Mito-TEMPO (Figure 4B). These data demonstrate that mitochondria are a major source of ROS in DOX-treated platelets.

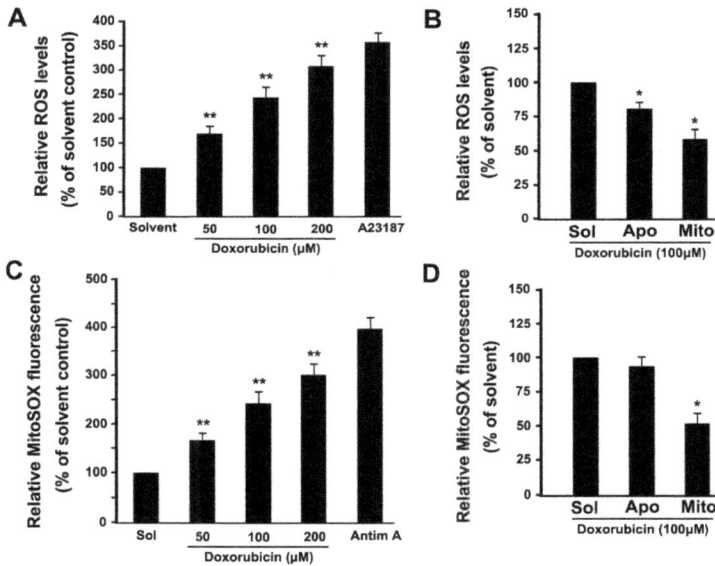

Figure 4. DOX increased intracellular ROS and mitochondrial ROS production. (A–D) Platelets were loaded with 2′7′-dichlorofluorescin diacetate (DCFDA) (A,B) or MitoSOX™ Red (C,D), and incubated with various concentrations of DOX or solvent (A,C), or pre-incubated with apocynin, Mito-TEMPO, and then incubated with DOX (B,D). As a positive control, loaded platelets were incubated with A23187 (3 μM) or antimycin A (50 μM) at 37 °C for 30 min. Treated platelets were analyzed for intracellular ROS or mitochondrial ROS levels by flow cytometry. The relative ROS levels are expressed as a percentage of platelets, which were incubated with solvent. Data are expressed as a percentage of platelets that were incubated with solvent control. Percentage is presented as means ± SEM from three independent experiments. ** $p < 0.017$ (after Bonferroni correction) as compared with solvent control. * $p < 0.025$ (after Bonferroni correction) as compared with solvent control. Solvent, apocynin, Mito-TEMPO and Antimycin A are labeled as Sol, Apo, Mito and Antim A, respectively.

To assist in confirming that mitochondria were a major site of ROS production in DOX-treated platelets, we used MitoSOX™ Red fluorescence, which detects superoxide synthesis, to quantify mitochondrial ROS [7]. We found that DOX dose-dependently induced mitochondrial superoxide production (Figure 4C). In addition, Mito-TEMPO significantly inhibited DOX-induced mitochondrial ROS generation as compared with the solvent control (Figure 4D). Together, these observations further confirm that DOX can induce mitochondrial ROS production in platelets. As a positive control, we found that antimycin A markedly induced mitochondrial ROS production in platelets (Figure 4C).

2.5. DOX Dose-Dependently Increases Malonyldialdehyde (MDA) Production and Cardiolipin Peroxidation in Platelets

Phospholipids are rich in unsaturated fatty acids that are particularly susceptible to ROS attack, which promotes lipid peroxidation. In order to demonstrate whether DOX treatment induces lipid peroxidation in platelets, we detected the production of MDA, which is a sensitive indicator of ROS-mediated lipid peroxidation. Production of MDA was increased in a dose-dependent manner (Figure 5A), suggesting that DOX induces platelet lipid peroxidation. Mito-TEMPO partly inhibited DOX-induced MDA production (Figure 5B). Mitochondria are the primary site of ROS generation and the major target of ROS. Cardiolipin, a unique phospholipid located at the level of the inner mitochondrial membrane, contains polyunsaturated fatty acid residues, and are thus highly prone to oxidation. In order to demonstrate whether DOX treatment induces cardiolipin peroxidation in platelets, we used the fluorescent dye NAO to estimate cardiolipin peroxidation. NAO binds to cardiolipin with high affinity, and the fluorochrome loses its affinity for peroxidized cardiolipin [7]. As shown in Figure 5C, cardiolipin peroxidation was increased in a dose-dependent manner. Mito-TEMPO obviously inhibited DOX-induced cardiolipin peroxidation (Figure 5D).

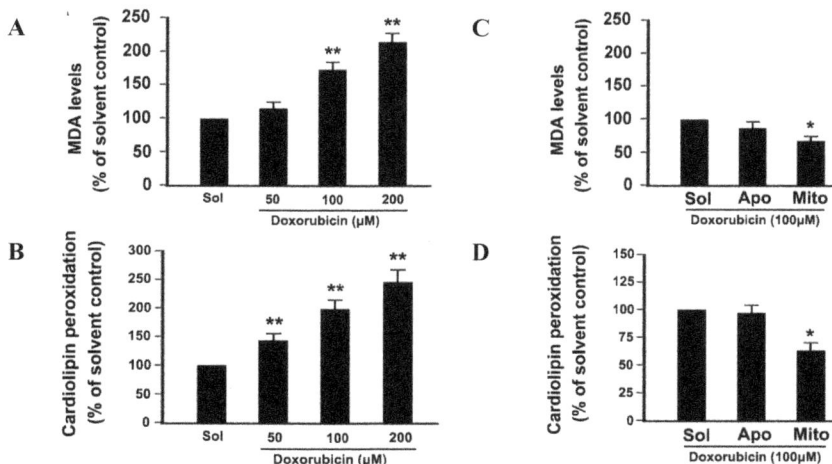

Figure 5. DOX increased malonyldialdehyde (MDA) production and cardiolipin peroxidation. (**A,B**) Platelets were incubated with different concentrations of DOX or solvent (**A**), or pre-incubated with apocynin, Mito-TEMPO, and then incubated with DOX (**B**). MDA levels were measured using an MDA assay kit. The MDA levels are expressed as a percentage of platelets that were incubated with solvent; (**C,D**) Platelets were incubated with different concentrations of DOX or solvent (**C**), or pre-incubated with apocynin, Mito-TEMPO, and then incubated with DOX (**D**). Cardiolipin peoxidation was detected as described in Methods. Data are expressed as a percentage of platelets that were incubated with solvent. Percentage is presented as means $\pm$ SEM from three independent experiments. ** $p < 0.017$ (after Bonferroni correction) as compared with solvent. * $p < 0.025$ (after Bonferroni correction) as compared with solvent control. Solvent, apocynin and Mito-TEMPO and are labeled as Sol, Apo and Mito, respectively.

2.6. Mitochondrial ROS Mediates DOX-Induced Platelet Apoptosis

The above observations confirmed that DOX treatment enhanced mitochondrial ROS levels in platelets. To investigate whether mitochondria-derived ROS were involved in DOX-induced platelet apoptotic events, Mito-TEMPO was pre-incubated with platelets before to DOX treatment. We found that Mito-TEMPO significantly inhibited DOX-induced platelets apoptosis, including $\Delta\Psi$m dissipation, PS exposure, caspase-3 activation, mitochondrial translocation of Bax, and cytochrome C release (Figure 6A–E). Together, these data indicate that mitochondrial-derived ROS play a pivotal role in DOX-induced platelet apoptosis. In addition, Mito-TEMPO also partly inhibited DOX-induced GPIbα shedding (Figure 6F).

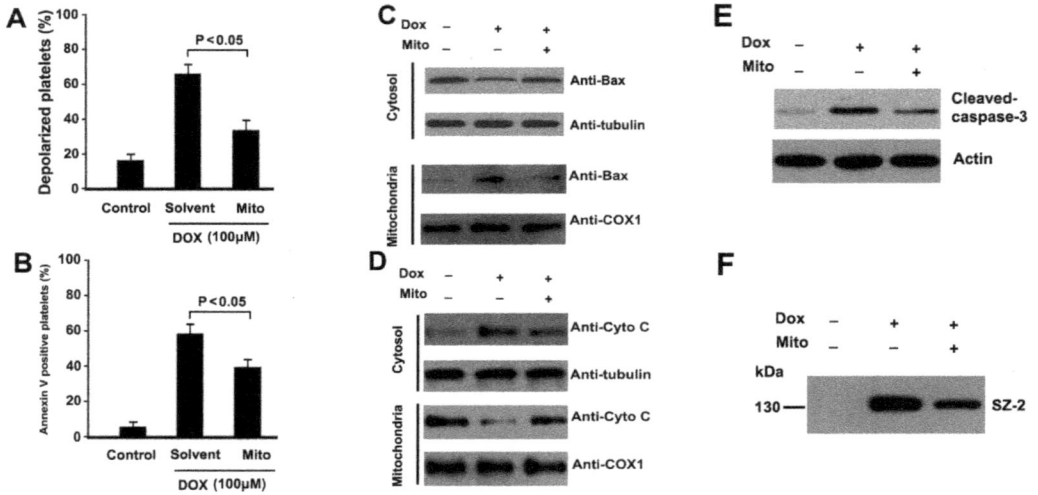

Figure 6. Mitochondria-targeted ROS scavenger attenuated DOX- induced platelet apoptosis. (**A,B**) Platelets were pre-incubated with Mito-TEMPO or solvent, and then incubated with DOX; Treated platelets were incubated with TMRE (**A**), or annexin V-FITC (**B**), and analyzed by flow cytometry. ΔΨm depolarization and PS exposure was quantified as the percentage of depolarized platelets. Means ± SEM from three independent experiments are shown; (**C,D**) Platelets were pre-incubated with Mito-TEMPO or solvent, and then incubated with DOX. Treated platelets were lysed, and cytosol and mitochondrial fractions were isolated and analyzed by Western blot with anti-Bax (**C**), and anti-cytochrome C antibodies (**D**); Representative results of three independent experiments are presented; (**E**) Platelets were pro-incubated with Mito-TEMPO or solvent, and then incubated with DOX. Treated platelets were lysed and analyzed by Western blot with anti-caspase-3 antibody. Actin levels were assayed to demonstrate equal protein loading. Representative results of three independent experiments are presented; (**F**) Platelets were pre-incubated with Mito-TEMPO or solvent, and then incubated with DOX. Treated platelets were centrifuged, and supernatants were analyzed by Western blot with SZ-2. Representative results of three independent experiments are presented. Mito-TEMPO is labeled as Mito.

3. Discussion

DOX is a highly effective chemotherapeutic agent that is widely used to treat a variety of cancers, however, its use is limited by some side effects, such as cardiotoxicity and thrombocytopenia [1,2,11,12]. Although it has been generally accepted that DOX exerts its anticancer effect by inducing different kinds of malignant cells apoptosis, it still remains unclear whether DOX incurs platelet apoptosis. In the current observation, DOX dose-dependently induces ΔΨm depolarization, PS exposure, mitochondrial translocation of Bax, cytochrome C release and caspase-3 activation, providing sufficient evidence to indicate that DOX incurs mitochondria-mediated intrinsic platelet apoptosis. We also found that DOX did not induce platelet activation through examining P-selectin expression and PAC-1 binding (data not shown). In addition, we have tried to explore the signaling cascades leading to DOX-induced platelet apoptosis, and the data indicate mitochondrial ROS is involved in the apoptotic process.

DOX induces ROS generation and apoptosis in various cell types, and the identities of the cellular sources of ROS remain controversial. Several studies have shown that NADPH oxidase is a major source of ROS in DOX-treated cells [15,16]. In our studies, we found that NADPH inhibitor apocynin did not significantly inhibited DOX-induced platelet apoptosis (data not shown). Recently, several studies have shown that mitochondria are major source of ROS in DOX-treated cells [12,13]. We found that mitochondria are the primary source of ROS in DOX-treated platelets based on our observations that (1) the mitochondria-targeted ROS scavenger inhibited DOX-induced ROS production; and (2) DOX-induced ROS was detected by the mitochondrial ROS probe MitoSOX™ Red. Therefore, different sources of DOX-induced ROS generation are likely to be dependent on cell type.

The precise mechanisms responsible for how DOX causes increased levels of mitochondrial ROS remain undetermined. The reasons may be manifold. On the one hand, DOX might increase mitochondrial ROS production. It has been reported that DOX could induce mitochondrial dysfunction, and thus augment mitochondrial ROS generation [14]. On the other hand, DOX might provoke decreased antioxidant capacity in mitochondria. Li *et al.* reported that myocardial MnSOD mRNA was not significantly changed, but its protein levels were significantly decreased in rats treated with DOX [24].

The functional role of mitochondrial ROS in DOX-induced platelet apoptosis was determined by pre-treating platelets with a mitochondria-targeted ROS scavenger before DOX treatment and then analyzing apoptotic markers. The mitochondria-targeted ROS scavenger was found to be effective in inhibiting DOX-induced platelet apoptosis. These observations indicate that mitochondrial ROS are key mediators of DOX-induced platelet apoptosis. However, the question remains, how does mitochondrial ROS triggers platelet apoptosis? It has been previously reported that mitochondria-derived ROS plays a pivotal role in triggering apoptosis in various cell types. Several studies have shown that mitochondrial ROS easily oxidizes cardiolipin, and oxidized cardiolipin appears to be essential for mitochondrial membrane permeabilization and releases of pro-apoptotic factors into the cytosol [25]. Conversely, prevention of cardiolipin peroxidation leads to inhibition of apoptosis [26]. These findings suggest that cardiolipin might be a crucial molecule that regulates the initiation of apoptosis. Our recent data demonstrated that hyperthermia increased cardiolipin peroxidation and that mitochondrial ROS plays an important role in hyperthermia-induced cardiolipin peroxidation [7]. Future work will be necessary to define how mitochondrial ROS regulates platelet apoptosis in DOX-treated platelets.

The reasons of Dox-induced thrombocytepenia may be manifold. On the one hand, DOX might increase platelet clearance. On the other hand, DOX might inhibit megakaryocyte function and decrease platelet generation. Up to now, the effect of DOX on megakaryocyte has not been clearly elucidated. Future work will be necessary to explore how DOX influence megakaryocyte function by *in vitro* or *in vivo* experiment.

In summary, our study provides direct evidence that DOX increases mitochondria-derived ROS generation in platelets, which in turn, induces platelet apoptosis. DOX does not incur platelet activation, whereas, it impairs platelet function. These findings may reveal a mechanism for platelet clearance and dysfunction *in vivo* or *in vitro*, and also suggest a possible pathogenesis of thrombocytopenia in some patients treated with DOX.

4. Experimental Section

4.1. Reagents and Antibodies

Anti-cleaved p17 fragment of caspase-3 antibody was obtained from Millipore (Billerica, MA, USA). Mito-TEMPO was obtained from Enzo Life Sciences (Plymouth Meeting, PA, USA). A23187, DOX, adenosine diphosphate (ADP), tetramethylrhodamine ethyl ester (TMRE), apocynin, 2', and 7'-dichlorofluorescin diacetate (DCFDA) were obtained from Sigma (St. Louis, MO, USA). Monoclonal antibodies against Bax, cytochrome C, tubulin, cytochrome C oxidase subunit 1 (COX1), actin, SZ-2, and HRP-conjugated goat anti-mouse IgG were obtained from Santa Cruz Biotechnology (Santa Cruz, CA, USA). FITC-conjugated annexin V was obtained from Bender Medsystem (Vienna, Austria). MitoSOX™ Red was obtained from Invitrogen/Molecular Probes (Eugene, OR, USA). Mitochondria isolation kit was obtained from Pierce (Rockford, IL, USA).

4.2. Preparation of Platelet-Rich Plasma (PRP) and Washed Platelets

For studies involving human subjects, approval was obtained from the Huashan Hospital institutional review board, China. Informed consent was provided in accordance with the Declaration of Helsinki. PRP and washed platelets were prepared as described previously [7]. Briefly, fresh blood from healthy volunteers (7 males and 5 females; age range: 24–35 years) was anti-coagulated with one-seventh volume of acid-citratedextrose (ACD, 2.5% trisodium citrate, 2.0% D-glucose and 1.5% citric acid). Anti-coagulated blood was separated by centrifuging, and the supernatant was PRP. Platelets were washed twice with CGS buffer (123 mM NaCl, 33 mM D-glucose, 13 mM trisodium citrate, pH 6.5) and re-suspended in modified Tyrode's buffer (MTB) (2.5 mM Hepes, 150 mM NaCl, 2.5 mM KCl, 12 mM NaHCO$_3$, 5.5 mM D-glucose, pH 7.4) to a final concentration of 3×10^8/mL, and incubated at room temperature (RT) for 1 h to recover to resting state.

4.3. Measurement of Mitochondrial Inner Transmembrane Potential ($\Delta\Psi m$)

Washed platelets were incubated with different concentrations of DOX (50, 100, 200 µM) or solvent control at 37 °C for indicated time. TMRE was added according to a previously described method [7]. For the inhibition experiments, washed platelets were pre-incubated with Mito-TEMPO (10 µM) or solvent control at 37 °C for 15 min, and then incubated with DOX at 37 °C for 3 h.

4.4. Phosphatidylserine (PS) Externalization Assay

Washed platelets were incubated with different concentrations of DOX or solvent control at 37 °C for indicated time. Annexin V binding buffer was mixed according to a previously described method [7]. For the inhibition experiments, washed platelets were pre-incubated with Mito-TEMPO (10 µM) or solvent control at 37 °C for 15 min, and then incubated with DOX at 37 °C for 3 h.

4.5. Measurement of Intracellular ROS and Mitochondrial ROS Levels

Intracellular ROS and mitochondrial ROS levels were examined using DCFDA and MitoSOX™ Red, respectively, according to a previously described method [7]. Briefly, washed platelets were loaded with DCFDA (10 µM) or MitoSOX™ Red (5 µM) at 37 °C for 20 min in the dark and washed three times with modified Tyrode's buffer (MTB). Pre-loaded platelets were incubated with different concentrations of DOX or solvent control at 37 °C for different time. For the inhibition experiments, pre-loaded platelets were incubated with apocynin (100 µM), Mito-TEMPO (10 µM), or solvent control at 37 °C for 15 min, and then treated with DOX at 37 °C for 3 h. A23187-treated and antimycin A-treated platelets were used as positive controls for intracellular cellular ROS and mitochondrial ROS levels, respectively.

4.6. Assessment of Malonyldialdehyde (MDA) Levels

Washed platelets were incubated with different concentrations of DOX or solvent control at 37 °C for 3 h. Samples were treated according to a previously described method [7]. For the inhibition experiments, washed platelets were pre-incubated with Mito-TEMPO, apocynin or solvent control at 37 °C for 15 min and then incubated with DOX at 37 °C for 3 h.

4.7. Assessment of Cardiolipin Peroxidation

Washed platelets were incubated with different concentrations of DOX or solvent control at 37 °C for 3 h, and then loaded with NAO according to a previously described method [7]. For the inhibition experiments, platelets were pre-incubated with Mito-TEMPO, apocynin or solvent control at 37 °C for 15 min and then incubated with DOX at 37 °C for 3 h.

4.8. Subcellular Fractionation

Washed platelets were incubated with different concentrations of DOX or solvent control at 37 °C for 3 h. Samples were suspended according to a previously described method [7]. For inhibition experiments, platelets were pre-incubated with Mito-TEMPO or solvent control at 37 °C for 15 min, and then further incubated with DOX at 37 °C for 3 h.

4.9. Western Blot Analysis

After subcellular fractionation Bax and cytochrome C were detected by Western blot using anti-Bax, and anti-cytochrome C antibodies. COX1 and tubulin were used as mitochondrial and cytosolic internal controls, respectively. Caspase-3 activation and GPIbα shedding was assessed with platelet whole lysates and supernatant, respectively. Washed platelets were incubated with different concentrations of DOX or solvent control at 37 °C for 3 h. One part treated platelets were lysed with an equal volume of lysis buffer on ice for 30 min. The samples were subjected to Western blot analysis using anti-cleaved p17 fragment of caspase-3 antibody. Anti-actin antibody was used as an equal protein loading control. Another part treated platelets were centrifuged at 4000 rpm for 5 min, and the supernatants were analyzed by Western blot with anti-GPIbα N-terminal antibody SZ-2. In the inhibition experiments, platelets were pre-incubated with Mito-TEMPO or solvent control at 37 °C for 15 min, and incubated with DOX at 37 °C for 3 h.

4.10. Platelet Aggregation

PRP or washed platelets were incubated with different concentrations of DOX or solvent control at 37 °C for 3 h. Platelet aggregation was induced by addition of ADP or thrombin at 37 °C with a stirring speed of 1000 rpm.

4.11. Platelet Adhesion under Flow Condition

The glass capillary was coated according to a previously described method [7]. Washed platelets were incubated with different concentrations of DOX or solvent control at 37 °C for 3 h, then perfused into the glass capillary by a syringe pump at a flow shear rate of 250 s^{-1} for 5 min, and then washed with MTB for 5 min. The number of adherent platelets was counted in 10 randomly selected fields of 0.25 mm^2 and at randomly selected time points.

4.12. Statistical Analysis

The experimental data were expressed as means $\pm$ SEM. Each experiment was carried out at least three times. Statistical analysis for multiple group comparisons were performed by one-way analysis of variance (ANOVA), followed by *post-hoc* Dunnett's test. A *p*-value of less than 0.05 was considered statistically significant.

5. Conclusions

In the current study, the data show that DOX induces mitochondria-mediated intrinsic apoptosis and GPIbα shedding. Dox does not incur platelet activation, however, it obviously impair platelet aggregation and adhesion. Meanwhile, DOX induces mitochondrial ROS generation, and mitochondria-targeted ROS scavenger obviously reduces DOX-induced platelet apoptosis and GPIbα shedding. Thus, mitochondria-targeted ROS scavenger would have potential clinical utility in platelet-associated disorders involving mitochondrial oxidative damage.

Acknowledgments: This study was supported by grants from the National Natural Science Foundation of China (NSFC 81270650), the Natural Science Foundation of Shanghai (12ZR1429900, 15ZR1438300).

Author Contributions: Zhicheng Wang and Yuan Lu contributed to the manuscript concept and design, and critical review of the literature; Zhicheng Wang and Jie Wang contributed to perform the experiments and data analysis; Rufeng Xie and Ruilai Liu contributed to interpretation of results and language revision. All authors wrote the manuscript.

Conflicts of Interest: The authors declare no conflict of interest.

References

1. Wasle, I.; Gamerith, G.; Kocher, F.; Mondello, P.; Jaeger, T.; Walder, A.; Auberger, J.; Melchardt, T.; Linkesch, W.; Fiegl, M.; *et al.* Non-pegylated liposomal DOX in lymphoma: Patterns of toxicity and outcome in a large observational trial. *Ann. Hematol.* **2015**, *94*, 593–601.
2. Kim, E.J.; Lim, K.M.; Kim, K.Y.; Bae, O.N.; Noh, J.Y.; Chung, S.M.; Shin, S.; Yun, Y.P.; Chung, J.H. DOX-induced platelet cytotoxicity: A new contributory factor for DOX-mediated thrombocytopenia. *J. Thromb. Haemost.* **2009**, *7*, 1172–1183.
3. Kim, S.H.; Lim, K.M.; Noh, J.Y.; Kim, K.; Kang, S.; Chang, Y.K.; Shin, S.; Chung, J.H. DOX-induced platelet procoagulant activities: An important clue for chemotherapy-associated thrombosis. *Toxicol. Sci.* **2011**, *124*, 215–224.
4. Mendivil-Perez, M.; Velez-Pardo, C.; Jimenez-Del-Rio, M. DOX induces apoptosis in Jurkat cells by mitochondria-dependent and mitochondria-independent mechanisms under normoxic and hypoxic conditions. *Anticancer Drugs* **2015**. in press.

5. Wang, H.; Lu, C.; Li, Q.; Xie, J.; Chen, T.; Tan, Y.; Wu, C.; Jiang, J. The role of Kif4A in DOX-induced apoptosis in breast cancer cells. *Mol. Cells* **2014**, *37*, 812–818.

6. Leytin, V. Apoptosis in the anucleate platelet. *Blood Rev.* **2012**, *26*, 51–63.

7. Wang, Z.; Cai, F.; Chen, X.; Luo, M.; Hu, L.; Lu, Y. The role of mitochondria-derived reactive oxygen species in hyperthermia-induced platelet apoptosis. *PLoS ONE* **2013**, *8*, e75044.

8. Zhang, J.; Chen, M.; Zhang, Y.; Zhao, L.; Yan, R.; Dai, K. Carmustine induces platelet apoptosis. *Platelets* **2014**, *23*, 1–6.

9. Thushara, R.M.; Hemshekhar, M.; Kemparaju, K.; Rangappa, K.S.; Devaraja, S.; Girish, K.S. Therapeutic drug-induced platelet apoptosis: An overlooked issue in pharmacotoxicology. *Arch. Toxicol.* **2014**, *88*, 185–198.

10. Zhang, W.; Liu, J.; Sun, R.; Zhao, L.; Du, J.; Ruan, C.; Dai, K. Calpain activator dibucaine induces platelet apoptosis. *Int. J. Mol. Sci.* **2011**, *12*, 2125–2137.

11. Danz, E.D.; Skramsted, J.; Henry, N.; Bennett, J.A.; Keller, R.S. Resveratrol prevents DOX cardiotoxicity through mitochondrial stabilization and the Sirt1 pathway. *Free Radic. Biol. Med.* **2009**, *46*, 1589–1597.

12. Cheung, K.G.; Cole, L.K.; Xiang, B.; Chen, K.; Ma, X.; Myal, Y.; Hatch, G.M.; Tong, Q.; Dolinsky, V.W. SIRT3 attenuates DOX-induced oxidative stress and improves mitochondrial respiration in H9c2 cardiomyocytes. *J. Biol. Chem.* **2015**. in press.

13. Min, K.; Kwon, O.S.; Smuder, A.J.; Wiggs, M.P.; Sollanek, K.J.; Christou, D.D.; Yoo, J.K.; Hwang, M.H.; Szeto, H.H.; Kavazis, A.N.; *et al.* Increased mitochondrial emission of reactive oxygen species and calpain activation are required for DOX-induced cardiac and skeletal muscle myopathy. *J. Physiol.* **2015**. in press.

14. Shokoohinia, Y.; Hosseinzadeh, L.; Moieni-Arya, M.; Mostafaie, A.; Mohammadi-Motlagh, H.R. Osthole attenuates DOX-induced apoptosis in PC12 cells through inhibition of mitochondrial dysfunction and ROS production. *Biomed. Res. Int.* **2014**, *2014*.

15. Zhao, Y.; McLaughlin, D.; Robinson, E.; Harvey, A.P.; Hookham, M.B.; Shah, A.M.; McDermott, B.J.; Grieve, D.J. Nox2 NADPH oxidase promotes pathologic cardiac remodeling associated with DOX chemotherapy. *Cancer Res.* **2010**, *70*, 9287–9297.

16. Gilleron, M.; Marechal, X.; Montaigne, D.; Franczak, J.; Neviere, R.; Lancel, S. NADPH oxidases participate to DOX-induced cardiac myocyte apoptosis. *Biochem. Biophys. Res. Commun.* **2009**, *388*, 727–731.

17. Balaban, R.S.; Nemoto, S.; Finkel, T. Mitochondria, oxidants, and aging. *Cell* **2005**, *120*, 483–495.

18. Schulz, E.; Wenzel, P.; Munzel, T.; Daiber, A. Mitochondrial redox signaling: Interaction of mitochondrial reactive oxygen species with other sources of oxidative stress. *Antioxid. Redox Signal.* **2014**, *20*, 308–324.

19. Quinlan, C.L.; Orr, A.L.; Perevoshchikova, I.V.; Treberg, J.R.; Ackrell, B.A.; Brand, M.D. Mitochondrial complex II can generate reactive oxygen species at high rates in both the forward and reverse reactions. *J. Biol. Chem.* **2012**, *32*, 27255–27264.

20. Block, K.; Gorin, Y.; Abboud, H.E. Subcellular localization of Nox4 and regulation in diabetes. *Proc. Natl. Acad. Sci. USA* **2009**, *106*, 14385–14390.

21. Koziel, R.; Pircher, H.; Kratochwil, M.; Lener, B.; Hermann, M.; Dencher, N.A.; Jansen-Durr, P. Mitochondrial respiratory chain complex I is inactivated by NADPH oxidase Nox4. *Biochem. J.* **2013**, *452*, 231–239.

22. Renault, T.T.; Manon, S. Bax: Addressed to kill. *Biochimie* **2011**, *93*, 1379–1391.

23. Wang, Z.; Shi, Q.; Yan, R.; Liu, G.; Zhang, W.; Dai, K. The role of calpain in the regulation of ADAM17-dependent GPIbα ectodomain shedding. *Arch. Biochem. Biophys.* **2010**, *495*, 136–143.

24. Li, T.; Danelisen, I.; Singal, P.K. Early changes in myocardial antioxidant enzymes in rats treated with adriamycin. *Mol. Cell. Biochem.* **2002**, *232*, 19–26.

25. Kagan, V.E.; Bayir, A.; Bayir, H.; Stoyanovsky, D.; Borisenko, G.G.; Tyurina, Y.Y.; Wipf, P.; Atkinson, J.; Greenberger, J.S.; Chapkin, R.S.; *et al.* Mitochondria-targeted disruptors and inhibitors of cytochrome c/cardiolipin peroxidase complexes: A new strategy in anti-apoptotic drug discovery. *Mol. Nutr. Food Res.* **2009**, *53*, 104–114.

26. Tyurina, Y.Y.; Tyurin, V.A.; Kaynar, A.M.; Kapralova, V.I.; Wasserloos, K.; Li, J.; Mosher, M.; Wright, L.; Wipf, P.; Watkins, S.; *et al.* Oxidative lipidomics of hyperoxic acute lung injury: Mass spectrometric characterization of cardiolipin and phosphatidylserine peroxidation. *Am. J. Physiol. Lung Cell Mol. Physiol.* **2010**, *299*, L73–L85.

Abnormal Mitochondrial Function and Impaired Granulosa Cell Differentiation in Androgen Receptor Knockout Mice

Ruey-Sheng Wang, Heng-Yu Chang, Shu-Huei Kao, Cheng-Heng Kao,
Yi-Chen Wu, Shuyuan Yeh, Chii-Reuy Tzeng and Chawnshang Chang

Abstract: In the ovary, the paracrine interactions between the oocyte and surrounded granulosa cells are critical for optimal oocyte quality and embryonic development. Mice lacking the androgen receptor ($AR^{-/-}$) were noted to have reduced fertility with abnormal ovarian function that might involve the promotion of preantral follicle growth and prevention of follicular atresia. However, the detailed mechanism of how AR in granulosa cells exerts its effects on oocyte quality is poorly understood. Comparing *in vitro* maturation rate of oocytes, we found oocytes collected from $AR^{-/-}$ mice have a significantly poor maturating rate with 60% reached metaphase II and 30% remained in germinal vesicle breakdown stage, whereas 95% of wild-type AR ($AR^{+/+}$) oocytes had reached metaphase II. Interestingly, we found these $AR^{-/-}$ female mice also had an increased frequency of morphological alterations in the mitochondria of granulosa cells with reduced ATP generation (0.18 ± 0.02 *vs.* 0.29 ± 0.02 µM/mg protein; $p < 0.05$) and aberrant mitochondrial biogenesis. Mechanism dissection found loss of AR led to a significant decrease in the expression of peroxisome proliferator-activated receptor γ (PPARγ) co-activator 1-β (PGC1-β) and its sequential downstream genes, nuclear respiratory factor 1 (NRF1) and mitochondrial transcription factor A (TFAM), in controlling mitochondrial biogenesis. These results indicate that AR may contribute to maintain oocyte quality and fertility via controlling the signals of PGC1-β-mediated mitochondrial biogenesis in granulosa cells.

Reprinted from *Int. J. Mol. Sci.* Cite as: Wang, R.-S.; Chang, H.-Y.; Kao, S.-H.; Kao, C.-H.; Wu, Y.-C.; Yeh, S.; Tzeng, C.-R.; Chang, C. Abnormal Mitochondrial Function and Impaired Granulosa Cell Differentiation in Androgen Receptor Knockout Mice. *Int. J. Mol. Sci.* **2015**, *16*, 9831–9847.

1. Introduction

Androgens classically mediate their genomic effects via the androgen receptor (AR), a protein encoded by an X chromosome gene, which exerts its biological function through activation of target gene expression via a sequence of processes [1,2]. The earlier studies indicated that androgens act directly on the development of ovarian follicles via AR, apart from serving as a substrate of aromatase (Cyp19a1) for estrogen

494

synthesis. Androgens have been implicated to have a role in promoting follicular development [3,4], by up-regulating follicle-stimulating hormone (FSH) receptor (FSHR) expression and augmenting FSH-stimulated follicular differentiation [5–8]. In addition, AR is expressed in all cell types of the ovarian follicle, including granulosa cells, theca cells and the oocytes [9–11].

By using Cre/LoxP system, we and others generated the global $AR^{-/-}$ [12–14] and granulosa cell-specific $AR^{-/-}$ female mice [15,16]. Those data showed that female mice lacking AR have a reduced fertility, fewer oocytes were recovered after superovulation with gonadotropins, and fewer corpora lutea were observed. However, ablation of AR in oocytes had no effect on female fertility [15]. All together, these animal examples suggested that androgen, functions through the AR in granulosa cells, plays a primary and essential role for normal follicle development and optimal fertility. Earlier studies have shown that AR antagonists slow down mouse follicle growth [3,17] and prevent primary to secondary follicle transition in bovine [18]. Moreover, granulosa cell-specific $AR^{-/-}$ female mice [15,16] have been found to contain more preantral follicles, more atretic follicles, with fewer antral follicles and corpora lutea in their ovaries. These animal examples suggest that androgen/AR signaling in granulosa cells regulates normal follicular growth, mainly by controlling preantral follicle growth and development to antral follicles, meanwhile preventing follicular atresia by inhibition of granulosa cell apoptosis. However, the definitive mechanisms of the subfertility and the potential pathways by which granulosa cell AR exerts its effects on follicle development and follicle atresia are still ill-defined.

Earlier studies suggested that mitochondria dysfunction in granulosa cells leads to poor oocyte quality and contribute to female subfertility [19–21]. In the current study, the prophase I arrested oocyte was collected from the ovary and the *in vitro* maturation rate of oocytes was evaluated as the preliminary indication for oocyte quality. In addition, oocyte maturation often required the energy sources from the surrounding nursing cells, such as granulose cells, as we reported here. Thus, the objective of this study was to determine whether lack of AR in granulosa cells has impacts on the oocyte quality by changing mitochondrial status and differentiation status in granulosa cells. We found alterations in mitochondrial morphology, biogenesis and metabolism in the granulosa cells of $AR^{-/-}$ mice, suggesting that ablation of AR leads to mitochondrial dysfunction. In addition, the results in this study demonstrated that pregnant mare's serum gonadotropin (PMSG)-induced granulosa cell differentiation was impaired in $AR^{-/-}$ ovaries as indicated by decreasing the expression of genes involved in the granulosa cell differentiation. These consequences may be interrelated to significantly reduce *in vitro* maturation rate of $AR^{-/-}$ oocytes.

2. Results

2.1. Confirmation of Knockout Androgen Receptor (AR) in the Ovaries in the $AR^{-/-}$ Mice

Three primers ("select", "2–3" and "2–9", for the relative position of each primer in the AR gene see Figure S1) were synthesized to amplify mouse genomic DNA to distinguish the floxed AR, $AR^{-/-}$ and $AR^{+/+}$ mice. We were able to identify $AR^{-/-}$ mice by using "select" and "2–9" primers to PCR amplify the 238-bp DNA (Figure 1, upper panel). In contrast, $AR^{+/+}$ mice can produce 580-bp DNA fragments by using "select" and "2–9" primers (Figure 1, upper panel). In this study, we were using mice with a ubiquitous deletion of the AR, so we further confirmed that AR was knocked out in ovaries by Western blot (Figure 1, lower panel).

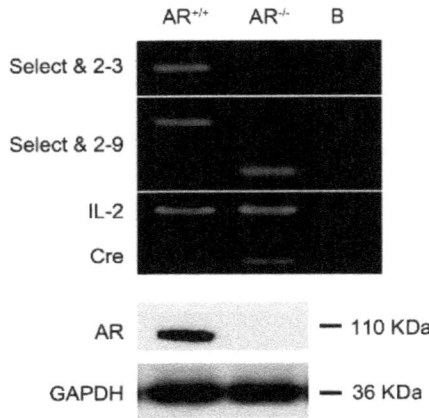

Figure 1. Genotyping of androgen receptor knockout ($AR^{-/-}$) female mice. We used two pairs of primers (i) "select" and "2–3"; and (ii) "select" and "2–9" to identify AR wild type ($AR^{+/+}$) and AR knockout ($AR^{-/-}$) female mice in our study. Genotyping was performed by using PCR on the genomic DNA isolated from the tails of 3-week-old mice. "Select" is a forward primer which is located in the intron 1 of AR gene with sequence 5'-GTTGATACCTTAACCTCTGC-3'. "2–3" is a reverse primer which is located at the 3' end of the exon 2 with the sequence 5'-CTTCAGCGGCTCTTTTGAAG-3'. "2–9" is a reverse primer which is located in intron 2 with the sequence 5'-CTTACATGTACTGTGAGAGG-3'. Using the "select" and "2–3" primers, we amplified a product with ~460 bp for $AR^{+/+}$ allele, and with no product for $AR^{-/-}$ allele. Using "select" and "2–9", we amplified a DNA fragment with 580-bp, which represents $AR^{+/+}$ allele and ~270 bp, which represents $AR^{-/-}$ allele (**upper panel**); The expression of Cre and internal control interleukin 2 (IL-2) were confirmed by PCR (bottom of upper panel). The lane marked B indicates blank control. Western blot shown that $AR^{-/-}$ ovaries do not express AR protein (**lower panel**).

2.2. Reduced Oocyte in Vitro Maturation Rate in $AR^{-/-}$ Mice

The *in vitro* oocyte maturation rate was examined to determine the potential AR roles in the oocyte maturation. The oocytes were collected from 4.5 weeks old $AR^{-/-}$ and $AR^{+/+}$ female mice that were previously treated with PMSG for 48 h. The result revealed that 95% of $AR^{+/+}$ oocytes had reached to metaphase II, whereas a significantly lower maturation rate (60%) was observed in the $AR^{-/-}$ oocytes (Figure 2, right panel). In the meantime, about 30% of $AR^{-/-}$ oocytes were arrested at prophase I and remained at germinal vesicle breakdown (GVBD) stage (Figure 2, left panel).

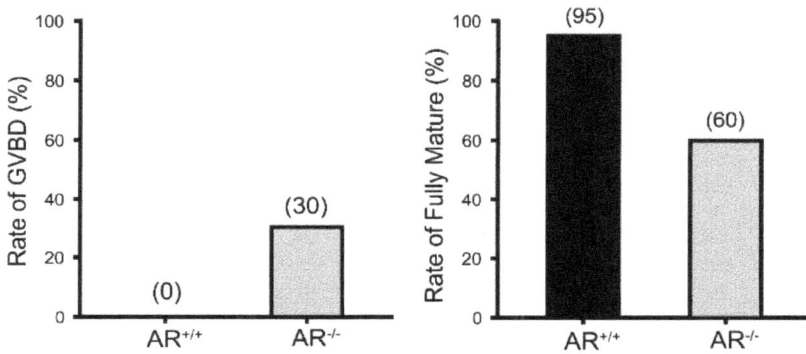

Figure 2. Reduced *in vitro* maturation rate of $AR^{-/-}$ oocytes. Histogram shows the *in vitro* maturation rate of $AR^{+/+}$ and $AR^{-/-}$ oocytes. After 18 h of *in vitro* culture, fewer $AR^{-/-}$ oocytes could reach to metaphase II (60%) compared with $AR^{+/+}$ oocytes (95%) (**right panel**); In the meantime, about 30% of $AR^{-/-}$ oocytes were remained in germinal vesicle breakdown (GVBD) stage (**left panel**), $n = 100$ follicles per genotype.

2.3. Change of Ovarian Follicle Morphology in $AR^{-/-}$ Mice

In order to elucidate whether poorer oocyte maturation rate in $AR^{-/-}$ mice was linked with morphological change of ovarian follicles, the morphology of ovarian follicles were checked with HE staining. Figure 3A shows follicle morphology after superovulation treatment. The largest follicles had reached to Pedersen class 6 (incipient antral) or Pedersen class 7 (early antral) stages in $AR^{+/+}$ ovaries, whereas in $AR^{-/-}$ ovaries, the majority of follicles did not show antrum formation in this time period. In agreement with our previous data [12], $AR^{-/-}$ ovaries contained considerably few antral follicles as compared to $AR^{+/+}$ littermates (Figure 3B).

(A)

(B)

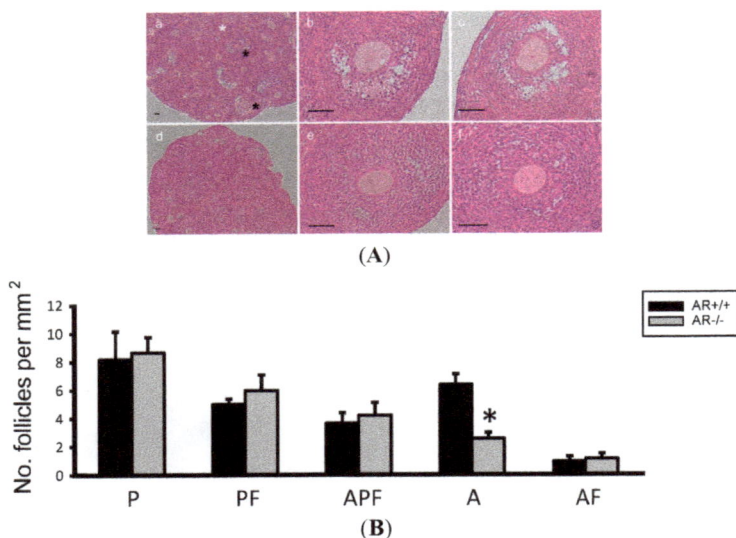

Figure 3. Morphological changes of ovarian follicle in $AR^{-/-}$ mice. (A) Comparison of morphology in ovaries of 4.5-week-old $AR^{+/+}$ and $AR^{-/-}$ female mice after PMSG-priming for 48 h ($n = 4$ mice per genotype). Representative hematoxylin and eosin-stained ovarian sections (**a**: $AR^{+/+}$, **b,c**: $AR^{+/+}$, **d**: $AR^{-/-}$ at 100× magnification; **e,f**: $AR^{-/-}$ at 400× magnification). The largest follicles had reached to Pedersen class 6 (incipient antral, white asterisk) or Pedersen class 7 (early antral, black asterisk) stages in $AR^{+/+}$ ovaries (**a**–**c**), whereas in $AR^{-/-}$ ovaries, the majority of follicles did not show antrum formation in this time period (**d**–**f**). Scale bar, 50 μm; (B) Statistical analysis of the number of the follicular compartments in $AR^{+/+}$ and $AR^{-/-}$ ovaries. The $AR^{-/-}$ female mice have few antral follicles relative to $AR^{+/+}$ mice. P, Primordial and primary follicle; PF, Preantral follicle; APF, Atretic primordial, primary, and preantral follicle; A, Antral follicle; AF, Atretic antral follicle. * $p < 0.05$, by Student's t test.

2.4. Alteration of Mitochondrial Morphology and Ultrastructure in Granulosa Cells from $AR^{-/-}$ Mice

To reveal the molecular mechanisms of ovarian follicles morphology changes, and impact on oocyte quality, we assayed the damage on the mitochondria of granulosa cells since the mitochondria are important for steroidogenisis and energy production [22,23]. We first examined the mitochondrial ultrastructure in granulosa cells using TEM on the ovaries of both $AR^{+/+}$ and $AR^{-/-}$ mice and found that most mitochondria of $AR^{+/+}$ granulosa cells contained highly folded inner membrane forming mitochondrial cristae, which were enveloped by an intact outer membrane. In contrast, the mitochondria of $AR^{-/-}$ granulosa cells showed increase electron density

498

of the matrix, generally more round in appearance with fewer, disarrayed cristae, and contained vacuoles (white arrowhead in Figure 4A). Immunofluorescent staining of mitochondria exhibited a normal perinuclear distribution in $AR^{+/+}$ granulosa cells. In contrast, the mitochondria in $AR^{-/-}$ granulosa cells displayed aggregating distribution pattern and small "donut-like" vesicles appearance (Figure 4B).

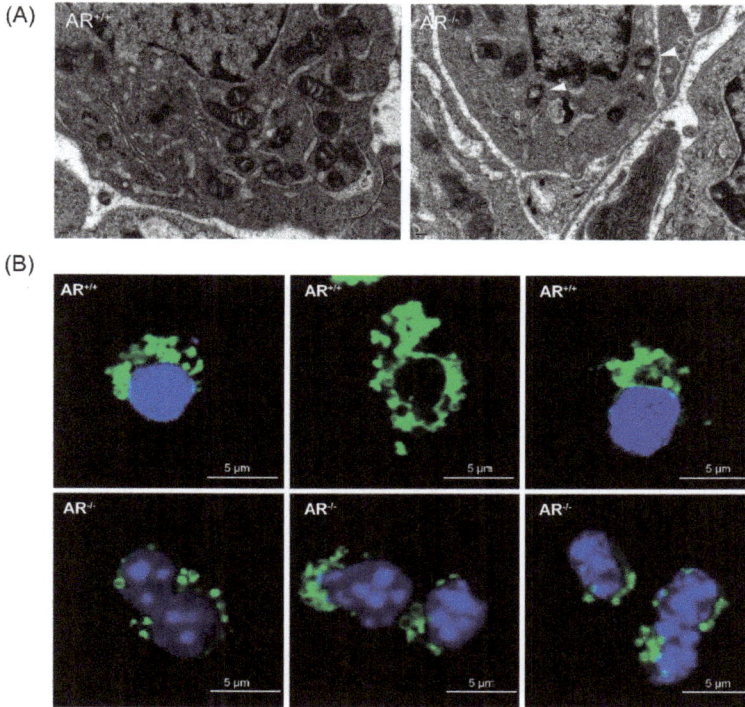

Figure 4. Alteration of mitochondrial morphology in granulosa cells of $AR^{-/-}$ mice. (**A**) Mitochondrial morphology in granulosa cells of $AR^{-/-}$ mice was visualized by transmission electron microscopy (TEM). Representative electron micrograph of mitochondria from control ($AR^{+/+}$) granulosa cells, showing bean-shaped structures with numerous transversely orientated cristae enveloped by an intact outer membrane; $AR^{-/-}$ granulosa cells displayed small spherical structures with fewer and disarrayed cristae, and the presence of large vacuoles in mitochondria (white arrowhead); (**B**) Mitochondrial morphology in granulosa cells of $AR^{-/-}$ mice was visualized by immunofluorescence. Granulosa cells collected from $AR^{+/+}$ and $AR^{-/-}$ mice were labeled with MitoTracker Green to visualize mitochondrial localization and co-stained with 4',6'-diamidino-2-phenylindole (DAPI) to visualize nuclei. Representative confocal sections of granulosa cells were shown. Mitochondria in granulosa cells from $AR^{-/-}$ mice displayed aggregating distribution pattern and small spherical structures.

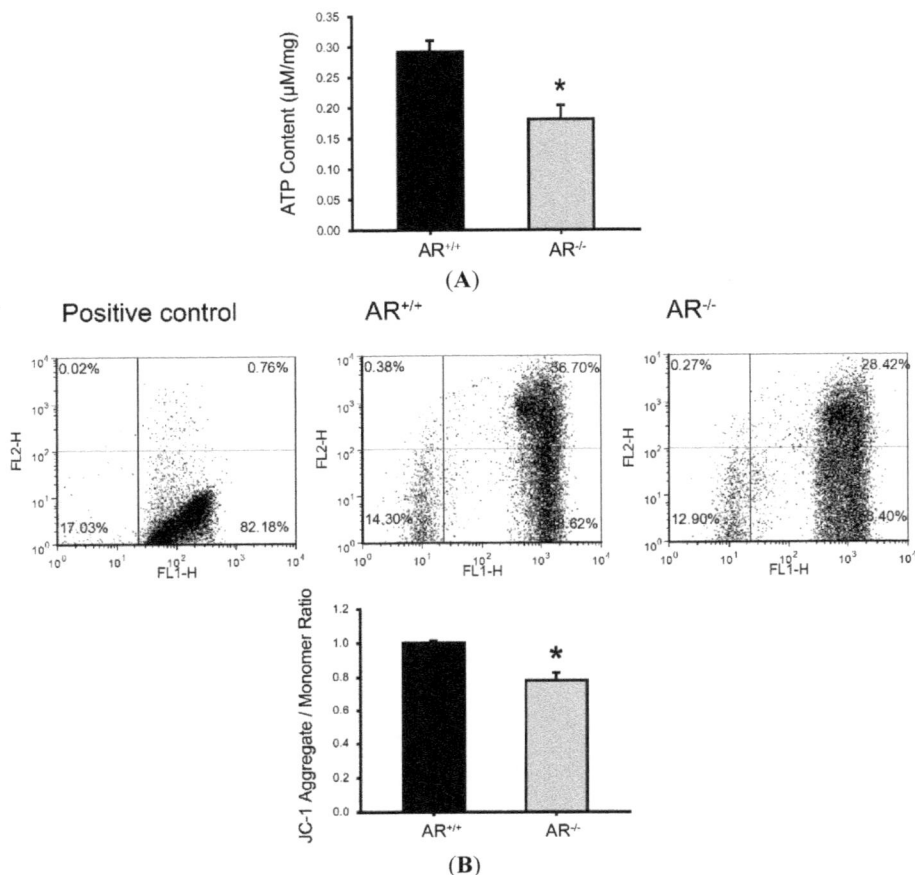

Figure 5. Metabolic dysfunction of mitochondria and reduced mitochondrial membrane potential in granulosa cells of $AR^{-/-}$ mice. (A) Reduced ATP content in granulosa cells of $AR^{-/-}$ mice. Granulosa cells removed from $AR^{+/+}$ and $AR^{-/-}$ mice were collected to determine the levels of ATP. Values are expressed as µM per mg protein. Histogram shows the average ATP content in granulosa cells from $AR^{+/+}$ and $AR^{-/-}$ mice. * $p < 0.05$, $n = 3$ mice per genotype; (B) Detection of mitochondrial membrane potential by flow cytometry analysis of JC-1 staining. Granulosa cells were collected from 4-week-old female mice ($AR^{+/+}$ and $AR^{-/-}$) injected with 48 h of 7.5 IU pregnant mare's serum gonadotropin (PMSG) treatment, and then stained with 5,5',6,6'-tetrachloro-1,1',3,3'-tetraethylbenzimidazolylcarbocyanine iodide (JC-1). JC-1 exists as a monomer in the cytosol (green fluorescent) and also accumulates as aggregates in the mitochondria (red fluorescent). In apoptotic and necrotic cells, JC-1 exists in monomeric form and stains the cytosol green. Results are expressed as the ratio of the aggregate to monomeric form of JC-1. The percentage is expressed as the mean ± SD ($n = 3$ mice per genotype). * $p < 0.05$.

500

2.5. Metabolic Dysfunction of Mitochondria and Reduced Mitochondrial Membrane Potential in Granulosa Cells of $AR^{-/-}$ Mice

Based on the above findings, we determine whether mitochondrial metabolism and mitochondria membrane potential were disturbed in granulosa cells of $AR^{-/-}$ mice. ATP synthesis by oxidative phosphorylation is the primary function associated with mitochondrial function [24]. Thus, the granulosa cells of $AR^{+/+}$ and $AR^{-/-}$ mice were processed to measure ATP content. As shown in Figure 5A, the ATP content were markedly decreased in the granulosa cells of $AR^{-/-}$ mice as compared to $AR^{+/+}$ mice (0.18 ± 0.02 μM/mg protein *vs.* 0.29 ± 0.02 μM/mg protein; $p < 0.05$). To detect the mitochondria membrane potential, we used flow cytometry analysis of JC-1 staining. JC-1 exists as a monomer in the cytosol (green fluorescent) and also accumulates as aggregates in the mitochondria (red fluorescent). In apoptotic and necrotic cells, JC-1 exists in monomeric form and stains the cytosol green. Results are expressed as the ratio of the aggregate to monomeric form of JC-1. As shown in Figure 5B, the mitochondria membrane potential were markedly reduced in the granulosa cells of $AR^{-/-}$ mice as compared to $AR^{+/+}$ mice, which suggesting a decline of mitochondrial function.

2.6. Decreased Mitochondrial Biogenesis in Granulosa Cells of $AR^{-/-}$ Mice

To investigate the potential effects of loss of AR in granulosa cells on mitochondrial biogenesis, we evaluated mtDNA content in granulosa cells from $AR^{+/+}$ and $AR^{-/-}$ mice by quantitative real-time PCR. Data are expressed as the ratio of mtDNA to nuclear DNA, as shown inFigure 6A (lower panel). We found that mtDNA content in granulosa cells were significantly lower in $AR^{-/-}$ mice than in $AR^{+/+}$ mice (0.2 ± 0.12 *vs.* 1.1 ± 0.2; $p < 0.05$). Figure 6A (upper panel) shows the mtDNA content by RT-PCR. Furthermore, we measured the mRNA levels of genes implicated in mitochondrial biogenesis, such as PGC-1α, PGC-1β, NRF1 and TFAM [25]. We found the granulosa cells of $AR^{-/-}$ mice exhibited decreased expression of PGC-1β, TFAM and NRF1 mRNA compared with $AR^{+/+}$ mice (Figure 6B). No significant difference was detected for PGC-1α transcripts (Figure 6B). Together these results revealed a deterioration of mitochondrial biogenic response in granulosa cells of $AR^{-/-}$ mice.

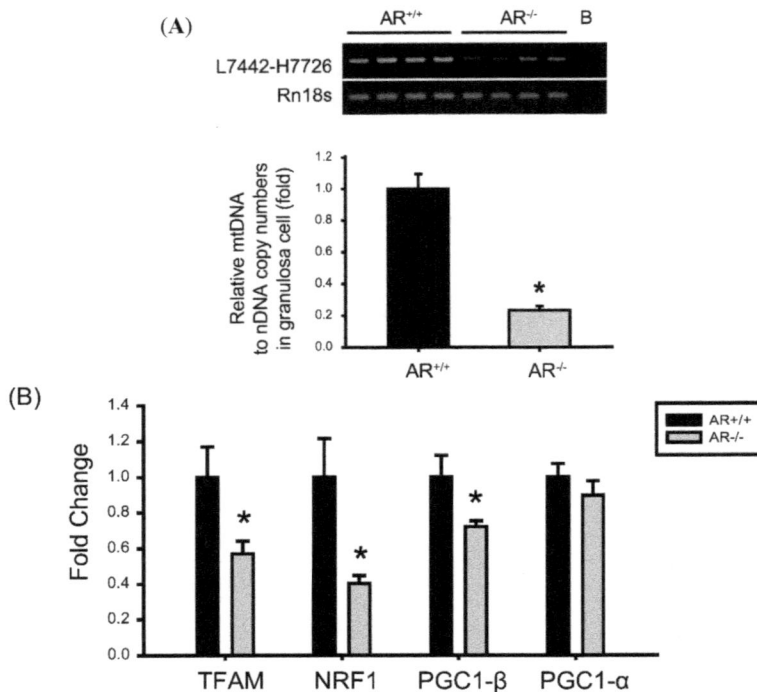

Figure 6. Decreased mitochondrial biogenesis in granulosa cells of $AR^{-/-}$ mice. (**A**) Decreased mitochondrial DNA (mtDNA) copy number in granulosa cells of $AR^{-/-}$ mice. mtDNA copy number was calculated using reverse transcription-PCR and quantitative real-time PCR by measuring the ratio of COX 2 (mtDNA) to Rn18s (nuclear DNA) DNA levels in granulosa cells of $AR^{+/+}$ and $AR^{-/-}$ mice. * $p < 0.05$, $n = 4$ mice per genotype; the lane marked B indicates blank control; (**B**) Decreased mitochondrial biogenesis in granulosa cells of $AR^{-/-}$ mice. The mRNA levels of genes implicated in mitochondrial biogenesis determined by real-time RT-PCR in granulosa cells of $AR^{+/+}$ and $AR^{-/-}$ mice. At least three experiments were performed and data are presented as the mean $\pm$ SEM. * $p < 0.05$, $n = 4$ mice per genotype. PGC-1: peroxisome proliferator-activated receptor γ coactivator-1. NRF1: nuclear respiratory factor 1. TFAM: mitochondrial transcription factor A.

2.7. Molecular Changes in Granulosa Cells of $AR^{-/-}$ Mice and the Effect of AR Deficiency on Serum E2 Levels

To understand the defects in granulosa cells during gonadotropin stimulation, we examined expression of genes that were important granulosa cell proliferation and differentiation markers during folliculogenesis. Granulosa cell differentiation is manifested during the process allowing the progression of preantral follicle to preovulatory follicle, which is dependent on sufficient FSH stimulation [26–28] and is marked by the acquisition of increased Cyp19a1 activity and luteinizing hormone (LH) receptor (LHCGR). Meanwhile, since androgen is known to augment the actions of FSH [3,4,7], we are specifically interested in evaluating the expression pattern of the genes encoding FSHR, LHCGR, and Cyp19a1 in $AR^{+/+}$ and $AR^{-/-}$ ovaries. A number of genes showed significant changes. The FSHR, LHCGR, Cyp11a1, Cyp19a1 and progesterone receptor (PR) expression were significantly decreased in granulosa cells of $AR^{-/-}$ mice (Figure 7A). These data indicate that preovulatory granulosa cells do not properly differentiate and develop sufficient aromatase activity in response to PMSG. Concomitantly, the level of serum estradiol (E2) was significantly decreased in $AR^{-/-}$ female mice as compared with $AR^{+/+}$ littermates (Figure 7B).

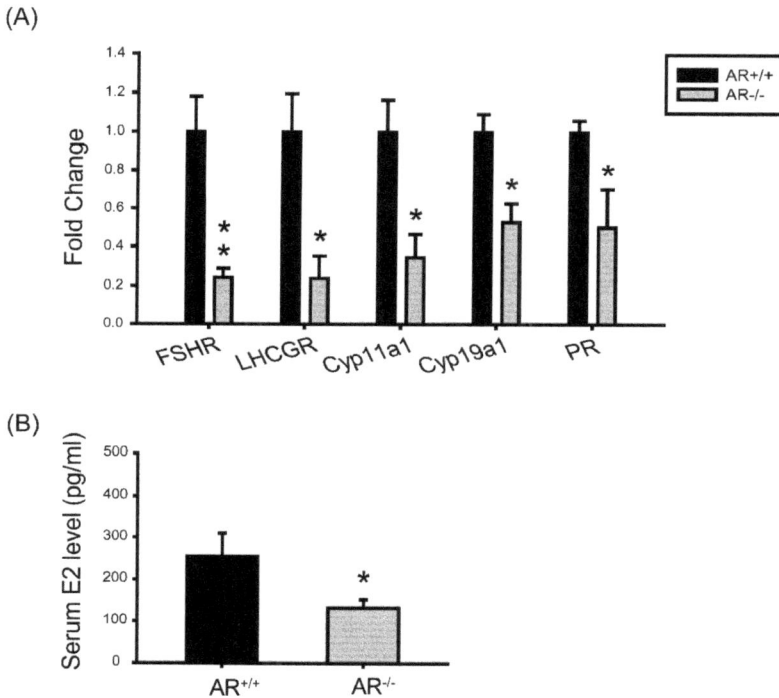

Figure 7. Molecular changes in granulosa cells of $AR^{-/-}$ mice and the effect of AR deficiency on serum estradiol (E2) levels. (**A**) The mRNA levels of genes that are important in granulosa cell differentiation during folliculogenesis and luteinization. At least three experiments were performed and data are presented as the mean $\pm$ SEM (n = 4 mice per genotype) of the fold changes; (**B**) Serum E2 level was measured by ELISA. (n = 4 mice per genotype). Data was expressed as the mean $\pm$ SEM. * $p < 0.05$, ** $p < 0.01$. FSHR: FSH receptor; LHCGR: Luteinizing hormone receptor; Cyp11a1: Cytochrome P450 side-chain-cleavage; Cyp19a1: Cytochrome P450 aromatase; PR: Progesterone receptor.

3. Discussion

The primary cause of reduced fecundity in $AR^{-/-}$ female mice is infrequent and inefficient ovulation [12–16]. In this study, we found a high frequency of morphological anomalies in the mitochondria of $AR^{-/-}$ granulosa cells. Mitochondria displayed an increased electron density of the matrix, small spherical structures with fewer, disarrayed cristae, and containing vacuoles under TEM (Figure 4A). It also displayed an aggregated distribution and a small "donut-like" vesicles appearance under immunofluorescence staining (Figure 4B) in the granulosa cells of $AR^{-/-}$ mice. These morphological changes are frequently referred to as "fragmented mitochondria" [29]. Evidences have suggested that fusion and fission of mitochondria affects the ability of cells to distribute their mitochondria to specific subcellular locations. It has been stated that trafficking of mitochondria are important for cellular function by placing them in appropriate locations relative to energy requiring processes [30]. Therefore, those fragmented and misplaced mitochondria in the granulosa cells of $AR^{-/-}$ mice might result in sub-cellular energy deficiency and even cell death [31,32]. Mitochondrial biogenesis requires the symphonious expression of mtDNA and nuclear genes that encode both mitochondrial proteins and their regulatory factors [33]. It has been proved that the PGC-1 family regulates mitochondrial biogenesis by serving as a coactivator of multiple transcription factors, such as NFR1 and TFAM [34,35]. Our quantitative real-time PCR analysis revealed that mtDNA copy number was significantly decreased in granulosa cells from $AR^{-/-}$ mice when compared to $AR^{+/+}$ mice. In agreement, PGC-1β, NRF1 and TFAM mRNAs were down-regulated in the $AR^{-/-}$ granulosa cells (Figure 6B). In the mean time, the average ATP contents and mitochondrial membrane potential were markedly reduced in the granulosa cells of $AR^{-/-}$ mice as compared to $AR^{+/+}$ mice (Figure 5A,B), suggesting a decline of mitochondrial function.

Within the follicle, cumulus cells surround the oocyte and directly contact with the oocyte via gap junctions [36]. The paracrine interactions between the oocyte and granulosa cells are bi-directional and critical for the development of granulosa cells, optimal oocyte quality and embryonic development [37,38]. Furthermore, earlier study has shown that oocytes enclosed by cumulus cells have higher ATP levels than those lacking cumulus cells [39], suggesting that cumulus cells provide ATP for oocyte development. Therefore, mitochondrial function is critical for the energy production in granulosa cells and then to participate as an energy source for oocyte maturation. The role of mitochondria in oocyte, in particular in the immature oocyte has not yet been identified. However, it is generally believed that mitochondria in the either oocyte or egg are in a transmitted state and undifferentiated. Therefore during oocyte maturation, the energy sources are believed from the surrounding cells [40]. Thus, in the current study, we demonstrated that the AR insufficiency that impact on mitochondria function will result in poor oocyte maturation rate.

We found that 95% of $AR^{+/+}$ oocytes had reached to metaphase II, whereas a significantly lower maturation rate (60%) was observed in the $AR^{-/-}$ oocytes. This is consistence with earlier studies that such a declined ATP content in oocytes has been suggested to significantly affect oocyte quality, embryonic development and even the implantation process [41,42]. We speculated that the decreased ATP levels and reduced mitochondria membrane potential in the granulosa cells of $AR^{-/-}$ mice might contribute to the decreased oocyte competence to undergo final maturation, as observed in this study.

Our results also showed that PMSG-stimulated $AR^{-/-}$ granulosa cells exhibit reduced FSHR, Cyp11a1 and Cyp19a1 mRNA expressions (Figure 7A). It has been reported, in the both of hypophysectomized rat [43], as well as mutant mice that lacking either gonadotrophins [44], FSH [26], or FSHR [27,28], causes follicular development arrest at the preantral stage. Androgens, via the activation of AR, can increase the expression of FSHR and augment FSH-stimulated follicular differentiation [5–8]. Cyp11a1 and Cyp19a1 are stimulated by FSH in granulosa cells as part of the differentiation program induced by this hormone [45]. Earlier study by Wu [46] nicely showed that androgens have a direct stimulatory effect on Cyp19a1 and Cyp11a1 expression in rat ovarian granulosa cells. In agreement with those earlier studies, our results provide the clear evidence that AR is required for maximum FSH stimulation of Cyp11a1 and Cyp19a1 activity. Our results also showed that the level of serum E2 was significantly decreased in $AR^{-/-}$ mice as compared with $AR^{+/+}$ littermates (Figure 7B). Another important characteristic of preovulatory granulosa cells is the acquisition of LHCGR and PR, which provide the mechanism of follicle selection by which follicles can respond to the LH surge and ovulate [47]. Induction of LHCGR [48,49] and PR [50,51] expression in preovulatory granulosa cells is primarily activated by FSH. Herein, our data demonstrated that AR mediates this effect of androgen on FSHR expression further onwards regulation of LHCGR and PR levels in preovulatory granulosa cells. Granulosa cells of $AR^{-/-}$ mice exhibit a reduced LHCGR and PR mRNA levels relative to $AR^{+/+}$ controls (Figure 7A). We proposed that the reduced LHCGR levels in the granulosa cells of $AR^{-/-}$ mice might contribute to compromised response to LH surge and oocyte competence to undergo final maturation.

Combining our previous findings [12] with the results presented here, we conclude that $AR^{-/-}$ mice exhibit a reduced fertility with defective folliculogenesis and reduced corpus luteum formation might have been contributed by the following reasons. Mitochondrial dysfunction in granulosa cells may compromise the competence of oocytes in $AR^{-/-}$ ovaries. These data indicate that the intraovarian actions of AR and androgen are critical for enhancing the FSH action and the generation of fully differentiated follicles that exhibit the proper cellular organization (*i.e.*, antrum formation and cumulus-oocyte complex), optimal mitochondrial function, the necessary enzymatic activity, and the essential receptor signaling pathways. All of those factors converge to provide the follicle with the normal capacity to respond to the LH surge and expel a healthy oocyte that is fully competent for fertilization.

4. Materials and Methods

All the culture media and chemicals were purchased from Sigma-Aldrich Chemical Company, unless otherwise stated.

4.1. Generation of Female $AR^{-/-}$ Mice

All mouse studies were approved by the Animal Studies Committee at Taipei Medical University and conformed to the Guide for the Care and Use of Laboratory Animals published by the National Institutes of Health. $AR^{-/-}$ female mice and heterozygous female mice ($AR^{+/-}$) were generated by crossing AR-flox mice (exon 2 flanked by loxP sites) with transgenic β-actin-Cre (ACTB-Cre) mice, as previously described [52,53]. Briefly, male mice carrying floxed AR were mated with females genotyped with *AR/ar* ACTB-Cre ($AR^{+/-}$) to produce female $AR^{-/-}$ mice carrying the genotype *ar/ar* ACTB-Cre. Genotype of offspring was confirmed by PCR. Based on the sequence of the AR genomic DNA, three primers have been designed to distinguish the $AR^{+/+}$, $AR^{-/-}$, and floxed AR X chromosome on mice. The primer "select" is the 5' primer which is located in the intron 1 and its sequence is 5'-GTTGATACCTTAACCTCTGC-3'. The primer "2–9" is the 3' end primer which is located in intron 2 and its sequence is 5'-CTTACATGTACTGTGAGAGG-3'. The "2–3" primer is the 3' end primer which is located in the exon 2 and its sequence is 5'-CTTCAGCGGCTCTTTTGAAG-3'. If the mouse contains $AR^{+/+}$, the primer pairs (select and "2–9") will generate a product with 580 bp and the other pair of primers (select and "2–3") will generate a product with ~460 bp. If the mice carry $AR^{-/-}$, the primer pairs (select and "2–9") will generate a product with ~270 bp and the other pair of primers (select and "2–3") will not have a product. The expression of Cre and internal control interleukin 2 (IL-2) were confirmed by PCR during genotyping. The primer design and PCR conditions of Cre and IL-2 follow The Jackson Laboratory's suggestions as previously described [52,53].

4.2. Western Blot Analysis

To determine the expression levels of AR, the protein extracts of ovaries were subjected to sodium dodecyl sulfatepolyacrylamide denaturing gel electrophoresis. Proteins were transferred to hybond-P PVDF membranes (GE Healthcare, Little Chalfont, UK) and processed by routine procedures. Immunoreactive bands were visualized by blotting with primary antibodies against AR (1:1000, LS-C137965, LifeSpan Biosciences, Seattle, WA, USA), followed by incubation with horseradish peroxidase-conjugated secondary antibodies and detection with enhanced chemiluminescence (Thermo Scientific, Rockford, IL, USA).

4.3. Tissue Sampling, Oocyte in Vitro Maturation Assay, RNA Extraction and Analysis

To collect fully grown germinal vesicle (GV) oocytes and granulosa cells, a minimum of six mice per genotype (4.5 weeks old; $AR^{-/-}$ and $AR^{+/+}$) were treated with 7.5 IU PMSG by intraperitoneal injection. After 48 h, the ovaries were removed and placed in a dish containing M2 medium. For the oocyte collection, cumulus-oocyte complexes (COCs) and granulosa cells were harvested by manually puncturing of large antral follicles with a sterile needle. Oocytes were denuded with gentle pipetting with pulled-glass Pasteur pipettes as previously described [54]. Oocytes were then cultured in M2 media on the 37 °C heated stage for another 18 h in order to score maturation rates. For collection of granulosa cells, ovarian debris was removed by filtered cells through a 40 μm nylon mesh filter. Granulosa cells were collected by centrifugation and processed for total RNA extraction using an RNeasy kit (Qiagen, Chatsworth, CA, USA). The final concentration of granulosa cells is approximately about 1.5 million per mL. Granulosa cells represent a mixture of both mural and cumulus cells and are referred to as "granulosa cells" in the text. One μg of total RNA was reverse transcribed and subjected to real-time PCR using LightCycler 2.0 instrument (Roche Applied Science, Mannheim, Germany). In general, the real-time PCR was performed with SYBR Green PCR Master Mix (Roche Applied Science). Each sample was run in triplicate. Data were analyzed by a LightCycler Software 4.0 (Roche Applied Science). The mRNA levels of genes of interest in $AR^{-/-}$ mice were compared with those of their $AR^{+/+}$ littermates. Each gene expression pattern was confirmed using at least three pairs of $AR^{-/-}$ and $AR^{+/+}$ mice. Primer sequences used for studying gene expression were designed by Beacon Designer II software (Bio-Rad Lab., Irvine, CA, USA) and are listed below: PPARγ co-activator 1-α (PGC-1α)-forward (f): GACATAGAGTGTGCTGCTCTGGT; PGC-1α-reverse (r): GTTCGCAGGCTCATTGTTGT; PGC-1β-f: CGCTCCAGGAGA CTG AATCCAG; PGC-1β-r: CTTGACTACTGTCTGTGAGGC; NRF1-f: TTACTCT GCTGTGGCTGATGG; NRF1-r: CCTCTGATGCTTGCGTCGTCT; TFAM-f: AGTT CATACCTTCGATTTTC; TFAM-r: TGACTTGGAGTTAGCTGC; FSHR-f: GAACG CCATTGAACTGAGATT; FSHR-r: CGGAGACTGGGAAGATTCTG; LHCGR-f: AGTCCATCACGCTGAAACTGT; LHCGR-r: GGCCTGCAATTTGGTGGAAG; Cytochrome P450 side-chain-cleavage (Cyp11a1)-f: GGTGGACACGACCTCCATGA; Cyp11a1-r: TGCTGGCTTTGAGGAGTGGA; Cyp19a1-f: TTCGCTGAGAGACGTG GAGA; Cyp19a1-r: AGGATTGCTGCTTCGACCTC; Progesterone receptor (PR)-f: CTCATGAGTCGGCCAGAGAT; PR-r: CACTGTCCTCTTCCACCTCC; β2-microglobulin (β2m)-f: ACCCTCATCAATGGCCTGTGGA; β2m-r: CATGGGCTTTGACCCTTGGG. Mouse β2m was used as a housekeeping gene.

4.4. Histology

Ovaries from 4.5-week-old $AR^{-/-}$ and $AR^{+/+}$ littermates ($n = 4$ per genotype) were obtained from PMSG treated animals. Ovaries were fixed in 4% paraformaldehyde at 4 °C and then paraffin embedded. Thereafter, 5-μm sections were taken at 30-μm intervals, mounted on slides, and subjected to hematoxylin and eosin staining for histological examination by light microscopy. All follicles with a visible oocyte nucleus were subsequently classified by stage of development. The follicle classification system was based on Pedersen's system [55]: (1) Primordial follicles—identified by an oocyte surrounded by a layer of flattened granulose progenitor cells; (2) Pedersen class 3 (primary)—identified by an oocyte surrounded by a layer of cuboid granulosa cells. There are 21 to 60 granulosa cells on the largest cross-section; (3) Pedersen classes 4–5 (secondary)—identified by an oocyte surrounded by two or more layers of granulosa cells. There are 61 to 400 granulosa cells on the largest cross-section; (4) Pedersen class 6 (incipient antral)—identified by a large oocyte with many layers of granulosa cells. The granulosa cells are separated by scattered areas of fluid. There are 401 to 600 granulosa cells on the largest cross-section; (5) Pedersen class 7 (early antral)—a follicle with a single cavity containing follicle fluid. There are more than 600 cells on the largest cross-section. The cumulus oophorus has formed; (6) Pedersen class 8 (Graafian follicle)—a large follicle with a single cavity with follicle fluid and a well-defined cumulus stalk. There are more than 600 cells on the largest cross-section. Follicles were considered as atretic based on any of the two morphometric criteria within a single cross section: A degenerated oocyte, three or more pyknotic nuclei or atretic bodies in granulosa cell layers or follicular antrum, disorganized granulosa cell layers, granulosa cells pulling away from basement membrane, and broken basement membrane as previously described [15,56].

4.5. Transmission Electron Microscopy (TEM)

For ultrastructural analysis of mitochondria, the ovaries from $AR^{-/-}$ and $AR^{+/+}$ mice were collected followed by 48 h PMSG treatment then fixed in a mixture of glutaraldehyde (1.5%) and paraformaldehyde (1.5%) in phosphate buffer at pH 7.3. They were then post-fixed in 1.0% osmium tetroxide, 1.5% potassium hexanoferrate, before being rinsed in cacodylate and 0.2 M sodium maleate buffers (pH 6.0) followed by block-staining with 1% uranyl acetate. Following dehydration, the ovaries were embedded in Epon and sectioned for TEM. Thin sections were cut, mounted on 200-mesh grids, stained with uranyl acetate and lead citrate, and examined using a H7100 Hitachi electron microscope (H7100, Hitachi High-Technologies Corporation, Tokyo, Japan). Digital images were captured using a MegaView III digital camera (OSIS Pro Software, Olympus Soft Imaging Solutions, Lakewood, CO, USA). The morphology of mitochondria in granulosa cells was examined.

4.6. Immunofluorescence

For mitochondrial localization, granulosa cells were collected and cultured overnight in μ-Slide 8 well plates (ibidi GmbH, Martinsried, Germany) to allow cell attachment. Granulosa cells were stained with 200 nM MitoTracker Green (Molecular Probes, Eugene, OR, USA) for 45 min at 37 °C and cells were then fixed with 4% paraformaldehyde for 20 min and treated with 0.5% Triton X-100 for 20 min, the liquid removed and blocked by 0.5% BSA in PBS 1 h. After empty the well, granulosa cells were counterstained with 4',6'-diamidino-2-phenylindole (DAPI; Sigma, St. Louis, MO, USA) for 10 min at 37 °C, samples were observed by *Delta Vision Elite* Microscope (Applied Precision, GE Healthcare, Olive Branch, MS, USA).

4.7. Determination of Mitochondrial DNA (mtDNA) Content in Granulosa Cells

Total DNA was extracted from granulosa cells using a DNeasy kit (Qiagen). mtDNA content was calculated using quantitative real-time PCR as described previously [57,58] by measuring the threshold cycle ratio (ΔC_t) of a mitochondria-encoded gene cytochrome c oxidase subunit 2 (COX2) (L7447, 5'-AATAGAACTT CCAATCCGTA-3', H7726, 5'-AAGGTTAACGCTCTTAGCTT-3') *vs.* a nuclear-encoded gene (18s-rRNA-f, 5'-CGGCTACCACATCCAAGGAA-3', 18s-rRNA-r, 5'-GCTGGAAT TACCGCGGCT-3'). Data were expressed as mtDNA/nuclear DNA. All measurements were performed in triplicate.

4.8. Metabolite Analytic Assays

Granulosa cells were measured for ATP content by ATPlite™ Luminescence ATP Detection Assay System (Perkin Elmer Life and Analytical Sciences B.V., Groningen, The Netherlands). Granulosa cells were homogenized in 50 μL mammalian cell lysis buffer and then 50 μL substrates solution were added. Adapt the plate to darkness for 10 min and then measure the luminescence. Reactions were normalized to total protein, and metabolite contents were expressed as μM/mg protein.

4.9. Detection of Mitochondrial Membrane Potential by Flow Cytometry

Granulosa cells were collected from 4-week-old female mice ($AR^{+/+}$ and $AR^{-/-}$) injected with 7.5 IU PMSG 48 h later. To get positive control from cells treated with CCCP (carbonyl cyanide 3-chlorophenylhydrazone, Sigma Immunochemicals, St. Louis, MO, USA) to depolarized the mitochondrial membrane potential. Cells were incubated with 500 nM JC-1 (5,5',6,6'-tetrachloro-1,1',3, 3'-tetraethylbenzimidazolylcarbocyanine iodide, Molecular Probes, Eugene, OR, USA) in the PBS for 15 min at 37 °C in the dark and then analyzed by flow cytometry (BD Biosciences, San Jose, CA, USA). The acquired data were further analyzed using FCS express V4 analysis software.

4.10. Assessment of Serum Hormone Levels

To determine the level of estradiol (E2) in the experimental animals, sera were collected from both $AR^{-/-}$ and $AR^{+/+}$ female mice that were previously treated with 7.5 IU PMSG for 48 h and stored at $-20\ ^\circ$C before further analysis were carried out. Total E2 level was measured by ELISA kits following the manufacture's instruction (USCN Life Science, Wuhan, China). The volume of serum used for ELISA was 50 μL, the minimal detectable concentration was 4.38 pg/mL, and the intra- and interassay coefficients of variation were 8.7% and 9.8%, respectively.

4.11. Statistical Analysis

Data are presented as mean $\pm$ standard error mean (SEM), unless otherwise indicated. Statistical comparisons were made with Student's t test. $p \leqslant 0.05$ as considered to be significant.

Supplementary Materials: Supplementary materials can be found at http://www.mdpi.com/1422-0067/16/05/9831/s1.

Acknowledgments: We wish to thank the Laboratory of Reproductive Medicine in the department of Gynecology and Obstetrics at the Taipei Medical University Hospital for the laboratory procedures. This work was supported by National Science Council, Taiwan (NSC 96-2314-B-038-018-MY2) and by Taipei Medical University (TMU100-AE1-B02).

Author Contributions: Ruey-Sheng Wang and Heng-Yu Chang did the writing of paper and oocyte experiments as well as molecular biology experiments; Cheng-Heng Kao did the TEM experiments; Yi-Chen Wu and Shuyuan Yeh bred the knockout mice; Shu-Huei Kao, Chii-Ruey Tzeng and Chawnshang Chang supervised the project.

Conflicts of Interest: The authors declare no conflict of interest.

References

1. Chang, C.S.; Kokontis, J.; Liao, S.T. Molecular cloning of human and rat complementary DNA encoding androgen receptors. *Science* **1988**, *240*, 324–326.
2. Heinlein, C.A.; Chang, C. Androgen receptor in prostate cancer. *Endocr. Rev.* **2004**, *25*, 276–308.
3. Murray, A.A.; Gosden, R.G.; Allison, V.; Spears, N. Effect of androgens on the development of mouse follicles growing *in vitro*. *J. Reprod. Fertil.* **1998**, *113*, 27–33.
4. Vendola, K.A.; Zhou, J.; Adesanya, O.O.; Weil, S.J.; Bondy, C.A. Androgens stimulate early stages of follicular growth in the primate ovary. *J. Clin. Investig.* **1998**, *101*, 2622–2629.
5. Tetsuka, M.; Hillier, S.G. Differential regulation of aromatase and androgen receptor in granulosa cells. *J. Steroid Biochem. Mol. Biol.* **1997**, *61*, 233–239.
6. Weil, S.; Vendola, K.; Zhou, J.; Bondy, C.A. Androgen and follicle-stimulating hormone interactions in primate ovarian follicle development. *J. Clin. Endocrinol. Metab.* **1999**, *84*, 2951–2956.

7. Daniel, S.A.; Armstrong, D.T. Enhancement of follicle-stimulating hormone-induced aromatase activity by androgens in cultured rat granulosa cells. *Endocrinology* **1980**, *107*, 1027–1033.

8. Sen, A.; Prizant, H.; Light, A.; Biswas, A.; Hayes, E.; Lee, H.J.; Barad, D.; Gleicher, N.; Hammes, S.R. Androgens regulate ovarian follicular development by increasing follicle stimulating hormone receptor and microRNA-125b expression. *Proc. Natl. Acad. Sci. USA* **2014**, *111*, 3008–3013.

9. Hirai, M.; Hirata, S.; Osada, T.; Hagihara, K.; Kato, J. Androgen receptor mRNA in the rat ovary and uterus. *J. Steroid Biochem. Mol. Biol.* **1994**, *49*, 1–7.

10. Tetsuka, M.; Whitelaw, P.F.; Bremner, W.J.; Millar, M.R.; Smyth, C.D.; Hillier, S.G. Developmental regulation of androgen receptor in rat ovary. *J. Endocrinol.* **1995**, *145*, 535–543.

11. Gill, A.; Jamnongjit, M.; Hammes, S.R. Androgens promote maturation and signaling in mouse oocytes independent of transcription: A release of inhibition model for mammalian oocyte meiosis. *Mol. Endocrinol.* **2004**, *18*, 97–104.

12. Hu, Y.C.; Wang, P.H.; Yeh, S.; Wang, R.S.; Xie, C.; Xu, Q.; Zhou, X.; Chao, H.T.; Tsai, M.Y.; Chang, C. Subfertility and defective folliculogenesis in female mice lacking androgen receptor. *Proc. Natl. Acad. Sci. USA* **2004**, *101*, 11209–11214.

13. Shiina, H.; Matsumoto, T.; Sato, T.; Igarashi, K.; Miyamoto, J.; Takemasa, S.; Sakari, M.; Takada, I.; Nakamura, T.; Metzger, D.; *et al.* Premature ovarian failure in androgen receptor-deficient mice. *Proc. Natl. Acad. Sci. USA* **2006**, *103*, 224–229.

14. Walters, K.A.; Allan, C.M.; Jimenez, M.; Lim, P.R.; Davey, R.A.; Zajac, J.D.; Illingworth, P.; Handelsman, D.J. Female mice haploinsufficient for an inactivated androgen receptor (AR) exhibit age-dependent defects that resemble the AR null phenotype of dysfunctional late follicle development, ovulation, and fertility. *Endocrinology* **2007**, *148*, 674–684.

15. Sen, A.; Hammes, S.R. Granulosa cell-specific androgen receptors are critical regulators of ovarian development and function. *Mol. Endocrinol.* **2010**, *24*, 1393–1403.

16. Walters, K.A.; Middleton, L.J.; Joseph, S.R.; Hazra, R.; Jimenez, M.; Simanainen, U.; Allan, C.M.; Handelsman, D.J. Targeted loss of androgen receptor signaling in murine granulosa cells of preantral and antral follicles causes female subfertility. *Biol. Reprod.* **2012**, *87*, 151.

17. Spears, N.; Murray, A.A.; Allison, V.; Boland, N.I.; Gosden, R.G. Role of gonadotrophins and ovarian steroids in the development of mouse follicles *in vitro*. *J. Reprod. Fertil.* **1998**, *113*, 19–26.

18. Yang, M.Y.; Fortune, J.E. Testosterone stimulates the primary to secondary follicle transition in bovine follicles *in vitro*. *Biol. Reprod.* **2006**, *75*, 924–932.

19. Seifer, D.B.; DeJesus, V.; Hubbard, K. Mitochondrial deletions in luteinized granulosa cells as a function of age in women undergoing *in vitro* fertilization. *Fertil. Steril.* **2002**, *78*, 1046–1048.

20. Wang, Q.; Frolova, A.I.; Purcell, S.; Adastra, K.; Schoeller, E.; Chi, M.M.; Schedl, T.; Moley, K.H. Mitochondrial dysfunction and apoptosis in cumulus cells of type I diabetic mice. *PLoS ONE* **2010**, *5*, e15901.

21. Gannon, A.M.; Stampfli, M.R.; Foster, W.G. Cigarette smoke exposure elicits increased autophagy and dysregulation of mitochondrial dynamics in murine granulosa cells. *Biol. Reprod.* **2013**, *88*, 63.

22. Newmeyer, D.D.; Ferguson-Miller, S. Mitochondria: Releasing power for life and unleashing the machineries of death. *Cell* **2003**, *112*, 481–490.

23. Miller, W.L. Steroid hormone synthesis in mitochondria. *Mol. Cell. Endocrinol.* **2013**, *379*, 62–73.

24. Ramalho-Santos, J.; Varum, S.; Amaral, S.; Mota, P.C.; Sousa, A.P.; Amaral, A. Mitochondrial functionality in reproduction: From gonads and gametes to embryos and embryonic stem cells. *Hum. Reprod. Update* **2009**, *15*, 553–572.

25. Scarpulla, R.C. Transcriptional paradigms in mammalian mitochondrial biogenesis and function. *Physiol. Rev.* **2008**, *88*, 611–638.

26. Kumar, T.R.; Wang, Y.; Lu, N.; Matzuk, M.M. Follicle stimulating hormone is required for ovarian follicle maturation but not male fertility. *Nat. Genet.* **1997**, *15*, 201–204.

27. Dierich, A.; Sairam, M.R.; Monaco, L.; Fimia, G.M.; Gansmuller, A.; LeMeur, M.; Sassone-Corsi, P. Impairing follicle-stimulating hormone (FSH) signaling *in vivo*: Targeted disruption of the FSH receptor leads to aberrant gametogenesis and hormonal imbalance. *Proc. Natl. Acad. Sci. USA* **1998**, *95*, 13612–13617.

28. Danilovich, N.; Babu, P.S.; Xing, W.; Gerdes, M.; Krishnamurthy, H.; Sairam, M.R. Estrogen deficiency, obesity, and skeletal abnormalities in follicle-stimulating hormone receptor knockout (FORKO) female mice. *Endocrinology* **2000**, *141*, 4295–4308.

29. Detmer, S.A.; Chan, D.C. Functions and dysfunctions of mitochondrial dynamics. *Nat. Rev. Mol. Cell Biol.* **2007**, *8*, 870–879.

30. Frazier, A.E.; Kiu, C.; Stojanovski, D.; Hoogenraad, N.J.; Ryan, M.T. Mitochondrial morphology and distribution in mammalian cells. *Biol. Chem.* **2006**, *387*, 1551–1558.

31. Martinou, J.C.; Youle, R.J. Mitochondria in apoptosis: Bcl-2 family members and mitochondrial dynamics. *Dev. Cell* **2011**, *21*, 92–101.

32. Chan, D.C. Mitochondria: Dynamic organelles in disease, aging, and development. *Cell* **2006**, *125*, 1241–1252.

33. Diaz, F.; Moraes, C.T. Mitochondrial biogenesis and turnover. *Cell Calcium* **2008**, *44*, 24–35.

34. Kelly, D.P.; Scarpulla, R.C. Transcriptional regulatory circuits controlling mitochondrial biogenesis and function. *Genes Dev.* **2004**, *18*, 357–368.

35. Lin, J.; Handschin, C.; Spiegelman, B.M. Metabolic control through the PGC-1 family of transcription coactivators. *Cell Metab.* **2005**, *1*, 361–370.

36. Gittens, J.E.; Barr, K.J.; Vanderhyden, B.C.; Kidder, G.M. Interplay between paracrine signaling and gap junctional communication in ovarian follicles. *J. Cell Sci.* **2005**, *118*, 113–122.

37. Eppig, J.J. Intercommunication between mammalian oocytes and companion somatic cells. *Bioessays* **1991**, *13*, 569–574.

38. Matzuk, M.M.; Burns, K.H.; Viveiros, M.M.; Eppig, J.J. Intercellular communication in the mammalian ovary: Oocytes carry the conversation. *Science* **2002**, *296*, 2178–2180.

39. Downs, S.M. The influence of glucose, cumulus cells, and metabolic coupling on ATP levels and meiotic control in the isolated mouse oocyte. *Dev. Biol.* **1995**, *167*, 502–512.

40. Sugiura, K.; Pendola, F.L.; Eppig, J.J. Oocyte control of metabolic cooperativity between oocytes and companion granulosa cells: Energy metabolism. *Dev. Biol.* **2005**, *279*, 20–30.

41. Quinn, P.; Wales, R.G. The relationships between the ATP content of preimplantation mouse embryos and their development *in vitro* during culture. *J. Reprod. Fertil.* **1973**, *35*, 301–309.

42. Van Blerkom, J.; Davis, P.W.; Lee, J. ATP content of human oocytes and developmental potential and outcome after *in vitro* fertilization and embryo transfer. *Hum. Reprod.* **1995**, *10*, 415–424.

43. Richards, J.S. Maturation of ovarian follicles: Actions and interactions of pituitary and ovarian hormones on follicular cell differentiation. *Physiol. Rev.* **1980**, *60*, 51–89.

44. Mason, A.J.; Hayflick, J.S.; Zoeller, R.T.; Young, W.S., 3rd.; Phillips, H.S.; Nikolics, K.; Seeburg, P.H. A deletion truncating the gonadotropin-releasing hormone gene is responsible for hypogonadism in the hpg mouse. *Science* **1986**, *234*, 1366–1371.

45. Hunzicker-Dunn, M.; Maizels, E.T. FSH signaling pathways in immature granulosa cells that regulate target gene expression: Branching out from protein kinase A. *Cell Signal.* **2006**, *18*, 1351–1359.

46. Wu, Y.G.; Bennett, J.; Talla, D.; Stocco, C. Testosterone, not 5α-dihydrotestosterone, stimulates LRH-1 leading to FSH-independent expression of Cyp19 and P450scc in granulosa cells. *Mol. Endocrinol.* **2011**, *25*, 656–668.

47. McGee, E.A.; Hsueh, A.J. Initial and cyclic recruitment of ovarian follicles. *Endocr. Rev.* **2000**, *21*, 200–214.

48. Segaloff, D.L.; Wang, H.Y.; Richards, J.S. Hormonal regulation of luteinizing hormone/chorionic gonadotropin receptor mRNA in rat ovarian cells during follicular development and luteinization. *Mol. Endocrinol.* **1990**, *4*, 1856–1865.

49. Kessel, B.; Liu, Y.X.; Jia, X.C.; Hsueh, A.J. Autocrine role of estrogens in the augmentation of luteinizing hormone receptor formation in cultured rat granulosa cells. *Biol. Reprod.* **1985**, *32*, 1038–1050.

50. Natraj, U.; Richards, J.S. Hormonal regulation, localization, and functional activity of the progesterone receptor in granulosa cells of rat preovulatory follicles. *Endocrinology* **1993**, *133*, 761–769.

51. Park-Sarge, O.K.; Mayo, K.E. Regulation of the progesterone receptor gene by gonadotropins and cyclic adenosine 3',5'-monophosphate in rat granulosa cells. *Endocrinology* **1994**, *134*, 709–718.

52. Yeh, S.; Tsai, M.Y.; Xu, Q.; Mu, X.M.; Lardy, H.; Huang, K.E.; Lin, H.; Yeh, S.D.; Altuwaijri, S.; Zhou, X.; *et al.* Generation and characterization of androgen receptor knockout (ARKO) mice: An *in vivo* model for the study of androgen functions in selective tissues. *Proc. Natl. Acad. Sci. USA* **2002**, *99*, 13498–13503.

53. Yeh, S.; Hu, Y.C.; Wang, P.H.; Xie, C.; Xu, Q.; Tsai, M.Y.; Dong, Z.; Wang, R.S.; Lee, T.H.; Chang, C. Abnormal mammary gland development and growth retardation in female mice and MCF7 breast cancer cells lacking androgen receptor. *J. Exp. Med.* **2003**, *198*, 1899–1908.

54. Chang, H.Y.; Minahan, K.; Merriman, J.A.; Jones, K.T. Calmodulin-dependent protein kinase γ 3 (CamKIIγ3) mediates the cell cycle resumption of metaphase II eggs in mouse. *Development* **2009**, *136*, 4077–4081.

55. Pedersen, T.; Peters, H. Proposal for a classification of oocytes and follicles in the mouse ovary. *J. Reprod. Fertil.* **1968**, *17*, 555–557.

56. Britt, K.L.; Drummond, A.E.; Cox, V.A.; Dyson, M.; Wreford, N.G.; Jones, M.E.; Simpson, E.R.; Findlay, J.K. An age-related ovarian phenotype in mice with targeted disruption of the *Cyp 19* (aromatase) gene. *Endocrinology* **2000**, *141*, 2614–2623.

57. Weng, S.W.; Lin, T.K.; Liou, C.W.; Chen, S.D.; Wei, Y.H.; Lee, H.C.; Chen, I.Y.; Hsieh, C.J.; Wang, P.W. Peripheral blood mitochondrial DNA content and dysregulation of glucose metabolism. *Diabetes Res. Clin. Pract.* **2009**, *83*, 94–99.

58. Bonnard, C.; Durand, A.; Peyrol, S.; Chanseaume, E.; Chauvin, M.A.; Morio, B.; Vidal, H.; Rieusset, J. Mitochondrial dysfunction results from oxidative stress in the skeletal muscle of diet-induced insulin-resistant mice. *J. Clin. Investig.* **2008**, *118*, 789–800.

MDPI AG
St. Alban-Anlage 66
4052 Basel, Switzerland
Tel. +41 61 683 77 34
Fax +41 61 302 89 18
http://www.mdpi.com

IJMS Editorial Office
E-mail: ijms@mdpi.com
http://www.mdpi.com/journal/ijms

www.ingramcontent.com/pod-product-compliance
Lightning Source LLC
Chambersburg PA
CBHW051926190326

41458CB00026B/6424